Seafloor Geomorphology as Benthic Habitat

Seafloor Geomorphology as Benthic Habitat

GeoHAB Atlas of Seafloor Geomorphic Features and Benthic Habitats

Edited by

Peter T. Harris
Geoscience Australia
Canberra, Australia

Elaine K. Baker
UNEP/GRID-Arendal
School of Geosciences
University of Sydney
Australia

AMSTERDAM • BOSTON • HEIDELBERG • LONDON • NEW YORK • OXFORD
PARIS • SAN DIEGO • SAN FRANCISCO • SINGAPORE • SYDNEY • TOKYO

Elsevier
32 Jamestown Road, London NW1 7BY
225 Wyman Street, Waltham, MA 02451, USA

First edition 2012

British Library Cataloguing-in-Publication Data
A catalogue record for this book is available from the British Library

Library of Congress Cataloging-in-Publication Data
A catalog record for this book is available from the Library of Congress

ISBN: 978-0-12-385140-6

Contents

Part II   Case Studies   157

Foreword

In 1967, while working with David Hopkins of the US Geological Survey (USGS) in Alaska, I had the privilege of flying all over the coastal plain of Nome in a helicopter to revisit the many gold exploration pits dug into the gravels of the plain during the gold rush of the mid to late 1800s. Dave was a Quaternary geologist and archeologist who spent his entire career studying alluvial and fluvial sedimentation, and he had a passion for understanding the migration of humans from Asia to the Americas. He wrote a book, *The Bering Land Bridge* (1967, Stanford University Press) on the subject. This was the first year that the helicopter, a relatively new technology to geology at that time, was used by the USGS in field mapping of the region. In that short summer season, Dave was able to visit every single pit he had examined and mapped on foot in the preceding two decades. Through this technological advancement, the introduction of the helicopter, geologic field mapping was rapidly advanced to the point that what once took 20 years or more to do could now be done in 3 months time. We are still benefiting from technological advancements.

Technology brought about a revolution in geology during that same time I was working in Nome. With the evolution of marine geophysical instruments during World War II that could measure magnetic anomalies, echosounders that could electronically measure and record depths of the oceans, and other such marine instruments, the understanding of tectonic processes advanced from the geosynclinal theory to the global plate tectonics theory, now generally accepted by the scientific community.

I spent my career at the USGS studying the seafloor as a marine geologist using the most sophisticated geophysical instruments at the time: high-precision echosounders, seismic reflection profilers, sidescan sonars, gravimeters, and magnetometers. Evaluation and interpretation of the data provided by this suite of instruments primarily produced data in two dimensions that could be used to develop three-dimensional interpretive products, often called at the time "geo-imagination." About 20 years ago, new technology was introduced in the form of wide-swath mulitibeam echosounders that could provide near 100% coverage of the seafloor at a resolution unobtainable before. These new instruments, along with sophisticated processing tools, now make it possible to computer generate three-dimensional images of the seafloor and thus advance seafloor mapping both in time expenditure and accuracy. What took me 20 years to imagine can now be accurately produced within a year.

Another evolution that has taken place in the sciences is the swing from a mainly discipline approach to the study of the marine environment to a multidiscipline, or systems approach, with geologists, biologists, ecologists, physicists, and oceanographers coming together to study the seafloor and its environs as a whole. This has led to the establishment of marine benthic habitat characterization and mapping. Out of this evolution, the organization GeoHab was born.

GeoHab spawned from a group of geologists and biologists concerned about the lack of attention being given by the oceanographic community to marine benthic habitat characterization and mapping, and to the technological advancements associated with these activities. This embryonic, unstructured, and loosely organized organization initiated its first meeting as a special session at the annual conference of the Geological Association of Canada in Saint Johns, Newfoundland, Canada, in 2001. In the past decade, GeoHab has evolved into a major international forum that has convened conferences every year, sequentially at: Moss Landing, California, USA; Hobart, Tasmania, Australia; Galway, Ireland; Sidney, British Columbia, Canada; Edinburgh, Scotland, UK; Noumea, New Caledonia; Sitka, Alaska, USA; Trondheim, Norway; Wellington, New Zealand; and Helsinki, Finland. Although the organization is still loosely structured, it has grown in numbers of participants, from about 100 strong in the early years to now over 400, and has combined its activities with the Circum-Pacific Council to establish a formal financial arm to the organization.

Another major activity of GeoHab, other than organizing and convening the GeoHab Conferences, is organizing and publishing the results of recent scientific advancements associated with the geological and biological mapping of marine benthic habitats, including the development of new applied mapping technologies and methodologies. All of the contributions published to date come from GeoHab participants, with many works presented at GeoHab conferences. This effort was initiated with a book published by the Geological Association of Canada, *Mapping the Seafloor for Habitat Characterization* (2007, Geological Association of Canada, Special Paper 47, Edited by Brian J. Todd and H. Gary Greene), followed by a Special Issue of *Continental Shelf Research* "Geological and Biological Mapping and Characterization of Benthic Marine Environments" (2011, Continental Shelf Research, Vol. 32, Issue 2, Edited by Andrew D. Heap and Peter T. Harris). This *Atlas*, the third major publication of GeoHab, represents the initiative and enthusiasm of GeoHab participants to contribute to the scientific literature being progressed by GeoHab as an organization.

This *Atlas* is a work in process. While not covering all geographical locations of the world's oceans, it represents a start to defining and illustrating marine benthic habitats that are critical to the management and sustainability of all marine resources, living and nonliving, on a worldwide basis. Although major gaps in geography occur offshore of Africa and Latin America, the major geomorphologic features such as submarine canyons, seamounts, rocky banks on continental shelves, and inland seas are well covered and represent the latest thinking on how these features as habitats should be mapped. There is no doubt in my mind that this work needs to continue, and hopefully GeoHab will support future efforts to do so.

A major goal in the production of this *Atlas* is to provide information useful to scientists endeavoring to map marine benthic habitats, to assist managers and policy makers concerned with the health and sustainability of ocean resources, and to educate the general populace on the state of the marine environment in regard to habitat uses and conditions.

With ever-increasing population, especially in regard to the migration and concentration of population along the coasts of continents and islands, and the demand for resources to sustain these populations, knowledge about the condition of the oceans' resources is increasing, but not keeping pace with the pressures to learn. Fortunately, as experienced in the past decades, technology has been advancing at a rate that can assist evaluation of these resources in a timely manner. To keep up to date with advancing mapping methodologies and imaging capabilities that present-day technologies provide, and in anticipation of the accelerated advancement in technology, publications such as this will be developed through GeoHab. The next decade promises to be as productive as the past, and hopefully those areas and unknown habitats not covered by this *Atlas* will be highlighted in a sequel.

I hope you find this *Atlas* useful and constructive. It represents a sincere dedication by GeoHab participants to further our knowledge on a major surface area of earth, the ocean floor. It illustrates how technology can be used to "map once, use many ways": a phrase coined by the late Roger Parson of NOAA/NOS and meant to say that many products (maps) can result from a single data set if effort is put forth to do so.

H. Gary Greene

Preface

This book started as an idea at the GeoHab meeting held in Noumea, New Caledonia, in May 2007. We noticed that multibeam bathymetry maps of geomorphic features, sometimes shown as 3D fly-through movies, followed by detailed sampling and photographic data (including underwater videos) illustrating the substrate conditions and associated biota, was a consistent theme of many papers. The presentations often also included a mathematical–statistical analysis that attempted to quantify the relationships between physical and biological variables. The idea of putting these studies together into a book grew during a discussion held at the subsequent GeoHab meeting held in Sitka, Alaska, in 2008; since so many maps are produced by GeoHab scientists, why not build our own "atlas" of different benthic habitats? An "e-book" was suggested, as this would allow for all maps and images to be published in color via online access; plus it would include the option of movies to illustrate the chapters. At the 2009 GeoHab meeting held in Trondheim, Norway, we sent out a call for papers to be submitted in time for the 2010 GeoHab meeting in Wellington, New Zealand. The response from the community was tremendous, with over 70 expressions of interest received for contributions.

The main content of this book comprises 57 "case studies" contributed by GeoHab authors from around the world. The case studies represent the "state of the art" in habitat-mapping science; they provide concise reports based on a template that all authors were asked to follow, with references included for readers to follow-up as needed. The book is divided into three parts:

1. Introduction
2. Case studies
3. Synthesis

The end-users of habitat-mapping science are drawn from a wide range of experiences and backgrounds. They are environmental managers, ecologists, fishermen, seabed mining and petroleum explorers, conservationists, and other marine scientists. Our aim is for the technical descriptions to be self-explanatory to the non-expert. The introductory Chapters 1–6 (Part 1) provide the reader with broad context and a basic understanding of the main concepts presented in the case studies; they might be thought of as an introduction to the science of benthic habitat mapping. A second purpose of the introductory and synthesis chapters (Parts 1 and 3) is to link general concepts and geomorphic features with cross-references to specific case studies. The Glossary is another important tool intended to assist our readers. It contains over 200 definitions of many technical terms used throughout the book, many of which were contributed by the case-study authors.

The 64 chapters comprising this book were all peer reviewed, which involved over 100 fellow scientists to whom we owe a considerable debt of gratitude. Quite simply, we could not have done it without them! We gratefully acknowledge the support and valuable contribution of peer reviewers listed below. A number of other people assisted us in preparing the final volume. In particular, we thank Jean-Nicolas Poussart (UNEP/GRID Arendal) and Tanya Whiteway (Geoscience Australia) for assistance with figures and maps.

We hope this book will provide a useful reference for students, scientists, managers, and industry specialists working in the field of habitat mapping, and that it will inspire further research and development of new methods and technologies, to explore, better understand, and appreciate the seafloor and the creatures that inhabit it.

Peter T. Harris
Elaine K. Baker

Peer Reviewers

Becky Allee
Franziska Althaus
Tara Anderson
Roy Armstrong
Vaughn Barrie
Nic Bax
Rob Beaman
Yannick Beaudoin
Brendan Brooke
Eleanor Bruce
Cleo Brylinsky
Mihai Burca
Dietmar Bürk
Alan Butler
Miguel Canals
Malcolm Clarke
Julia Clemons
Guy Cochrane
Roger Coggan
Jenny Collier
Antoine Collin
Tanya Compton
Alison Copeland
Emily Corcoran
Mark Costello
Silvana D'Angelo
Norbert Dankers
Pete Dartnell

Anna Maria De Biasi
Lies De Mol
Victor Díaz-del-Rio
Markus Diesing
Margaret Dolan
Federica Donda
Anna Downie
Isabelle du Four
Daniel Dunn
Piers Dunstan
Axel Ehrhold
Elena Ezhova
Andrea Fiorentino
Andre Freiwald
Ibon Galparsoro
Jonathan Gardner
Ann Gibbs
Miriam Sayago Gil
Michelle Greenlaw
Janine Guinan
Ralf Haese
Sarah Hamilton
Andrew Heap
Judi Hewitt
Will Heyman
Veerle Huvenne
Ceri James
Alan Jordan

Kathryn Julian
Anu Kaskela
Kaylene Keller
Ingi Klaucke
Shin Kobara
Vladimir Kostylev
Aarno Kotilainen
Tiina Kurvits
Geoffroy Lamarche
Yvonne Leahy
Claudio Lo Iacono
Vanessa Lucieer
Matthew McArthur
Andy Mackie
Mark Morrison
Lene Buhl-Mortensen
Scott Nichol
Scott Nodder
Phil O'Brien
Alan Orpin
Kim Picard
Alix Post
Lynda Radke
Jennifer Reynolds
Eli Rinde

Karen Robinson
Chris Roelfsema
Ashley Rowden
José Luis Rueda
Kathy Scanlon
Thierry Schmitt
Jody Smith
Samantha Smith
Alan Stevenson
Heather Stewart
Thomas Stieglitz
Fernando Tempera
Terje Thorsnes
Brian Todd
Page Valentine
Thaiënne A.G.P. van Dijk
Katrien Van Landeghem
Vera Vanlancker
Andy Wheeler
Alan Williams
Colin Woodroffe
Dawn Wright
Richard Wysoczanski
Mary Yoklavich
Vincent Zintzen

Contributors

Juan Acosta Instituto Español de OceanografíaC/Corazon de Maria, Madrid, Spain.

Rebecca J. Allee National Oceanic and Atmospheric Administration, Gulf Coast Services Center, Stennis Space Center, MS, USA.

Franziska Althaus CSIRO Wealth from Oceans Flagship, Hobart Marine Laboratories, Hobart, Tasmania, Australia.

German Alvarez GRC Geociències Marines, Parc Científic de Barcelona, Departament d'Estratigrafia, Paleontologia i Geociències Marines, Facultat de Geologia, Universitat de Barcelona, Campus de Pedralbes, Barcelona, Spain.

David Amblas GRC Geociències Marines, Parc Científic de Barcelona, Departament d'Estratigrafia, Paleontologia i Geociències Marines, Facultat de Geologia, Universitat de Barcelona, Campus de Pedralbes, Barcelona, Spain.

Tara J. Anderson Marine and Coastal Environment Group, Geoscience Australia, Canberra, ACT, Australia.

Philippe Archambault Institute of Marine Sciences, University du Québec à Rimouski, QC, Canada.

Roy A. Armstrong Department of Marine Sciences, University of Puerto Rico, Mayaguez, Puerto Rico.

Saara Bäck Ministry of Environment, Government, Finland.

Elaine K. Baker UNEP/GRID- Arendal School of Geosciences, University of Sydney, Australia.

Martin Baptist Institute for Marine Resources and Ecosystem Studies (IMARES-Wageningen UR), Department of Ecosystems, Texel, The Netherlands.

Neville Barrett Institute for Marine and Antarctic Studies, University of Tasmania, Hobart, Tasmania, Australia.

J. Vaughn Barrie Geological Survey of Canada, Pacific, Institute of Ocean Sciences, Sydney, British Columbia, Canada.

Rafael Bartolomé Unidad de Tecnología Marina, Consejo Superior de Investigaciones Científicas (UTM-CSIC), Barcelona, Spain.

Igor Bashmachnikov Departamento de Oceanografia e Pescas, Universidade dos Açores, Rua Prof. Dr. Frederico Machado, 4, Horta, Açores, Portugal; Instituto de Oceanografia, Faculdade de Ciências da Universidade de Lisboa, Campo Grande, Lisboa, Portugal.

Richard Bates School of Geography and Geosciences, University of St. Andrews, St. Andrews, Fife, Scotland, UK.

Chris Battershill Australian Institute of Marine Science, Townsville, QLD, Australia.

Nicholas J. Bax CSIRO Wealth from Oceans Flagship, Hobart Marine Laboratories, Hobart, Tasmania, Australia.

Robin J. Beaman School of Earth and Environmental Sciences, James Cook University, Cairns, QLD, Australia.

Yannick C. Beaudoin UNEP/GRID- Arendal School of Geosciences, University of Sydney, Australia.

Trevor Bell Geography Department, Memorial University, St. John's, NL, Canada.

Reidulv Bøe Geological Survey of Norway, Trondheim, Norway.

Ángel Borja AZTI-Tecnalia, Marine Research Division, Herrera Kaia, Portualdea s/n, Pasaia, Spain.

David A. Bowden National Institute of Water and Atmospheric Research (NIWA) Ltd, Kilbirnie, Wellington, New Zealand.

Andreia Braga Henriques Departamento de Oceanografia e Pescas, Universidade dos Açores, Rua Prof. Dr. Frederico Machado, 4, Horta, Açores, Portugal.

Thomas Bridge School of Earth and Environmental Sciences, James Cook University, Townsville, QLD, Australia.

Brendan P. Brooke Marine and Coastal Environment Group, Geoscience Australia, Canberra, ACT, Australia.

Lene Buhl-Mortensen Institute of Marine Research, Nordnes, Bergen, Norway.

Pål Buhl-Mortensen Institute of Marine Research, Nordnes, Bergen, Norway.

Pere Busquets GRC Geociències Marines, Parc Científic de Barcelona, Departament d'Estratigrafia, Paleontologia i Geociències Marines, Facultat de Geologia, Universitat de Barcelona, Campus de Pedralbes, Barcelona, Spain.

Antonio Calafat GRC Geociències Marines, Parc Científic de Barcelona, Departament d'Estratigrafia, Paleontologia i Geociències Marines, Facultat de Geologia, Universitat de Barcelona, Campus de Pedralbes, Barcelona, Spain.

Aldino S. Campos Portuguese Task Group for the Extension of the Continental Shelf (EMEPC), Rua Costa Pinto, Paço de Arcos, Portugal.

Miquel Canals GRC Geociències Marines, Parc Científic de Barcelona, Departament d'Estratigrafia, Paleontologia i Geociències Marines, Facultat de Geologia, Universitat de Barcelona, Campus de Pedralbes, Barcelona, Spain.

Diana Catarino Departamento de Oceanografia e Pescas, Universidade dos Açores, Rua Prof. Dr. Frederico Machado, 4, Horta, Açores, Portugal.

J.W. Ceri James British Geological Survey, Keyworth, Nottingham, England, UK; School of Ocean Sciences, Bangor University, Menai Bridge, Wales, UK.

Francesco L. Chiocci Department of Earth Sciences, University of Rome "La Sapienza", Roma, Italy.

Malcolm R. Clark National Institute of Water and Atmospheric Research (NIWA), Wellington, New Zealand.

Susan A. Cochran US Geological Survey, Santa Cruz, CA, USA.

Guy R. Cochrane USGS Pacific Coastal and Marine Science Center, Santa Cruz, CA, USA.

Roger A. Coggan Centre for Environment, Fisheries and Aquaculture Science (Cefas), Lowestoft, UK.

Enrique Coiras NATO Undersea Research Centre (NURC), La Spezia, Italy.

Ana Colaço Departamento de Oceanografia e Pescas, Universidade dos Açores, Rua Prof. Dr. Frederico Machado, 4, Horta, Açores, Portugal.

Jenny S. Collier Department of Earth Science and Engineering, Imperial College London, London, UK.

Antoine Collin Insular Research Center and Environment Observatory, Papetoai, French Polynesia.

Kim W. Conway Pacific Geoscience Centre, Sydney, BC, Canada.

Alison Copeland Geography Department, Memorial University, St. John's, NL, Canada.

Jenny Cremer Institute for Marine Resources and Ecosystem Studies (IMARES-Wageningen UR), Department of Ecosystems, Texel, The Netherlands.

Silvana D'Angelo Department for Soil Defense, Geological Survey of Italy-ISPRA via Curtatone, Rome, Italy.

Norbert Dankers Institute for Marine Resources and Ecosystem Studies (IMARES-Wageningen UR), Department of Ecosystems, Texel, The Netherlands.

Teresa Darbyshire Amgueddfa Cymru—National Museum Wales, Cathays Park, Cardiff, Wales, UK.

Andrew W. David National Oceanic and Atmospheric Administration, National Marine Fisheries Service, Panama City, FL, USA.

Steven Degraer Management Unit of the North Sea Mathematical Models, Royal Belgian Institute of Natural Sciences, Brussels; Marine Biology Section, Ghent University, Gent, Belgium.

Ben De Mol GRC Geociències Marines, Parc Científic de Barcelona, Departament d'Estratigrafia, Paleontologia i Geociències Marines, Facultat de Geologia, Universitat de Barcelona, Campus de Pedralbes, Barcelona, Spain.

Lies De Mol Renard Centre of Marine Geology (RCMG), Department of Geology and Soil Science, Ghent University, Gent, Belgium.

Laura De Santis Instituto Nazionale di Oceanografia e Geofisica Sperimentale, Sgnoico, Trieste, Italy.

Rodolphe Devillers Geography Department, Memorial University, St. John's, NL, Canada.

Víctor Díaz-del-Río Instituto Español de Oceanografía, Centro Oceanográfico de Málaga, Puerto Pesquero s/n, Fuengirola, Málaga, Spain.

Markus Diesing Centre for Environment, Fisheries and Aquaculture Science (Cefas), Lowestoft, UK.

Elze Dijkman Institute for Marine Resources and Ecosystem Studies (IMARES-Wageningen UR), Department of Ecosystems, Texel, The Netherlands.

Margaret F.J. Dolan Geological Survey of Norway, Trondheim, Norway.

Federica Donda Istituto Nazionale di Oceanografia e di Geofisica Sperimentale–OGS Trieste, Italy.

Terry Done Australian Institute of Marine Sciences, Townsville, QLD, Australia.

Pieter J. Doornenbal Deltares, Department of Marine and Coastal Systems, Delft, The Netherlands.

Dmitry Dorokhov P.P. Shirshov Institute of Oceanology, Atlantic Branch (ABIORAS), Kaliningrad, Russia.

Dayton Dove British Geological Survey, Edinburgh, UK.

Isabelle Du Four Renard Centre of Marine Geology, Ghent University, Gent, Belgium; International Polar Foundation, Brussels, Belgium.

Ruth Duran GRC Geociències Marines, Parc Científic de Barcelona, Departament d'Estratigrafia, Paleontologia i Geociències Marines, Facultat de Geologia, Universitat de Barcelona, Campus de Pedralbes, Barcelona, Spain.

Pablo Durán-Muñoz Instituto Español de Oceanografía, Centro Oceanográfico de Vigo, Subida al RadiofaroVigo, Pontevedra, Spain.

Evan Edinger Geography Department, Memorial University, St. John's, NL, Canada; Biology Department, Memorial University, St. John's, NL, Canada.

Sigrid Elvenes Geological Survey of Norway, Trondheim, Norway.

Lisa Etherington Cordell Bank National Marine Sanctuary, Olema, CA, USA.

Elena Ezhova P.P. Shirshov Institute of Oceanology, Atlantic Branch (ABIORAS), Kaliningrad, Russia.

Annalisa Falace Dipartimento di Scienze della Vita, Università degli Studi di Trieste, Italy.

Douglas Fenner Department of Marine and Wildlife Resources, American Samoa Government, Pago Pago, AS, USA.

Luis M. Fernández-Salas Instituto Español de Oceanografía, Centro Oceanográfico de Málaga, Puerto Pesquero s/n, Fuengirola (Málaga), Spain.

Andrea Fiorentino Department for Soil Defense, Geological Survey of Italy-ISPRA via Curtatone, Rome, Italy.

Robert Flemming Pacific Biological Station, Nanaimo, BC, Canada.

Thomas Furey Marine Institute, Renville, Oranmore, Co. Galway, Ireland.

Ibon Galparsoro AZTI-Tecnalia, Marine Research Division, Herrera Kaia, Portualdea s/n, Pasaia, Spain.

H. Gary Greene Tombolo Habitat Institute, Eastsound, WA, USA.

J. Germán Rodríguez AZTI-Tecnalia, Marine Research Division, Herrera Kaia, Portualdea s/n, Pasaia, Spain.

Julia E.R. Getsiv-Clemons NOAA Fisheries, Fishery Resource Analysis and Monitoring Division, Northwest Fisheries Science Center, Newport, OR, USA.

Eva Giacomello Departamento de Oceanografia e Pescas, Universidade dos Açores, Rua Prof. Dr. Frederico Machado, 4, Horta, Açores, Portugal.

Ann E. Gibbs US Geological Survey, Santa Cruz, CA, USA.

Josep Maria Gili Instituto de Ciencias del Mar, Consejo Superior de Investigaciones Científicas (ICM-CSIC), Barcelona, Spain.

João Gonçalves Departamento de Oceanografia e Pescas, Universidade dos Açores, Rua Prof. Dr. Frederico Machado, 4, Horta, Açores, Portugal.

Emiliano Gordini Istituto Nazionale di Oceanografia e di Geofisica Sperimentale– OGS Trieste, Italy.

Andrea Gori Instituto de Ciencias del Mar, Consejo Superior de Investigaciones Científicas (ICM-CSIC), Barcelona, Spain.

Eulàlia Gràcia Unidad de Tecnología Marina, Consejo Superior de Investigaciones Científicas (UTM-CSIC), Barcelona, Spain.

Janine Guinan INFOMAR Integrated Mapping for the Sustainable Development of Ireland's Marine Resource, Marine and Geophysics Programme, Geological Survey of Ireland, Beggars Bush, Haddington Road, Dublin 4, Ireland.

Annelise B. Hagan Cambridge Coastal Research Unit, Department of Geography, University of Cambridge, Cambridge, UK.

Sarah Hamylton Cambridge Coastal Research Unit, Department of Geography, University of Cambridge, Cambridge, UK.

Jodi Harney ENTRIX, Riverview, FL, USA.

Peter T. Harris Marine and Coastal Environment Group, Geoscience Australia, Canberra, ACT, Australia.

Andrew D. Heap Marine and Coastal Environment Group, Geoscience Australia, Canberra, ACT, Australia.

Jonathan Heifetz National Oceanic and Atmospheric Administration, National Marine Fisheries Service, Alaska Fisheries Science Center, Auke Bay Laboratories, Juneau, Alaska.

Jean-Pierre Henriet Renard Centre of Marine Geology (RCMG), Department of Geology and Soil Science, Ghent University, Gent, Belgium.

William D. Heyman Department of Geography, Texas A&M University, College Station, TX, USA.

Ana Hilário CESAM and Departamento de Biologia, Campus Universitário de Santiago, Aveiro, Portugal.

Nicole Hill Institute for Marine and Antarctic Studies, University of Tasmania, Hobart, Tasmania, Australia.

Emily R. Hirsch Geospatial Consulting Group International, Alexandria, VA, USA.

Hanne Hodnesdal Norwegian Hydrographic Service, Stavanger, Norway.

Kyle R. Hogrefe US Geological Survey, Alaska Science Center, Anchorage, AK, USA.

Stuart R. Humber Department of Earth Science and Engineering, Imperial College London, London, UK.

Veerle A.I. Huvenne National Oceanography Centre, Southampton, UK.

Eduardo J. Isidro Departamento de Oceanografia e Pescas, Universidade dos Açores, Rua Prof. Dr. Frederico Machado, 4, Horta, Açores, Portugal.

Glenn Johnstone Australian Antarctic Division, Channel Highway, Kingston, TAS, Australia.

Juan Jose Dañobeitia Unidad de Tecnología Marina, Consejo Superior de Investigaciones Científicas (UTM-CSIC), Barcelona, Spain.

Sara Kaleb Dipartimento di Scienze della Vita, Università degli Studi di Trieste, Italy.

Anu M. Kaskela Geological Survey of Finland (GTK), Espoo, Finland.

Rudy J. Kloser CSIRO Wealth from Oceans Flagship, Hobart Marine Laboratories, Hobart, Tasmania, Australia.

Shinichi Kobara Department of Oceanography, Texas A&M University, College Station, TX, USA.

Olga Kocheshkova P.P. Shirshov Institute of Oceanology, Atlantic Branch (ABIORAS), Kaliningrad, Russia.

Anthony A.P. Koppers College of Oceanic and Atmospheric Sciences, Oregon State University, Corvallis, OR, USA.

Vladimir E. Kostylev Geological Survey of Canada, Dartmouth, NS, Canada.

Aarno T. Kotilainen Geological Survey of Finland (GTK), Espoo, Finland.

Geoffroy Lamarche National Institute of Water and Atmospheric Research, Wellington, New Zealand.

Caroline Lavoie GRC Geociències Marines, Parc Cientific de Barcelona, Departament d'Estratigrafia, Paleontologia i Geociències Marines, Facultat de Geologia, Universitat de Barcelona, Campus de Pedralbes, Barcelona, Spain.

Yvonne Leahy Department of Arts, Heritage and the Gaeltacht, National Parks and Wildlife Service, Custom House, Flood Street, Galway, Ireland.

Philippe LeBlanc Geography Department, Memorial University, St. John's, NL, Canada.

Irati Legorburu AZTI-Tecnalia, Marine Research Division, Herrera Kaia, Portualdea s/n, Pasaia, Spain.

Jouni Leinikki Alleco Oy, Mekaanikonkatu, Helsinki, Finland.

Charles Lindenbaum Countryside Council for Wales, Maes y Ffynnon, Ffordd Penrhos, Bangor, Gwynedd, Wales, UK.

Michelle Linklater School of Earth and Environmental Science, University of Wollongong, NSW, Australia.

Claudio Lo Iacono Unidad de Tecnología Marina, Consejo Superior de Investigaciones Científicas (UTM-CSIC), Barcelona, Spain.

Bernard Long Department of Geosciences, University du Québec, QC, Canada.

Nieves López-González Instituto Español de Oceanografía, Centro Oceanográfico de Málaga, Puerto Pesquero s/n, Fuengirola (Málaga), Spain.

Vanessa Lucieer Institute for Marine and Antarctic Studies, University of Tasmania, Hobart, Tasmania, Australia.

Matthew A. McArthur Marine and Coastal Environment Group, Geoscience Australia, Canberra, ACT, Australia.

Kevin Mackay National Institute of Water and Atmospheric Research (NIWA) Ltd, Kilbirnie, Wellington, New Zealand.

Andrew S.Y. Mackie Amgueddfa Cymru—National Museum Wales, Cathays Park, Cardiff, Wales, UK.

Ruggero Marocco Dipartimento di Geoscienze, Università degli Studi di Trieste, Italy.

Ana Martins Departamento de Oceanografia e Pescas, Universidade dos Açores, Rua Prof. Dr. Frederico Machado, 4, Horta, Açores, Portugal.

Eleonora Martorelli Department of Earth Sciences, University of Rome "La Sapienza", Roma, Italy.

Douglas G. Masson National Oceanography Centre, Southampton, UK.

Monique MacKenzie Centre for Research into Ecological and Environmental Modeling (CREEM), University of St. Andrews, The Observatory, St. Andrews, Fife, Scotland, UK.

Ana Mendonça Departamento de Oceanografia e Pescas, Universidade dos Açores, Rua Prof. Dr. Frederico Machado, 4, Horta, Açores, Portugal.

Gui Menezes Departamento de Oceanografia e Pescas, Universidade dos Açores, Rua Prof. Dr. Frederico Machado, 4, Horta, Açores, Portugal.

L. Miguel Fernández-Salas Instituto Español de Oceanografía, Centro Oceanográfico de Málaga, Puerto Pesquero s/n, Fuengirola, Málaga, Spain.

Neil C. Mitchell School of Earth, Atmospheric and Environmental Sciences, University of Manchester, Manchester, UK.

Richard Mleczko Marine and Coastal Environment Group, Geoscience Australia, Canberra, ACT, Australia.

Geert Moerkerke Renard Centre of Marine Geology, Ghent University, Gent, Belgium; G-Tec NV, Deinze, Belgium.

Angela Morando British Geological Survey, Keyworth, Nottingham, UK.

Telmo Morato Departamento de Oceanografia e Pescas, Universidade dos Açores, Rua Prof. Dr. Frederico Machado, 4, Horta, Açores, Portugal.

Joshu Mountjoy National Institute of Water and Atmospheric Research, Wellington, New Zealand.

Araceli Muñoz Tragsa-SGM C/Nuñez de balboa, Madrid, Spain.

F. Javier Murillo Instituto Español de Oceanografía, Centro Oceanográfico de Vigo, Subida al RadiofaroVigo, Pontevedra, Spain.

Iñigo Muxika AZTI-Tecnalia, Marine Research Division, Herrera Kaia, Portualdea s/n, Pasaia, Spain.

David F. Naar College of Marine Science, University of South Florida, St. Petersburg, FL, USA.

Scott L. Nichol Marine and Coastal Environment Group, Geoscience Australia, Canberra, ACT, Australia.

Scott D. Nodder National Institute of Water and Atmospheric Research (NIWA) Ltd, Kilbirnie, Wellington, New Zealand.

Brenda L. Norcross University of Alaska Fairbanks, School of Fisheries and Ocean Sciences, Fairbanks, Alaska.

Philip E. O'Brien Geoscience Australia, Canberra, ACT, Australia.

Covadonga Orejas Instituto Español de Oceanografía (IEO), Santander, Spain.

Arne Pallentin National Institute of Water and Atmospheric Research (NIWA) Ltd, Kilbirnie, Wellington, New Zealand.

Marta Pascual AZTI-Tecnalia, Marine Research Division, Herrera Kaia, Portualdea s/n, Pasaia, Spain.

Abigail D.C. Pattenden National Oceanography Centre, Southampton, UK.

Bryony Pearce Marine Ecological Surveys Ltd, Bath, UK.

Kim Picard Geological Survey of Canada, Pacific, Institute of Ocean Sciences, Sydney, British Columbia, Canada; Pacific Geoscience Centre, Sydney, BC, Canada.

Mário R. Pinho Departamento de Oceanografia e Pescas, Universidade dos Açores, Rua Prof. Dr. Frederico Machado, 4, Horta, Açores, Portugal.

Jennifer Pinnion Marine Ecological Surveys Limited, Bath, UK.

Oscar Pizarro Australian Centre for Field Robotics, the University of Sydney, Sydney, NSW, Australia.

Filipe M. Porteiro Departamento de Oceanografia e Pescas, Universidade dos Açores, Rua Prof. Dr. Frederico Machado, 4, Horta, Açores, Portugal.

Alexandra L. Post Marine and Coastal Environment Group, Geoscience Australia, Canberra, ACT, Australia.

Pere Puig Instituto de Ciencias del Mar, Consejo Superior de Investigaciones Científicas (ICM-CSIC), Barcelona, Spain.

Marji Puotinen School of Earth and Environmental Sciences, University of Wollongong, Wollongong, NSW, Australia.

Marijn Rabaut Marine Biology Section, Ghent University, Gent, Belgium.

E. Ivor S. Rees School of Ocean Sciences, Bangor University, Menai Bridge, Wales, UK.

Susana Requena Instituto de Ciencias del Mar, Consejo Superior de Investigaciones Científicas (ICM-CSIC), Barcelona, Spain.

Jennifer R. Reynolds University of Alaska Fairbanks, School of Fisheries and Ocean Sciences, Fairbanks, Alaska.

Marta Ribó Instituto de Ciencias del Mar, Consejo Superior de Investigaciones Científicas (ICM-CSIC), Barcelona, Spain.

Martin J. Riddle Environmental Protection and Change, Australian Antarctic Division, Channel Highway, Kingston, TAS, Australia.

Stephen R. Rintoul CSIRO Marine and Atmospheric Research, Hobart, TAS, Australia.

Jesus Rivera Instituto Español de Oceanografía, C/ Corazon de Maria, Madrid, Spain.

Jed T. Roberts Oregon Department of Geology and Mineral Industries, Portland, OR, USA.

Karen A. Robinson Countryside Council for Wales, Maes y Ffynnon, Ffordd Penrhos, Bangor, Gwynedd, Wales, UK.

Sean C. Rooney University of Alaska Fairbanks, School of Fisheries and Ocean Sciences, Fairbanks, Alaska.

Ashley A. Rowden National Institute of Water and Atmospheric Research, Wellington, New Zealand.

José L. Rueda Instituto Español de Oceanografía, Centro Oceanográfico de Málaga, Puerto Pesquero s/n, Fuengirola (Málaga), Spain.

Daria Ryabchuk A.P. Karpinsky Russian Geological Research Institute (VSEGEI), St. Petersburg, Russia.

Stephen Sagar Marine and Coastal Environment Group, Geoscience Australia, Canberra, ACT, Australia.

William G. Sanderson Countryside Council for Wales, Maes y Ffynnon, Ffordd Penrhos, Bangor, Gwynedd, Wales, UK.

Ricardo S. Santos Departamento de Oceanografia e Pescas, Universidade dos Açores, Rua Prof. Dr. Frederico Machado, 4, Horta, Açores, Portugal.

Miriam Sayago-Gil Instituto Español de Oceanografía, Centro Oceanográfico de Málaga, Puerto Pesquero s/n, Fuengirola (Málaga), Spain.

Jan Seiler CSIRO Wealth from Oceans Flagship, Hobart Marine Laboratories, Hobart, Tasmania, Australia; University of Tasmania, Tasmania, Australia.

Alberto Serrano Instituto Español de Oceanografía, Centro Oceanográfico de Santander, Promontorio San Martín, Santander, Spain.

S. Kalei Shotwell National Oceanic and Atmospheric Administration, National Marine Fisheries Service, Alaska Fisheries Science Center, Auke Bay Laboratories, Juneau, Alaska.

Jodie Smith Geoscience Australia, Canberra, ACT, Australia.

John R. Smith Hawaii Undersea Research Laboratory, University of Hawaii, Honolulu, HI, USA.

Samantha Smith Nautilus Minerals Inc., Milton, QLD 4064, Australia.

Stephen J. Smith Fisheries and Oceans Canada, Dartmouth, NS, Canada.

Tom Spencer Cambridge Coastal Research Unit, Department of Geography, University of Cambridge, Cambridge, UK.

Hanumant Singh Woods Hole Oceanographic Institution, Woods Hole, MA, USA.

Vadim Sivkov P.P. Shirshov Institute of Oceanology, Atlantic Branch (ABIORAS), Kaliningrad, Russia.

Jonathan S. Stark Australian Antarctic Division, Channel Highway, Kingston, TAS, Australia.

Ian J. Stewart NOAA Fisheries, Fishery Resource Analysis and Monitoring Division, Northwest Fisheries Science Center, Seattle, Washington, USA.

Thomas C. Stieglitz AIMS@JCU, Townsville, QLD, Australia; School of Engineering and Physical Sciences, James Cook University, Townsville, QLD, Australia; Australian Institute of Marine Science, Townsville, QLD, Australia.

David R. Tappin British Geological Survey, Keyworth, Nottingham, UK.

Fernando Tempera Departamento de Oceanografia e Pescas, Universidade dos Açores, Rua Prof. Dr. Frederico Machado, 4, Horta, Açores, Portugal; School of Geography and Geosciences, University of St. Andrews, St. Andrews, Fife, Scotland, UK.

Terje Thorsnes Geological Survey of Norway, Trondheim, Norway.

Brian J. Todd Geological Survey of Canada, Dartmouth, NS, Canada.

Luke Trusel Clark University Graduate School of Geography, Worcester, MA, USA.

Giorgio Tunis Dipartimento di Geoscienze, Università degli Studi di Trieste, Italy.

Paul A. Tyler School of Ocean and Earth Science, University of Southampton, European Way, Southampton, UK.

Page C. Valentine US Geological Survey, Woods Hole, MA, USA.

Jan A. van Dalfsen Deltares, Department of Marine and Coastal Systems, Delft, The Netherlands.

Thaiënne A.G.P. van Dijk Deltares, Department of Applied Geology and Geophysics, Utrecht, The Netherlands; Department of Water Engineering and Management, University of Twente, Enschede, The Netherlands.

Willem van Duin Institute for Marine Resources and Ecosystem Studies (IMARES-Wageningen UR), Department of Ecosystems, Texel, The Netherlands.

Sytze van Heteren TNO Built Environment and Geosciences, Department of Geo-Modelling, Geological Survey of the Netherlands, Utrecht, The Netherlands.

Vera Van Lancker Management Unit of the North Sea Mathematical Models, Gulledelle 100, Brussels, Belgium; Management Unit of the North Sea Mathematical Models, Royal Belgian Institute of Natural Sciences, Brussels, Belgium; Renard Centre of Marine Geology, Ghent University, Gent, Belgium.

Katrien J.J. van Landeghem School of Ocean Sciences, Bangor University, Menai Bridge, Anglesey, Wales, UK.

Ronnie A. van Overmeeren TNO Built Environment and Geosciences, Department of Geo-Modelling, Geological Survey of the Netherlands, Utrecht, The Netherlands.

David Van Rooij Renard Centre of Marine Geology (RCMG), Department of Geology and Soil Science, Ghent University, Gent, Belgium.

Juan T. Vázquez Instituto Español de Oceanografía, Centro Oceanográfico de Málaga, Puerto Pesquero s/n, Fuengirola (Málaga), Spain.

Koen Verbruggen INFOMAR Integrated Mapping for the Sustainable Development of Ireland's Marine Resource, Marine and Geophysics Programme, Geological Survey of Ireland, Beggars Bush, Haddington Road, Dublin 4, Ireland.

Anne-Laure Verdier National Institute of Water and Atmospheric Research, Wellington, New Zealand.

Els Verfaillie Renard Centre of Marine Geology, Ghent University, Gent, Belgium; Department of Geography, Carto-GIS cluster, Ghent University, Gent, Belgium.

W. Waldo Wakefield NOAA Fisheries, Fishery Resource Analysis and Monitoring Division, Northwest Fisheries Science Center, Newport, OR, USA.

Jody M. Webster School of Geosciences, the University of Sydney, Sydney, NSW, Australia.

Leslie Whaylen Clift Coastal Marine Resource Associates, Honolulu, HI, USA.

Curt E. Whitmire NOAA Fisheries, Fishery Resource Analysis and Monitoring Division, Northwest Fisheries Science Center, Newport, OR, USA.

Alan Williams CSIRO Wealth from Oceans Flagship, Hobart Marine Laboratories, Hobart, Tasmania, Australia.

Stefan Williams Australian Centre for Field Robotics, the University of Sydney, Sydney, NSW, Australia.

Colin D. Woodroffe School of Earth and Environmental Science, University of Wollongong, NSW, Australia.

Dawn J. Wright Department of Geosciences, Oregon State University, Corvallis, OR, USA.

Joseph Wroblewski Ocean Sciences Centre, Memorial University, St. John's, NL, Canada.

Richard J. Wysoczanski National Institute of Water and Atmospheric Research (NIWA), Wellington, New Zealand.

K. Lynne Yamanaka Pacific Biological Station, Nanaimo, BC, Canada.

Mary Yoklavich National Marine Fisheries Service, Southwest Fisheries Science Center, Santa Cruz, CA, USA.

Vladimir Zhamoida A.P. Karpinsky Russian Geological Research Institute (VSEGEI), St. Petersburg, Russia.

Part I

Introduction

1 Why Map Benthic Habitats?

Peter T. Harris[1], Elaine K. Baker[2]

[1]Marine and Coastal Environment Group, Geoscience Australia, Canberra, ACT, Australia, [2]UNEP/GRID- Arendal School of Geosciences, University of Sydney, Australia

Habitat is the property that inherently integrates many ecosystem features, including higher and lower trophic level species, water quality, oceanographic conditions and many types of anthropogenic pressures. Thus, strengthening assessments of status and trends in habitat quality and extent will be an important priority in the development of a global marine assessment.

(Assessment of Assessments Report, UNEP and IOC-UNESCO [1])

Abstract

This introductory chapter provides an overview of this book's contents and definitions of key concepts, including benthic habitat, potential habitat, and seafloor geomorphology. It concludes with a summary of commonly used habitat mapping technologies. Benthic (seafloor) habitats are physically distinct areas of seabed that are associated with particular species, communities, or assemblages that consistently occur together. Benthic *habitat maps* are spatial representations of physically distinct areas of seabed that are associated with particular groups of plants and animals. Habitat maps can illustrate the nature, distribution, and extent of distinct physical environments and, importantly, they can predict the distribution of the associated species and communities.

The data sets collected for constructing habitat maps provide fundamental information that can be used for a range of management and industry applications, including the management of fisheries, spatial marine environmental management, design of marine reserves, supporting offshore oil and gas infrastructure development, port and shipping channel construction, maintenance dredging, tourism, and seabed aggregate mining. Seafloor habitat mapping provides fundamental baseline information for decision makers working in these sectors.

GeoHab (www.geohab.org) is an international association of marine scientists conducting research using a range of mapping technologies in the study of biophysical (i.e., geologic and oceanographic) indicators of benthic habitats and ecosystems as proxies for biological communities and species diversity. Using this approach, combinations of physical attributes of the seabed identify habitats that have been demonstrated to be effective surrogates for the benthic communities that they typically support. Thus, management priorities can be identified using seabed habitat maps as a guide. The work of GeoHab demonstrates how knowledge of seabed properties can be employed to guide marine environmental management, marine resource management, and conservation efforts.

Seafloor Geomorphology as Benthic Habitat. DOI: 10.1016/B978-0-12-385140-6.00001-3

Seafloor geomorphology is one of the most useful physical attributes of the seabed mapped and measured by GeoHab scientists. Different geomorphic features (e.g., submarine canyons, seamounts, atolls, and fjords) are commonly associated with particular suites of habitats. Knowledge of the geomorphology and biogeography of the seafloor has improved markedly over the past 10 years. Using multibeam sonar, submarine features such as fjords, sand banks, coral reefs, seamounts, canyons, and spreading ridges have been revealed in unprecedented detail. The 57 case studies presented in this book represent a range of seabed geomorphic features where detailed bathymetric maps have been combined with seabed video and sampling to yield an integrated picture of the benthic communities that are associated with different types of benthic habitat.

Key Words: Benthic habitats; geomorphic features; physical surrogates; biodiversity; spatial marine planning; environmental management; habitat mapping technology

General Outline of the Content of This Book

This book provides a synthesis of seabed geomorphology and benthic habitats based on up-to-date information contained in the case studies. Part 1 of the book provides an introduction in which the drivers that underpin the need for benthic habitat maps are examined, including threats to benthic habitats. The habitat mapping approach and classification schemes, based on principles of biogeography and benthic ecology, are reviewed, and the use of biophysical surrogates for habitats and benthic biodiversity are surveyed. Part 1 ends with a brief summary of seafloor geomorphology and geomorphic features that are the subject of the case studies.

Part 2 of the book includes 57 separate case studies representing a diverse range of geomorphic features and their associated habitats, from the coast to the abyss (Figure 1.1). The case studies are cross-referenced throughout Part 1 to provide the reader with a broad overview and context for the detailed information they contain.

Spatial mapping is one of the most important tools used by GeoHab scientists to convey information and demonstrate relationships among different variables. The content of the case studies, combined with the review of information provided in Part 1, forms an atlas comprised of a collection of maps that represent a range of different geomorphic features and habitats.

To be accepted, case studies had to conform to a template. They are required to contain both geomorphic and biologic data, provide a clear description of at least one geomorphic feature type, describe the oceanographic setting, and provide an assessment of the naturalness of the environment. The spatial comparison of biological data with spatial physical data is a key element of every case study. Authors were given the opportunity to describe surrogacy relationships and methods used to identify and quantify them.

Part 3 attempts to synthesize the content of the case studies and is partly based on responses to a questionnaire that was completed by the authors; responses to the questionnaire are also considered in the introductory chapters (Part 1). Part 3 includes headings such as attributes of the case study areas (depth range, naturalness, and geomorphic feature types), surrogates and classification systems used, the

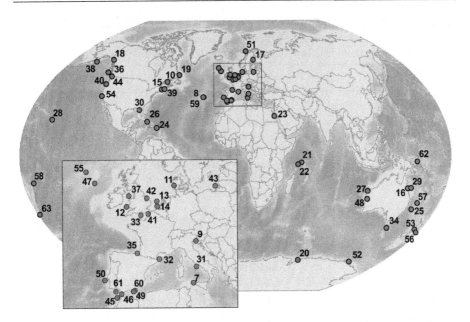

Figure 1.1 Distribution of 57 case studies presented in Part 2 of this book. Chapter numbers are indicated for each case study.

socioeconomic aspects underpinning habitat mapping (main clients for habitat maps and funding sources), gap analysis (i.e., geographic areas, geomorphic features, and environmental variables not included in the case studies), and finally what constitutes best practices for habitat mapping.

The book's Glossary provides definitions of key terms used in habitat mapping, description, and classification.

What Are the Main Purposes of Habitat Mapping?

When asked this question, the case study authors nominated a number of purposes for mapping benthic habitats (Table 1.1). Among these, four are preeminent: (1) to support government spatial marine planning, management, and decision making; (2) to support and underpin the design of marine-protected areas (MPAs); (3) to conduct scientific research programs aimed at generating knowledge of benthic ecosystems and seafloor geology; and (4) to conduct living and nonliving seabed resource assessments for economic and management purposes, including the design of fishing reserves.

An important point is that many authors nominated more than one purpose for their case study (Table 1.1). This highlights another particular benefit of habitat mapping: the data collected to manage one sector can be applied to others, since most of the information required about habitats is essentially the same for all applications.

Table 1.1 List of Different Purposes of Habitat Mapping in Relation to Case Studies Included in This Book (Purpose Nominated by Case Study Authors). Many Authors Nominated More than One Purpose for Their Study. See Figure 1.1 for Location of Case Studies

Purpose of Study	Case Study Reference	Chapter Number
Spatial marine planning and management (*n* = 19)	Althaus et al. [2]	48
	Barrie et al. [3]	44
	Beaman et al. [4]	29
	Cochrane et al. [5]	18
	Coggan and Diesing [6]	33
	Collier and Humber [7]	21
	Copeland et al. [8]	19
	Dankers et al. [9]	11
	Ezhova et al. [10]	43
	Hamylton [11]	23
	Hamylton et al. [12]	22
	James et al. [13]	12
	James et al. [14]	41
	Lamarche et al. [15]	53
	O'Brien et al. [16]	20
	Post et al. [17]	52
	Sayago-Gil et al. [18]	55
	Tempera et al. [19]	8
	van Dijk et al. [20]	13
MPA design (*n* = 17)	Allee et al. [21]	30
	Althaus et al. [2]	48
	Brooke et al. [22]	25
	Buhl-Mortensen et al. [23]	51
	Dankers et al. [9]	11
	Galparsoro et al. [24]	35
	Gordini et al. [25]	9
	Guinan et al. [26]	47
	Harris et al. [27]	57
	James et al. [14]	41
	Lo Iacono et al. [28]	49
	Lo Iacono et al. [29]	32
	Nichol et al. [30]	27
	Rueda et al. [31]	61
	Tempera et al. [32]	59
	Tempera et al. [19]	8
	Van Lancker et al. [33]	14
Scientific research and knowledge (*n* = 9)	Brooke et al. [22]	25
	De Mol et al. [34]	45
	De Mol et al. [35]	60
	Huvenne et al. [36]	50
	Lucieer et al [37]	34

(*Continued*)

Table 1.1 (Continued)

Purpose of Study	Case Study Reference	Chapter Number
	Nichol et al. [30]	27
	Robinson et al. [38]	37
	Stieglitz [39]	16
	Wright et al. [40]	58
Assessment of nonliving resources ($n = 5$)	Beaudoin and Smith [41]	62
	Harris et al. [27]	57
	Lamarche et al. [15]	53
	Pearce et al. [42]	42
	Wright et al. [40]	58
Fisheries resource assessment, management, and planning ($n = 8$)	Armstrong and Singh [43]	24
	Getsiv-Clemmons et al. [44]	40
	Nodder et al. [45]	56
	Rueda et al [31]	61
	Tempera et al. [32]	59
	Todd et al. [46]	39
	Todd and Valentine [47]	15
	Yoklavich and Greene [48]	54
Fisheries reserve design ($n = 5$)	Allee et al. [21]	30
	Copeland et al. [8]	19
	Heyman and Kobara [49]	26
	Tempera et al. [32]	59
	Yamanaka et al. [50]	36
Baseline mapping and information for management decisions ($n = 3$)	Collier and Humber [7]	21
	Gibbs and Cochran [51]	28
	Wysoczanski and Clark [52]	63
Input to UNESCO world heritage application ($n = 1$)	Kotilainen et al. [53]	17
Geological mapping ($n = 2$)	D'Angelo and Fiorentino [54]	7
	Martorelli et al. [55]	31
Hydrographic charting ($n = 1$)	O'Brien et al. [16]	20
Monitoring MPAs and environments ($n = 2$)	Cochrane et al. [5]	18
	Collin et al. [56]	10
Equipment testing ($n = 1$)	Getsiv-Clemmons et al. [44]	40
Assessment of the cold water coral distribution ($n = 2$)	De Mol et al. [35]	60
		46
Testing existing national bioregions using invertebrate data ($n = 1$)	Althaus et al. [2]	48
Hazard assessment ($n = 1$)	Lamarche et al. [15]	53

The goal to "map once, use many ways" underpins and justifies most government-funded seafloor mapping programs as well as the creation of national and regional databases and information systems containing essential marine environmental data.

What Are Benthic Habitats?

A habitat (Latin for "it inhabits") is an ecological or environmental area that is inhabited by a particular species of animal, plant, or other type of organism. *Benthic habitats* are physically distinct areas of seabed that are associated with the occurrence of a particular species. More broadly, habitats are often utilized by communities or assemblages that consistently occur together (e.g., shallow, wave-influenced rocky seabed, kelps, mollusks, and fish occur in a kelp forest habitat [57]). The term *biotope* is commonly used to refer to both the abiotic and biotic elements (physical habitats and their associated biota). The benthic habitat includes the natural environment in which an organism or community lives, or the physical environment that surrounds (influences and is utilized by) a species or community.

The classification of habitats may be structured in a hierarchy to reflect degrees of similarity (e.g., biotopes, biotope complexes, and broad habitats). *Seascapes* (the marine version of "landscapes") are comprised of suites of habitats that consistently occur together. Chapter 4 contains more detailed descriptions of the fundamental concepts of biogeography and habitat classifications arising throughout this book.

Potential Habitat Mapping

In order to truly understand the spatial relationships between the occurrence of organisms and their preferred habitats, information should be collected about both. However, the available mapping technologies generally reveal only the physical aspects of the marine environment, and at broad spatial scales they do not provide much information about the occurrence of individual organisms. In other words, our ability to map the physical spaces that organisms might utilize far exceeds our ability to measure the extent to which those spaces are actually occupied.

Mapping the physical habitats is commonly known as the "potential habitat mapping" approach [58]. The data contained in the Ocean Biogeographic Information System (Figure 1.2) makes a clear point. Although the database is already extensive and contains over 30 million records, there are still large gaps in the species record. We will never possess perfect knowledge of the existence of species or their spatial distribution. It is impossible to map the ocean's true species biodiversity. However, using potential habitat maps based on relationships that have been tested in different settings, we can at least estimate biodiversity and make predictions about its spatial distribution.

The underlying tenet of potential habitat mapping is that mapping the spatial distribution of habitats provides a means of estimating the occurrence of biota which commonly utilize that habitat type [58]. From the perspective of management and conservation, if the potential habitats are protected, then the biodiversity associated

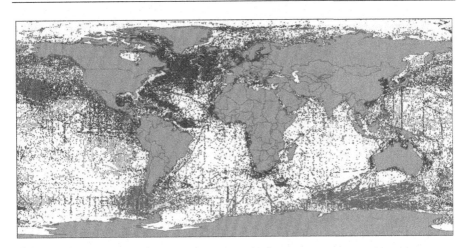

Figure 1.2 Map showing nearly 30 million Ocean Biogeographic Information System (OBIS) records of 120,000 species. Colors represent data collected prior to the Census of Marine Life (in blue) and data collected during the program (yellow and red). This database provides global coverage with an average of one data point per every 12 km². The map also illustrates broad areas of seafloor where no samples have been collected [59]. (For interpretation of the references to color in this figure legend, the reader is referred to the web version of this book.)

with them will also be protected (at least to some extent). Furthermore, it follows that an area that supports a diversity of habitats (high habitat heterogeneity) can be expected to support a greater biodiversity than an area which contains only a few (or one) habitat types; this is the so-called habitat heterogeneity hypothesis, a cornerstone of ecological theory [60].

Habitats are a shorthand way of describing and integrating biophysical and ecosystem information. To nominate tropical coral reef habitat, temperate kelp forest habitat, or abyssal seamount habitat, for example, immediately specifies particular associated biota plus the accompanying environmental attributes. It follows that there is a clear role for using environmental attributes that we can map and which exert control over biodiversity. In other words, we study and map habitats and other surrogates for biodiversity and use these to design our marine environmental management measures. The objective for marine scientists tasked with conserving biodiversity is therefore to identify and make use of measurable attributes or indicators of biodiversity [61]. Understanding the different measurable environmental parameters that exert control over marine biodiversity underlies much of the content of this book.

Geomorphology and Habitats

Among the physical attributes mapped and measured in detail in recent times using multibeam sonar equipment is the geomorphology of the seafloor. Temperate rocky reefs on the continental shelf, seamounts, submarine canyons, rocky ridges, pinnacles, ledges, escarpments, and muddy basins are examples of different geomorphic features that might each be expected to be associated with particular types of benthic habitat. The organization of this book (in terms of geomorphic features) is designed to advance our understanding of the different habitats associated with particular geomorphic features and allow examples to be compared and contrasted between different regions of the earth.

It might also be argued that the diversity of seabed geomorphic features has an intrinsic value of its own. The natural diversity of geological features has been termed *geodiversity* by some scientists, and the conservation of such diversity can be included as a criterion in making management decisions [62]. This concept is not unfamiliar to conservationists because many iconic terrestrial parks are defined on the basis of a prominent physical feature (e.g., the Grand Canyon and Mount Rainier in the United States or Uluru in Australia) and similarly some MPAs are defined by the presence of a particular reef, island, or rocky promontory. However, biological aspects of habitats are emphasized by most government agencies and nature conservation organizations, and in many cases there is little if any acknowledgment of the geological aspects of habitats [62].

Habitat Mapping Technologies and Approaches

The case studies in this book present examples of habitat maps that have been produced using a range of technologies, including satellite and airborne remote sensing, laser-light detection and ranging (LIDAR), multibeam swath sonar, sidescan sonar, ship-deployed remotely operated vehicles (ROVs), ship-deployed underwater cameras and videos, autonomous underwater vehicles (AUV), manned submersibles, and direct sampling of the seafloor (Table 1.2). A key point is that habitat mapping surveys use several complimentary technologies to map and sample the environment; determining the optimal combination of technologies to be deployed on a survey is a challenging task for habitat mapping scientists. The different systems have different applications for mapping different habitats, and the terminology may be confusing for some readers. Briefly reviewed here are seabed mapping technologies divided into four broad groups: (1) acoustic, sonar technology; (2) remote sensing based on natural or transmitted light; (3) underwater photography and video; and (4) direct sampling of sediment and biota.

Sonar Systems

Gauging water depth using acoustic (sonar) technology involves measuring the time taken for sound waves to travel between the vessel and the seafloor and back again. Transducers are used to transmit and receive sound pulses from a vessel. The most advanced technology is "multibeam sonar," which uses multiple (>100) sound beams to map the depth of water in a swath of the seabed across the track of the ship (Figure 1.3), in contrast to a single-beam sonar, which maps only a single row of points located directly below the ship, modern multibeam systems are coupled with the global positioning system (GPS) to create accurate bathymetric maps (seabed topographic maps)

Table 1.2 List of Seafloor Mapping Technologies Used in the Case Studies Presented in This Book, in Relation to Mapping Effort (Area Mapped per Hour), Typical Data Resolution, and Remarks About Applications (Based Partly on Kenny et al. [61]). Note that Most Case Studies Employed Multiple Technologies and that the Technology Listed Here Was Featured as a Unique and Major Part of the Case Study

Technology	Mapping Effort (km²/h)	Resolution (m)						Case Study Examples (This Volume)	Remarks
		10^3	10^2	10	1	10^{-1}	10^{-2}	10^{-3}		
Satellite remote sensing	>1,000	X	X	X	X				Dankers et al. [9]	Restricted to operational coverage, mainly shallow seas <10 m
Aircraft remote sensing (LIDAR, hyperspectral, CASI, etc.)	>10		X	X	X				Collin et al. [56] Galparsoro et al. [24] Gibbs and Cochran [51] Hamylton [11] Hamylton et al. [12]	Generally restricted to depths <30 m
12–30 kHz multibeam sonar	>100		X	X					Rueda et al. [31] Tempera et al. [32]	Backscatter plus depth data, low mapping effort trade-off with lower resolution, expensive system
30–100 kHz multibeam sonar	~30			X	X				Barrie et al. [3] Beaman et al. [4] Getsiv-Clemons et al. [44] Harris et al. [27] Lamarche et al. [15] Pearce et al. [42] Todd and Valentine [47]	Backscatter plus depth data, intermediate mapping effort and resolution

(Continued)

Table 1.2 (Continued)

Technology	Mapping Effort (km²/h)	Resolution (m)							Case Study Examples (This Volume)	Remarks
		10^3	10^2	10	1	10^{-1}	10^{-2}	10^{-3}		
>200 kHz multibeam sonar	~2				X	X			Brooke et al. [22] Copeland et al. [8] Huvenne et al. [36] James et al [13] Nichol et al. [30] O'Brien et al. [16] Robinson et al. [38] Stieglitz [39] van Dijk et al. [20]	Backscatter plus depth data; greater mapping effort required but finer resolution data collected; moderate cost for (semiportable) system
Side scan sonar (~100 –300 kHz)	~3				X	X			Collier and Humber [7] D'Angelo and Fiorentino [54] Martorelli et al. [55] Van Lancker et al. [33]	Backscatter data only; inexpensive
Single-beam echo sounder	~1	X	X						Coggan and Diesing [6] Ezhova et al. [10] Heyman et al. [49] James et al. [14] Kotilainen et al. [53]	Can use seabed classification software (QTC, Roxann, etc.); inexpensive
Subbottom profiler, shallow seismic profiler	~1	X	X						Gordini et al. [25] De Mol et al. [35, 63] Sayago-Gil et al. [18] Todd et al. [46]	Able to quantify thickness of unconsolidated sediments

Method	Resolution				References	Notes
Towed video, ROV, and other ship-deployed underwater cameras	0.001		X		Allee et al. [21] Althaus et al. [2] Beaudoin et al. [41] Cochrane et al [5] Guinan et al. [26] Lo Iacono et al. [28] Nodder et al. [45] Post et al. [17] Tempera et al. [19] Wysoczanski and Clark [52] Yamanaka et al. [50]	Megabenthos and geological feature identification
AUV	0.001		X	X	Armstrong et al. [43] Lucier et al. [37]	Able to replicate exact survey line for habitat monitoring
Manned submersible	0.0001		X	X	Lo Iacono et al. [29] Reynolds et al. [64] Wright et al. [40] Yoklavich and Greene [48]	ID and limited sampling of benthos and geology; expensive
Grab and core samples	$<1 \times 10^{-6}$	X	X		Buhl-Mortensen [23] De Mol et al. [34]	Quantitative data on fauna and sediments

Figure 1.3 Multibeam sonar is used to map the depth of water in a swath across the ship's track, allowing a map of the seafloor to be constructed. (Figure used with permission from the New Zealand Institute of Water and Atmospheric Sciences.)

that are presented in many case studies in this book. Different frequencies (measured in kilohertz, kHz) are used to map different water depths: higher frequencies (>100 kHz) are used in water depths of 10–100 m, frequencies of less than around 30 kHz are used in water depths of 100–2,000 m, and a frequency of around 12 kHz is used to map the abyssal depths of the ocean. Lower frequency (<30 kHz) systems utilize large (expensive) arrays of transducers that must be mounted on the hull of a ship, whereas higher frequency (>100 kHz) systems are smaller in size and can be deployed from smaller research vessels (often as portable systems). For different frequencies, there is also a trade-off in resolution: higher frequency, shallow water systems provide finer spatial resolution than lower frequency, deep-water systems (Table 1.2).

When the sound pulses bounce off the seafloor, the strength of the echo depends on the roughness and hardness of the seafloor; rougher/harder surfaces produce a stronger echo. Because of this, the strength of the sonar reflection (the backscatter) provides information on the seafloor topography and presence of rock or sediment on the bottom. An older technology employing transducers located in a "fish" towed behind the survey vessel, known as sidescan sonar, collects only acoustic backscatter data. Sidescan sonars are still used mainly because the technology is easy to deploy from small vessels and is less expensive than multibeam sonar. The resolution of towed sidescan sonar systems can exceed that of multibeam systems, but the exact location of the tow-fish behind the vessel is difficult to detect. This means that the data cannot easily be accurately located, which introduces errors when the backscatter data are combined with existing bathymetric data. A significant advantage of multibeam sonar is that it generates both accurate (geo-referenced) water depth and backscatter data simultaneously.

Continuous seismic reflection profiling ("seismic profiling") is another acoustic method used in some case studies (Table 1.2). This method is based on a sound source that generates acoustic pulses generally of a low frequency (usually <3–4 kHz, depending on the source) and having much higher energy than conventional echo sounders. Some of the energy from a seismic acoustic pulse is reflected from the seafloor directly back to the ship (as in an echo sounder), but part of the energy is able to penetrate into the seabed and reflect back off different rock layers beneath the seafloor. In this way, a single vertical profile, showing the thicknesses of different rock and sediment layers, is created as the vessel traverses an area.

Large (more powerful) seismic systems commonly use a separate sound source in which the sound pulses are generated by high-pressure "air guns" or electric "sparkers" or "boomers." The return signal is received by a second towed array of receivers (contained in a "seismic eel"). Smaller, less powerful subbottom profiler systems (basically large echo sounders) have transducers that send and receive the acoustic pulse from a single unit. Most modern research vessels have subbottom profilers built in. Smaller portable systems can be towed behind, or deployed over the side of, smaller research vessels.

Remote Sensing Based on Natural or Transmitted Light

Remotely sensed images of the shallow marine environment can be collected from satellites or aircraft to generate snapshots and time series of chlorophyll, ocean temperature [65], wave climate [66], and a number of other properties [9]. Satellite passive sensors rely on natural solar radiation reflected from the surface of the earth. Systems deployed from aircraft not only use natural light but also use active radar sensors or laser sources to create images and gather information.

LIDAR technology utilizes the reflective and transmissive properties of water and the seafloor to measure water depth using a laser, usually deployed from an aircraft. When an airborne laser beam is aimed vertically at the sea surface, the infrared is reflected while the blue-green light is transmitted through the water column. The blue-green light reflects off the seafloor (in shallow water), and water depth is calculated from the time difference between the surface and bottom returns. LIDAR systems are useful for mapping shallow water areas, to a maximum depth of around 30 m (depending on the clarity of the water).

Underwater Cameras

In order to map the occurrence of plants and animals on the seafloor, scientists collect underwater images using still and video cameras. Video data can be overlain on acoustic data such as multibeam bathymetry and backscatter to examine the relationships between seafloor depth, shape, composition, and plant and animal distributions. Directly observing the seafloor geology, plants, and animals not only allows for rapid characterization [67] but also provides the foundation to monitor future changes. Cameras can be lowered on a wire to the seabed, towed behind the vessel on a sled, deployed from submersibles, or mounted in ROVs or AUVs. In most ship-deployed systems, the digital camera images are sent via a cable to a recorder and TV screen on the ship, allowing biota and habitats to be viewed and assessed in real time [67].

Seafloor Sampling

In order to build our understanding of habitats, all imagery and mapping data must be correlated with samples obtained from the seabed. In particular, sediment properties (grain size, mineralogy, etc.) and the taxonomy of most species can only be accurately determined from physical samples. To lower on a wire, some device to the seafloor in order to obtain a sediment sample or biological specimen involves technologies that have been developed since the Challenger expedition of 1872–1876, and there are literally hundreds of different kinds of seafloor sampling devices in existence. Different devices were used in many case studies presented in this book. Some good textbooks describing different marine geological and biological sampling methods and techniques are Seibold and Berger [68], Ericson [69], and Levinton [70].

A drawback of these older seabed sampling technologies is that the samples are collected from random locations on the seabed—the spatial context of the biological specimen returned is poorly constrained. Using modern satellite navigation systems coupled with acoustic telemetry, the location of sampling devices can now be accurately calculated, but even this technology has its limitations in depths of more than a few hundred meters. One alternative is to use manned submersibles, which allow scientists to collect samples using a robotic arm, with the advantage that the samples are collected from known locations and the surrounding environment that can be imaged and measured at the same time the samples are collected. The trade-off is that manned submersibles are expensive to build and operate, and they can cover only small areas of seabed during each deployment (Table 1.2).

Acknowledgments

Thanks to Emily Corcoran (OSPAR), Marcus Diesing (CEFAS), and Alix Post (Geoscience Australia) for helpful comments and suggestions on an earlier draft of this chapter. This work was produced with the support of funding from the Australian Government's Commonwealth Environment Research Facilities (CERF) program and is a contribution of the CERF Marine Biodiversity Hub and UNEP/GRID Arendal. This chapter is published with the permission of the Chief Executive Officers of Geoscience Australia and UNEP/GRID Arendal.

References and further reading

[1] IOC-UNESCO, An Assessment of Assessments, Findings of the Group of Experts. Start-up Phase of a Regular Process for Global Reporting and Assessment of the State of

the Marine Environment, Including Socio-economic Aspects, United Nations, UNEP, and IOC-UNESCO, Valetta, Malta, 2009, p. 208.

[2] F. Althaus, A. Williams, R.J. Kloser, J. Seiler, N.J. Bax, Evaluating geomorphic features as surrogates for benthic biodiversity on Australia's western continental margin, in: P.T. Harris, E.K. Baker, (Eds.), Seafloor Geomorphology as Benthic Habitat: GeoHab Atlas of Seafloor Geomorphic Features and Benthic Habitats, Elsevier, Amsterdam, The Netherlands, 2011 (Chapter 48).

[3] J.V. Barrie, H.G. Greene, K.W. Conway, K. Picard, Inland Tidal Sea of the Northeastern Pacific, in: P.T. Harris, E.K. Baker (Eds.), Seafloor Geomorphology as Benthic Habitat: GeoHab Atlas of Seafloor Geomorphic Features and Benthic Habitats, Elsevier, Amsterdam, The Netherlands, 2011 (Chapter 44).

[4] R.J. Beaman, T. Bridge, T. Done, J.M. Webster, S. Williams, O. Pizarro, Habitats and benthos at Hydrographers Passage, Great Barrier Reef, Australia, in: P.T. Harris, E.K. Baker (Eds.), Seafloor Geomorphology as Benthic Habitat: GeoHab Atlas of Seafloor Geomorphic Features and Benthic Habitats, Elsevier, Amsterdam, The Netherlands, 2011 (Chapter 29).

[5] G.R. Cochrane, L. Trusel, J. Harney, L. Etherington, Habitats and benthos of an evolving fjord, Glacier Bay, Alaska, in: P.T. Harris, E.K. Baker (Eds.), Seafloor Geomorphology as Benthic Habitat: GeoHab Atlas of Seafloor Geomorphic Features and Benthic Habitats, Elsevier, Amsterdam, The Netherlands, 2011 (Chapter 18).

[6] R. Coggan, M. Diesing, Rock ridges in the central English channel, in: P.T. Harris, E.K. Baker (Eds.), Seafloor Geomorphology as Benthic Habitat: GeoHab Atlas of Seafloor Geomorphic Features and Benthic Habitats, Elsevier, Amsterdam, The Netherlands, 2011 (Chapter 33).

[7] J.S. Collier, S.R. Humber, Fringing reefs of the Seychelles inner granitic islands, Western Indian Ocean, in: P.T. Harris, E.K. Baker (Eds.), Seafloor Geomorphology as Benthic Habitat: GeoHab Atlas of Seafloor Geomorphic Features and Benthic Habitats, Elsevier, Amsterdam, The Netherlands, 2011 (Chapter 21).

[8] A. Copeland, E. Edinger, T. Bell, P. LeBlanc, J. Wroblewski, R. Devillers, Geomorphic features and benthic habitats of a subarctic fjord: Gilbert Bay, Southern Labrador, Canada, in: P.T. Harris, E.K. Baker (Eds.), Seafloor Geomorphology as Benthic Habitat: GeoHab Atlas of Seafloor Geomorphic Features and Benthic Habitats, Elsevier, Amsterdam, The Netherlands, 2011 (Chapter 19).

[9] N. Dankers, W. van Duin, M. Baptist, E. Dijkman, J. Cremer, The Wadden Sea in the Netherlands: ecotopes in a World Heritage barrier island system, in: P.T. Harris, E.K. Baker (Eds.), Seafloor Geomorphology as Benthic Habitat: GeoHab Atlas of Seafloor Geomorphic Features and Benthic Habitats, Elsevier, Amsterdam, The Netherlands, 2011 (Chapter 11).

[10] E. Ezhova, D. Dorokhov, V. Sivkov, V. Zhamoida, D. Ryabchuk, O. Kocheshkova, Benthic habitats and benthic communities in South-Eastern Baltic Sea, Russian sector, in: P.T. Harris, E.K. Baker (Eds.), Seafloor Geomorphology as Benthic Habitat: GeoHab Atlas of Seafloor Geomorphic Features and Benthic Habitats, Elsevier, Amsterdam, The Netherlands, 2011 (Chapter 43).

[11] S. Hamylton, Hyperspectral remote sensing of the geomorphic features and habitats of the Al Wajh Bank reef system, Saudi Arabia, Red Sea, in: P.T. Harris, E.K. Baker (Eds.), Seafloor Geomorphology as Benthic Habitat: GeoHab Atlas of Seafloor Geomorphic Features and Benthic Habitats, Elsevier, Amsterdam, The Netherlands, 2011 (Chapter 23).

[12] S. Hamylton, T. Spencer, A.B. Hagan, Coral reefs and reef islands of the Amirantes Archipelago, western Indian Ocean, in: P.T. Harris, E.K. Baker (Eds.), Seafloor

Geomorphology as Benthic Habitat: GeoHab Atlas of Seafloor Geomorphic Features and Benthic Habitats, Elsevier, Amsterdam, The Netherlands, 2011 (Chapter 22).

[13] J.W.C. James, A.S.Y. Mackie, E.I.S. Rees, T. Darbyshire, Sand wave field: the OBel Sands, Bristol Channel, U.K., in: P.T. Harris, E.K. Baker (Eds.), Seafloor Geomorphology as Benthic Habitat: GeoHab Atlas of Seafloor Geomorphic Features and Benthic Habitats, Elsevier, Amsterdam, The Netherlands, 2011 (Chapter 12).

[14] J.W.C. James, B. Pearce, R.A. Coggan, A. Morando, Open shelf valley system, Northern Palaeovalley, English Channel, U.K., in: P.T. Harris, E.K. Baker (Eds.), Seafloor Geomorphology as Benthic Habitat: GeoHab Atlas of Seafloor Geomorphic Features and Benthic Habitats, Elsevier, Amsterdam, The Netherlands, 2011 (Chapter 41).

[15] G. Lamarche, A.A. Rowden, J. Mountjoy, V. Lucieer, A.-L. Verdier, The Cook Strait Canyon, New Zealand: a large bedrock canyon system in a tectonically active environment, in: P.T. Harris, E.K. Baker (Eds.), Seafloor Geomorphology as Benthic Habitat: GeoHab Atlas of Seafloor Geomorphic Features and Benthic Habitats, Elsevier, Amsterdam, The Netherlands, 2011 (Chapter 53).

[16] P.E. O'Brien, J. Stark, G. Johnston, J. Smith, M.J. Riddle, Sea bed character and habitats of a rocky Antarctic coastline: a preliminary view of the Vestfold Hills, East Antarctica, in: P.T. Harris, E.K. Baker (Eds.), Seafloor Geomorphology as Benthic Habitat: GeoHab Atlas of Seafloor Geomorphic Features and Benthic Habitats, Elsevier, Amsterdam, The Netherlands, 2011 (Chapter 20).

[17] A.L. Post, P.E. O'Brien, R.J. Beaman, M.J. Riddle, L. De Santis, S.R. Rintoul, Distribution of hydrocorals along the George V slope, East Antarctica, in: P.T. Harris, E.K. Baker (Eds.), Seafloor Geomorphology as Benthic Habitat: GeoHab Atlas of Seafloor Geomorphic Features and Benthic Habitats, Elsevier, Amsterdam, The Netherlands, 2011 (Chapter 52).

[18] M. Sayago-Gil, P. Durán-Muñoz, F.J. Murillo, V. Díaz-del-Río, A. Serrano, L.M. Fernández-Salas, A study of geo-morphological features of the sea bed and the relationship to deep sea communities on the western slope of Hatton Bank (NE Atlantic Ocean), in: P.T. Harris, E.K. Baker (Eds.), Seafloor Geomorphology as Benthic Habitat: GeoHab Atlas of Seafloor Geomorphic Features and Benthic Habitats, Elsevier, Amsterdam, The Netherlands, 2011 (Chapter 55).

[19] F. Tempera, M. McKenzie, I. Bashmachnikov, M. Puotinen, R.S. Santos, R. Bates, Predictive modelling of dominant macroalgae abundance on temperate island shelves (Azores, northeast Atlantic), in: P.T. Harris, E.K. Baker (Eds.), Seafloor Geomorphology as Benthic Habitat: GeoHab Atlas of Seafloor Geomorphic Features and Benthic Habitats, Elsevier, Amsterdam, The Netherlands, 2011 (Chapter 8).

[20] T.A.G.P. van Dijk, J.A. van Dalfsen, R. van Overmeeren, V. Van Lancker, S. van Heteren, P.J. Doornenbal, Benthic habitat variations over tidal ridges, North Sea, Netherlands, in: P.T. Harris, E.K. Baker (Eds.), Seafloor Geomorphology as Benthic Habitat: GeoHab Atlas of Seafloor Geomorphic Features and Benthic Habitats, Elsevier, Amsterdam, The Netherlands, 2011 (Chapter 13).

[21] R.J. Allee, A.W. David, D.F. Naar, Two shelf edge marine protected areas in the Eastern Gulf of Mexico, in: P.T. Harris, E.K. Baker (Eds.), Seafloor Geomorphology as Benthic Habitat: GeoHab Atlas of Seafloor Geomorphic Features and Benthic Habitats, Elsevier, Amsterdam, The Netherlands, 2011 (Chapter 30).

[22] B.P. Brooke, M.A. McArthur, C.D. Woodroffe, M. Linklater, S.L. Nichol, T.J. Anderson, et al., Geomorphic features and infauna diversity of a subtropical mid-ocean carbonate shelf: Lord Howe Island, Southwest Pacific Ocean, in: P.T. Harris, E.K. Baker (Eds.), Seafloor Geomorphology as Benthic Habitat: GeoHab Atlas of Seafloor Geomorphic Features and Benthic Habitats, Elsevier, Amsterdam, The Netherlands, 2011 (Chapter 25).

[23] L. Buhl-Mortensen, R. Bøe, M.F.J. Dolan, P. Buhl-Mortensen, T. Thorsnes, S. Elvenes, et al., Banks, troughs and canyons on the continental margin off Lofoten, Vesterålen, and Troms, Norway, in: P.T. Harris, E.K. Baker (Eds.), Seafloor Geomorphology as Benthic Habitat: GeoHab Atlas of Seafloor Geomorphic Features and Benthic Habitats, Elsevier, Amsterdam, The Netherlands, 2011 (Chapter 51).

[24] I. Galparsoro, Á. Borja, J.G. Rodríguez, I. Muxika, M. Pascual, I. Legorburu, Rocky reef and sedimentary habitats within the continental shelf of the southeastern Bay of Biscay, in: P.T. Harris, E.K. Baker (Eds.), Seafloor Geomorphology as Benthic Habitat: GeoHab Atlas of Seafloor Geomorphic Features and Benthic Habitats, Elsevier, Amsterdam, The Netherlands, 2011 (Chapter 35).

[25] E. Gordini, A. Falace, S. Kaleb, F. Donda, R. Marocco, G. Tunis, Methane-related carbonate cementation of marine sediments and related macroalgal coralligenous assemblages in the Northern Adriatic Sea, in: P.T. Harris, E.K. Baker (Eds.), Seafloor Geomorphology as Benthic Habitat: GeoHab Atlas of Seafloor Geomorphic Features and Benthic Habitats, Elsevier, Amsterdam, The Netherlands, 2011 (Chapter 9).

[26] J. Guinan, Y. Leahy, K. Verbruggen, T. Furey, Habitats at the Rockall Bank slope failure features, Northeast Atlantic Ocean, in: P.T. Harris, E.K. Baker (Eds.), Seafloor Geomorphology as Benthic Habitat: GeoHab Atlas of Seafloor Geomorphic Features and Benthic Habitats, Elsevier, Amsterdam, The Netherlands, 2011 (Chapter 47).

[27] P.T. Harris, S.L. Nichol, T.J. Anderson, A.D. Heap, Habitats and benthos of a deep sea marginal plateau, Lord Howe Rise, Australia, in: P.T. Harris, E.K. Baker (Eds.), Seafloor Geomorphology as Benthic Habitat: GeoHab Atlas of Seafloor Geomorphic Features and Benthic Habitats, Elsevier, Amsterdam, The Netherlands, 2011 (Chapter 57).

[28] C. Lo Iacono, E. Gràcia, R. Bartolomé, E. Coiras, J.J. Dañobeitia, J. Acosta, Habitats of the Chella Bank, Eastern Alboran Sea (Western Mediterranean), in: P.T. Harris, E.K. Baker (Eds.), Seafloor Geomorphology as Benthic Habitat: GeoHab Atlas of Seafloor Geomorphic Features and Benthic Habitats, Elsevier, Amsterdam, The Netherlands, 2011 (Chapter 49).

[29] C. Lo Iacono, C. Orejas, A. Gori, J.M. Gili, S. Requena, P. Puig, et al., Habitats of the Cap de Creus continental shelf and Cap de Creus canyon, north-western Mediterranean, in: P.T. Harris, E.K. Baker (Eds.), Seafloor Geomorphology as Benthic Habitat: GeoHab Atlas of Seafloor Geomorphic Features and Benthic Habitats, Elsevier, Amsterdam, The Netherlands, 2011 (Chapter 32).

[30] S.L. Nichol, T.J. Anderson, C. Battershill, B.P. Brooke, Submerged reefs and aeolian dunes as inherited habitats, Point Cloates, Carnarvon Shelf, Western Australia, in: P.T. Harris, E.K. Baker (Eds.), Seafloor Geomorphology as Benthic Habitat: GeoHab Atlas of Seafloor Geomorphic Features and Benthic Habitats, Elsevier, Amsterdam, The Netherlands, 2011 (Chapter 27).

[31] J.L. Rueda, V. Díaz-del-Río, M. Sayago-Gil, N. López, L.M. Fernández, J.T. Vázquez, Fluid venting through the seabed in the Gulf of Cadiz (SE Atlantic Ocean, western Iberian Peninsula): geomorphic features, habitats and associated fauna, in: P.T. Harris, E.K. Baker (Eds.), Seafloor Geomorphology as Benthic Habitat: GeoHab Atlas of Seafloor Geomorphic Features and Benthic Habitats, Elsevier, Amsterdam, The Netherlands, 2011 (Chapter 61).

[32] F. Tempera, E. Giacomello, N. Mitchell, A.S. Campos, A.B. Henriques, A. Martins, et al., Mapping the Condor seamount seafloor environment and associated biological assemblages (Azores, NE Atlantic), in: P.T. Harris, E.K. Baker (Eds.), Seafloor Geomorphology as Benthic Habitat: GeoHab Atlas of Seafloor Geomorphic Features and Benthic Habitats, Elsevier, Amsterdam, The Netherlands, 2011 (Chapter 59).

[33] V. Van Lancker, G. Moerkerke, I. Du Four, E. Verfaillie, M. Rabaut, S. Degraer, Fine-scale geomorphological mapping for the prediction of macrobenthic occurrences in shallow marine environments, Belgian part of the North Sea, in: P.T. Harris, E.K. Baker (Eds.), Seafloor Geomorphology as Benthic Habitat: GeoHab Atlas of Seafloor Geomorphic Features and Benthic Habitats, Elsevier, Amsterdam, The Netherlands, 2011 (Chapter 14).

[34] L. De Mol, A. Hilário, D. Van Rooij, J.-P. Henriet, Habitat mapping of a cold-water coral mound on Pen Duick Escarpment (Gulf of Cadiz), in: P.T. Harris, E.K. Baker (Eds.), Seafloor Geomorphology as Benthic Habitat: GeoHab Atlas of Seafloor Geomorphic Features and Benthic Habitats, Elsevier, Amsterdam, The Netherlands, 2011 (Chapter 46).

[35] B. De Mol, D. Amblas, A. Calafat, M. Canals, R. Duran, C. Lavoie, Alboran Seamounts, Western Mediterranean sea: cold-water Coral colonization of knolls, in: P.T. Harris, E.K. Baker (Eds.), Seafloor Geomorphology as Benthic Habitat: GeoHab Atlas of Seafloor Geomorphic Features and Benthic Habitats, Elsevier, Amsterdam, The Netherlands, 2011 (Chapter 60).

[36] V.A.I. Huvenne, A.D.C. Pattenden, D.G. Masson, P.A. Tyler, Habitat heterogeneity in the Nazaré deep-sea canyon offshore Portugal, in: P.T. Harris, E.K. Baker (Eds.), Seafloor Geomorphology as Benthic Habitat: GeoHab Atlas of Seafloor Geomorphic Features and Benthic Habitats, Elsevier, Amsterdam, The Netherlands, 2011 (Chapter 50).

[37] V. Lucieer, N. Barrett, N. Hill, S.L. Nichol, Characterisation of shallow inshore coastal reefs on the Tasman Peninsula, South Eastern Tasmania, Australia, in: P.T. Harris, E.K. Baker (Eds.), Seafloor Geomorphology as Benthic Habitat: GeoHab Atlas of Seafloor Geomorphic Features and Benthic Habitats, Elsevier, Amsterdam, The Netherlands, 2011 (Chapter 34).

[38] K.A. Robinson, A.S.Y. Mackie, C. Lindenbaum, T. Darbyshire, K.J.J. van Landeghem, W.G. Sanderson, Seabed habitats of the Southern Irish Sea, in: P.T. Harris, E.K. Baker (Eds.), Seafloor Geomorphology as Benthic Habitat: GeoHab Atlas of Seafloor Geomorphic Features and Benthic Habitats, Elsevier, Amsterdam, The Netherlands, 2011 (Chapter 37).

[39] T.C. Stieglitz, The Yongala's Halo of Holes—systematic bioturbation close to a shipwreck, in: P.T. Harris, E.K. Baker (Eds.), Seafloor Geomorphology as Benthic Habitat: GeoHab Atlas of Seafloor Geomorphic Features and Benthic Habitats, Elsevier, Amsterdam, The Netherlands, 2011 (Chapter 16).

[40] D.J. Wright, J.T. Roberts, D. Fenner, J.R. Smith, A.A.P. Koppers, D.F. Naar, et al., Seamounts, ridges, and reef habitats of American Samoa, in: P.T. Harris, E.K. Baker (Eds.), Seafloor Geomorphology as Benthic Habitat: GeoHab Atlas of Seafloor Geomorphic Features and Benthic Habitats, Elsevier, Amsterdam, The Netherlands, 2011 (Chapter 58).

[41] Y.C. Beaudoin, S. Smith, Habitats of the Su Su Knolls hydrothermal site, eastern Manus Basin, Papua New Guinea, in: P.T. Harris, E.K. Baker (Eds.), Seafloor Geomorphology as Benthic Habitat: GeoHab Atlas of Seafloor Geomorphic Features and Benthic Habitats, Elsevier, Amsterdam, The Netherlands, 2011 (Chapter 62).

[42] B. Pearce, D.R. Tappin, D. Dove, J. Pinnion, Benthos supported by the tunnel-valleys of the southern North Sea, in: P.T. Harris, E.K. Baker (Eds.), Seafloor Geomorphology as Benthic Habitat: GeoHab Atlas of Seafloor Geomorphic Features and Benthic Habitats, Elsevier, Amsterdam, The Netherlands, 2011 (Chapter 42).

[43] R.A. Armstrong, H. Singh, Mesophotic coral reefs of the Puerto Rico Shelf, in: P.T. Harris, E.K. Baker (Eds.), Seafloor Geomorphology as Benthic Habitat: GeoHab Atlas of Seafloor Geomorphic Features and Benthic Habitats, Elsevier, Amsterdam, The Netherlands, 2011 (Chapter 24).

[44] J.E.R. Getsiv-Clemons, W.W. Wakefield, I.J. Stewart, C.E. Whitmire, Using meso-habitat information to improve abundance estimates for West Coast groundfish: a test case at Heceta Bank, Oregon, in: P.T. Harris, E.K. Baker (Eds.), Seafloor Geomorphology as Benthic Habitat: GeoHab Atlas of Seafloor Geomorphic Features and Benthic Habitats, Elsevier, Amsterdam, The Netherlands, 2011 (Chapter 40).

[45] S.D. Nodder, D.A. Bowden, A. Pallentin, K. Mackay, Seafloor habitats and benthos of a continental ridge: Chatham Rise, New Zealand, in: P.T. Harris, E.K. Baker (Eds.), Seafloor Geomorphology as Benthic Habitat: GeoHab Atlas of Seafloor Geomorphic Features and Benthic Habitats, Elsevier, Amsterdam, The Netherlands, 2011 (Chapter 56).

[46] B.J. Todd, V.E. Kostylev, S.J. Smith, Seabed habitat of a glaciated shelf, German Bank, Atlantic Canada, in: P.T. Harris, E.K. Baker (Eds.), Seafloor Geomorphology as Benthic Habitat: GeoHab Atlas of Seafloor Geomorphic Features and Benthic Habitats, Elsevier, Amsterdam, The Netherlands, 2011 (Chapter 39).

[47] B.J. Todd, P.C. Valentine, Large submarine sand features and gravel lag substrates on Georges Bank, Gulf of Maine, in: P.T. Harris, E.K. Baker (Eds.), Seafloor Geomorphology as Benthic Habitat: GeoHab Atlas of Seafloor Geomorphic Features and Benthic Habitats, Elsevier, Amsterdam, The Netherlands, 2011 (Chapter 15).

[48] M. Yoklavich, H.G. Greene, The ascension-monterey canyon system—habitats of demersal fishes and macro-invertebrates along the Central California Coast of the USA, in: P.T. Harris, E.K. Baker (Eds.), Seafloor Geomorphology as Benthic Habitat: GeoHab Atlas of Seafloor Geomorphic Features and Benthic Habitats, Elsevier, Amsterdam, The Netherlands, 2011 (Chapter 54).

[49] W.D. Heyman, S. Kobara, Geomorphology of reef fish spawning aggregations in Belize and the Cayman Islands (Caribbean), in: P.T. Harris, E.K. Baker (Eds.), Seafloor Geomorphology as Benthic Habitat: GeoHab Atlas of Seafloor Geomorphic Features and Benthic Habitats, Elsevier, Amsterdam, The Netherlands, 2011 (Chapter 26).

[50] K.L. Yamanaka, K. Picard, K.W. Conway, R. Flemming, Rock reefs of British Columbia, Canada: inshore rockfish habitats, in: P.T. Harris, E.K. Baker (Eds.), Seafloor Geomorphology as Benthic Habitat: GeoHab Atlas of Seafloor Geomorphic Features and Benthic Habitats, Elsevier, Amsterdam, The Netherlands, 2011 (Chapter 36).

[51] A.E. Gibbs, S.A. Cochran, Coral habitat variability in Kaloko-Honokohau National Park, Hawaii, USA, in: P.T. Harris, E.K. Baker (Eds.), Seafloor Geomorphology as Benthic Habitat: GeoHab Atlas of Seafloor Geomorphic Features and Benthic Habitats, Elsevier, Amsterdam, The Netherlands, 2011 (Chapter 28).

[52] R. Wysoczanski, M. Clark, Southern Kermadec Arc—Havre Trough Geohabitats, in: P.T. Harris, E.K. Baker (Eds.), Seafloor Geomorphology as Benthic Habitat: GeoHab Atlas of Seafloor Geomorphic Features and Benthic Habitats, Elsevier, Amsterdam, The Netherlands, 2011 (Chapter 63).

[53] A.T. Kotilainen, A.M. Kaskela, S. Bäck, J. Leinikki, Submarine De Geer moraines in the Kvarken Archipelago, the Baltic Sea, in: P.T. Harris, E.K. Baker (Eds.), Seafloor Geomorphology as Benthic Habitat: GeoHab Atlas of Seafloor Geomorphic Features and Benthic Habitats, Elsevier, Amsterdam, The Netherlands, 2011 (Chapter 17).

[54] S. D'Angelo, A. Fiorentino, Phanerogam meadows: a characteristic habitat of the Mediterranean shelf. Examples from the Tyrrhenian Sea, in: P.T. Harris, E.K. Baker (Eds.), Seafloor Geomorphology as Benthic Habitat: GeoHab Atlas of Seafloor Geomorphic Features and Benthic Habitats, Elsevier, Amsterdam, The Netherlands, 2011 (Chapter 7).

[55] E. Martorelli, S. D'Angelo, A. Fiorentino, F.L. Chiocci, Non-tropical carbonate shelf sedimentation. The Archipelago Pontino (Central Italy) case history, in: P.T. Harris, E.K. Baker (Eds.), Seafloor Geomorphology as Benthic Habitat: GeoHab Atlas

of Seafloor Geomorphic Features and Benthic Habitats, Elsevier, Amsterdam, The Netherlands, 2011 (Chapter 31).

[56] A. Collin, B. Long, P. Archambault, Coastal kelp forest habitat in the Baie des Chaleurs, Gulf of St. Lawrence, Canada, in: P.T. Harris, E.K. Baker (Eds.), Seafloor Geomorphology as Benthic Habitat: GeoHab Atlas of Seafloor Geomorphic Features and Benthic Habitats, Elsevier, Amsterdam, The Netherlands, 2011 (Chapter 10).

[57] D.W. Connor, J.H. Allen, N. Golding, K.L. Howell, L.M. Lieberknecht, K.O. Northen, et al., Marine Habitat Classification for Britain and Ireland Version 04.05, Joint Nature Conservation Committee, Peterborough, UK, 2004.

[58] H.G. Greene, J.J. Bizzarro, V.M. O'Connell, C.K. Brylinsky, Construction of digital potential benthic habitat maps using a coded classification scheme and its application, in: B.J. Todd, H.G. Greene (Eds.), Mapping the Seafloor for Habitat Characterisation, Geological Association of Canada, St. John's, NL, 2007, pp. 141–156, (Special Paper 47).

[59] J.H. Ausubel, D.T. Crist, P.E. Waggoner (Eds.), First Census of Marine Life 2010: Highlights of a Decade of Discovery, Census of Marine Life, Washington, DC. Available from: <http://www.coml.org/>, 2010.

[60] J. Tews, U. Brose, V. Grimm, K. Tielbörger, M.C. Wichmann, M. Schwager, et al., Animal species diversity driven by habitat heterogeneity/diversity: the importance of keystone structures, J. Biogeogr. 31 (2004) 79–92.

[61] A.J. Kenny, I. Cato, M. Desprez, G. Fader, R.T.E. Schüttenhelm, J. Side, An overview of seabed-mapping technologies in the context of marine habitat classification, ICES J. Mar. Sci. 60 (2003) 411–418.

[62] M. Gray, Geodiversity—Valuing and Conserving Abiotic Nature, John Wiley & Sons, Chichester, UK, 2004.

[63] B. De Mol, D. Amblas, G. Alverez, P. Busquets, A. Calafat, M. Canals, et al., Erosional environment at the gateway between the Atlantic and Mediterranean Sea: Strait of Gibraltar cold-water coral distribution, in: P.T. Harris, E.K. Baker (Eds.), Seafloor Geomorphology as Benthic Habitat: GeoHab Atlas of Seafloor Geomorphic Features and Benthic Habitats, Elsevier, Amsterdam, The Netherlands, 2011 (Chapter 45).

[64] J.R. Reynolds, S.C. Rooney, J. Heifetz, H.G. Greene, B.L. Norcross, Habitats and benthos in the vicinity of Albatross Bank, Gulf of Alaska, in: P.T. Harris, E.K. Baker (Eds.), Seafloor Geomorphology as Benthic Habitat: GeoHab Atlas of Seafloor Geomorphic Features and Benthic Habitats, Elsevier, Amsterdam, The Netherlands, 2011 (Chapter 38).

[65] C.R. McClain, A decade of satellite ocean color observations, Annu. Rev. Mar. Sci. 1 (2009) 19–42.

[66] M.A. Hemer, J.A. Church, J.R. Hunter, Variability and trends in the directional wave climate of the Southern Hemisphere, Int. J. Climatol. 30 (2009) 475–491.

[67] T.J. Anderson, G.R. Chochrane, D.A. Roberts, H. Chezar, G. Hatcher, A rapid method to characterise seabed habitats and associated macro-organisms, in: G. Greene, B.J. Todd (Eds.), Mapping the Seafloor for Habitat Characterisation, Geological Association of Canada, St. John's, NL, 2007, pp. 75–83.

[68] E. Seibold, W.H. Berger, The Sea Floor: An Introduction to Marine Geology, Springer Verlag, Berlin, 1996.

[69] J. Ericson, Marine Geology: Exploring the New Frontiers of the Ocean, Facts on File, New York, NY, 2003.

[70] J.S. Levinton, Marine Biology: Function, Biodiversity, Ecology, Oxford University Press, New York. NY, 2001.

2 Habitat Mapping and Marine Management

Elaine K. Baker[1], Peter T. Harris[2]

[1]UNEP/GRID- Arendal School of Geosciences, University of Sydney, Australia, [2]Marine and Coastal Environment Group, Geoscience Australia, Canberra, ACT, Australia

Abstract

Demands are being made of the marine environment that threaten to erode the natural, social, and economic benefits that human society derives from the oceans. Expanding populations ensure a continuing increase in the variety and complexity of marine-based activities—fishing, power generation, tourism, mineral extraction, shipping, and so on.

The two most commonly acknowledged purposes for habitat mapping in the case studies contained in this book are to support government spatial marine planning, management, and decision making, and to support and underpin the design of marine protected areas (MPAs; see Chapter 64).

Key Words: Ecosystem based management, benthic habitat, fisheries management, marine-protected areas, marine environmental assessments

Marine Habitat Mapping: A Vital Component in Ecosystem-Based Management

Introduction

The UNEP review publication *In Dead Water* [1] paints a bleak picture of the future of marine resources if the current situation of overexploitation, habitat loss, pollution, invasive species, and the stress of climate change continue. At present, the majority of human activities are still localized within the exclusive economic zones (EEZs) of coastal states (e.g., 100% of oil and gas extraction, 100% of aggregate mining, and 87% of the global fish catch [2]) and are therefore managed by local, regional, or national bodies under the governance of the sovereign state. Often, numerous organizations are involved, each responsible for a sector, activity, or location.

Analysis of the shortcomings of marine management often points to the fragmented sectoral perspective (e.g., shipping, fishing, and oil and gas development) that addresses activities separately [3]. A more holistic management approach is advocated, one that considers the interaction between people and the environment and the resultant impact on ecosystem function and resilience, rather than managing one issue or resource in isolation.

Adoption of the EBM Approach

Many organizations (Regional Seas Conventions, governments, etc.) have recognized the potential for an ecosystem-based approach to deliver more effective management [4,5], and it is an underpinning concept in the Convention for the Conservation of Antarctic Marine Living Resources (CCAMLR, [6]) and the Convention on Biological Diversity (CBD) [7]. The United Nations Convention on the Law of the Sea (UNCLOS) [8] provides the legal framework for the protection and preservation of the marine environment. It contains a general obligation for states to protect and preserve the marine environment, both within and beyond national jurisdiction (Article 192). For example, it requires states to consider the effects of fishing on species associated with targeted species—that is, to go beyond the consideration of a single species and recognize the connectivity of elements comprising a functioning ecosystem. However, while ecosystem-based management (EBM) is advocated in international agreements like these and by national and local authorities, it has often been difficult to translate into operational management [9].

Marine Habitat Mapping to Support EBM

Cogan et al. [10] examined the essential components of EBM in order to illustrate how marine habitat mapping can be incorporated into EBM and help facilitate the adoption of this form of management. They note that one of the initial steps in EBM should be to characterize the habitat features of the ecosystem. Accurate descriptions of marine habitats are central to spatial marine management. As discussed in Chapter 1, habitat mapping utilizing different techniques can be scaled to provide information on a single habitat (e.g., a hydrothermal vent) or expanded to national or international levels, such as a large marine ecosystem (e.g., the UNEP Transboundary Waters Assessment Programme, which includes LMEs as an assessment unit [11]). The important point is that EBM requires that the scale of governance is matched to the scale of the ecosystem, and it is largely through marine mapping that ecological boundaries are defined. Spatial mismatches often occur when jurisdictional boundaries are too small for effective management (e.g., in cases where species forage or migrate across national boundaries [3]).

The drivers, pressures, state, impact, response (DPSIR) framework ([12]; Figure 2.1) is an effective way of illustrating the role of habitat mapping in deriving indicators to describe the state of the environment within the EBM context. The "state" (Figure 2.1) is defined as the condition of the system at a specific time and is represented by a set of descriptors of system attributes that are affected by pressures. Examples of state descriptors could be sediment type, species composition, habitat structure, and the like. The DPSIR framework relates large-scale drivers of change (e.g., trawling) to the pressures they exert (e.g., habitat loss), which cause changes in the state of the environment (e.g., an increase in the area of modified habitat), resulting in impacts on biodiversity and human well-being (e.g., loss of fisheries income),

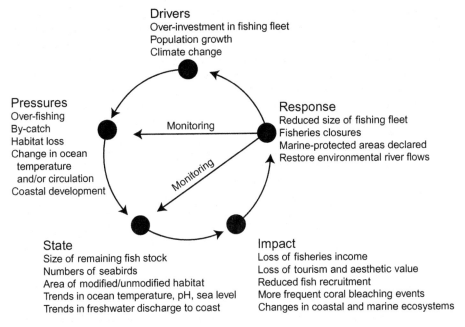

Drivers
Over-investment in fishing fleet
Population growth
Climate change

Pressures
Over-fishing
By-catch
Habitat loss
Change in ocean
  temperature
  and/or circulation
Coastal development

Monitoring

Monitoring

Response
Reduced size of fishing fleet
Fisheries closures
Marine-protected areas declared
Restore environmental river flows

State
Size of remaining fish stock
Numbers of seabirds
Area of modified/unmodified habitat
Trends in ocean temperature, pH, sea level
Trends in freshwater discharge to coast

Impact
Loss of fisheries income
Loss of tourism and aesthetic value
Reduced fish recruitment
More frequent coral bleaching events
Changes in coastal and marine ecosystems

Figure 2.1 The DPSIR tool is used for organizing information about the state of the environment and describing interactions between society and environment. The figure illustrates where habitat mapping can fit into this process (i.e., assessment of the "state" of the habitat/ecosystem and predicting impacts (adapted from [13]).

thereby leading to institutional responses, policy development, and target setting (e.g., establishment of an MPA or other zoning framework). The fundamental concept is that the species abundance and composition of any ecosystem are determined by the physical environment in combination with impacts from human management.

Habitat Mapping to Support National, Transboundary, and High Seas Marine Management Planning

Many national governments have adopted marine spatial planning (MSP), incorporating habitat mapping, in the development of marine management plans. The Great Barrier Reef Marine Park is an often cited example of a successful multiuse marine environment where MSP has been applied. It is referred to as a place-based management system [9,14], the key element of which is a zoning plan that defines what activities can occur in specific locations in order to protect the environment and to separate potentially conflicting activities. The zoning plan was updated in 2004 based on the results of investigations to determine the major habitat types of the Great Barrier Reef region [15]. The new zoning plan aims to protect "representative" examples of each habitat type within a network of no-take or highly protected areas. Representative areas were determined from habitat maps compiled using all available biophysical, biological, and oceanographic data sets and surrogates developed to approximate different habitat types. This information was used to define 70 different habitat types or "bioregions" (30 reef bioregions and 40 nonreef bioregions) across the Great Barrier Reef. The Great Barrier Reef is an example of EBM and MSP, underpinned by comprehensive habitat mapping that provides for conservation, tourism, fishing (including dredging and trawling), and other activities within a World Heritage area.

The Geological Survey of Canada Atlantic (GSCA) has been a pioneer in the development of habitat mapping as a tool for fish stock assessment, maximizing fishing efficiency and sustainable ocean management [16]. GSCA developed a habitat template that integrates data that can be associated with the life history traits of organisms. The approach is based on the theory that relates species life history traits to particular properties of the environment. In this case, the template utilizes selective forces related to physical disturbance of the environment and "scope for growth" (related to food availability, temperature, temperature variability, and oxygen saturation). In contrast to many habitat mapping techniques, the template predicts which life history traits will be best suited to a particular physical location and not which organisms will live there [17]. In documenting the role of habitat mapping in managing the Canadian marine environment, major improvements to the Canadian scallop fishery, identification of sensitive sponge habitats for incorporation into MPAs on the Pacific Coast, and improved decision making related to offshore infrastructure development are cited [18].

Transboundary marine habitat mapping initiatives have been developed both to provide an EBM approach to a shared marine environment and to produce standardized MSP tools and templates for management planning and decision making of shared resources [19–21]. The EU Marine Strategy Framework Directive [22] requires member states to manage their seas to achieve or maintain good environmental status (GES) by 2020. The GES assessment includes biological diversity and seafloor integrity, both of which will require habitat mapping to develop indicators and set quality targets.

Ardron et al. [23], in their chapter on MSP in the high seas, noted that while not explicitly specified in UNCLOS or the CBD, MSP is a practical way for states to fulfill their obligations related to the protection and preservation of the marine environment. However, due to the difficulty and expense of direct observation and biological sampling in the deep sea, there have been suggestions that more predicative modeling is required [24,25]. Habitat mapping based on seafloor geomorphology is an alternative approach to defining benthic habitats in the absence of data. This information can be used as proxies for biological communities and species diversity [26–30]. A global seafloor geomorphology map would provide spatial information on the potential distribution of ecosystems for management of the deep-sea region. Impacts to deep-sea habitats are mostly related to deep-sea trawling and disposal of marine debris.

Benthic Habitat Mapping Applications

Effective management requires decisions on what, where, and how much of a particular activity is sustainable. Habitat maps are effective tools that provide science-based information to help make these decisions. The majority of case studies contained in this volume were funded by government organizations and carried out specifically to support industry or conservation management and decision making.

Under the broad heading of MSP, there are numerous specific examples of where benthic habitat mapping has been applied, including:

- Fisheries management
- Supporting the design of marine reserves
- Site selection and environmental impact of offshore development
- Dredge spoil disposal management.

Habitat Mapping and Fisheries Management

Poor fisheries management is one of the biggest threats to marine biodiversity and benthic habitats. It is universally recognized that many target species are overexploited, that impacts on target species impact the whole ecosystem, and that these impacts are cumulative [2,31]. The World Summit on Sustainable Development called for the restoration of depleted fish stocks to maximum sustainable yields where possible by 2015 and the elimination of destructive fishing practices. In a recent analysis of fisheries management effectiveness across 209 EEZs, it was found that only 5% scored in the top quarter of the defined effectiveness scale [32].

The adverse effects of fishing on the environment have been comprehensively reviewed ([33]; Chapter 3), but in summary include, reduction in abundance of the target species, reduced spawning potential, and evolutionary changes such as smaller sizes, earlier maturity, and elevated reproductive effort. Organisms associated with the target species are also affected, as bycatch or through changes in predator–prey dynamics, competitive interactions, relative species abundance, and other ecological relationships ([34,35] and references therein). There are also impacts from habitat disturbance. Fishing practices can alter or destroy seabed habitats, which can lead to loss of species diversity as well as reduced populations and spawning ability. The degree and severity of these adverse effects on biodiversity and the seabed depend on a variety of factors, including the spatial extent of fishing, the level of fishing effort, and the fishing method used. Benthic habitat mapping can provide critical information relevant to these factors, which aids both environmental and economic sustainability.

In fact, habitat mapping is part of the commitment to an ecosystem approach to fisheries (EAF) that has been gaining ground for the last decade or more [36]. To implement EAF, managers need to make decisions based on an understanding of how these decisions will affect both the fishery and the ecosystem. Improving fisheries management was cited as the second most common reason (after conservation) for carrying out the studies in this volume (Chapter 64). Habitat maps are, for example, compiled to assess a fishery resource, including determining areas of suitable or essential habitat of a particular species or life stage, to assess the vulnerability of seafloor features to fishing impacts, and to aid in the design of fisheries reserves. The ability to track and predict the spatial dynamics of fish using key environmental indicators like benthic habitat may become increasingly important as climate change alters the geographical distribution patterns of many marine populations [37].

Many fish species depend on specific habitats to complete their life cycle, so identifying and protecting these essential habitats also benefits fisheries. The destructive impact of some fishing methods, such as bottom trawls and dredges, on benthic habitats is widely acknowledged [38]. Numerous regulatory examples focus on habitat protection. For example, the United States Sustainable Fisheries Act (SFA) requires fisheries managers to identify and describe essential fish habitat (EFH) and evaluate the effects of all fishing practices on seafloor habitat.

Assessment of Fishery Resources

Fisheries management has in the past relied on stock assessments models to set catch allowances or maximum sustainable yields, which do not include habitat data [39]. In the traditional single-species stock assessment, catch, abundance, and life history data are used to construct models that are used to establish allowable harvest quotas [40]. But as the ecosystem approach becomes an important part of management, determining catch levels and auditing the effectiveness of management needs to include managing the spatial distribution of effort and developing indicators to evaluate the effect of the fishery on target species, bycatch, and habitat features [41]. The EAF reverses the order of management priorities to start with the ecosystem rather than the target species, and because EAF emphasizes habitat and ecosystem function, management models need to incorporate spatial structure and environmental processes [42].

Examples of the use of habitat mapping in stock assessment can be found in numerous studies, and with improved mapping technology and data coverage, it is becoming more widely applied to fisheries management (see Chapters 9–11, 15, 18, 19, 24, 26, 30, 32, 34–36, 38, 42, 43, 49, 51, 54–56, 59, and 63). For example, Kostylev et al. [43] found links between scallop abundance, sediment type, and habitat structure that allowed multibeam backscatter data to be used to greatly improve stock abundance estimates of a scallop fishery on the Scotian Shelf. In southeast Alaska, yellow-eye rockfish stocks are estimated directly from habitat maps of rugged rocky seafloor terrain [44]. Galparaoro et al. [45] used seafloor morphological characteristics in the Bay of Biscay to predict the most suitable habitat characteristics for the European lobster. A similar study was carried out to predict rockfish distribution on Cordell Bank, California [9]. Various habitat mapping applications that accurately characterize seabed habitats can be used to maximize fishing effort, making it more cost-effective by locating the habitat where the resource is more likely to be found. This also has the added benefit of limiting environmental damage, as habitats where the resource is less likely to be found can be avoided.

A workshop on integrating seafloor mapping and benthic ecology into fisheries management in the Gulf of Maine [39] reported that future stock assessments needed to include EFH information that identified habitats where spawning, growth, and survival are high. To understand ecosystem dynamics in order to maximize productivity, fisheries managers need to know what habitat features support increased productivity, where they are located, and how they are affected by different kinds of human-induced (e.g., trawling) and natural disturbance.

Assessing the Vulnerability of Ecosystems to Fishing Impacts

Identifying areas that are vulnerable to fishing impacts has been important in the conservation efforts behind the delineation of MPAs or no-catch zones. Criteria for identifying vulnerable marine ecosystems (VMEs) include uniqueness or rarity of species or habitats, functional significance of the habitat, fragility, and ecosystems that are structurally complex or have life-history traits that hinder the chance of recovery (e.g., slow growth rates and late maturity [2]). VMEs are generally associated with specific undersea morphology, including topographically abrupt features, the summits and flanks of seamounts, submarine canyons, hydrothermal vents, and cold seeps [46]. For this reason habitat mapping is an effective tool in identifying and predicting the location of VMEs. There is a continuum between EFH and VMEs, both require protection, and when taking an ecosystem management approach, similar concerns exist.

Fishing affects marine habitats and ecosystems in a number of ways depending on the type of fishing gear employed and the spatial and temporal extent of fishing. Trawls and dredges are traditionally dragged across the seabed to catch demersal fish, some semipelagic species, and shellfish. This method of fishing modifies the seabed habitat; some trawls act like plows that can turn over the top 30 cm of sediment [47]. In disturbing the structure of the seafloor, towed fishing gear changes the composition of the biological community and disrupts the food web [38].

In 2005, the General Fisheries Commission for the Mediterranean banned the use of any towed fishing gear below 1,000 m, which equates to approximately 58% of the Mediterranean Basin. This area contains ecosystems associated with distinctive benthic habitats such as mud volcanoes and thermal vents [48], which are associated with high biodiversity and are also vulnerable to structural damage.

Seamounts have been identified as areas of enhanced biodiversity and productivity compared to the surrounding ocean. The occurrence of hard substrate provides for habitat-building organisms such as corals and sponges. Seamounts are known as aggregating locations for tuna, and as a consequence many are heavily exploited by fisheries. A recent study examining the association between long-line tuna catch and seamounts identified higher catch rates associated with some seamounts in the Pacific Ocean [49]. If we can clarify the relationship between seamounts and tuna, this kind of information can have significant management applications related to both fishing effort and conservation of these VMEs. The authors suggest that, to understand the factors driving tuna aggregation on specific seamounts, habitat mapping—incorporating both detailed oceanographic and improved seamount morphological data—is required.

Mapping techniques can be used in environmental monitoring—comparing the habitat structure and ecosystem between fished and unfished areas and monitoring recovery rates of degraded systems. For example, sessile fauna that are particularly vulnerable to damage from trawling, like the deep-water corals that form structural habitats on the tops and upper sides of seamounts, can be identified by their acoustic signature [18]. Similarly, seabed disturbance mapping based on trawl marks is one component of habitat vulnerability analysis that can be used to select management areas [39].

Habitat Mapping and the Design of Marine Reserves (MPAs)

Declaring MPAs Is a Priority for Nations

Given the perilous state of many marine species, a particularly urgent question is, what can be done to conserve the diversity of life in the oceans? In answering that question, the global consensus clearly and specifically highlights the need for nations to proclaim sanctuaries, or protected areas, for marine life within their marine jurisdictions. A global target of placing 10% of the oceans into MPAs by the year 2010 has been endorsed by a large majority of the international community (www.seaaroundus.org).

An MPA is defined by the IUCN as "any area of intertidal or subtidal terrain, together with its overlying water and associated flora, fauna, historical and cultural features, which has been reserved by law or other effective means to protect part or all of the enclosed environment" [50]. MPAs are therefore a spatial management tool, as opposed to regulations or laws which prohibit or limit specific activities like fishing for a certain species during certain periods. As summarized by Zacharias and Roff [51], past efforts in protecting marine environments have focused on the conservation of species but, over the past 20 years, this has been supplemented by the conservation of spaces (MPAs).

In the establishment of MPAs, managers often focus on ecological areas that are well understood by the public, such as coral reefs or seagrass beds. Value judgments are made about the importance of protecting some areas, but the development of a *representative* system of MPAs requires that all habitats be considered. Although conservation has been the main driver for the establishment of MPAs, they are also being delineated for the benefit of fisheries. MPAs are an important component of EBM [52], and habitat mapping plays a major role by providing a spatial context for understanding ecosystems and dependent benthic species and communities.

How Does Habitat Mapping Assist in Design of MPAs?

Although it is possible to use mainly biological information to select MPAs in the case of some small areas, such as restricted coastal or coral reef settings where there is sufficient biological information available [53,54], in many cases—particularly within large planning bioregions—the necessary biological data sets on species distributions simply do not exist at the spatial scales that are needed to design MPAs. In such cases, a number of workers advocate supplementing biological information with abiotic (i.e., geologic and oceanographic) indicators of benthic habitats and ecosystems as proxies for biological communities and species diversity [24,26–30]. In studies such as these, applications of spatially more complete, abiotic information have been employed to systematically map different habitats to support MPA design. Indeed, Greene et al. [55] have devised a benthic marine habitat classification scheme that is strongly dependent upon seabed geology, while in Canada, Roff and Taylor [27] and Zacharias and Roff [51] used primarily bottom physiography and oceanographic information in their hierarchical geophysical approach to classify and map marine environments (Chapter 11).

From a precautionary perspective, it is more valid to use the available information to identify and protect all of the physical variability that occurs between different habitats in an area that may or may not host a particular species or community than it is to attempt to map actual biodiversity using species richness or assemblages. As stated by Day and Roff [56], "to best conserve biodiversity, we should be identifying and conserving representative spaces in conjunction with preserving individual species. If we can identify the appropriate representative spaces to be protected, then these will contain species we wish to conserve, as well as a suite of factors necessary for the health of those species, such as habitat and community structure."

The combination of physical variables that define different habitats can be mapped if we know what variables to measure and over what spatial–temporal scales to map and measure them. The point is that communities will always exploit the availability of any given habitat, and although the species comprising that community will vary depending on biological factors (e.g., predator–prey relationships), the overall community types (as opposed to communities of specific species) are recognizable. Different species occupy the same ecological niche in different occurrences of the same habitat [56].

This has been called a *potential habitats mapping approach* [57] and is comparable to the geophysical approach advocated by Day and Roff [56], Zacharius and Roff [51], and Roff et al. [58] (Chapter 11). In this approach, biological data are used to determine those physical parameters that are most important for habitat characterization or are the best surrogates for mapping habitats (e.g., using a probability distribution function) rather than attempting to directly map the diversity of species or communities. This is clearly a sensible approach for mapping biodiversity when dealing with large areas (e.g., at the scale of continental margins), but even for smaller areas, geophysical information should be included along with biological data to assist with the identification of different habitats.

A good example of this approach is the recent study by Clark et al. [59] of the distribution of deep-sea stony coral communities associated with seamounts, which are potentially threatened by deep benthic trawl fishing. The actual distribution of deep-sea stony coral communities is unknown. In order to estimate the *potential* distribution of stony corals, Clark et al. [59] used surrogates including known depth range, bottom water temperature, and seamount occurrence (i.e., the potential habitat) to predict stony coral distribution in the South Pacific Ocean. This information can be used by managers to design MPAs and to put in place conservation measures to protect them, even without a map of known coral distribution.

The part that habitat mapping plays in MPA design involves identifying critical habitat for threatened, endangered, or protected species (TEPS), mapping the location of known biodiversity "hotspots," or iconic features (Figure 2.2), and identifying measurable ecological indicators that can be used to gauge the MPA's performance. Biophysical information may be relevant to helping to identify iconic features such as submarine valleys, seamounts, or reef habitats, but a major role also arises for spatial information in understanding ecosystem processes and the distribution of representative habitats (Figure 2.2).

Principles applied by many countries specify that MPAs must be comprehensive, adequate, and representative [61]. In applying these principles to the design of a national MPA network, "comprehensive" means that MPAs must contain the full range of biodiversity, measured at an appropriate scale (habitat, ecosystem, species,

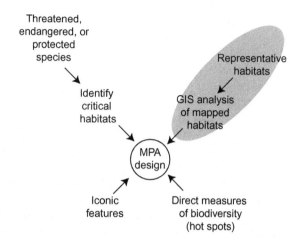

Figure 2.2 Four different ways that an MPA might be designed: identifying TEPS, critical TEPS habitat, the location of known biodiversity "hotspots," or iconic features (from Harris et al. [60]). Habitat information may be relevant to helping to identify iconic features such as submarine valleys, seamounts, or reef habitats, but a major role arises for spatial information in understanding ecosystem processes and the distribution of representative habitats.

or genetic) that occurs in the region of interest. This will include unique biological communities or habitats, but only as one part of the full spectrum of biodiversity that occurs in an area. The MPA network will be "adequate" if it contains a large enough area, with sufficiently redundant representation of different habitats, and is protected to a large enough extent (if not a fully protected MPA) that the ecosystems it contains will remain viable. Adequate replication of ecosystems is considered to be insurance against loss or damage caused by either natural events or anthropogenic activities. Finally, the MPA network should contain examples of habitats that are "representative," meaning they are typical of that region at some scale (e.g., in terms of population or community composition). A representative habitat should contain similar geophysical features, oceanographic conditions, and ecosystem processes in the same proportions as other habitats in that area, and it is assumed that they all contain similar communities and species [56].

Application of these principles is dependent upon reliable habitat maps for the areas of interest; without them it is not possible to determine if a proposed MPA design contains comprehensive, adequate, and representative areas of the habitats present in a given area [60].

Habitat Mapping in Support of Offshore Development

Spatial and temporal habitat information is required for making strategic planning decisions related to development applications, such as construction of oil and gas installations (Chapters 11, 42, 43, and 57), wind farms (Chapters 12 and 35), aquaculture sites, port dredging, coastal development (Chapter 45), aggregate mineral extraction (Chapters 12, 41, 42, and 56), and deep-sea mining (Chapters 62 and 63). The benefits of habitat maps have been demonstrated for determining both the acceptable location and the scale of developments. They support decision making related to site selection from an environmental (e.g., potential loss of biodiversity, EFH, and iconic feature) and socioeconomic (e.g., loss/gain of income and loss of other ecosystem service) perspective. Habitat maps provide an inventory of seafloor assets that include biological and physical resources, allowing valuations to be made that can be used to help decide whether to protect or exploit these resources. For example, a coral reef can be used for making concrete, as a food source, or as a recreational and tourist resource. Habitat mapping, including sensitivity mapping and impact modeling, can define areas of the seabed sensitive to physical disturbance from development and also identify resilient or already degraded ecosystems where development may have little negative impact.

Habitat Mapping Contribution to Marine Environmental Assessments

State of marine environment condition assessments and reporting are carried out by many nations and intergovernmental agencies, as reviewed in the Assessments of Assessments report [9]. Such assessments, viewed within the DPSIR framework (Figure 2.1), require information on the state of the environment, where habitat mapping has a clear role. It should be noted, however, that fully integrated assessments contain more than simple environmental condition assessments (i.e., more than a report on the state); rather, they include an integration of the multiple pressures and the policy responses. The United Nations Regular Process goes even further to include the social and economic aspects of human interactions with the marine environment (e.g., the numbers of people employed in shipping, marine-based tourism, and fisheries, and the wealth generated for dependent communities).

Viewing the pressures, state, and response components of the DPSIR diagram as three separate limbs of a triangle, Figure 2.3 illustrates some important aspects of how these components interact. The pressures located along the base of the triangle are cumulative, as summarized by Halpern et al.'s [62] global study. The pressures also summarize the social and economic aspects of human interactions with the marine environment, as many of the pressures relate to an industry sector (fishing, tourism, mining, etc.).

The right-hand limb of the triangle refers to the state of the environment, which is measured and observed by scientists in relation to one or all of the anthropogenic pressures. The reduction of data collected is illustrated by progressing up the side of the triangle, so that observations can be analyzed to create maps or time series showing trends in condition. Such analyses can sometimes reveal new insights into ecosystem processes and habitats. Habitat mapping is included explicitly within the scope of marine environmental condition assessments and is useful for monitoring

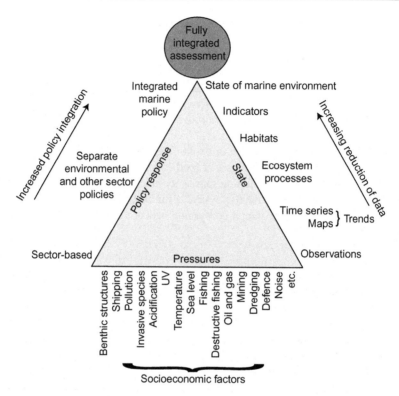

Figure 2.3 The relationships between anthropogenic pressures on the marine environment, state, and policy response as a subset of the DPSIR framework. See text for explanation.

changes in ecosystem health. Habitat mapping can identify indicators that are also useful for environmental monitoring, which is a crucial component of the DPSIR framework (Figure 2.1).

The left side of the triangle includes the response of governments in terms of developing policies (and implementing legislation), taking into account and informed by both the pressures (including social–economic aspects) and the state of the environment. The growth and development of policies are represented by a vertical progression toward the apex of the triangle. Policies may simply be based on sectors (e.g., policy on fisheries, tourism, oil, and gas) or broad groups of pressures. Development of a fully integrated marine environmental policy remains a challenge for most countries.

Concluding Remarks

Marine habitat mapping provides the spatial framework for EBM. Its increasing prominence in international policy and regional and national management organizations highlights its growing acceptance, especially as a tool for designing and delineating MPAs. The management priorities of protecting biodiversity and developing sustainable use solutions, while managing stakeholder expectations and avoiding competing use conflicts need to be implemented with the use of credible scientific information. As we face the reality of increasing ocean use, and with human activities moving into deeper and deeper water, we need to advocate for the continued coevolution of habitat mapping technology, predictive modeling techniques, and applications to management. At present there are wide gaps between the essential policy requirements, especially in the area beyond national jurisdiction, and the science to underpin this policy. If we do not know what ocean habitats exist or where they are located, we risk undervaluing and mismanaging them. Measuring the true value of benthic ecosystems will become increasingly important in the developing green economy, where they are sure to play an important role.

Acknowledgments

Thanks to Emily Corcoran (OSPAR), Tiina Kurvits (UNEP/GRID Arendal), and Norbert Dankers (IMARES) for helpful comments and suggestions on an earlier draft of this chapter. This work was produced with the support of UNEP/GRID Arendal and the Australian Government's Commonwealth Environment Research Facilities (CERF) program. This chapter is published with the permission of the Chief Executive Officers of UNEP/GRID Arendal and Geoscience Australia.

References

[1] C. Nellemann, S. Hain, J. Alder, In Dead Water—Merging of Climate Change with Pollution, Over-Harvest, and Infestations in the World's Fishing Grounds, United Nations Environment Programme, GRID-Arendal, Norway, 2008.

[2] FAO, The State of World Fisheries and Aquaculture: 2008, FAO Food and Agriculture Organization of the United Nations, Rome, 2009, p. 176.

[3] L.B. Crowder, G. Osherenko, O.R. Young, S. Airame, E.A. Norse, N. Baron, et al., Sustainability—resolving mismatches in US ocean governance, Science 313 (2006) 617–618.

[4] R. O'Boyle, G. Jamieson, Observations on the implementation of ecosystem-based management: experiences on Canada's east and west coasts, Fish. Res. 79 (2006) 1–12.

[5] G. Ottersen, E. Olsen, G.I. van der Meeren, A. Dommasnes, H. Loeng, The Norwegian plan for integrated ecosystem-based management of the marine environment in the Norwegian Sea, Mar. Policy 35 (2011) 389–398.

[6] Convention for the Conservation of Antarctic Marine Living Resources (CCAMLR), Report of the Twenty-seventh Meeting of the Commission for the Conservation of Antarctic Marine Living Resources, Hobart, TAS.

[7] Convention on Biological Diversity (CBD). Available from http://www.cbd.int/.

[8] United Nations Convention on the Law of the Sea. Available from: http://www.un.org/Depts/los/convention_agreements/convention_overview_convention.htm.

[9] O.R. Young, G. Osherenko, J. Ogden, L.B. Crowder, J.A. Wilson, J.C. Day, et al., Solving the crisis in ocean governance: place-based management of marine ecosystems, Environment 49 (2007) 20–32.

[10] C.B. Cogan, B.J. Todd, P. Lawton, T.T. Noji, The role of marine habitat mapping in ecosystem-based management, ICES J. Mar. Sci. 66 (2009) 2033–2042.

[11] L. Jeftic, P. Glennie, L. Talaue-McManus, J.A. Thornton (Eds), Methodology and Arrangements for the GEF Transboundary Waters Assessment Programme, vol. 1, United Nations Environment Programme, Nairobi, 2011, pp. 59.

[12] EEA, State and Pressures of the Marine and Coastal Mediterranean Environment, (Environment Assessment Series 5) European Environment Agency, Copenhagen, Denmark, 1999.

[13] UNEP and IOC-UNESCO, An Assessment of Assessments, Findings of the Group of Experts. Start-up Phase of the Regular Process for Global Reporting and Assessment of the State of the Marine Environment Including Socio-Economic Aspects, UNEP and IOC/UNESCO, Malta, 2009.

[14] F. Douvere, C.N. Ehler, New perspectives on sea use management: initial findings from European experience with marine spatial planning, J. Environ. Manage. 90 (2009) 77–88.

[15] GBRMPA, Zoning Plan 2003, Great Barrier Reef Marine Park Authority, Townsville, QLD, 2004.

[16] V.E. Kostylev, B.J. Todd, J. Shaw, Benthic habitat mapping in Canada—a perspective, J. Ocean Technol. 3 (2008) 7–12.

[17] V.E. Kostylev, C.G. Hannah, Process-driven characterization and mapping of sea-bed habitats, in: B.J. Todd, H.G. Greene (Eds.), Mapping the Seafloor for Habitat Characterization, Geological Association of Canada, St. Johns, NL, 2007, pp. 171–186, (Special Paper 47).

[18] R.A. Pickrill, V.E. Kostylev, Habitat mapping and national seafloor mapping strategies in Canada, in: B.J. Todd, H.G. Greene (Eds.), Mapping the Seafloor for Habitat Characterization, Geological Association of Canada, St. Johns, NL, 2007, pp. 483–495, (Special Paper 47).

[19] K.A. Robinson, K. Ramsay, C. Lindenbaum, N. Frost, J. Moore, A.P. Wright, et al., Predicting the distribution of seabed biotopes in the southern Irish Sea, Cont. Shelf Res. 31 (2011) S120–S131.

[20] T.T. Noji, S.A. Snow-Cotter, B.J. Todd, M.C. Tyrell, P.C. Valentine, Gulf of Maine Mapping Initiative: a Framework for Ocean Management, Gulf of Maine Council on the Marine Environment, Portland, ME, p. 22.

[21] BALANCE, Towards Marine Spatial Planning in the Baltic Sea, Balance Technical Summary Report 4/4, 2008.

[22] EU, Marine Strategy Framework Directive. Available from http://ec.europa.eu/environment/water/marine/index_en.htm, 2008.

[23] J.A. Ardron, G.S. Jamieson, D. Hangaard, Spatial identification of closures to reduce the bycatch of corals and sponges in the groundfish trawl fishery, British Columbia, Canada, Bulletin of Marine Science, Volume 81, Supplement 1, November 2007, pp. 157–167.

[24] P.T. Harris, T. Whiteway, High seas marine protected areas: benthic environmental conservation priorities from a GIS analysis of global ocean biophysical data, Ocean Coastal Manage. 52 (2009) 22–38.

[25] ICES, Report of the Working Group on Marine Habitat Mapping (WGMHM), Calvi, Corsica, France, 3–7 May 2010.

[26] P.A.R. Hockey, G.M. Branch, Criteria, objectives and methodology for evaluating marine protected areas in South Africa, S. Afr. J. Mar. Sci. 18 (1997) 369–383.

[27] J.C. Roff, M.E. Taylor, National frameworks for marine conservation—a hierarchical geophysical approach, Aquat. Conserv. Mar. Freshwater Ecosyst. 10 (2000) 209–223.

[28] S.A. Banks, G.A. Skilleter, Mapping intertidal habitats and an evaluation of their conservation status in Queensland, Australia, Ocean Coastal Manage. 45 (2002) 485–509.

[29] C. Roberts, S. Andelman, G. Branch, R. Bustamante, J.C. Castilla, J. Dugan, et al., Ecological criteria for evaluating candidate sites for marine reserves, Ecol. Appl. 13 (2003) 199–214.

[30] C. Roberts, S. Andelman, G. Branch, R. Bustamante, J.C. Castilla, J. Dugan, et al., Application of ecological criteria in selecting marine reserves and developing reserve networks, Ecol. Appl. 13 (2003) 215–228.

[31] B. Worm, R. Hilborn, J.K. Baum, T.A. Branch, J.S. Collie, C. Costello, et al., Rebuilding Global Fisheries, Science 325 (2009) 578–585.

[32] C. Mora, R.A. Myers, M. Coll, S. Libralato, T.J. Pitcher, R.U. Sumaila, et al., Management effectiveness of the world's marine fisheries, PLoS Biol. (2009) 7.

[33] P.K. Dayton, S. Thrush, F.C. Coleman, Ecological Effects of Fishing in Marine Ecosystems of the United States, Pew Oceans Commission, Arlington, VA, 2002, pp. 45.

[34] S.M. Garcia, K.L. Cochrane, Ecosystem approach to fisheries: a review of implementation guidelines, ICES J. Mar. Sci. 62 (2005) 311–318.

[35] C. Jorgensen, K. Enberg, E.S. Dunlop, R. Arlinghaus, D.S. Boukal, K. Brander, et al., Ecology—managing evolving fish stocks, Science 318 (2007) 1247–1248.

[36] FAO, The ecosystem approach to fisheries, FAO Fisheries Technical Paper, p. 443, 2003.

[37] B. Planque, J.M. Fromentin, P. Cury, K.F. Drinkwater, S. Jennings, R.I. Perry, et al., How does fishing alter marine populations and ecosystems sensitivity to climate? J. Mar. Syst. 79 (2010) 403–417.

[38] NRC, Effects of Trawling and Dredging on Seafloor Habitat, A report by the Committee on Ecosystem Effects of Fishing, the Ocean Studies Board of the National Research Council, 2002.

[39] T.E. Hart, J.H. Grabowski. 2009. Integrating seafloor mapping and benthic ecology into fisheries management in the Gulf of Maine, Workshop Proceedings, 15-16 June 2009, Portland, ME.

[40] S.L. Copps, M.M. Yoklavich, G. Parkes, W.W. Wakefield, A. Bailey, H.G. Greene, et al., Applying marine habitat data to fishery management on the US west coast: initiating a policy-science feedback loop, in: B.J. Todd, H.G. Greene (Eds.), Mapping the Seafloor for Habitat Characterization, Geological Association of Canada, St. Johns, NL, 2007, pp. 439–450, (Special Paper 47).

[41] J.C. Rice, D. Rivard, The dual role of indicators in optimal fisheries management strategies, ICES J. Mar. Sci. 64 (2007) 775–778.

[42] E.K. Pikitch, C. Santora, E.A. Babcock, A. Bakun, R. Bonfil, D.O. Conover, et al., Ecosystem-based fishery management: reversing the means to an end, Science 305 (2004) 346–347.

[43] V.E. Kostylev, R.C. Courtney, G. Robert, B.J. Todd, Stock evaluation of giant scallop (*Placopecten magellanicus*) using high-resolution acoustics for seabed mapping, Fish. Res. 60 (2003) 479–492.

[44] V.M. O'Connell, C.K. Brylinsky, H.G. Greene, The use of geophysical data in fisheries management: a case history from southeast Alaska, in: B.J. Todd, H.G. Greene (Eds.), Mapping the Seafloor for Habitat Characterization, Geological Association of Canada, St. Johns, NL, 2007, pp. 319–328, (Special Paper 47).

[45] I. Galparsoro, A. Borja, J. Bald, P. Liria, G. Chust, Predicting suitable habitat for the European lobster (*Homarus gammarus*), on the Basque continental shelf (Bay of Biscay), using Ecological-Niche Factor Analysis, Ecol. Modell. 220 (2009) 556–567.

[46] P.J. Auster, K. Gjerde, E. Heupel, L. Watling, A. Grehan, A.D. Rogers, Definition and detection of vulnerable marine ecosystems on the high seas: problems with the "move-on" rule, ICES J. Mar. Sci. (2010).

[47] J.F. Caddy, Underwater observations on tracks of dredges and trawls and some effects of dredging on a scallop ground, J. Fish. Res. Board Can. 30 (1973) 173.

[48] P Dando, D Stben, S. Varnavas, Hydrothermalism in the Mediterranean sea, Prog. Oceanogr. 44 (1999) 333–367.

[49] T. Morato, S.D. Hoyle, V. Allain, S.J. Nicol, Tuna longline fishing around West and Central Pacific seamounts, PLoS ONE 5 (12) (2010) e14453.

[50] S. Gubbay, Marine protected areas—past, present and future, in: S. Gubbay (Ed.), Marine Protected Areas: Principles and Techniques for Management, Chapman & Hall, London, 1995, pp. 1–14.

[51] M.A. Zacharias, J.C. Roff, A hierarchical ecological approach to conserving marine bio-diversity, Conserv. Biol. 14 (2000) 1327–1334.

[52] B. Halpern, S.E. Lester, K.L. McLeod, Placing marine protected areas onto the ecosystem based management seascape, PNAS 107 (2010) 18312–18317.

[53] G.J. Edgar, J. Moverley, N.S. Barrett, D. Peters, C. Reed, The conservation-related benefits of a systematic marine biological sampling programme: the Tasmanian reef bio-regionalisation as a case study, Biol. Conserv. 79 (1997) 227–240.

[54] W. Gladstone, The potential value of indicator groups in the selection of marine reserves, Biol. Conserv. 104 (2002) 211–220.

[55] H.G. Greene, M.M. Yoklavich, R.M. Starr, V.M. O'Connell, W.W. Wakefield, D.E. Sullivan, et al., A classification scheme for deep seafloor habitats, Oceanolog. Acta 22 (1999) 663–678.

[56] J.C. Day, J.C. Roff, Planning for Representative Marine Protected Areas: A Framework for Canada's Oceans, World Wildlife Fund Canada, Toronto, ON, 2000, p. 148.

[57] H.G. Greene, J.J. Bizzarro, V.M. O'Connell, C.K. Brylinsky, Construction of digital potential benthic habitat maps using a coded classification scheme and its application, in: B.J. Todd, H.G. Greene, (Eds.), Mapping the Seafloor for Habitat Characterisation, Geological Association of Canada, St. Johns, NL, 2007, pp. 141–156.

[58] J.C. Roff, M.E. Taylor, J. Laughren, Geophysical approaches to the classification, delineation and monitoring of marine habitats and their communities, Aquat. Conserv. Mar. Freshwater Ecosyst. 13 (2003) 77–90.

[59] M.R. Clark, L. Watling, C. Smith, A. Rowden, J.M. Guinotte, A global seamount classification to aid the scientific design of marine protected area networks, Ocean Coastal Manage. 54 (2011) 19–36.

[60] P.T. Harris, A.D. Heap, T. Whiteway, A.L. Post, Application of biophysical information to support Australia's representative marine protected area program, Ocean Coastal Manage. 51 (2008) 701–711.

[61] UNEP-WCMC, National and Regional Networks of Marine Protected Areas: A Review of Progress, UNEP-WCMC, Cambridge, 2008.

[62] B.S. Halpern, S. Walbridge, K.A. Selkoe, C.V. Kappel, F. Micheli, C. D'Agrosa, et al., A global map of human impact on marine ecosystems, Science 319 (2008) 948–952.

3 Anthropogenic Threats to Benthic Habitats

Peter T. Harris

Marine and Coastal Environment Group, Geoscience Australia, Canberra, ACT, Australia

Abstract

Anthropogenic threats to benthic habitats do not pose an equal risk, nor are they uniformly distributed over the broad depth range of marine habitats. Deep-sea benthic environments have, by and large, not been heavily exploited and most are in relatively good condition. In contrast, shelf and coastal habitats, and deep-ocean pelagic fisheries, have been exploited extensively, and human impacts here are locally severe. A critical point is that anthropogenic threats do not act in isolation: They are cumulative, and their impacts are compounded for every affected habitat. In general, human impacts on benthic habitats are poorly understood.

Habitat mapping provides condition assessments and establishes baselines against which changes can be measured. GeoHab scientists ranked the impacts on benthic habitats from fishing as the greatest threat, followed by pollution and litter, aggregate mining, oil and gas, coastal development, tourism, cables, shipping, invasive species, climate change, and construction of wind farms. The majority of authors (84%) reported that monitoring changes in habitat condition over time was a planned or likely outcome of the work carried out. In this chapter, the main anthropogenic threats to benthic habitats are reviewed in relation to their potential impacts on benthic environments.

Key Words: Benthic habitat condition; cumulative impacts; naturalness; fishing; oil and gas; mining; shipping; introduced marine pests; noise; pollution; global climate change

Measuring and Mapping Human Impacts

It is important at the outset to emphasize that human impacts are cumulative; overfishing, pollution, noise, mining, and so on often affect the same areas, and individual species are simultaneously affected by more than one stressor ([1–3]; Figure 3.1). The cumulative impacts of human activities have affected all parts of the oceans to greater or lesser degrees, but the greatest impacts have been in the coastal and shelf environments. Separate impacts ranked in order of priority for percentage of species affected [4], for example, shows that a simple list of threats does not predict the combined effect of multiple stressors.

Seafloor Geomorphology as Benthic Habitat. DOI: 10.1016/B978-0-12-385140-6.00003-7

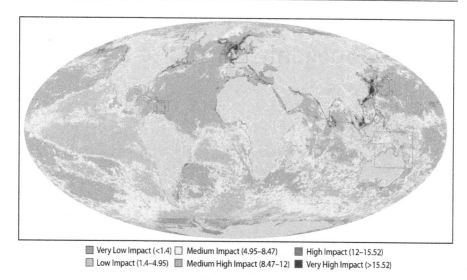

Very Low Impact (<1.4) ☐ Medium Impact (4.95–8.47) ▨ High Impact (12–15.52)
Low Impact (1.4–4.95) ▨ Medium High Impact (8.47–12) ■ Very High Impact (>15.52)

Figure 3.1 Map showing the cumulative human impacts on ocean health based on the synthesis of 17 global data sets of anthropogenic drivers of ecological change (after Halpern et al. [3]). Note that only a few areas of the Arctic and Antarctic oceans are assessed as having suffered very low impact. (For interpretation of the references to color in this figure legend, the reader is referred to the web version of this book.)

In order to detect anthropogenic impacts, measurements must first account for the natural variability that affects any signal. Baseline observations of dynamic systems must be referenced to a long time series of measurements in order to account for natural fluctuations that occur in many cases. For example, Gregg and Conkright [5] estimate that global primary productivity has decreased by 6% between the 1980s and 1990s based on a comparative analysis of SeaWiffs satellite imagery. These authors attribute this decline to natural variability of ocean productivity over decadal time scales, and not to anthropogenic climate change. This study is an example of the complexity involved with measuring change in marine systems.

The case studies presented in Part 2 of this book describe the relative *naturalness* of the studied habitats, that is, the extent to which a habitat or environment has been modified from its natural state by human activities and impacts [1]. When asked to list the threats they believed posed the most immediate risk to the habitats they studied, more authors nominated fishing, followed by pollution, aggregate mining, oil and gas, coastal development, tourism, submarine cables, shipping, invasive pests, climate change, and wind farms (Table 3.1).

Perhaps one of the greatest challenges facing the management of shelf and coastal marine environments is that pristine, benchmark sites are rare or absent, making the assessment of human impacts difficult. The goals for shelf and coastal conservation efforts therefore frequently deal with restoration of previously existing ecosystems, rather than with maintenance of current status. A corollary of this observation is that the identification of remaining pristine sites is a priority for conservation, in order to

Table 3.1 List of Threats Nominated by Case Study Authors

Threat	Case Study Chapter
Fishing	8–12, 14, 15, 18, 19, 22–24, 26, 29, 30, 32–46, 48–61, 63
Pollution including litter	7, 11, 17, 20, 28, 32, 34, 35, 43, 50, 53
Aggregate mining	8, 12, 13, 33, 37, 41, 42
Other mining, oil, and gas	11, 40, 43, 48, 56, 62, 63
Coastal development, coastal inputs	7, 8, 11, 14, 28, 37
Tourism	11, 16, 18, 25–27
Cables	12, 33, 41, 53, 57
Shipping	12, 17, 33, 37, 41, 43
Invasive species	8, 43, 45
Climate change	21, 24, 34
Wind farms	12, 13, 40

The case study chapters (in Part 2 of this book) are listed against each threat they nominated. Multiple stressors are represented where case study chapters are cited against more than one threat.

help establish benchmarks and control sites for condition assessments and monitoring the performance of conservation measures.

Of the 57 case studies included in this book, habitat mapping was intended to be part of an ongoing monitoring program in 24 cases (two case studies specified that an ongoing monitoring program was part of their purpose; Chapters 10 and 18). A further 24 case studies reported that their habitat map would form the baseline for monitoring future changes. Thus, 48 out of 57 studies (84%) reported that monitoring changes in habitat condition over time was a planned or likely outcome of the work carried out.

Fishing

Compared with all other human activities, fishing is, without a doubt, the most immediate and pervasive threat to marine ecosystems ([6,7]; see also the chapters listed in Table 3.1). In the modern industrial age, and especially over the past 50 years, fishing has expanded into almost every part of the world's coastal and shelf seas (Figure 3.2). The consequences have been severe. Many fisheries have been overexploited to the point of permanent closure. Of particular significance to benthic habitats is the impact of towed fishing gear (trawls and dredges), which plows through soft sediments, destroying biological structures such as worm tubes and burrows, but also overturns rocks, levels bedforms, and dislodges sessile and colonial benthic animals such as sponges, corals, and bryozoans. An entire recent symposium volume of the American Fisheries Society is devoted to the subject [9], and in Norse and Crowder's [1] recent textbook, five of the nine chapters devoted to human impacts on marine environments are concerned with the impacts of fishing.

The most productive fisheries are concentrated in approximately 10% of the ocean, mainly on the continental shelf, while vast areas of the oceans are relatively

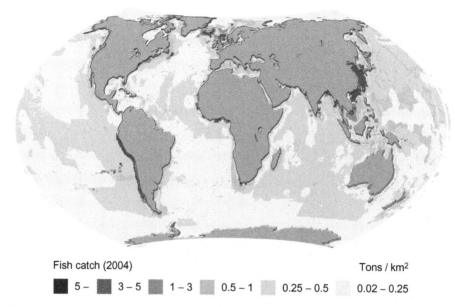

Fish catch (2004)   Tons / km²

■ 5 –   ■ 3 – 5   ■ 1 – 3   ░ 0.5 – 1   ░ 0.25 – 0.5   ░ 0.02 – 0.25

Figure 3.2 Global fish catch in 2004 [8]. Most fish are caught in approximately 10% of the ocean area, mainly on the continental shelves where productivity is high. High productivity is associated with freshwater runoff from land and upwelling onto the shelf of cold, nutrient-rich waters from deep-ocean basins. (For interpretation of the references to color in this figure legend, the reader is referred to the web version of this book.)

unproductive (Figure 3.2). The direct damage caused by fishing is most apparent on reefs, seamounts, ridges, and other rocky habitats where bottom trawling techniques are used. Trawl gear not only captures the target species (fish and shrimp) but doors, weights, and chains used to hold the nets open impact the bottom and kill or injure nontarget benthic epifauna [10], such as cold-water coral and sponge communities (Figure 3.3A and B). Seamount trawl fisheries developed over the past 50 years have a very poor track record for being managed sustainably. Unregulated fishing occurs on seamounts and ridges located in the high seas; the amount of damage caused is unknown but is probably widespread and significant. The amount of biodiversity at risk is probably large, based on biogeographic patterns from the few seamounts that have been studied [11].

The sheer quantity of commercial oceanic fishing has greatly reduced the uppermost sections of trophic levels. Fishing of pelagic top predators has reduced their numbers by around 90% of what existed prior to 1950 [12]. With the higher predators removed, effort has switched to lower trophic levels [13]. For example, the extermination of the majority of baleen whales between 1950 and 1970 resulted in a so-called krill surplus of around 150 million tons/year, representing the amount of krill that was formerly consumed by whales and that was now available for human use (note the current catch of Antarctic krill is only around 120,000 tons/year). The effect of these dramatic changes on populations at the higher trophic levels and on populations at lower levels (such as krill) are unknown [14]. Their effects on the ocean ecosystem as a whole are incredibly complex and may never be fully understood.

(A)

(B)

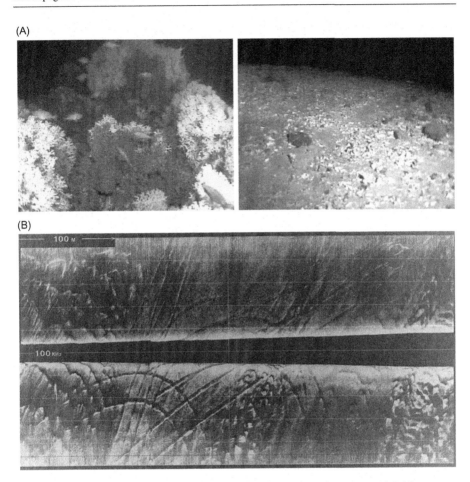

Figure 3.3 (A) Bottom photographs showing seafloor before (left) and after (right) bottom trawling has occurred on deep-sea coral gardens on the continental slope off Norway. Note in photo on right the elongate trawl mark on the seafloor, resulting from dragging trawl doors. (B) Side scan sonograph showing elongate and curved tracks made by bottom trawl boards on the seabed of Moreton Bay, Australia. Light-toned areas are elongate trains of sand dunes. (A) Photos from UNEP/GRID Arendal, Norway, and (B) image from Geoscience Australia.

Oil and Gas Exploration and Production

The threats to marine life from mining of hydrocarbons (not including the effects of climate change) are generally lower than for either fishing or mining [10]. Only two case studies nominated oil and gas development as a threat (Chapters 43 and 48). The risks of impact from oil spills are greatest during transport, from pipeline rupture, vessel loss, or spillage, when large volumes of oil can be released suddenly. Accidental (anthropogenic) spills are ecologically damaging because they result in unnatural concentrations of oil at a particular site that is incompatible with local marine life. However, crude oil is a naturally occurring substance on the earth and an

Table 3.2 Estimates of Global Inputs of Oil into the Marine Environment (Based on 15–18)

Source	1970s	1980s	1990s
Land-based sources: Urban runoff and discharges	2,500	1,080 (500–1,250)	1,175
Coastal refineries	200	100 (60–600)	–
Other coastal effluents	–	50 (50–200)	–
Oil transportation and shipping: Operational discharges from tankers	1,080	700 (400–1,500)	564
Tanker accidents (ITOPF, 2007)	314	118	114
Losses from nontanker shipping	750	320 (200–600)	–
Offshore production discharges	80	50 (40–60)	47
Atmospheric fallout	600	300 (50–500)	306
Natural seeps (Kvenvolden and Cooper, 2003)	600 (200–2,000)	600 (200–2,000)	600 (200–2,000)
Total discharges	**6,1240**	**3,318**	**2,806**

Units are thousands tons/year with error range in brackets where available ("–" indicates no data available).

amount of oil equal to or larger than that spilled accidentally by humans enters the oceans each year through natural seepage [15]. Natural seepage is a gradual, ongoing process, and ecosystems have evolved that use it as a food source. Also it should be noted that accidental oil spills account for only a small percentage of the total volume of oil that enters the oceans due to humans. Most oil enters the ocean mixed with sewage and urban stormwater runoff (Table 3.2), but such diffuse sources do not have the same dramatic impact on ecosystems as a spill because the oil is delivered continuously in low concentrations over a broad area. The long-term effect of low-level oil pollution from diffuse sources is unknown [16].

Environmental impacts arise throughout petroleum exploration–drilling–production development operations, although the nature and degree of impact varies [17]. Seismic surveys, oil and gas production, and transportation all have associated environmental impacts; these are described briefly in the following section.

Seismic Surveys

Marine acoustic survey equipment is used by the oil and gas industry as well as by the military and other marine industries to map the seafloor and to study seafloor geology and the water column. These systems are generally far more powerful than equipment used for marine research or normal ship navigation (see also section "Noise" regarding marine noise). Equipment that is most likely to affect marine mammals and other life are airgun (seismic) arrays and low-frequency, high-power transducers with wide-beam angles (deep-water multibeam sonar systems). Cetaceans have been observed avoiding powerful, low-frequency sound sources, and there is now a documented case of injury to whales and whale stranding caused by multiple, midfrequency (2.6–8.2 kHz) military echo sounders [18]. Although there is a historical record of cetaceans stranding themselves prior to the industrial age (i.e., cetacean stranding occurs under natural environmental conditions), an increase in the number of stranding events may be an indicator of human impact. Some whale populations coexist with commercial seismic exploration surveys. In general, it seems that high-powered military type sonar systems are likely to have the greatest impact on marine life [19]. In the case of marine animals other than cetaceans, there is some evidence for short-term displacement of seals and fish by seismic surveys, but there is little literature available [20].

Drilling and Production Activities

Drilling activities are carried out from ships or fixed platforms during exploration and to extract oil once it has been found. In 2004, there were some 8,300 offshore oil production platforms in the world producing a total of around 11 million tons (80 million barrels) of oil per day. A further approximately 3,500 rotary drill rigs are used for exploration globally. Direct damage to the seafloor is caused by the anchors used to hold the rig in place, as well as by the impact of the drill itself. Drilling requires the use of lubricant (drilling mud) and the disposal of drill cuttings onto the seabed at the drill site. Drilling mud and some of the drill cuttings can be toxic, so the environmental impacts of drilling may involve burial together with toxicity affects [17]. The initial impact is generally confined to the immediate surrounds, typically within 150 m of the drill site. However, Olsgard and Gray [21] reported barium, total hydrocarbons, zinc, copper, cadmium, and lead contamination sourced from production platforms on the Norwegian shelf had spread after a period of 6–9 years, so that evidence of contamination was found 2–6 km away from the platforms.

During the production of oil and gas, water from the hydrocarbon reservoir is also brought to the surface. This so-called produced formation water (PFW) is a by-product of oil production and is disposed of in the ocean [17]. Compared with ambient seawater, PFW may contain elevated concentrations of heavy metals (e.g., arsenic, mercury, barium, copper, lead, and zinc), radium isotopes, and hydrocarbons. The proportion of oil/water varies between locations, but generally the proportion of water increases over time as the oil deposit is depleted (i.e., older wells discharge more PFW than new wells). The proportion of water produced per barrel of oil typically ranges from around 3:1 to 7:1, although in some extreme cases the fluid pumped from a well might be 98% water and only 2% oil [22]. At the production platform, most of the PFW is separated from the oil, treated (typically to around 30 mg/l hydrocarbon), and disposed of in the ocean.

PFW forms a buoyant plume because it is typically 40–80°C warmer (and therefore less dense) than ambient seawater. Thus, it will be dispersed by wind and currents away from the production platform. Mixing and dilution with seawater results in toxic effects of PFW being generally confined to within 1 km of production platforms, although PFW plumes may be detected in surface waters for distances exceeding 10 km from the point source [23]. Cases of coral discoloration (coral bleaching) have been attributed to dilute (~12%) PFW concentrations [24]. Hence, the situation of production platforms in relation to prevailing winds and currents and to the proximity of sensitive habitats is a consideration for offshore petroleum development.

Oil Spills

Between 1970 and 2005, some 2.5 million tons of oil was accidentally spilled into the ocean from oil tankers and barges. It is significant that most of this oil was spilled during a few catastrophic events resulting from tankers running aground, colliding with other ships or breaking up in stormy seas. Of the 9,309 accidents documented since 1967 by ITOPF [25], 4,987 (about 50%) occurred during routine loading/unloading operations but resulted in very small volume spills. The worst year for the tanker industry was 1979, when some 670,000 tons of oil was spilled, of which 287,000 tons was from one ship (MV Atlantic Express). In the 2010, Gulf of Mexico oil spill, it is estimated that around 4.4 million barrels (about 600,000 tons, assuming a specific gravity of 0.88) was discharged into the sea before the well was capped—more oil than was spilled by the Exxon Valdez in 1989 [26]. The impacts of this huge volume of oil on deep-water habitats in the Gulf of Mexico are unknown at the time of this writing (March 2011).

Accidents that occur in coastal waters have the most severe environmental impact because of the fact that most oil floats on the sea surface and its effects are concentrated at the shoreline (Figure 3.4). The coast is also a habitat for a diversity of species of birds, mammals, invertebrates, and marine plants. For this reason, spills that impact the coast, such as the Exxon Valdez spill that occurred in Alaska in 1989, have the greatest impact on the ecosystem [27].

The impacts of oil spills range from the immediate effects of oiling to longer term consequences of habitats being modified by the presence of oil and tar balls. Traces

Figure 3.4 On March 24, 1989, the supertanker *Exxon Valdez* grounded on a rocky reef in Prince William Sound, Alaska, spilling about 100,000 tons of crude oil into the sea. (USGS Photo)

of hydrocarbons can remain in coastal sediments for many years after an oil spill [28]. Some species exhibit reduced abundance associated with the timing of spills [29], although direct causal evidence is not always available [30]. Some opportunistic species are able to take advantage of the changed habitat conditions and the attendant reduced abundance of some species, giving rise to a short-term *increase* in local biodiversity [31,32]; this is an example of why biodiversity statistics alone are not a reliable indicator of environmental health. Recovery time for sites varies as a function of the type of oil spilled, the biological assemblage impacted, substrate type, climate, wave/current regime, and coastal geomorphology, ranging from years to decades depending on these and other factors [33–35].

Seabed Mining

Mining of the seabed is carried out in mostly shallow shelf and coastal waters to extract heavy minerals, gold, diamonds, tin, sand, and gravel to nourish beaches and as a source of shell and aggregate for use in making concrete [35,36]. Such activities have the potential to impact benthic flora and fauna, planktonic ecosystems, fisheries, and marine mammals that utilize the area being mined. Where mining activities result in the removal of large volumes of bed material, changes in wave transformation, storm surge, bottom currents, and the dynamics of the shoreline can occur [37].

Of all seabed mining activities, aggregate extraction is by far the largest and most widespread (see the seven case studies listed in Table 3.1). Statistics are difficult to obtain from most countries. However, in the United Kingdom, around 20 million tons of sand and gravel is extracted from an area of seabed equal to about $1,500 \, km^2$ each year [38]. Studies on the impacts of sand/gravel extraction have been carried out in several countries including Japan [39], Australia [40,41], the United States [42], and the United Kingdom [43]. Recovery from the effects of aggregate mining at three sites in the United Kingdom was studied by Boyd et al. [44]. These workers found that evidence of physical disturbance of the seabed and perturbation of benthic fauna was detectable from 3 to 7 years after mining had ceased. Disturbance of the sediments from mining can make available organic matter previously trapped among sediment particles, thus driving a temporary increase in the abundance of benthos [40].

Mining in the deep sea has, at the present time, not progressed beyond a few pilot studies to extract manganese nodules. Mining of metal-rich sulfide deposits associated with an extinct hydrothermal vent system in the Manus Basin area of Papua New Guinea (http://www.nautilusminerals.com/s/Home.asp) is being planned but is yet to commence (see Chapter 62). In a recent publication, Hein et al. [45] evaluate seamount associated manganese crusts as a source of high-tech minerals such as tellurium, cobalt, bismuth, thallium, and others. They predict that seabed mining on seamounts will become a reality sooner rather than later. If commercial-scale mining for manganese nodules and/or metalliferous hydrothermal vent deposits were to proceed on a broad scale, there would potentially be significant environmental impacts. These would include (1) direct impacts on the benthic communities where nodules/ore deposits are removed; (2) impacts on the benthos due to deposition of mobilized sediment; and

(3) impacts in the water column in cases where mining vessels discharge a plume of sediment near the sea surface, thus affecting photosynthesizing biota and pelagic fish [46,47]. In general, environmental impact studies have been inconclusive. For example, in their review of the potential environmental effects of manganese nodule mining, Morgan et al. [46] found that none of the studies completed up to that time had been able to establish quantitative relationships between burial depth and faunal succession.

Pollution

Throughout the twentieth century, there has been widespread use of the oceans as a global waste repository, for dredge spoil, sewage sludge, industrial chemical waste, worn-out ships, unwanted military hardware, and radioactive waste [1,48–50]. To this we must add the waste and litter that is inadvertently washed or blown into the oceans from land. One remotely operated vehicle (ROV) survey on the mid-Atlantic ridge reported the "frequent occurrence of garbage (e.g., plastic bags and other objects) at all depths over very wide areas" ([49]; see also Chapter 50). Seafloor observations taken on 27 marine surveys carried out in European waters between 50 and 2,700 m water depth were compiled by Galgani et al. [49], who reported maximum concentrations of garbage (mostly plastic bags and bottles) of over 100,000 pieces of debris per square kilometer. The effects of litter on marine biota are poorly understood.

Since the London Convention on ocean dumping came into force in 1975, most countries have gradually reduced waste disposal in the ocean, although much (mostly inert) material is still disposed of in the ocean today. For example, in the United States, about 52 million m^3 of dredged material from ports and shipping channels is disposed of at sea each year [48]. The disposal of all radioactive waste into the oceans was finally banned in 1994 under amendments made to the London Convention. Disposal has generally been proximal to the point of origin (closest deep-water area to major ports), although deep-ocean trenches have been proposed for the disposal various kinds of waste. The effects of waste disposal on marine biota are poorly understood (see case studies listed in Table 3.1).

Coastal Development

A major cause of habitat loss in coastal and inner-shelf environments derives from coastal development, the discharge of untreated sewage into coastal waters, and other land-based discharges from various human activities (see the six case studies listed in Table 3.1). In their assessment of 12 once diverse and productive estuaries and coastal seas, Lotze et al. [51] found that human impacts have depleted 90% of formerly important species, destroyed 65% of seagrass and wetland habitat, and degraded water quality. In the Caribbean, Africa, and Latin America, 80–90% of sewage discharged into coastal waters is untreated, leading to widespread eutrophication, harmful algal blooms, and oxygen depletion [52]. Apart from sewage, other land-derived pollutants are introduced as a result of agriculture (fertilizers and pesticides), logging (sediment), and mining (sediment and heavy metals). These effects can be locally cumulative, resulting in significant impacts on marine ecosystems and habitats.[1] The degradation and destruction of coastal habitats resulting from human activities means a reduction in the available habitat for the maintenance of biodiversity [52].

Submarine Cables

Submarine cable installation is perceived as a possible threat by five case study authors (Table 3.1). Cables are laid in trenches where waves or strong currents prevail, but otherwise they are simply draped across the seabed. Apart from damage caused by trenching, subsequent vibration or movement of the cable can damage the seabed. On the remote Lord Howe Rise in the South Pacific, the only human impact on the benthic environment detected was the identification of elongate furrows in bottom photographs attributed to telecommunication cable-laying activities, presumably caused by simply lowering the cable onto the seabed [53]. Two studies on the impacts of submarine cables published from sites in the Baltic Sea [54] and off the coast of California [55] showed few changes in the abundance or distribution of benthic fauna based on video observations (epifaunal) and sediment core samples (infauna). Overall, the results indicate that the biological impacts of cables are minor at most. The California study [55] found that Actiniarians (sea anemones) colonized the cable when it was exposed on the seafloor and were therefore generally more abundant on the cable than in surrounding, sediment-dominated seafloor habitats. Some fish were also more abundant near the cable, apparently due to the higher habitat complexity provided.

Shipping

Shipping is listed as a threat by six case studies (Table 3.1). The impacts of shipping include the introduction of exotic species and ship-generated noise in the environment.

Introduction of Exotic Species

The impact of shipping to the natural distribution of species has been locally dramatic. Some commercially valuable species have been deliberately transported to different locations for the production of food and ornaments (fish for aquaculture, aquarium specimens, pearls or pearl shell, etc.; [56]). Most exotic species have, however, been accidentally introduced from one continent to another by ships that unwittingly carry plants and animals attached to their hulls or else contained in ballast water [57]. Mollusk species simply attach themselves to the hull. Planktonic algae, dinoflagellates, fish, spores, and larvae can be pumped aboard when the ship takes on ballast. When the ship arrives at a port, it discharges its ballast to take on cargo.

---

[1] The literature on the effects of coastal pollution on the marine environment is vast; good reviews can be found in the volumes edited by Norse and Crowder [1] and by Nellemann et al. [8].

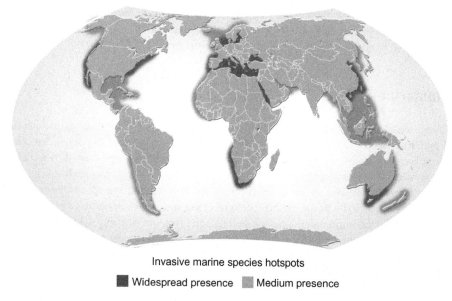

Invasive marine species hotspots

■ Widespread presence   ■ Medium presence

Figure 3.5 Global impact of invasive species, from Nellemann et al. [8].

Animals and plants that attached themselves to a ship's hull in one port can be dislodged and dropped off in another. In these ways, an estimated 15,000 exotic species are carried around the world (Figure 3.5) and dispersed between different ports *every week* [55]. In some cases, the introduced species can outcompete and replace the native species occupying a similar ecological niche.

Canal digging provides another pathway for exotic species to be introduced to new areas. When the Suez Canal was opened in 1869, the shipping route between Europe and Asia was shortened by about 4,400 km, but a link was also formed between two previously separate bodies of water, the Red Sea and the Mediterranean Sea. Since the sea level in the Red Sea is slightly higher than the Mediterranean (due to evaporation exceeding precipitation in the Mediterranean), water flows in only one direction through the canal toward the Mediterranean. The water has carried some 300 species of plants and animals from the Red Sea that have begun to colonize the Mediterranean, including some deep-water species. The new arrivals, comprising the so-called Lessepsian migration [58], have replaced many of the natives, such that today 30% of fish caught in Israel are Red Sea migrants. The breach in the barrier at Suez has allowed a trans-Indian/Mediterranean/Atlantic invasion to occur.

In this context, the threat of breaching land bridges between seas in at least one other location, Panama, has major implications for the mixing of separate populations. The two sides of Panama have been separated for at least 3 million years, and species once common to both sides have now evolved into different species. The present Panama Canal uses locks to lift vessels 25.9 m into a large freshwater lake, which is an effective barrier to marine species. Although the possible construction of a sea-level canal in Panama is not presently on the agenda for the Panamanian government, the idea has been put forward in the past [59].

Noise

Anthropogenic noise in the oceans is another direct impact from shipping, but it is also caused by commercial seismic surveys for oil and gas (see above), military sonar operations, and other human activities (e.g., jet skis, motor yachts, and wind turbines). However, noise from shipping is the largest and most widely dispersed source in the ocean. Anthropogenic ocean noise interferes with breeding and feeding, displaces animals from their habitat, and may injure or even kill animals. Species as disparate as fish, giant squid, dolphins, and whales are affected.

Sound in the ocean travels as vibrations of water molecules that exert push–pull pressure on objects in their path. Sound is heard by push–pull of the ear or hearing mechanism of an animal and it similarly exerts pressure on the rest of the animal. Sound frequency is the rate of oscillation or vibration measured in cycles per second, known as Hertz (Hz). Ultrasonic frequencies are too high to be heard by humans (>20,000 Hz) but may be heard by dolphins. Infrasound is too low to be heard by humans (<20 Hz) but can be heard by baleen whales [60]. Commercial container ships, tankers, and other large freighters generate noise in the frequency range far from 10 to 1,000 Hz, which coincides with frequencies used by marine mammals for communication and navigation.

Sound energy dissipates as a function of the distance squared, so the impact of noise on benthic communities is greater for shallow-water habitats closer to the sound source (at the sea surface) than for habitats in greater (off-shelf) water depths. Not only large marine mammals are affected; one recent study has documented a response of invertebrates to noise [61]. However, few papers have been published on the impacts of noise on benthic ecosystems [62].

Climate Change

Predicting the impacts on benthic habitats caused by global climate change is difficult because ocean ecosystems are complex and poorly understood. This may explain why only three case studies cited climate change as an immediate threat (Chapters 21, 24, and 34; Table 3.1). Humans have increased the concentration of greenhouse gases in the global atmosphere, particularly carbon dioxide, through the combustion of fossil fuels, cement production, agriculture, and deforestation. The concentration of CO_2 in the atmosphere has been increasing from its preindustrial level of about 280 parts per million (ppm) to about 390 ppm today. Increased greenhouse gas concentrations are expected to cause acidification and warming of the ocean, changes to ocean circulation patterns, sea-level rise, changed evaporation and precipitation patterns, and changes to ocean wave patterns, among a range of other effects. Climate change impacts will combine with other pressures, exacerbating their effects in many cases [63–69].

Acidification

One impact of increased atmospheric CO_2 concentration is the acidification of the oceans. The surface waters of the oceans are slightly alkaline, with an average pH of about 8.2, although this varies across the oceans by $\pm 0.3\,pH$ units because of local, regional, and seasonal variations. When CO_2 dissolves in seawater, it forms a weak acid, called *carbonic acid*. Part of this acidity is neutralized by the buffering effect of seawater, but the overall impact is to increase the acidity (lower the pH). The dissolution of CO_2 has already lowered the average pH of the oceans by about 0.1 unit from preindustrial levels [64].

These small changes in ocean chemistry could have major impacts on some pelagic calcifying organisms. For example, ocean acidification is likely to affect calcification of foraminifera, a group of calcite-producing protists. As ocean pH declines, it might be expected that calcifying organisms in near-polar areas, where pH is the lowest, will be affected first. Owing to the higher solubility of aragonite, corals and pteropods that produce $CaCO_3$ in its aragonite form will be more strongly affected than the coccolithophores and foraminifera which produce the calcite form [65]. It has been predicted that if CO_2 emissions continue on current trends, the aragonite saturation horizon will rise to the surface of the oceans before the end of this century, making aragonite skeletons unstable throughout the water column of the entire ocean [64–66].

Ocean Warming

The most obvious affect of the rising CO_2 concentrations in the atmosphere is an overall warming of the oceans (attendant with greenhouse warming of the atmosphere). Warming of the surface ocean has already caused coral bleaching events in tropical areas, with almost 30% of the world's corals having disappeared since the beginning of the 1980s [66]. The combination of thermal stress and acidification of the oceans will affect most corals, and the prognosis is that many coral reefs will not survive to the end of the twenty-first century [67,68].

Coastal and shelf species will also have to adjust to the gradual poleward shift of climatic belts as global warming proceeds. Human land uses often create "islands" of habitat surrounded by developed land, which precludes species from migrating as climate changes, thus causing local extinctions. Local variations in the response to global warming will mean that the generalized pattern of poleward migrating eco-systems may not occur everywhere; some habitats may disappear locally, with differ-ent habitats being created in their place. Warming may not affect locations that have strong buffering processes (e.g., where ocean temperature is governed by upwelling of cold ocean water masses). A poleward retreat path is not available to coastal or shelf species at the southern ends of South Africa, South America, or Tasmania (to name a few locations) owing to geography. The global response to warming is there-fore complex and difficult to predict at a local or even regional level.

Changes to Ocean Circulation

The oceanic responses to global warming include the possible slowdown of ocean thermohaline circulation (Figure 3.6), deduced mainly from studies of coupled ocean-atmospheric computer models. Observations of the extent of winter sea ice

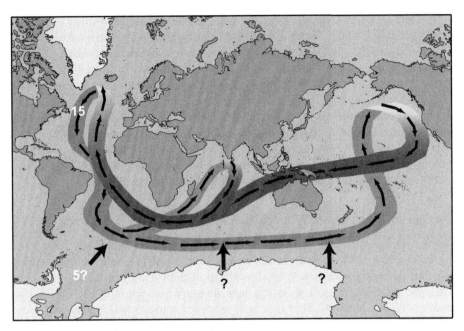

Figure 3.6 The global ocean "conveyor" thermohaline circulation [70–72]. Bottom water is formed in the polar seas via sea-ice formation in winter, which rejects cold, salty (dense) water. This sinks to the ocean floor and flows into the Indian and North Pacific Oceans before returning to complete the loop in the North Atlantic. Numbers indicate estimated volumes of bottom water production in "Sverdrups" (1 Sverdrup = 1 million m^3/s), which may be reduced by global warming because less sea ice will be formed during winter.

cover in both the northern and southern hemispheres show significant reductions over the past 50 years, and some model results suggest that this might reduce, or even shut down, thermohaline circulation in the ocean over the next century [69–71]. Such changes could profoundly affect surface ocean productivity, the strength of major ocean surface currents like the Gulf Stream, and the ventilation of the deep oceans, leading to less food and oxygen for deep-sea benthic communities [72–74]. Limits in the ability of models to accurately predict ocean response to climate change means such scenarios are speculative. However, the consequences of a slowing or shutdown of thermohaline circulation is potentially catastrophic for oceanic benthic ecosystems.

Global Sea-Level Rise

Global warming is predicted to cause a rise in mean sea level by as much as 1 m by the year 2100 [72]. Rising sea level will change the configuration and habitat composition of the coastline by inundating wetlands and other low-lying coasts, inducing erosion to beaches, and increasing the salinity of estuaries, bays, and groundwater tables. As sea level rises, low-lying areas that are not directly inundated will experience more frequent and higher storm surges.

The human response to sea-level rise will include armoring coastlines to protect real estate, thus cutting off the natural (landward) retreat path of coastal and intertidal organisms. Coastal development that has occurred on low-gradient, sandy coastlines is the most vulnerable, since the natural response of such systems is to retreat landward as sea level rises. Although a positive net sediment supply will ensure that many atolls will maintain their freeboard and possibly continue to expand in spite of rising sea levels [75], the demise of coral reefs from bleaching (caused by warmer ocean surface temperatures), and acidification will counteract natural reef-growth processes. The result is that many inhabited coral atolls that have only a few meters of freeboard are under direct threat of becoming uninhabitable when their water tables disappear or they become submerged when sea level rises.

Changes to Evaporation/Precipitation Patterns

Changes in rainfall obviously impact terrestrial ecosystems, but there are consequences for the estuarine and coastal ecosystems as well. Habitats within estuaries and river deltas are intimately related to fluvial systems and their delivery of water, nutrients, and sediment to the coastal zone. Primary production is often related to freshwater (and nutrient) discharge, for example. Animals and plants that have evolved to tolerate a certain salinity range may be unable to survive in coastal areas impacted by severe changes in precipitation and evaporation patterns [76].

Changes to Storm Intensity and Ocean Swell Waves

Another apparent response to the warming of the Earth's atmosphere is a change in the ocean wave climate, manifested as increased wave heights associated with more intense storm events [77–81]. Changes in wave regime may affect the stability of sandy shorelines and potentially dramatic changes in coastal geomorphology may occur locally. For example, the transformation from tide-dominated to wave-dominated coastal systems is possible in some locations [82]. On the continental shelf, increased wave height and period translates into an increase in the water depth at which sediments may be mobilized, thereby fundamentally changing the character of the seabed habitat. For example, areas of sandy seafloor previously stable under the prevailing wave and current regime may become transformed into a different habitat type, subject to mobilizing forces of episodic storms [83].

Wind Farms

Three of the case studies specified the construction of wind farms as a threat to habitats (Chapters 12, 13, and 40; Table 3.1). Inger et al. [62] reviewed the positive and negative ecological effects of offshore renewable energy installations, including wind turbines. The negative aspects include the displacement of animals from their habitat, noise generated by the machinery, and electromagnetic fields generated along the cables that join the turbine to the onshore electricity grid. Positive aspects are that the structures may form artificial reefs and fish aggregation points, and they act as small marine protected areas because of the exclusion zones needed around each installation.

From a habitat perspective, wind turbines are preferentially built on hard substrates to provide for a secure foundation. Multibeam sonar mapping work is commonly undertaken to find suitable rocky substrate for installation sites. This may mean that rocky habitats are preferentially targeted for wind farm construction. Rocky habitat types are by far the smallest in proportional area (compared with soft sediment habitat) in most bioregions. They provide critical habitat for many species, which increases their conservation value, while at the same time they are targeted for other human uses, such as fishing and tourism (scuba diving).

Concluding Remarks

Anthropogenic threats to benthic habitats are a key driver underpinning the need for building accurate and comprehensive habitat maps. Habitat mapping provides condition assessments and establishes baselines against which changes can be measured. Repeated mapping and sampling are essential for quantifying change, measuring impacts, and tracking recovery of affected habitats. Habitat mapping is viewed by many government agencies as an essential component of ecosystem-based management.

Destructive fishing practices are viewed by GeoHab scientists and many other experts as the greatest single threat to the marine environment (an order of magnitude greater than the total extent of all other activities, according to Benn et al.; [84]). However, it is important to recognize that human impacts are cumulative. Destructive fishing, pollution, mining, climate change, and so on often affect the same habitats, and the benthos are simultaneously affected by more than one stressor. One final theme that emerges from the literature (and from the case studies) is that the responses of benthic habitats to anthropogenic pressures (singly or cumulatively) are poorly understood. It is therefore often difficult to give reliable, scientifically defensible answers to questions as to how a habitat or ecosystem will respond to any particular pressure (or combination of pressures). This situation is exacerbated by the ocean floor being out of sight from the public view so that we are not aware of the full extent of the impacts and damage we cause. The greatest threat to the oceans from humans may simply be ignorance.

Acknowledgments

Thanks to Emily Corcoran (OSPAR), Brendan Brooke (Geoscience Australia), and Kathy Scanlon (USGS) for helpful comments and suggestions on an earlier draft of this chapter. This work was produced with the support of funding from the Australian Government's Commonwealth Environment Research Facilities (CERF) program and is a contribution of the CERF Marine Biodiversity Hub. This chapter is published with the permission of the Chief Executive Officer, Geoscience Australia.

References

[1] E.A. Norse, L.B. Crowder (Eds.), Marine Conservation Biology, Island Press, Washington, DC, 2005.

[2] E. Hoyt, Marine Protected Areas for Whales, Dolphins and Porpoises, Earthscan, London, 2005.

[3] B.S. Halpern, S. Walbridge, K.A. Selkoe, C.V. Kappel, F. Micheli, C. D'Agrosa, et al., A global map of human impact on marine ecosystems, Science 319 (2008) 948–952.

[4] C.V. Kappel, Losing pieces of the puzzle: threats to marine, estuarine and diadromous species, Front. Ecol. Environ. 3 (2005) 275–282.

[5] W.W. Gregg, M.E. Conkright, Decadal changes in global ocean chlorophyll, Geophys. Res. Lett. 29 (2002) 1730–1733.

[6] J.B.C. Jackson, Ecological extinction and evolution in the brave new ocean, Proc. Natl. Acad. Sci. 105 (2008) 11458–11465.

[7] D. Pauly, R. Watson, J. Alder, Global trends in world fisheries: impacts on marine ecosystems and food security, Philos. Transact. Royal Soc. B Biol. Sci. B 360 (2005) 5–12.

[8] C. Nellemann, S. Hain, J. Alder, (Eds.), Dead Water—Merging of Climate Change with Pollution, Over-Harvest, and Infestations in the World's Fishing Grounds, United Nations Environment Programme, GRID-Arendal, Arendal, Norway, 2008.

[9] P.W. Barnes, J.P. Thomas (Eds.), Benthic Habitats and the Effects of Fishing, American Fisheries Society Symposium, 2005.

[10] L. Watling, The global destruction of bottom habitats by mobile fishing gear, in: E.A. Norse, L.B. Crowder (Eds.), Marine Conservation Biology, Island Press, Washington, DC, 2005, pp. 198–210.

[11] T. Pitcher, T. Morato, P. Hart, M. Clark, N. Haggan, R. Santos (Eds.), Seamounts: Ecology, Fisheries and Conservation, Blackwell, Oxford, U.K.

[12] R.A. Myers, B. Worm, Extinction, survival, or recovery of large predatory fishes, Proc. Royal Soc. B 360 (2005) 13–20.

[13] D. Pauly, V. Christensen, J. Dalsgaard, R. Froese, F. Torres, Fishing down marine food webs, Science 279 (1998) 860–863.

[14] S. Nicol, Y. Endo, Review. Krill fisheries: development, management and ecosystem implications, Aquat. Living Resour. 12 (1999) 105–120.

[15] K.A. Kvenvolden, C.K. Cooper, Natural seepage of crude oil into the marine environment, Geo Mar. Lett. 23 (2003) 140–146.

[16] S. Pantin, Oil Pollution of the Sea, Copyright EcoMonitor Publishing, East Northport, NY, http://www.offshore-environment.com/oilpollution.html.

[17] J.M. Swan, J.M. Neff, P.C. Young (Eds.), Environmental Implications of Offshore Oil and Gas Development in Australia—The Findings of an Independent Scientific Review, Australian Petroleum Exploration Association, Sydney, 1994.

[18] K.C. Balcomb, D.E. Claridge, A mass stranding of cetaceans caused by naval sonar in the Bahamas, Bahamas J. Sci. 5 (2001) 2–12.

[19] O. Boebel, P. Clarkson, R. Coates, R. Larter, P.E. O'Brien, J. Ploetz, et al., Risks posed to the Antarctic marine environment by acoustic instruments: a structured analysis— SCAR action group on the impacts of acoustic technology on the Antarctic marine environment, Antarct. Sci. 17 (2005) 533–540.

[20] P.E. O'Brien, W. Arnt, I. Everson, K. Gohl, J.C.D. Gordon, K. Goss, et al., Impacts of marine acoustic technology on the Antarctic environment, SCAR Ad Hoc Group on marine acoustic technology and the environment, 2002.

[21] F. Olsgard, J.S. Gray, A comprehensive analysis of the effects of offshore oil and gas exploration and production on the benthic communities of the Norwegian continental shelf, Mar. Ecol. Progr. 122 (1995) 277–306.

[22] Z. Khatib, P. Verbeek, Water to value—produced water management for sustainable field development of mature and green fields, J. Petrol. Tech. 55 (2003) 26–28.

[23] D. Holdway, D.T. Heggie, Direct hydrocarbon detection of produced formation water discharge on the Northwest Shelf, Australia, Estuar. Coast. Shelf Sci. 50 (2000) 387–402.

[24] R.J. Jones, A.J. Hayward, The effects of produced formation water (PFW) on coral and isolated symbiotic dinoflagellates of coral, Mar. Freshwat. Res. 54 (2003) 153–162.

[25] ITOPF, International Tanker Owners Pollution Federation, http://www.itopf.com/index.html, 2007.

[26] T.J. Crone, M. Tolstoy, Magnitude of the 2010 Gulf of Mexico oil leak, Science 330 (2010) 634.

[27] D.G. Shaw, The Exxon Valdez oil-spill: ecological and social consequences, Environ. Conservat. 19 (1992) 253–258.

[28] M.W. Hester, I.A. Mendelssohn, Long-term recovery of a Louisiana brackish marsh plant community from oil-spill impact: vegetation response and mitigating effects of marsh surface elevation, Mar. Environ. Res. 49 (2000) 233–254.

[29] F. Sánchez, F. Velasco, J.E. Cartes, I. Olaso, I. Preciado, E. Fanelli, et al., Monitoring the Prestige oil spill impacts on some key species of the Northern Iberian shelf, Mar. Pollut. Bull. 53 (2006) 332–349.

[30] M.G. Carls, G.D. Marty, J.E. Hose, Synthesis of the toxicological impacts of the Exxon Valdez oil spill on Pacific herring (*Clupea pallasi*) in Prince William Sound, Alaska, U.S.A., Can. J. Fish. Aquat. Sci. 59 (2002) 153–172.

[31] G.J. Edgar, L. Kerrison, S.A. Shepherd, M.V. Toral-Granda, Impacts of the Jessica oil spill on intertidal and shallow subtidal plants and animals, Mar. Pollut. Bull. 47 (2003) 276–283.

[32] T. Yamamoto, M. Nakaoka, T. Komatsu, H. Kawai, K. Ohwada, Impacts by heavy-oil spill from the Russian tanker Nakhodka on intertidal ecosystems: recovery of animal community, Mar. Pollut. Bull. 47 (2003) 91–98.

[33] W. Ritchie, The short-term impact of the Braer oil spill in Shetland and the significance of coastal geomorphology, Scot. Geogr. Mag. 109 (1993) 50–56.

[34] S.C. Jewett, T.A. Dean, R.O. Smith, A. Blanchard, "Exxon Valdez" oil spill: impacts and recovery in the soft-bottom benthic community in and adjacent to eelgrass beds, Mar. Ecol. Progr. 185 (1999) 59–83.

[35] D.P. French-McCay, Oil spill impact modeling: development and validation, Environ. Toxicol. Chem. 23 (2004) 2441–2456.

[36] D.V. Ellis, A review of some environment issues affecting marine mining, Marine Georesour. Geotechnol. 19 (2001) 51–63.

[37] C.H. Hobbs, III, An investigation of potential consequences of marine mining in shallow water: an example from the mid-Atlantic coast of the United States, J. Coast. Res. 18 (2002) 94–101.

[38] Marinet, Dredging Statistics for 2005, http://www.marinet.org.uk/mad/stats05.html, 2007.

[39] K. Tsurusaki, T. Iwasaki, M. Arita, Seabed sand mining in Japan, Mar. Min. 7 (1988) 49–67.

[40] I.R. Poiner, R. Kennedy, Complex patterns of change in the macrobenthos of a large sandbank following dredging, Mar. Biol. 78 (1984) 335–352.

[41] C.B. Pattiaratchi, P.T. Harris, Hydrodynamic and sand transport controls on *en echelon* sandbank formation: an example from Moreton Bay, eastern Australia, J. Mar. Res. 53 (2002) 1–13.

[42] B.S. Drucker, W. Waskes, M.R. Byrnes, The U.S. Minerals Management Service outer continental shelf sand and gravel program: environmental studies to assess the potential effects of offshore dredging operations in Federal waters, J. Coast. Res. 20 (2004) 1–5.

[43] D.R. Hitchcock, S. Bell, Physical impacts of marine aggregate dredging on seabed resources in coastal deposits, J. Coast. Res. 20 (2004) 101–114.

[44] S.E. Boyd, K.M. Cooper, D.S. Limpenny, R. Kilbride, H.L. Rees, M.P. Dearnaley, et al., Assessment of the Rehabilitation of the Seabed Following Marine Aggregate Dredging, Centre for Environment, Fisheries and Aquaculture Science, Lowestoft, 2004 , p. 154.

[45] J.R. Hein, T.A. Conrad, H. Staudigel, Seamount mineral deposits—a source of rare metals for high-technology industries, Oceanography 23 (2010) 184–189.

[46] C.L. Morgan, N.A. Odunton, A.T. Jones, Synthesis of environmental impacts of deep seabed mining, Marine Georesour. Geotechnol. 17 (1999) 307–356.

[47] R. Sharma, Indian deep-sea environment experiment (INDEX): an appraisal, Deep Sea Res. II Top. Stud. Oceanogr. 48 (2001) 3295–3307.

[48] USCAE, US Core of Army Engineers Ocean Disposal Database, http://el.erdc.usace.army.mil/odd/, 2007.

[49] F. Galgani, J.P. Leaute, P. Moguedet, A. Souplet, Y. Verin, A. Carpentier, et al., Litter on the sea floor along European coasts, Mar. Pollut. Bull. 40 (2000) 516–527.

[50] MAR-ECO Expedition "Deep sea search finds species surprises", http://www.msnbc.msn.com/id/5610557/.

[51] H.K. Lotze, H.S. Lenihan, B.J. Bourque, R.H. Bradbury, R.G. Cooke, M.C. Kay, et al., Depletion, Degradation, and Recovery Potential of Estuaries and Coastal Seas, Science 312 (2006) 1806–1809.

[52] N.N. Rabalais, The potential for nutrient overenrichment to diminsh marine biodiversity, in: E.A. Norse, L.B. Crowder (Eds.), Marine Conservation Biology, Island Press, Washington, DC, 2005, pp. 105–108 (Chapter 7).

[53] P.T. Harris, S.L. Nichol, T.J. Anderson, A.D. Heap, Habitats and benthos of a deep sea marginal plateau, Lord Howe Rise, Australia, in: P.T. Harris, E.K. Baker (Eds.), Seafloor Geomorphology as Benthic Habitat: GeoHab Atlas of Seafloor Geomorphic Features and Benthic Habitats, Elsevier, Amsterdam, 2011 (Chapter 57).

[54] I. Kogan, C.K. Paull, L.A. Kuhnz, E.J. Burton, S. Von Thun, H. Gary Greene, et al., ATOC/Pioneer Seamount cable after 8 years on the seafloor: observations, environmental impact, Continent. Shelf Res. 26 (2006) 771–787.

[55] E. Andrulewicz, D. Napierska, Z. Otremba, The environmental effects of the installation and functioning of the submarine SwePol Link HVDC transmission line: a case study of the Polish Marine Area of the Baltic Sea, J. Sea Res. 49 (2003) 337–345.

[56] J.T. Carlton, G.M. Ruiz, The magnitude and consequences of bioinvasions in marine ecosystems—implications for conservation biology, in: E.A. Norse, L.B. Crowder (Eds.), Marine Conservation Biology, Island Press, Washington, DC, 2005, pp. 123–148.

[57] J.C. Briggs, Marine biogeography and ecology: invasions and introductions, J. Biogeaogr. 34 (2007) 193–198.

[58] F.D. Por, Lessepsian migration—the influx of Red Sea biota into the Mediterranean Sea by way of the Suez Canal, Springer, Berlin, 1978.

[59] D. McCullough, The Path Between the Seas: The Creation of the Panama Canal, Simon and Schuster, New York, NY, 1977.

[60] W.J. Richardson, C.R.J. Greene, C.I. Malme, D.H. Thomson, Marine Mammals and Noise, Academic Press, San Diego, CA, 1995.

[61] S.D. Simpson, A.N. Radford, E.J. Tickle, M.G. Meekan, A.G. Jeffs, Adaptive avoidance of reef noise, PLoS One 6 (2011) e16625.

[62] R. Inger, M.J. Attrill, S. Bearhop, A.C. Broderick, W.J. Grecian, D.J. Hodgson, et al., Marine renewable energy: potential benefits to biodiversity? An urgent call for research, J. Appl. Ecol. 46 (2009) 1145–1153.

[63] F.C. Coleman, C.C. Koenig, The effects of fishing, climate change, and other anthropogenic disturbances on red grouper and other reef fishes in the Gulf of Mexico, Integr. Comp. Biol. 50 (2010) 201–212.

[64] K. Caldeira, M.E. Wickett, Anthropogenic carbon and ocean pH, Nature 425 (2003) 365.

[65] R.A. Feely, C.L. Sabine, K. Lee, W. Berelson, J. Kleypas, V.J. Fabry, et al., Impact of anthropogenic CO_2 on the $CaCO_3$ system in the ocean, Science 305 (2004) 362–366.

[66] O. Hoegh-Guldberg, Low coral cover in a high CO_2 world, J. Geophys. Res. 110 (2005) C09S06.

[67] T.P. Hughes, A.H. Baird, D.R. Bellwood, M. Card, S.R. Connolly, C. Folke, et al., Climate change, human impacts, and the resilience of coral reefs, Science 301 (2003) 929–933.

[68] J.M. Pandolfi, R.H. Bradbury, E. Sala, T.P. Hughes, Global trajectories of the long-term decline of coral reef ecosystems, Science 301 (2003) 955.

[69] S. Rahmstorf, Risk of sea-change in the Atlantic, Nature 388 (1997) 825–826.

[70] X. Wu, W.F. Budd, Modelling global warming and Antarctic sea-ice changes over the past century, Ann. Glaciol. 29 (1999) 413–419.

[71] D. Bi, W.F. Budd, A.C. Hirst, X. Wu, Collapse and reorganisation of the Southern Ocean overturning under global warming in a coupled model, Geophys. Res. Lett. 28 (2001) 3927–3930.

[72] IPCC Fourth Assessment Report: Climate Change (AR4), http://www.ipcc.ch/publications_and_data/publications_and_data_reports.htm#1, 2007.

[73] F. Joos, G.-K. Plattner, T. Stocker, A. Körtzinger, D.W.R. Wallace, Trends in marine dissolved oxygen: implications for ocean circulation changes and the carbon budget, EOS Tans. AGU 84 (2003) 197–204.

[74] G. Shaffer, S.M. Olsen, J.O.P. Pedersen, Long-term ocean oxygen depletion in response to carbon dioxide emissions from fossil fuels, Nature Geosci. 2 (2009) 105–109.

[75] A.P. Webb, P. Kench, The dynamic response of Reef Islands to sea level rise: evidence from multi-decadal analysis of Island change in the Central Pacific, Global Planet. Change 72 (2010) 234–246.

[76] A.H. Baldwin, I.A. Mendelssohn, Effects of salinity and water level on coastal marshes: an experimental test of disturbance as a catalyst for vegetation change, Aquat. Bot. 61 (1998) 255–268.

[77] D.J.T. Carter, L. Draper, Has the northeast Atlantic become rougher? Nature 332 (1988) 494.

[78] WASA, Changing waves and storms in the northeast Atlantic? The WASA group, Bull. Am. Meteorol. Soc. 79 (1998) 741–760.

[79] S.K. Gulev, L. Hasse, Changes of wind-waves in the North Atlantic over the last 30 years, Int. J. Climatol. 19 (1999) 1091–1117.

[80] J. Allan, P. Komar, Are ocean wave heights increasing in the eastern North Pacific? EOS Trans. Am. Geophy. Union 81 (2000) 561.

[81] I. Grevemeyer, R. Herber, H.H. Essen, Microseismological evidence for a changing wave climate in the northeast Atlantic Ocean, Nature 408 (2000) 349–352.

[82] P.T. Harris, A. Heap, S. Bryce, R. Smith, D. Ryan, D. Heggie, Classification of Australian clastic coastal depositional environments based upon a quantitative analysis of wave, tidal and fluvial power, J. Sediment. Res. 72 (2002) 858–870.

[83] M.G. Hughes, P.T. Harris, B.P. Brooke, Seabed exposure and ecological disturbance on Australia's continental shelf: potential surrogates for marine biodiversity, Geoscience Australia Record 2010/43, Canberra, 2010, 76 pp.

[84] A.R. Benn, P.P. Weaver, D.S.M. Billet, S. van den Hove, A.P. Murdock, G.B. Doneghan, et al., Human activities on the deep seafloor in the north east Atlantic: an assessment of spatial extent, PLoS ONE 5 (2010) 1–15.

4 Biogeography, Benthic Ecology, and Habitat Classification Schemes

Peter T. Harris

Marine and Coastal Environment Group, Geoscience Australia, Canberra, ACT, Australia

Abstract

A number of terms used in this book are derived from the fields of biogeography and benthic ecology and these are defined in the glossary; the reader is also referred to the works cited at the end of this chapter for further information. Many of the case studies presented in this book refer to habitat classification schemes that have been developed based on principles of biogeography and ecology. For these reasons, a brief overview is provided here to explain the concepts of biodiversity, biogeography, and benthic ecology that are most relevant to habitat mapping and classification. Of particular relevance is that these concepts underpin classification schemes employed by GeoHab scientists in mapping habitats and other bioregions. A selection of published schemes, from both deep and shallow water environments, are reviewed and their similarities and differences are examined.

Key Words: Biodiversity, endemism, biogeographic classification schemes, patch theory, hierarchy, bioregionalisation

Biodiversity and Biogeography

Biodiversity

The term "biodiversity" can have different meanings, and it is important that these are clearly understood. Biodiversity is, put simply, the richness and variety of life in the natural world. The term "biodiversity" (short for "biological diversity") can be at the level of habitat or community diversity, species diversity, or at genetic diversity. Biodiversity is important according to the International Convention on Biological Diversity (CBD)[1] because "At least 40 per cent of the world's economy and 80 per cent of the needs of the poor are derived from biological resources. In addition, the richer the diversity of life, the greater the opportunity for medical discoveries, economic development, and adaptive responses to such new challenges as climate change."

---

[1] The International Convention on Biodiversity, http://www.cbd.int/.

Seafloor Geomorphology as Benthic Habitat. DOI: 10.1016/B978-0-12-385140-6.00004-9

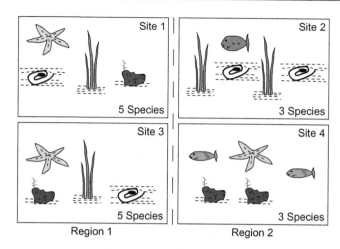

Figure 4.1 Diagram illustrating the difference between alpha, beta, and gamma biodiversity (based on Perlman and Adelson [1]). Alpha diversity is measured locally at a single site. Site 1 has more species than Site 2, so it has a higher alpha diversity. Beta diversity measures the amount of change between two places or along a gradient, as in Regions 1 and 2. Region 2 has higher beta diversity than Region 1 because the difference in species composition is greater between sites within the region. Gamma diversity is similar to alpha diversity but measured over broad spatial scales; Region 2 has more species than Region 1, so it has higher gamma diversity.

A crucial point is that species are interdependent components of ecosystems; removing any one species will reduce ecosystem resilience and functioning. Finally, biodiversity has an intrinsic value to humans—we value the diversity of life on Earth and wish to preserve it for future generations to enjoy.

Most commonly, biodiversity is conceptualized in terms of the diversity of species (species richness, alpha diversity), although other measures are possible (beta and gamma diversity) when viewed at different spatial scales (Figure 4.1). Species are distinguished based on strict rules of taxonomy in the classification hierarchy of species, genera, families, orders, classes, and phyla. At higher taxonomic levels, marine ecosystems are far more diverse than terrestrial ones; for example, of the 33 animal phyla, only 11 occur on land while 28 occur in the oceans. The international Census of Marine Life program [2] reported that approximately 250,000 marine species have been described by scientists, and there are probably at least 1 million marine species in total (and probably hundreds of millions, including bacteria and other groups of microscopic taxa).

Humans do not value all species equally. There is generally a clear anthropocentric bias in any discussion of biodiversity conservation. The organisms targeted for conservation are often those to which we have an aesthetic or emotional attachment; most species listed as threatened or endangered by human activities are large plants and animals. Humans are less aware of their impact on small, inconspicuous plants, animals, or microbial life—and there are more than a billion microorganisms living in each liter of seawater. In as much as human impacts in the ocean extend beyond the species targeted for food, anthropocentric bias can be an obstacle for the conservation of species diversity.

The number of species on Earth has increased over geologic time, from zero in the pre-Cambrian (>600 million years ago) up to the present time [3]. The rise in the numbers of species has not been a linear progression, since major extinctions have occurred several times over the past 600 million years (divisions in geological time are commonly defined on the basis of an increased rate of extinction). A key lesson from the study of geology is that all species, including humans, are transient; ultimately, they go extinct. The species alive today are a small subset of all those that have ever existed, perhaps only 1%. On average, a species typically survives between 1 and 10 million years. Fossils are the only testimony of species that have gone extinct over geologic time, and the geologic (fossil) record is far from perfect. Nevertheless, it points to the existence of an immense diversity of life spanning Earth's history; for example, the Natural History Museum in London houses 6 million fossils of marine snails. The possible natural causes of mass extinctions range from climate change to the impact of large meteors, such as the one blamed for the demise of the dinosaurs 65 million years ago. It is a dubious honor for humans to have laid claim to being the cause of a major extinction event at the present time (about 10–50 times higher than the natural rate [4]) that will be recorded as a divide in the geological record with the onset of the so-called Anthropocene [5].

Species diversity can be measured as simply the number of species (S) that occur in an area (also known as the species "richness"). However, this measure overlooks the relative abundance of different species; a sample comprised of 50 individuals of species A and 50 of species B is more *diverse* than one comprised of 99 individuals of species A and one of species B (even though both samples of 100 individuals contain two species). Therefore, other measures, such as the Shannon index (H), are used:

$$H = -\sum_{i=1}^{S} p_i \ln p_i \qquad (4.1)$$

where p_i is the proportion of the ith species [6,7]. The Shannon index is sensitive to both the total number of species and their relative abundance.

Within a population of a single given species, there may be considerable variation in genetic composition. Genetically distinct populations of the same species can occur at different geographic locations. The preservation of this genetic diversity is another conservation goal [8]. Genetic diversity is important for the survival of a species because it can be a factor in a species' resistance to diseases and its ability to adapt to different environmental stresses.

Endemism

The geographic range of species is limited by physical and environmental barriers; no species occurs worldwide. The occurrence of species found only in a restricted geographic area is known as *endemism*. Habitat mapping is often employed to manage the sustainable harvest of commercial species and the conservation of protected endemic species, but also to measure and quantify the biophysical conditions associated with the occurrence of a particular species or group of species, in order to attempt to explain spatial relationships between the physical environment and the occurrence of a particular species.

Endemic species are important ecologically because, over different spatial scales, they signal significant changes in communities and populations. Acknowledging the uncertainty of the known distribution of species, scientists may refer to a species having a restricted known geographic range [9]. Nevertheless, the term endemism is widely used (even with imperfect knowledge) with the meaning given above.

There are two main reasons why species are endemic to a particular geographic area. First, most species can only survive and reproduce in a limited range of habitats. For example, bivalves that require a soft sediment substrate are limited spatially to a particular type of seabed. Species that have a life stage that requires shallow, warm water are unable to disperse to deep oceanic habitats. Species with a pelagic larvae stage are dependent upon ocean currents for their dispersal. The second factor is that there are different natural barriers to the dispersal of species. Deep ocean basins comprise vast expanses of relatively homogenous mud habitat. But ocean basins are divided by rocky, mid-ocean spreading ridges, which are a barrier to any species that can only live in mud. Similarly, estuaries and rocky peninsulas are barriers for the dispersal of some coastal species. Latitudinal temperature changes are also a primary barrier because species are often only able to survive within a specific temperature range. The configuration of dispersal barriers divides the world into geographic provinces, each having distinct assemblages of endemic species.

Many of the barriers that divide the world into the provinces we see today are geologically recent features. Nearly all of the world's continental shelves were emergent 20,000 years ago during the last ice age, when global sea level was around 120 m below its present position, so all species found on continental shelves today are relatively recent arrivals; they were able to colonize the shelf as sea level rose and flooded it, creating new habitat. Similarly, ocean currents and the Earth's climate were quite different during the ice ages than they are today. Over longer time scales, plate tectonic processes have moved some continents into different climate regimes, created new barriers, and changed the course of ocean currents. For example, the uplift of the isthmus of Panama between 1 and 3 million years ago separated the populations located on the Caribbean and Pacific sides [10]. Species that were once common to both sides have evolved into two distinct species. Another example is the northward movement of the Australian plate after it separated from Gondwanaland approximately 60 million years ago; northward movement has created habitat suitable for coral reefs to become established at locations progressively further south, thus exerting a fundamental control over the evolution of the Great Barrier Reef [11]. The present distribution of species is the result of millions of years of speciation, dispersal, and extinction.

Biogeographic Provinces

Biogeographic "provinces," defined based on their geologic history and endemic species, are divided based on their ecology into the intertidal, neritic, and bathypelagic biomes. *Biomes* are spatially recognizable units that are nested within provinces having high levels of associations between organisms [12]. Biogeographic concepts used to distinguish different provinces and biomes are important for the habitat classification schemes that we will discuss later in this chapter.

Interestingly, the deep ocean (slope and abyssal regions) tends to have more species (both planktonic and benthic) than shelf and coastal habitats. For a wide variety of animal groups, diversity increases with increasing water depth reaching a maximum at around 2,000 m [13]; below this depth diversity decreases with increasing depth. There is no single explanation for this pattern, though there are several plausible theories: the interaction of competition, predation, and disturbance; the "center of origin" theory, which states that the tropical, high-diversity regions are the sources of species to regions of low diversity; and the "*vicariance* hypothesis," which argues that once a geographic barrier arises that separates a species, it will evolve into two separate species (or groups of species). It is paradoxical that the deep ocean basins comprise the largest ecosystem on Earth and yet it is also the least well understood [13]; deep ocean biogeography is a relatively new science and much work has yet to be done in describing species and mapping their distribution [14,15].

The diversity of species generally increases from the polar oceans to the equator. In the Pacific Ocean, species diversity is highest in the Philippines, Indonesia, and Northeast Australia. In the Atlantic Ocean, the highest species diversity is found in the Caribbean [2,16]. Thus, the areas of highest biodiversity in the oceans are associated with tropical oceans and coral reefs.

Relative Species Abundance and Rarity

It is important not to overlook differences in the relative *abundances* of different species. Since biodiversity is more than simply the number of species (e.g., the Shannon Index; Eq. (4.1)), managers working in the fields of fisheries management or biodiversity conservation require knowledge of the fauna composition, including relative species abundance (as well as diversity).

Observations and measurements of species density and abundance demonstrate that numerically dominant species are far less common than species that are few in number. There is a great number of rare species. The explanation for this phenomenon is the intimate link between species and their habitats. Species have evolved to specialize in exploiting resources within a restricted range of environmental parameters. As environments change, the range of any given species also changes. Meanwhile, new species evolve and compete with previously existing species. Small changes in the environment can thus lead to the extinction of one species and a simultaneous population explosion in another [16].

Three concepts that together go a long way toward explaining the commonness and rarity of individual species are the geographic range, habitat specificity, and total population. Species that have a broad geographic range may be rare in any one location but still have a large population because of their broad distribution. Some species can survive only in specific habitats; their geographic range may be large, but the population size is constrained by the available, specialized habitat. Other species may be able to occupy a diversity of habitats over a broad geographic range, but the population is never large in any one area. The rarest species (and those most vulnerable to human impacts) are those that have a small geographic range, high habitat specificity, and low population [16].

The distinction between rarity and endemism is fundamental for understanding and interpreting habitat maps. Rare species that are widely distributed across a given habitat (e.g., a species that occurs on many seamounts) require quite different management approaches than spot endemics that are spatially restricted to an individual feature (e.g., a species found only on one particular seamount).

Landscape Ecology and the Theory of Island Biogeography

The subject of benthic habitat mapping has many close associations with the field of landscape ecology. Both are concerned with the interactions between spaces within landscape complexes and their spatial–temporal variability. While it is true that much can be learned from landscape ecology for understanding benthic habitats, it is important to emphasize that there are distinct differences between terrestrial and marine ecosystems. Chief among these differences is the transport capacity of water, which has 50 times the density of air. Nevertheless, landscape ecology has much to say that is relevant to mapping and understanding benthic habitats.

Patch Theory

Landscape ecology is based on the presumption that discrete landscape elements (patches) exert control over key aspects of ecosystems and biological assemblages [16]. Habitats are generally defined with this assumption in mind (or at least implied). The relationship between patch size and species richness was first recognized by biologists studying the communities inhabiting oceanic islands. Islands adjacent to a mainland coast have lower species diversity and generally hold only a subset of the continental species. The further offshore the island, the poorer the island species richness. Also, small islands generally have lower species richness than larger islands. In their pioneering study, MacArthur and Wilson [17] found that on a logarithmic plot, the number of species present on an island (S) is proportional to island area (A), given by:

$$S = CA^Z \tag{4.2}$$

where "C" is the intercept of the y-axis and "Z" is the slope of the line. These concepts form the basis of the *Theory of Island Biogeography* (Figure 4.2).

The explanation for the relationship between island area and species richness is that the number of species present on an island is governed by the rates of colonization and extinction. The probability of any species being able to colonize an island from another landmass is a function of the island size and distance offshore.

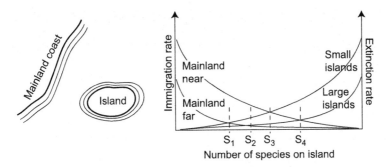

Figure 4.2 Diagram showing the effects of island distance from the mainland and island size relative to species immigration rate, extinction rate, and overall number of species [17,18]. Small islands that are far away from the mainland have the smallest number of species (S_1), followed by large, far-away islands (S_2), small, nearby islands (S_3). Large, nearby islands have the largest number of species (S_4). The downward bowing of the curves represents effects of competition.

The chances of an individual washing ashore or being carried by wind are greater for smaller islands and greater for islands that are closer to other land (sources). On the other hand, extinction is more likely to occur on small islands, where the diversity of habitats is less and resources are more limited than on larger islands. On islands where the colonization and extinction rates are equal, the species number is stable and the community is in equilibrium (indicated by the "S" values in Figure 4.2). The theory of island biogeography seems to hold true for most islands, but how is it relevant to benthic habitats?

To answer this question, let us conduct a thought experiment. Consider an imaginary seabed characterized by abundant, upright habitat-forming fauna in an area that is subject to bottom trawling. Repeated trawling removes the large, upright, habitat-forming species in the trawled areas, but they persist in patches that the fishermen avoid for some reason (e.g., the presence of local rocky substrate that might damage the trawl gear). The patches of surviving, intact habitat will initially contain all of the species formerly present (Figure 4.3). But over time, the species numbers will decline as more extinctions than colonizations take place (e.g., via pelagic larvae). The time required for species and populations to adjust to the changed habitat conditions is known as the *relaxation time*. The theory of island biogeography predicts that untrawled patches that are small and isolated from other patches will (after the relaxation time) only support a subset of the original species diversity.

After trawling, the original continuous benthos is divided into discrete areas having different species composition and different physical conditions [19]; this is an example of habitat fragmentation (Figure 4.3B). The borders of the patches of untrawled seabed will suffer what in landscape ecology are known as "edge effects." Along the edges of the patch, the environment is different from conditions in the interior of the patch (Figure 4.3C). For example, currents that were previously slowed by epibenthic sponges and other protruding forms are able to reach the seabed unimpeded in

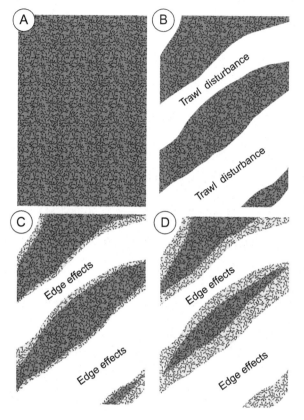

Figure 4.3 Example showing (A) undisturbed habitat; (B) effects of trawling disturbance removing parts of the original habitat; (C) changed conditions altering the composition of original habitat along exposed edges; (D) expansion of edge effects causing smaller patches (or elongate patches with a large *P/A* ratio) to contain only a subset of the original community.

the disturbed areas. Current speeds near the bed are thus greater, and sediments are mobilized more frequently and with greater intensity in disturbed areas compared with undisturbed areas. Thus, seabed sediments that were previously stable under the prevailing current regime may become unstable, resulting in higher water turbidity and bedload sediment movement, forming ripples and dunes. The disturbed sediment may also have different physical properties than the original sediment (e.g., sediment porosity is higher and compaction is lower in undisturbed, bioturbated sediments than in the disturbed sediments). Species that are unable to tolerate the altered conditions present at the habitat edges will be restricted to the interior (Figure 4.3C and D).

The average shape and size of patches can be assessed using the perimeter/area (*P/A*) ratio, a standard technique in landscape ecology [16]. The ratio is minimized for circle-shaped patches and increases in value for elongate patches. Obviously, smaller patches have proportionately more edges per unit area than do larger patches. In some cases, the patch may be too small to provide any significant internal areas where the edge effects are not felt, and such environments may not be tolerable to some of the original inhabitants (Figure 4.3D). Wide-ranging foraging species that feed preferentially within the original habitat will focus on the remaining patches in the fragmented habitat; this will tend to increase competition and predation within the patches of original habitat.

Application of the theory of island biogeography to the above "trawled seabed" example is not exact, because the trawled areas of seabed are not equivalent to the ocean surrounding an island (i.e., the difference between an island and the surrounding ocean is far greater than the difference between a patch of untrawled seabed and the surrounding trawled seabed). The disturbed seabed will still support some benthos, favoring those species least impacted by the disturbance caused by benthic trawling. Importantly, the disturbed seabed may also provide habitat for some pioneer species that were not originally present.

This leads to another key point: the species found in the original habitat may be supplemented by new species that are able to colonize the disturbed trawled habitat, adding up to an increase in total species richness (i.e., seabed disturbance caused by trawling can actually result in an increase in biodiversity). In this case, the heterogenous, disturbed, patchy habitat has a greater biodiversity than homogenous, undisturbed habitat. This is one reason why measure of species diversity, on its own, is not an exact measure of ecosystem health or pristine condition. Rather, information on habitat condition and naturalness is also required to make an assessment.

Despite its limitations, the theory of island biogeography provides a conceptual framework for understanding the interactions between habitat patches and the species they are able to sustain. In some naturally occurring, isolated ecosystems (e.g., isolated rocky reefs on the continental shelf, deep ocean trenches, seamounts, hydrothermal vent communities), the theory may have direct applications, since these habitats have strong similarities to isolated "islands." The patch sizes and their relative degrees of isolation from each other may have ecological attributes that can best be understood from the theory of island biogeography.

Metapopulations and Source-Sink Population Models

In the above example of a trawled seabed, a complex of discrete habitat patches is formed. Some patches will be too small to support all of the species originally present because of edge effects. In the case of isolated patches, species that have a limited migration range (and therefore a limited capacity to recolonize more remote patches) may become locally extinct. Theoretically, other species may be able to persist in a network of patches where populations within individual patches regularly go extinct, but which are then recolonized (i.e., by pelagic larvae dispersal); such connected populations, that could not persist in isolation but are viable as a whole, are termed *metapopulations* [20].

The importance of the concept of metapopulations is not so much that it directly explains the dynamics of a number of species in a disturbed environment (in fact, very few species exhibit true metapopulation characteristics; [16]), but that it highlights the consequences of habitat destruction and increased fragmentation. As more habitat patches are removed or diminished, the ability of species to recolonize is reduced while at the same time extinction rates within the remaining patches remain the same. A point is reached eventually where the patches are so small and isolated that the metapopulation becomes unviable; extinction may not be immediate but will occur once the relaxation time elapses.

A criticism of the metapopulation theory is that it assumes all patches are comprised of habitat having equal quality, which is not observed to be the case in the real world. In fact, some patches will invariably have areas of superior quality habitat supporting larger populations that produce a surplus of offspring, compared with other patches having poorer quality habitat that will require the arrival of immigrants to maintain their populations. In other words, for any given species, not all patches are equal; some patches act as population sources while others act as sinks. Over time, changing environmental conditions may mean that any given patch is transformed from being a source to a sink (or vice versa). Thus, the evaluation of individual patches as being either one or the other may not be possible based on a single set of observations; a time series of observations may be needed to make a reliable determination.

The distinction between high- and low-quality habitat is an important consideration for the application of habitat maps. For example, in the design of marine protected area (MPA) networks, habitat quality must be considered and the MPA network should aim to include as many source habitat patches as possible. In particular, Crowder et al. [21] point out that if the total fishing effort is not reduced in areas where MPAs are declared, then placing reserves in "sink" habitat patches can theoretically harm fish populations because all of the fishing effort will then be displaced onto the "source" patches. A metapopulation will survive if some sink patches are depleted, but might not if important source patches are overfished [22].

Causes of Patchiness in Marine Benthic Habitats

Patches of disturbed and undisturbed habitat are created by human actions (as in the example of seabed trawling discussed above). However, all natural systems exhibit patchiness when viewed at certain scales. In their study of patchiness in terrestrial ecosystems, Wu and Loucks [23] noted that patchiness depends on scale and the perception of dependent organisms; if the patch of seagrass, for example, is large compared with the foraging range of a particular animal, then the patchiness may not be perceived (even though the seagrass beds appear patchy to a human observer in an airplane).

Patchiness includes both biological and physical aspects that are interactive across a range of spatiotemporal and organizational scales (Figure 4.4). In other words, patches in nature are hierarchical, with smaller patches nested within larger ones.

In Wu and Loucks' [23] original conceptual framework of patchiness, derived for terrestrial ecosystems, biological causes of patchiness were divided into two groups related to consumers and effects of vegetation. In the benthic marine environment, the "vegetation" aspects of patchiness are limited because benthic plants are restricted to coastal and inner-shelf habitats where the seabed is within the photic zone. In this restricted shallow area, seagrasses, kelp beds, and other algae provide food sources and habitat for many animals. There is, however, no real comparison between terrestrial vegetation and benthic vegetation in the oceans. Oceanic primary

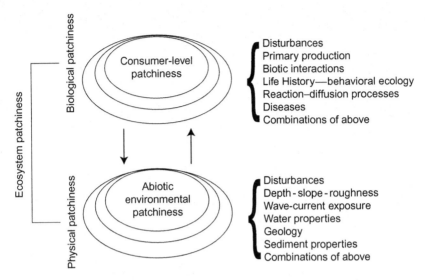

Figure 4.4 Conceptual framework of patchiness in the benthic environment, causes and mechanisms in ecological systems.
(adapted from Ref. [24].)

production occurs predominantly in the pelagic-photic zone inhabited by floating alga—an ecosystem that has no terrestrial equivalent. Hence, we have grouped benthic and pelagic primary production together for the purposes of this discussion on causes of marine benthic patchiness (Figure 4.4).

Other biological causes of patchiness in the benthic marine environment include interactions between animals (e.g., predation), life history, reaction to stimuli and resultant dispersal (e.g., food availability or physical stimuli), and the effects of diseases (Figure 4.4). Physical causes of patchiness in the marine environment include depth, slope, bed roughness, exposure to currents, physical disturbances (from storms, slumping, etc.; see below), water properties (temperature, salinity, dissolved oxygen content, etc.), seabed geology, and sediment properties (Figure 4.5). In each of these examples, a biological or physical attribute results an environment that excludes some plants or animals, which in turn gives rise to patchiness [23].

Disturbance as a Cause of Patchiness in Marine Benthic Habitats

Many ecosystems and species evolve in response to particular environmental disturbances that create patches of disturbed habitat and play a significant role in controlling things such as life cycles, food and nutrient supply, and habitat availability. Disturbances may be either biological or physical in nature (Figure 4.3). An environmental disturbance regime is a process or set of processes that characterizes and sustains an environment and that species have evolved to exploit to their own advantage.

A definition of disturbance was provided by Pickett and White [25] as "any discrete event in time that disrupts ecosystem, community or population structure and

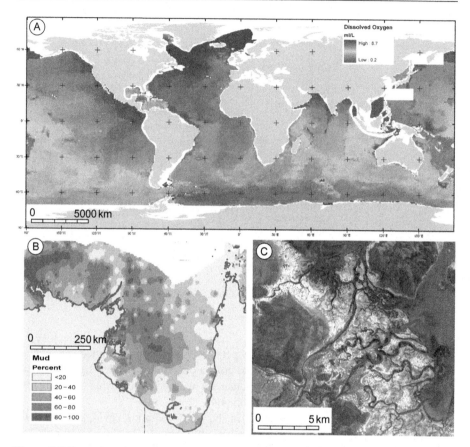

Figure 4.5 Examples of patchiness in the marine and coastal environment at different scales: (A) Ocean bottom water dissolved oxygen content based on data from the NOAA Ocean Atlas [24]; (B) percentage mud content in bottom sediments in the Gulf of Carpentaria, northern Australia; and (C) air photograph of the McArthur River delta in Australia's northern territory (white areas are salt flats, red is vegetation, and pink high ground is sediment flats). (For interpretation of the references to color in this figure legend, the reader is referred to the web version of this book.)
Source: Images (B) and (C) are from Geoscience Australia.

changes resources, substrate availability, or the physical environment." Patch-clearing processes are emphasized by landscape ecologists. Storms causing a large tree to fall in the forest, fire destroying an area of forest or scrubland, and a tree succumbing to drought are examples of important natural disturbances. The disturbance creates a patch of open space that is available for opportunistic species to colonize. An ecological succession ensues, with different species arriving over time, until the disturbed patch finally reverts to the original, undisturbed landscape [26]. Hence, landscapes that are subject to different disturbance regimes exhibit a degree of patchiness that relates to past disturbances, their colonization by opportunists, and gradual recovery.

Examples of marine environmental disturbances include El Nino-associated warming of sea surface temperature and attendant bleaching of coral reefs [27]; intertidal mud flats near a river mouth subject to broad variations in water salinity during extreme, annual, or interannual, flood events [28]; muddy seabeds of the continental shelf mobilized during extreme storm events [29]; submarine canyons and fans subject to pulses of sediment influx [30]; glacial-eustatic changes in sea level causing shelf areas to be exposed and flooded over 100,000-year cycles; and shelf seafloor episodically ploughed by drifting icebergs [29,30]. Physical processes may also exert a stress on organisms, tearing plants from their place of attachment [31], mobilizing sediment and burying plants and animals [32], damaging organisms by abrasion [33], or limiting light availability [34,35]. Biological disturbances can be caused by outbreaks of disease, invasions of predators such as sea urchins or starfish, or interruptions to food supply.

In each of these examples, a natural process gives rise to a disturbance that disrupts the ecosystem, community, or population structure and changes the availability of habitat or resources. Therefore, natural disturbances to marine environments may be physical, chemical, biological, periodic, or episodic in nature. They occur at multiple scales and are often contemporaneous, asynchronous, and heterogeneous. These processes play a critical role in the dynamic fluctuation of habitat availability and biotic diversity.

Habitat Recovery Rates

The time required for an ecosystem to recover from a disturbance will depend on the spatial extent and intensity of the disturbance. A localized disturbance, such as a few kelp fronds breaking free during a gale, will require less time to recover than if an entire kelp forest is affected by a severe storm [31]. The climatic regime and water depth are factors too, since ecological processes are generally slower in cold-deep than warm-shallow ecosystems [1]; cold-deep ecosystems might generally be expected to take longer to recover from a disturbance than warm-shallow ecosystems. Habitats characterized by larger, long-lived species will require more time to recover than habitats containing mostly smaller, short-lived species. Other factors might include proximity to source stock for larvae, availability of food, and connectivity [36]. Kostylev and Hannah [37] defined a species' "scope for growth," which they suggest is based on four factors: food availability (which combines stratification and surface chlorophyll as a measure of benthic–pelagic coupling), annual bottom temperature (as an indicator of metabolic rates), temperature variability (as an indicator of both thermal stress and temporal uncertainty for reproduction), and oxygen saturation (as a measure of metabolic stress).

There have been few studies published regarding recovery time of the benthos from natural disturbances. This is probably because, as noted by Hall [34], storms are by their very nature unpredictable and few studies have been funded to study benthic community dynamics for extended periods. Also, Berumen and Pratchett [27] note that it is often difficult to distinguish between the effects of multiple, unrelated but near-contemporaneous disturbances—such as crown of thorns starfish

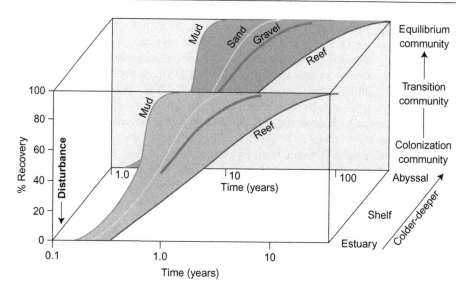

Figure 4.6 Rate of ecological succession: generalized relationship between elapsed time versus percentage of recovery for different benthic communities as a function of substrate type [36], depth, and temperature.

outbreaks, coral bleaching events, and cyclone events—which frustrates attempts to identify causal relationships.

A review of the rates of ecological successions following a natural disturbance event, typical of temperate to tropical coastal and shelf ecosystems [38], suggests full-recovery times of about 1 year for muddy substrates, to several years for sandy to gravely substrates, to 10 years or more for rocky, coral reef-type habitats (Figure 4.6).

Intermediate Disturbance Hypothesis

One of the more compelling possible explanations for the observed nonuniformity in the spatial distribution of biodiversity on Earth is the "intermediate disturbance hypothesis" (IDH; [27]). This hypothesis involves communities living in two end-member types of environment: highly dynamic environments versus highly stable environments, measured in terms of the length of time needed for ecological succession (Figure 4.6). In the former, diversity is low because few species can thrive under such stressful, ever-changing conditions. In stable environments, competitive exclusion takes its toll as weaker, less well-adapted species are eliminated. It is the intermediate zone of quasi-stable environments that allows the greatest diversity of species to exist.

The IDH predicts that biodiversity is controlled (in part) by the frequency of disturbance events, their spatial extent, and the amount of time required for ecological succession (Figure 4.6). These factors give rise to patchiness of the seafloor, with coexisting patches at different stages of ecological recovery. After a disturbance, propagules of a few species arrive and settle on the open space. Diversity is low at first because only the fastest reproducers or species that happen to be located in close proximity can colonize in the short time available. If the frequency of disturbance is too high, the community will consist only of the few early colonizers. With lower disturbance frequencies, the species number will increase with new arrivals until, at some intermediate frequency, maximum biodiversity is attained. If the disturbance frequency is too low, then species numbers are reduced by competition for resources and/or by interference [26]. Hence, an indicator that distinguishes communities that exist in an intermediate disturbance regime from other communities is the co-occurrence of habitats in various stages of postdisturbance impact and recovery.

Habitat Connectivity, Scale, and Hierarchy

Connectivity of Habitats

Connectivity exists where all the species occurring in an area have access to habitats and resources needed for life cycle completion and are able to recover from natural disturbances [39]. For example, access to seagrass habitat is crucial for juvenile tiger prawns as nursery areas [40]. Waves and currents generated by tropical cyclones locally remove seagrass beds, which may take months to years to recover, depending on the position of the disturbed patch in relation to potential seed-source areas [41].

Habitats have an inherent degree of "connectedness" in terms of their ability to be a source or sink of larvae and juvenile animals, but although superficially appearing to be a simple relationship related to the control of prevailing currents and the dispersal of larvae, it is actually very complex. This is due to the combination of different biological parameters (e.g., different life stages, mobility, animal life span, spawning intervals, predation) with different physical controls (e.g., temporally reversing flows, the nature of substrate within and between habitats) operating over different spatial and temporal scales. The concept of larvae being released at one location and being transferred by currents to a different location where a new colony is established is an oversimplification of what actually occurs in most cases [42–44]. Although the connectedness of habitats may be an important, if not essential, attribute, particularly where the application is for fisheries management or for the design of marine reserves, connectivity is a major gap in our knowledge and understanding of habitats.

Scale and Hierarchy of Habitats

How large and how small can a habitat be? The issue of scale for the identification and mapping of habitats is dependent upon their being viewed as nested within a hierarchy of spatial and temporal scales [39,45,46]. The ecosystem-based approach to planning uses habitats and other natural regions as planning units, but natural regions need to be identified on a range of hierarchically nested scales to design habitat mapping programs (e.g., selection of the right equipment) as well as for planning or management purposes.

Using a hierarchical scheme has several advantages. It provides context for the spatial information that is being considered, as well as a common reference framework for discussion and decision making. It also provides an agreed-upon set of environmental indicators that define different levels within the hierarchy that may become important for different applications of the habitat map.

The choice of a hierarchical system, and its application to mapping habitats in a particular area, carries a cost: The more refined and detailed the hierarchical system, the more expensive it will be in time and effort to gather the necessary information and apply it to a particular area. On the other hand, if information about the diversity of habitats is *not* collected at a scale that is relevant to its application or purpose, then the mapping product cannot be expected to inform managers or be useful in making decisions.

In the following sections, we will review some of the more commonly known hierarchical classification systems and compare them on the bases of information requirements and spatial application.

Hierarchical Classification Systems

Recalling from above, Wu and Loucks [23] noted that patchiness depends on scale and the perception of dependent organisms. The scale of a "patch" or habitat depends upon the observer, which leads us to recognize that habitats may be grouped together to form mosaics of like habitat types. Mosaics may also be grouped, and so on across a range of spatiotemporal scales.

As sentient beings, humans perceive order in nature, which leads us to classify and to try to describe systematically what can be observed. The properties used to distinguish different hierarchical levels in natural systems are themselves continuously varying functions; thus, sharp, unequivocal boundaries are the exception rather than the rule. Alan and Starr [47] observed that within any hierarchy, "discrete levels need to be recognized as convenience, not truth." Nevertheless, in most systems, real discontinuities can be recognized, and these have prompted the development of a number of classification schemes for different purposes. Hierarchical classification systems have the intrinsic predictive power of describing the relationships between physical habitats and their associated biological communities, and a good hierarchical classification system must be able to be modified when missing components are identified [46].

Although a simple Log_{10} scale can be used to describe marine environments, for example, macroscale (thousands of kilometers), mesoscale (hundreds of kilometers), microscale (tens of kilometers), and picascale ($<10\,km$) habitats [48], the levels in habitat classification schemes cannot usually be specified in terms of a fixed-length scale. This is because it is difficult to put even rough average lengths on the units being mapped, unless they are first placed in the context of the level above. For example, deep ocean muddy-habitats are likely to have greater spatial scales than coastal muddy-habitats, so an average-length scale for "muddy-habitat" depends first the biome (coastal or oceanic?) in question.

A number of hierarchical classification systems have been described in the literature, most of which are "rule based," meaning that different levels in the hierarchy are defined on the basis of a theory that explains the difference between levels; the process of subdivision is based on a set of rules derived from the theory [39]. "Rule-based" classifications can be contrasted against other more analytical approaches that use multivariate analysis techniques (discussed below).

Hayden et al. [49] presented a rule-based classification of the world's oceans and coasts into 40 provinces based mainly on wind and ocean current information. The analysis was an early attempt to identify the primary variables from which to derive a large-scale (global) biogeography for coasts and oceans. The work was inspired by Dietrich's [50] classification of the oceans based on current regime, and by Pielou's [51] earlier conclusion that "the evidence seems overwhelming that the boundaries of coastal biotic provinces are determined by modern abiotic factors."

Classification Scheme of Roff et al. (2003)

A widely known rule-based scheme proposed by Roff et al. [52] for application in Canada comprises eight nested "levels" which are considered to have effects at either global (Levels 1–4), regional (hundreds to thousands of kilometers; Levels 4–6), or local (tens to hundreds of kilometers; Levels 6–8) scales. Their approach, developed in a series of papers [46,52–55], is to apply mainly abiotic information in a hierarchy, inclusive of both pelagic and benthic environments, as follows:

- Level 1 places the area in either the marine or terrestrial environments.
- Level 2 is the major ocean basin (North Atlantic Ocean, Arctic Ocean, North Pacific Ocean, etc.).
- Level 3 refers to the atmospheric temperature regime (climate), which in broad terms places the area in polar, temperate, or tropical (or subdivisions of these) climate zones.
- Level 4 relates to sea-ice cover, which is clearly of importance for nations having polar climates within their jurisdictions.
- Level 5 divides the area into either the benthic or pelagic realms.
- Level 6a subdivides the pelagic realm, based on ocean stratification, into epipelagic, mesopelagic, bathypelagic, and abyssal/hadal zones. The benthic realm is subdivided based on light penetration into euphotic, dysphotic/aphotic, bathyal, and abyssal/hadal zones.
- Level 6b further subdivides the benthic realm based on bottom water temperature.
- Level 7 divides parts of pelagic and benthic realms based on mixing, oceanic fronts, wave action, and physiography. The epipelagic zone is thus either stratified or mixed, and it may or may not contain oceanic fronts. The benthic euphotic zone may or may not be exposed to strong swell-wave energy, and the whole of the benthic realm may have either high slopes (>2°) or low slopes (<2°).
- Level 8 applies only to the benthic realm and is based on sediment type (surface sediment gravel, sand and mud content).

The physical habitat approach used by Roff et al. [52] is supported by a number of independent studies. For example, in their review of ecosystem classification schemes, Snelder et al. [56] concluded that environmental classifications that use existing spatial data representing physical components of the environment

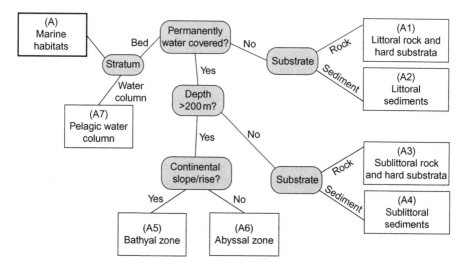

Figure 4.7 EUNIS Habitat univariate classification for marine habitats to Level 2 [57]. The shaded/enclosed text are deciding factors that separate seven habitat levels (A1–A7), shown as enclosed rectangles.

(e.g., climate, topography, and geology) have the potential to satisfy most requirements of environmental management applications.

EUNIS Classification Scheme

A different approach was applied to the European seas in the European Nature Information System (EUNIS) habitat classification system [57,58]. The EUNIS scheme was applied in four case studies (Chapters 8, 11, 33, and 35). The overall EUNIS classification is extremely complex because it applies to both terrestrial and marine environments. Of interest here is that part of the EUNIS scheme which relates to the marine environment (Figure 4.7). The EUNIS scheme broadly comprises six levels that are distinguished by a univariate assessment [58]:

- Level 1 places the area in either the marine or terrestrial environments.
- Level 2 defines seven major broad marine habitat types: (A1) littoral rock and other hard substrata; (A2) littoral sediments; (A3) sublittoral rock and other hard substrata; (A4) sublittoral sediments; (A5) slope and rise benthic habitats; (A6) abyssal benthic habitats; and (A7) the pelagic water column.
- Level 3 are "habitat complexes," which are frequently occurring combinations or mosaics of individual, possibly interdependent, habitat types. These are identified (in the littoral and sublittoral zone) based on sediment type (gravel, sand, mud, etc.), different degrees of wave exposure along the coast, and different depths of light penetration. Connor et al. [59] list 89 possible combinations of variables for habitat complexes in the littoral and sublittoral (A1–A4) habitats alone.
- Level 4 are biotope complexes, which are inferred to have the same physical information as for habitat complexes, but are subdivided based on community structure information.

- Level 5 are biotopes (habitats) having finer details of species and community structure within a similar physical environment.
- Level 6 are subbiotopes (microhabitats).

The contrast between the EUNIS scheme and that of Roff et al. [52] is therefore quite marked. Level 2 in the EUNIS scheme is based on depth and substrate type, which although not directly comparable to the Roff et al. [52] scheme is equivalent to a combination of their Level 5 and 6. Whereas the EUNIS scheme uses ecological terms, like habitat and biotope, to define levels, the Roff et al. [52] scheme is based on a range of physical parameters that apply over broad areas. Levels 2 and 3 in the Roff et al. [52] scheme are similar in scale to biogeographic provinces. Level 6 units are equal to biomes. In contrast, biomes in the EUNIS scheme occur at Level 2. Hence, there is little overlap between the two schemes at the very broad scales represented by Roff et al.'s [52] Levels 2–5. Another difference between the EUNIS scheme and that of Roff et al. [52] is the transformation from univariate decisions rules at Levels 1 and 2 in the EUNIS hierarchy to several variables defining Level 3 "habitat complexes"; in contrast, the Roff et al. [52] scheme is univariate throughout.

Classification Scheme of Greene et al. (1999)

The Roff et al. [52] and EUNIS schemes, with their broad regional focus and inclusion of both pelagic and benthic realms, may be contrasted with the benthic-only habitat classification of Greene et al. [45], whose largest units are physiographic provinces (continental shelf, slope, and abyssal plain; [51]). The Greene et al. [45] scheme was used in five case studies (Chapters 18, 38, 40, 44, and 54). Physiographic provinces are implied in EUNIS Level 2 and although not explicitly acknowledged in the Roff et al. [52] scheme, they coincide with parts of their Level 6. Greene et al. [45] developed their scheme for application to shelf and slope rockfish assemblages in 30–300 m water depth along the west coast of North America. They envisaged that it could be applied in other areas and expanded to include endofauna.

Apart from the scale and benthic versus pelagic differences mentioned above, the main contrast between the Roff et al. [52] and Greene et al. [45] schemes is the latter's focus on geomorphology and geomorphic features as mapable habitats. *Megahabitats* are defined by Greene et al. [45] as large geomorphic seabed features, 1–10 km in size, nested within major physiographic provinces. These are large geomorphic features, examples being submarine canyons, seamounts, lava fields, plateaus, large banks, reefs, terraces, and expanses of sediment-covered seafloor. *Mesohabitats* are smaller geomorphic features, from tens of meters to 1 km in size, and may include small seamounts, canyons, banks, coral reefs and other bioherms, caves, glacial moraines, gas-escape pock marks, mass wasting (slump) scars and deposits, and bedrock outcrops. *Macrohabitats* are 1–10 m in size and include seafloor materials such as boulders, rocky reefs, sediment waves (sandwaves or dunes), bars, cracks, caves, scarps, sink holes, gas-escape pock marks, feeding pits, and rock outcrops. Biogenic structures such as kelp beds, corals (solitary or reef building), and algal mats are also defined as macrohabitats. Lastly, *microhabitats* are centimeters or less in size, such as sediment grains or small cracks and crevices in a solid (rock) substrate. Greene et al. [45] go on to include a list of modifiers that qualify and add more detail to the broad habitat classification. The list of modifiers is extensive and includes the depth range, grain size, rock type (igneous, metamorphic, or sedimentary), seafloor slope, morphology type (regular, irregular, hummocky, etc.), sediment cover, bottom current regime, biological processes, and anthropogenic processes. Greene et al. [45] give several examples of how their scheme can be applied to describe seafloor habitats and make special mention of the use of multibeam sonar technology for seafloor mapping.

NOAA's Coastal/Marine Ecological Classification Standard

The United States' National Oceanic and Atmospheric Administration (NOAA) Coastal/Marine Ecological Classification Standard (CMECS) was used in two case studies (Chapters 18 and 30). It is based on eight nested hierarchical levels that span a range of bioregions from broad "ecological regions" that have length scales of $100-1,000 \, km^2$ to "biotopes" that have length scales of $1-100 \, m^2$ [60]. The scheme was developed for application in North America, but the authors suggest the scheme may be applied anywhere in the global marine environment.

The eight levels may be summarized as follows:

- *Level 1*: "Ecological regions" (scale 100 to $>1,000 \, km^2$) are broad regions whose boundaries are determined mainly by physiographic variables such as climate, water temperature, ocean currents, or ocean basins.
- *Level 2*: "Regimes" (scale 10 to $>1,000 \, km^2$) differentiate between freshwater and marine waters, defined as waters where the salinity is $>30 \, psu$ for more than 11 months of the year.
- *Level 3*: "Systems" (scale 1 to $>1,000 \, km^2$) are based on five types of coastal and marine environments: estuaries, estuary-influenced, marine, nearshore, neritic, and oceanic.
- *Level 4*: "Geoforms and hydroforms" (scale $10,000 \, m^2$ to $100 \, km^2$) are geomorphic or hydrologic features comprised of large elements approximately $10,000 \, m^2$ in area that influence ecological and biological processes. Examples might include small islands, sandbanks, or embayments (geoforms), or upwelling zones, oceanic gyres, and hydrothermal vents (hydroforms). It is noted that geoforms and hydroforms may exist at different levels in the hierarchy.
- *Level 5*: "Zones" (scale 100 to $10,000 \, m^2$) introduce a vertical dimension to the classification of different systems and geoforms and hydroforms (Levels 3 and 4). Three zones are defined: (1) littoral (intertidal) zone; (2) the water column zone; and (3) the seabed zone. These zones may be further subdivided; for example, the littoral zone may include supratidal and intertidal subzones, and the seabed may include shelf, slope, bathyal, and abyssal subzones.
- *Level 6*: "Macrohabitats" (scale 100 to $1,000 \, m^2$) are specific recognizable units of the physical environment having characteristic biophysical attributes, such as sediment type, temperature, salinity, current regime, and so on. These are subdivisions of geoforms and hydroforms, such as the flanks of sandbanks or the inner versus outer part of an oceanic gyre.
- *Level 7*: "Habitats" (scale 1 to $100 \, m^2$) are defined as "the physical biotic and abiotic features within the environment that are critical for biological and ecosystem health and function on a local scale." A habitat is a self-contained and physically distinct feature that is not differentiable into separate physical components.
- *Level 8*: "Biotopes" (scale 1 to $100 \, m^2$) are environmentally uniform in structure, environment, and biota. The physical habitat is characterized by recurring, predictable biological associations. The distinction between habitats and biotopes is the focus on physical aspects for habitats, whereas biotopes are determined by both biological and physical aspects.

Integrated Australian Classification Scheme

A scheme developed in Australia by Last et al. [12] for the selection of a national representative system of MPAs, and used in one case study (Chapter 48), integrates biological and physical criteria, with emphasis on different criteria at different levels in the hierarchy (Table 4.1). Hence, the scheme is a hybrid mix of the EUNIS, CMECS, Roff et al. [52] and Greene et al. [45] schemes. In the Australian scheme, biological criteria are emphasized at the highest level (provinces) and also at the lowest levels (biological facies and microcommunities). Physical variables like depth, geomorphology, sediments, and current strength play a dominant role in the middle levels (bathomes, geomorphological units, biotopes, and sub-biotopes; see Table 4.1).

Global Open Ocean and Deep Sea-habitats Classification

A committee formed under the auspices of the United Nations Ninth Meeting of the Convention on Biological Diversity (CBD) in May 2008 produced separate pelagic and benthic global bioregionalizations [61]. Of interest here is the benthic bioregionalization (Figure 4.8), which was created by dividing the ocean floor into nine bathyal provinces (800–3,500 m water depth), 13 abyssal provinces (3,500–6,500 m), and 13 hadal provinces (ocean trenches >6,500 m deep). The 35 different provinces were distinguished mainly by physical oceanographic variables extrapolated to the seafloor (i.e., dissolved oxygen, temperature, salinity, and organic matter flux) as well as published biogeographical information where it was available [61]. Separate hydrothermal vent provinces were also delineated based on biological data and other unpublished records from field sampling and observation [14].

The Global Open Ocean and Deep Sea-habitats (GOODS) classification is strictly not hierarchical since it contains only one level. It is presented here because it is the only benthic classification of its kind covering the entire world ocean (Figure 4.8). It is interesting to note that while the GOODS benthic bioregions are biomic (defined mainly on depth ranges), they are also provincial, since they subdivide ocean basins based on differences in biological assemblages.

Coastal Classification Schemes

The schemes discussed so far were devised either for the classification of broad-scale continental margins, or in the case of the GOODS classification, for the classification of global bioregions. By contrast, many other schemes have been devised for much smaller coastal and estuarine systems. For example, the Dethier [62] scheme is a modified version of an earlier scheme devised for the state of Washington, USA, and contains six main levels:

1. System, which divides fully marine from estuarine environments.
2. Subsystem, which divides subtidal from intertidal environments.
3. Class, based on the substratum type, which divides consolidated from unconsolidated substrates, rocky reefs, and artificial (anthropogenic) substrates. There is also a "subclass" in

Table 4.1 Hierarchical Scheme for Habitat Mapping and Classification in Southeast Australia [12]

Level	Names	Descriptions
1	Province	Large-scale biogeographic units. Evolutionary biogeography is the key process at this level, as reflected by the presence of regions of endemism. Provinces are typically of the order of ~1,000 km in extent.
2	2a Bathome	Marine version of "biomes," comprised of neritic and oceanic zones divided by the continental shelf break. The neritic zone has four primary bathomes: estuarine, coastal marine, demersal shelf, and pelagic shelf, whereas the oceanic zone consists of three primary demersal bathomes (continental slope, abyssal, and hadal), and five pelagic bathomes (epi-, meso-, bathy-, abysso-, and hadopelagic bathomes). These are nested within provincial units and are typically several hundreds of kilometers or more in extent.
	2b Sub-biomes	Based on recognizably distinct composition of the biota. For example, along-shelf subdivision of the inner-, mid-, and outer-shelf sub-biomes, some of which have quite narrow depth ranges.
3	Geomorphological units	Areas characterized by similar geomorphology. These may include (on the continental shelf) fields of sand waves, rocky outcrops, incised valleys, flat muddy seabeds, etc., and (on the slope and at abyssal depths) submarine canyons, seamounts, oceanic ridges, and troughs, etc. Such units may typically be about 100 km in extent.
4	Primary biotopes	Soft, hard, or mixed substrate-based units, together with their associated suites/collections of floral and faunal communities, modified by hydrological variables such as wave exposure, turbidity, and current speed.
5	Secondary biotopes	Rock types (e.g., fossiliferous limestone; granite); sediment types (e.g., poorly sorted shelly sands) or biota (e.g., seagrasses).
6	Biological facies	Defined by a biological indicator, or suite of indicator species, that identify a biological assemblage used as a surrogate for a biocoenosis or community. Includes, for example, a particular species of seagrass, or group of corals, sponges, or other macrofauna strongly adherent to the facies.
7	Microcommunities	Species that depend on facies (e.g., isopods on seagrass).

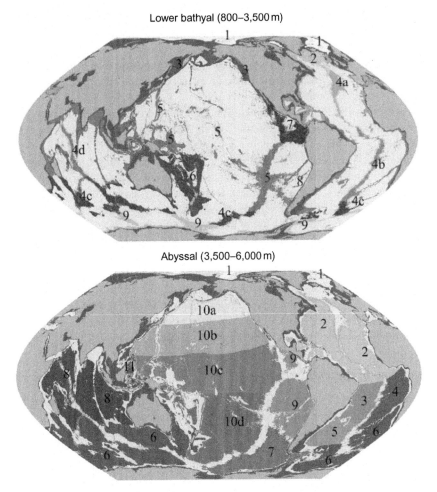

Lower bathyal (800–3,500 m)

Abyssal (3,500–6,000 m)

Figure 4.8 GOODS benthic classification [61] showing the lower bathyal and abyssal bioregions. Not shown is the hadal classification for 13 separate ocean trench bioregions.

the scheme, based on the dominant substrate type including bedrock, hardpan, boulder, cobble, gravel, sand, mud, and organic constituents.

4. Energy/enclosure is a relative measure of the wave and tidal current regime.
5. Modifiers include two categories of intertidal zone, three categories of subtidal zone based on depth, and five categories based on salinity.
6. Characteristic species, which may include the most abundant species, a combination of numerical and biomass measures, the most "obvious" species, and/or functionally important species that either provide habitat (e.g., eel grass or mussel beds) or strongly interact (e.g., key predators).

The Zacharias and Howes [63] classification was devised for application to British Columbia in western Canada. It contains two main levels: ecosections and ecounits. Ecosections are broad geographic spatial units based on differences in ocean circulation, mixing, salinity, temperature, primary and total productivity, continental slope and shelf influence, and freshwater influence. Ecounits are based on a combination of depth, current speed, swell-wave exposure, relief, and substrate. Zacharias and Howes [63] reported that 12 ecosections and 619 ecounits could be mapped in British Columbia.

Connor et al. [58,59] devised a classification for Britain and Ireland, based on a matrix of primary habitats nested within Levels 2 and 3 (broad habitat types and habitat complexes) of the EUNIS scheme described above. The matrix places substrate type along the x-axis and water depth along the y-axis. Two substrate types are recognized, rock and sediment, which are subdivided based essentially on wave energy (for rock) and grain size (for sediment). Depth on the y-axis is divided into littoral and sublittoral zones, with the sublittoral zone further subdivided into infra-littoral and circa-littoral subzones (see Connor et al. [58], their Table 3).

Thus, Connor et al. [58] consider rocky versus sediment substrate types found along the shore (intertidal) and in the subtidal or deep ocean to be Level 2 units, with Level 3 habitat complexes being different types of sediment (gravel, sand, mud, etc.), different degrees of wave exposure along the coast, and depth of light penetration. Combinations of these variables in the matrix show there are a total of 24 possible habitat complexes (EUNIS Level 3), 75 biotope complexes (Level 4), and 370 possible biotopes and sub-biotopes (Levels 5 and 6; [58]).

Comparison of Different Schemes

It is difficult to identify equivalence between the different schemes because they have been derived from different information sources and are intended for application to different environments. However, from the above, it is evident that there are some common elements to the hierarchical classifications. At the highest level, the distinction is made between marine and terrestrial environments in both the Roff et al. [52] and the EUNIS schemes; other schemes only include the marine environment (Figure 4.9). The largest regional divisions reflect global-scale oceanic basins, endemism, and atmospheric latitudinal climate variation. At its highest level, the Australian scheme [12] reflects broad patterns in biodiversity, such as the large marine ecosystems (LMEs) or biogeochemical provinces [64], which are based mainly on spatial patterns in surface ocean primary productivity. Evolutionary biogeography is a key process at this level and geological factors, such as the movement of continental plates over millions of years, are significant.

Nested within this highest level are broad physiographic divisions (*biomes* or "*bathomes*") that divide the seabed into neritic and oceanic zones, with the boundary between the two at the continental shelf break. The neritic zone has three primary biomes (estuarine, coastal marine, and demersal shelf), while the oceanic zone consists of three primary biomes (continental slope, abyssal, and hadal). This level of division is explicitly stated in the Australian scheme (Table 4.1) and is approximately equal to the EUNIS "habitat types," the physiographic provinces of Greene et al. [45], the "system" of Dethier [62], and the "ecosections" of Zacharias and Howes ([63]; see Figure 4.9).

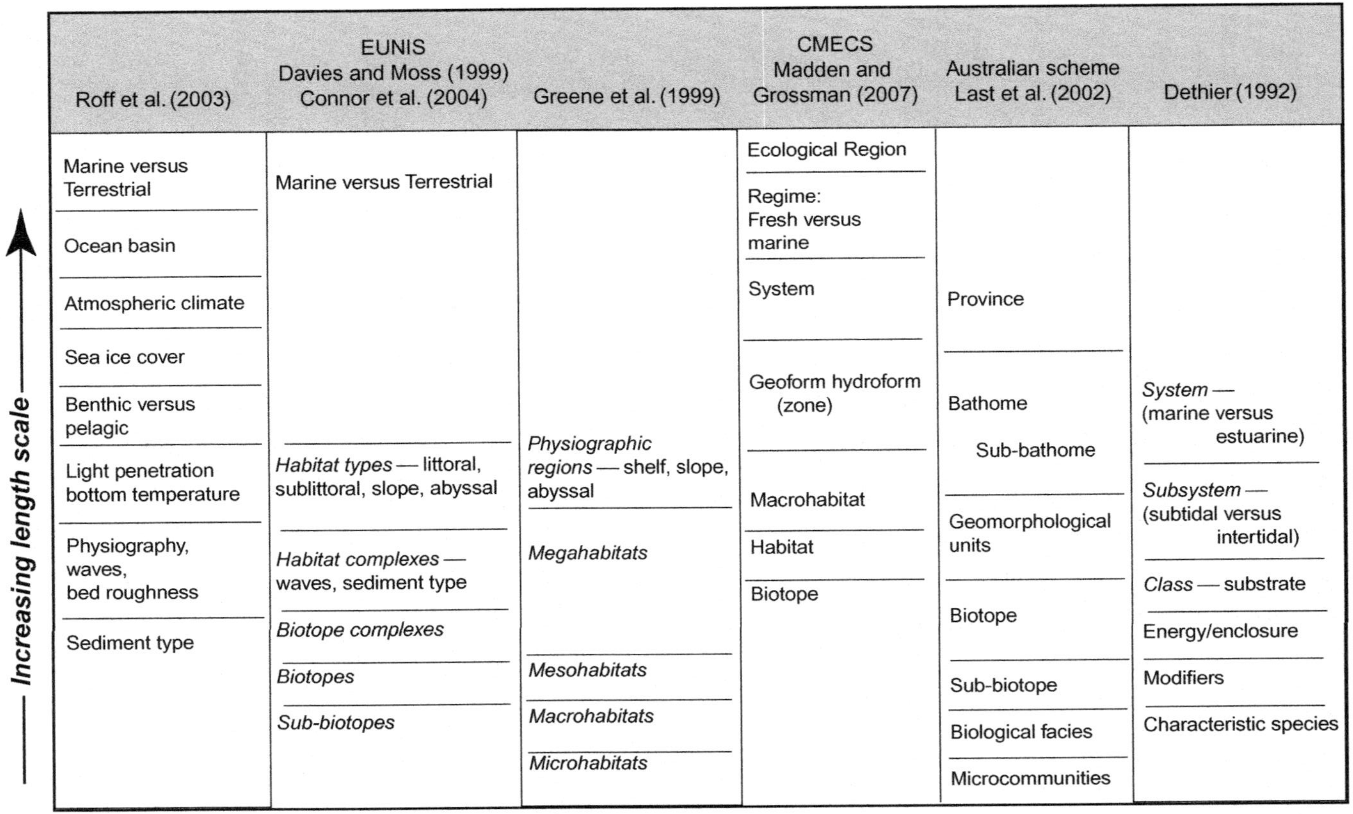

Roff et al. (2003)	EUNIS Davies and Moss (1999) Connor et al. (2004)	Greene et al. (1999)	CMECS Madden and Grossman (2007)	Australian scheme Last et al. (2002)	Dethier (1992)
Marine versus Terrestrial	Marine versus Terrestrial		Ecological Region		
Ocean basin			Regime: Fresh versus marine		
Atmospheric climate			System	Province	*System* — (marine versus estuarine)
Sea ice cover					
Benthic versus pelagic	*Habitat types* — littoral, sublittoral, slope, abyssal	*Physiographic regions* — shelf, slope, abyssal	Geoform hydroform (zone)	Bathome	*Subsystem* — (subtidal versus intertidal)
Light penetration bottom temperature				Sub-bathome	
Physiography, waves, bed roughness	*Habitat complexes* — waves, sediment type	*Megahabitats*	Macrohabitat	Geomorphological units	*Class* — substrate
Sediment type	*Biotope complexes*		Habitat		Energy/enclosure
	Biotopes	*Mesohabitats*	Biotope	Biotope	Modifiers
	Sub-biotopes	*Macrohabitats*		Sub-biotope	Characteristic species
		Microhabitats		Biological facies	
				Microcommunities	

Figure 4.9 Comparison, on the basis of length scale, of the hierarchical schemes.

Broad subdivisions of biomes might be comparable to the "megahabitats" of Greene et al. [45], geomorphological units of the Australian scheme, and "habitat complexes" of Connor et al. [58], based on physiography, wave and current regime, and seafloor roughness (Figure 4.8). These are large geomorphic features, such as seamounts or offshore plateaus in the deep sea, and coral reef complexes or large sand-wave fields on the continental shelf. They would include Dethier's [62] "subsystem," which is based on the distinction between intertidal and subtidal zones, and may form some of the larger "ecounits" of Zacharias and Howes [63].

The Australian and EUNIS schemes both use the terms "biotopes" (physical habitats and their associated biota) and "sub-biotopes," which are consistent with Levels 7 and 8 in the Roff et al. [52] scheme and include variables such as physiography, wave and current regime, bed roughness, and sediment type. These factors are also included in the other schemes. For example, Dethier's [62] "class," the "ecounits" of Zacharias and Howes [63], and the EUNIS "habitat complexes" all include these variables (see also Table 4.1). The biotope level is also where the scheme of Greene et al. [45] includes the greatest number of subdivisions ("meso-," "macro-," and "microhabitats").

Some hierarchies, such as those of Roff et al. [52] and Greene et al. [45], have no classifications below the "biotope" level. Schemes that contain hierarchical levels at this fine scale generally move away from physical, habitat-defining parameters and invoke biotic information to differentiate between levels. Examples are Dethier's [62] "characteristic species," the Australian scheme's "biological facies," and the EUNIS "sub-biotopes."

Bioregionalization—Mapping Biologically Distinctive Regions

Mapping Bioregions

Once a hierarchical classification has been chosen, maps can be generated showing the boundaries of the areas represented by each level (provinces, biomes, habitat complexes, etc.) forming what is commonly referred to as a *bioregionalization* (Figure 4.8). The spatial units are referred to generally as *bioregions*, regardless of their level in the hierarchy (i.e., bioregions can be either biomes or geomorphologic units). While it may be obvious that the information used to construct the bioregionalization is indicated by the content of the various hierarchical classification schemes, it should be noted that a hierarchy is an "ideal"—useful conceptually to provide a context and common terminology for users, but not necessarily easily applicable to real-world situations. The challenges involved with converting the idealized hierarchy into a map include the fact that many data sets are incomplete and do not provide 100% coverage of the area of interest. There are also some complex problems surrounding the interpretation of different data sets.

For example, a difficult question facing scientific teams tasked with building a bioregionalization is knowing what biological, geological, and oceanographic information is relevant to habitat mapping, and to understanding ecosystem processes, within their particular area of interest. This problem manifests itself in that mapping of bioregions at one level requires information at the next level down in the hierarchy. Boundaries between bathomes are based on geomorphology, whereas boundaries between geomorphic units are based at least in part on information about substrate type and sedimentary processes (e.g., defining the limits of a geomorphic unit comprising fields of sandwaves on the continental shelf requires information on sand transport and sediment type).

Another problem arises with attempting to simply apply the information contained in "rule-based" hierarchical classifications directly to generating a bioregionalization map. This is because the lower levels in a hierarchy cannot always be expected to nest wholly within the higher levels in every case. For example, in the Roff et al. [52] scheme, there is no logical reason why spatial areas defined on the bases of physiography, wave exposure, and bed roughness (Level 7) should nest wholly within larger areas defined by a combination of light penetration and bottom water temperature (Level 6). A single, large, physiographic feature such as a seamount (Level 7 unit) might rise from the cold abyssal depths up into warmer, near-surface water masses, and thus it cuts across the boundaries that define a higher level in the hierarchy.

One way around this dilemma is to assess *all* the data layers using multivariate analysis and clustering techniques [52,65]. Advantages of the multivariate approach are that: (i) it maximizes the differences between the physical components that characterize each level in the hierarchy; (ii) it can be applied to large and complex data sets; (iii) it lends itself to computer-based geographic information system (GIS) mapping tools; and (iv) it is scientifically defensible [65,66]. However, such an approach seems to contradict the conceptualization of the hierarchy as having distinct levels, with each being defined by a few particular dominant variables. A distinct disadvantage to the multivariate approach is that the classification provides little insight into the causes of differences between classes. The differences between hierarchical levels derived from a multivariate approach are difficult to communicate to the general public, which may limit their usefulness to managers. By comparison, the reasoning behind the identification of different levels within a "rule-based" hierarchy is much more obvious, which makes the maps of such hierarchical levels easy to understand and to communicate.

In practice, the mapping of different hierarchical levels to create a bioregionalization can proceed using both rule-based and multivariate analysis. For example, the rule-based approach can be used to map Level 3 geomorphological units, whereas mapping Level 4 biotopes or Connor et al.'s [58] "habitat complexes" might be more easily accomplished using a multivariate analysis approach.

Bioregion Boundaries—Fuzzy or Crisp?

Bioregions may be separated by distinct, unequivocal, "crisp" boundaries, but more often than not the division between bioregions is gradual and the exact location is best expressed as a "fuzzy" gradient. "Provinces," for example, are distinguished based on continental-scale geomorphology (e.g., oceanic basins), atmospheric latitudinal climate variation, and associated biological endemism. Province boundaries are consequently broad, gradational "ecotones" that are generally expected to trend normal to the continental margin (although there are exceptions). Ecotones, in turn, are associated with a gradient in environmental conditions, from one province to the next, and are themselves ecologically unique, possessing attributes of both provinces. It has been suggested that ecotone habitats are occupied by specialized species that have adapted to the edge effects and may contain a greater diversity of species than either of the adjacent provinces.

Acknowledgments

Thanks to Lene Buhl-Mortensen and Antoine Collin for helpful comments and suggestions on an earlier draft of this chapter. This work was produced with the support of funding from the Australian Government's Commonwealth Environment Research Facilities (CERF) program and is a contribution of the CERF Marine Biodiversity Hub. This chapter is published with the permission of the Chief Executive Officer, Geoscience Australia.

References

[1] D.L. Perlman, G. Adelson, Biodiversity: Exploring Values and Priorities in Conservation, Blackwell Science, Oxford, U.K.

[2] J.H. Ausubel, D.T. Crist, P.E. Waggoner (Eds.), First Census of Marine Life 2010: Highlights of a Decade of Discovery, Census of Marine Life, Washington, DC, 2010, http://www.coml.org/.

[3] R.V. Sole, M. Newman, Extinctions and biodiversity in the fossil record, in: H.A. Mooney, J.G. Canadell (Eds.), The Earth System: Biological and Ecological Dimensions of Global Environmental Change, John Wiley & Sons, Chichester, 2002, pp. 297–301.

[4] B. Lomborg, The Skeptical Environmentalist: Measuring the Real State of the World, Cambridge University Press, Cambridge, 2001.

[5] P.J. Crutzen, E.F. Stoermer, The "Anthropocene", Global Change Newsletter 41 (2000) 12–13.

[6] J.S. Gray, The Ecology of Marine Sediments, Cambridge University Press, Cambridge, 1981.

[7] R. Noss, Indicators for monitoring biodiversity: a hierarchical approach, Conserv. Biol. 4 (1990) 355–364. J.S. Levinton, Marine Biology Function, Biodiversity, Ecology, Oxford University Press, New York, NY, 2001.

[8] G.G. Kelleher, C.J. Bleakley, S. Wells, A Global Representative System of Marine Protected Areas, Great Barrier Reef Marine Park Authority, World Bank and IUCN, Washington, DC, 1995.

[9] P.V.R. Snelgrove, Getting to the bottom of marine biodiversity: sedimentary habitats, Bioscience 49 (1999) 129–138.

[10] N. Knowlton, L.A. Weigt, New dates and new rates for the divergence across the Isthmus of Panama, Proc. R. Soc. London B 265 (1998) 2257–2263.

[11] P.J. Davies, P.A. Symonds, D.A. Feary, C.J. Pigram, Horizontal plate motion: a key allocyclic factor in the evolution of the Great Barrier Reef, Science 238 (1987) 1697–1700.

[12] P.R. Last, V.D. Lyne, A. Williams, C.R. Davies, A.J. Butler, G.K. Yearsley, A hierarchical framework for classifying seabed biodiversity with application to planning and managing Australia's marine biological resources, Biol. Conserv. 143 (2010) 1675–1686.

[13] L.A. Levin, R.J. Etter, M.A. Rex, A.J. Gooday, C.R. Smith, J. Pineda, et al., Environmental influences on regional deep-sea species diversity, Annu. Rev. Ecol. Syst. 32 (2001) 51–93.

[14] A.V. Gebruk, E.C. Southward, P.A. Tyler (Eds.), The Biogeography of the Oceans, Academic Press, San Diego, CA, 1997

[15] J.D. Gage, Diversity in deep-sea benthic macrofauna: the importance of local ecology, the larger scale, history and the Antarctic, Deep-Sea Res. II 51 (2004) 1689–1708.

[16] A.S. Pullin, Conservation Biology, Cambridge University Press, Cambridge, UK, 2002.

[17] R.H. MacArthur, E.O. Wilson, The Theory of Island Biogeography, Princeton University Press, Princeton, NJ, 1967.

[18] S.P. Hubbell, The Unified Neutral Theory of Biodiversity and Biogeography, Princeton University Press, Princeton, NJ, 2001.

[19] J. Engel, R. Kvitek, Effects of otter trawling on a benthic community in Monterey Bay National Marine Sanctuary, Conserv. Biol. 12 (1998) 1204–1214.

[20] R. Levins, Some demographic and genetic consequences of environmental heterogeneity for biological control, Bull. Entemologic. Soc. Am. 15 (1969) 237–240.

[21] L.B. Crowder, S.J. Lyman, W.F. Figueira, J. Priddy, Source-sink population dynamics and the problem of siting marine reserves, Bull. Mar. Sci. 66 (2000) 799–820.

[22] F.C. Coleman, P.B. Baker, C.C. Koening, A review of Gulf of Mexico marine protected areas: successes, failures and lessons learned, Fisheries 29 (2004) 10–21.

[23] J. Wu, O.L. Loucks, From balance of nature to hierarchical patch dynamics: a paradigm shift in ecology, Q. Rev. Biol. 70 (1995) 439–466.

[24] NOAA. World Ocean Atlas data, 2005. Available from http://www.nodc.noaa.gov/OC5/ WOA05/pr_woa05.html.

[25] S.T. Pickett, P.S. White, The Ecology of Natural Disturbance and Patch Dynamics, Academic Press, London, 1985.

[26] J.H. Connell, Diversity in tropical rain forests and coral reefs, Science 199 (1978) 1302–1310.

[27] M. Berumen, M. Pratchett, Recovery without resilience: persistent disturbance and long-term shifts in the structure of fish and coral communities at Tiahura Reef, Moorea, Coral Reefs 25 (2006) 647–653.

[28] W.J. Kimmerer, Open water processes of the San Francisco Estuary: from physical forcing to biological responses, San Francisco Estuary Watershed Sci. (2004) Vol. 2, Issue 1, pp. 1–142, http://repositories.cdlib.org/jmie/sfews/vol2/iss1/art1.

[29] P.T. Harris, R. Coleman, Estimating global shelf sediment mobility due to swell waves, Mar. Geol. 150 (1998) 171–177.

[30] S. Hess, F.J. Jorissen, V. Venet, R. Abu-Zied, Benthic foraminiferal recovery after recent turbidite deposition in Cap Breton Canyon, Bay of Biscay, J. Foraminiferal Res. (2005) 35.

[31] M.S. Thomsen, T. Wernberg, G.A. Kendrick, The effect of thallus size, life stage, aggregation, wave exposure and substratum conditions on the forces required to break or dislodge the small kelp Ecklonia radiata, Botanica Marina 47 (2004) 454–460.

[32] J.Y. Aller, J.R. Todorov, Seasonal and spatial patterns of deeply buried calanoi copepods on the Amazon shelf: evidence for periodic erosional/depositional cycles, Estuarine Coastal Shelf Sci. 44 (1997) 57–66.

[33] A.G. Cheroske, S.L. Williams, R.C. Carpenter, Effects of physical and biological disturbances on algal turfs in Kaneohe Bay, Hawaii, J. Exp. Mar. Biol. Ecol. 248 (2000) 1–34.

[34] S.J. Hall, Physical disturbance and marine benthic communities: life in unconsolidated sediments, in: A.D. Ansell, R.N. Gibson, M. Barnes, (Eds.), Oceanography and Marine Biology: An Annual Review, UCL Press, London, pp. 179–239.

[35] T.J.B. Carruthers, W.C. Dennison, B.J. Longstaff, M. Waycott, E.G. Abal, L.J. McKenzie, et al., Seagrass habitats of north-east Australia: models of key processes and controls, Bull. Marine Sci. 71 (2002) 1153–1169.

[36] S. Condie, J. Waring, J.V. Mansbridge, M.L. Cahill, Marine connectivity patterns around the Australian continent, Environ. Model. Softw. 20 (2005) 1149–1157.

[37] V.E. Kostylev, C.G. Hannah, Process-driven characterization and mapping of seabed habitats, in: B.J. Todd, H.G. Greene, (Eds.), Mapping the Seafloor for Habitat Characterization, Geological Association of Canada, St. John's, NL, 2007, pp. 171–184.

[38] P.T. Harris, On seabed disturbance, marine ecological succession and applications for environmental management: a physical sedimentological perspective, in: M. Li, C. Sherwood, P. Hill (Eds.), Sediments, Morphology and Sedimentary Processes on Continental Shelves, International Association of Sedimentologists Special Publication, Blackwell Publishing Ltd, Oxford, 2011

[39] K.A. Poiani, B.D. Richter, M.G. Anderson, H.E. Richter, Biodiversity conservation at multiple scales: functional sites, landscapes, and networks, Bioscience 50 (2000) 133–146.

[40] W.J. Lee Long, J.E. Mellors, R.G. Coles, Seagrasses between Cape York and Hervey Bay, Queensland, Australia, Aust. J. Mar. Freshw. Res. 44 (1993) 19–31.

[41] T.J. Done, Effects of tropical cyclone waves on ecological and geomorphological structures on the Great Barrier Reef, Cont. Shelf Res. 12 (1992) 859–872.

[42] I.J. Dight, M.K. James, L. Bode, Modelling the larvel dispersal of Acanthaster planci: patterns of reef connectivity, Coral Reefs 9 (1990) 125–134.

[43] R.K. Cowan, K.M.M. Lwiza, S. Sponaugle, C.B. Paris, D.B. Olson, Connectivity of marine populations: open or closed?, Science 287 (2000) 857–859.

[44] R.A. Briers, Incorporating connectivity in reserve design, Biol. Conserv. 103 (2002) 77–83.

[45] H.G. Greene, M.M. Yoklavich, R.M. Starr, V.M. O'Connell, W.W. Wakefield, D.E. Sullivan, et al., A classification scheme for deep seafloor habitats, Ocenaologica Acta 22 (1999) 663–678.

[46] J.C. Roff, M.E. Taylor, National frameworks for marine conservation—a hierarchical geophysical approach, Aquatic Conserv. Mar. Freshw. Ecosyst. 10 (2000) 209–223.

[47] T.F.H. Allen, T.B. Starr, Hierarchy—Perspectives for Ecological Complexity, University of Chicago Press, Chicago, IL, 1982.

[48] T. Stevens, R.M. Connolly, Testing the utility of abiotic surrogates for marine habitat mapping at scales relevant to management, Biol. Conserv. 119 (2004) 351–362.

[49] B.P. Hayden, G.C. Ray, R. Dolan, Classification of coastal and marine environments, Environ. Conserv. 11 (1984) 199–207.

[50] D. Dietrich, General Oceanography: An Introduction, Interscience Publishers, New York, NY, 1963.

[51] E.C. Pielou, Biogeography, Wiley-Interscience, New York, NY, 1979.

[52] J.C. Roff, M.E. Taylor, J. Laughren, Geophysical approaches to the classification, delineation and monitoring of marine habitats and their communities, Aquatic Conserv. Mar. Freshw. Ecosyst. 13 (2003) 77–90.

[53] J.C. Day, J.C. Roff, Planning for Representative Marine Protected Areas: A Framework for Canada's Oceans, World Wildlife Fund Canada, Toronto, 2000 , p. 148.

[54] M.A. Zacharias, J.C. Roff, A hierarchical ecological approach to conserving marine biodiversity, Conserv. Biol. 14 (2000) 1327–1334.

[55] M.A. Zacharias, J.C. Roff, Zacharias and Roff vs. Salomon et al.: Who adds more value to marine conservation efforts?, Conserv. Biol. 15 (2001) 1456–1458.

[56] T. Snelder, J. Leathwick, B. Biggs, M. Weatherhead, Ecosystem classification: a discussion of various approaches and their application to environmental management, Report prepared for the Ministry for the Environment, NIWA, Christchurch, New Zealand, 2001.

[57] C.E. Davies, D. Moss, The EUNIS Classification, European Environment Agency p. 124, http://eunis.eea.eu.int/index.jsp.

[58] D.W. Connor, J.H. Allen, N. Golding, K.L. Howell, L.M. Lieberknecht, K.O. Northen, et al., Marine Habitat Classification for Britain and Ireland Version 04.05. Joint Nature Conservation Committee Peterborough, UK, www.jncc.gov.uk/page-1584.

[59] D.W. Connor, K. Hiscock, R.L. Foster-Smith, R. Covey, A classification system for benthic marine biotopes, in: A. Eleftheriou, A.D. Ansell, C.J. Smith, (Eds.), Biology and Ecology of Shallow Coastal Waters, Olsen and Olsen, Fredensberg, Denmark, 1995, pp. 155–166.

[60] C.J. Madden, D.H. Grossman, A framework for a coastal/marine ecological classification standard (CMECS), in: B.J. Todd, G. Greene (Eds.), Mapping the Seafloor for Habitat Characterisation, Geological Association of Canada, St. John's, NL, 2007, pp. 185–210.

[61] V. Agnostini, E. Escobar-Briones, I. Cresswell, K. Gjerde, D.J.A. Niewijk, A. Polacheck, et al., Global Open Oceans and Deep Sea-habitats (GOODS) bioregional classification, in: M. Vierros, I. Cresswell, E. Escobar-Briones, J. Rice, J. Ardron (Eds.), Report submitted to the Ninth Meeting of the Conference of the Parties to the Convention on Biological Diversity (CBD). document number UNEP/CBD/COP/9/INF/44. http://ioc-grame.grouphub.com/projects/1101472/file/15064932, pp. 94.

[62] M.N. Dethier, Classifying marine and estuarine natural communities, Nat. Areas J. 12 (1992) 90–100.

[63] M.A. Zacharias, D.E. Howes, An analysis of marine protected areas in British Columbia using a marine ecological classification, Nat. Areas J. 18 (1998) 4–13.

[64] A.R. Longhurst, Seasonal cycles of pelagic production and consumption, Prog. Oceanogr. 36 (1995) 77–167.

[65] P.T. Harris, T. Whiteway, High seas marine protected areas: benthic environmental conservation priorities from a GIS analysis of global ocean biophysical data, Ocean Coast. Manag. 52 (2009) 22–38.

[66] J. Rice, K.M. Gjerde, J. Ardron, S. Arico, I. Cresswell, E. Escobar, et al., Policy relevance of biogeographic classification for conservation and management of marine biodiversity beyond national jurisdiction, and the GOODS biogeographic classification, Ocean Coast. Manag. 54 (2011) 110–122.

5 Surrogacy

Peter T. Harris

Marine and Coastal Environment Group, Geoscience Australia, Canberra, ACT, Australia

Abstract

The term "surrogacy" is used in habitat mapping with reference to the biophysical variables that can be mapped with a quantifiable correspondence to the occurrence of benthic species and communities. Surrogacy research can be defined as an empirical method of determining which easily measured characteristics best describe the species assemblage in a particular space and at a particular time. These characteristics act as predictors (with some known probability and uncertainty) for the occurrence of species assemblages in unexplored areas. Abiotic variables are, in general, more easily and less expensively obtained than biological observations, which is a key driver for surrogacy research. However, the suite of abiotic factors that exert control over the occurrence of a species (its niche) is also a scientifically interesting aspect of ecology that provides important insights into species evolution and biogeography. This chapter provides a review of surrogates used by case study authors and the methods used to quantify relationships between variables.

Key Words: Niche theory; physical surrogates; biodiversity; analytical techniques; multivariate analysis

Niche Theory

City lights seen from space are a surrogate for the distribution and density of the Earth's human habitation, just as burrows on a tide flat are surrogates for the crustaceans living within. The idea that physical, chemical, and biological properties of the environment can act as surrogates for biodiversity is intuitive; it seems obvious that the spatial occurrence of a particular species will in some way be associated with the combination of physical characteristics defining its preferred habitat. In fact, every species has a certain range of abiotic (physical and chemical) variables within which it is able to survive and reproduce, a space known as the "fundamental niche" for that species. But most species are unable to fully exploit the entire volume of their fundamental niche because of competition with other species, diseases, and disturbances, all of which reduces their occurrence to a "realized niche." This imposes a limitation on the use of abiotic surrogates, which may be used to estimate the fundamental niche for a species but cannot predict the realized niche, because physical

Seafloor Geomorphology as Benthic Habitat. DOI: 10.1016/B978-0-12-385140-6.00005-0

and chemical surrogates will not account for biological processes such as competition with other species and diseases, and may not account for physical disturbances. Abiotic surrogates will therefore tend to overestimate the distribution of species (the fundamental rather than realized niches [1]).

Physical Surrogates

The relationships between physical variables and the occurrence and abundance of benthic biota have been analyzed by a number of workers (Table 5.1). The abiotic variables with greatest known influence over benthic organism distributions are temperature, salinity, oxygen concentration, light availability, and sediment composition [12,13]. However, the more studies that are undertaken to find physical surrogates for the occurrence of species, the less clear the relationships appear to be.

Consider the following two observations: (1) As much as a quarter of the variance in benthic diversity can be explained by depth alone [14]; (2) For soft-sediment habitats, patch structure is mainly based on sediment characteristics and geomorphological/topographic features [15]. Given that each of these observations is justified from the evidence available, why is it that they seem to contradict each other?

The answer is partly because there are different types of physical surrogates, and they relate to biodiversity in different ways that must be considered when interpreting the results of comparisons. These differences relate to niche theory (discussed earlier), direct versus indirect variables, and interpretation errors (e.g., false homogeneity and false heterogeneity).

Table 5.1 Key Results of Surrogacy Studies in Which Species Assembly Was Analyzed Against Abiotic Variables [2]

Study	Variables Best/Most Important	% Variability Accounted For	Statistical Process
Post et al. [3]	% Mud, % gravel disturbance, depth	59	BIOENV
Passlow et al. [4]	Depth, longitude	25	BIOENV
Sanders et al. [5]	Depth	65	BIOENV
Beaman et al. [6]	Slope, % gravel, % $CaCO_3$	75	BIOENV
Beaman and Harris [7]	Slope, % gravel, turbidity	62	BIOENV
Williams et al. [8]	Depth, latitude, gear type longitude	44	BIOENV
Stevens and Connolly [9]	% Mud, distance to ocean	30	Spearmans
Ellingsen et al. [10]	Depth	29	Linear regression
Gogina et al. [11]	Depth, TOC	50, 43	BIOENV (also Spearmans)

Direct and Indirect Variables

Some physical variables, while useful predictors of biodiversity, may not be the cause of the patterns they describe. A species may be observed to occur over a particular depth range, for example, so "depth" would seem to be a good predictor of the occurrence of that species. However, the species may actually be responding (for example) to the location of a thermal gradient in the water column where food is plentiful, which happens to intersect the seabed over that depth range. In this case, "depth" is an indirect surrogate for the species, whereas the gradient in bottom water temperature is a direct variable. Indirect variables may correlate with the occurrence of a species, but they are not the reason for the species to occur at a particular location.

Examples of indirect variables (other than depth) are latitude, distance from the shore, distance from a river mouth, and acoustic backscatter. Some indirect variables used by case study authors are listed in Table 5.2. Depth covaries with many physical attributes of the ocean including temperature, light availability, pressure, food availability, and aragonite solubility; thus, it is an indirect measure of all of these variables. Similarly, acoustic backscatter is affected by seabed slope, rugosity, sediment grain size, the presence of bedforms, and other factors that vary in importance depending on the sonar frequency and grazing angle [60].

Direct variables are attributes of the environment sought by species; they are actually the cause for the occurrence of a species at a particular location. Direct variables used by case study authors (Table 5.2) include substrate type, sediment grain size, seafloor rugosity, water temperature, salinity, and bed stress caused by waves and tidal currents. In order to understand causal relationships and ecosystem function, it is important to recognize the inherent differences between direct and indirect variables.

False Heterogeneity and False Homogeneity

In many studies, habitat mapping relies on the correlation of the occurrence of one or more species with a specific range of values for one or more abiotic surrogates. However, due to the complexity of marine ecosystems and ecosystem processes (including biological, physical, and chemical processes), these correlations are frequently confounded. In some cases, areas of the seafloor may appear to be different on the basis of abiotic variables but are biologically similar (false heterogeneity) or, conversely, areas of the seafloor may appear to be similar on the basis of abiotic variables but are biologically distinct (false homogeneity). For example, in an underwater video study of the macrobenthos in soft-sediment habitats of Moreton Bay, Australia, Stevens and Connolly [9] found that errors of false homogeneity were between 20% and 62%. In other words, soft-sediment habitats of Moreton Bay appeared to be physically similar, but subtle (or unmeasured) differences resulted in the occurrence of significantly different habitats occupied by different benthic communities.

Table 5.2 Most Important Physical Surrogacy Variables and Analytical Techniques Reported in the Papers Included in This Volume

Author	Chapter	Most Important Variables	Analysis Technique
D'Angelo and Fiorentino [16]	7	Backscatter	GIS
Tempera et al. [17]	8	Depth, slope, exposure, primary production (chl-a)	Ordered logistic regression
Collin and Archambault [18]	10	TPI	ENVI (GIS)
Dankers et al. [19]	11	Bed stress, emergence time	GIS
James et al. [20]	12	Grain size, depth	PRIMER
Van Dijk et al. [21]	13	Grain size, bed morphology	SIMPER, cluster analysis
Van Lancker et al. [22]	14	Grain size	GIS
Stieglitz [23]	16	Slope	GIS
Kotilainen et al. [24]	17	Substrate type, depth	GIS
Cochrane et al. [25]	18	Substrate type, exposure	GIS, ERDAS
Copeland et al. [26]	19	Grain size, backscatter, slope	ANOSIM, SIMER
Collier and Humber [27]	21	Backscatter	Cluster analysis, Mann–Whitney U-test
Hamylton et al. [28]	22	Depth, wave power, rugosity	Spatial ANOVA
Hamylton [29]	23	Depth	Spatially lagged autoregressive model
Armstrong and Singh [30]	24	Light attenuation	GIS
Brooke et al. [31]	25	Sediment type, rugosity	GIS
Heyman and Kobara [32]	26	Morphology	GIS
Nichol et al. [33]	27	Grain size, depth	GIS
Gibbs and Cochran [34]	28	Slope, depth, aspect	GIS
Beaman et al. [35]	29	Depth, backscatter, slope, rugosity, aspect, BTM zones	BTM, Maxent
Allee et al. [36]	30	Rugosity	CCA
Martorelli et al. [37]	31	Backscatter	GIS
Coggan and Diesing [38]	33	Substrate type	BTM
Lucieer et al. [39]	34	Depth, exposure, morphology	BTM, Landserf
Galparsoro et al. [40]	35	Sediment properties, wave stress	GIS, PRIMER
Yamanaka et al. [41]	36	TPI, depth, backscatter	GIS
Robinson et al. [42]	37	Grain size, sediment composition	PRIMER, CANOCO
Reynolds et al. [43]	38	Depth, grain size	GIS
Todd et al. [44]	39	Depth, oxygen, sediment type	BioENV, ANOSIM
Getsiv-Clemmons et al. [45]	40	Backscatter, slope, TPI	GIS, Geowizard
James et al. [46]	41	Grain size, current speed	PRIMER
Pearce et al. [47]	42	Sorting, current and wave speed, rugosity, % sand, temperature	PRIMER

(Continued)

Table 5.2 (Continued)

Author	Chapter	Most Important Variables	Analysis Technique
Ezhova et al. [48]	43	Grain size, T, S, current speed	GIS
De Mol et al. [49]	44	Rugosity, slope, BPI	Max likelihood, Max entropy
Althaus et al. [50]	48	Depth, backscatter	Bray–Curtis dissimilarity
Lo Iacono et al. [51]	49	Sediment type, backscatter, rugosity, bed stress	Linear classification
Huvenne et al. [52]	50	Depth, substrate type	MVA
Buhl-Mortensen et al. [53]	51	Depth, backscatter, sediment type, TPI	DCA ordination, GIS
Post et al. [54]	52	Hard/soft substrate, iceberg	MVA
Lamarche et al. [55]	53	Backscatter	Image analysis
Yoklavich and Greene [56]	54	Grain size, depth	GIS
Wright et al. [57]	58	Rugosity, TPI	GIS
Tempera et al. [58]	59	Substrate type, depth	GIS
De Mol et al. [59]	60	Backscatter, TPI, rugosity, slope	Max likelihood, PCA, GIS

GIS, geographic information system; BTM, benthic terrain modeler; CCA, canonical correspondence analysis; MVA, multivariate analysis.

Drivers of Some Biophysical Surrogates

In Chapter 4, it was shown that most classification schemes developed for marine environments acknowledge hard versus soft substrate as being a key habitat determining factor. Abiotic surrogates are useful if they can distinguish between hard and soft substrates, or if they allow for more detailed characterization and subdivision of one of these two categories. Many case studies presented in this volume focus on the development of approaches for distinguishing between hard and soft substrates [23,26,35,36,48,61–63].

Hard-Substrate Surrogates

There is a definite bias in studies of benthic habitats toward hard substrates (coral reefs, temperate rocky reefs, seamount peaks, rocky scarps at the heads of submarine canyons, rocky banks and shoals, etc.). This is because hard-substrate habitats are often associated with iconic features of interest to humans. They are also generally more biodiverse and support a higher biomass per unit area than soft substrates, which makes them of interest for fishing, recreational diving, tourism, and so on. (Figure 5.1).

From an ecological perspective, there are good reasons why biota compete for access to hard-substrate habitats [65]. Irregular rocky surfaces that project into the water column above the level of surrounding seabed provide advantageous habitat for passive filter-feeding invertebrates, because being located in an area of stronger currents means access to potentially more food. Plants that anchor to the seafloor via a holdfast and that require access to sunlight colonize rocky surfaces located in the

Figure 5.1 Underwater video compilation of submerged coral reefs and associated fishes in the Gulf of Carpentaria, Australia, reported by Harris et al. [64]. (Video is only available in the on-line version of this book.)

euphotic zone to maximize their stability. Small juvenile fish and invertebrates seek the complex surfaces that rocky habitats provide for shelter and hiding places.

Several case studies in this book are concerned with technologies used for distinguishing between hard and soft substrates, such as acoustic backscatter and geographic information system (GIS)-derived measures of seafloor roughness from gridded bathymetry data. Acoustic backscatter, seafloor roughness, and rugosity (estimated using ArcGIS tools applied to a gridded bathymetry data set) are commonly used surrogates. Rugosity, either on its own or combined with backscatter and/or other derived geomorphic indices (e.g., topographic position index (TPI) or bathymetric position index (BPI) slope), can be an indicator of hard substrates. Backscatter data are acquired as part of multibeam and sidescan sonar mapping surveys, and when added to the gridded values of seafloor depth, they provide an important suite of abiotic variables for surrogacy research.

Soft-Substrate Surrogates

Several studies have found sediment properties to be good predictors of benthic communities (Table 5.1). Among the direct surrogates listed in Table 5.1 are sediment grain size and total organic carbon (TOC) content of bottom sediments.

It should be noted that some parameters have been widely measured but do not seem to be particularly useful as surrogates for benthic organisms, and other parameters are useful surrogates only for particular habitats. Water salinity, for example, may be very important in the coastal and shelf biomes, where salinity exerts a direct control over the occurrence of biota: reduced or more variable salinity has a direct

Figure 5.2 Seafloor photograph taken in 554 m water depth on the outer Antarctic continental shelf off George Vth Land, showing numerous unidentified arthropods feeding on pelagic detritus deposited on the seabed.
Source: After Brancolini, Harris and shipboard party [68].

negative effect on species diversity. However, salinity varies by less than approximately 1 psu over most of the deep ocean [66] and does not appear to play any significant role in controlling species distribution in the deep sea [67].

A fundamental parameter that controls all benthic ecosystems is food supply. Fecal pellets are the most important source of particulate organic carbon (POC) flux to the seafloor, where animals have evolved to survive on this food source (Figure 5.2). Pelagic sediments, composed of calcareous or siliceous ooze, are supplied to the seabed as part of this POC flux. There are three main spatial gradients in ocean surface primary productivity that are known to affect POC supply to the deep ocean [69]. First, there is a lateral decrease in productivity from the shallow continental shelves seaward to the deep ocean basins. Second, there is a lateral decrease in productivity from the equator into the ocean current gyres that occupy the major ocean basins where the lowest surface productivity (oligotrophic) waters in the world ocean occur. Third, there is a general vertical decrease in POC flux with increasing water depth (i.e., the deeper the water, the food arriving from above is less).

Explanations for the lateral productivity gradients are related to the supply of river-derived nutrients supporting higher coastal and shelf productivity and the availability of sunlight, coupled with upwelling along the equator, supporting higher productivity in that region [69]. The vertical gradient is related to the degradation of organic matter as it falls through the water column, which introduces another variable into the discussion—dissolved oxygen.

Oxygen is consumed during the degradation process, particularly beneath areas of high productivity, at depths of between 100 and 1,200 m [69]. If the rate of oxygen consumption exceeds the rate at which it can be resupplied by mixing downward from the surface or upward from bottom waters, hypoxic zones referred to as "oxygen minimum zones" are formed. Such zones have dissolved oxygen concentrations of <0.5 ml/l and occur in the eastern Pacific Ocean, Gulf of Mexico, Arabian Sea, and off the Atlantic coast of Africa. Where an oxygen minimum zone makes contact with the seabed (e.g., in the case of a seamount whose summit pierces an oxygen minimum zone), studies have shown that the benthic macrofauna diversity is reduced [69–71]. The dissolved oxygen content of bottom water is also governed by the elapsed time since the water was last in contact with the sea surface. It is highest in areas of bottom water production (in the North Atlantic and adjacent to Antarctica) where bottom water is young, and it is lowest where the bottom water mass has been long out of contact with the sea surface.

Much of the POC reaching the seafloor is recycled by benthic animals. Seiter et al. [72] calculated that approximately 0.5 GtC/year of POC reaches the deep seafloor (>1,000 m water depth), whereas only a small part, around 0.002–0.12 GtC/year (equal to 0.01–0.4% of surface primary production), is buried in the sediments.

The thickness of pelagic sediments accumulating at a given location on the seafloor depends upon several factors, chiefly the rate of mass flux to the seafloor, whether the site is above or below the carbonate compensation depth (CCD; normally this occurs at a water depth of ~4,000 m), and the age of the ocean crust (sediments are thinnest on young crust adjacent to mid-ocean spreading ridges, since there has been less time for sediment to accumulate). Deep-sea red clay is commonly deposited at depths greater than approximately 4,000 m (below the CCD) and covers broad areas of seafloor, with calcareous and siliceous oozes draping over features at more shallow depths.

A second source of POC to the oceans is supplied by rivers, equal to approximately 0.54 GtC/year [73]. This carbon is recycled in the coastal zone, and a portion is deposited along continental margins in association with terrigenous sediments. The thickness of terrigenous sediments on the ocean floor is therefore correlated with the supply of terrigenous POC. In comparison to the high seas, sediment thickness (and food supply) is much greater adjacent to continents that have high sediment-yield river systems and/or that were glaciated during the Pleistocene ice ages [74].

The temperature of bottom water also exerts a direct control over the distribution of benthic animals (Figure 5.3); most benthic animals are confined to live within a specific temperature range and metabolic rates are lower with decreasing temperature [12,69,75]. The coldest bottom water is found at the poles, in the North Atlantic and adjacent to the Antarctic continental margin. However, bottom water temperature in the mid to lower latitudes below 2,000 m water depth is <4°C and does not vary significantly in the abyssal or hadal ocean environments [67].

Food and oxygen supply are crucial to the existence of animal life, but geophysical data sets also provide insights into the character of the substrate that, in turn, exerts control over the type of benthic communities that may exist. For example, hard-rocky substrates (that support sessile filter-feeding communities) are more likely to be associated with thin sediment cover and steep slopes, whereas soft, sediment-covered substrates (detritus-feeding communities) are associated with flat seabeds and thick

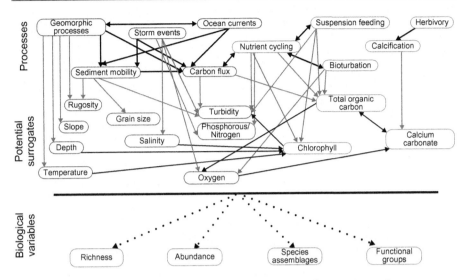

Figure 5.3 Conceptual map of the drivers of biodiversity in marine systems and some potential surrogates. Black lines indicate relationships between potential surrogates. Bold black lines represent relationships between processes. Gray lines show potential surrogates linked to a process. Dotted lines show potential relationships between surrogates and biological variables. Brown borders enclose abiotic factors. Green borders enclose biological or biophysical factors. (For interpretation of the references to color in this figure legend, the reader is referred to the web version of this book.)
Source: From McArthur et al. [2].

sediments. In this way, combinations of different data sets provide even greater insight into potential habitats than simply analyzing individual data layers in isolation.

Biological Surrogates

Biological surrogates are measures of species richness, endemism, rarity, and complementarity of taxonomic or functional groups that are presumed to be indicators of overall biodiversity [76]. Examples of marine biodiversity studies that have adopted this approach include those of Vanderklift et al. [77], who used fish, invertebrate, and macroalgal species richness data to estimate biodiversity in Jervis Bay, Australia (see also Ward et al. [78]). Olsgard et al. [79] used polychaetes as indicators of biodiversity in soft-sediment habitats from the North Atlantic. Hirst [80] investigated macroalgae and faunal subsets derived by aggregating the arthropod fauna at higher taxonomic levels to predict arthropod faunal biodiversity on subtidal rocky reefs.

These studies suggest that the use of biological surrogates involves four main sources of uncertainty: (1) uncertainty that there exists a reliable, known relationship between the surrogate and the species or groups of species whose distribution is being predicted; (2) spatial uncertainty in the sampling program; (3) temporal uncertainty in the surrogacy relationship and in the sampling program; and (4) uncertainty in the spatial and temporal variability in prevailing ecosystem processes. The latter two sources of temporal uncertainty are particularly significant in habitats subject to long-term changes to environmental conditions, such as estuaries and coastal environments. The problem stems from the fragmentary knowledge we have of marine ecosystems, which derives from marine surveys that can only take one-off "snapshots" when it is obvious that long-term monitoring of environments is required for a full understanding of their functioning.

Surrogacy Analytical Techniques

The case studies in this book have applied different techniques to further our understanding of surrogacy relationships. In most cases, the problem is one of identifying useful surrogates from multiple layers of abiotic spatial data that are tested against a limited number of biological observations made from sampling or video–photographic observations of the seabed. Hence, the problem requires multivariate analytical techniques to identify and quantify the relationships between variables.

Spatial analysis tools incorporated into ArcGIS spatial analyst software are the primary analytical tool used in 52% of case studies (Table 5.2). An advantage of using ArcGIS is that many of the input data layers are Raster files (e.g., multibeam sonar products) that are easily imported and manipulated spatially. Point data provided by seabed sampling and video observations may also be included in the analysis.

Biological associations are commonly investigated using *Plymouth Routines in Multivariate Ecological Research* (PRIMER) software to examine statistical relationships between variables [81]. PRIMER is a collection of specialist routines for analyzing species or sample abundance (biomass). Multivariate routines included in PRIMER (Table 5.2) are grouping (CLUSTER); sorting (MDS); principal component identification (PCA); hypothesis testing (ANOSIM); sample discrimination (SIMPER); trend correlation (BEST); comparisons (RELATE); and diversity, dominance, and distribution calculation. PRIMER allows nonparametric tests of hypotheses about multivariate data to test questions like: Is there a statistically significant difference in community composition between sites A and B?

Nonparametric (or distribution-free) inferential statistical methods are mathematical procedures for statistical hypothesis testing which, unlike parametric statistics, make no assumptions about the probability distributions of the variables being assessed. An example is the Mann–Whitney U-test used by Collier and Humber [27]. Other statistical methods used by case study authors are listed in Table 5.2.

Acknowledgments

Thanks to Sarah Hamylton for helpful comments and suggestions on an earlier draft of this chapter. This work was produced with the support of funding from the Australian Government's Commonwealth Environment Research Facilities (CERF) program and is a contribution of the CERF Marine Biodiversity Hub. This chapter is published with the permission of the Chief Executive Officer, Geoscience Australia.

References

[1] D.M. Moore, B.G. Lees, S.M. Davey, A new method for predicting vegetation distributions using decision tree analysis in a geographic information system, Environ. Manage. 15 (1991) 59–71.

[2] M.A. McArthur, B. Brooke, R. Przeslawski, D.A. Ryan, V.L. Lucieer, S. Nichol, et al., A review of abiotic surrogates for marine benthic biodiversity, Geoscience Australia Record 2009/42, 2009, 61 pp.

[3] A.L. Post, T. Wassenberg, V. Passlow, Physical surrogates for macrofaunal distributions and abundance in a tropical gulf, Mar. Freshw. Res. 57 (2006) 469–483.

[4] V. Passlow, T.D. O'Hara, T. Daniell , R.J. Beaman, L.M. Twyford, Sediments and benthic biota of Bass Strait: an approach to benthic habitat mapping, Geoscience Australia Record 2004/23, 2006, Canberra.

[5] J.L. Sanders, M.A. Kendall, A.J.S. Hawkins, J.I. Spicer, Can functional groups be used to indicate estuarine ecological status? Hydrobiologia 588 (2007) 45–58.

[6] R.J. Beaman, J. Daniell, P.T. Harris, Geology–benthos relationships on a temperate rocky bank, eastern Bass Strait, Australia, Mar. Freshw. Res. 56 (2005) 943–958.

[7] R.J. Beaman, P.T. Harris, Geophysical proxies as predictors of megabenthos assemblages from the northern Great Barrier Reef, Australia, in: B.J. Todd, H.G. Greene (Eds.), Mapping the Seafloor for Habitat Characterization: Geological Association of Canada, Special Paper 47, 2007, pp. 241–258.

[8] A. Williams, K. Gowlett-Holmes, F. Althaus (Eds.), Biodiversity Survey of Seamounts and Slopes of the Norfolk Ridge and Lord Howe Rise: Final Report to the Department of the Environment and Heritage (National Oceans Office), CSIRO, Hobart, 2006, 591.

[9] T. Stevens, R.M. Connolly, Testing the utility of abiotic surrogates for marine habitat mapping at scales relevant to management, Biol. Conserv. 119 (2004) 351–362.

[10] K.E. Ellingsen, K.R. Clarke, P.J. Somerfield, R.M. Warwick, Taxonomic distinctness as a measure of diversity applied over a large scale: the benthos of the Norwegian continental shelf, J. Anim. Ecol. 74 (2005) 1069–1079.

[11] M. Gogina, M. Glockzin, M.L. Zettler, Distribution of benthic macrofaunal communities in the western Baltic Sea with regard to near-bottom environmental parameters. 2. Modelling and prediction, J. Mar. Syst. 79 (2010) 124.

[12] P.V.R. Snelgrove, Getting to the bottom of marine biodiversity: sedimentary habitats, Bioscience 49 (1999) 129–138.

[13] V.E. Kostylev, C.G. Hannah, Process-driven characterization and mapping of seabed habitats, in: B.J. Todd, H.G. Greene (Eds.), Mapping the Seafloor for Habitat Characterization, Geological Association of Canada, St. Johns, Newfoundland, 2007, pp. 171–184.

[14] C. Olabarria, Faunal change and bathymetric diversity gradient in deep-sea prosobranchs from Northeastern Atlantic, Biodivers. Conserv. 15 (2006) 3685–3702.

[15] R.N. Zajac, Challenges in marine, soft-sediment benthoscape ecology, Landsc. Ecol. 23 (2007) 7–18.

[16] S. D'Angelo, A. Fiorentino, Phanerogam meadows: a characteristic habitat of the Mediterranean shelf. Examples from the Tyrrhenian Sea, in: P.T. Harris, E.K. Baker (Eds.), Seafloor Geomorphology as Benthic Habitat: GeoHab Atlas of Seafloor Geomorphic Features and Benthic Habitats, Elsevier, Amsterdam, 2011 (Chapter 7).

[17] F. Tempera, M. McKenzie, I. Bashmachnikov, M. Puotinen, R.S. Santos, R. Bates, Predictive modelling of dominant macroalgae abundance on temperate island shelves (Azores, northeast Atlantic), in: P.T. Harris, E.K. Baker (Eds.), Seafloor Geomorphology

as Benthic Habitat: GeoHab Atlas of Seafloor Geomorphic Features and Benthic Habitats, Elsevier, Amsterdam, 2011 (Chapter 8).

[18] A. Collin, B. Long, P. Archambault, Coastal kelp forest habitat in the Baie des Chaleurs, Gulf of St. Lawrence, Canada, in: P.T. Harris, E.K. Baker (Eds.), Seafloor Geomorphology as Benthic Habitat: GeoHab Atlas of Seafloor Geomorphic Features and Benthic Habitats, Elsevier, Amsterdam, 2011 (Chapter 10).

[19] N. Dankers, W. van Duin, M. Baptist, E. Dijkman, J. Cremer, The Wadden Sea in the Netherlands: ecotopes in a World Heritage barrier island system, in: P.T. Harris, E.K. Baker (Eds.), Seafloor Geomorphology as Benthic Habitat: GeoHab Atlas of Seafloor Geomorphic Features and Benthic Habitats, Elsevier, Amsterdam, 2011 (Chapter 11).

[20] J.W.C. James, A.S.Y. Mackie, E.I.S. Rees, T. Darbyshire, Sand wave field: The OBel Sands, Bristol Channel, U.K., in: P.T. Harris, E.K. Baker (Eds.), Seafloor Geomorphology as Benthic Habitat: GeoHab Atlas of Seafloor Geomorphic Features and Benthic Habitats, Elsevier, Amsterdam, 2011 (Chapter 12).

[21] T.A.G.P. van Dijk, J.A. van Dalfsen, R. van Overmeeren, V. Van Lancker, S. van Heteren, P.J. Doornenbal, Benthic habitat variations over tidal ridges, North Sea, Netherlands, in: P.T. Harris, E.K. Baker (Eds.), Seafloor Geomorphology as Benthic Habitat: GeoHab Atlas of Seafloor Geomorphic Features and Benthic Habitats, Elsevier, Amsterdam, 2011 (Chapter 13).

[22] V. Van Lancker, G. Moerkerke, I. Du Four, E. Verfaillie, M. Rabaut, S. Degraer, Fine-scale geomorphological mapping for the prediction of macrobenthic occurrences in shallow marine environments, Belgian part of the North Sea, in: P.T. Harris, E.K. Baker (Eds.), Seafloor Geomorphology as Benthic Habitat: GeoHab Atlas of Seafloor Geomorphic Features and Benthic Habitats, Elsevier, Amsterdam, 2011 (Chapter 14).

[23] T.C. Stieglitz, The Yongala's Halo of Holes—systematic bioturbation close to a shipwreck, in: P.T. Harris, E.K. Baker (Eds.), Seafloor Geomorphology as Benthic Habitat: GeoHab Atlas of Seafloor Geomorphic Features and Benthic Habitats, Elsevier, Amsterdam, 2011 (Chapter 16).

[24] A.T. Kotilainen, A.M. Kaskela, S. Bäck, J. Leinikki, Submarine De Geer moraines in the Kvarken Archipelago, the Baltic Sea, in: P.T. Harris, E.K. Baker (Eds.), Seafloor Geomorphology as Benthic Habitat: GeoHab Atlas of Seafloor Geomorphic Features and Benthic Habitats, Elsevier, Amsterdam, 2011 (Chapter 17).

[25] G.R. Cochrane, L. Trusel, J. Harney, L. Etherington, Habitats and benthos of an evolving fjord, Glacier Bay, Alaska, in: P.T. Harris, E.K. Baker (Eds.), Seafloor Geomorphology as Benthic Habitat: GeoHab Atlas of Seafloor Geomorphic Features and Benthic Habitats, Elsevier, Amsterdam, 2011 (Chapter 18).

[26] A. Copeland, E. Edinger, T. Bell, P. LeBlanc, J. Wroblewski, R. Devillers, Geomorphic features and benthic habitats of a subarctic fjord: Gilbert Bay, Southern Labrador, Canada, in: P.T. Harris, E.K. Baker (Eds.), Seafloor Geomorphology as Benthic Habitat: GeoHab Atlas of Seafloor Geomorphic Features and Benthic Habitats, Elsevier, Amsterdam, 2011 (Chapter 19).

[27] J.S. Collier, S.R. Humber, Fringing reefs of the Seychelles inner granitic islands, Western Indian Ocean, in: P.T. Harris, E.K. Baker (Eds.), Seafloor Geomorphology as Benthic Habitat: GeoHab Atlas of Seafloor Geomorphic Features and Benthic Habitats, Elsevier, Amsterdam, 2011 (Chapter 21).

[28] S. Hamylton, T. Spencer, A.B. Hagan, Coral reefs and reef islands of the Amirantes Archipelago, western Indian Ocean, in: P.T. Harris, E.K. Baker (Eds.), Seafloor Geomorphology as Benthic Habitat: GeoHab Atlas of Seafloor Geomorphic Features and Benthic Habitats, Elsevier, Amsterdam, 2011 (Chapter 22).

[29] S. Hamylton, Hyperspectral remote sensing of the geomorphic features and habitats of the Al Wajh Bank reef system, Saudi Arabia, Red Sea, in: P.T. Harris, E.K. Baker (Eds.), Seafloor Geomorphology as Benthic Habitat: GeoHab Atlas of Seafloor Geomorphic Features and Benthic Habitats, Elsevier, Amsterdam, 2011 (Chapter 23).

[30] R.A. Armstrong, H. Singh, Mesophotic coral reefs of the Puerto Rico Shelf, in: P.T. Harris, E.K. Baker (Eds.), Seafloor Geomorphology as Benthic Habitat: GeoHab Atlas of Seafloor Geomorphic Features and Benthic Habitats, Elsevier, Amsterdam, 2011 (Chapter 24).

[31] B.P. Brooke, M.A. McArthur, C.D. Woodroffe, M. Linklater, S.L. Nichol, T.J. Anderson, et al., Geomorphic features and infauna diversity of a subtropical mid-ocean carbonate shelf: Lord Howe Island, Southwest Pacific Ocean, in: P.T. Harris, E.K. Baker (Eds.), Seafloor Geomorphology as Benthic Habitat: GeoHab Atlas of Seafloor Geomorphic Features and Benthic Habitats, Elsevier, Amsterdam, 2011 (Chapter 25).

[32] W.D. Heyman, S. Kobara, Geomorphology of reef fish spawning aggregations in Belize and the Cayman Islands (Caribbean), in: P.T. Harris, E.K. Baker (Eds.), Seafloor Geomorphology as Benthic Habitat: GeoHab Atlas of Seafloor Geomorphic Features and Benthic Habitats, Elsevier, Amsterdam, 2011 (Chapter 26).

[33] S.L. Nichol, T.J. Anderson, C. Battershill, B.P. Brooke, Submerged reefs and aeolian dunes as inherited habitats, Point Cloates, Carnarvon Shelf, Western Australia, in: P.T. Harris, E.K. Baker (Eds.), Seafloor Geomorphology as Benthic Habitat: GeoHab Atlas of Seafloor Geomorphic Features and Benthic Habitats, Elsevier, Amsterdam, 2011 (Chapter 27).

[34] A.E. Gibbs, S.A. Cochran, Coral habitat variability in Kaloko-Honokohau National Park, Hawaii, USA, in: P.T. Harris, E.K. Baker (Eds.), Seafloor Geomorphology as Benthic Habitat: GeoHab Atlas of Seafloor Geomorphic Features and Benthic Habitats, Elsevier, Amsterdam, 2011 (Chapter 28).

[35] R.J. Beaman, T. Bridge, T. Done, J.M. Webster, S. Williams, O. Pizarro, Habitats and benthos at Hydrographers Passage, Great Barrier Reef, Australia, in: P.T. Harris, E.K. Baker (Eds.), Seafloor Geomorphology as Benthic Habitat: GeoHab Atlas of Seafloor Geomorphic Features and Benthic Habitats, Elsevier, Amsterdam, 2011 (Chapter 29).

[36] R.J. Allee, A.W. David, D.F. Naar, Two shelf edge marine protected areas in the eastern Gulf of Mexico, in: P.T. Harris, E.K. Baker (Eds.), Seafloor Geomorphology as Benthic Habitat: GeoHab Atlas of Seafloor Geomorphic Features and Benthic Habitats, Elsevier, Amsterdam, 2011 (Chapter 30).

[37] E. Martorelli, S. D'Angelo, A. Fiorentino, F.L. Chiocci, Non-tropical carbonate shelf sedimentation. The Archipelago Pontino (Central Italy) case history, in: P.T. Harris, E.K. Baker (Eds.), Seafloor Geomorphology as Benthic Habitat: GeoHab Atlas of Seafloor Geomorphic Features and Benthic Habitats, Elsevier, Amsterdam, 2011 (Chapter 31).

[38] R. Coggan, M. Diesing, Rock Ridges in the central English Channel, in: P.T. Harris, E.K. Baker (Eds.), Seafloor Geomorphology as Benthic Habitat: GeoHab Atlas of Seafloor Geomorphic Features and Benthic Habitats, Elsevier, Amsterdam, 2011 (Chapter 33).

[39] V. Lucieer, N. Barrett, N. Hill, S.L. Nichol, Characterisation of shallow inshore coastal reefs on the Tasman Peninsula, South Eastern Tasmania, Australia, in: P.T. Harris, E.K. Baker (Eds.), Seafloor Geomorphology as Benthic Habitat: GeoHab Atlas of Seafloor Geomorphic Features and Benthic Habitats, Elsevier, Amsterdam, 2011 (Chapter 34).

[40] I. Galparsoro, Á. Borja, J.G. Rodríguez, I. Muxika, M. Pascual, I. Legorburu, Rocky reef and sedimentary habitats within the continental shelf of the southeastern Bay of Biscay, in: P.T. Harris, E.K. Baker (Eds.), Seafloor Geomorphology as Benthic Habitat: GeoHab Atlas of Seafloor Geomorphic Features and Benthic Habitats, Elsevier, Amsterdam, 2011 (Chapter 35).

[41] K.L. Yamanaka, K. Picard, K.W. Conway, R. Flemming, Rock reefs of British Columbia, Canada: inshore rockfish habitats, in: P.T. Harris, E.K. Baker (Eds.), Seafloor

Geomorphology as Benthic Habitat: GeoHab Atlas of Seafloor Geomorphic Features and Benthic Habitats, Elsevier, Amsterdam, 2011 (Chapter 36).

[42] K.A. Robinson, A.S.Y. Mackie, C. Lindenbaum, T. Darbyshire, K.J.J. van Landeghem, W.G. Sanderson, Seabed habitats of the Southern Irish Sea, in: P.T. Harris, E.K. Baker (Eds.), Seafloor Geomorphology as Benthic Habitat: GeoHab Atlas of Seafloor Geomorphic Features and Benthic Habitats, Elsevier, Amsterdam, 2011 (Chapter 37).

[43] J.R. Reynolds, S.C. Rooney, J. Heifetz, H.G. Greene, B.L. Norcross, Habitats and benthos in the vicinity of Albatross Bank, Gulf of Alaska, in: P.T. Harris, E.K. Baker (Eds.), Seafloor Geomorphology as Benthic Habitat: GeoHab Atlas of Seafloor Geomorphic Features and Benthic Habitats, Elsevier, Amsterdam, 2011 (Chapter 38).

[44] B.J. Todd, V.E. Kostylev, S.J. Smith, Seabed habitat of a glaciated shelf, German Bank, Atlantic Canada, in: P.T. Harris, E.K. Baker (Eds.), Seafloor Geomorphology as Benthic Habitat: GeoHab Atlas of Seafloor Geomorphic Features and Benthic Habitats, Elsevier, Amsterdam, 2011 (Chapter 39).

[45] J.E.R. Getsiv-Clemons, W.W. Wakefield, I.J. Stewart, C.E. Whitmire, Using meso-habitat information to improve abundance estimates for West Coast groundfish: a test case at Heceta Bank, Oregon, in: P.T. Harris, E.K. Baker (Eds.), Seafloor Geomorphology as Benthic Habitat: GeoHab Atlas of Seafloor Geomorphic Features and Benthic Habitats, Elsevier, Amsterdam, 2011 (Chapter 40).

[46] J.W.C. James, B. Pearce, R.A. Coggan, A. Morando, Open shelf valley system, Northern Palaeovalley, English Channel, U.K., in: P.T. Harris, E.K. Baker (Eds.), Seafloor Geomorphology as Benthic Habitat: GeoHab Atlas of Seafloor Geomorphic Features and Benthic Habitats, Elsevier, Amsterdam, 2011 (Chapter 41).

[47] B. Pearce, D.R. Tappin, D. Dove, J. Pinnion, Benthos supported by the tunnel-valleys of the southern North Sea, in: P.T. Harris, E.K. Baker (Eds.), Seafloor Geomorphology as Benthic Habitat: GeoHab Atlas of Seafloor Geomorphic Features and Benthic Habitats, Elsevier, Amsterdam, 2011 (Chapter 42).

[48] E. Ezhova, D. Dorokhov, V. Sivkov, V. Zhamoida, D. Ryabchuk, O. Kocheshkova, Benthic habitats and benthic communities in South-Eastern Baltic Sea, Russian sector, in: P.T. Harris, E.K. Baker (Eds.), Seafloor Geomorphology as Benthic Habitat: GeoHab Atlas of Seafloor Geomorphic Features and Benthic Habitats, Elsevier, Amsterdam, 2011 (Chapter 43).

[49] B. De Mol, D. Amblas, G. Alverez, P. Busquets, A. Calafat, M. Canals, et al., Erosional environment at the gateway between the Atlantic and Mediterranean Sea: Strait of Gibraltar cold-water coral distribution, in: P.T. Harris, E.K. Baker (Eds.), Seafloor Geomorphology as Benthic Habitat: GeoHab Atlas of Seafloor Geomorphic Features and Benthic Habitats, Elsevier, Amsterdam, 2011 (Chapter 45).

[50] F. Althaus, A. Williams, R.J. Kloser, J. Seiler, N.J. Bax, Evaluating geomorphic features as surrogates for benthic biodiversity on Australia's western continental margin, in: P.T. Harris, E.K. Baker (Eds.), Seafloor Geomorphology as Benthic Habitat: GeoHab Atlas of Seafloor Geomorphic Features and Benthic Habitats, Elsevier, Amsterdam, 2011 (Chapter 48).

[51] C. Lo Iacono, E. Gràcia, R. Bartolomé, E. Coiras, J.J. Dañobeitia, J. Acosta, The Chella Bank, Eastern Alboran Sea (Western Mediterranean), in: P.T. Harris, E.K. Baker (Eds.), Seafloor Geomorphology as Benthic Habitat: GeoHab Atlas of Seafloor Geomorphic Features and Benthic Habitats, Elsevier, Amsterdam, 2011 (Chapter 49).

[52] V.A.I. Huvenne, A.D.C. Pattenden, D.G. Masson, P.A. Tyler, Habitat heterogeneity in the Nazaré deep-sea canyon offshore Portugal, in: P.T. Harris, E.K. Baker (Eds.), Seafloor Geomorphology as Benthic Habitat: GeoHab Atlas of Seafloor Geomorphic Features and Benthic Habitats, Elsevier, Amsterdam, 2011 (Chapter 50).

[53] L. Buhl-Mortensen, R. Bøe, M.F.J. Dolan, P. Buhl-Mortensen, T. Thorsnes, S. Elvenes, et al., Banks, troughs and canyons on the continental margin off Lofoten, Vesterålen, and Troms, Norway, in: P.T. Harris, E.K. Baker (Eds.), Seafloor Geomorphology as Benthic Habitat: GeoHab Atlas of Seafloor Geomorphic Features and Benthic Habitats, Elsevier, Amsterdam, 2011 (Chapter 51).

[54] A.L. Post, P.E. O'Brien, R.J. Beaman, M.J. Riddle, L. De Santis, S.R. Rintoul, Distribution of hydrocorals along the George V slope, East Antarctica, in: P.T. Harris, E.K. Baker (Eds.), Seafloor Geomorphology as Benthic Habitat: GeoHab Atlas of Seafloor Geomorphic Features and Benthic Habitats, Elsevier, Amsterdam, 2011 (Chapter 52).

[55] G. Lamarche, A.A. Rowden, J. Mountjoy, V. Lucieer, A.-L. Verdier, The Cook Strait Canyon, New Zealand: a large bedrock canyon system in a tectonically active environment, in: P.T. Harris, E.K. Baker (Eds.), Seafloor Geomorphology as Benthic Habitat: GeoHab Atlas of Seafloor Geomorphic Features and Benthic Habitats, Elsevier, Amsterdam, 2011 (Chapter 53).

[56] M. Yoklavich, H.G. Greene, The Ascension-Monterey Canyon system—habitats of demersal fishes and macro-invertebrates along the central California coast of the USA, in: P.T. Harris, E.K. Baker (Eds.), Seafloor Geomorphology as Benthic Habitat: GeoHab Atlas of Seafloor Geomorphic Features and Benthic Habitats, Elsevier, Amsterdam, 2011 (Chapter 54).

[57] D.J. Wright, J.T. Roberts, D. Fenner, J.R. Smith, A.A.P. Koppers, D.F. Naar, et al., Seamounts, ridges, and reef habitats of American Samoa, in: P.T. Harris, E.K. Baker (Eds.), Seafloor Geomorphology as Benthic Habitat: GeoHab Atlas of Seafloor Geomorphic Features and Benthic Habitats, Elsevier, Amsterdam, 2011 (Chapter 58).

[58] F. Tempera, E. Giacomello, N. Mitchell, A.S. Campos, A.B. Henriques, A. Martins, et al., Mapping the Condor seamount seafloor environment and associated biological assemblages (Azores, NE Atlantic), in: P.T. Harris, E.K. Baker (Eds.), Seafloor Geomorphology as Benthic Habitat: GeoHab Atlas of Seafloor Geomorphic Features and Benthic Habitats, Elsevier, Amsterdam, 2011 (Chapter 59).

[59] B. De Mol, D. Amblas, A. Calafat, M. Canals, R. Duran, C. Lavoie, Alboran Seamounts, Western Mediterranean sea: Cold-water Coral colonization of knolls, in: P.T. Harris, E.K. Baker (Eds.), Seafloor Geomorphology as Benthic Habitat: GeoHab Atlas of Seafloor Geomorphic Features and Benthic Habitats, Elsevier, Amsterdam, 2011 (Chapter 60).

[60] J.V. Gardner, M.E. Field, H. Lee, B.E. Edwards, D.G. Masson, N. Kenyon, et al., Ground-truthing 6.5 kHz side scan sonographs: what are we really imaging? J. Geophys. Res. 96 (1991) 5955–5974.

[61] Y.C. Beaudoin, S. Smith, Habitats of the Su Su Knolls hydrothermal site, eastern Manus Basin, Papua New Guinea, in: P.T. Harris, E.K. Baker (Eds.), Seafloor Geomorphology as Benthic Habitat: GeoHab Atlas of Seafloor Geomorphic Features and Benthic Habitats, Elsevier, Amsterdam, 2011 (Chapter 62).

[62] P.T. Harris, S.L. Nichol, T.J. Anderson, A.D. Heap, Habitats and benthos of a deep sea marginal plateau, Lord Howe Rise, Australia, in: P.T. Harris, E.K. Baker (Eds.), Seafloor Geomorphology as Benthic Habitat: GeoHab Atlas of Seafloor Geomorphic Features and Benthic Habitats, Elsevier, Amsterdam, 2011 (Chapter 57).

[63] J.L. Rueda, V. Díaz-del-Río, M. Sayago-Gil, N. López, L.M. Fernández, J.T. Vázquez, Fluid venting through the seabed in the Gulf of Cadiz (SE Atlantic Ocean, western Iberian Peninsula): geomorphic features, habitats and associated fauna, in: P.T. Harris, E.K. Baker (Eds.), Seafloor Geomorphology as Benthic Habitat: GeoHab Atlas

of Seafloor Geomorphic Features and Benthic Habitats, Elsevier, Amsterdam, 2011 (Chapter 61).

[64] P.T. Harris, A.D. Heap, J.F. Marshall, M.T. McCulloch, A new coral reef province in the Gulf of Carpentaria, Australia: colonisation, growth and submergence during the early Holocene, Mar. Geol. 251 (2008) 85–97.

[65] P.K. Dayton, Competition, disturbance, and community organization: the provision and subsequent utilization of space in a rocky intertidal community, Ecol. Monogr. 41 (1971) 351–389.

[66] V. Agnostini, E. Escobar-Briones, I. Cresswell, K. Gjerde, D.J.A. Niewijk, A. Polacheck, et al., Global Open Oceans and Deep Sea-Habitats (GOODS) Bioregional Classification, in: M. Vierros, I. Cresswell, E. Escobar-Briones, J. Rice, J. Ardron (Eds.), United Nations Conference of the Parties to the Convention on Biological Diversity (CBD), 2008, p. 94.

[67] P.A. Tyler, Conditions for the existence of life at the deep sea floor: an update, Oceanogr. Mar. Biol. Annu. Rev. 33 (1995) 221–244.

[68] G. Brancolini, P.T. Harris, and shipboard party, Post-Cruise Report: Joint Italian/Australian marine geoscience expedition aboard the *R.V. Tangaroa* to the George Vth Land region of East Antarctica during February–March, 2000, Australian Geological Survey Organisation Record 2000/19, 2000, 181 pp.

[69] L.A. Levin, J.D. Gage, Relationships between oxygen, organic matter and the diversity of bathyal macrofauna, Deep Sea Res. Part II 45 (1998) 129–163.

[70] K. Wishner, L.A. Levin, M. Gowing, L. Mullineaux, Involvement of the oxygen minimum zone in benthic zonation on a deep sea mount, Nature 346 (1990) 57–59.

[71] C.R. Smith, L.A. Levin, D.J. Hoover, G. McMurtry, J.D. Gage, Variations in bioturbation across the oxygen minimum zone in the northwest Arabian Sea, Deep Sea Res. Part II 47 (2000) 227–257.

[72] K. Seiter, C. Hensen, M. Zabel, Benthic carbon mineralization on a global scale, Global Biogeochem. Cycles 19 (2005), GB1010, doi:10.1029/2004GB002225.

[73] M. Meybeck, Riverine transport of atmospheric carbon: sources, global typology and budget, Water Air Soil Pollut. 70 (1993) 443–463.

[74] D. Divins, Total Sediment Thickness of the World's Oceans & Marginal Seas, NOAA National Geophysical Data Center, Boulder, CO. http://www.ngdc.noaa.gov/mgg/sedthick/sedthick.html, 1998.

[75] L.A. Levin, R.J. Etter, M.A. Rex, A.J. Gooday, C.R. Smith, J. Pineda, et al., Environmental influences on regional deep-sea species diversity, Annu. Rev. Ecol. Syst. 32 (2001) 51–93.

[76] Y. Carmel, L. Stoller-Cavari, Comparing environmental and biological surrogates for biodiversity at a local scale, Isr. J. Ecol. Evol. 52 (2006) 11–27.

[77] M.A. Vanderklift, T.J. Ward, J.C. Phillips, Use of assemblages derived from different taxonomic levels to select areas for conserving marine biodiversity, Biol. Conserv. 86 (1998) 307–315.

[78] T.J. Ward, M.A. Vanderklift, A.O. Nicholls, R.A. Kenchington, Selecting marine reserves using habitats and species assemblages as surrogates, Ecol. Appl. 9 (1999) 691–698.

[79] F. Olsgard, T. Brattegard, T. Holthe, Polychaetes as surrogates for marine biodiversity: lower taxonomic resolution and indicator groups, Biodivers. Conserv. 12 (2003) 1033–1049.

[80] A.J. Hirst, Surrogate measures for assessing cryptic faunal biodiversity on macroalgal-dominated subtidal reefs, Biol. Conserv. 141 (2008) 211–220.

[81] K.R. Clarke, Non-parametric multivariate analyses of changes in community structure, Aust. J. Ecol. 18 (1993) 117–143.

6 Seafloor Geomorphology—Coast, Shelf, and Abyss

Peter T. Harris

Marine and Coastal Environment Group, Geoscience Australia,
Canberra, ACT, Australia

Abstract

The overarching theme of this book (and for the GeoHab organization in general) is that mapping seafloor geomorphic features is useful for understanding benthic habitats. Many case studies in this volume demonstrate that geomorphic feature type is a powerful surrogate for associated benthic communities. Here, we provide a brief overview of the major geomorphic features that are described in the detailed case studies (which follow in Part 2 of this book). Starting from the coast we will consider sandy temperate coasts, rocky temperate coasts, estuaries and fjords, barrier islands, and glaciated coasts. Moving offshore onto the continental shelf we will consider sand banks, sandwaves, rocky ridges, shallow banks, coral reefs, shelf valleys, and other shelf habitats. Finally, on the continental slope and deep ocean environments, we will review the general geomorphology and associated habitats of escarpments, submarine canyons, seamounts, plateaus, and deep-sea vent communities.

Key Words: geomorphic features; estuaries; fjords; continental shelf; coral reefs; rocky reefs; seagrass; submarine canyons; seamounts; escarpments; plateaus; hydrothermal vents

Geomorphology and Geomorphic Features

Broadly speaking, seafloor geomorphology is the scientific study of the formation, alteration, and configuration of seabed features and their relationship with the underlying geology. The geomorphologic classification of any area of seabed is a fundamental, first-order descriptor containing information about an area's relief, geology, geologic history, and formative processes. Geomorphic classification provides a synthesis of these attributes and information relevant for characterizing habitats. For example, geomorphic features and substrate type are interlinked in as much as some geomorphic features are inherently "rocky" in character (e.g., pinnacles, reefs, and ridges), whereas others are inherently "sediment covered" (e.g., basins, abyssal plains, and sandwaves). The case studies included in Part 2 of this book provide information on a range of different geomorphic feature types, located around the world (Figure 6.1; Table 6.1).

Seafloor Geomorphology as Benthic Habitat. DOI: 10.1016/B978-0-12-385140-6.00006-2

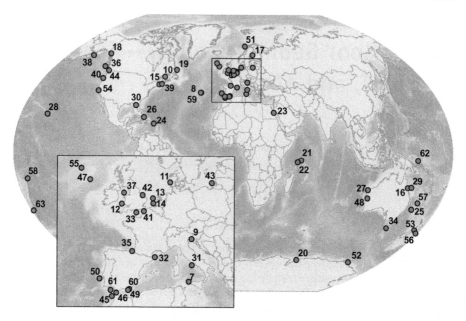

Figure 6.1 Locations of case studies contained in Part 2 (numbers refer to chapter numbers; see Table 6.1).

Table 6.1 List of Geomorphic Features Described in the Case Studies of This Book with the Relevant Chapter Numbers and References (Listed in Same Order as Chapters)

Geomorphic Feature (Habitat)	Chapters	References
Seagrass and macroalgae	7–10	D'Angelo and Fiorentino [1]; Tempera et al. [2]; Gordini et al. [3]; Colin et al. [4]
Barrier island	11	Dankers et al. [5]
Tidal inlet	11	Dankers et al. [5]
Estuary	11, 12	Dankers et al. [5]; James et al. [6]
Sandwave–sandbank	12–15, 29, 37–39, 42	James et al. [6]; van Dijk et al. [7]; Van Lanker et al. [8]; Todd and Valentine [9]; Beaman et al. [10]; Robinson et al. [11]; Reynolds et al. [12]; Todd et al. [13]; Ezhova et al. [14]
Holes	16	Stieglitz [15]
Glaciated shelf	17, 20, 39, 42, 43	Kotilainen et al. [16]; O'Brien et al. [17]; Todd et al. [13]; Ezhova et al. [14]; Barrie et al. [18]
Fjord	18–20	Cochrane et al. [19]; Copeland et al. [20]; O'Brien et al. [17]

(Continued)

Table 6.1 (Continued)

Geomorphic Feature (Habitat)	Chapters	References
Coral reef	21–28	Collier and Humber [21]; Hamylton et al. [22]; Hamylton [23]; Armstrong and Singh [24]; Brooke et al. [25]; Heyman and Kobara [26]; Nichol et al. [27]; Gibbs and Cochran [28]
Ridge	27, 30, 37	Nichol et al. [27]; Allee et al.[29]; Robinson et al. [11]
Platform	29	Beaman et al. [30]
Terrace	29	Beaman et al. [30]
Pinnacle	30, 38	Allee et al.; Reynolds et al. [12]
Mound	31	Martorelli et al. [31]
Embayment	32	Lo Iacono et al. [32]
Temperate rocky reef	33–36	Coggan and Diesing [33]; Lucieer et al. [34]; Galparsoro et al. [35]; Yamanaka et al. [36]
Shelf valley	37, 40, 42, 43	Robinson et al. [37]; James et al. [38]; Pearce et al. [39]; Ezhova et al. [14]
Sill	45	De Mol et al. [40]
Escarpment	46, 47	De Mol et al. [41]; Guinan et al. [42]
Channel	47	Guinan et al. [42]
Trough/trench	47, 63	Guinan et al. [42]; Wysoczanski and Clark [56]
Canyon	48, 50–54	Althaus et al. [43]; Huvenne et al. [157]; Buhl-Mortensen et al. [44]; Post et al. [45]; Lamarche et al. [46]; Yoklavich and Greene [47]
Peak	48	Althaus et al. [43]
Seamount/guyot	49, 58–60	Lo Iacono et al. [48]; Wright et al. [158]; Tempera et al. [49]; De Mol et al. [50]
Plateau	55–58	Sayago-Gil et al. [51]; Nodder et al. [52]; Harris et al. [53]; Wright et al. [158]
Cold seep	9, 61	Gordini et al. [53]; Rueda et al. [54]
Hydrothermal vents	62, 63	Beaudoin and Smith [55]; Wysoczanski and Clark [56]

Conceptually, the habitat mapping approach based on geomorphic features has the added advantage that detailed ecological models already exist for many geomorphically defined habitats [57]. As noted by Roff et al. [58], marine ecology textbooks are commonly organized into chapters having broad, geomorphically defined habitat types as titles (the ecology of estuaries, coral reefs, temperate rocky reefs, etc.). In this chapter, we briefly review each of the main geomorphic feature types described in more detail in the case studies that follow in Part 2.

Coastal Geomorphic Features and Habitats

The coastal biome contains estuaries, deltas, sandy–muddy coasts, and rocky coasts. These features occur in all climate zones, from polar to tropical. Coastal environments may be subdivided into soft-substrate, sedimentary coasts, and hard-substrate, rocky coasts, often characterized by eroding cliffs and rocky shores. From a geomorphic perspective, these two broad categories are the result of sediment supply to the coast and coastal evolution over geologic time. Coasts that receive a large sediment supply and are actively prograding seaward are characterized by deltas, strand plains, and tidal flats. Coasts that have received only a limited (or zero) sediment supply exhibit geomorphic features associated with coastal erosion such as rocky headlands, cliffs, and unfilled river valleys that have become estuaries or lagoons (Figure 6.2).

Coasts are geologically recent features that formed following the postglacial sea-level rise, less than 10,000 years ago. Many glaciated coasts are still undergoing isostatic adjustment related to the melting of continental ice sheets; for example, the Kvarken Archipelago in Finland is undergoing glacioisotatic uplift at a rate of approximately 8.0–8.5 mm/year (Chapter 17). Consequently, marine plants and animals living on the coasts today are adaptable colonists, able to move into habitat as it becomes available.

The gross geomorphology of coasts is also affected by the relative importance of waves and tides in controlling an area's exposure as well as the amount, nature, distribution, and transport of sediment. Large swell waves generate significant along-shore sediment transport that produces coast-parallel sedimentary features such as spits, barriers, sand bars, and barrier islands. In contrast, large tidal ranges (>4 m) and strong tidal currents generally produce coast-normal sedimentary features, including elongate tidal sand banks, wide-mouthed estuaries, funnel-shaped (in plan view) deltaic distributary channels, and broad intertidal flats (Figure 6.2).

Estuaries

There are numerous definitions of "estuary" in the literature, but the geological definition ("a drowned river valley that receives sediment from both landward and seaward sources" [60]) is more closely linked to geomorphology than other definitions. Wave- and tide-dominated estuarine environments both receive sediment from rivers and from the adjacent sea, but they exhibit contrasting arrangements in habitats. In tide-dominated estuaries, the characteristic "funnel" shape in map view is evident in the distribution of salt marsh and intertidal mudflat habitats (Figure 6.2). In the center of tide-dominated estuaries, tidal sand banks are aligned with their long axes normal to the overall trend of the coast (onshore–offshore alignment).

By contrast, wave-dominated estuaries have a tripartite distribution of habitats, arranged as follows in a progressively seaward direction: salt marsh with bay-head delta; central muddy basin; and coastal sandy barrier. Salt marsh habitat is largely confined to the landward head of the estuary, colonizing fluvial deposits associated with the bay-head delta. The central muddy basin is the deepest part of the estuary, bounded on its seaward side by a sandy barrier composed of marine-derived sediments. The barrier is cut by a tidal inlet, which supports ebb and flood-tidal deltas on its seaward and landward sides, respectively [60].

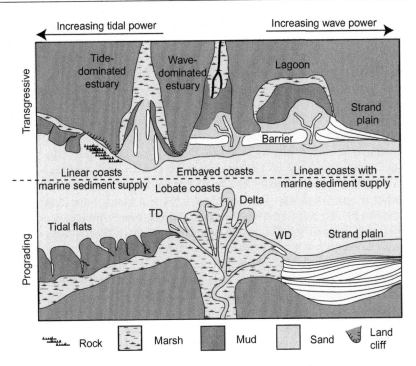

Figure 6.2 Diagram contrasting prograding and transgressive coastal geomorphic features, in terms of relative wave and tide energy and marine versus fluvial sediment supply. Prograding and transgressive coasts dominated by tides and marine sediment supply (left side of diagram) exhibit tidal flats backed by salt marsh. For lobate coasts receiving a fluvial sediment supply (center of diagram), prograding coasts exhibit deltas, ranging from funnel-shaped tide-dominated (TD) to coast-parallel wave-dominated (WD) morphotypes, with the classic "bird's foot" morphotype shown in between. Transgressive, lobate coasts exhibit tide- and wave-dominated estuaries along with lagoons, which all exhibit distinctive geomorphic features [60]. Broad intertidal flats and elongate tidal sand banks oriented normal to the coast characterize tide-dominated estuaries. Sandy barriers with tidal inlets and muddy basins backed by salt marsh are found for both lagoons and wave-dominated estuaries. The latter also contain river channels with bay-head deltas, which are absent from lagoons. Transgressive coasts may also be eroding locally, giving rise to cliffs and offshore rocky reefs (for both tide- and wave-dominated cases). Prograding and transgressive coasts dominated by waves and marine sediment supply (right side of diagram) exhibit strand plains.
Source: Modified after Boyd et al. [59].

While it has been shown that different estuarine habitats support different benthic assemblages, studies have seldom been able to correlate the occurrence of particular communities with the physical variables or gradients that are so characteristic of estuaries (sediments, salinity, temperature, or turbidity). The main reason is that these variables are interdependent, as sediment type, for example, covaries with wave energy or tidal current energy and salinity often fluctuates with temperature (see review by Dethier and Schoch [61]). Some generalizations are possible, nevertheless.

Halophytic vegetation (i.e., mangroves and salt marshes), occurs in many coastal environments and provides a unique coastal habitat that supports a wide range of marine species at different parts of their life cycle (Chapter 11). The spatial patterns of vegetation are different in wave- and tide-dominated systems. In their summary of wave-dominated estuaries of southeastern Australia, Roy et al. [62] recognized four zones (marine flood-tidal delta, central mud basin, fluvial delta, and riverine channel/alluvial plain), which possess characteristic water quality, nutrient cycling/primary productivity signatures, and ecosystems. Each zone is also characterized by different values of species richness and commercial fisheries production. Sandy beaches characteristic of wave-dominated coasts exhibit a trend for species richness of macro-invertebrates to decrease along a morphodynamic gradient from the dissipative to the reflective condition [63]. On beaches in Spain studied by Rodil and Lastra [64], for example, the number of species increases linearly with relative tidal range and decreases with increasing average grain size, with the same trend occurring on exposed to very exposed beaches; biomass decreases exponentially with increasing average grain size.

In the outer part of the macrotidal Bristol Channel, the case study by James et al. (Chapter 12) found that species richness on sandwaves reaching over 10 m in height above the level of surrounding seabed was lower than that found on the intervening coarse lag sediments between sandwaves. Similarly, North Sea sand ridges contain very poor communities (low density/low diversity) on the well-sorted sandy tops of ridges contrasting with comparatively rich communities (high density/high diversity) located in the poorly sorted sediments (clay to gravel) found in the adjacent troughs (Chapters 13 and 14).

Barrier islands are a feature of many wave-dominated coasts (Figure 6.2) and case studies in this volume include reference to a drowned barrier island system in the Gulf of Mexico (Chapter 30) and a modern barrier island system in the Wadden Sea, along the North Sea coast of the Netherlands (Chapter 11). Lagoons are also a feature of many wave-dominated coasts (Figure 6.2) as described in the case study of Ezhova et al. (Chapter 43) for the Curonian and Vistula lagoons located on the southeast coast of the Baltic Sea.

Fjords

Fjords are the drowned portions of glacially eroded valleys. They are long, narrow, deep inlets of the sea bounded by steep slopes. Fjords typically reach a maximum depth midway along their length and shoal at their mouth, forming a sill, or in more complex examples, a series of basins and sills (Chapter 19). This morphology has a profound influence on water circulation, limiting water exchange with the ocean and commonly resulting in oxygen depletion in water masses located in basins below the sill depth [65]. Because of the restricted water circulation and mixing in fjords, the salinity gradient, and temporal variations in salinity, exert significant control over benthic species. At high latitudes, a tidewater glacier may empty into the head of the fjord (Chapter 18); a glacially fed river is more commonly found at the head of the fjords in warmer climates (Chapter 44). The low salinity and extreme environmental conditions of the Baltic Sea results in a very low species diversity reported in the case studies of Kotilainen et al. (Chapter 17) and Ezhova et al. (Chapter 43).

Continental Shelf Geomorphic Features and Habitats

Continental Shelf Geomorphology

Continental shelves extend from beach (*foreshore*) environments, across the *shoreface* to an offshore location where the seaward dipping, low-gradient (~1:2,000) shelf gives way at the *shelf break* to a steeper gradient continental *slope*. In a global context, the depth of the shelf break varies from 20 to 550 m water depth, though it is commonly assumed to be 200 m, and the width of the shelf varies from 2 to 1,500 km. Continental shelves cover an area of about 27 million km^2, equal to about 7% of the surface area of the oceans.

All plants and animals living today on the continental shelf are colonists that arrived in the last 10,000 years or less. During the last ice age, which reached its peak approximately 18,000 years ago, global sea level was approximately 120 m below its present position (Figure 6.3) and most of the world's shelf area was exposed.[1] Terrestrial plants and animals lived on the continental shelf during the Pleistocene[2] ice ages. The shelf has only recently become a marine environment and in some locations the process of colonization may still be underway. Interglacial, high sea-level conditions that which exist at present have occurred for only around 12% of the time during the past 150,000 years (Figure 6.3).

The major controls on shelf morphology include tectonic setting, sea-level history, glaciation, rate and type of sediment supply, and energy available to erode, rework, and disperse sediment. Different controls may overlap, forming a matrix of possible combinations of shelf type. Changes in sea level have profoundly affected the biodiversity and distribution of marine life on continental shelves. At the peak of the last ice age, the area of shallow continental shelf, where benthic photosynthesis is possible, was approximately 80% less than its present extent [68]. Shallow coral reef habitat was reduced by a similar proportion during the ice age [69].

During past lower sea-level stands, coastlines would have occupied positions on what is now the outer continental shelf or upper slope. Therefore, the morphology, sedimentology, and benthic habitats of the present outer shelf and upper slope are partially the product of past, low sea-level terrestrial and coastal sedimentary processes and partially the product of modern sea-level shelf processes.

Glaciated Coasts and Shelves

Glaciation of the continents during the last ice age extended across what are now the continental shelves of Antarctica, western and northeastern North America, western Europe, Greenland, Iceland, South America, and New Zealand. U-shaped glacial valleys that exist as fjords along the coast extend in places across the full width of

---

[1] The continental shelf of Antarctica is a notable exception. It has an average depth of around 350 m and therefore it was not exposed during the last ice age.

[2] The Pleistocene epoch extends from about 2 million years ago up to the beginning of the Holocene, 11,700 years ago.

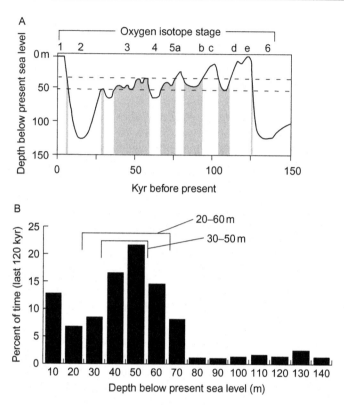

Figure 6.3 (A) Global eustatic sea-level curve and oxygen isotope stages for the last 150,000 years [66]. (B) Histogram showing percentage of time that sea level has been within 10 m depth bands (i.e., 0–10 m and 10–20 m) over that past 120,000 years (isotope stages 1 to 5d, inclusive, equal to one full glacial cycle), based on the curve shown above [67]. The graphs show that sea level was within the 30–50 m depth range for approximately 38% of the time (46,400 years) over the past 120,000 years. For comparison, sea level has been within the 20–60 m depth range for approximately 60% of the time (74,500 years), and in the 0–10 m range for only 12.8% of the time (15,500 years).

the shelf, with middle-shelf, overdeepened troughs rising upward near the outer shelf (Figure 6.4).

Around Antarctica, the shelf is suppressed by the weight of the continental ice sheet, and the shelf has remained submerged below sea level throughout the Pleistocene. During ice ages, the ice sheet extended out across the continental shelf, incising a trough in many locations (Figure 6.4). Characteristic of shallow coastal marine benthic communities around Antarctica are high levels of endemism, gigantism, slow growth, longevity, and late maturity, as well as adaptive radiations that have generated considerable biodiversity in some taxa [70].

In northern latitudes, glacially incised valleys were flooded during the postglacial sea-level rise when glaciers retreated and the continental ice sheets disintegrated, leaving marine basins that are perched on the shelf; the Alaskan, Canadian, and Norwegian shelves are examples of this type of morphology (Chapters 15, 18, 39,

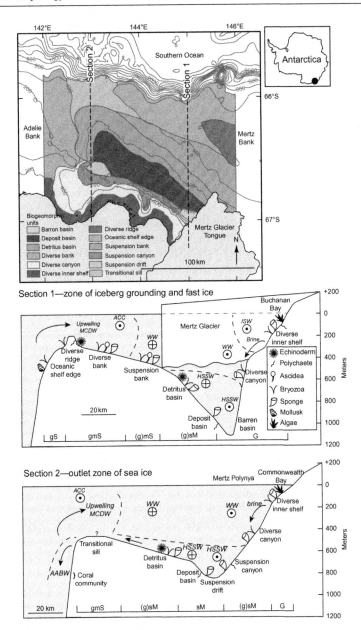

Figure 6.4 Bathymetry (contours) and biotopes (colors) of the George V Land shelf, Antarctica [10]. Profiles shown in Sections 1 and 2 show the occurrence of different biotopes in relation to depth, currents (ACC = Antarctic Coastal Current), and water masses (WW = Winter Water; HSSW = High-Salinity Shelf Water; MCDW = Modified Circumpolar Deep Water; AABW = Antarctic Bottom Water). The coral community at 800 m depth in Section 2 is based on Post et al. [45]. Note the profiles exhibit an outer shelf sill with an overdeepened glacially incised trough on the inner shelf. (For interpretation of the references to color in this figure legend, the reader is referred to the web version of this book.)

and 44). The unique habitat provided by such basins is characterized by rapid, fine-grained sedimentation, restricted water circulation, and a tendency toward anoxic bottom water and sediment conditions [71].

Many relict glacial features formed by ice sheets during the Pleistocene are apparent today and form unique benthic habitats. These features include tunnel valleys formed by subglacial meltwater (Chapter 42), glacial palaeovalleys (Chapters 10, 33, 39, and 41), and glacial till moraine deposits (Chapter 15). Benthic communities associated with the English Channel palaeovalley system, for example, reach a peak in abundance and diversity in association with the palaeovalley margins (Chapter 41; see also Table 6.1).

A distinguishing process of polar shelves is the effect of ice-turbation on benthos. Icebergs calved from glaciers run aground on the seafloor to depths of up to 350 m, killing the benthos and plowing the seabed sediments [72]. Where sea-ice intersects the coastline, it causes disturbance in the intertidal zone, resulting in a distinct depth zonation of benthos ([73]; Chapter 20). In their comparative analysis, Brey and Gerdes [74] reported that there is no significant difference of macrobenthic biomass in the 0–10 m depth range between Antarctic versus nonpolar regions. However, Antarctic macrobenthic biomass between 10 and 1,000 m water depth is significantly greater than the biomass of nonpolar continental margins [74].

Communities mapped by Beaman and Harris [10] on the George V Land glacially incised shelf of Antarctica comprise sponges, echinoderms, and mollusks that occur in association with different depths, currents, and water masses (Figure 6.4). Below the effects of iceberg scour (depths >500 m) in the basin, the broad-scale distribution of macrofauna is largely determined by substrate type, specifically the mud content of bottom sediments [10]. Dense, high-salinity shelf water formed on the shelf in winter by brine rejection during sea-ice formation flows across the shelf and cascades down the George V slope, contributing to the production of Antarctic Bottom water (Figure 6.4). Cold-water corals viewed in underwater video are described by Post et al. (Chapter 52) at the site of the bottom-water cascade, presumably benefiting from the suspended food particles carried by the flow.

Rocky habitat is common on glaciated shelves where glacial erosion has left bedrock exposed on the seabed. This is the case at German Bank, eastern Canada, where Todd et al. (Chapter 39) observed that hard substrates are commonly overgrown with dense mats of hydrozoa and bryozoa, and fish (flounder, hake, skate, and cod) are common. On stable lag gravel, calcareous polychaete tubes and sponges are common, while fauna abundance is low on the patches of mobile sand and shell hash, characterized by spider crabs (*Hyas araneaus*), urchins (*Strongylocentrotus* sp.), and infrequent occurrence of flatfish. Coggan et al. (Chapter 33) described an extensive system of rock ridges, located 30 km south of the Isle of Wight (UK) in water depths of 40–80 m, supporting a diverse fauna including sponges, bryozoans, hydroids, and anemones.

Tropical Coral Reefs

Ginsburg and James [75] suggested that shelves can be grouped into two broad categories: (i) open shelves and (ii) rimmed shelves where a shelf-edge barrier reef has accreted over geologic time. The barrier reef (rim) acts to restrict the propagation

Figure 6.5 Three-dimensional color bathymetry image showing an example of the northern Great Barrier Reef (GBR) "rimmed" continental shelf (Geoscience Australia). Note how the rimmed GBR shelf morphology contrasts with the nonrimmed Gulf of Papua. Elongate coral reefs and incised valleys adjacent to Torres Strait are the product of strong tidal flows. Ashmore reef is a coral atoll. Colors are related to depth and elevation (red = mountains; yellow = low-lying coastal plains; green-aqua = shelf depths 5–50 m; light blue = deep shelf depths 50–200 m; and dark blue = continental slope depths 200–2,000 m). (For interpretation of the references to color in this figure legend, the reader is referred to the web version of this book.)

of surface waves and water circulation (Figure 6.5). In contrast, open shelves have the profile of a relatively smooth seaward-dipping ramp. The initiation and growth of coral reefs forming the major barrier reef systems on Earth occurred over several sea-level cycles, with new coral limestone being deposited during each interglacial period. The carbonate rim is a high-energy, barrier reef environment where coral reefs flourish and on which there is a vast scientific literature.

The total area of coral reefs in the world's oceans is estimated to be about 255,000 km^2 [76,77], comprised of shallow reef crests and reef flats close to the sea surface (Figure 6.6). But there also exists an unknown number of submerged coral reefs below the sea surface and often invisible to satellite imagery or aerial photography. Such submerged reefs were difficult to detect prior to the advent of multibeam sonar systems and consequently they have been documented in only a few studies [78].

Armstrong et al. (Chapter 24) found that mesophotic (submerged) coral ecosystem occurrence between 30 and 100 m on low-gradient platforms, and high-gradient slopes of the Puerto Rico Shelf, is determined by a combination of suitable hard substrates and physical factors such as temperature, light availability, and low sedimentation. Nichol et al. (Chapter 27) found that the abundance of corals and sponges was greatest on the tops of approximately 15 m high limestone ridges interpreted as drowned

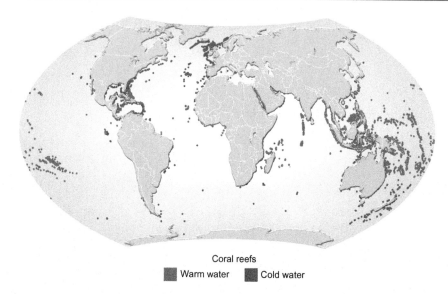

Coral reefs
■ Warm water   ■ Cold water

Figure 6.6 Known global distribution of tropical warm-water and cold-water (deep) coral reefs. *Source*: From UNEP-GRID Arendal.

fringing reefs and relict aeolian dunes on the Carnarvon shelf of Western Australia. Low densities of epibenthic organisms occurred on the surrounding, 60 m deep, sandy seafloor. Allee et al. (Chapter 30) describe a rocky limestone ridge located off the Florida coast in the Gulf of Mexico in 60–140 m of water, which is thought to be the remnant of a 14,000+ year old (relict) coral reef. The ridge supports living coral reefs, octocorals, sponges, nonhermatypic corals, and other encrusting invertebrates (Chapter 30). The fact that the submerged relict reefs still support live hard corals is evidence for a "give-up" growth history [79] relative to sea-level rise. A "give-up" growth history involves slow reef growth following submergence of the reef pedestal during transgression, followed by limited to zero growth under present high sea-level conditions.

In other locations, coral colonization and growth occurs on different (noncoral limestone) hard surfaces. Gibbs and Cochran (Chapter 28) note that in contrast to coral habitat elsewhere in the Hawaiian Islands, where coral typically populates relict carbonate platforms, coral cover in the Kaloko-Honokohau National Park is typically only a thin veneer of live coral and rubble on exposed volcanic pavement. Hamylton et al. (Chapter 22) describe coral reef lagoons, reef flats, and sand cays associated with platform reefs, atolls, and submerged atolls in the Amirantes archipelago, Seychelles, western Indian Ocean. Collier and Humber (Chapter 21), who also worked in the Seychelles, found corals growing on gently seaward-dipping granitic terraces in 8–10 m water depth, although bare sediment substrate comprises more than 70% of the seabed that they surveyed.

The morphology of coral reefs itself exerts control over the occurrence of different species. Heyman and Kobara (Chapter 26) note that multispecies spawning aggregation sites are generally located near the inflection points of convex-shaped reefs, near the shelf edge in 20–40 m water depth in Belize and the Cayman Islands

(Caribbean), adjacent to the shelf break where water depth drops to several hundred meters. Hamylton (Chapter 23) studied lagoonal reef communities inside the Al Wajh Bank, Red Sea and found that coral larval recruitment is strongly controlled by the availability of suitable antecedent topography for settlement. Brooke et al. (Chapter 25) report that the morphology of submerged relict coral reefs adjacent to Lord Howe Island in the southwest Pacific Ocean influences the distribution of benthic fauna. The island is surrounded by a mid-shelf relict coral reef in water depths of 30–40 m inboard of which is a sandy basin (mean depth 45 m). Infauna species abundance and richness are similar for sediment samples collected on the outer shelf and relict reef, while samples from the basin have significantly lower infauna abundance and richness (Chapter 25).

Temperate Reefs, Seagrass Meadows, and Other Features

The buildup of biogenic sediment produces a variety of mound-shaped, tropical, and also nontropical deposits formed mainly of biologically produced sediment (known as bioherms) that occur on temperate to polar continental shelves. Nontropical bioherms are produced by many different organisms including sponges, bryozoans, bivalves, and cold-water corals. Such communities usually develop in highly productive waters, such as areas of upwelling, and may be associated with sills, escarpments, seamounts, and the heads of submarine canyons, as described elsewhere in this chapter. For example, in the Irish Sea, Robinson et al. (Chapter 37) describe horse-mussel (*Modiolus modiolus*) bioherms, inhabited by rich macroinfaunal and epifaunal assemblages, which form part of a Special Area of Conservation off the Welsh coast. In the Archipelago Pontino in the eastern central Tyrrhenian Sea, Martorelli et al. (Chapter 31) describe "coralligène" (coralline algal buildups) formed mainly by coralline algae, foraminifers, bryozoans, and undifferentiated bioclasts. In a cool-temperate environment, Barrie et al. (Chapter 44) describe sponge banks in the Salish Sea of western Canada.

Seagrass meadows occur globally (Figure 6.7) and are described in Chapters 7–10. Seagrass foliage and roots trap sediments that accrete over time to form local deposits. In the Tyrrhenian Sea, D'Angelo and Fiorentino (Chapter 7) describe "mattes," bioherms that have formed from sediments trapped in the roots of seagrasses.

Seafloor features may also form by indirect biological processes such as the production of methane in bottom sediments, forming unique habitats. For example, Gordini et al. (Chapter 9) describe rocky seabed features associated with 112 different macroalgal taxa in the Adriatic Sea. The rocky hardgrounds are produced by seepage of methane-rich fluids, which causes the precipitation of biogenic methane-derived calcium carbonate cement. Methane seepage is also implicated in the formation of abundant pockmarks observed on the seabed of the Great Barrier Reef lagoon (Chapter 16) and in the Irish Sea (Chapter 37).

Temperate Rocky Reefs

Rocky habitats occur on most shelves, particularly transgressive shelves that have received relatively limited fluvial sediment input during the Holocene (Figure 6.2). Rocky reefs are often targeted by fishermen and conservation groups because they

Figure 6.7 Global distribution of seagrass meadows.
Source: From UNEP-GRID Arendal.

support relatively diverse and abundant fauna compared with the surrounding sedimented shelf areas. Rocky shelf habitats are naturally protected from trawl fishing as they are commonly "untrawlable" by benthic trawl gear. Instead, temperate rocky reefs are exploited using other commercial and recreational fishing technologies, such as fish and crustacean traps, scuba-diver fishing, and line fishing.

Several of the case studies presented in this book related to temperate to subpolar rocky reef habitat were conducted to support commercial fisheries management. Getsiv-Clemons et al. (Chapter 40; [156]) were able to estimate the abundance of commercially important groundfish on Heceta Bank, Oregon, one of the largest rocky banks along the US west coast, by mapping the area of available habitat. Reynolds et al. (Chapter 38) reported on commercial rockfish habitat on Albatross Bank, Gulf of Alaska, comprised of a series of flat banks of sedimentary bedrock located in 50–100 m water depth. Yamanaka et al. (Chapter 36) found that different commercial bottom fish habitats corresponded with bedrock outcrop, cobbles, gravel, and muddy substrates. Rockfish habitat equates with bedrock areas; inshore rockfish habitat is well described by bedrock and cobble substrata; and spotted ratfish habitat is described by gravel and mud substrata (Chapter 36).

On islands that have a surrounding shelf, it is often the case that rocky habitats are found because the island produces an insufficient sediment supply to drape and bury the rocky shelf exposures. Rocky infralittoral biotopes in the Azores archipelago investigated by Tempera et al. (Chapter 8) used ordered logistic regression models to find the combinations of major environmental variables (depth, slope, swell exposure, maximum tidal currents, sea surface temperature, and chlorophyll-a concentration) that best explain spatial variations observed in macroalgal communities. Off the eastern coast of the island of Tasmania, Australia, Lucieer et al. (Chapter 34) found that the degrees of change of seabed slope over a distance of 6 m and the seabed morphology were useful for explaining the distribution of biological communities.

The likelihood of occurrence of rocky reefs is increased in exposed shelf locations where wave energy limits sediment deposition and erodes the substrate to expose underlying bedrock. Galparsoro et al. (Chapter 35) studied rocky reefs and sedimentary habitats in the Bay of Biscay. These workers found that rocky bottoms are dominant along the shore, and that wave energy and sedimentary characteristics are the main environmental factors explaining the composition and spatial distribution of sedimentary benthic communities. Wave energy and light availability also affect the distribution of biota on rocky substrates off eastern Tasmania, where Lucieer et al. (Chapter 34) found that while brown and red algae dominated communities to a depth of 40 m, no algae are recorded below 50 m and sponges became more evident below this depth. Collin et al. (Chapter 10) describe kelp forest at depths of up to 16 m on the north shore of the Baie des Chaleurs, Gulf of St. Lawrence (Canada), growing on ridges and channels.

Shelf Sediment Banks and Bedforms

Continental shelves where the transport of sediment is controlled mainly by tidal currents are typified by the west European shelf seas (Figure 6.8). Here, shelf sand (derived mainly from the reworking of Pleistocene glacial deposits) is dispersed from discrete zones of seabed scouring and erosion in a divergent pattern, such that an increasing supply of sand of decreasing grain size occurs moving away from the scour zone. This gives rise to a succession of habitats, ranging from a high-energy rocky seabed inhabited by sessile filter feeders to low-energy soft-substrate habitat inhabited by deposit feeding communities [80]. Tidal sand banks (shown as black lines in Figure 6.8) support a particular macrobenthos, as documented for shallow sandbank areas of the Belgian part of the North Sea by Van Lancker et al. (Chapter 14), the Bristol Channel (Chapter 12), and the English Channel (Chapter 33). Other locations where the benthos of shelf sand banks is documented in this volume include the Salish Sea of western Canada (Chapter 44) and the Great Barrier Reef, Australia (Chapter 29).

Todd and Valentine (Chapter 15) describe lag-gravel and sandy-bedform habitats on Georges Bank off eastern Canada. Bedforms here consist of large, mobile, asymmetrical sandwaves up to 19 m in height, whose dynamic surfaces provide habitat for a low diversity of infaunal bivalves (e.g., razor clams, *Ensis directus*, and surf clams, *Spisula solidissima*) and epifaunal gastropods that prey on them (e.g., moon snails, *Lunatia heros*). In contrast, the static surfaces of the lag-gravel deposits provide habitat for attached sponges, bryozoans, tube worms, and anemones as well as infaunal bivalves and worms, epifaunal bivalves (e.g., sea scallops, *Placopecten magellanicus*), and arthropod scavengers (Chapter 15).

Sills

Sills are geomorphic features of particular significance for oceanography and biogeography. From an oceanographic perspective, sills separate basins that may contain particular water masses of different properties. Sills have already been mentioned in relation to fjords and glaciated shelves (Figure 6.4), but they also occur on nonglaciated shelves. Sills are commonly a biogeographic barrier for species located in deep

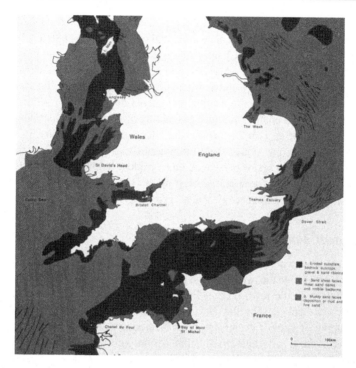

Figure 6.8 Distribution of three main seabed types of the tidally dominated west European shelf seas in order of decreasing tidal energy: (1) scoured rock and lag gravel (blue); (2) sand with superimposed bedforms (red area) and linear sand banks (black lines); and (3) muddy sand (green). (For interpretation of the references to color in this figure legend, the reader is referred to the web version of this book.)
Source: Modified from Harris et al. [81].

basins on either side. The basin environments are vastly different from the shallow water depths, high current energy, and warmer water temperature of the sill environment.

The Strait of Gibraltar is a type-case for geomorphic sills. Here, ocean circulation is characterized by a two-layer system: a surface eastward Atlantic water inflow and a deep westward outflow of saline Mediterranean water, with variable interface depth of around 100 m and a sill depth of around 200 m. De Mol et al. (Chapter 45) document reef-forming cold-water coral deposits up to 40 m in thickness in the deepest part of the Strait of Gibraltar between 180 and 330 m water depth, which have developed to exploit the unique oceanographic conditions. Coral buildups on the tops of N-S orientated rocky crests are the most common form in the Strait of Gibraltar.

Geomorphic Features and Habitats of Continental Slopes

The continental slope is located seaward from the shelf edge and extends to water depths typically of around 3,000–4,000 m. It may be bounded on its seaward margin

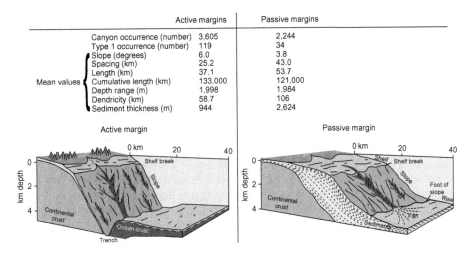

		Active margins	Passive margins
	Canyon occurrence (number)	3,605	2,244
	Type 1 occurrence (number)	119	34
	Slope (degrees)	6.0	3.8
	Spacing (km)	25.2	43.0
	Length (km)	37.1	53.7
Mean values	Cumulative length (km)	133,000	121,000
	Depth range (m)	1,998	1,984
	Dendricity (km)	58.7	106
	Sediment thickness (m)	944	2,624

Figure 6.9 Geomorphic aspects contrasting submarine canyons incising active and passive continental margins.
Source: After Harris and Whiteway [84].

by thick deposits comprising the continental rise or basin-filling deposits of the (essentially flat) abyssal plains. The seaward limit of the continental slope is denoted by a general reduction in slope, tending toward the horizontal.

A first-order control on continental slope morphology is the tectonic setting of the margin. Emery [82] and Uchupi and Emery [83] noted that active and passive continental margin types exhibit differences in morphology that can be attributed to the processes governing their formation. Passive margin morphology is controlled by deposition and erosion processes, whereas active margin morphology is controlled by tectonic/magmatic processes. Thus, passive margins generally are less steep and are likely to have adjacent sedimentary continental rises and abyssal plains. In contrast, active margins are thinly draped by sediments and may have an adjacent ocean trench or trough (Figure 6.9).

Depth is found in many studies to be a fundamental (biome-defining) parameter that correlates with the occurrence of biota (see Chapter 5), and continental slopes span a great range of depths, from 0 to >4,000 m. For these reasons, associations between benthic communities and geomorphic features on the continental slope must be considered in the context of their depth of occurrence. Althaus et al. (Chapter 48) conclude that "while some (slope) geomorphic features have high potential to act as surrogates for biodiversity at intermediate spatial scales, a hierarchical context is necessary to define and validate them within a larger, biogeographical context."

Submarine Canyons

Submarine canyons are perhaps the most conspicuous features occurring on continental slopes. They are incised into all continental margins and act as conduits for the transfer of sediment from the continents to the deep sea [85]. Shepard [86,87] recognized that submarine canyons may have several origins and restricted his definition to "steep-walled, sinuous valleys with V-shaped cross sections, axes sloping outward as continuously as river-cut land canyons and relief comparable to even the largest of land canyons." This definition therefore excludes other seafloor valleys including: delta-front troughs (located on the prograding slope of large deltas); fan valleys (the abyssal, seaward continuation of submarine canyons, some of which are remarkably long [88,89]); slope gullies (incised into prograding slope sediments); fault valleys (structural-related, trough-shaped valleys, generally with broad floors); shelf valleys (incised into the shelf by rivers during the ice ages, generally <120 m deep); and glacial troughs incised into the continental shelf by glacial erosion during the ice ages, generally U-shaped in profile and having a raised sill at their seaward terminus [87]. The heads of some submarine canyons terminate on the slope, making the so-called "blind" or "headless" canyons. The largest canyons, however, commonly incise into the continental shelf and may even continue as shelf valleys that have a direct connection to modern terrestrial fluvial systems (Figure 6.9).

A global assessment of canyons compiled by Harris and Whiteway [84] based on an analysis of ETOPO-1 bathymetry data identified 5,849 separate large submarine canyons in the world ocean. Active continental margins contain over 50% more canyons (3,605) than passive margins (2,244), and the canyons are steeper, shorter, more dendritic, and are more closely spaced on active than on passive continental margins (Figure 6.10). River-associated, shelf-incising canyons are more numerous on active continental margins ($n = 119$) than on passive margins ($n = 34$). They are most common on the western margins of South and North America, where they comprise 11.7% and 8.6% of canyons, respectively; but they are absent from the margins of Australia and Antarctica. Finally, steeper canyons (active margins) are more closely spaced than gently sloping canyons (passive margins), and the greatest canyon spacing occurs in the Arctic and the Antarctic. Canyons are more closely spaced in the Mediterranean than in other areas [84].

Subaerial erosion can be an important factor in canyon evolution. Most shelves were subaerially exposed at the peak of the last ice age when global eustatic sea level was approximately 120 m below its present position and rivers incised valleys across what is today the continental shelf. The delivery of sediments to the shelf break during the Pleistocene ice ages provided a sediment source for down-slope turbidity flows and canyon incision—a process that occurs in only a few canyon systems during interglacial high-sea-level periods (e.g., Congo River in Africa, described by Shepard and Emery [90] and Sepik River in Papua New Guinea, described by Kineke et al. [91]). Margin exposure has been especially significant in isolated marine basins like the Mediterranean Sea, where extensive sea-level lowering and desiccation occurred during the late Miocene "Messinian Salinity Crisis" [92] and the Pleistocene evaporation of the Black Sea [93,94].

Interest in the evolution, occurrence, and distribution of canyons in the oceans has been driven by the need to lay cables and pipelines across the seafloor [95], to support naval submarine operations, to understand the geological evolution of continental margins, and to understand oceanographic and ecological processes associated with canyons [96–98]. At the down-slope terminus of canyons may be found depositional

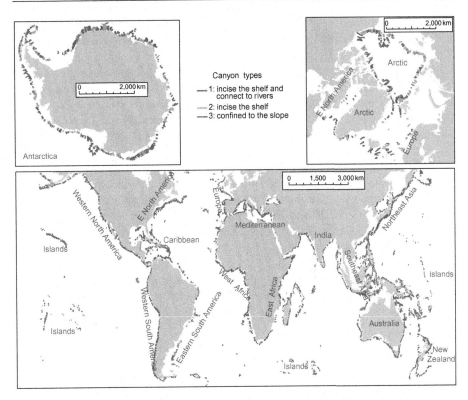

Figure 6.10 Global distribution of large submarine canyons, as classified by Harris and Whiteway [84]. Active continental margins are labeled in red, passive margins in green. (For interpretation of the references to color in this figure legend, the reader is referred to the web version of this book.)

submarine fans with their often extensive fan valley complexes, which have been studied in detail as analogs for ancient deposits of economic significance [99,100].

Oceanographic processes such as internal waves, coastally trapped waves, the modification (e.g., bathymetric steering) of outer-shelf and upper-slope currents and internal tides cause the mixing of canyon waters and upwelling of cold, nutrient-rich waters to the sea surface [101–102]. For example, ocean mixing rates inside Monterey Canyon are as much as 1,000 times greater than rates measured in the open ocean [103]. The upwelling and mixing associated with canyons enhances local primary productivity and the effects extend up the food chain to include birds and mammals [101]. Consequently, commercially important pelagic and demersal fisheries, as well as cetacean feeding grounds [104], are commonly located at the heads of submarine canyons [105]. Buhl-Mortensen et al. (Chapter 51) found that submarine canyons have a significant influence on the distribution of biological communities through their modification of current patterns.

Recent interest has focused on the benthic habitats associated with submarine canyons, particularly the heads of shelf-incising canyons that are characterized by steep (vertical to overhanging) bedrock exposures upon which biologically diverse communities may occur ([106]; Chapter 54). In particular, canyon heads are commonly found to be the preferred habitat for cold-water coral communities (Chapters 32 and 52).

Interestingly, the reverse pattern is observed in a shelf-incising canyon from New Zealand, where Lamarche et al. (Chapter 53) estimate that taxa richness is lowest for the presumably most disturbed habitats-angular and smoothed gullies, and the canyon walls affected by mass failure. Taxa richness is estimated to be highest for the communities of sand banks located on the canyon margins and the continental slope. The communities of the predominantly soft sediment canyon floor and continental shelf are estimated to have intermediate levels of taxa richness (Chapter 53).

Submarine canyons that extend across the continental shelf and approach the coast are known to intercept organic-matter-rich sediments being transported along the inner shelf zone [86,107–109]. This process causes organic-rich material to be supplied to the head of Scripps Canyon, for example (Chapter 54), and transported down slope, where it provides nourishment to feed a diverse and abundant macrofauna [110,111]. A similar situation occurs in the Nazaré deep-sea canyon offshore Portugal (Chapter 50; [157]). Canyons that having no significant landward extension would presumably not intercept littoral sediments and would not be expected to contain such a rich fauna.

Escarpments

On carbonate-depositional continental margins, or margins undergoing erosion and/or tectonic faulting, and also along the fractures of mid-ocean ridges, vertical cliff-like escarpment features may occur. Examples of some well-known carbonate escarpments include the Florida escarpment, Bahama escarpment, and the Mazagan escarpment off central Morocco [112,113]. Escarpments formed through faulting and tectonism are typified by those found along the margins of the Mediterranean and Red Seas, and also along the margins of California [114] and Chile [115]. Some escarpments, such as the Blake Escarpment near the Bahamas, are believed to have formed mainly by gravitational, slump or slide, mass-wasting processes [116], although erosion plays a role in the development of all escarpments. Escarpments are characteristically many kilometers in length and >100 m in vertical relief.

Interest in the oceanography and ecology of escarpments has increased in recent years because of the discovery of deep-water coral communities associated with them [117,118]. The benthos colonizing an escarpment on the flank of the Rockall Plateau, located in the North Atlantic at water depths from 200 m and extending to 3,000 m, was described by Guinan et al. (Chapter 47). They reported scarps, overhangs and horizontal ledges, vertical rock walls, and pinnacles colonized by framework reef-building cold-water coral *Lophelia pertusa*, gorgonians, encrusting sponges, soft corals, black corals, corallimorphs, desmospongiae, and encrusting sponges.

De Mol et al. (Chapter 46) studied the benthos occurring at the summit of the Pen Duick Escarpment on the Moroccan continental margin in the southern Gulf of Cadiz. Using a remotely operated underwater vehicle (ROV), these workers investigated a mound-shaped bioherm comprised mostly of dead coral with one living *Dedrophyllia cornigera*. The coral rubble consists mainly of dead *Dendrophyllia*,

Lophelia pertusa, and *Madrepora oculata*, and it was colonized by crinoids, sponges, squat lobsters, echinoids, holothurians, and the fish *Helicolenus dactylopterus* (Chapter 46).

Geomorphic Features of Ocean Basins

The world ocean has an area of about 361 million km^2, an average depth of about 3,730 m, and a total volume of about 1,347 million km^3. Although 70.78% of the Earth is covered by oceans, only about 5% of that area relates to the shallow continental shelf seas; the remaining 65% of the Earth is characterized by the deep continental slopes, ocean basins, and mid-ocean spreading ridges that encircle the planet [119]. The hypsometric curve for the Earth shows that 53.5% of Earth's surface (75.3% of the oceans) is between depths of 3,000 and 6,000 m. The study of geomorphic features of the ocean floor, therefore, covers most of the surface of the Earth. It is incongruous that this environment is also the least well studied or understood by marine scientists. In fact, the deep ocean floor is the most poorly studied and understood environment on Earth.

Geomorphic Provinces Map of the World's Deep Ocean

Agapova et al. [120] interpreted the existing bathymetric data that was available to them at the time to create a global geomorphic provinces map of the oceans (Figure 6.11). Some key features mapped by Agapova et al. [120] may be visualized in three dimensions, as depicted in Figure 6.12. An analysis of Agapova et al.'s [120] map (Figure 6.11) provides an estimate of the surface areas of the different geomorphic provinces and their proportional occurrence on Earth (Table 6.2). The results show that about 40% of the ocean floor is abyssal plain, 19% is mid-ocean ridge, and less than 0.5% is deep ocean trench environment. Of the total area of mid-ocean ridges, only about 5.6% is associated with the central rift zone.

Ocean Basins

The ocean basins are partially bounded by the continents, but they are interconnected; this is why marine scientists refer to a single "world ocean." The world ocean is divided into the North and South Pacific, Atlantic and Indian oceans, and the Arctic Ocean. Oceanographers also recognize the Southern Ocean, which encircles Antarctica and includes the southernmost parts of the Pacific, Atlantic and Indian oceans.[3] Ocean basins are formed geologically of oceanic (basalt) crust, in contrast to continental (granite) crust that forms the Earth's major land masses. Thus, some smaller "seas" that are underlain by ocean crust and are isolated from the world ocean as separate basins, such as the Mediterranean, Tasman, Coral, Caribbean, and

---

[3] The United Nations Food and Agriculture Organization (FAO) reporting ocean areas are commonly used: see www.fao.org.

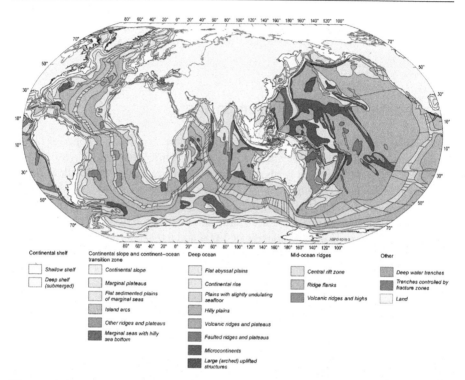

Figure 6.11 Global geomorphic provinces map of the oceans (based on Agapova et al. [120]). Note: the Arctic Ocean is not included in the mapped area.

Red Seas, may be considered oceanic. By contrast, other shelf seas and gulfs are perched on continental crust (e.g., the North Sea, Arafura Sea, Baltic Sea, and Yellow Sea) and are consequently distinguished on a geological basis from oceanic basins.

Abyssal Plain

Sediment deposited adjacent to the continents forms the continental rise. Seaward of this, the land-derived sediment wedge becomes thinner and the morphology gives way to the flat abyssal plains that are underlain by basaltic ocean crust (Figure 6.11). Abyssal plains are remarkably flat, having a slope of less than 1:1,000 (or <1 m change in height over a distance of 1 km), because of the thick sediment drape that covers and subdues most of the underlying basement topography. Ocean basins that receive the greatest sediment input have the best developed abyssal plains (e.g., the Atlantic and Indian oceans and the Gulf of Mexico). Abyssal plains are less well developed in the North Pacific and Southern Ocean basins because these areas do not receive as much land-derived sediment. This, in turn, is because of the lack of large river systems draining into them, and/or because of large ocean trench systems that intercept and trap land-derived sediments.

Hilly abyssal plains are much more common and cover over 100 million km^2 of the seafloor (Figure 6.11; Table 6.2), comprising the Earth's largest geomorphic feature type

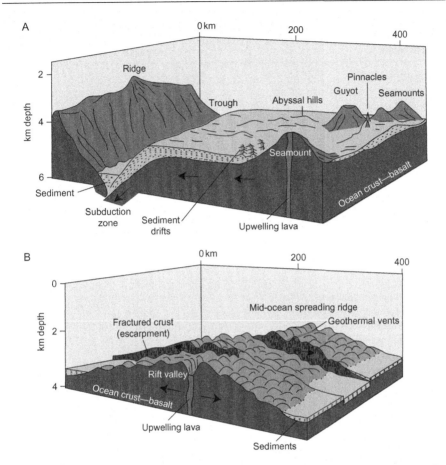

Figure 6.12 Three-dimensional representations of deep-sea geomorphic features: (A) subduction of ocean crust beneath other ocean crust, forming a trench/trough and ridge, with flanking hilly abyssal plain and volcanogenic seamounts; (B) mid-ocean spreading ridge with rift valley, fractures, and draping sediments that thicken with increasing distance from the ridge axis.

by surface area. The hilly morphology results from the expression of underlying subsided but hilly basaltic ocean crust, which is evident through the thin sediment cover.

Mid-Ocean Ridges

Mid-ocean ridges are created by the upwelling of basaltic lava and lateral rifting of ocean crust (Figure 6.12). They form a rift valley system that encircles the Earth along a total length of over 75,000 km (Figure 6.11). The mid-ocean ridges are the earth's largest volcanic system, accounting for >75% of all volcanic activity on the planet. The heat from this volcanism is dispersed by hydrothermal circulation of seawater. Hot seawater venting from the seafloor supports exotic benthic communities that have evolved to survive by using the hydrogen sulfide dissolved in the hot fluid

Table 6.2 Geomorphic Provinces of the World Ocean and Their Surface Areas

Feature Name	Area (10^6 km²)	Percent
Continental shelf		
Shallow shelf	39.63	11.4
Deep shelf (submerged)	5.27	1.52
Continental slope and continent–ocean transition zone		
Slope	22.28	6.41
Marginal plateaus	5.38	1.55
Flat sedimented plains of marginal seas	7.60	2.19
Island arcs	3.96	1.14
Other ridges and plateaus	3.22	0.926
Marginal seas with hilly sea bottom	2.87	0.825
Deep ocean		
Hilly abyssal plains	106.99	30.8
Flat abyssal plains	29.11	8.38
Continental rise	20.75	5.97
Undulating plains	3.88	1.12
Volcanic ridges + plateaus	5.42	1.56
Faulted ridges and plateaus	5.83	1.68
Microcontinents	3.04	0.875
Ridges and fore-arc bulges	15.33	4.41
Mid-ocean ridges		
Central rift zone	19.49	5.61
Ridge flanks	44.53	12.8
Vol ridges + highs	0.99	0.285
Ocean trenches	1.59	0.457

Source: Based on the map of Agapova et al. [120].
The areas listed cover the whole of the ocean floor, excluding the Arctic Ocean. The percent area is a proportion of the sum of all areas mapped, which is equal to 347 million km².

(discussed later). The mid-ocean ridges are flanked on either side by abyssal hills and hilly abyssal plains (Figure 6.12).

Troughs, Trenches, and Back-Arc Basins

Along active plate margins, the ocean crust collides with and is overridden by continental crust (or in some cases by less-dense ocean crust) in a region called the *subduction zone*, where ridge and trench complexes are created. This occurs because the denser ocean crust is subducted beneath the lighter continental crust (forming a V-shaped trench) and the lateral pressure pushes the continental crust upward (forming a ridge; Figure 6.12). In some locations, ocean crust is subducted beneath other ocean crust, so not all subduction involves continental crust. Sediment that is eroded from areas of surrounding seafloor slumps down into the trench as debris flows or turbidity currents. Over time, sediment may partially infill the valley to form a flat-floored *trough* (Figure 6.12). Along passive plate margins there is no subduction zone (or trench), and the oceanic and continental crusts simply abut one another.

Ocean trenches are the deepest parts of the ocean, commonly 6–10 km in depth[4] and have a highly specialized fauna. Trenches are separated from each other by comparatively shallow ocean floor over which any trench-associated animal would have to pass, and so trenches are isolated from each other ecologically. This isolation has given rise to a high degree of endemism for trench fauna. Oceanographic conditions vary between trenches, but generally species diversity decreases with increasing depth, with a high percentage of species endemic to individual trenches [121]. Most of what we know about ocean trench fauna was derived from Danish and Russian expeditions carried out in the 1950s and 1960s, during the "heroic" era of deep-sea exploration [121], that culminated in the 10,911 m decent of Jacque Piccard and Don Walsh aboard the bathyscaphe *Trieste* on 23 January, 1960, to the bottom of the Challenger Deep in the Mariana Trench. Looking out of the *Trieste's* porthole, Piccard observed a solitary flat bottom fish,[5] confirming the existence of life at even the greatest ocean depths. Species most common in the hadal community are mollusks, polychaete worms, and particularly holothurians.

In some regions, two separate oceanic crustal plates collide and the overriding plate is raised up to form a concave "bulge" with a basin located behind (Figure 6.12; [122]). These are termed *back-arc basins* because they are bounded by volcanic island arcs and can occur in association with both ocean–ocean crust and ocean–continental crust collision zones. An example of ocean–ocean back-arc basin is the Kermadec Arc, described by Wysoczanski and Clark (Chapter 63). Ocean–continental examples are the Kuril, Japan, Ryukyu, Banda, and Hellenic Arcs. Back-arc basins are associated with volcanism and hydrothermal vent communities. For example, in the Tonga-Kermadec arc, comprising one of the fastest moving pieces of ocean crust on Earth [123], about 30 submarine volcanoes are hydrothermally active out of 70 that have been investigated in the region.

Abyssal Hills and Sediment Drapes

Sediments raining down through the water column blanket the seafloor, forming thick deposits that drape over the topography. These deposits are thinnest near the mid-ocean ridge, because, for one reason, the ocean crust here is youngest and sediments have had less time to accumulate. With increasing distance from the center of seafloor spreading, the ocean crust is older and the overlying sediment blanket is thicker. Sediment thickness increases away from mid-ocean ridges in proportion to the amount of time that has elapsed since that particular area of seafloor was created and moved laterally away from the spreading center, cooled and subsided (Figure 6.12). The oldest marine sediment occurs at the far western edge of the Pacific and is around 200 million years old (middle Jurassic). *Abyssal hills* <1,000 m in vertical relief occur in the region between the mid-ocean ridge and the comparatively flat abyssal plains. These hills have their origins as prominent peaks of the mid-ocean ridge, but subsidence coupled with thick blankets of sediment obscures their original relief.

---

[4] The Mariana Trench in the Pacific Ocean is the deepest place in the world ocean at 11,034 m water depth.
[5] Piccard's "fish" was more likely to have been a holothurian because flatfish are uncommon below approximately 2,000 m depth.

Plateaus

Areas of flat ocean floor that are raised above the level of surrounding seafloor, delimited by a steep slope, are known as submarine plateaus. Plateaus mapped by Agapova et al. ([120]; Figure 6.11) include "marginal plateaus," "other ridges and plateaus," and "microcontinents," which together have a total area of about 11,630,000 km² (Table 6.2). The largest plateaus were formed from fragments of continental crust that were left stranded offshore following the rifting apart of continents (Figure 6.13). Other plateaus are produced from large volcanic eruptions or from tectonic uplift of ocean crust. Plateaus are normally draped by sediments, but since the tops of plateaus are cut off from receiving any sediment from the continents because of their raised profile, the only sediment (and food source to the benthos) is derived from the rain of pelagic organic matter from the overlying water column, and sedimentation rates are very slow.

The Lord Howe Rise, located in the Tasman Sea between Australia and New Zealand, is a fragment of continental crust that separated from Australia around 40 million years ago. It has subsided to a depth of around 1,500 m and because the overlying ocean is oligotrophic, pelagic sediments are accumulating on its flat top at a sluggish rate of only a few centimeters per thousand years [124]. It is characterized mostly (95%) by sediment infauna, with isolated volcanic cones providing hard substrate for filter-feeding epifauna (Chapter 57).

In the northeast Atlantic, Sayago-Gil et al. (Chapter 55) describe multibeam bathymetry and benthic samples collected from Hatton Bank, between 600 and 2,000 m water depth. Benthic communities are sparse where the seabed comprises

Figure 6.13 Bathymetric image from Geoscience Australia showing a three-dimensional view of the South Tasman Rise (submarine plateau, lower left) located off the south coast of Tasmania (southeast Australia), and part of the Tasmantid Seamount chain. The continental shelf (in red and yellow) drops dramatically to abyssal depths of 3,000–4,000 m (in green and blue). (For interpretation of the references to color in this figure legend, the reader is referred to the web version of this book.)

mobile sediments (contourite drift deposits) in deeper water, whereas cold-water corals are common on rocky outcrops closer to the top of the bank (Chapter 55).

Nodder et al. (Chapter 56) describe the benthic communities of the Chatham Rise, a 160,000 km^2 submarine continental ridge that extends eastward from New Zealand into the Southwest Pacific Ocean in water depths of <50 to >2,000 m. Isolated groups of volcanic seamounts are present on the flanks of the rise. Most of the Chatham Rise consists of sediment substrates populated by mobile fauna, with conspicuous examples including scampi (*Metanephrops challengeri*), squat lobsters (*Munida gracilis*), several crab species, quill worms (*Hyalinoecia tubicola*), urchins (*Parametia peloria*), and other spatangids and asteroids (Chapter 56).

Seamounts

Beneath the ocean crust, local hotspots of upwelling lava erupt to build submarine volcanoes over 1,000 m high, called *seamounts* (Figure 6.12); smaller volcanoes between 500 and 1,000 m in elevation above the level of surrounding seabed are sometimes called *knolls*, and those <500 m are *abyssal hills*. When basaltic ocean crust is melted, the lava flows relatively easily, unlike melted granitic continental crust, which is much more viscous. The great pressure that occurs at abyssal depths precludes the explosive, violent eruptions associated with continental (subaerial) volcanoes. The eruptions of lava underwater that form seamounts are comparatively quiet, characterized by the accretion of talus slopes comprised of basaltic boulders and rubble. This loose structure provides a porous environment within seamounts that is believed to provide habitat for sulfur-reducing bacteria and also gives rise to unstable slopes subject to massive slumping of debris flows [125]. Volcanic seamounts, knolls, and abyssal hills commonly have locally steep rocky slopes that give way with depth to the low-relief abyssal plains upon which they rest.

Estimates of the numbers of seamounts in the world ocean and their distribution have been published by Agapova et al. [120], Craig and Sandwell [126], and Kitchingham and Lai [127]. Our understanding of the distribution of seamounts is very much dependent upon the available data. As the quality of exiting bathymetric data improves, more seamounts have been discovered and mapped. Agapova et al. [120], basing their assessment on bathymetric data available in the 1970s, identified 7,080 seamounts over 1,000 m in elevation, including 302 flat-topped guyots. Using satellite altimetry data, Craig and Sandwell [126] identified 8,556 seamounts having a diameter of at least 15 km. The most recent estimate of 14,287 seamounts published by Kitchingman and Lai [127][6] was based on an analysis of the US National Oceanographic and Atmospheric Agency (NOAA) ETOPO2 raster bathymetric data set[7] (Figure 6.14). In the South Pacific near Eastern Samoa, Wright et al. (Chapter 58) present multibeam data that reveals 51 previously undocumented seamounts.

---

[6] A new estimate of over 33,000 seamounts was recently published [128].
[7] ETOPO2 Global 2 minute elevations, 2001, National Geophysical Data Center, NOAA/NGDC, USA, www.ngdc.noaa.gov.

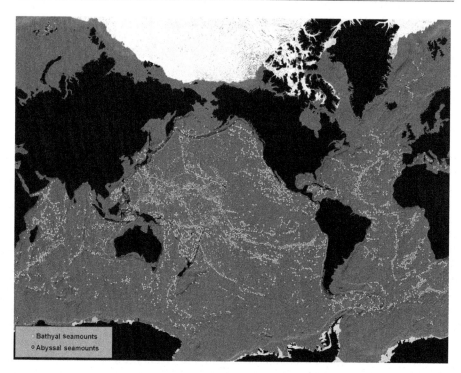

Figure 6.14 Distribution of seamounts in the ocean classified by Clark et al. [129] to highlight those occurring at abyssal and bathyal depths.
Source: Based on Kitchingman and Lai [127].

Table 6.3 Number and Frequency of Seamounts in Ocean Basins

Ocean Basin	Number of Seamounts	Seamounts per Million km²
Pacific	8,952	49.7
Atlantic	2,763	25.8
Indian	1,651	22.3
Southern	883	4.3

Source: After Kitchingman and Lai [127].

There are more seamounts in the Pacific Ocean than in the Atlantic Ocean (Table 6.3). Their distribution can be described as comprising several elongate chains of seamounts superimposed on a more or less random background distribution [126]. Seamount chains occur in all three major ocean basins, with the Pacific having the most numerous and most extensive seamount chains. These include the Hawaiian (Emperor), Mariana, Gilbert, Tuomotu, and Austral Seamounts (and island groups) in the north Pacific, and the Louisville and Sala y Gomez ridges in the southern Pacific Ocean. In the North Atlantic Ocean, the New England Seamount Chain extends from the eastern coast of the US to the mid-ocean ridge. Craig and Sandwell [126] noted that clusters of larger Atlantic seamounts tend to be associated with other evidence of hotspot activity, such as on the Walvis Ridge, Bermuda Islands, and Cape Verde Islands. The mid-Atlantic ridge and spreading ridges in the Indian Ocean are also associated with abundant seamounts [127]. Otherwise, seamounts tend not to form distinctive chains in the Indian and Southern Oceans; rather, their distribution appears to be more or less random (Figure 6.13).

Ecologically, there may be no significant difference between seamounts and some other rocky seabed features that have a prominent vertical relief, such as knolls, ridges, pinnacles, or escarpments. These features all have locally steep slopes with rocky outcrop, and the habitats might also be similar among them. The isolated nature of seamounts that plays a role in the evolution of endemic species [130] is not represented by ridges or escarpments, but knolls and pinnacles do have this factor in common with seamounts. However, the smaller size of knolls and pinnacles has significant implications for how they interact with the ocean currents flowing over them. This affects sedimentation on their flanks and, presumably, the dispersal of fauna as well as providing a smaller area for larvae to settle and colonize. Thus, the comparisons that can be made between seamounts and other rocky seabed features are limited, and many ecological processes will be unique to different classes of feature.

A key factor that controls the occurrence and particularly the abundance of benthic animals is the water depth at the seamount's summit [129]. Seamounts that reach to within approximately 1,500 m of the ocean surface have much higher density of faunal coverage than deeper seamounts. Faunal density reached 90% coverage at 1,000–1,100 m depth on 13 unfished seamounts surveyed by Williams et al. [131], but coverage was <10% at 1,400–1,500 m. In their investigation of the Condor seamount, Tempera et al. (Chapter 59) reported coral gardens on the seamount summit at 287 m depth, locally dominated by gorgonians and hydrarians. Corals colonize the top and flanks of the Chella Bank in the western Mediterranean (Chapters 49 and 60). This flat-topped knoll rises from approximately 600 m to a depth of only 70 m and has steep slopes of up to 70°, with stepped ridges and near-vertical walls colonized with corals.

The acceleration of currents over obstacles on the seabed is a function of the object's height in relation to the depth of water, as well as to its orientation with respect to the current for nonsymmetrical features. Therefore, the larger a feature is with respect to the depth of water in which it occurs, the more it will interact with currents and cause local flow acceleration (the absolute flow speed varying as a function of the speed of the ocean current being deflected). Wind-driven, surface currents flowing around seamounts are affected by seamount size, shape, the proximity of other seamounts or continental margins, the Coriolis force, and stratification. Seamounts may generate coastally trapped waves and internal waves, and may amplify the ocean tide. Mixing in the waters over a seamount can therefore be 100–10,000 times greater than mixing rates in the surrounding ocean [132]. The resulting strong currents carry food to filter feeders, remove waste, and inhibit sedimentation.

Seamounts rising to within a few hundred meters of the sea surface interact with waves and surface currents, deflecting flow around the obstacle and forming local eddies. In some cases, an eddy may become trapped over the seamount, forming a closed circulation cell known as a "Taylor column." These oceanographic features have been observed to persist over an individual seamount for 6 weeks. The turbulence generated by a Taylor column may induce upwelling and locally enhanced primary productivity in an otherwise oligotrophic oceanic regime [130].

Seamounts are isolated habitats that have evolved slowly over millions of years, and they support communities having a high degree of endemism [130,133]. For example, 40–51% of the species found at the Nasca/Sala y Gomez chain off western South America [134] and 29–34% of those from seamounts in the Coral and Tasman Seas were new to science [135].

The biology of hard substrata has been studied on shallow seamount tops (within scuba-diving range) and to a lesser extent in deeper waters using submersibles and ROVs. Due to clear water conditions that often occur in the open ocean, photosynthesis is possible at great depths. For example, live coralline algae was recovered at 268 m from the San Salvador Seamount in the Caribbean Sea [136]. At greater depths the dominant species are suspension feeders: corals, gorgonians, actinarians, pennatulids, and hydroids (Figure 6.15). These species require a hard substrate to anchor themselves, and strong currents to provide food, remove waste products and, most importantly, keep sediment from accumulating. The densities of attached filter feeders is greatest on the peaks and upper flanks of seamounts, with decreasing abundances found at greater depths [130]. Submersible observations on Axial Seamount on the Juan de Fuca ridge in the northeast Pacific revealed dense communities of organisms (up to 100 individuals per square meter) on the vertical walls of a caldera where current speeds were measured at 25 cm/s [137]. Suspension feeder abundance is inversely correlated with soft sediment cover, and so their numbers decline around the base of seamounts, where sediment deposits and weaker bottom currents are also found.

Enhanced primary productivity over seamounts attracts migratory birds and pelagic species such as sharks, tuna, billfishes, and cetaceans. The relatively abundant benthos attracts and supports a range of demersal and epipelagic fish species, which in turn makes them the preferred aggregation and spawning grounds for deep-sea species such as orange roughy [138]. These factors compound to make the abundance of life as much as 100 times greater on seamounts than occurs on the adjacent sediment-covered abyssal plain.

Factors that might influence the dispersal of larvae and colonization of one seamount from another include the distance between seamounts, seamount size, the speed and direction of prevailing currents, the occurrence of Taylor columns, and the depth of seamount peaks. In the Kermadec Arc region of the South Pacific, Wysoczanski and Clark (Chapter 63) found that the diversity of invertebrate assemblages on the hard substrate of seamounts, characterized by the presence of gorgonian corals, echinoids, ophiuroids, alcyonacea, gastropods, and asteroids, varies between seamounts by as much as a factor of three.

The current regime plays a major role in larvae dispersal and hence the orientation of seamounts relative to the prevailing direction of flow determines whether one seamount is effectively downstream of another (and hence a potential site for colonization). The

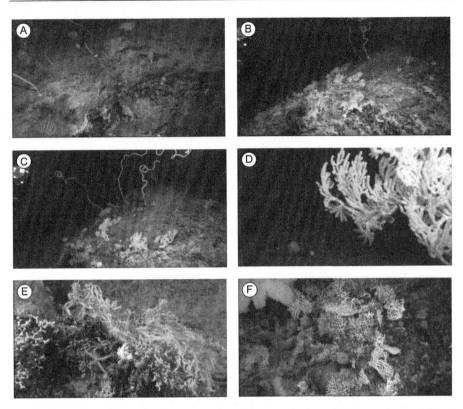

Figure 6.15 Photographs taken by ROV from the flanks of the Manning Seamount, one of the New England Seamounts located in the North Atlantic Ocean. Photos taken at around 2,500 m water depth show: (A) black coral, whip-like bamboo corals, sponges, crinoids, and sea stars; (B) yellow Enallopsammia stony coral, whip-like bamboo corals, sponges and crinoids; (C) Enallopsammia stony coral, whip-like bamboo corals, sponges, and crinoids; (D) close-up view of Keratoisis sea fan with crinoids; (E) Lophelia, Candidella, solitary cup corals, brittle stars, crinoids, and sponges; and (F) Lophelia, Enallopsammia stony coral, dead coral skeleton, and ruffled sponges with purple Trachythela octocoral growing on dead coral skeleton. (For interpretation of the references to color in this figure legend, the reader is referred to the web version of this book.)
Source: Photos courtesy of Mountains in the Sea Research Team, the IFE Crew, and NOAA: http://www.oceanexplorer.noaa.gov/explorations/04mountains/welcome.html.

speed of flow determines how long larvae will have to be able to survive as it is carried along passively with the plankton from one seamount to the next. For these reasons we might expect recruitment to be infrequent and episodic, punctuated with hiatuses of nonrecruitment periods.

In some situations, an eddy may become established over the top of a seamount, establishing Taylor columns (as described earlier). The persistence of Taylor columns over seamounts implies that it may be difficult for larvae to be transported by currents to another seamount site within the normal larval lifespan. Seamounts prone to having Taylor columns are therefore not only geographically isolated, but also oceanographically disconnected from surrounding areas, limiting the seamount's ability to send or receive colonizers.

Hydrothermal Vents

One of the most remarkable discoveries in oceanographic research during the twentieth century is the existence of thriving colonies of animals located on the mid-ocean spreading ridges, feeding on sulfide-rich fluids escaping from hydrothermal vents. Deep-sea vent communities comprise entirely separate ecosystems, decoupled from solar-powered life on the Earth's surface, having evolved to utilize organic matter synthesized by hydrogen sulfide-reducing bacteria. Over 300 endemic species have been found near the vents, including corals, clams, shrimps, crabs, and the now famous giant, red-tipped tube worms, 4 m tall creatures that flourish in waters close to the hot springs (Figure 6.16).

The sulfides, sulfates, and oxides precipitated from the hot (over 350°C) fluids escaping from deep-sea vents form constructional, chimney-like features over 10 m in height. Vent chimneys are most commonly <20 m in height, but one over 45 m in height, named "Godzilla," occurs on the Juan de Fuca Ridge. On exiting the chimney, the minerals in the hot fluid precipitate rapidly, forming what are described as black or white "smokers." Mounds of precipitated pyrite-chalcopyrite several meters high cover the seafloor around the vents [140].

The vents and their associated chemosynthetic communities are most commonly located on the axial ridge of the rift valley, where the geothermal heat flux is highest. Typically only a few kilometers in width, axial ridges generally comprise a rocky substrate devoid of sediment cover, although ridge morphology varies depending upon whether the flanks of the Atlantic-type rift valley are far apart (5–15 km) and the valley depth is large (1–3 km deep). In contrast, fast-spreading, Pacific-type ridges do not exhibit a well-defined rift valley; rather, the ridge supports a linear caldera or eruptive fissure 50–1,000 m wide and only a few tens of meters deep [141]. The East Pacific Rise has the fastest spreading ocean crust, with rates of from 90 to 170 mm/year.

Of significance for conservation is the correlation between the number of vent sites on slow versus fast-spreading ridges; fast-spreading ridges on the East Pacific Rise may have one active site for every 5 km of ridge crest, whereas vents along the slow-spreading mid-Atlantic ridge occur only once in every 100–350 km [142]. Based on these figures, we can estimate that, whereas the fast-spreading ridges in the Pacific, Indian, and Southern oceans may support as many as 12,000 vent sites, the slow-spreading mid-Atlantic ridge supports only about 40 or so vents (i.e., there are perhaps only ~40 hydrothermal vents along the entire length of the mid-Atlantic ridge in the North and South Atlantic Ocean).

The communities that inhabit hydrothermal vents exhibit high degrees of endemism and diversity. The average biomass associated with vents is an order of magnitude at least larger than that associated with the surrounding deep-sea environment. Beaudoin and Smith (Chapter 62) describe vent sites from the Manus Basin, Bismarck Sea, Papua New Guinea, which typically support tube worms, gastropods, bivalves, and crustaceans that

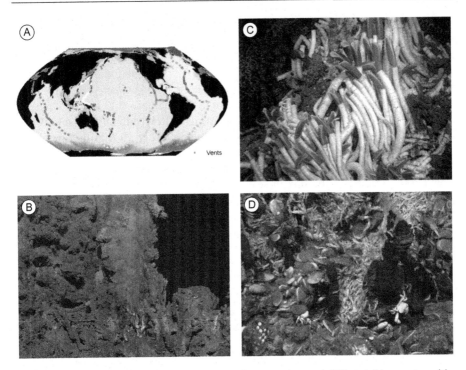

Figure 6.16 Hydrothermal vents: (A) map showing vent sites and different ridge communities along the mid-ocean ridges of the world (from GOODS bioregionalization [139]); (B) photograph taken by Charles Fisher, Penn State University, of a vent-plume at the Mariner vent field on the Valu Fa Ridge near Fiji, South Pacific Ocean; (C) giant tube worms *Riftia pachyptila* at 13°N on the East Pacific Rise, Copyright Ifremer–Phare Expedition, communication@ifremer.fr; (D) swarming vent fauna, mid-Atlantic ridge (University of Bremen, akluegel@uni-bremen.de).

are dependent on the microorganisms that break down hydrogen sulfide in the venting fluid. Many animals host symbiotic, chemosynthetic bacteria that convert sulfides into organic matter. Other chemosynthetic organisms make filamentous mats and biofilms that provide food for grazers and deposit feeders. At the apex of the food web are predators and scavengers such as spider crabs, who move in from the surrounding abyssal areas. Some species of predators and scavengers have evolved to become vent specialists and are found only on vents [142].

In the Kermadec Arc of the South Pacific, Wysoczanski and Clark (Chapter 63) described black-smoker vent communities on active submarine volcanoes. Vent assemblages differ between sites, with regional trends of increased abundance of tube worms from southern to northern seamounts and a reversed pattern in the case of stalked barnacles, which are not found in northern vent sites, but which are common in southern sites.

Factors that play a part in the connectivity between vents are the rate of vent formation, vent longevity, vent distribution, spacing between active vents, the temperature of venting seawater, and bottom current regime at the vent site. The first factors are all related to the rate at which new ocean crust is formed. Thus, a fundamental factor is the spreading rate of the ocean ridge system and the amount of volcanic activity. As noted earlier, volcanic activity along the mid-Atlantic ridge is generally much less than in the Pacific, and the spacing between individual vent communities is much larger in the Atlantic than in the Pacific.

All the water in the world ocean is estimated to flow through the mid-ocean ridges and escape through hydrothermal vents once in every 10–100 million years. For comparison, the rivers refill the oceans via the hydrological cycle about once every 30,000 years. Therefore, the absolute flux of water via river inflow is equal to around 300–3,000 times the flux of seawater through hydrothermal vents. Although the total amount of water escaping from deep-sea vents is not large, the fact that vents may occur along the entire length of the mid-ocean ridge, and also on the flanks of erupting undersea volcanoes, suggests this process is of global significance.

Larvae spawning from a vent point source must rely upon bottom currents for dispersal, rather than the flow of water escaping from the vent. Axial ridges typically rise over 1,000 m in height above the level of the abyssal plains and thus interact with the flow of bottom water. Their linear geometry tends to deflect bottom currents to flow parallel to the ridge, enhancing larvae dispersal and colonization of vent sites along the axial ridge. In contrast, rift valleys and other basins perched between the parallel axial ridges may contain isolated volumes of water that have sluggish flow rates limiting larvae dispersal and colonization of vent sites. For these reasons the distribution patterns of today's vent fauna display the strong imprint of the timing and geometry of plate boundaries [143].

Cold Seeps

Species taxonomically related to hydrothermal vent sites have been found to occur in association with cold-water seeps (deep-sea springs), where their high biomass is due to methane-rich fluids having a chemoautotrophic basis. Like the hydrothermal vent communities, cold-seep communities rely on sulfate-reducing, free-living microorganisms. Cold-seep communities are characterized by invertebrate species (typically bivalves and tube worms) that are commonly host to sulfide-oxidizing endosimbionts. Five different geological settings have been generally related to the occurrence of cold-seep habitats: (1) carbonate platform seeps; (2) hydrocarbon seeps; (3) subduction zone seeps; (4) slump scarp seeps; and (5) seeps related to locations of rapid sediment accumulation, such as submarine fans [142].

Water moving through porous sedimentary rocks comprising carbonate platforms escapes at the base of steep escarpments in the form of brine fluids. The brines are rich in methane and hydrogen sulfide, which are believed to have a chemosynthetic origin. Brine seep communities were first discovered at the base of the Florida escarpment in the Caribbean Sea [144], where communities of mussels and tube worms have been studied using submersibles. Such brine seeps and associated communities could be found in similar carbonate platform, coral reef provinces of the southwest Pacific and Indian oceans.

Over geologic time scales, organic matter that has been deeply buried beneath sediments is transformed into oil and gas hydrocarbon deposits. Thermogenic hydrocarbons (so-called because geothermal heat affects their formation during burial) are usually trapped within geological structures, such as faulted sedimentary rocks or salt domes. However, these structures may develop leaks that allow the hydrocarbons to escape; for example, natural oil seeps have been observed as ocean surface slicks in satellite images [145].

Hydrocarbon-seep communities were first documented from the continental slope off Louisiana, in the Gulf of Mexico at water depths of 600–900 m [146]. Seep communities have since been documented at greater depths of up to 2,750 m in the Gulf [129] and up to 2,220 m depth in association with submarine canyon systems [37]. Hydrocarbon liquids and methane gas migrate vertically upward through overlying sediments to escape into the ocean. The escaping hydrocarbons can be diffused through unconsolidated sediments or may be vented via anoxic "brine pools" of methane-rich hypersaline seawater [147,148]. Within the near-surface sediments, microbial sulfate reduction of the organic matter produces sulfide that supports vestimentiferan tube worms, vesticomyid clams [146], microbial matts, and mussel beds.

In the Gulf of Cádiz, compressional tectonic stress exerted upon the >2 km thick marginal sediment wedge produces hydrocarbon-rich fluid venting that generates different hard-substrate habitats (Chapter 61). The fauna from stable hard bottoms is the most biodiverse, including porifera (*Asconema setubalense*), cold-water corals (*M. oculata*), gorgonians (*Callogorgia verticillata*), and mollusks (*Asperarca nodulosa, Neopycnodonte zibrowii*).

Where ocean crust is subducted beneath overriding continental crust, the unconsolidated sediments that overlie the ocean crust are subducted along with it. As the ocean crust descends, the soft, unconsolidated sediments are subject to compaction under great pressure, and the pore water is expelled. The pore water escapes via diapirs that give rise to "mud volcanoes" on the seafloor, some as much as 250 m in height [149]. The expelled pore water is rich in methane, which forms the basis for unique communities characterized by cladorhizid sponges clustered around the "caldera" of the mud volcano, with clams hosting sulfide-oxidizing endosymbionts located around the rim. In other locations, pore water escapes via elongate fissures in the seabed. Subduction-related seeps were first described at the Cascadia subduction zone in the Pacific Ocean off the US west coast [150] and have since been described at a number of locations including the Barbados Trench [151], Japan Trench, off western Peru, and in the Gulf of Cadiz [152].

In some locations, thermogenic or biogenic hydrocarbon accumulations are exposed as the result of seabed slumps or massive submarine landslides. In such cases, the previously trapped hydrocarbons are uncovered and able to escape, giving rise to seep communities. An example is the Laurentide submarine fan, which was mobilized during the 1929 Grand Banks earthquake. The earthquake generated a submarine landslide with associated bottom currents that produced 1–3 m high submarine dunes comprised of porous gravel. Hydrocarbons previously trapped in deeper sediment layers escape through the sediments, focused along the crests of the coarse gravel waves. Chemoautotrophic communities dominated by clams (reaching numbers of 10–100 individuals per m^2) occur along the crests of the gravel waves and at other sites along the slopes of the Eastern Valley in approximately 3,800 m water depth [153]. An example at more shallow (200–400 m) water depths is from the Skagerrak near Denmark. Slumping along the margins of the central Skagerrak basin has exposed underlying, gas-saturated sediment layers, where seeps occur and seep communities are established [154].

If the sedimentation rate is large compared with the rate of consumption of organic matter in the sediments, then the potential exists to bury excess organic matter that will generate biogenic gas. Conditions favorable for the burial of organic carbon in sediments occur in areas where surface primary productivity is high, bottom waters have low-oxygen content, and sedimentation rates are rapid. These conditions exist adjacent to continents where upwelling zones and/or river input of nutrients drives high primary production and rapid sedimentation rates; an example is the Adriatic Sea where Gordini et al. (Chapter 9) document methane-rich fluids, which causes the precipitation of methane-derived calcium carbonate cement. Organic-rich layers of sediment (called sapropels) deposited under anoxic conditions are found in the Black Sea, on the continental slope of the Arabian Peninsula and in the Santa Barbara Basin and Monterey Bay off California. Where active gas or fluid expulsion occurs in association with such deposits, mud volcanoes, "pockmark" gas-escape features, and chimneys composed of authigenic carbonate are found [155].

Acknowledgments

Thanks to Brian Todd (Natural Resources, Canada), Miriam Sayago-Gil (Instituto Español de Oceanografía, Spain), and Andrew Heap (Geoscience Australia) for helpful comments and suggestions on an earlier draft of this chapter. This work was produced with the support of funding from the Australian Government's Commonwealth Environment Research Facilities (CERF) programme and is a contribution of the CERF Marine Biodiversity Hub. This chapter is published with the permission of the Chief Executive Officer, Geoscience Australia.

References

[1] S. D'Angelo, A. Fiorentino, Phanerogam meadows: a characteristic habitat of the Mediterranean shelf. Examples from the Tyrrhenian Sea, in: P.T. Harris, E.D. Baker (Eds.), Seafloor Geomorphology as Benthic Habitat: GeoHab Atlas of Seafloor Geomorphic Features and Benthic Habitats, Elsevier, Amsterdam, 2011 (Chapter 7).

[2] F. Tempera, M. McKenzie, I. Bashmachnikov, M. Puotinen, R.S. Santos, R. Bates, Predictive modelling of dominant macroalgae abundance on temperate island shelves (Azores, northeast Atlantic), in: P.T. Harris, E.D. Baker (Eds.), Seafloor Geomorphology as Benthic Habitat: GeoHab Atlas of Seafloor Geomorphic Features and Benthic Habitats, Elsevier, Amsterdam, 2011 (Chapter 8).

[3] E. Gordini, A. Falace, S. Kaleb, F. Donda, R. Marocco, G. Tunis, Methane-related carbonate cementation of marine sediments and related macroalgal coralligenous assemblages in the Northern Adriatic Sea, in: P.T. Harris, E.D. Baker (Eds.), Seafloor Geomorphology as

Benthic Habitat: GeoHab Atlas of Seafloor Geomorphic Features and Benthic Habitats, Elsevier, Amsterdam, 2011 (Chapter 9).

[4] A. Collin, B. Long, P. Archambault, Coastal kelp forest habitat in the Baie des Chaleurs, Gulf of St. Lawrence, Canada, in: P.T. Harris, E.D. Baker (Eds.), Seafloor Geomorphology as Benthic Habitat: GeoHab Atlas of Seafloor Geomorphic Features and Benthic Habitats, Elsevier, Amsterdam, 2011 (Chapter 10).

[5] N. Dankers, W. van Duin, M. Baptist, E. Dijkman, J. Cremer, The Wadden Sea in the Netherlands: ecotopes in a World Heritage barrier island system, in: P.T. Harris, E.D. Baker (Eds.), Seafloor Geomorphology as Benthic Habitat: GeoHab Atlas of Seafloor Geomorphic Features and Benthic Habitats, Elsevier, Amsterdam, 2011 (Chapter 11).

[6] J.W.C. James, A.S.Y. Mackie, E.I.S. Rees, T. Darbyshire, Sand wave field: the OBel Sands, Bristol Channel, U.K., in: P.T. Harris, E.D. Baker (Eds.), Seafloor Geomorphology as Benthic Habitat: GeoHab Atlas of Seafloor Geomorphic Features and Benthic Habitats, Elsevier, Amsterdam, 2011 (Chapter 12).

[7] T.A.G.P. van Dijk, J.A. van Dalfsen, R. van Overmeeren, V. Van Lancker, S. van Heteren, P.J. Doornenbal, Benthic habitat variations over tidal ridges, North Sea, Netherlands, in: P.T. Harris, E.D. Baker (Eds.), Seafloor Geomorphology as Benthic Habitat: GeoHab Atlas of Seafloor Geomorphic Features and Benthic Habitats, Elsevier, Amsterdam, 2011 (Chapter 13).

[8] V. Van Lancker, G. Moerkerke, I. Du Four, E. Verfaillie, M. Rabaut, S. Degraer, Fine-scale geomorphological mapping for the prediction of macrobenthic occurrences in shallow marine environments, Belgian part of the North Sea, in: P.T. Harris, E.D. Baker (Eds.), Seafloor Geomorphology as Benthic Habitat: GeoHab Atlas of Seafloor Geomorphic Features and Benthic Habitats, Elsevier, Amsterdam, 2011 (Chapter 14).

[9] B.J. Todd, P.C. Valentine, Large submarine sand features and gravel lag substrates on Georges Bank, Gulf of Maine, in: P.T. Harris, E.D. Baker (Eds.), Seafloor Geomorphology as Benthic Habitat: GeoHab Atlas of Seafloor Geomorphic Features and Benthic Habitats, Elsevier, Amsterdam, 2011 (Chapter 15).

[10] R.J. Beaman, P.T. Harris, Bioregionalization of the George V Shelf, East Antarctica, Cont. Shelf Res. 25 (2005) 1657–1691.

[11] K.A. Robinson, A.S.Y. Mackie, C. Lindenbaum, T. Darbyshire, K.J.J. van Landeghem, W.G. Sanderson, Seabed habitats of the Southern Irish Sea, in: P.T. Harris, E.D. Baker (Eds.), Seafloor Geomorphology as Benthic Habitat: GeoHab Atlas of Seafloor Geomorphic Features and Benthic Habitats, Elsevier, Amsterdam, 2011 (Chapter 37).

[12] J.R. Reynolds, S.C. Rooney, J. Heifetz, H.G. Greene, B.L. Norcross, Habitats and benthos in the vicinity of Albatross Bank, Gulf of Alaska, in: P.T. Harris, E.D. Baker (Eds.), Seafloor Geomorphology as Benthic Habitat: GeoHab Atlas of Seafloor Geomorphic Features and Benthic Habitats, Elsevier, Amsterdam, 2011 (Chapter 38).

[13] B.J. Todd, V.E. Kostylev, S.J. Smith, Seabed habitat of a glaciated shelf, German Bank, Atlantic Canada, in: P.T. Harris, E.D. Baker (Eds.), Seafloor Geomorphology as Benthic Habitat: GeoHab Atlas of Seafloor Geomorphic Features and Benthic Habitats, Elsevier, Amsterdam, 2011 (Chapter 39).

[14] E. Ezhova, D. Dorokhov, V. Sivkov, V. Zhamoida, D. Ryabchuk, O. Kocheshkova, Benthic habitats and benthic communities in South-Eastern Baltic Sea, Russian sector, in: P.T. Harris, E.D. Baker (Eds.), Seafloor Geomorphology as Benthic Habitat: GeoHab Atlas of Seafloor Geomorphic Features and Benthic Habitats, Elsevier, Amsterdam, 2011 (Chapter 43).

[15] T.C. Stieglitz, The Yongala's Halo of Holes—systematic bioturbation close to a ship-wreck, in: P.T. Harris, E.D. Baker (Eds.), Seafloor Geomorphology as Benthic Habitat: GeoHab Atlas of Seafloor Geomorphic Features and Benthic Habitats, Elsevier, Amsterdam, 2011 (Chapter 16).

[16] A.T. Kotilainen, A.M. Kaskela, S. Bäck, J. Leinikki, Submarine De Geer moraines in the Kvarken Archipelago, the Baltic Sea, in: P.T. Harris, E.D. Baker (Eds.), Seafloor Geomorphology as Benthic Habitat: GeoHab Atlas of Seafloor Geomorphic Features and Benthic Habitats, Elsevier, Amsterdam, 2011 (Chapter 17).

[17] P.E. O'Brien, J. Stark, G. Johnston, J. Smith, M.J. Riddle, Sea bed character and habitats of a rocky Antarctic coastline: a preliminary view of the Vestfold Hills, East Antarctica, in: P.T. Harris, E.D. Baker (Eds.), Seafloor Geomorphology as Benthic Habitat: GeoHab Atlas of Seafloor Geomorphic Features and Benthic Habitats, Elsevier, Amsterdam, 2011 (Chapter 20).

[18] J.V. Barrie, H.G. Greene, K.W. Conway, K. Picard, Inland Tidal Sea of the Northeastern Pacific, in: P.T. Harris, E.D. Baker (Eds.), Seafloor Geomorphology as Benthic Habitat: GeoHab Atlas of Seafloor Geomorphic Features and Benthic Habitats, Elsevier, Amsterdam, 2011 (Chapter 44).

[19] G.R. Cochrane, L. Trusel, J. Harney, L. Etherington, Habitats and benthos of an evolving fjord, Glacier Bay, Alaska, in: P.T. Harris, E.D. Baker (Eds.), Seafloor Geomorphology as Benthic Habitat: GeoHab Atlas of Seafloor Geomorphic Features and Benthic Habitats, Elsevier, Amsterdam, 2011 (Chapter 18).

[20] A. Copeland, E. Edinger, T. Bell, P. LeBlanc, J. Wroblewski, R. Devillers, Geomorphic features and benthic habitats of a subarctic fjord: Gilbert Bay, Southern Labrador, Canada, in: P.T. Harris, E.D. Baker (Eds.), Seafloor Geomorphology as Benthic Habitat: GeoHab Atlas of Seafloor Geomorphic Features and Benthic Habitats, Elsevier, Amsterdam, 2011 (Chapter 19).

[21] J.S. Collier, S.R. Humber, Fringing reefs of the Seychelles inner granitic islands, Western Indian Ocean, in: P.T. Harris, E.D. Baker (Eds.), Seafloor Geomorphology as Benthic Habitat: GeoHab Atlas of Seafloor Geomorphic Features and Benthic Habitats, Elsevier, Amsterdam, 2011 (Chapter 21).

[22] S. Hamylton, T. Spencer, A.B. Hagan, Coral reefs and reef islands of the Amirantes Archipelago, Western Indian Ocean, in: P.T. Harris, E.D. Baker (Eds.), Seafloor Geomorphology as Benthic Habitat: GeoHab Atlas of Seafloor Geomorphic Features and Benthic Habitats, Elsevier, Amsterdam, 2011 (Chapter 22).

[23] S. Hamylton, Hyperspectral remote sensing of the geomorphic features and habitats of the Al Wajh Bank reef system, Saudi Arabia, Red Sea, in: P.T. Harris, E.D. Baker (Eds.), Seafloor Geomorphology as Benthic Habitat: GeoHab Atlas of Seafloor Geomorphic Features and Benthic Habitats, Elsevier, Amsterdam, 2011 (Chapter 23).

[24] R.A. Armstrong, H. Singh, Mesophotic coral reefs of the Puerto Rico Shelf, in: P.T. Harris, E.D. Baker (Eds.), Seafloor Geomorphology as Benthic Habitat: GeoHab Atlas of Seafloor Geomorphic Features and Benthic Habitats, Elsevier, Amsterdam, 2011 (Chapter 24).

[25] B.P. Brooke, M.A. McArthur, C.D. Woodroffe, M. Linklater, S.L. Nichol, T.J. Anderson, et al., Geomorphic features and infauna diversity of a subtropical mid-ocean carbonate shelf: Lord Howe Island, Southwest Pacific Ocean, in: P.T. Harris, E.D. Baker (Eds.), Seafloor Geomorphology as Benthic Habitat: GeoHab Atlas of Seafloor Geomorphic Features and Benthic Habitats, Elsevier, Amsterdam, 2011 (Chapter 25).

[26] W.D. Heyman, S. Kobara, Geomorphology of reef fish spawning aggregations in Belize and the Cayman Islands (Caribbean), in: P.T. Harris, E.D. Baker (Eds.), Seafloor

Geomorphology as Benthic Habitat: GeoHab Atlas of Seafloor Geomorphic Features and Benthic Habitats, Elsevier, Amsterdam, 2011 (Chapter 26).

[27] S.L. Nichol, T.J. Anderson, C. Battershill, B.P. Brooke, Submerged reefs and aeolian dunes as inherited habitats, Point Cloates, Carnarvon Shelf, Western Australia, in: P.T. Harris, E.D. Baker (Eds.), Seafloor Geomorphology as Benthic Habitat: GeoHab Atlas of Seafloor Geomorphic Features and Benthic Habitats, Elsevier, Amsterdam, 2011 (Chapter 27).

[28] A.E. Gibbs, S.A. Cochran, Coral habitat variability in Kaloko-Honokohau National Park, Hawaii, USA, in: P.T. Harris, E.D. Baker (Eds.), Seafloor Geomorphology as Benthic Habitat: GeoHab Atlas of Seafloor Geomorphic Features and Benthic Habitats, Elsevier, Amsterdam, 2011 (Chapter 28).

[29] R.J. Allee, A.W. David, D.F. Naar, Two shelf edge marine protected areas in the eastern Gulf of Mexico, in: P.T. Harris, E.D. Baker (Eds.), Seafloor Geomorphology as Benthic Habitat: GeoHab Atlas of Seafloor Geomorphic Features and Benthic Habitats, Elsevier, Amsterdam, 2011 (Chapter 30).

[30] R.J. Beaman, T. Bridge, T. Done, J.M. Webster, S. Williams, O. Pizarro, Habitats and benthos at Hydrographers Passage, Great Barrier Reef, Australia, in: P.T. Harris, E.D. Baker (Eds.), Seafloor Geomorphology as Benthic Habitat: GeoHab Atlas of Seafloor Geomorphic Features and Benthic Habitats, Elsevier, Amsterdam, 2011 (Chapter 29).

[31] E. Martorelli, S. D'Angelo, A. Fiorentino, F.L. Chiocci, Non-tropical carbonate shelf sedimentation. The Archipelago Pontino (Central Italy) case history, in: P.T. Harris, E.D. Baker (Eds.), Seafloor Geomorphology as Benthic Habitat: GeoHab Atlas of Seafloor Geomorphic Features and Benthic Habitats, Elsevier, Amsterdam, 2011 (Chapter 31).

[32] C. Lo Iacono, C. Orejas, A. Gori, J.M. Gili, S. Requena, P. Puig, et al., Habitats of the Cap de Creus continental shelf and Cap de Creus canyon, north-western Mediterranean, in: P.T. Harris, E.D. Baker (Eds.), Seafloor Geomorphology as Benthic Habitat: GeoHab Atlas of Seafloor Geomorphic Features and Benthic Habitats, Elsevier, Amsterdam, 2011 (Chapter 32).

[33] R. Coggan, M. Diesing, Rock Ridges in the central English Channel, in: P.T. Harris, E.D. Baker (Eds.), Seafloor Geomorphology as Benthic Habitat: GeoHab Atlas of Seafloor Geomorphic Features and Benthic Habitats, Elsevier, Amsterdam, 2011 (Chapter 33).

[34] V. Lucieer, N. Barrett, N. Hill, S.L. Nichol, Characterization of shallow inshore coastal reefs on the Tasman Peninsula, South Eastern Tasmania, Australia, in: P.T. Harris, E.D. Baker (Eds.), Seafloor Geomorphology as Benthic Habitat: GeoHab Atlas of Seafloor Geomorphic Features and Benthic Habitats, Elsevier, Amsterdam, 2011 (Chapter 34).

[35] I. Galparsoro, Á. Borja, J.G. Rodríguez, I. Muxika, M. Pascual, I. Legorburu, Rocky reef and sedimentary habitats within the continental shelf of the southeastern Bay of Biscay, in: P.T. Harris, E.D. Baker (Eds.), Seafloor Geomorphology as Benthic Habitat: GeoHab Atlas of Seafloor Geomorphic Features and Benthic Habitats, Elsevier, Amsterdam, 2011 (Chapter 35).

[36] K.L. Yamanaka, K. Picard, K.W. Conway, R. Flemming, Rock reefs of British Columbia, Canada: Inshore rockfish habitats, in: P.T. Harris, E.D. Baker (Eds.), Seafloor Geomorphology as Benthic Habitat: GeoHab Atlas of Seafloor Geomorphic Features and Benthic Habitats, Elsevier, Amsterdam, 2011 (Chapter 36).

[37] C.A. Robinson, J.M. Bernhard, L.A. Levin, G.F. Mendoza, J.K. Blanks, Surficial hydrocarbon seep infauna from the Blake Ridge (Atlantic Ocean, 2150m) and the Gulf of Mexico (690–2240m), Mar. Ecol. 25 (2004) 313–336.

[38] J.W.C. James, B. Pearce, R.A. Coggan, A. Morando, Open shelf valley system, Northern Palaeovalley, English Channel, U.K., in: P.T. Harris, E.D. Baker (Eds.), Seafloor Geomorphology as Benthic Habitat: GeoHab Atlas of Seafloor Geomorphic Features and Benthic Habitats, Elsevier, Amsterdam, 2011 (Chapter 41).

[39] B. Pearce, D.R. Tappin, D. Dove, J. Pinnion, Benthos supported by the tunnel-valleys of the southern North Sea, in: P.T. Harris, E.D. Baker (Eds.), Seafloor Geomorphology as Benthic Habitat: GeoHab Atlas of Seafloor Geomorphic Features and Benthic Habitats, Elsevier, Amsterdam, 2011 (Chapter 42).

[40] B. De Mol, D. Amblas, G. Alverez, P. Busquets, A. Calafat, M. Canals, et al., Erosional environment at the gateway between the Atlantic and Mediterranean Sea: Strait of Gibraltar cold-water coral distribution, in: P.T. Harris, E.D. Baker (Eds.), Seafloor Geomorphology as Benthic Habitat: GeoHab Atlas of Seafloor Geomorphic Features and Benthic Habitats, Elsevier, Amsterdam, 2011 (Chapter 45).

[41] L. De Mol, A. Hilário, D. Van Rooij, J.-P. Henriet, Habitat mapping of a cold-water coral mound on Pen Duick Escarpment (Gulf of Cadiz), in: P.T. Harris, E.D. Baker (Eds.), Seafloor Geomorphology as Benthic Habitat: GeoHab Atlas of Seafloor Geomorphic Features and Benthic Habitats, Elsevier, Amsterdam, 2011 (Chapter 46).

[42] J. Guinan, Y. Leahy, K. Verbruggen, T. Furey, Habitats at the Rockall Bank slope failure features, Northeast Atlantic Ocean, in: P.T. Harris, E.D. Baker (Eds.), Seafloor Geomorphology as Benthic Habitat: GeoHab Atlas of Seafloor Geomorphic Features and Benthic Habitats, Elsevier, Amsterdam, 2011 (Chapter 47).

[43] F. Althaus, A. Williams, R.J. Kloser, J. Seiler, N.J. Bax, Evaluating geomorphic features as surrogates for benthic biodiversity on Australia's western continental margin, in: P.T. Harris, E.D. Baker (Eds.), Seafloor Geomorphology as Benthic Habitat: GeoHab Atlas of Seafloor Geomorphic Features and Benthic Habitats, Elsevier, Amsterdam, 2011 (Chapter 48).

[44] L. Buhl-Mortensen, R. Bøe, M.F.J. Dolan, P. Buhl-Mortensen, T. Thorsnes, S. Elvenes, et al., Banks, troughs and canyons on the continental margin off Lofoten, Vesterålen, and Troms, Norway, in: P.T. Harris, E.D. Baker (Eds.), Seafloor Geomorphology as Benthic Habitat: GeoHab Atlas of Seafloor Geomorphic Features and Benthic Habitats, Elsevier, Amsterdam, 2011 (Chapter 51).

[45] A.L. Post, P.E. O'Brien, R.J. Beaman, M.J. Riddle, L. De Santis, S.R. Rintoul, Distribution of hydrocorals along the George V slope, East Antarctica, in: P.T. Harris, E.D. Baker (Eds.), Seafloor Geomorphology as Benthic Habitat: GeoHab Atlas of Seafloor Geomorphic Features and Benthic Habitats, Elsevier, Amsterdam, 2011 (Chapter 52).

[46] G. Lamarche, A.A. Rowden, J. Mountjoy, V. Lucieer, A.-L. Verdier, The Cook Strait Canyon, New Zealand: a large bedrock canyon system in a tectonically active environment, in: P.T. Harris, E.D. Baker (Eds.), Seafloor Geomorphology as Benthic Habitat: GeoHab Atlas of Seafloor Geomorphic Features and Benthic Habitats, Elsevier, Amsterdam, 2011 (Chapter 53).

[47] M. Yoklavich, H.G. Greene, The Ascension-Monterey Canyon system—habitats of demersal fishes and macro-invertebrates along the central California coast of the USA, in: P.T. Harris, E.D. Baker (Eds.), Seafloor Geomorphology as Benthic Habitat: GeoHab Atlas of Seafloor Geomorphic Features and Benthic Habitats, Elsevier, Amsterdam, 2011 (Chapter 54).

[48] C. Lo Iacono, E. Gràcia, R. Bartolomé, E. Coiras, J.J. Dañobeitia, J. Acosta, The Chella Bank, Eastern Alboran Sea (Western Mediterranean), in: P.T. Harris, E.D. Baker (Eds.), Seafloor Geomorphology as Benthic Habitat: GeoHab Atlas of Seafloor Geomorphic Features and Benthic Habitats, Elsevier, Amsterdam, 2011 (Chapter 49).

[49] F. Tempera, E. Giacomello, N. Mitchell, A.S. Campos, A.B. Henriques, A. Martins, et al., Mapping the Condor seamount seafloor environment and associated biological assemblages (Azores, NE Atlantic), in: P.T. Harris, E.D. Baker (Eds.), Seafloor Geomorphology as Benthic Habitat: GeoHab Atlas of Seafloor Geomorphic Features and Benthic Habitats, Elsevier, Amsterdam, 2011 (Chapter 59).

[50] B. De Mol, D. Amblas, A. Calafat, M. Canals, R. Duran, C. Lavoie, Alboran seamounts, Western Mediterranean sea: cold-water coral colonization of knolls, in: P.T. Harris, E.D. Baker (Eds.), Seafloor Geomorphology as Benthic Habitat: GeoHab Atlas of Seafloor Geomorphic Features and Benthic Habitats, Elsevier, Amsterdam, 2011 (Chapter 60).

[51] M. Sayago-Gil, P. Durán-Muñoz, F.J. Murillo, V. Díaz-del-Río, A. Serrano, L.M. Fernández-Salas, A study of geo-morphological features of the sea bed and the relationship to deep sea communities on the western slope of Hatton Bank (NE Atlantic Ocean), in: P.T. Harris, E.D. Baker (Eds.), Seafloor Geomorphology as Benthic Habitat: GeoHab Atlas of Seafloor Geomorphic Features and Benthic Habitats, Elsevier, Amsterdam, 2011 (Chapter 55).

[52] S.D. Nodder, D.A. Bowden, A. Pallentin, K. Mackay, Seafloor habitats and benthos of a continental ridge: Chatham Rise, New Zealand, in: P.T. Harris, E.D. Baker (Eds.), Seafloor Geomorphology as Benthic Habitat: GeoHab Atlas of Seafloor Geomorphic Features and Benthic Habitats, Elsevier, Amsterdam, 2011 (Chapter 56).

[53] P.T. Harris, S.L. Nichol, T.J. Anderson, A.D. Heap, Habitats and benthos of a deep sea marginal plateau, Lord Howe Rise, Australia, in: P.T. Harris, E.D. Baker (Eds.), Seafloor Geomorphology as Benthic Habitat: GeoHab Atlas of Seafloor Geomorphic Features and Benthic Habitats, Elsevier, Amsterdam, 2011 (Chapter 57).

[54] J.L. Rueda, V. Díaz-del-Río, M. Sayago-Gil, N. López, L.M. Fernández, J.T. Vázquez, Fluid venting through the seabed in the Gulf of Cadiz (SE Atlantic Ocean, western Iberian Peninsula): geomorphic features, habitats and associated fauna, in: P.T. Harris, E.D. Baker (Eds.), Seafloor Geomorphology as Benthic Habitat: GeoHab Atlas of Seafloor Geomorphic Features and Benthic Habitats, Elsevier, Amsterdam, 2011 (Chapter 61).

[55] Y.C. Beaudoin, S. Smith, Habitats of the Su Su Knolls hydrothermal site, eastern Manus Basin, Papua New Guinea, in: P.T. Harris, E.D. Baker (Eds.), Seafloor Geomorphology as Benthic Habitat: GeoHab Atlas of Seafloor Geomorphic Features and Benthic Habitats, Elsevier, Amsterdam, 2011 (Chapter 62).

[56] R. Wysoczanski, M. Clark, Southern Kermadec Arc—Havre Trough geohabitats, in: P.T. Harris, E.D. Baker (Eds.), Seafloor Geomorphology as Benthic Habitat: GeoHab Atlas of Seafloor Geomorphic Features and Benthic Habitats, Elsevier, Amsterdam, 2011 (Chapter 63).

[57] H.G. Greene, M.M. Yoklavich, R.M. Starr, V.M. O'Connell, W.W. Wakefield, D.E. Sullivan, et al., A classification scheme for deep seafloor habitats, Ocenaologica Acta 22 (1999) 663–678.

[58] J.C. Roff, M.E. Taylor, J. Laughren, Geophysical approaches to the classification, delineation and monitoring of marine habitats and their communities, Aquat. Conserv.: Mar. Freshwater Ecosyst. 13 (2003) 77–90.

[59] R. Boyd, R. Dalrymple, B.A. Zaitlin, Classification of clastic coastal depositional environments, Sediment. Geol. 80 (1992) 139–150.

[60] R.W. Dalrymple, B.A. Zaitlin, R. Boyd, Estuarine facies models: conceptual basis and stratigraphic implications, J. Sediment. Petrol. 62 (1992) 1130–1146.

[61] M.N. Dethier, G.C. Schoch, The consequences of scale: assessing the distribution of benthic populations in a complex estuarine fjord, Estuar. Coast. Shelf Sci. 62 (2005) 253–270.

[62] P.S. Roy, R.J. Williams, A.R. Jones, I. Yassini, P.J. Gibbs, B. Coates, et al., Structure and function of southeast Australian estuaries, Estuar. Coast. Shelf Sci. 53 (2001) 351–384.

[63] A. Brazeiro, Relationship between species richness and morphodynamics in sandy beaches: what are the underlying factors? Mar. Ecol. Progr. 224 (2001) 35–44.

[64] I.F. Rodil, M. Lastra, Environmental factors affecting benthic macrofauna along a gradient of intermediate sandy beaches in northern Spain, Estuar. Coast. Shelf Sci. 61 (2004) 37–44.

[65] R.J. Diaz, R. Rosenberg, Marine benthic hypoxia: A review of its ecological effects and the behavioural responses of benthic macrofauna, Oceanogr. Mar. Biol. Ann. Rev. 33 (1995) 245–303.

[66] J. Chappell, N.J. Shackleton, Oxygen isotopes and sea level, Nature 324 (1986) 137–140.

[67] P.T. Harris, A. Heap, V. Passlow, M. Hughes, J. Daniell, M. Hemer, et al., Tidally-incised valleys on tropical carbonate shelves: an example from the northern Great Barrier Reef, Australia, Mar. Geol. 220 (2005) 181–204.

[68] G. Tassinari, D.J.H.C. Campara, M.P. Kothiyal, H.J. Tiziani, A. Schaaf, Sea level changes, continental shelf morphology, and global paleoecological constraints in the shallow benthic realm: a theoretical approach, Palaeogeogr. Palaeoclimatol. Palaeoecol. 121 (1996) 259–271.

[69] J.E.N. Veron, A Reef in Time: The Great Barrier Reef from Beginning to End, Harvard University Press, Cambridge, MA, 2008.

[70] A. Brandt, A.J. Gooday, S.N. Brandao, S. Brix, W. Brokeland, T. Cedhagen, et al., First insights into the biodiversity and biogeography of the Southern Ocean deep sea, Nature 447 (2007).

[71] M.J. Hambrey, Glacial Environments, UCL Press, London, 1994.

[72] C.M.T. Woodworth-Lynas, H.W. Josenhans, J.V. Barrie, C.F.M. Lewis, D.R. Parrott, The physical processes of seabed disturbance during iceberg grounding and scouring, Cont. Shelf Res. 11 (1991) 939–961.

[73] J. Gutt, On the direct impact of ice on marine benthic communities, a review, Polar Biol. 24 (2001) 553–564.

[74] T. Brey, D. Gerdes, Is Antarctic benthic biomass really higher than elsewhere? Antarct. Sci. 9 (1997) 266–267.

[75] R.N. Ginsburg, N.P. James, Holocene carbonate sediments of continental shelves, in: C.A. Burk, C.L. Drake (Eds.), The Geology of Continental Margins, Springer-Verlag, Berlin, 1974, pp. 137–155.

[76] M.D. Spalding, A.M. Greenfell, New estimates of global and regional coral reef areas, Coral Reefs 16 (1997) 225–230.

[77] M.D. Spalding, C. Ravilious, E.P. Green, World Atlas of Coral Reefs. Prepared at the UNEP World Conservation Monitoring Centre, University of California Press, Berkeley, USA, 2001.

[78] P.T. Harris, A.D. Heap, J.F. Marshall, M.T. McCulloch, A new coral reef province in the Gulf of Carpentaria, Australia: colonisation, growth and submergence during the early Holocene, Mar. Geol. 251 (2008) 85–97.

[79] A.C. Neumann, I.G. Macintyre (1985) Reef response to sea level rise: keep-up, catch-up or give-up. Proceedings 5th International Coral Reef Congress, International Coral Reef Society, Tahiti, pp. 105–110.

[80] A.H. Stride (Ed.), Offshore Tidal Sands—Processes and Deposits, Chapman and Hall, London, 1982.

[81] P.T. Harris, C.B. Pattiaratchi, M.B. Collins, R.W. Dalrymple, What is a bedload parting? in: B.W. Flemming, A. Bartholoma (Eds.), Tidal signatures in modern and ancient sediments. IAS Special Publication No. 24, Blackwell, Oxford, 1995, pp. 1–18.

[82] K.O. Emery, Continental margins; classification and petroleum prospects, AAPG Bull. 64 (1980) 297–315.

[83] E. Uchupi, K.O. Emery, Genetic global geomorphology: a prospectus, in: R.H. Osborne (Ed.), From Shoreline to Abyss: Contributions in Marine Geology in Honor of Francis Parker Shepard, SEPM Special Publication, Tulsa, OK, 1991, pp. 273–290.

[84] P.T. Harris, T. Whiteway, Global distribution of large submarine canyons: geomorphic differences between active and passive continental margins, Mar. Geol. (in press).

[85] C.A. Nittrouer, L.D. Wright, Transport of particles across continental shelves, Rev. Geophys. 32 (1994) 85–113.

[86] F.P. Shepard, Submarine Geology, Harper & Row, New York, NY, 1963.

[87] F.P. Shepard, Submarine canyons: multiple causes and long-time persistence, AAPG Bull. 65 (1981) 1062–1077.

[88] K.I. Skene, D.J.W. Piper, Late Cenozoic evolution of Laurentian Fan: development of a glacially-fed submarine fan, Mar. Geol. 227 (2006) 67–92.

[89] J. Bourget, S. Zaragosi, T. Garlan, I. Gabelotaud, P. Guyomard, B. Dennielou, et al., Discovery of a giant deep-sea valley in the Indian Ocean, off eastern Africa: the Tanzania channel, Mar. Geol. 255 (2008) 179–185.

[90] F.P. Shepard, K.O. Emery, Congo submarine canyon and Fan Valley, AAPG Bull. 57 (1973) 1679–1691.

[91] G.C. Kineke, K.J. Woolfe, S.A. Kuehl, J.D. Milliman, T.M. Dellapenna, R.G. Purdon, Sediment export from the Sepik River, PNG: evidence for a divergent sediment plume, Cont. Shelf Res. 20 (2000) 2239–2266.

[92] J. Lofi, C. Gorini, S. Berné, G. Clauzon, A. Tadeu Dos Reis, W.B.F. Ryan, et al., Erosional processes and paleo-environmental changes in the Western Gulf of Lions (SW France) during the Messinian Salinity Crisis, Mar. Geol. 217 (2005) 1–30.

[93] W.B.F. Ryan, W.C. Pitman, C.O. Major, K. Shimkus, V. Moskalenko, G.A. Jones, et al., An abrupt drowning of the Black Sea shelf, Mar. Geol. 138 (1997) 119–126.

[94] I. Popescu, G. Lericolais, N. Panin, A. Normand, C. Dinu, E. Le Drezen, The Danube submarine canyon (Black Sea): morphology and sedimentary processes, Mar. Geol. 206 (2004) 249–265.

[95] D.J.W. Piper, P. Cochonate, M.L. Morrison, The sequence of events around the epicentre of the 1929 Grand Banks earthquake: initiation of debris flows and turbidity current inferred from side scan sonar, Sedimentology 46 (1999) 79–97.

[96] B.C. Heezen, R.J. Menzies, E.D. Schneider, W.M. Ewing, N.C.L. Granelli, Congo submarine canyon, AAPG Bull. 48 (1964) 1126–1149.

[97] F.P. Shepard, R.F. Dill, Submarine Canyons and Other Sea Valleys, Rand McNally, Chicago, IL, 1966.

[98] D.J.W. Piper, Late Cenozoic evolution of the continental margin of eastern Canada, Norw. J. Geol. 85 (2005) 305–318.

[99] R.G. Walker, Turbidites and submarine fans, in: R.G. Walker, N.P. James (Eds.), Facies Models—Response to Sea Level Change, Geological Association of Canada, St. John's, Newfoundland, Canada, 1992, pp. 239–263 (Chapter 13).

[100] J.D. Clark, N.H. Kenyon, K.T. Pickering, Quantitative analysis of the geometry of submarine channels: implications for the classification of submarine fans, Geology 20 (1992) 633–636.

[101] B.M. Hickey, Coastal submarine canyons: Topographic Effects in the Ocean: 'Aha Huliko'a Hawaiian Winter Workshop, University of Hawaii at Manoa, Honolulu, Hawaii, 1995, pp. 95–110.

[102] M. Sobarzo, M. Figueroa, L. Djurfeldt, Upwelling of subsurface water into the rim of the Biobío submarine canyon as a response to surface winds, Cont. Shelf Res. 21 (2001) 279–299.

[103] G.S. Carter, M.C. Gregg, Intense, variable mixing near the head of Monterey Canyon, J. Phys. Oceanogr. 32 (2002) 3145–3165.

[104] S.J. Rennie, C.B. Pattiaratchi, R.D. McCauley, Numerical simulation of the circulation within the Perth submarine canyon, Western Australia, Cont. Shelf Res. 29 (2009) 2020–2036.

[105] S.K. Hooker, H. Whitehead, S. Gowans, Marine protected area design and the spatial and temporal distribution of cetaceans in a submarine canyon, Conser. Biol. 13 (1999) 592–602.

[106] L. De Mol, D. Van Rooij, H. Pirlet, J. Greinert, N. Frank, F. Quemmerais, et al., Coldwater coral habitats in the Penmarc'h and Guilvinec Canyons (Bay of Biscay): deepwater versus shallow-water settings, Mar. Geol. (2010).

[107] B.L. Mullenbach, C.A. Nittrouer, P. Puig, D.L. Orange, Sediment deposition in a modern submarine canyon: Eel Canyon, northern California, Mar. Geol. 211 (2004) 101–119.

[108] D.J.W. Piper, W.R. Normark, Processes that initiate turbidity currents and their influence on turbidites: a marine geology perspective, J. Sediment. Res. 79 (2009) 347–362.

[109] J.P. Walsh, C.A. Nittrouer, Understanding fine-grained river-sediment dispersal on continental margins, Mar. Geol. 263 (2009) 34–45.

[110] E.W. Vetter, P.K. Dayton, Macrofaunal communities within and adjacent to a detritus-rich submarine canyon system, Deep Sea Res. II 45 (1998) 25–54.

[111] E.W. Vetter, P.K. Dayton, Organic enrichment by macrophyte detritus, and abundance patters of megafaunal populations in submarine canyons, Mar. Ecol. Prog. Ser. 186 (1999) 137–148.

[112] C.K. Paull, B. Hecker, R. Commeau, R.P. Freeman-Lynde, A.C. Neumann, W.P. Corso, et al., Biological communities at the Florida Escarpment resemble hydrothermal vent taxa, Science 226 (1984) 965–967.

[113] W. Schlager, O. Camber, Submarine slope angles, drowning unconformities, and self-erosion of limestone escarpments, Geology 14 (1986) 762–765.

[114] B. Leitner, A.M. Tréhu, N.J. Godfrey, Crustal structure of the northwestern Vizcaino block and Gorda Escarpment, offshore northern California, and implications for post-subduction deformation of a paleoaccretionary margin, J. Geophys. Res. 103 (1998) 23795–23812.

[115] R. Von Huene, W. Weinrebe, F. Heeren, Subduction erosion along the North Chile margin, J. Geodyn. 27 (1999) 345–358.

[116] C.K. Paull, W.P. Dillon, Erosional origin of the Blake Escarpment: an alternative hypothesis, Geology 8 (1980) 538–542.

[117] J.K. Reed, D.C. Weaver, S.A. Pomponi, Habitat and fauna of deep-water *Lophelia pertusa* coral reefs off the southeastern U.S.: Blake plateau, Straits of Florida, and Gulf of Mexico, Bull. Mar. Sci. 78 (2006) 343–375.

[118] A. Freiwald, J.M. Roberts, M. Taviani, A. Freiwald, H. Zibrowius, Deep coral growth in the Mediterranean Sea: an overview, in: A. Freiwald (Ed.), Cold-Water Corals and Ecosystems, Springer, Berlin Heidelberg, 2005, pp. 137–156.

[119] A.C. Duxbury, A.B. Duxbury, An Introduction to the World's Oceans, W.C. Brown, Dubuque, IN, 1991.

[120] G.V. Agapova, L.Y. Budanova, N.L. Zenkevich, N.I. Larina, V.M. Litvin, N.A. Marova, et al., Geomorphology of the ocean floor: Geofizika okeana. Geofizika okeanskogo dna, Neprochnov, Izd, Nauka, Moscow, 1979, pp. 150–205.

[121] J.D. Gage, P.A. Tyler, Deep-Sea Biology: A Natural History of Organisms at the Deep-Sea Floor, Cambridge University Press, Cambridge, 1991.

[122] R.J. Arculus, Aspects of magma genesis in arcs—a review, Lithos 33 (1994) 189–208.

[123] M. Bevis, F.W. Taylor, B.E. Schultz, J. Recy, B.L. Isacks, S. Helu, et al., Geodetic observations of convergence and back-arc spreading at the Tonga Island arc, Nature 374 (1995) 249–251.

[124] K.M. Grant, G.R. Dickens, Coupled productivity and carbon isotope records in the southwest Pacific Ocean during the late Miocene—early Pliocene biogenic bloom, Palaeogeogr. Palaeoclimatol. Palaeoecol. 187 (2002) 61–82.

[125] A. Malahoff, Summit construction, caldera formation cone growth and hydrothermal processes on submarine volcanoes of the southern Kermadec arc, in: D. Denham (Ed.), Australian Earth Sciences Convention, Geological Society of Australia, Melbourne, 2006, p. 138.

[126] C.H. Craig, D.T. Sandwell, Global distribution of seamounts from Seasat profiles, J. Geophys. Res. 93 (1988) 10,408–10,420.

[127] A. Kitchingman, S. Lai, Inferences on potential seamount locations from mid-resolution bathymetric data, in: T. Morato, D. Pauly (Eds.), FCRR Seamounts: Biodiversity and Fisheries, University of British Columbia, Vancouver, BC, 2004, pp. 7–12. Fisheries Centre Research Reports.

[128] C. Yesson, M.R. Clark, M.L. Taylor, A.D. Rogers, The global distribution of seamounts based on 30 arc seconds bathymetry data, Deep Sea Res. I 58 (2011) 442–453.

[129] M.R. Clark, L. Watling, C. Smith, A. Rowden, J.M. Guinotte, A global seamount classification to aid the scientific design of marine protected area networks, Ocean Coast. Manag. 54 (2011) 19–36.

[130] A.D. Rogers, The biology of seamounts, Adv. Mar. Biol. 30 (1994) 305–350.

[131] A. Williams, N.J. Bax, R.J. Kloser, F. Althaus, B. Barker, G. Keith, Australia's deep-water reserve network: implications of falsehomogeneity for classifying abiotic surrogates of biodiversity, ICES J. Mar. Sci. 66 (2009) 214–224.

[132] R.G. Lueck, T.D. Mudge, Topographically induced mixing around a shallow seamount, Science 276 (1997) 1831–1833.

[133] P.A. Tyler, Conditions for the existence of life at the deep sea floor: an update, Oceanogr. Mar. Biol. Annu. Rev. 33 (1995) 221–244.

[134] N.V. Parin, A.N. Mironov, K.N. Nesis, Biology of the Nazca and Sala y Gomez submarine ridges, an outpost of the Indo-West Pacific fauna in the eastern Pacific Ocean: composition and distribution of the fauna, its communities and history, Adv. Mar. Biol. 32 (1997) 145–252.

[135] B. Richer de Forges, J.A. Koslow, G.C.B. Poore, Diversity and endemsism of the benthic seamount macrofauna in the southwest Pacific, Nature 405 (2000) 944–947.

[136] M.M. Littler, D.S. Littler, S.M. Blair, J.N. Norris, Deep-water plant communities from an uncharted seamount off San Salvador Island, Bahamas: distribution, abundance, and primary productivity, Deep Sea Res. I 33 (1986) 881–892.

[137] V. Tunnicliffe, S.K. Juniper, M.E. de Burgh, The hydrothermal vent community on Axial Seamount, Juan de Fuca Ridge, Bull. Biol. Soc. Wash. 6 (1985) 453–464.

[138] B. Bull, I. Doonan, D. Tracey, A. Hart, Diel variation in spawning orange roughy (*Hoplostethus atlanticus, Trachichthyidae*) abundance over a seamount feature on the northeast Chatham Rise, New Zeal. J. Mar. Freshwater Res. 35 (2001) 435–444.

[139] V. Agnostini, E. Escobar-Briones, I. Cresswell, K. Gjerde, D.J.A. Niewijk, A. Polacheck, et al. Global Open Oceans and Deep Sea-habitats (GOODS) bioregional classification, in: M. Vierros, I. Cresswell, E. Escobar-Briones, J. Rice, J. Ardron (Eds.), United Nations Conference of the Parties to the Convention on Biological Diversity (CBD) (2008), p. 94.

[140] J.B. Corliss, J. Dymond, L.I. Gordon, J.M. Edmond, R.P. von Herzen, R.D. Ballard, et al., Submarine thermal springs on the Galapagos Rift, Science 203 (1979) 1073–1083.

[141] K.S. Macdonald, D.S. Scheirer, S.M. Carbotte, Mid-ocean ridges: discontinuities, segments and giant cracks, Science 253 (1991) 986–994.

[142] C. Van Dover, The Ecology of Deep-Sea Hydrothermal Vents, Princeton University Press, Princeton, NJ, 2000.

[143] V. Tunnicliffe, M.R. Fowler, Influence of sea-floor spreading on the global hydrothermal vent fauna, Nature 379 (1996) 531–533.

[144] C.K. Paull, A.C. Neumann, Continental margin brine seeps: their geological consequences, Geology 15 (1987) 545–548.

[145] I.R. MacDonald, N.L. Guinasso, Jr., S.G. Ackleson, J.F. Amos, R. Duckworth, R. Sassen, et al., Natural oil slicks in the Gulf of Mexico visible from space, J. Geophys. Res. 98 (1993) 16351–16364.

[146] J.M. Brooks, M.C. Kennicutt, II, C.R. Fisher, S.A. Macko, K. Cole, J.J. Childress, et al., Deep-sea hydrocarbon seep communities: evidence for energy and nutritional carbon sources, Science 238 (1987) 1138–1142.

[147] S.B. Joye, I.R. MacDonald, J.P. Montoya, M. Peccini, Geophysical and geochemical signatures of Gulf of Mexico seafloor brines, Biogeosci. Discuss. 2 (2005) 637–671.

[148] E.E. Cordes, D.C. Bergquist, C.R. Fisher, Macro-ecology of Gulf of Mexico cold seeps, Annu. Rev. Mar. Sci. 1 (2009) 143–168.

[149] A.V. Milkov, Worldwide distribution of submarine mud volcanoes and associated gas hydrates, Mar. Geol. 167 (2000) 29–42.

[150] L.D. Kulm, E. Seuss, J.C. Moore, B. Carson, B.T. Lewis, S.D. Riter, et al., Oregon subduction zone: venting, fauna and carbonates, Science 231 (1986) 561–566.

[151] K. Olu, S. Lance, M. Sibuet, P. Henry, A. Fiala-Medioni, A. Dinet, Cold seep communities as indicators of fluid expulsion patterns through mud volcanoes seaward of the Barbados accretionary prism, Deep Sea Res. 44 (1997) 811–841.

[152] D. Van Rooij, D. Depreiter, I. Bouimetarhan, E. De Boever, K. De Rycker, A. Foubert, et al., First sighting of active fluid venting in the Gulf of Cadiz, EOS Transact. Am. Geophys. Union (Suppl. 86) (2005) 509–511.

[153] L.A. Mayer, A.N. Shore, J.E. Hughes Clarke, D.J.W. Piper, Dense biological communities at 3850 m on the Laurentian Fan and their relationship to the deposits of the 1929 Grand Banks earthquake, Deep Sea Res. 35 (1988) 1235–1246.

[154] P.R. Dando, I. Bussman, S.J. Niven, S.C.M. O'Hara, R. Schmaljohann, L.J. Taylor, A methane seep area in the Skagerrak, the habitat of the pogonophore *Siboglinum poseidoni* and the bivalve mollusc *Thyasira sarsi*, Mar. Ecol. Progr. Ser. 107 (1994) 157–167.

[155] D.L. Orange, H.G. Greene, D. Maher, D. Stakes, J.P. Barry, Widespread fluid expulsion on a translational continental margin: mud volcanoes, fault zones, headless canyons, and organic-rich substrate in Monterey Bay, California, Geol. Soc. Am. Bull. 111 (1999) 992–1009.

[156] J.E.R. Getsiv-Clemons, W.W. Wakefield, I.J. Stewart, C.E. Whitmire, Using meso-habitat information to improve abundance estimates for West Coast groundfish: a test case at Heceta Bank, Oregon, in: P.T. Harris, E.D. Baker (Eds.), Seafloor Geomorphology as Benthic Habitat: GeoHab Atlas of Seafloor Geomorphic Features and Benthic Habitats, Elsevier, Amsterdam, 2011 (Chapter 40).

[157] V.A.I. Huvenne, A.D.C. Pattenden, D.G. Masson, P.A. Tyler, Habitat heterogeneity in the Nazaré deep-sea canyon offshore Portugal, in: P.T. Harris, E.D. Baker (Eds.), Seafloor Geomorphology as Benthic Habitat: GeoHab Atlas of Seafloor Geomorphic Features and Benthic Habitats, Elsevier, Amsterdam, 2011 (Chapter 50).

[158] D.J. Wright, J.T. Roberts, D. Fenner, J.R. Smith, A.A.P. Koppers, D.F. Naar, et al., Seamounts, ridges, and reef habitats of American Samoa, in: P.T. Harris, E.D. Baker (Eds.), Seafloor Geomorphology as Benthic Habitat: GeoHab Atlas of Seafloor Geomorphic Features and Benthic Habitats, Elsevier, Amsterdam, 2011 (Chapter 58).

Part II

Case Studies

7 Phanerogam Meadows: A Characteristic Habitat of the Mediterranean Shelf—Examples from the Tyrrhenian Sea

Silvana D'Angelo, Andrea Fiorentino

Department for Soil Defense, Geological Survey of Italy–ISPRA
via Curtatone, Rome, Italy

Abstract

This study examines seagrass habitat in the Tyrrhenian Sea, focusing on the Egadi Islands. Phanerogam meadows, particularly *Posidonia oceanica* and *Cymodocea nodosa*, have been mapped using sidescan sonar, which indicates a different backscatter signature for each type.

P. oceanica represents a marine ecobiological system (biocoenosis) crucial for the Mediterranean Sea and its biota; *Posidonia* meadows accommodate an abundant epiphyte community and offer recovery to many other organisms. They originate specific sedimentary settings and constitute peculiar geomorphic features. *Posidonia* meadows are considered habitats of community interest (site of community importance (SCI)) and have priority in the protection strategies.

Key Words: Tyrrhenian, phanerogam meadow, sidescan sonar, *Posidonia*, *Cymodocea*, habitat mapping

Introduction

This study describes the occurrence and seagrass species of phanerogam meadows characteristic of the Tyrrhenian Sea continental shelf, located off the western coast of Italy (Figure 7.1). The shelf width varies from 5 to over 100 km, and the coast is mainly flat and sandy (78%), but parts are also steep and rocky (22%). Phanerogam meadows have developed in all shelf settings.

The Tyrrhenian Sea is a deep back-arc basin originated by the Alpine and Apenninic orogenesis. It is characterized by a microtidal regime, which has only a secondary influence on the coastal dynamics. Average surface salinity is about 37.5–38.5 PSU; surface temperatures span from a minimum of 13°C to a maximum of 25°C. A seasonal thermocline is present at a depth between 15 and 40 m, caused by the

Seafloor Geomorphology as Benthic Habitat. DOI: 10.1016/B978-0-12-385140-6.00007-4

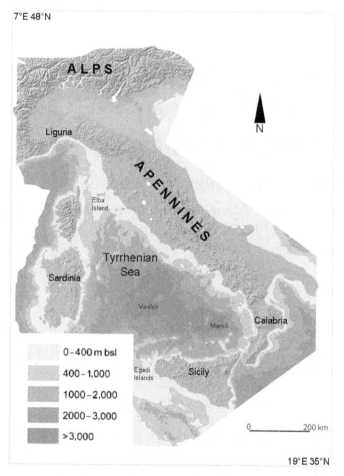

7°E 48°N

ALPS

Liguria

APENNINES

Elba
Island

Tyrrhenian
Sea

Sardinia

Vavilov

Marsili

Calabria

0–400 m bsl

400–1,000   Egadi   Sicily
                  Islands

1000–2,000

2000–3,000

>3,000   0_____200 km

                                                    19°E 35°N

Figure 7.1 Map of the seas surrounding Italy: bathymetry represented by blue tones, emerged areas by shaded relief.

summer heating of surface waters. Winter stirring of surface water causes the progressive deepening and dissipation of the thermocline. Below a depth of 100–200 m, the water temperature is almost constantly around 13–14°C [1]. The sedimentary setting of each area varies according to the geomorphology of its land counterpart.

Posidonia oceanica (L.) Delile and *Cymodocea nodosa* (Ucria) Asch. are the most common species found in the Mediterranean Sea. *P. oceanica* is endemic to the Mediterranean Sea, whereas *C. nodosa* is also present in the northeastern Atlantic Ocean. These species characterize environments ranging from 0 to 50 m and are widespread along the Italian coast. Factors affecting the health of *Posidonia* meadows include the following:

• Variation of sedimentary equilibrium (a high sedimentation rate may cover and suffocate the rhizomes; a very low sedimentation rate may be insufficient to support the development of a meadow).

Figure 7.2 Location (dots) of the main *Posidonia* meadows mapped and monitored within the Italian Ministry of Environment program (http://www.minambiente.it/).

- Increase of water turbidity due to anthropogenic factors or eutrophication.
- Anchoring and trawling.
- Competition of other plants, as is the case of *Caulerpa racemosa* (Forsskal) and *Caulerpa taxifolia* (Vahl).
- Pollution.

The Italian Ministry of Environment started a mapping and monitoring program in the 1990s of *Posidonia* meadows on a 100,000 ha surface (http://www.minambiente.it/) (Figure 7.2). *Posidonia* meadows are included among the SCI defined by the European Community "Habitat" Directive (92/43/EEC), which refers to natural habitats, because they contribute significantly to the preservation of biodiversity.

In terms of the naturalness of the study area, *Posidonia* meadows represent the climax of the shelf biocoenosis; therefore, their presence indicates pristine environmental conditions [2,3]. The *Posidonia* ecosystem is very delicate and sensitive to pollution and is thus an important biological and environmental indicator. Seagrass (*Posidonia*) monitoring is an effective method for the early identification of coastal erosion processes, and observations have shown evidence of anthropogenic disturbance of seagrass beds along several Mediterranean coasts [3]. The study area contains examples of intact *Posidonia* meadows as well as sites impacted by human activities.

Geomorphic Features and Habitats

The examples presented here come from the shelf surrounding the Egadi Islands (Figure 7.3); these islands are located west of Sicily at about 12°E and 38°N. Physical parameters are the same as for the Tyrrhenian; geomorphologically, they are surrounded by a shelf extending mainly southeastwardly down to −120 m, passing to a slope cut by canyons. The basement is constituted predominantly by dolomites,

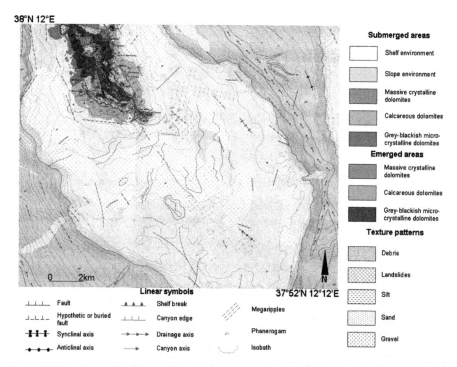

Figure 7.3 Egadi Islands (detail of the geological map of Italy); the distribution of phanerogams varies according to the substrate. Some of them are following megaripple alignment, others are bordering the rocky outcrops. (For interpretation of the references to colour in this figure legend, the reader is referred to the web version of this book.)

whereas sediments are mainly composed of biogenic carbonates. Sediment texture varies depending on local wave and current energy; sand and gravel characterize the shelf, except for a few small patches of silt, whereas the slope areas are covered exclusively by silt. In Figure 7.3, phanerogams are represented in the area with megaripples, as well as close to the coast on a rocky substrate. However, in the geological map of Italy, when the sediment cover is less than 2 m thick, the underlying substrate is directly represented. Therefore, in the rocky area, phanerogams indicate the presence of a thin sediment cover.

P. oceanica and *C. nodosa* have different geomorphic expressions and different acoustic (sonar) responses (Figure 7.4), and they produce different habitats. *P. oceanica* grows on sandy sea bottoms. The roots and rhizomes capture sediment, forming a carpet-like structure, called a "matte," which is elevated above the level of surrounding seafloor, representing a characteristic geomorphic feature of the Italian shelf. *Posidonia* meadows and *mattes* reduce coastal erosion by fixing unconsolidated seabed sediments and by decreasing the wave energy reaching the shoreface [2].

The other common phanerogam of the Mediterranean is *C. nodosa*, which is also present in the eastern Atlantic Ocean ranging from the Bay of Biscay to the coast of Senegal. It lives in the upper photic zone on low-energy sandy or muddy sea bottoms, between 5 and 20 m depth. It is a pioneer species that can tolerate anoxia and the presence of hydrogen sulfide, preparing the substrate for *P. oceanica*, or colonizing dead *P. oceanica* matte. Unlike *P. oceanica*, *C. nodosa* does not form a matte.

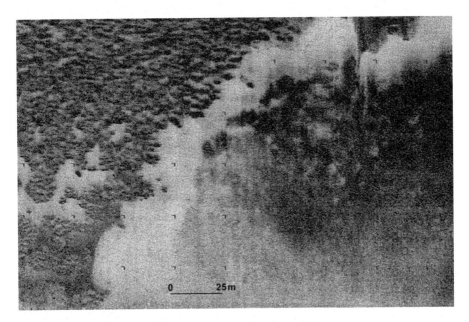

Figure 7.4 Egadi Islands: SSS 100–500 kHz image showing the different response of *Posidonia* (on the left) and *Cymodocea* (on the right) separated by a channel full of sediment.

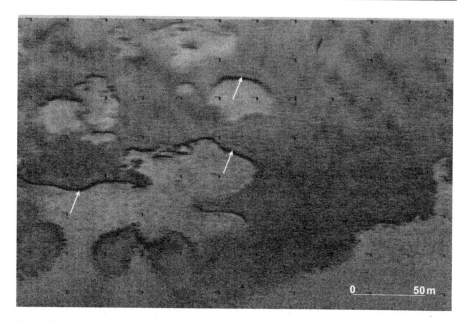

Figure 7.5 Egadi Islands: SSS 100–500 kHz image showing *Posidonia* matte. The darkest outline indicated by the arrows represents the elevation (1–1.5 m) above the surrounding sediments.

The acoustic response of seagrass beds to sidescan sonar (SSS) is affected by the anatomical structure of the plants and by the sonar frequency used (100–500 kHz, in this study). Surveying was performed within the Italian Geological Cartography Project (CARG) at a scale of 1:50,000; mapped areas are completely covered by SSS mosaics. Ground-truthing of acoustic maps was achieved by grab sampling, in which sample spacing depends upon local characteristics but is never less than one sample per square kilometer.

P. oceanica returns a strong reflection due to its leaves and complex rhizome structure; the characteristic morphological structure of *Posidonia* matte beds appear on sonographs as well-defined topographic features (Figure 7.5). Consequently, the spatial coverage of the meadows can be detected by SSS, and repeated surveys can provide spatial information on human impacts and recovery/deterioration trends [4].

Cymodocea communities appear in sonographs as dark areas with a fine and homogeneous smooth texture, with a high contrast against the soft sediment (characterized by a weak acoustic response; Figure 7.4). The extremely high acoustic response of this vegetation contrasts with its external morphology, as the plant presents spatially separated shoots of thin leaves on the surface, while the rhizomes and root system remain buried between 5 and 40 cm depth (see section "Biological Communities). What can be observed in a sonograph is due to the presence of air canals in the very dense rhizome and root system [5].

The echo level produced by a submerged object is a consequence of its size and the relationship between its acoustic impedance and that of the water [6]. Thus, while a rock has acoustic impedance 10 times that of water, an air pocket has a value

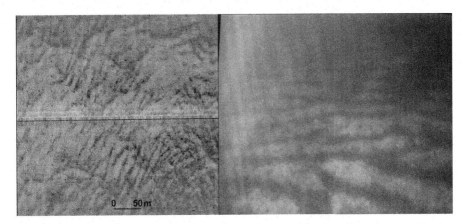

Figure 7.6 Egadi Islands: *Posidonia* accumulating in alignments along megaripples (on the left, sonograph; on the right, photograph by ROV).

in the order of 4,000 times, meaning that it will produce a much stronger echo than an equally sized rock. Therefore, aeriferous canals found in the leaves, rhizomes, and the root system—which grow in a very dense mesh, forming a carpet just below the sediment surface—explain the strong acoustic response [7].

P. oceanica grows on the seafloor, following different patterns according to the local geomorphology. It can colonize extended areas along the coast where the submerged beach forms a regular wide and flat bottom. In other cases, it follows the irregular outline of a rocky substrate growing in the troughs filled by sediments. Finally, following differential sediment accumulation and erosion, it can create alignments, as happens in association with megaripples (Figure 7.6) or irregular-shaped patches, in the case of meadows affected by human impacts (Figure 7.7).

In summary, phanerogam meadows are associated with particular sedimentary settings and give rise to specific geomorphic features (matte mounds) that can be detected during SSS surveying and are consequently represented on geological maps. Furthermore, geological mapping places phanerogam habitat within a geomorphological context (Figure 7.3).

Biological Communities

Marine phanerogams, as with any angiosperm, are constituted by roots, stems, leaves, flowers, and fruits. They have vertical rhizomes attached to horizontal rhizomes. In *P. oceanica*, rhizome internodes are short (0.5–2 mm); roots are branched, attaching the plant to the substratum. Leaf bundles consisting of 5 to 10 leaves are attached to the vertical rhizomes. The leaves are broad (5–12 mm) and the length usually varies from 20 to 40 cm, but may be up to 1 m. *C. nodosa* develops mainly horizontally, with more spaced internodes and buried rhizomes.

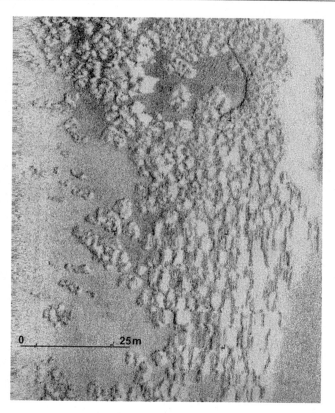

Figure 7.7 Egadi Islands: SSS image of an impacted *Posidonia* meadow forming scattered patches.

P. oceanica grows a few centimeters per year, can tolerate a temperature range of 10–28°C, is sensitive to salinity variations, and is less common in brackish waters. It lives in shallow water depths to a maximum depth of 30–50 m, depending on turbidity and light penetration. When sediment supply is scarce, rhizomes develop horizontally; when sediment input increases, rhizomes grow vertically [2].

The anatomical structure of phanerogams is the effect of their adaptation to the marine environment. It is characterized by a network of air channels in leaves, rhizomes, and roots. Differently from other marine plants, phanerogams store gases within their tissues. A network of lacunae allows transport of sufficient oxygen from leaves to roots in order to maintain aerobic respiration and to guarantee oxygenated conditions among the roots, even in anoxic substrates.

Posidonia meadows play a fundamental role for Mediterranean coastal environments affecting benthic communities. *Posidonia* is one of the main oxygen (*ca.* 14 l/m²/year) and organic matter (*ca.* 20 tons/ha/year) primary producers; it forms an essential habitat for several fishes, mollusks, echinoderms, and crustaceans, providing an important site for their reproduction and a source of their nutrition. Many epiphyte taxa live on rhizomes, others on the leaves. Approximately 30% of a meadow's organic matter production is exported to other ecosystems (http://www.minambiente.it/).

Organisms living on leaves include diatoms, bacteria, and red and brown algae. These organisms are predated by mollusks, gastropods, crustaceans, amphipods, and polychaetes [8]. About 70% of the total living population of a typical meadow is represented by herbivores, mainly echinoderms [9–11]. The infaunal community is constituted mainly of detritivores, living within the matte.

Remnants of the leaves, transported to the supralittoral plain by waves, offer shelter and food to insects, amphipods, and isopods.

P. oceanica is suffering from the competition of the two tropical species, *C. taxifolia* and *C. racemosa*, which penetrated into the Mediterranean between the 1950s and the 1980s. They adapted to the new environment, negatively affecting *Posidonia* meadows, especially *C. taxifolia*, which produces toxic secondary metabolites, inducing regression and necrotization [12,13].

Surrogacy

No statistical analyses have been carried out on this data set to examine relationships between physical surrogates and benthos.

Acknowledgments

We want to thank Profs R. Catalano and M. Agate, from the University of Palermo, for their cooperation in the production of the geological map of the Egadi Islands. We are grateful to all reviewers who helped us improve our chapter.

References

[1] E. Martorelli, S. D'Angelo, A. Fiorentino, F.L. Chiocci, Non-tropical carbonate shelf sedimentation. The Archipelago Pontino (Central Italy) case history, this volume pp. 441–447.

[2] R. Molinier, J. Picard, Récherches sur les herbiers de phanerogames marines du littoral méditerranéen francais, Ann. Inst. Océanogr. Paris 27 (3) (1952) 127–324.

[3] G. Pergent, C. Pergent-Martini, Ch.F. Boudouresque, Utilisation de l'herbier à *Posidonia oceanica* comme indicateur biologique de la qualité du milieu littoral en Méditerranée: état des connaissances, Mésogée 54 (1995) 3–29.

[4] P.S. Ribed, Submarine vegetation studies using remote sensing: importance in beach management strategies, J. Environ. Protect. Ecol. 3 (4) (2002) 953–958.

[5] P. Siljeström, J. Rey, A. Moreno, Characterization of phanerogam communities (*Posidonia oceanica* and *Cymodocea nodosa*) using side-scan-sonar images, J. Photogram. Remote Sens. 51 (6) (1996) 308–315.

[6] C.S. Clay, H. Medwin, Acoustical Oceanography: Principles and Applications, John Wiley & Sons, New York, NY, 1977, p. 544.

[7] P. Siljeström, A. Moreno, R. Carbó, J. Rey, J. Cara, Selectivity in the acoustic response of *Cymodocea nodosa* (Ucria) Ascherson, Int. J. Rem. Sens. 23 (14) (2002) 2869–2876.

[8] M.C. Gambi, M.C. Buia, E. Casola, M. Scardi, Estimates of water movement in *Posidonia oceanica* beds: first approach, GIS Posidonie Publ 2 (1989) 101–112.

[9] J.A. Ott, Growth and production in *Posidonia oceanica* (L.) Delile, Mar. Ecol. 1 (1980) 47–64.

[10] C. Pergent-Martini, V. Rico-Raimondino, G. Pergent, Primary production of *Posidonia oceanica* in the Mediterranean Basin, Mar. Biol. 120 (1994) 9–15.

[11] C.F. Boudouresque, A. Meinesz, Découverte de l'herbier de Posidonies, Cahier Parc Nat. Port-Cros 4 (1982) 1–79.

[12] A. Meinesz, J. de Vaugelas, B. Hesse, X. Mari, Spread of the introduced tropical green alga *Caulerpa taxifolia* in northern Mediterranean waters, J. Appl. Phycol. 5 (2) (1993) 141–147 (Springer, The Netherlands).

[13] E. Ballesteros, O. Delgado, E. Gacia, C. Rodríguez-Prieto, Lack of severe nutrient limitation in *Caulerpa taxifolia* (Vahl) C. Agardh, an introduced seaweed spreading over the oligotrophic northwestern Mediterranean, Bot. Mar. 39 (1–6) (1996) 61–68.

8 Predictive Modeling of Dominant Macroalgae Abundance on Temperate Island Shelves (Azores, Northeast Atlantic)

Fernando Tempera[1,2], Monique MacKenzie[3], Igor Bashmachnikov[1,4], Marji Puotinen[5], Ricardo S. Santos[1], Richard Bates[2]

[1]Departamento de Oceanografia e Pescas, Universidade dos Açores, Rua Prof. Dr. Frederico Machado, 4, Horta, Açores, Portugal, [2]School of Geography and Geosciences, University of St. Andrews, St. Andrews, Fife, Scotland, UK, [3]Centre for Research into Ecological and Environmental Modeling (CREEM), University of St. Andrews, The Observatory, St. Andrews, Fife , Scotland, UK, [4]Instituto de Oceanografia, Faculdade de Ciências da Universidade de Lisboa, Campo Grande, Lisboa, Portugal, [5]School of Earth and Environmental Sciences, University of Wollongong, Wollongong, NSW, Australia

Abstract

Volcanic oceanic islands typically rise steeply from the ocean floor and are surrounded by narrow shelves produced by swell erosion on the islands' flanks. This study focuses on mapping the distribution of six macroalgae that dominate infralittoral on-shelf hard substrate biotopes around the island of Faial (Azores, northeast Atlantic): articulated Corallinaceae, *Codium elisabethae*, *Dictyota* spp., *Halopteris filicina*, *Padina pavonica*, and *Zonaria tournefortii*. Semiquantitative data on their abundance, collected by SCUBA diving, ROV, and drop-down camera surveys, are intersected with a series of gemorphological and oceanographical explanatory variables collated from various sources that include multibeam surveys, satellite imagery, oceanographic modeling, and GIS analysis. Ordered logistic regression models are used to find the combinations of major environmental variables that best explain the abundance variations observed. The predictive distribution maps obtained for the six macroalgae are combined to produce the first predictive map of macroalgal facies on an island shelf in the Azores. Depthwise general and sectoral macroalgal zonation are also presented.

Key Words: Macroalgae, temperate rocky reefs, ordered logistic regression, facies, predictive map, zonation, Faial Island, Azores

Seafloor Geomorphology as Benthic Habitat. DOI: 10.1016/B978-0-12-385140-6.00008-6

Introduction

The Azores is a volcanic archipelago located in the northeast Atlantic approximately 1,600 km westward of Portugal's mainland coast. Its islands are typically volcanic and rise steeply from the ocean floor, which in the region is on average located at *ca.* 2,000 m depth. All the Azores islands typically evolved from seamounts that eventually broke through the sea surface and experienced the shelf-carving processes induced by swells [1].

Faial and Pico, two neighboring volcanic islands located east of the Mid-Atlantic Ridge, are estimated to have emerged during the Pleistocene (800 kyr and 270 kyr BP, respectively). They are united by an 8-km-wide shelf and are elsewhere surrounded by narrow shelves ranging between approximately 4 km and 140 m wide and exhibiting depths at shelf edge between 14 and 183 m depth [2].

Despite their limited extension, the studied island shelves encompass a wide range of environmental conditions. West- and north-facing sectors are typically exposed to the full force of prevailing swell waves, while small embayments on the southern and eastern shores experience much more sheltered conditions [3]. In the passage separating the two islands, tides are an additional hydrodynamic force of considerable importance due to vertical and horizontal current funneling [4].

Owing to the volcanic constitution of the islands, hard substrates are major components of the island shelves. Since the macroalgae recorded for the Azores archipelago develop primarily on consolidated surfaces, this seafloor type generally contains the algal-dominated biotopes, besides providing an abundance of niches for other epibenthic species. In turn, this attracts benthic feeders like many of the species targeted by commercial and recreational fisheries, making hard substrates of great significance to both biodiversity conservation and fishery resource management. In view of this, the EU Habitats and Species Directive [5] classifies *reefs* (code 1170 in Annex I) as a marine habitat requiring designation of Special Areas of Conservation (SACs).

This case study concentrates on the shelves surrounding Faial Island and western Pico Island (Figure 8.1), which hold important temperate reefs in good conservation status. This fact is attested by the multiple designations received under (i) the European Natura 2000 network (5 SACs), (ii) the Azorean system of protected areas (two nature parks), and (iii) the OSPAR network of marine-protected areas (MPAs).

The seafloor of the area has been mapped in high resolution using swath sonars [6], providing an opportunity to infer the distribution of benthic species associated with the shelf habitats. In this perspective, semiquantitative data on infralittoral macroalgae living on hard substrates were brought together from a number of biological surveys and related to geomorphological and oceanographic variables (namely, depth, slope, swell exposure, maximum tidal currents, sea surface temperature, and chlorophyll-*a* concentration). Ordered logistic regression models were used to find the combinations of major environmental variables that best explain the abundance variations observed and to map the distribution of the selected macroalgae. This provided the basis for producing the first predictive map of macroalgal facies on Azores island shelves and assessing general and sectoral depth-wise infralittoral zonations.

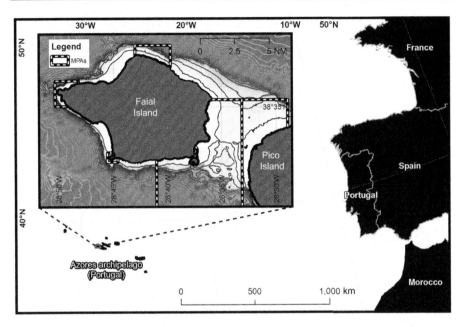

Figure 8.1 Location of Faial Island and neighboring passage to Pico Island. The study area is demarcated in the inset as a white area. Designated MPAs are delimited by the dashed lines.

Seafloor Surveys

The first swath survey of the study area was conducted in the autumn of 2003 using a Reson 8160 beamforming multibeam mounted on RV *Arquipélago*. A complementary survey was carried out in the summer of 2004 to cover the nearshore gaps left by the previous survey, using a Submetrix 2000 phase-measuring sonar installed on RL *Águas Vivas*. A total of 1,227 km of survey tracks were completed around the island of Faial and neighboring passage to Pico, corresponding to a surveyed area of approximately 537.5 km². Apart from bathymetry post-processing, backscatter from both surveys was also mosaiced and its patterns were used in interpreting seafloor nature and texture aspects that were not apparent in the bathymetric image (see [2] for details).

Overall, a total perimeter of 84 km was studied using seafloor digital terrain models of 5 m resolution and backscatter mosaics at 1 m resolution. Delineation of substrate nature was conducted in a GIS environment using bathymetry texture and tonal and textural properties of the backscatter, together with seafloor nature ground-truthing information provided by video imagery, grab samples, and historical nautical chart point data. Seafloor video imagery was collected using a VideoRay Explorer ROV (summer 2004) and a Tritech MD4000 drop-down camera (summer 2005). A minimum of two locations representative of each physiographic/acoustic signature was sampled. In addition, geo-referenced data from 50 locations sampled using a Duncan box corer and a Van Veen grab provided validation and established the nature of the seafloor and grain size of the sediments mapped acoustically down to a maximum depth of 80 m [1]. Publicly available hydrographical charts—despite being based on an armed sounding lead survey conducted as long ago as the 1940s—provided additional information on surficial bottom nature, particularly near the coast and around harbor entrances (1,612 sites sampled).

Biological Surveys

Since the belt dominated by frondose algae around the Azores islands frequently extends to depths beyond access by compressed air SCUBA diving, carrying out visual sampling along the full extent of the infralittoral horizon required not only SCUBA divers working down to 40 m depth [7] but also remote video-based platforms, like ROVs [8], and a drop-down camera [9] that was operated down to the shelf edge.

The sampling was conducted using a visual survey method shown to produce valid semiquantitative data for conspicuous benthic species [10–12]. Observations were collected along depth-transects oriented approximately perpendicular to the shore, except in ROV deployments, where only point locations could be surveyed. Along each transect, observers annotated the presence and abundance (by visually estimating the counts per area or percentage cover) of conspicuous sessile or sedentary species within depth-delimited biological facies that presented a homogeneous appearance. Abundance estimates were recorded using the Marine Nature Conservation Review percentage cover/density categories, also denoted SACFOR [11]. Only the data pertaining to horizontal to subhorizontal rocky surfaces (0–45° inclination) were of interest for the analysis, as they contained the most conspicuous biologic growth for which comparable abundance estimates could be derived from the different sampling techniques used. Comparability was further facilitated by the fact that the species analyzed in this study are all easily identifiable *in situ* and are conspicuous videographically.

Sampling stations were distributed across the study area to guarantee a broad coverage of the range of environmental conditions. In order to do this, the study area was first divided in sectors that were more or less homogeneous in their degree of exposure (Figure 8.2A). Within each of these zones, a minimum of three sampling stations were then randomly located that covered the full depth range of macroalgae distribution.

In total, 85 SCUBA dives, 21 ROV deployments, and 11 drop-down camera deployments executed between 1999 and 2005 are included in the analyzed data set (Figure 8.2A).

Geomorphic Features and Habitats

The shelf seafloor revealed complex tectonic, volcanic, and erosional processes that are detailed in Ref. [2].

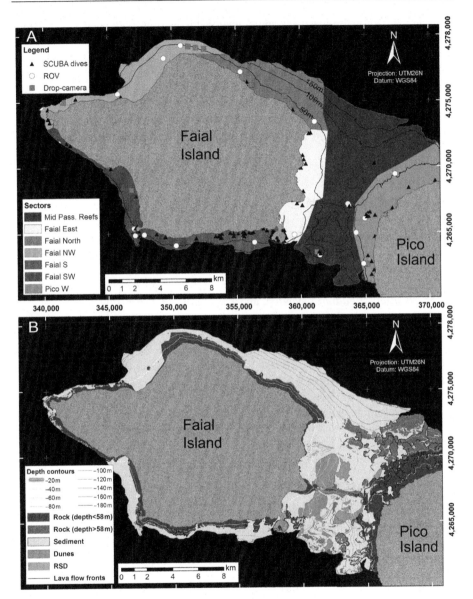

Figure 8.2 (A) Sectors used in distributing sampling effort and location of the SCUBA dives, ROV, and drop-down camera deployments. (B) Distribution of on-shelf geomorphic features on hard substrates and unconsolidated sediments.

Sediments represented 67% of the shelf area and were prevailingly found on the wider shelf sectors (Figure 8.2B). Large sediment wave fields and rippled scour depressions were the main geomorphic features of note, particularly in the inter-island passage. Variations of backscatter amplitude were observed on this type of substrate, but radiometric artifacts and gradational transitions precluded a systematic interpretation of grain sizes (Figure 8.3).

The remaining 33% of the shelf areas were interpreted as hard substrates (Figure 8.2B), mostly comprised of sloping boulder fields adjacent to tall subaerial cliffs, lava flows, and submerged volcanic cones and extrusion centers, as well as underwater cliffs associated to faults and palaeoshorelines (Figure 8.3). Backscatter texture was insufficiently clear to consistently inform on the type of rocky seafloor; therefore, seafloor was classified into two general classes: unconsolidated sediments and hard substrates.

Biological Communities

The statistical modeling of biological distributions focused on six *taxa* belonging to the three groups of macroalgae dominating infralittoral hard substrates in the Azores: calcareous red algae, brown algae and, to a lesser extent, green algae. The species chosen were articulated Corallinaceae, *Codium elisabethae*, *Dictyota* spp., *Halopteris filicina*, *Padina pavonica*, and *Zonaria tournefortii* (Figure 8.4A–H).

Response Variable

For modeling purposes, the ordered SACFOR categories were assigned ranked numerical values of 6 (superabundant) to 1 (rare) according to their natural order from high to low. Zero (0) was used to numerically represent the absence of the species. The resulting response variable therefore comprises seven mutually exclusive, collectively exhaustive, and inherently ranked outcomes as supposed for ordered categorical variables [13].

Explanatory Variables

Potential explanatory variables were collated from multibeam surveys (seafloor nature, depth, slope), satellite imagery (SST and chlorophyll-*a* concentration), a GIS-based fetch analysis (swell exposure), and an oceanographic model (maximum tidal currents). These variables are described in Table A (supplementary material) and their fields are shown in Fig. A (supplementary material) at their original resolution.

A full description of the methods used to obtain each covariate field can be found in Ref. [2].

Data Intersection and Visualization

ArcGIS® ESRI was used to (i) integrate the geo-referenced vector and raster information; (ii) overlay and visualize sampling locations and environmental variable maps; (iii) extract matrices for statistical analysis by spatially intersecting the

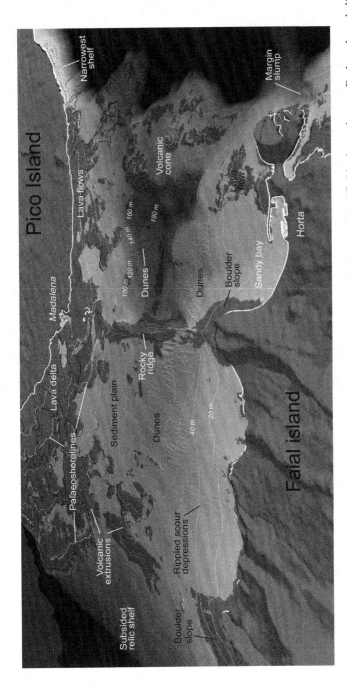

Figure 8.3 Perspective of the Faial–Pico passage highlighting examples of the geomorphic features identified in the study area. Darkened on-shelf patches represent hard substrates. Vertical exaggeration is 2×.

Figure 8.4 Aspect of rocky surfaces covered by the different species analyzed. (A) Articulated Corallinaceae turf. (B) *Codium elisabethae* plants. (C) Climatic population of *Codium elisabethae*. (D) Rocky surface dominated by *Dictyota dichotoma*. (E) *Halopteris filicina*. (F) Climatic population of *Padina pavonica*. (G) Well-developed *Zonaria tournefortii* fronds. (H) Climatic population of *Zonaria tournefortii* on deep infralittoral boulders.

biologic data with the colocated values of the environmental variables; and (iv) produce predictive and composite maps.

Model Fitting

The ordered categorical nature of the response variable determined that ordered logistical methods [14] were most appropriate to investigate the relationship between the suite of colocated environmental parameters (explanatory variables) and the raw abundance (response variable). The STATA™ (v.10 from StataCorp LP) *ologit* module was used to fit the models, which estimate the parameters of the logistic regression using a maximum-likelihood approach.

The number of observations per species varied between 926 and 1,020. In order to tackle spatial autocorrelation of observations performed in neighboring sites, where major environmental variables were supposedly very similar, *clusters* (between 104 and 113, depending on the species) were defined to aggregate observations. The correlation of the observations within clusters was computed using Generalized Estimating Equations (robust clustering option in STATA™). To accommodate nonlinearities between the response variable and the explanatory variables on the scale of the link function (an assumption of the modeling framework used), multivariable fractional polynomials transformations were applied to the explanatory variables ([15]; *mfp* command in STATA™).

The final model for each species was obtained by discarding the *mfp*-transformed explanatory variables that showed nonsignificant coefficients at the 5% level (p-value >0.05). When this nonsupervised backward selection procedure resulted in only one statistically significant explanatory variable (excessive parsimony), alternative models with supervised combinations of more than one explanatory variable were tested, aiming at obtaining a model providing a higher goodness-of-fit (as measured by the MacFadden pseudo-R^2 [16]).

Surrogacy

A summary of the combination of covariates found to be statistically significant for each species is presented in Table 8.1, including MacFadden pseudo-R^2 goodness-of-fit values for the final model. The results show that the geomorphological covariates *depth* (<58 m) and *slope* were consistently significant in all but one case. Water-column covariates such as *exposure to swell*, *exposure to currents*, and chlorophyll-*a* were significant for part of the species, and SST was nonsignificant for all species. Goodness-of-fit levels can be classified as reasonably good for articulated Corallinaceae and *Codium elisabethae* and as intermediate for *Halopteris filicina*, *Padina pavonica*, and *Zonaria tournefortii*. Unsatisfactory fit levels were obtained for *Dictyota* spp. (for thresholds see Refs. [17,18]).

A detailed description of model parametrizations and spatialization approach is provided in Ref. [2].

Table 8.1 Summary of Final Model Configurations (i.e., Keeping Variables with Coefficients Significantly Different from 0 at $p > 0.05$) and Fit Measures

Species	Depth	Slope	Exposure to Swell	Exposure to Current	SST	Chlorophyll-*a*	MacFadden Pseudo-R^2
Articulated Corallinaceae (Rhodophyta, Corallinales)	X_{mfp}	X_{mfp}	X_{mfp}	X_{mfp}		X_{mfp}	0.2202
Codium elisabethae (Chlorophyta, Bryopsidales)	X_{mfp}	X_{mfp}		X_{mfp}			0.1875
Dictyota spp. (Ochrophyta, Dictyotales)	X_{mfp}	X_{mfp}					0.0755
Halopteris filicina (Ochrophyta, Sphacelariales)	X_{mfp}	X_{mfp}	X_{mfp}				0.1537
Padina pavonica (Ochrophyta, Dictyotales)	X_{mfp}			X_{mfp}		X_{mfp}	0.1177
Zonaria tournefortii (Ochrophyta, Dictyotales)	X_{mfp}	X_{mfp}					0.1115

X_{mfp} indicates variable integration in the final model after *multivariable fractional polynomial* transformation.

Benthic Communities

A map of the spatial distribution of the different *facies* in the study area is presented in Figure 8.5. In pixels where more than one species held the same maximum score, codominance was assigned and multispecific classes were composed.

Species succession along depth-wise profiles is summarized in Table 8.2 for the general study area and for the original sectors employed in allocating minimum sampling effort (Figure 8.2A). A general succession emerges that is dominated by (i) articulated Corallinacea in the shallowest strata down to 15 m depth, followed by (ii) *Zonaria tournefortii*, prevailing in the depth strata between 17 and 47 m, and (iii) *Halopteris filicina* taking over in the deepest strata between 48 and 57 m depth. In a comparative sectoral analysis, the major differences to highlight are (i) the occurrence of a *Dictyota* spp. belt on the NW, N, and E sectors and (ii) the distinct succession of species on the mid-passage reefs. In the latter, *Padina pavonica* is dominant or codominant in the depth strata down to 22 m, and *Halopteris filicina* occupies most strata below 22 m depth, noticeably taking over ecological space that is typically occupied by *Z. tournefortii*. The reemergence of *Dictyota* spp. observed in the deepest strata is interpreted as a modeling artifact induced by the poor performance of the model fit for this species.

Evidence collected by previous studies focusing on assemblage structure and definition of algae-based biotopes for the region [19,20] suggests that the selected

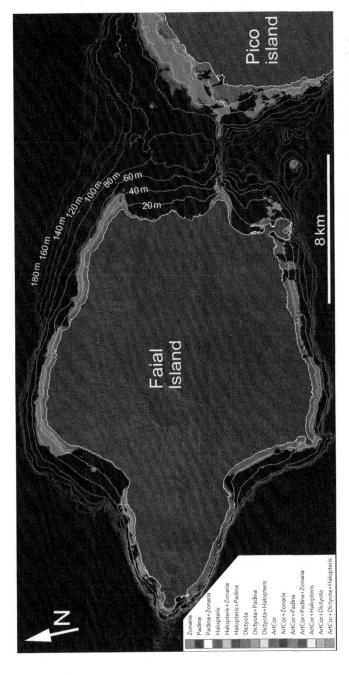

Figure 8.5 Faial Island shelf and neighboring passage to Pico with the modeled distribution of the different rocky infralittoral algal facies.

Table 8.2 General and Sectoral Zonation of the Dominant Macroalgae around Faial and Western Pico Islands

	Faial					Passage	Pico	General
	North	East	South	SW	NW	Reefs	West	All sectors
0m	Art. Cor.	Art. Cor.	Art. Cor.	Art. Cor.			Art. Cor.	Art. Cor.
	Art. Cor.	Art. Cor.	Art. Cor.	Art. Cor.	Art. Cor.		Art. Cor.	Art. Cor.
	Art. Cor.	Art. Cor.	Art. Cor.	Art. Cor.	Art. Cor.		Art. Cor.	Art. Cor.
	Art. Cor.	Art. Cor.	Art. Cor.	Art. Cor.	Art. Cor.		Art. Cor.	Art. Cor.
	Art. Cor.	Art. Cor.	Art. Cor.	Art. Cor.	Art. Cor.		Art. Cor.	Art. Cor.
5m	Art. Cor.	Art. Cor.	Art. Cor.	Art. Cor.	Art. Cor.		Art. Cor.	Art. Cor.
	Art. Cor.	Art. Cor.	Art. Cor.	Art. Cor.	Art. Cor. + Dict. spp.		Art. Cor.	Art. Cor.
	Art. Cor.	Art. Cor. + Dict. spp.	Art. Cor.	Art. Cor.	Art. Cor. + Dict. spp.	Art. Cor. + Pad. pavo.	Art. Cor.	Art. Cor.
	Art. Cor.	Art. Cor. + Pad. spp.	Art. Cor.	Art. Cor.	Art. Cor. + Dict. spp.	Art. Cor. + Pad. pavo.	Art. Cor.	Art. Cor.
	Art. Cor.	Art. Cor. + Dict. spp.	Art. Cor.	Art. Cor.	Art. Cor. + Pad. spp.	Pad. pavo.	Art. Cor.	Art. Cor.
10m	Art. Cor.	Art. Cor. + Pad. spp.	Art. Cor.	Art. Cor.	Art. Cor. + Dict. spp.	Pad. pavo.	Art. Cor.	Art. Cor.
	Art. Cor.	Art. Cor. + Pad. spp.	Art. Cor.	Art. Cor.	Art. Cor. + Dict. spp.	Pad. pavo.	Art. Cor.	Art. Cor.
	Art. Cor.	Art. Cor. + Dict. spp.	Art. Cor.	Art. Cor.	Art. Cor. + Dict. spp.	Pad. pavo.	Art. Cor.	Art. Cor.
	Art. Cor.	Art. Cor. + Dict. spp.	Art. Cor.	Art. Cor.	Art. Cor. + Dict. spp.	Pad. pavo.	Art. Cor.	Art. Cor.
15m	Art. Cor. + Dict. spp.	Hal. fili.	Art. Cor.	Art. Cor.	Art. Cor. + Dict. spp.	Pad. pavo.	Art. Cor.	Art. Cor. + Dict. spp.
	Art. Cor. + Dict. spp.	Hal. fili.	Art. Cor. + Dict. spp.	Zon. tourn.	Art. Cor. + Dict. spp.	Pad. pavo.	Art. Cor.	Art. Cor. + Dict. spp.
	Art. Cor. + Dict. spp.	Hal. fili.	Hal. fili.	Zon. tourn.	Zon. tourn.	Pad. pavo.	Zon. tourn.	Zon. tourn.
	Zon. tourn.	Zon. tourn.	Zon. tourn.	Zon. tourn.	Zon. tourn.	Pad. pavo.	Zon. tourn.	Zon. tourn.
	Zon. tourn.	Zon. tourn.	Zon. tourn.	Zon. tourn.	Zon. tourn.	Pad. pavo.	Zon. tourn.	Zon. tourn.
20m	Zon. tourn.	Zon. tourn.	Zon. tourn.	Zon. tourn.	Zon. tourn.	Pad. pavo.	Zon. tourn.	Zon. tourn.
	Zon. tourn.	Zon. tourn.	Zon. tourn.	Zon. tourn.	Zon. tourn.	Pad. pavo.	Zon. tourn.	Zon. tourn.
	Zon. tourn.	Zon. tourn.	Zon. tourn.	Zon. tourn.	Zon. tourn.	Hal. fili.	Zon. tourn.	Zon. tourn.
	Zon. tourn.	Zon. tourn.	Zon. tourn.	Zon. tourn.	Zon. tourn.	Hal. fili.	Zon. tourn.	Zon. tourn.
25m	Zon. tourn.	Zon. tourn.	Zon. tourn.	Zon. tourn.	Zon. tourn.	Hal. fili.	Zon. tourn.	Zon. tourn.
	Zon. tourn.	Zon. tourn.	Zon. tourn.	Zon. tourn.	Zon. tourn.	Hal. fili.	Zon. tourn.	Zon. tourn.
	Zon. tourn.	Zon. tourn.	Zon. tourn.	Zon. tourn.	Zon. tourn.	Hal. fili.	Zon. tourn.	Zon. tourn.
	Zon. tourn.	Zon. tourn.	Zon. tourn.	Zon. tourn.	Zon. tourn.	Hal. fili.	Zon. tourn.	Zon. tourn.
30m	Zon. tourn.	Zon. tourn.	Zon. tourn.	Zon. tourn.	Zon. tourn.	Zon. tourn.	Zon. tourn.	Zon. tourn.
	Zon. tourn.	Zon. tourn.	Zon. tourn.	Zon. tourn.	Zon. tourn.	Zon. tourn.	Zon. tourn.	Zon. tourn.
	Zon. tourn.	Hal. fili. + Zon. tourn.	Zon. tourn.	Zon. tourn.	Zon. tourn.	Zon. tourn.	Zon. tourn.	Zon. tourn.
	Zon. tourn.	Hal. fili. + Zon. tourn.	Zon. tourn.	Zon. tourn.	Zon. tourn.	Zon. tourn.	Zon. tourn.	Zon. tourn.
	Zon. tourn.	Hal. fili. + Zon. tourn.	Zon. tourn.	Zon. tourn.	Zon. tourn.	Zon. tourn.	Zon. tourn.	Zon. tourn.
35m	Zon. tourn.	Hal. fili. + Zon. tourn.	Zon. tourn.	Zon. tourn.	Zon. tourn.	Hal. fili.	Zon. tourn.	Zon. tourn.
	Zon. tourn.	Hal. fili. + Zon. tourn.	Zon. tourn.	Zon. tourn.	Zon. tourn.	Zon. tourn.	Zon. tourn.	Zon. tourn.
	Zon. tourn.	Hal. fili. + Zon. tourn.	Zon. tourn.	Zon. tourn.	Zon. tourn.	Hal. fili.	Zon. tourn.	Zon. tourn.
	Zon. tourn.	Hal. fili. + Zon. tourn.	Zon. tourn.	Zon. tourn.	Zon. tourn.	Hal. fili.	Zon. tourn.	Zon. tourn.
40m	Zon. tourn.	Hal. fili. + Zon. tourn.	Zon. tourn.	Zon. tourn.	Zon. tourn.	Hal. fili.	Zon. tourn.	Zon. tourn.
	Zon. tourn.	Zon. tourn.	Zon. tourn.	Zon. tourn.	Zon. tourn.	Hal. fili.	Zon. tourn.	Zon. tourn.
	Zon. tourn.	Zon. tourn.	Zon. tourn.	Zon. tourn.	Zon. tourn.	Hal. fili.	Zon. tourn.	Zon. tourn.
	Zon. tourn.	Zon. tourn.	Zon. tourn.	Zon. tourn.	Zon. tourn.	Hal. fili.	Zon. tourn.	Zon. tourn.
45m	Zon. tourn.	Zon. tourn.	Zon. tourn.	Zon. tourn.	Zon. tourn.	Hal. fili.	Zon. tourn.	Zon. tourn.
	Hal. fili.	Zon. tourn.	Zon. tourn.	Hal. fili.	Hal. fili.	Hal. fili.	Zon. tourn.	Zon. tourn.
	Hal. fili.	Hal. fili.	Hal. fili.	Hal. fili.	Hal. fili.	Hal. fili.	Hal. fili.	Hal. fili.
	Hal. fili.	Hal. fili.	Hal. fili.	Hal. fili.	Hal. fili.	Hal. fili.	Hal. fili.	Hal. fili.
50m	Hal. fili.	Hal. fili.	Hal. fili.	Hal. fili.	Hal. fili.	Hal. fili.	Hal. fili.	Hal. fili.
	Hal. fili.	Hal. fili.	Hal. fili.	Hal. fili.	Hal. fili.	Hal. fili.	Hal. fili.	Hal. fili.
	Hal. fili.	Hal. fili.	Hal. fili.	Hal. fili.	Hal. fili.	Hal. fili.	Hal. fili.	Hal. fili.
	Hal. fili.	Hal. fili.	Hal. fili.	Hal. fili.	Hal. fili.	Hal. fili.	Hal. fili.	Hal. fili.
55m	Hal. fili.	Hal. fili.	Hal. fili.	Hal. fili.		Hal. fili.	Hal. fili.	Hal. fili.
	Hal. fili.	Hal. fili.	Hal. fili.	Dict. spp.	Dict. spp.	Hal. fili.	Hal. fili.	Hal. fili.
	Dict. spp.	Hal. fili.	Dict. spp.	Dict. spp.	Dict. spp.	Dict. spp.		Dict. spp.

Legend:

- Art. Cor. — Art. Corallinacea
- Art. Cor. + Dict. spp. — Art. Corallinacea + *Dictyota* spp.
- Art. Cor. + Hal. fili. — Art. Corallinacea + *Halopteris filicina*
- Art. Cor. + Pad. pavo. — Art. Corallinacea + *Padina pavonica*
- Dict. spp. — *Dictyota* spp.
- Hal. fili. — *Halopteris filicina*
- Hal. fili. + Zon. tourn. — *H. filicina* + *Zonaria tournefortii*
- Pad. pavo. — *Padina pavonica*
- Zon. tourn. — *Zonaria tournefortii*
- No data

species are major components of the algal biotopes of the Azores. The results show that statistical models can be developed from existing biological and environmental data sets that explain the spatial variability of most of these dominant macroalgae at an island scale and can be used to produce maps of biological facies distribution. As marine spatial planning is implemented in the Azores, such products will be increasingly required to inform decisions regarding the conservation of both biodiversity and essential habitats for species of commercial importance.

Acknowledgments

IMAR-DOP/UAz is Research and Development Unit No. 531 and LARSyS-Associated Laboratory No. 9 funded by the Portuguese Foundation for Science and Technology (FCT) through pluriannual and programmatic funding schemes (OE, FEDER, POCI2001, FSE) and by the Azores Directorate for Science and Technology (DRCT). This study was carried out while FT was financed by a Ph.D. grant from the Fundação para a Ciência e a Tecnologia (ref: SFRH/BD/12885/2003). The field research was funded by the projects MARÉ (Life-Nature B4-3200/98/509), MAROV—(PDCTM/P/MAR/15249/1999), OGAMP (INTERREG IIIb—MAC/4.2/A2 2001), MARMAC (INTERREG IIIb—03/MAC/4.2/A2 2004), MAYA (AdI/POSI/2003), GEMAS (DROTRH contract), and MARINOVA (INTERREG IIIb—MAC/4.2/M11). The writing of this paper was further supported by project MeshAtlantic (AA-10/1218525/BF). Acknowledgments are also due to DOP/UAç divers and vessel crews for all the hard work at sea and to Dr Jack Jarvis (University of St. Andrews) for his support with setting up the TELEMAC current model.

References

[1] R. Quartau, A.S. Trenhaile, N.C. Mitchell, F. Tempera, Development of volcanic insular shelves: insights from observations and modelling of Faial Island in the Azores Archipelago, Mar. Geol. 275 (2010) 66–83.

[2] F. Tempera, Benthic habitats of the extended Faial Island Shelf and their relationship to geologic, oceanographic and infralittoral biologic features, Ph.D. Thesis, University of St. Andrews, Scotland, UK, 2008. Available from http://hdl.handle.net/10023/726.

[3] F. Carvalho, Elementos do clima de agitação marítima no Grupo Central dos Açores (Mar Alto), Instituto de Meteorologia, Lisboa, Portugal, 2003.

[4] A. Simões, R. Duarte, M. Alves, A pilot ocean monitoring site at Azores islands, Elsev. Oceanogr. Ser. 62 (1997) 444–451.

[5] European Council, Council Directive 92/43/EEC of 21 May 1992 on the conservation of natural habitats and of wild fauna and flora, Off. J. Eur. Commun. L 206 (1992) 7–50.

[6] N.C. Mitchell, T. Schmitt, E. Isidro, F. Tempera, F. Cardigos, J.C. Nunes, et al., Multibeam sonar survey of the central Azores volcanic islands, InterRidge News 12 (2) (2003) 30–32.

[7] R. Holt, B. Sanderson, Procedural guideline no. 3-3: *in situ* survey of subtidal (epibiota) biotopes and species using diving techniques, in: J. Davies, J. Baxter, M. Bradley, D. Connor, J. Khan, E. Murray, et al. (Eds.), Marine Monitoring Handbook, NHBS, Totnes, 2001, pp. 233–239.

[8] C.M. Howson, A. Davison, Trials of Monitoring Techniques Using Divers and ROV in
     Loch Maddy cSAC, North Uist, Scottish Natural Heritage, Edinburgh, 2000.
[9] R. Holt, B. Sanderson, Procedural guideline no. 3-5: identifying biotopes using video
     recordings, in: J. Davies, J. Baxter, M. Bradley, D. Connor, J. Khan, E. Murray, et al.
     (Eds.), Marine Monitoring Handbook, NHBS, Totnes, 2001, pp. 241–251.
[10] J.C. Gamble, Diving, in: N.A. Holme, A.D. McIntyre (Eds.), Methods for the Study of
     Marine Benthos, second ed., Blackwell Scientific Publications, Oxford, 1984, pp. 99–
     139. International Biological Programme Handbook No. 16
[11] D. Connor, K. Hiscock, Data collection methods, in: K. Hiscock (Ed.), Marine Nature
     Conservation Review: Rationale and Methods, Joint Nature Conservation Committee,
     Peterborough, 1996, pp. 51–65.
[12] M. Kingsford, C. Battershill, Studying Marine Temperate Environments: A Handbook
     for Ecologists, Canterbury University Press, Christchurch, 1998.
[13] V.K. Borooah, Logit and probit: ordered and multinomial models, Sage University Papers
     Series on Quantitative Application in the Social Sciences, 07–138, Sage, Thousand Oaks,
     CA, 2001.
[14] P. McCullagh, Regression models for ordinal data (with discussion), J. R. Stat. Soc. B42
     (1980) 109–142.
[15] W. Sauerbrei, C. Meier-Hirmer, A. Benne, P. Royston, Multivariable regression model
     building by using fractional polynomials: description of SAS, STATA and R programs,
     Comput. Stat. Data Anal. 50 (12) (2006) 3464–3485.
[16] W.M. Hanemann, B. Kanninen, The statistical analysis of discrete-response CV data,
     in: I.J. Bateman, K.G. Willis (Eds.), Valuing Environmental Preferences: Theory and
     Practice of the Contingent Valuation Method in the US, EU and Developing Countries,
     Oxford University Press, Oxford, 1999, pp. 302–442.
[17] D.A. Hensher, L.W. Johnson, Applied Discrete Choice Modelling, Wiley, New York, NY,
     1981.
[18] J.J. Louviere, D.A. Hensher, J.D. Swait, Stated Choice Methods, Cambridge University
     Press, Cambridge, 2000.
[19] I. Tittley, A.I. Neto, A provisional classification of algal-characterised rocky shore bio-
     topes in the Azores, Hydrobiologia 440 (2000) 19–25.
[20] F.F.M.M. Wallenstein, A.I. Neto, N.V. Álvaro, C.I. Santos, Algae-based biotopes of the
     Azores (Portugal): spatial and seasonal variation, Aquat. Ecol. 42 (4) (2007) 547–559.

Supplementary Material

Table A Summary of the Variables Used in the Statistical Models

Response Variable	Type	Source and Raster Resolution Used for Spatialization	Time Scale of Observations
Abundance of selected infralittoral algae	Ordered categorical	SACFOR score attributed at every 1 m depth interval along the transects	Summers 1999, 2000, 2003, 2004, and 2005
Explanatory variable			
Depth	Continuous	Raster compiled from multibeam surveys. Resolution: 5 m	Surveys in autumn 2003 and summer 2004
Zonal slope	Continuous	Raster derived from multibeam bathymetry. Resolution: 20 m	Combination of bathymetric surveys in autumn 2003 and summer 2004
Exposure to swell action	Continuous	Raster derived from GIS-computed fetch windows weighted by mean swell statistics by compass interval (15°). Resolution: 200 m	Multiannual average 1989–2002
Exposure to currents	Continuous	Raster derived from shallow-water oceanographic current model. Resolution: 500 m	Maximum spring-tide model
Sea surface temperature	Continuous	Raster field of average SST computed from AVHRR satellite imagery. Resolution: 0.819′ × 0.684′ (lat. × long.)	Multiannual weighted average from April 2001 to April 2006
Surface chlorophyll-*a* concentration	Continuous	Raster field of average surface chlorophyll-*a* concentration computed from SeaWiFS satellite imagery. Resolution: 0.412′ × 1.130′ (lat. × long.)	Multiannual weighted average from January 1999 to October 2004
Analysis mask			
Seafloor nature	Categorical	Raster layer representing rocky seafloor areas and blanking out sediment surfaces interpreted from backscatter and bathymetric texture validated by seafloor observations conducted by SCUBA diving, ROV, and drop-down camera deployments. Resolution: 5 m	Combination of seafloor surveys in autumn 2003 and summers 2004 and 2005

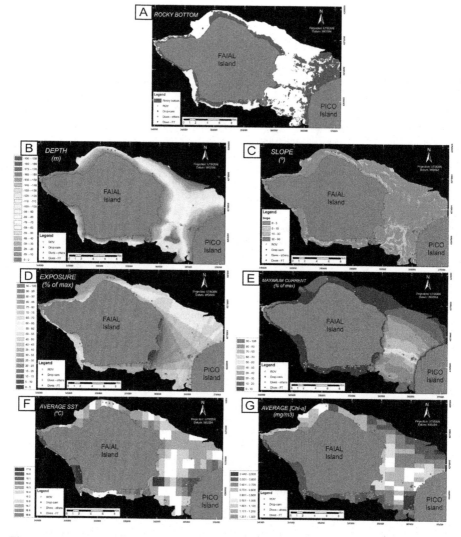

Figure A Covariate fields. A. Seafloor nature and location of the SCUBA dives, ROV and drop-down camera deployments; B. Depth. C. Slope; D. Swell exposure (percentage of maximum); E. Maximum tidal currents (percentage of maximum); F. SST (yearly-average); G. Chl-*a* concentration (yearly-average).

9 Methane-Related Carbonate Cementation of Marine Sediments and Related Macroalgal Coralligenous Assemblages in the Northern Adriatic Sea

Emiliano Gordini[1], Annalisa Falace[2], Sara Kaleb[2], Federica Donda[1], Ruggero Marocco[3], Giorgio Tunis[3]

[1]Istituto Nazionale di Oceanografia e di Geofisica Sperimentale–OGS Trieste, Italy, [2]Dipartimento di Scienze della Vita, Università degli Studi di Trieste, Italy, [3]Dipartimento di Geoscienze, Università degli Studi di Trieste, Italy

Abstract

This multidisciplinary study presents the major characteristics of a set of submarine rock outcrops in the Northern Adriatic Sea: the rock occurrence, embedded down to about 1 m from the seafloor in the bottom sediments; numerous gas accumulations in the proximity of the outcrops (e.g., bubbling gas from the sediment–water interface); small-scale mud volcanoes; and microbial mats around the seepage site. Geochemical analysis revealed that these features are related to the seepage of CH_4-rich fluids, which causes the precipitation of methane-derived calcium carbonate as a cement ($\delta^{13}C = -49.80‰$ V-PDB). ^{14}C data gave radiocarbon ages of about $21,700 \pm 2,265$ years BP for two cement samples. From the biological point of view, a total of 112 macroalgal taxa were recorded. The different number of taxa recorded at the San Pietro and Bardelli sampling sites has been related to the different distance from the coast and to water depth. Some species that characterize these outcrops are acknowledged as important bioconstructors distinctive of the Mediterranean area.

Key Words: Northern Adriatic Sea, rock outcrop, methane-related carbonate cementation, Quaternary, subbottom profiler Chirp, sidescan sonar, multibeam, CH_4 seepage, geochemistry, macroalgal taxa

Introduction

The investigated area is located in the Gulf of Trieste (Northern Adriatic Sea) (Figure 9.1), in a maximum water depth of 22 m, close to the Slovenian–Croatian coast. Due to the orientation of this portion of the Adriatic Sea (NE–SW), the study

Seafloor Geomorphology as Benthic Habitat. DOI: 10.1016/B978-0-12-385140-6.00009-8

Figure 9.1 Location map of the northern Adriatic Sea rock outcrops (red dots) where the position of **San Pietro** and **Bardelli** is highlighted. (For interpretation of the references to color in this figure legend, the reader is referred to the web version of this book.)

area is exposed to different winds and wave regimes. The prevailing winds, especially during the fall and winter, are mainly represented by an ENE wind called Bora [1]. The entire area is also affected by the northern winds (diurnal) caused by thermal variations; the Sirocco and Libeccio winds blow less frequently from the southwest [2,3].

The tidal variations observed along the coast are microtidal (<1 m). The greatest amplitude occurs during spring tide, with an average semidiurnal difference of 86 cm in the northeasternmost part of the Adriatic Sea, in the proximity of Trieste, and 100 cm in Venice (with peaks of 130 up to 200 cm). The minimum amplitude is noted during the mean neap tides, with diurnal differences of about 22 and 20 cm in Trieste and Venice, respectively [4,5]. Annual mean significant wave heights are lower than 0.5 m [6], while the highest offshore wave height, generated by the Bora- and Sirocco-related storms, is about 5 m [7]. The mean basin circulation is counterclockwise.

From the stratigraphical and depositional point of view, the study area is characterized by sedimentary deposits interpreted as late Quaternary transgressive systems [8–10]. They are represented by lowstand system tract (LST) alluvial clays and sands of continental origin, buried by sediments displaying different thicknesses that constitute the transgressive system tract (TST) deposits, mainly characterized by "barrier–lagoon–estuary" systems. The highstand system tract (HST) deposits are located in correspondence to the actual shoreline or sometimes even more inland [11]. The seismic stratigraphy and tectonic evolution of the Gulf of Trieste is described in detail [12,13].

The seafloor sedimentary deposits show a gradual increase of the fine fraction from the coast toward the offshore; then a progressive increase in the coarse fraction in the central part of the study area (residual sands of the TST) is recognized.

Geophysical (multibeam, singlebeam, sidescan sonar (SSS), and Chirp subbottom profiler) and underwater surveys carried out in the Gulf of Trieste allowed a more detailed understanding of the depositional and erosional features of this sea floor sector. It is characterized by a series of reliefs (called Trezze and Banco della Mula di Muggia), elongated and round depressions, and widespread subaqueous sand-dunes fields.

However, the most peculiar features of the Northern Adriatic Sea are submarine rock exposures, irregularly distributed on the seabed [14], whose origin is still largely debated. These rock formations have been initially interpreted as beachrocks [15–17], while the current hypothesis on their genesis suggests they are most likely related to seeping methane and cementation and lithification processes [18–23]. The migration of shallow gas through marine sediments usually causes the precipitation of methane-derived calcium carbonate as a cement within otherwise unconsolidated sediment [22,24–26]. The peculiarity of this cement is the remarkable carbon isotopic depletion in ^{13}C compared to normal marine carbonates. It has been interpreted as the result of coupled processes, that is, anaerobic methane oxidation and sulfate reduction, operated by *Archaea* and sulfate-reducing bacteria.

These rock outcrops are characterized by a rich community of associated flora and fauna and thus represent a unique hotspot of biodiversity on the rather monotonous seabed of the Northern Adriatic Sea. The coralligenous biogenic concretions that characterize these reefs play a fundamental role as habitat for reproduction and nurseries of demersal and pelagic species. These structures are very sensitive to human impacts. Scuba diving confirmed the widespread fragmentation of these features, probably as a result of fishing and overexploitation of *Callista chione* (Linnaeus), 1758, and *Venus verrucosa* (Linnaeus), 1758.

This chapter discusses the most relevant results concerning the study of two rock outcrops, San Pietro (45°36′191″N″ − 13°20′276″E″) and Bardelli (45°29′843″N″ − 13°14′662″E″).

Geomorphic Features and Habitats

The San Pietro outcrop lies at a distance of about 9 km from the coast, at 15 m depth. The study of this structure was first performed through an SSS survey (Edgetech DF-1000/DCI, 100–500 kHz; data processing: Coda Octopus Geokit Mosaics; cell size, 50 cm), which covered a 1,500 × 2,500 m area (*ca.* 4 km²; Figure 9.2). The SSS mosaic consists of 13 profiles, 100 m spaced and acquired with a 75 m wide swath, which allowed an overlap of 20%.

The data highlights that the seabed is almost flat, locally showing the occurrence of rock outcrops, with a range of sizes, shapes, and spatial orientations. On the SSS data, they appear as high backscatter features. The seabed is mainly composed of sands. The rock outcrops can be grouped into two blocks with 40 smaller (0.6 up

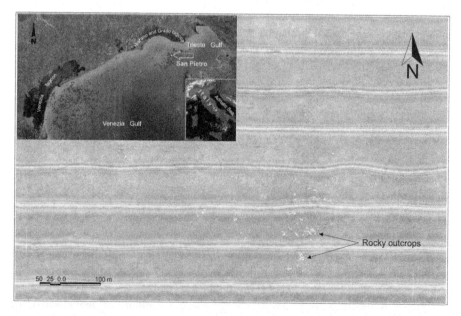

Figure 9.2 San Pietro SSS mosaic. Data reveal that the seabed, with a dominant sandy component, is almost flat, with local high backscatter features, indicating the occurrence of rock outcrops.

to 6 m in length) outcrops in the surroundings, covering an area of 7,000 m² (about 175 m × 75 m; E–W oriented). The larger block is 260 m² large (29 m × 11 m).

In order to obtain a better morphological characterization of the rock outcrops, a multibeam survey has also been performed (Multibeam Reson SeaBat 8125, 455 kHz; processing: Reson PDS2000; grid cell size: 20 cm) in an area of 1.6 km². The surveyed area shows a NW–SE, 1.5‰ dipping seafloor, with a depth ranging from 15 to 1 m. The rock outcrops have a pinnacle morphology and are 1.5–1.9 m high, and 1–5 m wide (Figure 9.3).

In addition, a Chirp survey (Edgetech 3200-XS/SB-216S, 2–16 kHz; data processing: Discover SB 3200-XS and Kingdom Suite; Figure 9.4) has been performed in the rock outcrop area.

Subbottom profiler data were characterized by a signal penetration of 10 m and a vertical resolution of about 20 cm; the seismostratigraphic setting is generally flat, and where four major sequences were recognizable. Based on a correlation with a 2 m long core collected in the area (GT1bis; see [8]), the uppermost sequence consists of medium to fine sand, with abundant organic matter (*Cymodocea nodosa* (Ucria) Ascherson), probably related to the TST system.

The lowermost sequence is interpreted as representing the LST top (silty sands with *Polmonata* spp. and *Pisidium* spp. and littoral organic matter).

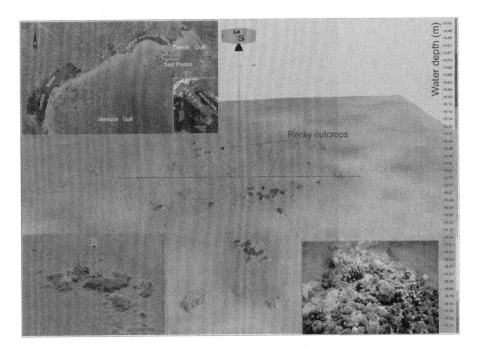

Figure 9.3 Multibeam sonar bathymetric map showing a 3D perspective view of San Pietro outcrop area, 1.6 km² wide. The surveyed area shows a NW–SE, 1.5‰ dipping seafloor, with a depth ranging from 15 to 17 m. The rock outcrops have a pinnacle morphology and are 1.5–1.9 m high.

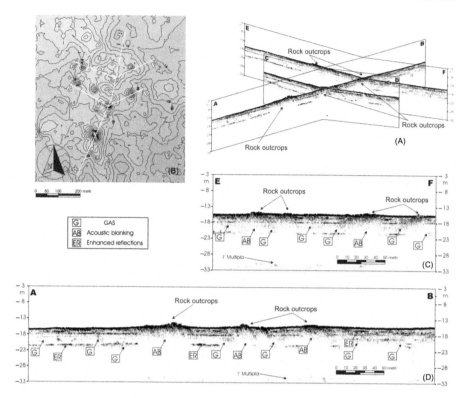

Figure 9.4 Subbottom profiler Chirp (2–7 kHz) survey on the San Pietro outcrop: (A) 3D sketch; (B) seafloor map (digital echosounder survey); (C, D) high-resolution Chirp, interpreted profiles performed along the direction of maximum development of the rock outcrops (A, B) and across them (E–F). The investigated outcrops, represented by a seismic acoustic blanking, are clearly recognizable on the seismic data and appear to be partially buried within the superficial sediments. The signal penetration is primarily influenced by the presence of sandy-silty sand superficial sediments.

The occurrence of gas has also been identified in both the stratigraphic sequence and the water column (Figure 9.7).

The rock outcrops appear on the seismic profiles as semitransparent *facies* embedded down to about 1 m from the seafloor within the sedimentary sequence.

SSS data acquired in the Bardelli outcrop area, which lies at about 19 km away from the coast, covers 600 m × 800 m (Figure 9.5). The SSS mosaic consists of five profiles, 100 m spaced, and acquired with a 75 m wide swath, which allowed an overlap of 20%. The mosaic shows a predominantly sandy seafloor; locally pelitic sediments are evidenced as lighter backscatter.

The Bardelli outcrop consists of two main groups, which have an E–W orientation (Figure 9.5). They cover an area of 250 m × 90 m (6,300 m²) and are separated by a flat sandy seafloor morphology, where minor rock outcrops are present. The larger group is 95 m × 80 m wide (4,100 m²), shows an almost round shape, and

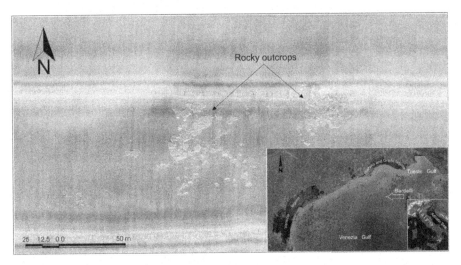

Figure 9.5 Bardelli SSS mosaic showing a predominant flat, sandy seafloor; locally pelitic sediments are evident as lighter backscatter, whereas the outcrops appear as high backscatter features.

is formed by smaller blocks, separated by sandy to gravelly deposits. The second group, composed of closely spaced rock slabs, reveals an ENE–WSW orientation and is 55 m × 38 m wide (2,200 m²). The slabs are separated by vertical fractures smaller than 0.2–0.3 m.

In order to obtain a better morphological characterization of this outcrop, a single-beam survey (Kongsberg-Simrad EA400, 38–200 kHz; processing: Reson PDS2000; grid cell size: 2 m; Figure 9.6) was performed over the Bardelli outcrop in an area of about 0.2 km². The analysis of 2D and 3D bathymetric data highlighted that the outcrop lies at a depth ranging from 20.2 to 22.5 m (dip, 4‰). Three morphological highs can be identified at 20.2, 20.4, and 20.6 m, the latter two being located in correspondence to the Bardelli outcrop. In the eastern sector of the surveyed area, a bathymetric low (maximum depth 22.5 m) is recognizable. The rock outcrops have a maximum elevation above the seafloor of 1.2 m.

A Chirp subbottom profiler survey was also conducted; one example is shown in Figure 9.7D. The maximum signal penetration was about 10 m. The high-resolution seismic profiles show the occurrence of a highly reflective *facies* on the seafloor indicative of the presence of the Bardelli outcrop. The sedimentary sequence consists of four depositional units (Figure 9.7D). The lowermost unit is composed of a tabular horizon geometry, locally affected by upward migrating gas.

The uppermost unit, which extends up to the seafloor, is onlapping against the outcrops. It can easily be noted that the two outcrop groups are not laterally connected and that two major plumes are recognizable just above the outcrops. They may represent fluids seeping from the sediments and the outcrops in the water column. Both the Chirp subbottom profiler investigations show, as regards the previous described outcrops, an embedding of only 1 m within the stratigraphic sequence and

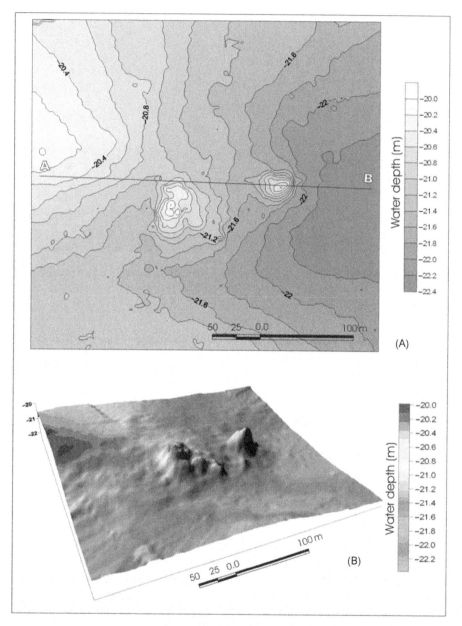

Figure 9.6 Singlebeam data above Bardelli outcrop. (A) Seafloor contour map where the
two blocks constituting the outcrop are recognizable; (A, B) Chirp subbottom profiler
profile. (B) 3D perspective view; the outcrop lies at a depth of 20.2–22.5 m, with a maximum
elevation above the seafloor of 1.2 m.

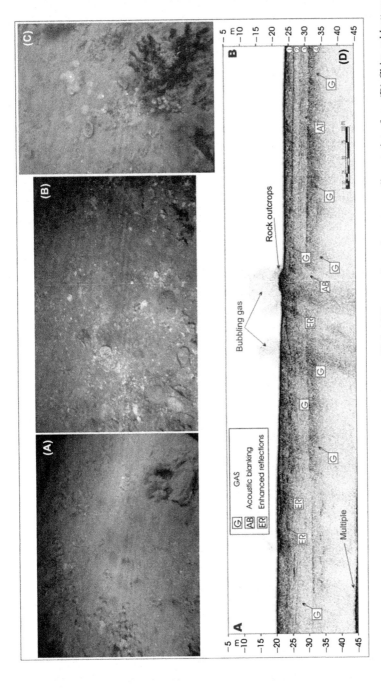

Figure 9.7 (A) Small-scale (cm) mud volcanoes, (B) microbial mats, (C) bubbling gas from the water–sediments interface, (D) Chirp subbottom profiler (2–7 kHz) survey on the Bardelli outcrop. High-resolution seismic profile showing the occurrence of four depositional units. The highly reflective facies on the seafloor is indicative of the presence of the Bardelli outcrop. Numerous gas accumulations in the proximity of the outcrop can also be recognized; their occurrence is testified by bubbling gas seeping in the water column from the sediments and from the outcrops.

numerous gas accumulations in the proximity of the outcrop, for example, bubbling gas from the water–sediments interface (Figure 9.7C), small-scale mud volcanoes (Figure 9.7A), and microbial mats (Figure 9.7B) around the seepage site.

Four outcrop samples have been collected in correspondence to each outcrop during several diver surveys, at the outcrop top and bottom, and in peculiar areas, such as the gas seepages. SEM-EDX image analysis of eight thin sections of such samples of both the San Pietro and Bardelli outcrops confirmed that their origin is related to marine precipitation cements. Microstructural and petrographical observations were fundamental for the distinction of the different types of authigenic carbonates. Aragonite characterizes the Bardelli outcrop and is primarily associated with shell accumulation in the sediments. Calcite was mostly observed in correspondence to the San Pietro outcrop, in the calcareous sandstones. The mineralogical composition of the rocks reveals a high percentage of carbonates (calcite, aragonite, dolomite, and fragments of carbonate rocks), which may sum up to 80% of the sample. The remaining 20% is made of quartz, feldspar, biotite, chlorite, glauconite, phosphates, and small inclusions of pyrite. The composition of the examined samples on the whole proves the alluvial and marine origin of the deposits, which commonly characterize the study area.

Geochemical analyses, performed through the method described in [27] with a "dual inlet" mass spectrometer (Europa Scientific GEO 20-20), were also carried out to investigate the possible relationships of the rocks with the seepage of CH_4 fluids. The geochemical signatures performed on 16 bulk samples at different sampling sites (mixed marine sediments, shells, and carbonate cements) showed that they are depleted in ^{13}C with $\delta^{13}C$ values ranging between -10.28‰ and -26.30‰ V-PDB. More specifically, two samples of authigenic, acicular, aragonite carbonate cement alone revealed strong negative $\delta^{13}C$ values (e.g., -49.80‰ to -39.40‰ V-PDB), thus showing that methane is the primary source of the carbonate cement carbon.

^{14}C data gave apparent radiocarbon ages of about $21,700 \pm 2,265$ and $15,940 \pm 360$ years BP for two aragonitic cement samples, and an age of about $4,990 \pm 45$ years BP for shells of gastropod and bivalve specimens embedded in the rock (Analyses performed at the Angstrom Laboratory of Uppsala University: the samples are ultrasonically cleaned in boiled distilled water [pH = 3]; each sample is leached with 0.5 M HCl, giving rise to separate fractions of CO_2; the desired fraction of CO_2 is graphitized using an Fe-catalyst reaction prior to the accelerator measurement). It would then suggest that the carbonate cement ages are compatible with the youngest ages for the fossil methane source, that is, the late Quaternary peat, which age is 16,000 to 22,000 years BP [8,28]. Instead, the ages from the shells would suggest that the cementation processes are recent (Holocene). However, the hypothesis on the age attribution to the outcrops remains at the moment speculative; in fact, we are going to perform an evaluation of the gas age also.

Biological Communities

Macroalgal assemblages were seasonally investigated from 2008 to 2009 on the San Pietro and Bardelli outcrops by means of a 30×30 cm frame arranged randomly

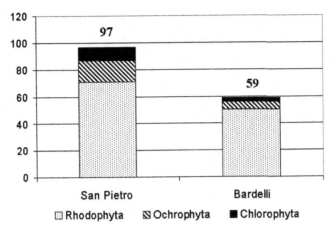

Figure 9.8 Number of taxa assessed on the two rock outcrops.

on the bottom, within which the substratum was scrubbed clean. Algae collection was supported by underwater photography [29]. Furthermore, on the nearby soft bottoms, SCUBA surveys were made along two 10 m long radial transects across the reef structure. Samples were collected on three quadrates (50 × 50 cm) randomly selected along each transect.

A total of 114 taxa were recorded: 87 Rhodophyta, 17 Ochrophyta, and 10 Chlorophyta. The different number of species recorded at the San Pietro and Bardelli sampling sites (Figure 9.8) may be related to the different distances from the coast (3 and 18 miles, respectively) and depth (16 and 22 m). In particular, at Bardelli, the dim-light conditions favored the development of a greater relative percentage of red algae (84.7% versus 73.1% at San Pietro). These observations are in accordance with [30,31] results on nearby similar outcrops. Among the most conspicuous bioconstructors are the crustose red algae, with 21 Corallinales and 5 Peyssonneliaceae, which at Bardelli showed a coverage up to 70–80%. Nevertheless, these rock outcrops are characterized by a high spatial and temporal variability, even on a small to medium scale of observation. As suggested by Casellato et al. [30], in the Northern Adriatic, the unpredictability of the oceanographical conditions drive the observed variability in benthic communities.

On both outcrops, the most frequent species is the nongeniculate coralline red alga [32] *Lithophyllum pustulatum* (J.V. Lamouroux) Foslie (Figure 9.9B), while *Lithothamnion philippii* Foslie and the genus *Peyssonnelia* showed the higher coverage. Bardelli was also characterized by *Pneophyllum confervicola* (Kützing) Y.M. Chamberlain (Figure 9.9C–D), *Mesophyllum alternans* (Foslie) Cabioch & Mendoza, and *Neogoniolithon mamillosum* (Hauck) Setchell & L.R. Mason. The latter two species are acknowledged as important corallinegous bioconstructors distinctive of the Mediterranean [33–35]. It has been suggested that the calcareous organisms had an important role in the origin of the rock outcrops [30]: the precipitation of carbonate due to methane seeps formed the hard substrate on which the benthic building organisms developed. Such calcareous organisms act as bioconstructors, providing a

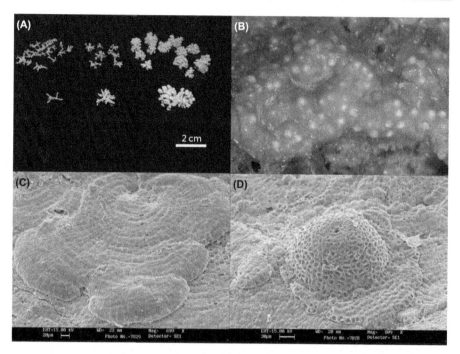

Figure 9.9 (A) Rhodoliths specimens. From left to right: *L. corallioides*, *P. calcareum*, and *L. racemus*; (B) thallus in reproduction with conceptacles of *L. pustulatum*; (C) habitus of *P. confervicola*; (D) conceptacle of *P. confervicola*.

secondary substratum for further colonization and promoting the construction of the coralligenous reefs [35]. Furthermore, calcareous organisms represent a major contributor to coastal sediments and fossil deposits of biogenic calcium carbonate [36,37].

Finally, on the seabed surrounding the outcrops, free-living, unattached nodules of coralline algae [38,39] were identified. These assemblages represent complex and heterogenous substrata on otherwise monotonous sedimentary bottoms, providing habitat for many associated organisms. They are recognized worldwide as one of the most important benthic communities dominated by marine algae [40].

In particular, the sandy seabed at San Pietro was characterized by the rhodolith *Lithophyllum racemus* (Lamarck) Foslie, while on the pelitic–sandy sediments at Bardelli, a maërl association (Figure 9.9A) with two species characteristic, *Lithothamnion corallioides* (P.L. Crouan & H.M. Crouan) P.L. Crouan & H.M. Crouan and *Phymatolithon calcareum* (Pallas) W.H. Adey & D.L. McKibbin, along with *Lithothamnion minervae* Basso and *Lithophyllum racemus* (Lamarck) Foslie, were detected. Living rhodoliths are crucial for the survival of beds and their associated biodiversity [41]. The assessed ratio of living (pigmented)/nonliving thalli was higher at San Pietro ($L/nL=10$) than at Bardelli ($L/nL=1.64$). Following [42–44], the rhodoliths' sphericity was also assessed. The analyzed specimens are mostly likely spheroidal and ellipsoidal (Figure 9.10). As far as the density of branching is concerned, in accordance with the scale proposed by [42], the dominance of thalli with

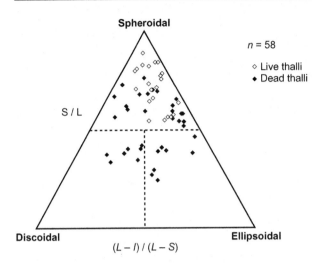

Figure 9.10 Ternary diagram of *L. racemus* shape (L = long axis; I = intermediate axis; S = short axis).

low-density branches (Classes II–III) was observed. In literature, the sphericity and branching density of rhodoliths have often been correlated to environmental factors, such as hydrodynamic features (currents, water turbulence, wave action) and sedimentation rate [39,42,45,46]. Sphericity has also been used as an indicator of frequency of movement and overturning [42,45]. On Bardelli, the lack of a dominant form, together with the almost total absence of fleshy epiphytic algae, seems to delineate a situation highly variable in local conditions. Nevertheless, the relationships linking the algal morphology and the environmental conditions are still questionable and mainly based on few correlative studies on living beds; therefore, more experimental data are imperative to better predict such correlations.

With reference to the EU Habitat Directive (Natura 2000, 92/43/EEC), the rock outcrops of the Northern Adriatic are regarded as "reef habitat" (Habitat: 1170), characterized by biogenic/geogenic concretions. None of the species until now reported are included in Annex II of the Habitat Directive, while two Mollusca (*Lithophaga lithophaga* Linnaeus, 1758, and *Pinna nobilis* Linnaeus, 1758), and two macroalgae (*L. corallioides* and *P. calcareum*) are respectively reported in Annexes IV and V.

The high biodiversity recorded, together with the occurrence of ecological gradients, represent valuable elements on the rather monotonous coastal bottoms of the Northern Adriatic Sea. In addition, the rock outcrops play a pivotal role as nurseries for demersal and pelagical species reproduction and spawning. The fragile biological equilibrium of these structures, the slow-growing calcareous organisms considered as nonrenewable resources, and the dredging and bottom trawling pressure that strongly threatens the physical characteristics of the habitat and negatively affect the associated biota, all motivate the need to protect the Trezze and their surrounding seabed.

References

[1] F. Carrera, M. Cerasuolo, A. Tomasin, P. Canestrelli, La nebbia a Venezia nel quarantennio (1951–1990), Commissione di studio dei provvedimenti per la conservazione e

difesa della laguna e della città di Venezia. Rapporti e studi, Istituto Veneto di Scienze Lettere e Arti, Venezia, vol. XII, 1995, pp. 235–271.

[2] G. Catani, R. Marocco, Considerazioni sulle caratteristiche mareografiche e anemografiche a San Nicolò di lido, Punta Tagliamento e Grado, Quaderni della Ricerca Scientifica 92 (1976) 21–25.

[3] A. Brambati, Studio sedimentologico e marittimo-costiero dei litorali del Friuli Venezia Giulia, Servizio Idraulica della Regione Friuli Venezia Giulia (1987) 67.

[4] L. Dorigo, La laguna di Grado e le sue foci. Ufficio Idrografico dle Magistrato delle Acque di Venezia. Ricerche e rilievi idrografici, Venezia (1965) 231 pp.

[5] S. Polli, Tabelle di previsione delle maree per Trieste e l'Adriatico Settentrionale per l'anno 1971, Istituto Talassografico Sperimentale, Trieste (1970) 20 pp.

[6] R. Dal Cin, U. Simeoni, A model for determining the classification, vulnerability and risk in the Southern coastal zone of the Marche (Italy), J. Coast. Res. 10 (1994) 18–29.

[7] L. Cavaleri, L. Bergamasco, L. Bertotti, L. Bianco, M. Drago, L. Iovenitti, et al., Wind and waves in the Northern Adriatic Sea, Il Nuovo Cimento 19 (1996) 1–36.

[8] E. Gordini, R. Marocco, E. Vio, Stratigrafia del sottosuolo della "Trezza Grande" (Golfo di Trieste, Adriatico Settentrionale), Gortania 24 (2002) 31–63.

[9] F. Trincardi, A. Correggiari, M. Roveri, Late Quaternary transgressive erosion and deposition in a modern epicontinental shelf: the Adriatic Semienclosed Basin, Geo. Mar. Lett. 14 (1994) 41–51.

[10] A. Cattaneo, F. Trincardi, The late-Quaternary transgressive record in the Adriatic epicontinental sea: basin widening and facies partitioning, in: K.M. Bergman, J.W. Snedden (Eds.), Isolated Shallow Marine Sand Bodies: Sequence Stratigraphic Analysis and Sedimentologic Interpretation, SEPM Special Publication (1999), vol. 64, pp. 127–146.

[11] R. Marocco, Considerazioni sedimentologiche sui sondaggi S19 e S20 (Delta del F. Tagliamento), Gortania 10 (1988) 101–120.

[12] M. Busetti, V. Volpi, E. Barison, M. Giustiniani, M. Marchi, R. Ramella, et al., Cenozoic seismic stratigraphy and tectonic evolution of the Gulf of Trieste (Northern Adriatic), GeoActa, Special Publication 3 (2010) 1–14.

[13] M. Busetti, V. Volpi, R. Nicolich, E. Barison, R. Romeo, L. Baradello, et al., Dinaric tectonic features in the Gulf of Trieste (Northern Adriatic) in: D. Slejko (Ed.), Novelties in Geophysics, Select paper from the 27th Annual Conference of the Italian Group for Solid Earth Geophysics, Trieste (Italy), October 6–8, 2008, Bollettino di Geofisica Teorica e Applicata, 51(2–3) (2010) 117–128.

[14] G. Olivi, Zoologia Adriatica, Reale Accademia Scienze Lettere Arti, Bassano (1792) 334 pp.

[15] G. Braga, A. Stefanon, Beachrock ed alto Adriatico: aspetti paleogeografici, climatici, morfologici ed ecologici del problema, Atti Ist. Ven. Sci. Lett. Art. CXXVII (1969) 351–359.

[16] A. Stefanon, The role of beachrock in the study of the evolution of the North Adriatic Sea, Mem. Biogeogr. Adriat 8 (1970) 79–99.

[17] R. Newton, A. Stefanon, The "Tegnue de Ciosa" area: patch reefs in the Northern Adriatic Sea, Mar. Geol. 46 (1975) 279–306.

[18] P. Colantoni, G. Gabbianelli, L. Ceffa, C. Ceccolini. Bottom features and gas seepages in the Adriatic Sea, Vth International Conference on Gas in Marine Sediments (1998) pp. 28–31.

[19] A. Stefanon, G.M. Zuppi, Recent carbonate rock formation in the northern Adriatic Sea, Hydroèologie 4 (2000) 3–10.

[20] A. Conti, A. Stefanon, G.M. Zuppi, Gas seeps and rock formation in the northern Adriatic Sea, Continent. Shelf Res. 22 (2002) 2333–2344.

[21] E. Gordini, R. Marocco, G. Tunis, R. Ramella, The cemented deposits of the Trieste Gulf (Northern Adriatic Sea): areal distribution, geomorphologic characteristics and high resolution seismic survey, Il Quaternario Ital. J. Quaternary Sci. 17 (2/2) (2004) 555–563.

[22] A. Mazzini, M.K. Ivanov, J. Parnell, A. Stadnitskaia, B.T. Cronin, E. Poludetkina, et al., Methane-related authigenic carbonates from the Black Sea: geochemical characterisation and relation to seeping fluids, Mar. Geol. 212 (1–4) (2004) 153–181.

[23] P. Jensen, I. Aagaard, R.A. Burke, Jr., P.R. Dando, N.O. Jorgensen, A. Kuijpers, et al., Bubbling reefs in the Kattegat: submarine landscapes of carbonate-cemented rocks support a diverse ecosystem at methane seeps, Mar. Ecol. Progr. 83 (2–3) (1992) 103–112.

[24] M. Hovland, M.R. Talbot, H. Qvale, S. Olaussen, L. Aasberg, Methane-related carbonate cements in pockmarks of the North Sea, J. Sediment. Petrol. 57 (5) (1987) 881–892.

[25] N.O. Jorgensen, Methane-derived carbonate cementation of marine sediments from the Kattegat, Denmark: geochemical and geological evidence, Mar. Geol. 103 (1992) 1–13.

[26] J. Peckmann, A. Reimer, U. Luth, Methane-derived carbonates and authigenic pyrite from the northwestern Black Sea, Mar. Geol. 177 (1–2) (2001) 129–150.

[27] J.M. McCrea, On the isotopic chemistry of carbonates and paleo temperature scale, J. Chem. Phys. 18 (1950) 849–857.

[28] R. Marocco, Evoluzione tardopleistocenica–olocenica del Delta del F. Tagliamento e delle lagune di Marano e Grado (Golfo di Trieste), Quaternario Ital. J. Quaternary Sci. 4 (1b) (1991) 223–232.

[29] R.J. Meese, P.A. Tomich, Dots on the rocks: a comparison of percent cover estimation methods, J. Exp. Mar. Biol. Ecol. 165 (1992) 59–73.

[30] S. Casellato, L. Masiero, E. Sichirollo, S. Soresi, Hidden secrets of the Northern Adriatic: "Tegnùe," peculiar reefs, Cent. Eur. J. Biol. 2 (1) (2007) 122–136.

[31] D. Curiel, A. Rismondo, C. Miotti, E. Checchin, C. Dri, G. Cecconi, et al., Le macroalghe degli affioramenti rocciosi (tegnùe) del litorale veneto, Soc. Ven. Sc. Nat. 35 (2010) 39–55.

[32] W.J. Woelkerling, The Coralline Red Algae: an analysis of the genera and subfamilies of nongeniculate Corallinaceae, British Museum (Natural History) & Oxford University Press, London & Oxford, 1988, pp. 1–268, 259.

[33] S. Sartoretto, M. Verlaque, J. Laborel, Age of settlement and accumulation rate of submarine "coralligène" (−10 to −60 m) of the northwestern Mediterranean Sea; relation to Holocene rise in sea level, Mar. Geol. 130 (1996) 317–331.

[34] C.F. Boudouresque, Groupes écologiques d'algues marines et phytocenoses benthiques en Méditerranée nord-occidentale: une revue, Giornale Botanico Italiano 118 (1985) 7–42.

[35] E. Ballesteros, Mediterranean coralligenous assemblages: a synthesis of present knowledge, Oceanogr. Mar. Biol. Annu. Rev. 44 (2006) 123–195.

[36] D.W. Bosence, J. Wilson, Maerl growth, carbonate production rates and accumulation rates in the northeast Atlantic, Aquat. Conservat. Mar. Freshwat. Ecosyst. 13 (2003) 21–31.

[37] M. Canals, E. Ballesteros, Production of carbonate particles by phytobenthic communities on the Mallorca–Menorca shelf, northwestern Mediterranean Sea, Deep Sea Res. II 44 (1997) 611–629.

[38] D.W. Bosence, Coralline algal reef frameworks, J. Geol. Soc. Lond. 140 (1983) 365–376.

[39] M.S. Foster, R. Riosmena-Rodríguez, D.L. Steller, W.J. Woelkerling, Living rhodolith beds in the Gulf of California and their implications for paleoenvironmental interpretation, Geol. Soc. Am. 318 (1997) 127–139.

[40] M.S. Foster, Rhodoliths: between rocks and soft places, J. Phycol. 37 (2001) 659–667.

[41] C. Barberá, C. Bordehore, J.A. Borg, M. Glémarec, J. Grall, J.M. Hall-Spencer, et al., Conservation and management of northeast Atlantic and Mediterranean maërl beds, Aquat. Conserv. Mar. Freshwater Ecosyst. 13 (2003) 65–76.

[42] D.W. Bosence, Ecological studies on two unattached coralline algae from western Ireland, Palaeontology 19 (1976) 365–395.

[43] E.D. Sneed, R.L. Folk, Pebbles in the Lower Colorado River, Texas: a study in particle morphogenesis, J. Geol. 66 (1958) 114–150.

[44] D.J. Graham, N.G. Midglay, Graphical representation of particle shape using triangular diagrams: an excel spreadsheet method, Earth Surf. Process. Landforms 25 (2000) 1473–1477.

[45] A. Bosellini, R.N. Ginsburg, Form and internal structure of recent algal nodules (Rhodolites) from Bermuda, J. Geol. 79 (1971) 669–682.

[46] R.S. Steneck, The ecology of coralline algal crusts: convergent patterns and adaptive strategies, Annu. Rev. Ecol. Syst. 17 (1986) 273–303.

10 Coastal Kelp Forest Habitat in the Baie des Chaleurs, Gulf of St. Lawrence, Canada

Antoine Collin[1], Bernard Long[2], Philippe Archambault[3]

[1]Insular Research Center and Environment Observatory, Papetoai, French Polynesia, [2]Department of Geosciences, University du Québec, QC, Canada, [3]Institute of Marine Sciences, University du Québec à Rimouski, QC, Canada

Abstract

Bonaventure and Paspébiac are two sites on the north shore of the Baie des Chaleurs (48°N, 65°W), Gulf of St. Lawrence (Canada), that encompass intertidal and nearshore zones. The two study areas together cover a total area of approximately 50 km^2 and reach water depths of up to 16 m. A seascape analysis conducted at broad and fine scales (100 m and 10 m horizontal resolution, respectively) classified the area into ridges, passages, and seachannels. Drop-down camera stations were carried out at 814 locations to characterize these geomorphic features in terms of biological cover and sediment composition. The distribution of fine sediment showed opposite trends at the two sites, decreasing in Bonaventure in the order: broad ridges > passages > seachannels, while increasing in Paspébiac. While broad ridges and passages constituted niches for *Laminaria* spp. (kelp) in both areas, broad seachannels were populated by *Laminaria* spp. in Bonaventure and covered by dead shells in Paspébiac. Biodiversity indices were higher in broad ridges and passages than in broad seachannels. Moderate correlations are shown for both areas between depth and biodiversity indices derived herein.

Key Words: sublittoral, macrobenthos, Light Detection and Ranging (LiDAR)

Introduction

Bonaventure and Paspébiac are two nearshore sites located in the Baie des Chaleurs (48°N, 65°W), southern Gulf of St. Lawrence, QC, Canada (Figure 10.1). The surveyed areas comprise two intertidal and subtidal zones approximately 10 m and 16 km long and 2 km wide. The two areas are comprised mainly of Quaternary fluvio-glacial sediments deposited during the Holocene transgression. Outcrops of Siluro-Devonian, coarse-grained, red conglomerate, sandstone, and mudstone occur

Seafloor Geomorphology as Benthic Habitat. DOI: 10.1016/B978-0-12-385140-6.00010-4

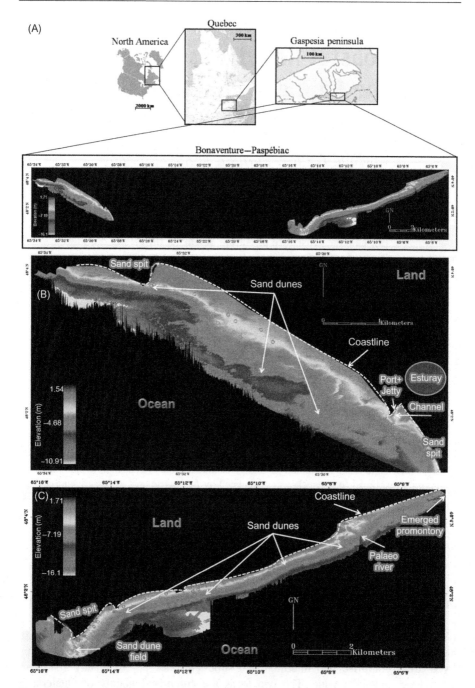

Figure 10.1 (A) Locations of Bonaventure's and Paspébiac's nearshores (QC, Canada) and the underlying LiDAR bathymetric maps (B and C, respectively) based on a 2-m grid. Survey stations are identified by red circles. (For interpretation of the references to color in this figure legend, the reader is referred to the web version of this book.)

locally [1]. This locality is mesotidal (2–4 m tidal range) [2] and experiences tidal currents of up to 0.25 m/s [3]. In both areas, fluvial sediments are dispersed by alongshore currents driven by east-southeast swells and deposited as sand spits. The interaction between hydrodynamics and sediment supply results in specific geomorphic features: ridges, passages, and seachannels. The surveyed depths range from 0 to 16.1 m below sea level and intertidal areas lie between 0.3 and 1.7 m in respect of the chart datum (Lowest Astronomical Tide). Both areas lie within the euphotic zone and are characterized by annual winter sea-ice cover. Summer water temperatures reach 16°C at 2 m depth and 14°C at 10 m depth. The mean water salinity is 25‰, and Secchi depth (turbidity) ranges from 2.4 to 5.3 m [4].

Average seabed slopes are 0.36° and 0.46° for the Bonaventure and Paspébiac study areas, respectively, and the orientations of the coast at Bonaventure and Paspébiac are 165° and 210°, respectively (relative to true north). The gentler slope at Bonaventure is consistent with that site being better protected from ocean swell and the eastward-flowing alongshore current that derives from the mouth of the Gulf of St. Lawrence. The deepest sounding within the Bonaventure survey area reached approximately 11 m, while the Paspébiac study area reached 16 m maximum depth. Both areas showed an average roughness (bathymetric variability within a 2-m raster resolution) of 0.1 m.

The Baie des Chaleurs harbors a rich variety of marine animals, from marine mammals (e.g., humpback whale, white-beaked dolphin, and harp seal) and seabirds (e.g., northern gannet), to benthic invertebrates (e.g., sand dollar), large crustaceans (e.g., snow crab and American lobster), and pelagic and demersal predators (e.g., Atlantic salmon and cod, respectively). Benthic habitats are impacted by the deployment of lobster pots during summer months, but ice cover hinders human activities during winter. The area is otherwise in its natural state and has not been significantly impacted by human activities.

Methods

The Bonaventure and Paspébiac study areas were surveyed between 1 and 3 July, 2006, using a high-resolution underwater camera and an airborne bathymetric Light Detection and Ranging (LiDAR). A Scanning Operational Airborne LiDAR System (SHOALS) was deployed to acquire depth measurements, operated at a rate of 3,000 Hz, which yielded 2 m horizontal resolution depth soundings over Bonaventure's area (~16 km²) and 4 m resolution over Paspébiac's area (~33 km²).

Sediment properties and the occurrence of benthos were derived from high-resolution underwater camcorder observations. The camcorder was mounted on a tetrapod frame that was dropped down and pulled up as soon as it made contact with the bottom. All 814 GPS-located images represented 0.16 m² of the seabed.

A total of 293 photographs (out of 300) in Bonaventure and 453 photographs (out of 514) in Paspébiac enabled four grain-size classes to be recognized (Figure 10.2). These are boulders (>256 mm), cobbles (256–64 mm), pebbles (64–4 mm), and finer sediments (<4 mm). The percentage cover of each sediment class was estimated by means of a grid of 100 uniformly distributed points superimposed on the geometrically corrected photographs [5].

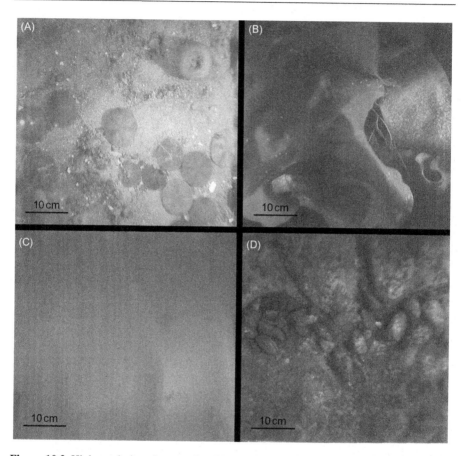

Figure 10.2 High-resolution photographs of bottoms of Paspébiac's nearshore, between 0 and 16 m depth. (A) Assemblage of *Laminaria* spp. and *Fucus* spp. (B) *Zostera marina* on finer sediments. (C) Mixture of pebbles and cobbles. (D) Mixture of cobbles and finer sediments with parts of *Laminaria* spp. (For interpretation of the references to color in this figure legend, the reader is referred to the web version of this book.)

Benthos composition was assessed using the same methodology presented for sediment characterization. A matrix of 300 stations by 12 biological variables, corresponding to the taxa-type components (Echinoidea (Figure 10.3A), Annelida (and inherent bioturbation marks), Gastropoda, Asteroidea, dead shells, *Fucus* spp. (Figure 10.2A), *Zostera marina* (Figure 10.2B), *Chondrus crispus*, *Laminaria* spp. (Figure 10.3B), *Chorda tomentosa*, Bivalvia (Figure 10.3D)) were Bonaventure's data set, and Paspébiac's data set comprised a matrix of 514 stations by 16 variables (12 previous ones + Anthozoa, Holothuroidea, Actinopterygii, *Polysiphonia* spp., and *Ulva lactuca*). The limitations of photograph-based identification only allowed macro-zoo-benthos to be identified down to the class rank, while the identification of macro-phyto-benthos reached either genus or species ranks.

The broad-scale geomorphic features were subdivided into smaller, fine-scale relief derived from LiDAR to characterize the larger features in terms of the smaller

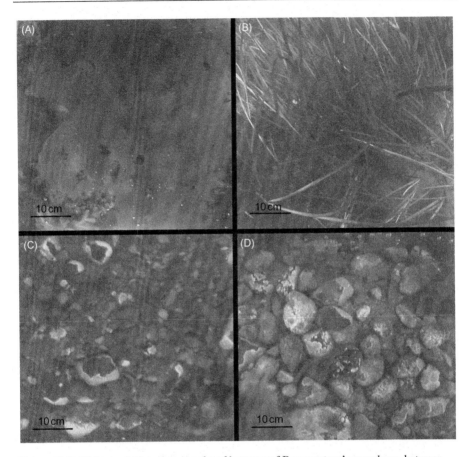

Figure 10.3 High-resolution photographs of bottoms of Bonaventure's nearshore, between 0 and 11 m depth. (A) Assemblage of Echinoidea (sand dollars, *Echinarachnius parma*) and an Anthozoa (sea anemone, *Metridium senile*) over pebbles and finer sediments. (B) A *Laminaria* spp. frond. (C) Sediment ripples. (D) Bivalvia and Echinoidea over boulders. (For interpretation of the references to color in this figure legend, the reader is referred to the web version of this book.)

features of which they are comprised. The thematic mapping of seabed geomorphic features in this study is based on terms and definitions standardized by the International Hydrographic Organization [6]. Geomorphic features were classified and mapped applying a decision tree to the morphometric variables of slope and curvature, using algorithms in the image analysis software ENVI 4.2 [7]. Three geomorphic features (ridges, passages, and seachannels) were identified in the two areas, both at broad scale and at fine scale (Figures 10.4 and 10.5). Ridges are defined by sloping surfaces that are convex in a cross section. Passages exhibit one convex curvature and one concave curvature. Seachannels are sloping surfaces that are concave in a cross section. In the following, the larger (broad-scale) geomorphic features are described in terms of mean depth, slope, and aspect values, as well as in terms of the occurrence of superimposed, small-scaled features.

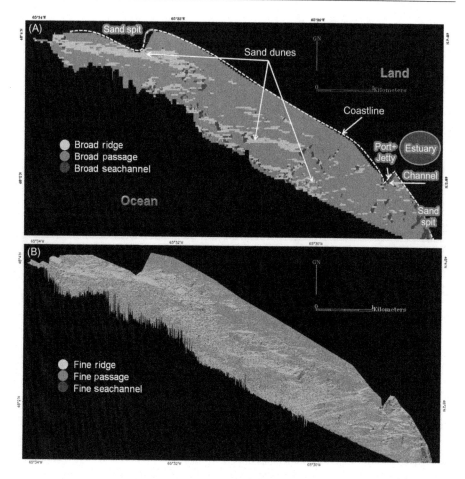

Figure 10.4 Broad- and fine-scale geomorphic features of Bonaventure's nearshore. (For interpretation of the references to color in this figure legend, the reader is referred to the web version of this book.)

Geomorphic Features and Habitats

Broad ridges: Broad ridges (Figures 10.4 and 10.5) cover 1.77 and 6.76 km^2, in Bonaventure and Paspébiac, comprising 17% and 39% of the total areas, respectively. The mean depth of broad ridges is 3.6 m at Bonaventure, and 5.8 m at Paspébiac. The slope of broad ridges was similar at both areas, being 0.0265° and 0.0338° at Bonaventure and Paspébiac, respectively. The aspect of broad ridges at Bonaventure was 132°, while it was 230° at Paspébiac. Within the two areas, the broad-ridge class was constituted of approximately the same percent of fine-scale features: 78% and 75% of fine ridges; 10% and 16% of fine passages; and 12% and 10% of fine seachannels, respectively. Broad ridges were constituted predominantly of fine-grained sediments in Bonaventure. Fine sediments also dominate in the Paspébiac study area, but supplemented with numerous boulders.

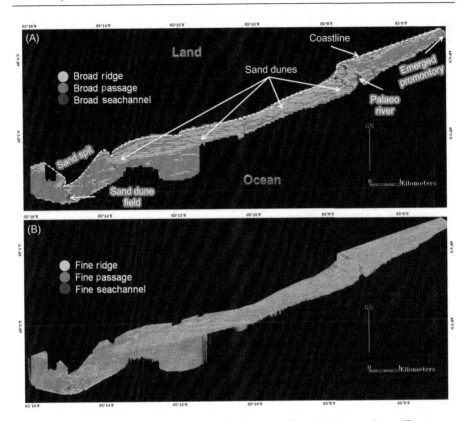

Figure 10.5 Broad- and fine-scale geomorphic features of Paspébiac's nearshore. (For interpretation of the references to color in this figure legend, the reader is referred to the web version of this book.)

Broad passages: Broad passages cover 8.22 and 10.41 km², in Bonaventure and Paspébiac, equal to 79% and 60% of each area, respectively. The mean slopes of broad passages are 0.0409° and 0.0573° in Bonaventure and Paspébiac, while the mean depths of these features are 2.3 and 6.3 m, respectively. Broad passages exhibit an aspect of 156.22° in Bonaventure and 222.32° in Paspébiac. Broad passages contained 11% fine seachannels in both sites, 40% and 34% fine passages, and 50% and 55% fine ridges at Bonaventure and Paspébiac, respectively.

Broad passages in Bonaventure (Figure 10.4) are situated mostly along the intertidal area both southeastward of the sand spit, located at the northwest end, and northwestward of the jetty. Broad passages at Paspébiac (Figure 10.5) occurred at the southwest and northeast ends of the area; that is to say, over deeper areas sheltered from currents by a sand spit in the southwest, and by an emerged sandstone promontory at the eastern boundary of the study area. Broad passages were mainly composed of pebbles in Bonaventure, and by fine-grained sediment in Paspébiac.

Broad seachannels: Broad seachannels cover only 0.44 and 0.27 km², equal to 4% and 2% of the area at Bonaventure and Paspébiac, respectively. The mean depth of broad seachannels is 3.4 m at Bonaventure, and 4.4 m at Paspébiac. The slope of broad seachannels was similar at both areas, being 0.0216° and 0.0247° at Bonaventure and Paspébiac, respectively. The aspect of broad seachannels at Bonaventure is 193°, while it is 164° at Paspébiac. Within the two areas, the broad-seachannel class was constituted of exactly the same percent of fine-scale features: 57% of fine ridges, 12% of fine passages, and 31% of fine seachannels, for both sites. At Bonaventure (Figure 10.4), occurrences of broad seachannels increased southeastward and were especially high within the channel of the port and along the massive sand spit at the southeast end. Sand dune fields occur in both study areas (Figures 10.4A and 10.5A). At Paspébiac, a submerged palaeo river channel crosses the study area. Broad seachannels are dominated by pebbles in Bonaventure, and by fine sediment in Paspébiac.

Biological Communities

All photographs (i.e., 300) in Bonaventure and 427 photographs (out of 514) in Paspébiac contained biological evidence. In terms of bottom surface observed, benthos covered 44% in Bonaventure and 30% in Paspébiac. Within Bonaventure, *Laminaria* spp., or kelp, was the most extensive macroalgae variable (48% of all observed biota), followed by the eelgrass *Zostera marina* (25%) and the wrack *Fucus* spp. (10%). *Laminaria* spp. and *Fucus* spp. were preferentially associated with a mixture of cobbles/pebbles (39%/46% and 45%/44%, respectively), and *Zostera marina* with finer sediments (49%). Dead shells were the most prominent macro-zoo-benthic feature (7%) and occurred frequently over pebbles (47%). In Paspébiac, *Laminaria* spp. (48%) was also the most widespread macroalgae, followed by the red algae *Polysiphonia* spp. (15%), *Fucus* spp. (5%), and *Chondrus crispus* (5%). They were most often found over boulders/pebbles (32% and 32%), boulders (36%), pebbles (48%), and finer sediments (49%). Echinoidea (10%) and Bivalvia (4%) constituted the most abundant class within macro-zoo-benthos and were monitored over pebbles/finer sediments (34% and 35%) and boulders (49%), respectively. Although occurrences of other taxa, such as Echinoidea, Annelida, and *Chorda tomentosa* in Bonaventure, as well as Actinopterygii in Paspébiac, were highly correlated with sediment type—that is, finer sediments (96%), finer sediments (65%), cobbles (51%), and pebbles (1%), respectively—they were never abundant.

Surrogacy

Sediment type, slope, and aspect did not emerge as useful surrogates to explain the occurrence of different biota. However, water depth is a good surrogate; for example, epimacrobenthos tended to show a higher biodiversity toward shallower depths. Species richness increased from 5 to 12 with decreasing depth at Bonaventure. In Paspébiac, species richness increased from 5 to 20 with decreasing depth. The Pielou (diversity index related to the normalized information entropy; that is to say, the normalized Shannon index) and Simpson (diversity index based upon the fraction of individuals with respect to category) indices increased with decreasing depth in both study areas (Figure 10.6).

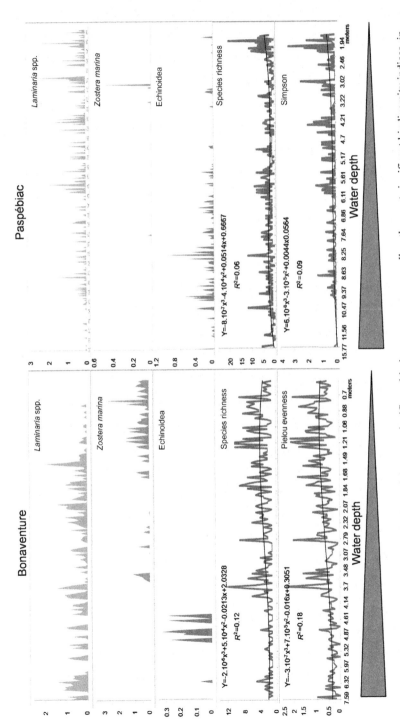

Figure 10.6 Area and line plots of distributions of the most significant biotic components, as well as the most significant biodiversity indices, in Bonaventure and Paspébiac.

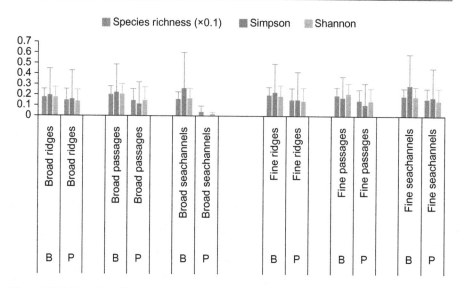

Figure 10.7 Bar plots of species richness (×0.1), Simpson index, and Shannon index values of the three broad-scale (100 m) geomorphic features in Bonaventure (B) and Paspébiac (P), and of the three fine-scale (10 m) geomorphic features in Bonaventure (B) and Paspébiac (P). Error bars show the standard deviations.

Several patterns can be highlighted in the relationships between biodiversity indices and the broad-scale geomorphic features. Species richness, Simpson index, and Shannon index values were slightly higher in broad passages than in broad ridges, but were significantly lower in broad seachannels (Figure 10.7). Regionally, there were important differences between Bonaventure's and Paspébiac's indices (Bonaventure has a higher biodiversity). Given the sampling gear used to collect the underwater photographs, biodiversity indices may be biased and underestimated when *Laminaria* spp. are abundant because of the device-induced horizontal spreading of thalli over seabed. Ancillary factors such as current speed, food supply, temperature range, predation pressure, and disturbance by fishing activities known to impact on structure and dynamics of epimacrobenthic communities [8] must be considered to properly model their spatial patterns.

Acknowledgments

Collin expresses deep appreciation to the Geoide project, Fisheries and Oceans Canada, and the Naval Oceanographic Office. The authors are grateful to anonymous reviewers and editorial board for providing useful comments on earlier drafts of the manuscript.

References

[1] P. Jutras, G. Prichonnet, S. McCutcheon, Alleghanian deformation in the eastern Gaspé Peninsula of Québec, Canada, Geol. Soc. Am. Bull. 115 (2003) 1538–1551.

[2] J.P.M. Syvitski, Marine geology of Baie des Chaleurs, Géographie physique et quaternaire 46 (3) (1992) 331–348.

[3] J.C. Bonardelli, K. Drinkwater, J.H. Himmelman, Current variability and upwelling along the north shore of Baie des Chaleurs, Atmos. Ocean 31 (4) (1993) 541–565.

[4] St. Lawrence Global Observatory. http://slgo.ca/.

[5] P. Archambault, K. Banwell, A.J. Underwood, Temporal variation in the structure of intertidal assemblages following the removal of sewage, Mar. Ecol. Prog. Ser. 222 (2001) 51–62.

[6] IHO. Standardization of undersea feature names: guidelines proposal form terminology, in: International Hydrographic Organisation and International Oceanographic Commission, Monaco, 2008.

[7] Research Systems. ENVI User's Guide, ENVI Version 4.2, Research Systems, Inc., Boulder, CO 80301, 2005.

[8] V.E. Kostylev, R.C. Courtney, G. Robert, B.J. Todd, Stock evaluation of giant scallop (*Placopecten magellanicus*) using high-resolution acoustics for seabed mapping, Fish. Res. 60 (2001) 479–492.

11 The Wadden Sea in the Netherlands: Ecotopes in a World Heritage Barrier Island System

Norbert Dankers, Willem van Duin, Martin Baptist, Elze Dijkman, Jenny Cremer

Institute for Marine Resources and Ecosystem Studies (IMARES-Wageningen UR), Department of Ecosystems, Texel, The Netherlands

Abstract

The Wadden Sea forms the coastal strip bordering the North Sea in NW Europe. Along most of its length it is separated from the North Sea by a string of barrier islands. The system consists of many recognizable geomorphic units (here called ecotopes), such as islands, high sandbanks, sand dunes, salt marshes, subtidal and intertidal sand and mudflats, and channels of different depths and widths. Within these units communities of biota such as mussel beds and seagrass fields can also be recognized and mapped. For management purposes, it is important that information on these units is available on different scales, depending on the management issue. For mapping it is therefore important to develop legends for different variables. The classes of the legend should be based on ecological considerations and reflect the habitat requirements of species, communities, or biotopes. The classes used are, for example, percentage emergence during a tidal cycle, sediment grainsize, and sediment dynamics based on shear stress and shear strength as well as potentiality of occurrence of bioengineers that are important as habitat or substrate for other organisms.

Key Words: Habitat, community, biotope, ecotope, Wadden Sea, barrier island, tidal flats

Introduction

The International Wadden Sea is located in northwestern Europe as a fringe along the North Sea, stretching along the coasts of the Netherlands, Germany, and Denmark between 53°00′N−4°40′E and 55°35′N−8°20′E (Figure 11.1). The area developed over the last 6,000 years, after the North Sea was formed in the millennia before when sea level rose 60–80 m during the last postglacial transgression. Since June 2009, the Wadden Sea has been listed as a UNESCO World Heritage site, and hence the area is very well described in the World Heritage nomination dossier http://www.waddenseasecretariat.org/management/whs/whs.html [1]. The Wadden Sea covers

Seafloor Geomorphology as Benthic Habitat. DOI: 10.1016/B978-0-12-385140-6.00011-6

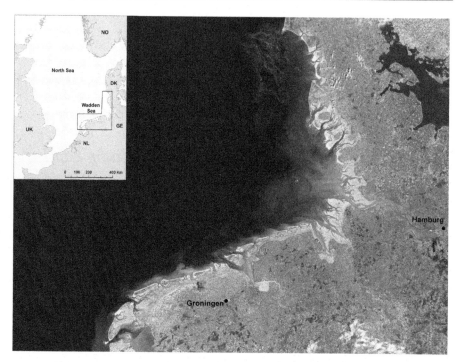

Figure 11.1 Satellite image of the Dutch, German, and Danish Wadden Sea.
© Eurimage 2003, Common Wadden Sea Secretariat, Brockmann Consult (with permission).

an area of 10,000 km², with 5,300 km² comprising intertidal flats and salt marshes. The permanently submerged areas consist mainly of gullies and channels with water depths up to 40 m in the tidal inlets. The area is characterized by a series of barrier islands behind which an intricate system of channels and intertidal flats has developed. The Wadden Sea intertidal flats are characterized by their location within tidal basins behind barrier islands. They contain a sequence of large to small ebb and flood gullies. The flats near the inlet are predominantly sandy and those near the coast are muddy. The Wadden Sea is a mesotidal barrier island system that only has minor river influences, fringing the flat and low-lying NW European coastal plain. In contrast to most mudflat systems in the world, the Wadden Sea area has not been formed as a river delta, nor is part of an estuary or bay, although there are some estuaries and bays within the Wadden Sea itself. The Wadden Sea consists of a number of very different habitats, ranging from islands to intertidal flats, salt marshes, and gullies and channels. Depending on the mapping scale, these can be further subdivided into smaller recognizable units. For management purposes, it is essential that the proper scale is adhered to when discussing specific issues. On a scale of 1:500,000, for example, individual salt marshes and tidal flats can be recognized, but for management of vegetation types in salt marshes maps with scales around 1:5,000 may be necessary. Many inventories are carried out in the region, most of them based on a specific management issue such as fisheries, nature conservation, and management of migratory birds. The scale of the available information is not always easy to match across the different surveys. Every 3–5 years, a Quality Status Report (http://www.waddensea-secretariat.org) is produced in which much information is brought together, some of which is used in constructing habitat or ecotope maps. The nature value of the Wadden Sea is unmistakably high. Natural processes that have developed the Wadden Sea are still fully operational and are responsible for maintaining the system. Ebb and flood tides are unrestricted and transport large amounts of sediment. These transports result in a net import on the order of 20 million m³/year. They cause the tidal flats to keep up with sea-level rise and to maintain the variety of sediment types. Organisms can find their preferred niches and bioengineers, such as mussels and oysters, from biogenic structures and reefs. Eolian transport is responsible for dune formation on extensive beach plains. Vegetation on salt marshes maintains these marshes by stimulating sedimentation and creek formation. However, the area is far from pristine. Part of the salt marshes are man-made; dune ridges on the barrier islands were extended by stimulating sedimentation of wind-blown sand; and beach nourishment is carried out for coastal protection every year. Shrimp and shellfish fisheries are still extensive and may influence the benthic communities. Along the mainland side of the Wadden Sea, most of the coastline is now protected by massive dikes. Much research and effort is expended in the Wadden Sea to find a balance between nature and human use, for example, fisheries (mainly shellfish and shrimps), gas extraction, coastal defenses, and recreation.

Geomorphic Features, Habitats, and Ecotopes

During the period of sea-level rise in the early Holocene, sandy sediment was transported toward the higher grounds along the North Sea, where ridges of sandbanks and dunes developed. When sea-level rise decreased to the present level of 20 cm per century, the dune islands remained more or less stationary and silty salt marshes developed leeward of the islands and along the mainland coast. During transgressions these marshes were flooded, and the Wadden Sea developed, most of which evolved around 1,000 years ago. During the last 500 years, much land was reclaimed. Along the coasts, salt marsh growth was promoted by developing shelter in the form of brushwood groynes and drainage. In these marshes, the intricate system of branched channels was replaced by a very regular pattern that can even be recognized from space. This type of reclamation has now ceased.

The diurnal tides enter and leave through the gaps between the barrier islands, forming flood and ebb channels. The flood currents meet behind the islands at the tidal divides that are characterized by their high altitude and silty sediments. Several mechanisms are responsible for the transport and deposition of sediment [2], resulting in tidal flats with coarse sediment near the tidal inlets to very silty ones near the mainland coast and the tidal divides. The very diverse physical processes operating in the Wadden Sea, such as tides, waves, wind, and freshwater inflow, have large natural ranges in magnitude. These mechanisms, together with biological processes that often involve "bioengineers" resulting in the formation of biogenic structures, lead to a variety of recognizable features, here called ecotopes (but often also called habitats, as defined in the following section).

Habitats and Ecotopes

In general, a habitat is defined as the characteristic space occupied by an individual, an organism, a species, a community, or a population. Populations, communities, or species occurring in the same geographic area often have clearly different habitats or parts of the environment in which they live. For sedentary organisms, the habitat is one specific place. In the case of organisms that are free to move, such as birds and fish, the habitat may be in very different locations depending on their life cycle or even time of day. A biotope is mostly defined as an area occupied by a community or, incorrectly in our view, an individual or an organism. The distinction between the two definitions is not always clear. When the term ecotope is used, it is meant to be almost synonymous with biotope, but in general relevant physical and geomorphologic parameters are also included in the definition. An ecotope can therefore be considered to be a distinct, recognizable part of an ecosystem that can be mapped easily.

For mapping, it is important that elements can be distinguished based on discernable characteristics. When a single organism uses different habitats for different purposes (e.g., salt marsh for resting, mussel bed for feeding, and a distant area for breeding), simple correlations between habitats and the occurrence of species are confounded. For the Wadden Sea maps, an approach was followed in which areas with similar (ecological) characteristics could be mapped. The system was based on a freshwater methodology developed by Verdonschot et al. [3]. They followed a hierarchical division of the ecosystem as used in the Netherlands in the terrestrial environment [4] that was based largely on a concept developed by Bailey [5,6]. New terminology was developed in order to discern different hierarchical scales. The highest category, called "Ecodistrict," was mapped at a scale of 1:50,000,000. The Wadden Sea would be called an "Ecosection" because it is maintained by morphodynamic processes. An "Ecotope" has a minimum size of 2,500–15,000 m² and can be mapped at scales of 1:5,000–1:25,000. For small specific units, such as a mussel bed, the term "Eco-element" can be used. The definition of an ecotope would therefore be:

a geographical unit homogeneous within limits for the most important hydraulic, morphological, and physico-chemical environmental factors that are relevant for biota. [3]

Protection of habitats or ecotopes is considered to be the best way to protect species [7]. The European Union therefore distinguished certain "habitat types" that are in need of protection. Several of these types occur in the Wadden Sea, resulting in its being given high-protection status. The protection status was laid down in international law (EU Habitats Directive, 92/43/EEC), and the interrelations between the habitat types were formally adopted in an European-wide regional network of protected sites known as NATURA 2000 (http://ec.europa.eu/environment/nature/index_en.htm). The Wadden Sea has been listed as a NATURA 2000 site. The European Union developed a habitat classification system called EUNIS (http://eunis.eea.europa.eu/habitats.jsp). The listing of the Wadden Sea was based on the occurrence of the following (EUNIS) habitats:

- 1110 sandbanks, which are slightly covered by seawater all the time
- 1130 estuaries
- 1140 mudflats and sandflats not covered by seawater at low tide
- 1310 *Salicornia* and other annuals colonizing mud and sand
- 1330 Atlantic salt meadows (*Glauco-Puccinellietalia maritimae*)

Also included are more terrestrial habitat types such as EUNIS 2110, 2120, and 2130 (embryonic shifting dunes, shifting dunes along the shoreline with *Ammophila arenaria* ("white dunes"), and fixed dunes with herbaceous vegetation ("gray dunes"), respectively).

The Ecotopes of the Wadden Sea

Along the landward side of the Wadden Sea, the majority of the coast consists of man-made dikes along low-lying land and reclaimed fresh and brackish marshes. In most subregions, there are (man-made) salt marshes along these dikes. A small part of the mainland coast consists of naturally sloping Pleistocene deposits that have been shaped by ice sheets and eolian processes during previous ice ages. Along the seaward side of the Wadden Sea, there is a chain of 25 vegetated and mostly inhabited larger islands and some permanently exposed sandbanks, resulting in 30 tidal inlets. Three of the islands have a Pleistocene nucleus deposited by glaciers 200,000 years ago. Because tidal (flood) flows converge behind the islands, tidal basins have developed that have limited water exchange between them. The tidal basins span the area between two tidal divides and vary in length between a few and more than 30 km. Each contains all the characteristics of the Wadden Sea, such as channels and intertidal flats. In the southeastern inner part of the Wadden Sea, there are a small number of islands (Halligen) with a different origin. They are remnants of larger salt marshes that were swallowed by the sea between 1,000 and 500 years ago. The tidal range across the Wadden Sea varies from 1.2 m in the southwestern and northeastern parts to more than 3.5 m in the southeastern bight. In areas with small tidal ranges (1.2–2.5 m), the tidal flats lie behind barrier islands; in contrast, in areas where the tidal range is 2.5–3.5 m, the flats are open to the sea.

The tidal prism in the inlets is up to 1 km^3 per tide. Altogether 15 km^3 of water is exchanged with the North Sea every 6 h. At low tide the Wadden Sea contains 15 km^3 of water and at high tide, 30 km^3.

There are three medium to large rivers entering the Wadden Sea: Ems (80 m³/s), Weser (327 m³/s), and Elbe (711 m³/s), resulting in small to medium estuaries. In the Dutch part, Lake IJssel discharges through sluices at an average of 500 m³/s.

Specific geomorphic units that can be delineated in the Wadden Sea are:

- Islands with coastal dunes (1,000 km²)
- Salt marshes (400 km²)
- Intertidal flats (4,900 km²)
- Subtidal regions, channels, and gullies (3,700 km²)

Ecotopes have been mapped in more detail in the Wadden Sea using several approaches. The first attempt was made by Dijkema and coworkers [8–10]. They used nautical charts, ordnance maps, aerial photography, vegetation maps, ground-truthing, and information from tide gauges. A large improvement from the pre-viously published charts was that they did not indicate the height of the tidal flats according to ordnance level (mean sea level, MSL). Dijkema et al. incorporated tidal ranges and their classification was based on the period of inundation (33%, 50%, and 67% of time covered by water), which is ecologically more relevant to intertidal ben-thic communities. For example, in the western part an area at 1 m below MSL hardly ever emerges; in the eastern part, that level emerges 2 h each tide.

A study by Bouma et al. [11] described the ecotopes of the Wadden Sea based on physical variables by considering water depth, dynamics (wave action), and emer-gence time as the major determining factors. Sediment composition was considered to be a result of wave and current dynamics. The presence or absence of certain organisms was assumed not to be related directly to sediment type.

For the Dutch Wadden Sea, a revision of the Dijkema map [10] was developed recently by Dankers et al. [12]. This new atlas gives a more detailed classification scheme than the EU typology. It is considered that the distribution and abundance of organisms in this dynamic environment are determined principally by abiotic fac-tors. The legend was therefore based on (combinations of) physical variables that were considered important. In contrast to the Bouma approach, Dankers et al. [12] included sediment type as a separate factor, and in their atlas the following variables were used and subdivided into a small number of classes:

- Depth (intertidal or subtidal)
- Emergence time (low intertidal ≤25% of the time emerged, middle intertidal = 25–75% of the time emerged, and high intertidal ≥75% of the time emerged)
- Sediment characteristics (percentage silt, fine sand, and coarse sand)

For the Wadden Sea as a whole, no reliable areal estimates of different (physically defined) ecotopes can be given because international agreement on mapping classi-fications has not yet been agreed upon. For the Dutch Wadden Sea, legend classes are agreed upon and more information can therefore be provided for the different tidal basins (Table 11.1). The table gives ecotopes at a scale that can be mapped at the level of flood basins. As an example, only the Dutch tidal basins are presented (Figure 11.2). For individual basins, more detailed information can be given and eco-elements, such as mussel beds, can be mapped (Figure 11.3). Examples of different ecotopes are shown in Figure 11.4.

Table 11.1 Area of Intertidal Ecotopes (in hectares, ha) in the Dutch Part of the Wadden Sea (For the Location of the Tidal Basins, See Figure 11.2. Intertidal Is Area Between Mean Spring Low Tide and Mean Spring High Tide)

Emergence Time	<25%			25–75%			>75%			Total per Tidal Basin
Sediment	Silty	Fine Sand	Coarse Sand	Silty	Fine Sand	Coarse Sand	Silty	Fine Sand	Coarse Sand	
TIDAL BASIN										
Marsdiep	832	8,168	357	1,284	2,784	31	279	64		13,799
Eierlandse Gat	306	4,024	245	165	5,709	384		599	91	11,522
Vlie	2,964	14,883	784	3,619	12,580	470	13	501	80	35,894
Borndiep	1,228	6,820	153	3,379	7,131	105	27	106	7	18,956
Ameland/Schiermonnikoog										
Pinkegat	85	1,020		734	2,443			47		4,330
Zoutkamperlaag	423	1,172	4	1,864	4,311	3	27	496		8,301
Rottum										
Eilanderbalg	86	314	1	549	1,856	12	35	232	2	3,087
Lauwers	104	1,823	2	1,350	5,260	1	9	452		9,001
Schild	23	317		149	1,761		4	342		2,596
Ems-Dollard	1,189	1,530	12	5,170	3,691	4	1,220	284		13,100
Total per ecotope	**7,239**	**40,070**	**1,558**	**18,263**	**47,527**	**1,010**	**1,614**	**3,124**	**180**	**120,586**

Characteristics Ecotope

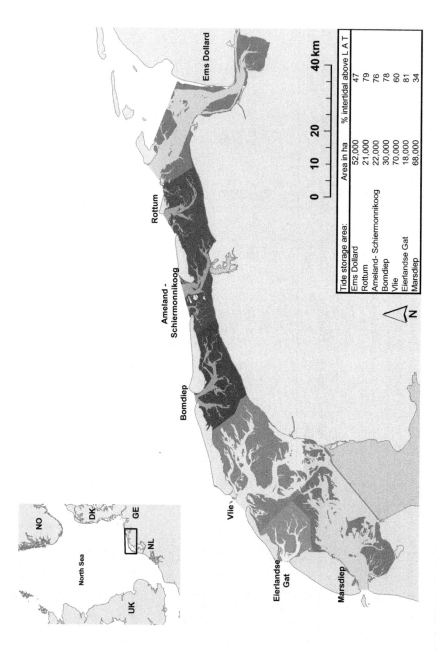

Tide storage area:	Area in ha	% intertidal above L A T
Ems Dollard	52,000	47
Rottum	21,000	79
Ameland- Schiermonnikoog	22,000	76
Bomdiep	30,000	78
Vlie	70,000	60
Eierlandse Gat	18,000	81
Marsdiep	68,000	34

Figure 11.2 Tidal basins in the Dutch part of the Wadden Sea. Area of each basin (in hectares, ha) and the percentage of intertidal mudflat above lowest astronomical tide (LAT) are indicated.

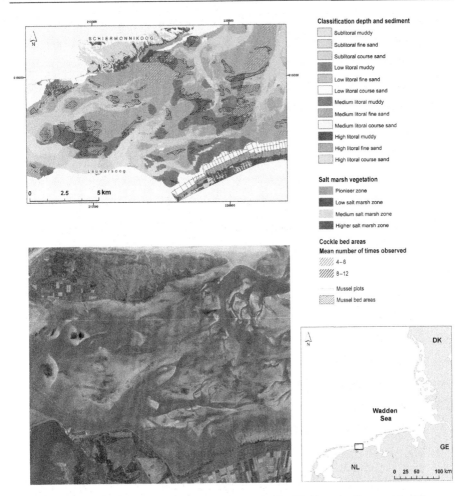

Figure 11.3 Map of Wadden Sea area south of the Island of Schiermonnikoog with different intertidal ecotopes, and a satellite photo from the same area from Google Earth.

Biological Communities

Biological communities that can be considered as ecotopes—or, when (almost) monospecific, eco-elements—and can therefore be mapped, include, for example, cockle beds, mussel beds, oyster reefs, seagrass beds, and different types of salt marshes and sand dunes.

In the ecological atlases of Dijkema et al. [8] and Dankers et al. [12], the majority of these features were mapped. Dijkema et al. gave areal estimates of mussel beds based on aerial photography. Dankers et al. have all maps in GIS and can therefore calculate areas for each eco-element. Mussel beds and cockle beds in the atlas of Dankers et al. are cumulative areas from multiple inventories over a number of

Figure 11.4 Examples of different ecotopes in the Wadden Sea: (A) mudflat with gullies; (B) man-made salt marsh with creek; (C) mussel (*Mytilus edulis*) bed; (D) Pacific oyster (*Crassostrea gigas*) reef.

years; they called them "mussel or cockle areas" following the approach developed by Herlyn and Millat [13] for the German Wadden Sea. Since mussel beds often develop in areas where they have been before, this is considered a useful approach for management, as not only the actually present beds are on the maps but also the areas where they often occur and may soon develop.

In the Wadden Sea, considerable areas of salt marshes occur. It is the largest stretch of salt marsh in Europe. These salt marshes, however, are a modest remainder

Figure 11.4 (Continued)

of the vast saline and brackish landscapes, peat regions, and lakes that were situated in the borderlands between the high lying Pleistocene land and the sea until about 1,000 years ago [14].

Salt marshes originate naturally on intertidal flats around the mean high tide level that are sheltered against waves and currents and have an adequate supply of sediment and plant parts or seeds. Interactions between physical and biological processes lead to the development of tidal flats with a few pioneer plants that may eventually form into a salt marsh, vegetated with halophytes. The salt marsh is typically situated above the mean high water mark and when secondary vegetation has developed becomes characterized by an accompanying geomorphologic pattern of creeks, levees, and basins within a period of 10–20 years. Purely natural development of the Dutch, German, and Danish salt marshes has become a rare phenomenon these days. Hard, man-made coastal defenses and extension of man-made salt marshes or coastal polders have little regard for natural processes, with the narrowing of the distance between mainland and islands as one of the consequences. This is known as "coastal squeeze" [15] and is considered to be responsible for an increase in dynamics because of wave reflection [15]. Moreover, sheltered bays suitable for salt marsh development have disappeared in favor of straight sea walls. The same physical processes that determine sediment habitat types in nearshore environments often also impose restrictions on the settlement and survival of certain benthic organisms. On the other hand, some species rely on specific environmental conditions for feeding, building burrows, or attachment. In the Wadden Sea, it has not been possible to develop robust methods for describing specific communities of macrozoobenthos associated with specific habitats, except for hard substrate communities in this environment of predominantly soft sediment. Many species show a distribution where highest biomass occurs in "average" conditions for tidal level, wave action, grain size, and so on [16]. The community as a whole, therefore, reaches highest biomass and diversity there. Few species have very specific requirements for sediment and only occur in specific sandy or silty environments.

If all environmental information is in Arc-GIS, the aerial cover for each specific ecotope can be presented and updated at regular intervals, especially in the case of dynamic ecotopes, such as mussel or cockle beds.

Surrogacy

Habitat or ecotope mapping is possible when general information is available for biotic and abiotic variables. We started with mapping the more important physical aspects, such as sediment type, emergence time, and shear stress. Sediment type was based on inventories in which samples were taken on a grid of 1 km×1 km. This information was collated and extended for specific management purposes by Zwarts et al. [17]. Emergence time is considered to be important, both for feeding possibilities for suspension feeders and for exposure to predation pressure by either fish or birds. Emergence time was calculated by overlaying bathymetry and tidal curves for specific locations in a 50 m×50 m grid.

On the basis of these variables, specific ecotopes were mapped. These maps were improved by adding information on the occurrence of mussel beds, cockle beds, salt marshes, and the like. Other biological attributes, such as seal haul outs and bird foraging areas, can be added when required for specific management issues. Several geostatistical techniques were available for developing areal coverage information (e.g., kriging) when only observations at point locations are available. When legends are developed with subdivisions based on ecological requirements, maps can be produced for different management questions. It is also possible to develop "potential" ecotope maps if the ecological requirements of these ecotopes are known. These maps are instrumental in developing nature protection areas where the aim is that specific species or ecotopes develop in the future. These maps can be added to the GIS files and can also be used in management decisions. In the Dutch Wadden Sea, this has been done for the potential development of mussel beds [18], and these maps have been used in the management of cockle, mussel, and lugworm fisheries.

References

[1] CWSS-World Heritage Nomination Group, Nomination of the Dutch-German Wadden Sea as a World Heritage Site, Wadden Sea Ecosystem No. 24, Common Wadden Sea Secretariat, Wilhelmshaven, Germany, 2008, p. 200.

[2] J. Dronkers, Import of marine sediment in tidal basins, Neth. J. Sea Res. Publ. Ser. 10 (1984) 83–107.

[3] P.F.M. Verdonschot, J. Runhaar, W.F. van der Hoek, Aanzet tot een ecologische indeling van het oppervlaktewater in Nederland, IBN Leersum/CML Leiden, 1992 (in Dutch).

[4] F. Klijn, U. de Haes, Hiërarchische ecosysteemclassificatie. Voorstel voor een eenduidig begrippenkader, Landschap 7 (1990) 215–233 (in Dutch).

[5] R.G. Bailey, Suggested hierarchy of criteria for multiscale ecosystem mapping, Landsc. Urban Plann. 14 (1987) 313–319.

[6] R.G. Bailey, Explanatory supplement to ecoregions map of the continents, Environ. Conserv. 16 (1989) 307–310.

[7] N. Dankers, Managing ecotopes: the best way to conserve habitats for species, Trends in Ecol. and Evol. 11 (1) (1996) 38.

[8] K.S. Dijkema, G. van Tienen, J.G. van Beek, Habitats of the Netherlands, German and Danish Wadden Sea 1:100,000. Veth Foundation/Research Institute for Nature Management, Leiden/Texel, 1989, 24 maps+6 pp.

[9] K.S. Dijkema, Towards a habitat map of the Netherlands, German and Danish Wadden Sea, Ocean Shoreline Manage. 16 (1991) 1–21.

[10] K.S. Dijkema, Habitats of the Netherlands, German and Danish Wadden Sea: an outline map 1:100,000, in: N. Dankers, C.J. Smit, M. Scholl (Eds.), Proceedings of the 7th International Wadden Sea Symposium, Ameland, The Netherlands, 22–26 October 1990, Neth. J. Sea Res. Publ. Ser., 20 (1992) 239–242.

[11] H. Bouma, D.J. de Jong, F. Twisk, K. Wolfstein, Zoute wateren EcotopenStelsel (ZES.1). Report RIKZ/2005.024, Rijkswaterstaat, Rijksinstituut voor Kust en Zee/RIKZ, Middelburg, 2005, pp. 156 (in Dutch) (English version: "Salt Water Ecotopes" also available from authors).

[12] N. Dankers, J. Cremer, E. Dijkman, S. Brasseur, K. Dijkema, F. Fey, et al., Ecologische Atlas Waddenzee, IMARES Wageningen UR, Texel, 2008, p. 32 (in Dutch).

[13] M. Herlyn, G. Millat, Wissenschaftliche Begleituntersuchungen zur Aufbauphase des Miesmuschelmanagements im Nationalpark "Niedersächsisches Wattenmeer". Niedersächsische Wattenmeerstiftung, Projekt Nr. 32/98, Abschlussbericht März, 2004, p. 226 (in German).

[14] K.S. Dijkema, Changes in salt-marsh area in the Netherlands Wadden Sea after 1600, in: A.H.L. Huiskes, C.W.P.M. Blom, J. Rozema (Eds.), Vegetation Between Land and Sea, Junk Publishers, Dordrecht, 1987, pp. 42–49.

[15] B.W. Flemming, Effects of climate and human interventions on the evolution of the Wadden Sea depositional system (southern North Sea), in: G. Wefer, W. Berger, K.-E. Behre, E. Jansen (Eds.), Climate Development and History of the North Atlantic Realm, Springer, Berlin, 2002, pp. 399–413.

[16] N. Dankers, J.J. Beukema, Distributional patterns of macrozoobenthic species in relation to some environmental factors, in: N. Dankers, H. Kühl, W.J. Wolff (Eds.), Invertebrates of the Wadden Sea, Balkema, Rotterdam, 1981, pp. 69–103.

[17] L. Zwarts, W. Dubbeldam, K. Essink, H. van de Heuvel, E. van de Laar, U. Menke, et al., Bodemgesteldheid en mechanische kokkelvisserij in de Waddenzee. RIZA rapport 2004.028. RIZA, Lelystad, 2004, p. 129 (in Dutch) (report including data is also available on CD).

[18] N. Dankers, A.G. Brinkman, A. Meijboom, E. Dijkman, Recovery of intertidal mussel beds in the Wadden Sea: use of habitat maps in the management of fishery, Hydrobiologia 465 (2001) 21–30.

12 Sand Wave Field: The OBel Sands, Bristol Channel, UK

J.W. Ceri James[1,3], Andrew S.Y. Mackie[2], E. Ivor S. Rees[3], Teresa Darbyshire[2]

[1]British Geological Survey, Keyworth, Nottingham, England, UK, [2]Amgueddfa Cymru—National Museum Wales, Cathays Park, Cardiff, Wales, UK, [3]School of Ocean Sciences, Bangor University, Menai Bridge, Wales, UK

Abstract

The OBel Sands, an area of sand waves up to 19 m high, cover an extensive area, >1,000 km² , in the Outer Bristol Channel off the coast of Wales. The sand wave field can be divided into a northern half with a dense concentration of bedforms on a sand substrate, and southern half with isolated sand waves on a coarse substrate. In both areas, the sand waves are generally asymmetric in cross profile, with steep west-facing lee slopes associated with the Channel's ebb tides. The sand waves commonly have abundant megaripples and secondary sand waves on their slopes; these dynamic environments maintain little or no epifauna. The infaunal assemblages are varied and primarily related to sediment composition, sediment stability, and depth. Species richness is highest in coarse sediment between isolated sand waves and on the nearby platform. These areas generally support a rich epifauna.

Key Words: marine habitat, sand wave, macrofauna, diversity, multivariate analyses, Bristol Channel

Introduction

The OBel Sands, an extensive area of sand waves up to 19 m high, cover over 1,000 km² in the Outer Bristol Channel off the Welsh coast (Figure 12.1). The sand wave field can be divided into a northern half, the NOBel Sands, with a dense concentration of bedforms on a sand substrate, and a southern half, the SOBel Sands, with isolated sand waves on a coarse sediment substrate. The sand wave field stretches west to east for about 40 km in its northern half; it narrows to the south to a width of about 12 km. Its north–south extent is over 37 km. The OBel Sands are surrounded by a sand sheet to the north, and elsewhere by a seabed predominantly of coarse sediment and rock; including these, the total area studied is about 2,400 km² (Figure 12.1).

Seafloor Geomorphology as Benthic Habitat. DOI: 10.1016/B978-0-12-385140-6.00012-8

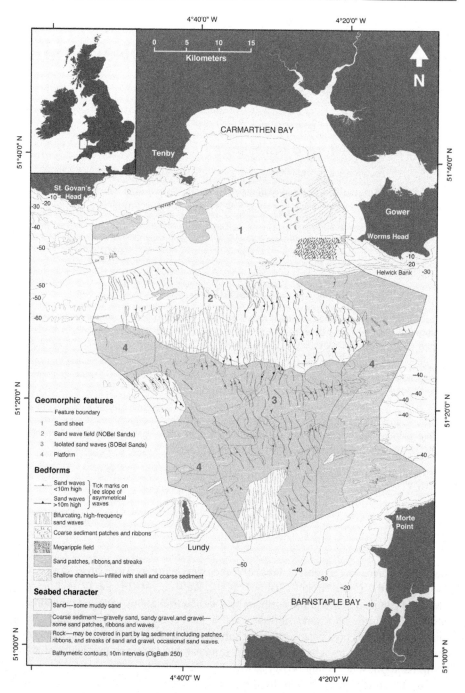

Figure 12.1 Geomorphic features, bedforms, and seabed character map of the Outer Bristol Channel study area.

Across the OBel Sands, the ambient seabed declines from east to west at a depth range of about 37 to >55 m. The Bristol Channel has the second highest tidal range in the world. Although not as large as further east, where the Bristol Channel becomes narrower, the tidal range in the Outer Bristol Channel is significant, with a mean spring tidal range varying from around 6.5 to 8 m. The tidal wave emanating from the Atlantic enters the study area from the southwest. Over the OBel Sands, the depth averaged mean spring tidal currents are relatively strong, with values of between 1.0 and 1.3 m/s. It is only within Carmarthen Bay and the lee of Lundy Island that tidal currents decrease to <0.5 m/s. The tidal streams are commonly rectilinear, with the western directed ebb tidal constituent being dominant. This tidal asymmetry is also matched by the dominance of west-facing asymmetry in the large sand waves of the area, commonly indicative of net sand transport to the west.

The Outer Bristol Channel is open to the west and southwest. This is the source direction of the strongest and most frequent winds. It also equates with the direction of the longest fetch out into the Atlantic—over hundreds of kilometers. However, modelling indicates that wave effects on the seabed in the Outer Bristol Channel are only likely to be significant in water depths <20 m [1].

The OBel Sands are in an area where human impact on the benthic environment is designated as high [2]. It is an area affected by shipping, some fishing, telecommunication cables, and recent initiation of aggregate extraction. It has also been designated as a potential area for wind farm development [3].

The area was surveyed and studied from 2003 to 2006, and five research cruises were conducted [4,5]. The geophysical survey strategy was based on 11 parallel corridors, 30–45 km long and about 5 km apart (Figure 12.2) covered with multibeam and sidescan sonar across a kilometer-wide swath and a single line of boomer sub-bottom profiler down the corridor center line. In total 2,177 line km of multibeam, 1,436 line km of sidescan, and 330 line km of boomer were collected. Singlebeam echo sounder data was also used to provide a seabed morphology model across most of the OBel Sands based on a 50-m grid (Figure 12.2) and enabled mapping and correlation of the major sand waves between the survey corridors.

Within the survey corridors, seabed samples were collected using a 0.1-m^2 modified Van Veen grab and provided 134 macrofaunal stations and 140 sediment samples. A 2-m beam trawl was deployed at 13 trawl box locations and completed 53 tows. The video and camera sledge was towed across the seabed at 24 locations and was particularly useful in ground truthing the multibeam and sidescan data and interpretations (Figure 12.3).

Geomorphic Features and Habitats

The present-day distribution of sediments and geomorphic features in the OBel Sands area appears to be the product of sedimentation processes associated with two major geological environments. The first is glacial and glacio-fluvial associated with the Quaternary Last Glaciation, when an ice lobe extended south out into Carmarthen Bay and the northern half of the OBel Sands and initially deposited

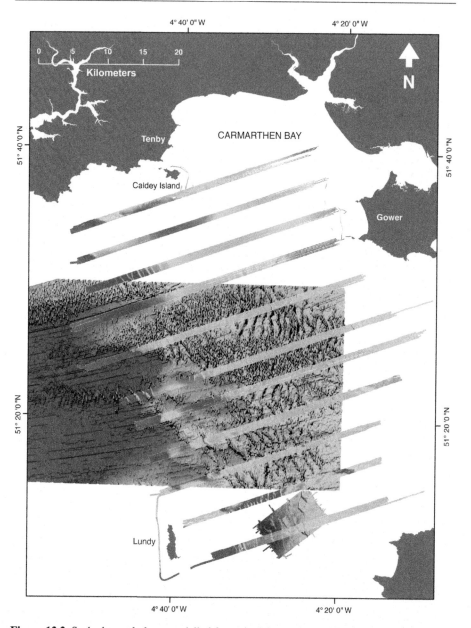

Figure 12.2 Seabed morphology modelled from singlebeam data (1977) overlain with multibeam data (corridors 1–11, south to north: 2003–2004; southeast rectangle: 2002). (Multibeam rectangle provided by UK Maritime and Coastguard Agency. Singlebeam data derived in part from material obtained from the UK Hydrographic Office with the permission of the Controller of Her Majesty's Stationery Office and UKHO. © British Crown & SeaZone Solutions Ltd. 2004. All rights reserved. Data License No. 112005.006.)

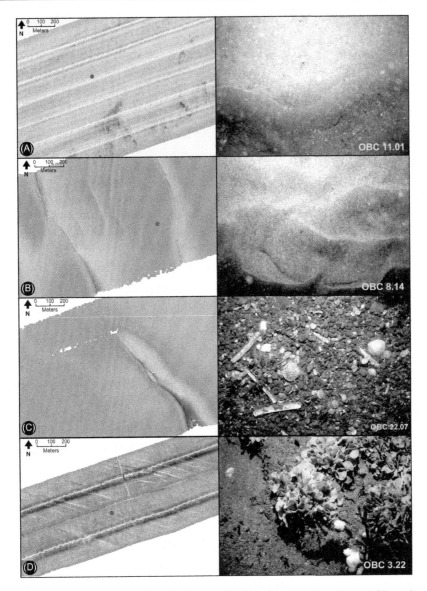

Figure 12.3 Examples of acoustic and photographic images of the Outer Bristol Channel seabed: (A) sidescan and photo showing minor sand ripples with mud/organic floc material from the sand sheet southeast of Caldey Island, off Tenby; (B) multibeam and photo showing minor sand ripples associated with a sand wave from the NOBel Sands; (C) example of multibeam (location not marked) and photo showing surface lag gravel and shells from between isolated sand waves in the SOBel Sands northeast of Lundy; (D) sidescan from platform in southeast, with photo showing gravelly mixed sediment and exposed rock ledge colonized by bryozoans *Flustra foliacea* and *Alcyonium digitata*, sponges, and the crab, *Cancer pagurus*.

glacial sediments subsequently overlain by sand and gravel. At the glacial maximum, around 22,000 years ago, sea level would have been at least 100 m lower than the present day, and the area would have been a marginal glaciated terrestrial environment. The subsequent amelioration of climate brought on the development of the second major geological environment with the gradual rise in sea level culminating in a fully marine environment. The morphology of the OBel Sands area underwent considerable metamorphosis as sea level rose, and wave and tidal currents began to fashion mobile sandy sediments to produce one of the most significant sand wave fields on the UK continental shelf.

The geomorphic features described here were identified from the interpretation of the corridor-based high-resolution seismic surveys, plus the seabed morphology model derived from singlebeam echo sounder data. Data from previous studies, particularly with regard to seabed sediment distribution based on sampling, have also been utilized. These have all been analyzed and mapped within a GIS system to produce a fourfold classification based on large and extensive regional scale features. The classification includes the OBel Sands subdivided into two areas, the NOBel Sands and the SOBel Sands.

Sand Sheet (Figure 12.1, 1)

Lies within the southwest-facing embayment of Carmarthen Bay, which is about 27 km wide and deepens gradually to the southwest and reaches maximum depths of 25–30 m across its mouth. It is dominantly a smooth seabed of fine to medium sand with some small rippled bedforms. There are some muddy (Figure 12.3A), coarse, and shelly patches and channels. Waves are important as a transport mechanism in shallow water as tidal current energy decreases. The inner bay acts as a store for sandy sediment. The western part of Helwick Bank, predominantly composed of medium sand, occurs in the southeastern part (Figure 12.1).

Sand Wave Field (NOBel Sands) (Figure 12.1, 2)

An extensive sand wave field covering an area of around 440 km². The crests of the sand waves in the eastern and central part of NOBel Sands lie in water depths between 25 and 40 m, whereas those further to the west have crests in water depths of 40–60 m. The distance between the individual waves varies on average from 1,000 to 1,500 m, though the lower and upper ranges observed are 600–3,000 m. The sand waves are laterally extensive and continuous, with crest lengths ranging from 1 to 7 km long. Although the sand waves themselves comprise medium- to coarse-grained sand, the seabed surrounding these features tends to be slightly coarser, consisting of gravelly sand or sandy gravel. Between the sand waves, the overall topography of the seabed tends to be relatively flat. The maximum sand wave height observed in the NOBel Sands is 19 m, though more commonly observed heights are 12–14 m. The sand wave crests are oriented with a regular sinuosity aligned on two principal trends: NNW (330°/340°) to SSE (150°/160°), and NNE (10°/20°) to SSW (190°/200°).

The sand waves display strong asymmetry, with lee slopes facing west to southwest. The angle of the lee slopes generally range between 5° and 10°, although upper crest sections of the lee slopes on some waves may be much steeper, with angles of up to 24°. The stoss slopes are much gentler and have angles generally less than 3°. The surfaces of the large sand waves are commonly covered by mobile sediment in the form of megaripples and ripples (Figure 12.3B). However, although their surfaces are mobile, the position and form of these large sand waves appears to have remained static in recent times given the resolution of the 1977 data set (Figure 12.2). Hence, the large sand waves may be in a state of *in situ* equilibrium. In the south of the sand wave field, there is an area of bifurcating high-frequency sand waves. These include primary waves that are generally 4–10 m high, with both west-facing asymmetrical sand waves and symmetrical sand waves, and secondary sand waves up to 4 m high.

Isolated Sand Waves (SOBel Sands) (Figure 12.1, 3)

In the SOBel Sands, the commonly isolated sand waves are generally less than 10 m high, with wavelengths ranging from 150 to 1,800 m. The majority of sand wave crests lie in water depths of 40 m, with a relatively flat seabed between the waves at around 45 m. The sand waves are oriented approximately normal to the peak tidal currents (ranging from NNW–SSE to N–S) and display strong asymmetry, with the lee slopes facing west to southwest. The angle of the lee slope ranges between 5° and 10°, although smaller sections of the lee slopes on some sand waves have steeper slopes, with angles of up to 18°. The stoss slopes are much gentler, with angles of less than 3°. Sidescan data indicate that megaripples occur on both the stoss and lee slopes of the >10- and <10-m sand waves (Figure 12.1). These megaripples are often oriented obliquely to parts of the sand wave crests, suggesting that the orientation of the megaripples is determined by the local flow conditions over the larger waves and not solely by the residual tidal currents. A number of sand waves have developed double crests that have an elliptical plan along sand wave crests. They are most common on individual waves in the SOBel Sands. They can be found in various stages of development with numerous individual ellipses aligned along crests. Some are not fully developed and may be an indication of restricted sediment supply. The seabed between the isolated sand waves in the SOBel Sands is covered by predominantly coarse sediment (Figure 12.3C) with thin sand, which can be seen in patches or as sand ribbons and streaks. These may have rippled or megarippled surfaces. In some areas, isolated outcrops of rock may appear at the seabed.

Platform (Figure 12.1, 4)

The platform which lies to the southwest and east of the OBel Sands is characterized by a coarse sediment substrate of gravelly sand, sandy gravel, and gravel with patches and streaks of thin sandy sediment. Rock exposed on the seabed in the southwest are mainly individual masses; there is some structure and lineation in the outcrops, but bedding is not well developed. The platform in the southeast is dominated

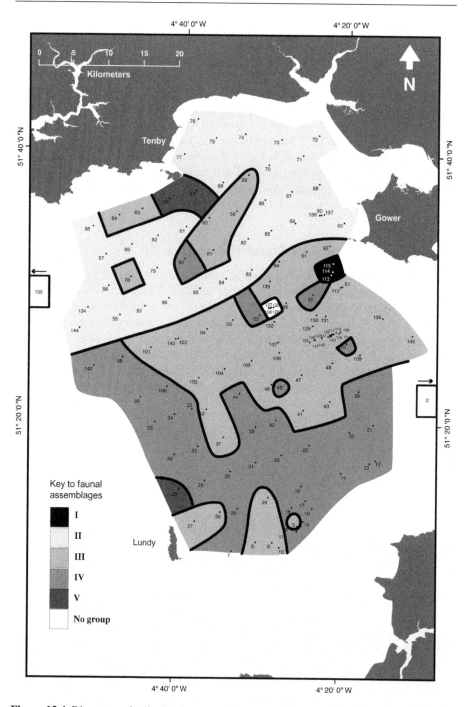

Figure 12.4 Diagrammatic distribution map of the benthic macrofaunal assemblages in the Outer Bristol Channel study area, determined from a semiquantitative cluster analysis of 127 stations (each 2 × 0.1 m² Van Veen grab samples; 0.5 mm mesh sieve).

by well-bedded rock outcrops, exposed at the seabed in water depths of 20 to less than 40 m in the center of the Channel. The rock outcrops have formed a very frequent dense series of small scarps and troughs up to a meter or two high; the majority are <0.5 m high (Figure 12.3D). The rocks have been subject to tectonic movement, and the bedding exposed on the seabed can be linear and sinuous, and disrupted by faults and folds. Sediment is commonly restricted to the troughs and can include gravel and sand. There are a few small isolated sand waves, as well as occasional sand ribbons and sand patches.

Biological Communities

Macrofaunal characterization was primarily done through analyses of quantitative data acquired through deployment of a modified $0.1\,m^2$ Van Veen grab at 127 stations. Seven more stations were sampled qualitatively using the grab or dredge, while three additional stations were from out of the study area. The larger epifauna and sediment surface were assessed at 13 beam trawl and 24 video locations. A total of 948 taxa (including 130 colonial species) were found, and an average of 1,665 individuals was enumerated per $0.2\,m^2$ quantitative station. Various analyses were carried out on the data, including a semiquantitative cluster analysis. In this analysis, frequency scale estimates of the colonial epifauna (hydroids and bryozoa) were assigned numerical equivalents and added to the quantitative infaunal data.

The semiquantitative cluster analysis (not shown) revealed five broad faunal clusters, and the distributions of each (Figure 12.4) were in general agreement with the major geomorphic features (Figure 12.1). Four subgroups were recognized within faunal assemblages II and III, while six subgroups were delineated within assemblage IV. The faunal characterizations within each main geomorphic feature are summarized below according to the habitat (biotope) classification manual for Britain and Ireland [6].

Sand Sheet

In Carmarthen Bay (Figures 12.1 and 12.4), the sand sheet supported the main assemblage II subgroup (IIb) "*Fabulina fabula* and *Magelona mirabilis* with venerid bivalves and amphipods in infralittoral compacted fine muddy sand." Areas of more mobile sand to the southwest and southeast (north of Helwick Bank) had assemblages with affinities to the sand wave field of the NOBel Sands to the south. The Helwick Bank fauna (assemblage I), associated with fine to medium sands, had only 45 taxa and was described as a "sparse fauna in infralittoral mobile clean sand." This is consistent with an earlier study of Helwick Bank [7].

The faunas (IIc and IId) of the stable sands to the south and southwest of the sand sheet exhibited affinities with other biotopes, characterized by bivalves "*Abra alba* and *Nucula nitidosa* in circalittoral muddy sand or slightly mixed sediment," or by polychaetes and brittle-stars—"*Owenia fusiformis* and *Amphiura filiformis* in offshore circalittoral sand or muddy sand." Stony areas southeast of Tenby were part of a loosely defined assemblage V. Juvenile mytilid mussels were abundant, and encrusting and colonial species common.

Sand Wave Field (NOBel Sands)

Dominated by two subgroups of faunal assemblage III (Figure 12.4). Subgroup IIIa was assigned to the "*Hesionura elongata* and *Microphthalmus similis* with other interstitial polychaetes in infralittoral mobile coarse sand" biotope. Subgroup IIIb was very closely related, but increased presence of tube-dwelling polychaetes such as *Lagis koreni* and *Spiophanes bombyx*, and reductions in small interstitial species, reflected the generally finer sediments and more stable conditions. Some gravelly areas had assemblage IV biotopes similar to those found in the SOBel Sands to the south. The biotopes in the northwest of the NOBel Sands were largely those attributed to assemblage subgroups IIc and IId, as found in the adjacent sand sheet area to the north.

Isolated Sand Waves (SOBel Sands)

Sand waves occurring within the SOBel Sands (Figures 12.1 and 12.4) had the same fauna as assemblage subgroup IIIa found in the sand waves of the NOBel Sands. However, the seabed of the SOBel Sands was mainly gravelly, and most of the area was characterized as assemblage subgroup IVa in terms of both its infauna and epifauna. The former, described as "*Mediomastus fragilis, Lumbrineris* spp. and venerid bivalves in circalittoral coarse sand or gravel," was overlain by the hydroids "*Sertularia cupressina* and *Hydrallmania falcata* on tide-swept sublittoral sand with cobbles or pebbles" with patches of the encrusting honeycomb worm "*Sabellaria spinulosa* on stable circalittoral mixed sediment." The fauna here closely resembles that found in gravelly areas of the southern Irish Sea [8,9].

Platform

Fauna within the platform areas was almost entirely attributable to assemblage subgroups IVa, IVb, and IVe. Subgroup IVb showed an increased presence of the cumacean *Pseudocuma similis*, echinoid *Echinocyamus pusillus*, and bivalve *Goodallia triangularis*, and was considered a sandier variant of subgroup IVa. The platform in the southeast, to the north of Morte Point (Figure 12.1), was characterized by either subgroup IVa or IVe. The fauna of the latter was indicative of a mosaic of bottom types—from bedrock to cobbles, sand veneers, and pockets of muddy mixed sediments. This area was described as having a "polychaete-rich deep *Venus* community in offshore mixed sediment" co-occurring with "*Sabellaria spinulosa* on stable circalittoral mixed sediment." The heterogeneity of the seabed was visible in videos of the area, and areas of exposed rock showed evidence of the "*Flustra foliacea* and *Hydrallmania falcata* on tide-swept circalittoral mixed sediment" biotope (Figure 12.3D).

Species Richness

Patterns in species richness were investigated in relation to geomorphic features and generalized sediment types: muddy, sandy, gravelly, and stony (Figure 12.5). The Helwick Sand Bank was clearly the most impoverished, with the more stable sand sheet of Carmarthen Bay supporting more species, while gravelly sediments were

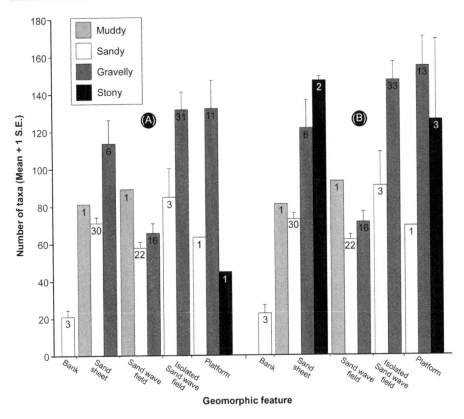

Figure 12.5 Macrofaunal species richness according to the broad sediment category for each geomorphic feature: (A) quantitative grab data (excluding colonial epifauna) and (B) qualitative data including colonial bryozoans and hydroids. Species richness calculated as the number of taxa per $0.2\,m^2$ ($2 \times 0.1\,m^2$ Van Veen Grab samples; 0.5 mm mesh sieve) station. Numbers on bars represent the number of stations involved.

the richest. Conversely, the gravelly sediments in the sand wave field (NOBel Sands) were not richer than the sands. The mobility of the sand was thought to be the controlling feature there, restricting the development of the rich infaunal and epifaunal assemblages usually found in more stable gravelly areas [8–10]. The gravelly sediments of the isolated sand wave field (SOBel Sands), platform, and stony areas had the highest species richness and the most epifaunal species.

Surrogacy

The relationships between the benthic macroinfauna were investigated [5] using the BIO-ENV and LINKTREE routines in PRIMER [11,12]. The strongest relationships were obtained with a combination of sediment parameters (primarily sand and mud content) and depth. The correlations ($\rho_s = 0.59-0.62$ for 3–6 variables) were similar to those found in studies carried out in the Celtic Sea/Irish Sea area to the west and northwest of the Bristol Channel [8,9]. The inclusion of hydrodynamic data from a current velocity model from the Proudman Oceanographic Laboratory did not improve the species–environment relationships, because the resolution of the model could not differentiate changes at the scale of the sand waves.

Acknowledgments

This paper is based on research undertaken for the Outer Bristol Channel Marine Habitat Study. The two principal funding bodies for the study were the Aggregate Levy Sustainability Fund for Wales, which is administered by the Welsh Assembly Government, and the Sustainable Land Won and Marine Dredged Aggregate Minerals Programme of the Aggregate Levy Sustainability Fund in England. The National Museum of Wales (NMW) and the British Geological Survey contributed data from their own research programs and made funding available for surveys. The Natural Environment Research Council provided ship time for one of the geophysical surveys. The Crown Estate and the British Marine Aggregate Producers Association (BMAPA) provided funding and contributed data. The Maritime and Coastguard Agency kindly provided multibeam survey data. Chris Howlett of the UKHO was instrumental in digitizing the 1977 singlebeam survey data. Sally Philpott, Angela Morando, and Gareth Jenkins of BGS and Kate Mortimer of NMW are thanked for their contribution to the marine habitat study. Ceri James publishes with the approval of the Director, British Geological Survey (NERC).

References

[1] H.R. Wallingford. Dredging on Nobel Banks (Area 476), Bristol Channel, Coastal Impact Study Report, EX 4597, Hydraulics Research Wallingford, Wallingford, 2002.

[2] B.S. Halpern, S. Walbridge, K.A. Selkoe, C.V. Kappel, F. Micheli, C. D'Agrosa, et al., A global map of human impact on marine ecosystems, Science 319 (2008) 948–952.

[3] J. McMahon, T. Golding. Atlantic Array: Environmental Impact Assessment Scoping Report, Bristol Channel Zone Ltd/RWE npower Renewables, Swindon, 2010, p. 309. http://infrastructure.independent.gov.uk/wp-content/uploads/2010/04/Atlantic-Array-Scoping-Report.pdf.

[4] J.W.C. James, S.L. Philpott, G. Jenkins, A.S.Y. Mackie, T. Darbyshire, E.I.S. Rees. Outer Bristol channel marine habitat study—2003 investigations and results, British Geological Survey Commissioned Report CR/04/054, 2004, p. 113.

[5] A.S.Y. Mackie, J.W.C. James, E.I.S. Rees, T. Darbyshire, S.L. Philpott, K. Mortimer, et al. (2006). The outer Bristol channel marine habitat study. Studies in Marine Biodiversity and Systematics from the National Museum of Wales, BIOMÔR Reports 4, 1-249 & A1-A227, +DVD-ROM, 2007.

[6] D.W. Connor, J.H. Allen, N. Golding, L.K. Howell, L.M. Lieberknecht, K.O. Northen, et al. (2004). The marine habitat classification for Britain and Ireland version 04.05. JNCC, Peterborough. ISBN 1 861 07561 8. (internet version) www.jncc.gov.uk/MarineHabitatClassification.

[7] T. Darbyshire, A.S.Y. Mackie, S.J. May, D. Rostron. A macrofaunal survey of Welsh sandbanks, Countryside Council for Wales CCW Report 539, 2002, p. 113.

[8] A.S.Y. Mackie, P.G. Oliver, E.I.S. Rees. Benthic biodiversity in the southern Irish Sea, Studies in Marine Biodiversity and Systematics from the National Museum of Wales, BIOMÔR Reports 1, 1995, p. 263.

[9] K.A. Robinson, T. Darbyshire, K. Van Landeghem, C. Lindenbaum, F. McBreen, S. Creavan, et al. Habitat Mapping for Conservation and Management of the Southern Irish Sea (HABMAP) I: Seabed Surveys, Studies in Marine Biodiversity and Systematics from the National Museum of Wales. BIOMÔR Reports 5(1), 2009, p. 148, Appendices p. 74.

[10] A.S.Y. Mackie, Macrofaunal assemblages and their sedimentary habitats: working toward a better understanding, Irish Sea Semin. Rep. 32 (2004) 30–38.

[11] K.R. Clarke, R.N. Gorley, PRIMERv5: User Manual/Tutorial, PRIMER-E Ltd, Plymouth, 2006.

[12] K.R. Clarke, R.M. Warwick, Change in marine communities: an approach to statistical analysis and interpretation, PRIMER-E Ltd, Plymouth, 2001.

13 Benthic Habitat Variations over Tidal Ridges, North Sea, the Netherlands

Thaiënne A.G.P. van Dijk[1,3], Jan A. van Dalfsen[2], Vera Van Lancker[4], Ronnie A. van Overmeeren[5], Sytze van Heteren[5], Pieter J. Doornenbal[2]

[1]Deltares, Department of Applied Geology and Geophysics, Utrecht, The Netherlands, [2]Deltares, Department of Marine and Coastal Systems, Delft, The Netherlands, [3]Department of Water Engineering and Management, University of Twente, Enschede, The Netherlands, [4]Management Unit of the North Sea Mathematical Models, Gulledelle 100, Brussels, Belgium, [5]TNO Built Environment and Geosciences, Department of Geo-Modelling, Geological Survey of the Netherlands, Utrecht, The Netherlands

Abstract

Marine ecosystems on continental shelves endure an increasing burden of human activity offshore, and the impacts on benthic habitats are not well known. An improved understanding of how benthic habitats vary in relation to substrate types and seabed features is therefore essential to both scientists and offshore developers. This case study shows that marine habitats over two tidal ridges in the North Sea vary from low-density/low-diversity communities on the well-sorted sandy crests of ridges to high-density/high-diversity communities in the poorly sorted muddy, gravelly sediments in the adjacent troughs.

Key Words: benthic habitat, tidal sand banks, tidal ridges, sand waves, seabed morphology, open marine ecosystems, North Sea, Netherlands Continental Shelf

Introduction

On sandy continental shelves, including the Netherlands Continental Shelf (NCS) in the North Sea, tidal bedforms occur of different spatial scales, such as sand banks (tidal ridges), sand waves, and megaripples [1]. Marine habitat maps reveal that benthic habitats vary spatially on continental shelves in relation to seabed morphology,

Seafloor Geomorphology as Benthic Habitat. DOI: 10.1016/B978-0-12-385140-6.00013-X

water depth, and sediment composition [2,3]. With this in mind, the different morphological elements of tidal ridges are expected to accommodate different benthic
habitats.

Some tidal ridge areas in the North Sea are nominated to become marine
protected areas, but are also attractive for their marine aggregates and may be designated in part as mining areas. Due to their composition and shallow water depths,
tidal ridges are also suitable locations for the construction of offshore wind farms. To
date, the characteristics of relatively inaccessible seabeds are too poorly understood
to explain the effects of physical parameters on benthic communities. Therefore, it is
important to expand our understanding of benthic habitat variations associated with
tidal ridges, for the benefit of both science and offshore development.

This case study describes the variation of benthic habitats within two tidal
ridge areas in the southern North Sea, Thornton Bank (TNT) and Brown Bank
(BNB), which lie 15 and 50 km, respectively, off the Netherlands coast. Bounding
coordinates are 51°20′ − 52°45′N and 2°30′ − 3°30′E (Figure 13.1).

Geomorphic Features and Habitats

The Thornton Bank is a coast-parallel tidal ridge approximately 20 m in height with
water depths ranging from 17 to 37 m below MLLWS at the study site. BNB is a
large, north–south-oriented offshore sand bank in the center of the southern North
Sea with a height of approximately 29 m and water depths ranging from 15 to 44 m
below MLLWS-level at the study site. Estimated (modelled) peak current velocities of the main tidal constituent (M2, semidiurnal lunar) are 0.72 m/s at TNT and
0.60 m/s at BNB. Both sites are open marine environments, although TNT may experience a minor influence of brackish water outflow of the Western Scheldt estuary.

Although the trough on the eastern side of BNB has been allocated since 1993
to be a test site for the equipment for aggregate extraction (Figure 13.1), this site
is undisturbed by human activities. The test site at BNB is only sporadically used
(not since 2002), and the occurrence of undisturbed megaripples on multibeam echo
sounder (MBES)-data sets of 2000 and 2006 surveys suggests that natural conditions
prevail at times of the surveys. Also the absence of beam trawl marks on the MBES
images suggests that the site was not disturbed by fishing activities in the otherwise
intensely fished southern North Sea. For TNT, the mining area within the survey site
(Figure 13.1) was last used in 2005 for 502,000 m³ sand extraction. Most mining areas
in the vicinity of the survey site have been abandoned since 2006; more extensive sand
extraction (1.6 Mm³ in 2006) occurred 7–10 km to the north-east.

Both sand banks are superimposed by sand waves and megaripples. Sand waves
at the study site on BNB are on average 190 m long and 2.6 m high, are symmetrical
in cross-sectional shape, and migrate toward the crest of the ridge at an average rate
of 2 m/year. Sand waves on TNT are on average 138 m long, 2.5 m high, and show an
ebb-tide asymmetry. Migration rates were not determined at TNT.

Hydrographic data were collected across the ridges and along their crests from
16 to 20 October 2006 for TNT and from 13 to 17 November 2006 for BNB, using
a hull-mounted Simrad EM 3000 multibeam echo sounder (300 kHz) and a towed

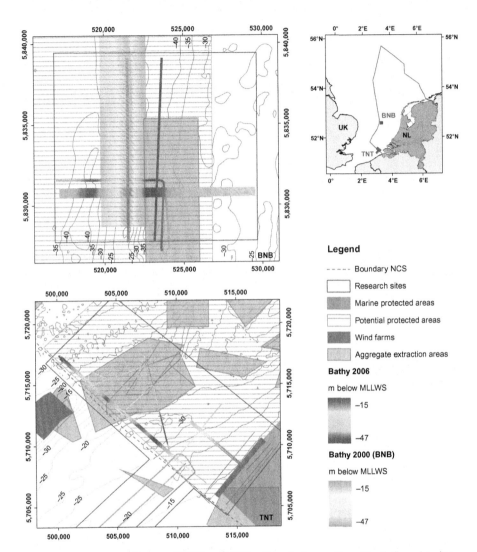

Figure 13.1 Bathymetric map of BNB and TNT regions in the southern North Sea, showing marine management areas. The locations of BNB (offshore site, water depth 15–44 m) and TNT (coastal site, water depth 17–37 m) are shown in the overview map. The blue line is the boundary of the NCS. Coordinates are in meters in a UTM 31N ED50 projection. The contour interval of the isobaths is 5 m.

EdgeTech 4300 MP sidescan sonar (410 kHz; Figure 13.1). Bathymetric digital elevation models (DEMs) were created with a Kriging interpolation algorithm with cell sizes of 1 m × 1 m. Unsupervised seabed classification of sidescan sonograms (SSS) was performed in QTC Side View, which uses a Principal Component Analysis

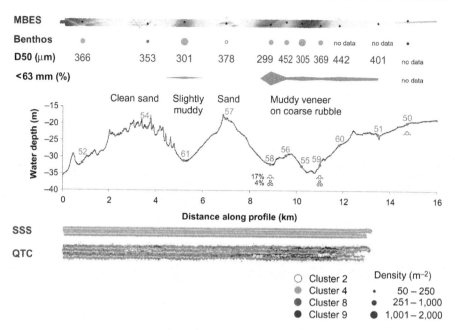

Figure 13.2 Overview of the analysis of TNT: MBES imagery and sample locations; benthos densities and clusters (see Figure 13.5); median grain size (D50, in micrometer) of the sand fraction; relative mud content (% smaller than 63 μm); a topographic cross section of the ridge with sample locations and descriptions of typical sediments (symbols: the three circles indicate a large gravel content and the three convex curves indicate a large content of large shells or shell fragments); sonogram (SSS); and a seabed classification map (QTC) based on sonograms.

for the identification of controlling factors in backscatter intensity and a Fourier analysis for the analysis of patterns. Seabed samples for the ground-truthing of biotic habitat characteristics and biological composition were collected within the same survey with a cylindrical box corer at the crests, in the troughs and on the slopes of both tidal ridges. Samples were also collected to identify acoustic facies that were observed on the sonograms. At each sample location (Figures 13.2 and 13.3), one undisturbed sediment core was collected for sedimentary bedding structure of the seabed sediments and subsampling for particle size analyses, and one benthos sample was collected for benthos analyses, which latter was sieved on board on a 1-mm sieve. No underwater video imagery was taken. Sediment grain sizes of the fractions smaller than 2 mm were determined using a laser diffraction method. The fraction larger than 2 mm was not analyzed further.

Grain size distributions of seabed sediments in the tidal ridge areas showed contrasts between crests, slopes, and troughs. Ridge crests are composed of well-sorted, clean medium sand (0% mud), whereas the troughs consist of muddy sediments on coarse

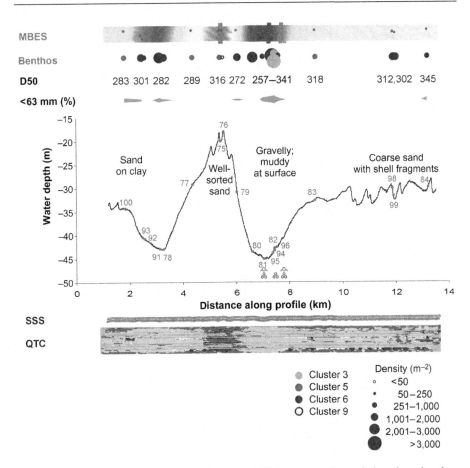

Figure 13.3 Overview of the analysis of BNB: MBES imagery and sample locations; benthos densities and clusters; median grain size (D50, in micrometer) of the sand fraction; relative mud content (% smaller than 63 μm); a topographic cross section of the ridge with sample locations and descriptions of typical sediments (symbols: the three circles indicate a large gravel content, and the three convex curves indicate a large content of large shells or shell fragments); SSS of a single track; and a seabed classification map (QTC).

rubble (gravel and large shells) (Figures 13.2–13.4). Ridge slopes consist of slightly muddy sand or thin veneers of mud and sand on the layered subsurface deposits. On both ridges, median grain sizes of the sand fraction range from 316 to 378 μm on the crests (medium sand) and from 252 to 452 μm in the troughs (medium sand).

Acoustic facies on the tidal ridges in the North Sea vary from sand wave and megaripple morphology to smooth and mottled facies where megaripples are locally absent (Figure 13.4).

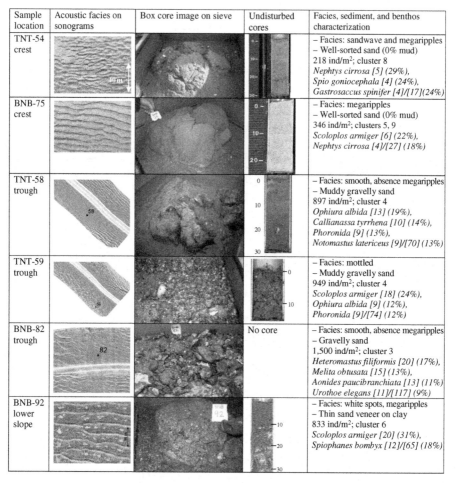

Figure 13.4 Acoustic facies on sonograms, box core images (benthos samples on sieve and undisturbed sediment cores), and the characterization of facies, sediments, and benthos (benthos density in ind/m², benthos clusters (see Figure 13.5) and dominant species) at sample locations on crests and slopes and in troughs of the BNB and TNT tidal ridges. The code *[4]/[17]* indicates the presence of four individuals of the species out of a total of 17 individuals per sample. Percentages of individuals of the species per sample are shown as (24%). Lengths of the cores are in centimeters. See Figures 13.2 and 13.3 for the location of the sample numbers shown here.

Biological Communities

Benthic assemblages were distinguished based on 35% similarity in a SIMPER cluster analysis (Figure 13.5). The analysis was performed on all samples that were collected during the campaign, including 13 samples across sand waves on the crest of the TNT. These 13 samples were taken just off the tidal ridge transect and are not indicated in Figure 13.2, but correspond in habitat to sample 54 of the transect.

Cluster 1 contains only one sample of low density and low diversity (11 individuals (ind) per sample, which corresponds to 141 ind/m^2, and 8 species per sample). The community of cluster 2 (two samples) has low densities (10–16 individuals per sample = 128–205 ind/m^2) of mainly single individuals (5–10 species per sample) and no statistically dominant species. Cluster 3 (two samples) is characterized by high densities (1,500–3,179 ind/m^2); the community contains annelids and crustaceans, dominated by *Heteromastus filiformis* ("capitellid thread worm"), and contains *Urothoe elegans* and *Spiophanes bombyx* (for quantifications, see Figure 13.4). Cluster 4 (six samples) has a moderate to high density (538–1,846 ind/m^2) and is dominated by annelids, ophiuroids, and crustaceans (*Callianassa tyrrhena, Ophiura albida, Phoronida* sp.). Cluster 5 (five samples) has a moderate density (346–679 ind/m^2) and contains some annelids and crustaceans (*Scoloplos armiger, Nephtys cirrosa, Bathyporeia elegans*). Cluster 6 (11 samples) has a high density (833–3,808 ind/m^2) and is dominated by annelids (*Scoloplos armiger, Spiophanes bombyx*), crustaceans (*Bathyporeia elegans, Urothoe elegans, U. poseidonis*), mollusks (*Tellina fabula*), and some echinoderms (*Echinocardium cordatum, Amphiura* sp.). Cluster 7 (four samples) has a low to moderate density (128–500 ind/m^2) dominated by annelids (*Pisione remota, Oligochaeta* sp., *Nephtys cirrosa*). Cluster 8 (eight samples) has a low density (115–231 ind/m^2) with mainly annelids and some crustaceans, dominated by *Nephtys cirrosa, Spiophanes goniocephala*, and *Glycera* sp. junvenile. Cluster 9 (two samples) has a very low density (26–77 ind/m^2) and species are represented by single individuals.

The community represented by cluster 7 is not displayed in the transect across the tidal ridge (Figure 13.2), because this cluster occurred in sand wave samples just off the transect, corresponding to the habitat of the crest of the TNT.

Surrogacy

The biological assemblages across the tidal ridges, as identified in the cluster analysis of seabed samples (Figure 13.5), display a systematic grouping of ridge crest and trough locations at the TNT. All samples in clusters 7, 8, and 2 represent the habitat on the crest of the ridge. All samples in cluster 4 represent the habitat in the trough of the ridge. Clusters 7 and 8 contain seabed samples collected across sand waves on the crest of TNT, just off the transect. Sand wave crests and troughs were not discriminated in the cluster analysis. The clustering across the BNB is slightly less systematic, but still habitats are identified in general. At BNB, cluster 5 represents both ridge slopes and crests, and clusters 3 and 6 represent ridge troughs and lower slopes, although a few samples in cluster 6 represent elevated seabed locations. The cluster analysis also shows that the contrasts between the two tidal ridges accounts for clustering.

Furthermore, a plot of the number of individuals per sample (density) versus the number of species per sample (diversity) (Figure 13.6) indicates that the

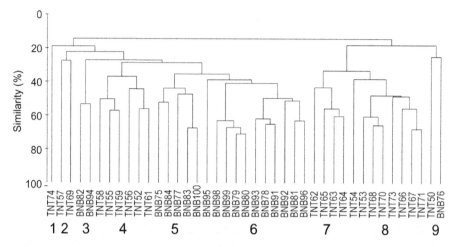

Figure 13.5 Cluster analysis result of both the TNT and the BNB, showing sample numbers and cluster numbers. The colors of sample numbers and cluster numbers correspond to cluster symbols in Figures 13.2 and 13.3. (For interpretation of the references to color in this figure legend, the reader is referred to the web version of this book.)

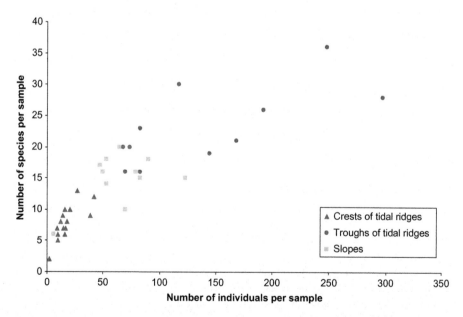

Figure 13.6 Density of individuals versus diversity of benthic species plot of samples from TNT and BNB, showing the identified groups that represent low-density, low-diversity communities on the ridge crests (red triangles), high-density, high-diversity communities in the troughs (blue circles), and moderate-density, moderate-diversity communities on the ridge slopes in between (green squares). (For interpretation of the references to color in this figure legend, the reader is referred to the web version of this book.)

morphological elements of the tidal ridges are characterized by (1) low-density, low-diversity communities on the sandy crests, (2) high-density, high-diversity communities in the troughs, and (3) moderate-density, moderate-diversity communities on the slopes.

The QTC Side View analyses of sonograms show large contrasts in seabed classification, suggesting that habitat types were identified across the ridges. At TNT, the deep trough at 10 km is clearly distinguished from the crests and slopes of the ridge (Figure 13.2). At BNB, the high crest is identified as contrasting habitat (Figure 13.3). The disadvantage of the QTC classification is that the diagnostic criteria for separating classes are not clarified, so it remains uncertain whether, for example, backscatter intensity or pattern was discriminative. In case of TNT, large contrasts in sediment grain size may be the criterion, whereas at BNB, large contrasts in the density of the benthic communities may be diagnostic.

References

[1] G.M. Ashley, Classification of large-scale subaqueous bedforms: a new look on an old problem, J. Sediment. Petrol. 60 (1) (1990) 160–172.

[2] M.J. Baptist, J. Van Dalfsen, A. Weber, S. Passchier, S. Van Heteren, The distribution of macrozoobenthos in the southern North Sea in relation to meso-scale bedforms, Estuar. Coast. Shelf Sci. 68 (3–4) (2006) 538–546.

[3] S.E. Holtmann, A. Groenewold, K.H.M. Schrader, J. Asjes, J.A. Craeymeersch, G.C.A. Duineveld, et al., Atlas of the Zoobenthos of the Dutch Continental Shelf, Ministry of Transport, Public Works and Water Management, Rijswijk, the Netherlands, 1996.

14 Fine-Scale Geomorphological Mapping of Sandbank Environments for the Prediction of Macrobenthic Occurrences, Belgian Part of the North Sea

Vera Van Lancker[1,2], Geert Moerkerke[2,3], Isabelle Du Four[2,4], Els Verfaillie[2,5], Marijn Rabaut[6], Steven Degraer[1,6]

[1]Management Unit of the North Sea Mathematical Models, Royal Belgian Institute of Natural Sciences, Brussels, Belgium, [2]Renard Centre of Marine Geology, Ghent University, Gent, Belgium, [3]G-Tec NV, Deinze, Belgium, [4]International Polar Foundation, Brussels, Belgium, [5]Department of Geography, Carto-GIS cluster, Ghent University, Gent, Belgium, [6]Marine Biology Section, Ghent University, Gent, Belgium

Abstract

Fine-scale geomorphological mapping is demonstrated to be able to predict macrobenthos occurrences within shallow sandbank areas of the Belgian part of the North Sea. Very high resolution sidescan sonar imagery was obtained covering a subtle grain size gradient from mud to coarse sands. Differences in reflectivity and texture patterns were discriminated. Interpretation in terms of bedforms and sedimentary characteristics led to 15 acoustic facies, which were further related to macrobenthic community occurrences. Successful validation over the entire Belgian coastal zone promotes its use for classifying sandbank environments. Furthermore, it is demonstrated that sidescan sonar and multibeam technology can be used to directly map reefs of tube-building polychaetes, such as *Lanice conchilega* and *Owenia fusiformis*. These ecosystem engineers are key species within hotspots of biodiversity among inter- and subtidal soft sediment environments of the North Sea.

Key Words: sandbank; subaqueous dunes; biogenic reefs; macrobenthos; ecosystem engineers; sidescan sonar; multibeam; fine scale; geomorphology

Seafloor Geomorphology as Benthic Habitat. DOI: 10.1016/B978-0-12-385140-6.00014-1

Introduction

The Belgian part of the North Sea (BPNS) (3,600 km^2) is a siliciclastic macrotidal environment (tidal range of 4.5 m) comprising several groups of sandbanks (Figure 14.1). The sandbanks represent a thin and patchy Holocene cover, which overlies Tertiary clayey sediments that outcrop locally in troughs. Sediment transport is mainly driven by tidal currents (maximum 1.18 m/s), though wind-induced currents and waves may have a direct effect on sediment resuspension and bedform morphology. Human activities are widespread and relate mainly to harbor infrastructure works, dredging and disposal of dredged material, marine aggregate extraction, and windmill farm construction.

To assess the relevance of fine-scale acoustic mapping to the prediction of macrobenthos, a sandbank area was chosen along the French–Belgian border (Location 1, Figure 14.1). The area includes coastal-attached and -detached tidal sandbanks (Trapegeer, Broers Bank, and Den Oever), as well as two sandbank troughs (Potje and Westdiep; Figures 14.1 and 14.4). Water depths in Locations 1 and 2 vary between 0 and 24 m relative to Mean Lowest Low-Water Springs (MLLWS). The area hosts a high biological diversity [1] and is designated as a EU Habitat Directive area. Similar mapping was performed along the eastern coastal zone (Location 2, Figure 14.1), 16 km offshore, and close to the Belgian–Dutch border (Figure 14.1). Depths of 24 and 15 m MLLWS in Location 2 characterize sandbank troughs and crests, respectively.

Sidescan sonar technology was used mainly for fine-scale mapping purposes. For Location 1, data were acquired using a GeoAcoustics sidescan sonar (410 kHz) in October 1999 and March 2000. The sonar, set at a range of 50 m, was towed at a speed of 4 knots, 3–4 m above the seafloor (*MV OostendeXI*). All data were recorded digitally using ISIS acquisition software (Triton-ELICS). Good weather and good DGPS positioning accuracy allowed for processing and mosaicing at a 10-cm resolution (ISIS and Delphmap). All of the sonar imagery was ground-truthed with a dense grid of 113 sediment and macrobenthos samples, collected using a Van Veen grab (sampling surface area of 0.1026 m^2). Sampling was at a spacing of 500 m during surveys conducted in October 1999 and March 2000, and additional stratified random sampling was applied in order to generate validation data sets.

At Location 2, multibeam sonar data were acquired aboard the *RV Belgica* in November 2007 using a Kongsberg EM1002 multibeam echosounder, which has 111 receive beams at a frequency of 95 kHz. Data are motion corrected and calibrated. Shallower than 30 m, depth accuracy is around 0.2% of the depth. Neptune (Kongsberg-Simrad) and Fledermaus were used for post-processing to produce digital terrain models (DTM) with a 1-m grid resolution. Following a random sampling approach, 117 ground-truth benthic samples were collected using a Van Veen grab.

Macrobenthos samples were sieved using a 1-mm mesh and fixed in an 8% formaldehyde–seawater solution. After staining with Bengal rose, all organisms were sorted and identified to species level when possible. Densities were expressed as the number of individuals per square meter (ind/m^2).

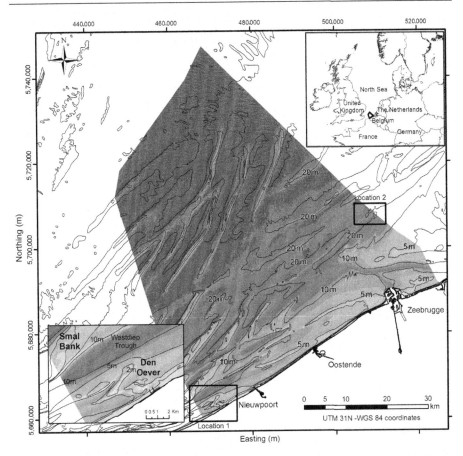

Figure 14.1 Sandbanks along the BPNS. Water depths vary from 0 to 55 m MLLWS in the BPNS. Locations 1 and 2 of detailed geo-acoustic mapping is indicated.

Geomorphic Features and Habitats

The shallow shoreface-attached sandbank complex near the French–Belgian border (Location 1, Figure 14.1) is used to describe fine-scale geomorphic features and habitats of sandbank areas. The following geomorphological features are discriminated and described:

> *Sandbank troughs*: Trough areas of sandbanks comprise very fine to fine sands with varying degrees of silt–clay percentages. In the deepest part of the troughs, mud may be deposited and ephemeral fluid mud layers have been observed. Still, sedimentary characteristics of trough areas vary with increasing complexity of the morphology. Tertiary clays may outcrop, giving rise to secondary relief, and gravel lag deposits also occur in sandbank troughs. Scour features such as sand ribbons and erosion terraces are observed in association with the exposed clay and lag gravel deposits, comprised of sediments enriched with shell debris.

Depending on current strengths and sand availability, small to medium dunes up to 0.75 m and, locally, even large (0.75–3 m) to very large dunes (>3 m) can occur in sandbank troughs. Along the slope of the terraces, small mound features are observed; they mostly occur in bands (200–400 m wide), parallel to the slope. In intertidal areas, these features are 7.5–11.5 cm high with patch sizes of 0.8–11.6 m² [2]. Increased seabed complexity is also seen where flood and ebb dominant channels meet. These areas are shallower, and increased current–bed interaction, with sediment trapping, may result in various bedforms. This is most striking at Location 2. Here, circular to elongated mounds 15–40 cm in height occur in the troughs of large dunes (Figure 14.2, inset), with patch sizes of 0.6–12 m².

Sandbank slopes: With decreasing depth, current–bed interaction increases and sediments tend to become coarser. Still, there are some flat, featureless areas where sediments become finer with decreasing depth; this is probably related to high sediment availability. Where sediments get coarser with decreasing depth, the bed exhibits small to medium dunes and/or large to very large dunes composed of medium sand with shell debris. Along the shoreface, mostly very fine to fine sands dominated; bedforms were hard to distinguish. Along the foot of the slope, in the transition zone from trough to slope, small mound features were observed, composed dominantly of fine sands with silt–clay (<10%), similar to those occurring in sandbank troughs described earlier.

Sandbank crests: In shallowest areas, fines are actively winnowed leaving shelly, coarse grained, symmetrical, wave ripples at a depth of 2 m. At depths shallower than 2 m, the bed is coarse grained and mostly featureless with some very large dunes corresponding with areas of active sand transport. Sediments in these areas are loosely packed and acoustically exhibit low reflectivity (on sidescan sonar images), due to active sediment reworking.

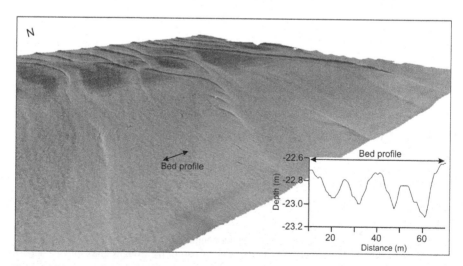

Figure 14.2 Large to very large dune features, offshore from the Belgian eastern coastal zone (Location 2, Figure 14.1). In the troughs, multibeam imagery (1 m grid) shows small mound features, corresponding to dense aggregations of the tube-building polychaete *Owenia fusiformis*. The bed profile indicates a height of 15–40 cm of these small mounds.

Biological Communities

Four subtidal communities were distinguished: (1) the *Macoma balthica* community (A), (2) the *Abra alba* community (B), (3) the *Nephtys cirrosa* community (C), and (4) the *Ophelia limacina* community (D). Next to these communities, several transitional species assemblages, connecting the four communities, were defined [1]. These communities are representative of the soft sediment macrobenthic community structure along the BPNS [1]. Detailed descriptions and photographs are available [3]. With respect to the main geomorphological entities, the following observations are made:

> *Sandbank troughs*: Generally, the *A. alba* community was observed in this habitat. They dominate the muddy sands, even though they normally tolerate only moderate silt–clay percentages (mostly <10%). Where sediments become finer and silt–clay percentages increase, the *M. balthica* community occurred. Where secondary relief is present in the troughs (e.g., terraces, extremities of flood/ebb channels), the thriving macrobenthic community will depend on the sedimentary characteristics: *N. cirrosa* community for the fine to medium sands, and *O. limacina* community for the coarser sands. Along slopes, the *A. alba* community dominated. In the areas with small mound features (see earlier), high densities of tube-building polychaetes, such as *Lanice conchilega* or sand mason, were often found. By building (semi-) permanent tubes, extending some centimeters above the seabed, and when occurring in dense aggregations they are able to induce local sediment accumulations [4]. Patches or mounds are formed that can be observed directly on very high resolution sonar imagery [2]. The larger mounds at Location 2 (Figure 14.2) correspond with high densities (e.g., 583 ± 57 ind/m²) of the more robust polychaete *Owenia fusiformis*; outside the mound area, only 3.8 ± 1.8 ind/m² were found.
>
> *Sandbank slopes*: Generally, the *N. cirrosa* community is observed in this habitat and is typical of more exposed sandy habitats. Still, very fine to fine sands, with silt–clay enrichment, were mostly present along the lower slope areas. At the transition from the troughs to the slopes, the *A. alba* community was most abundant. In these zones, the small mound (tube worm) features were most evident. Figure 14.3 shows a cross-shore transect over a sandbank where such mounds are observed at the foot of the slope of the more coastward sandbank. Where the slope gradient decreases, higher up the slope, and fines are more actively deposited, the *M. balthica* community is dominant.
>
> *Sandbank crests*: The *O. limacina* community was observed mostly with typically an impoverished fauna. In areas where fine to medium sands prevailed, the *N. cirrosa* community was more important.

Surrogacy

The statistical habitat suitability model of Degraer et al. [1] predicts the occurrence of the four macrobenthic communities based only on median grain size and silt–clay content. On the scale of the BPNS, marine landscapes have been mapped and translated into macrobenthic community preferences [5]. In this study, emphasis is placed upon relationships between the macrobenthos and the fine-scale geomorphological and sediment nature, for which sonar imagery is needed. Ideally,

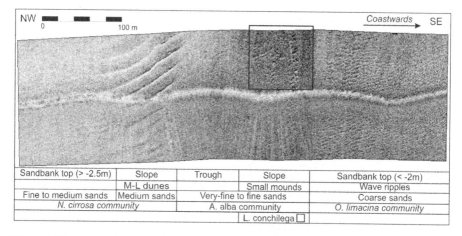

Sandbank top (> -2.5m)	Slope	Trough	Slope	Sandbank top (< -2m)
	M-L dunes		Small mounds	Wave ripples
Fine to medium sands	Medium sands	Very-fine to fine sands		Coarse sands
N. cirrosa community		A. alba community		O. limacina community
		L. conchilega ☐		

Figure 14.3 A cross-shore transect over a shallow sandbank area. The relatively low backscatter along the sandbank crest (left) is due to the loose packing of the sediments. The higher backscatter in the trough is due to compaction of silt–clay rich sediments. Coarse-grained wave ripples cause the high backscatter along the second sandbank (right). The pattern of the acoustic image at the foot of the slope of the second sandbank is typical for dense aggregations of *L. conchilega*. (Corrected sidescan sonar image, 0.10 × 0.10 m; for location, see Figure 14.4).

seafloor classification techniques would allow automated discrimination of acoustic facies, but results did not depict the fine-scale variation, as seen from the imagery. Therefore, a classification table (Table 14.1) was created, matching sonar imagery and its interpretation to macrobenthic community preferences.

Sonar images are classified in terms of reflectivity, texture, and patterns. The reflectivity was divided into low, medium, and high, related to the darkness or light-ness of the imagery (signal amplitude). Low reflectivity means that most of the acoustic energy is lost, absorbed by loose to loosely packed sediments. These can be distinguished easily and are mostly associated with fluid mud layers or strongly reworked sandy seafloors (e.g., sandbank crests). A higher reflectivity equates with harder substrates (e.g., coarse sands, shell debris, or gravel). Texture was classed as smooth, grainy, or rough. Smooth textures are featureless. Patterns relate to seabed features (e.g., bedforms). Mound features (patchy to mottled textures) were directly correlated with the presence of dense fields of tube-building polychaetes. Fifteen acoustic classes could be discriminated and linked to a macrobenthic community preference. The correlations are based on the ground-truthed 1999/2000 sonar data sets and were later validated along the entire Belgian coastal zone.

Figure 14.4 shows the distribution of the different acoustic facies along the Western Coastal Banks (Location 1). Based on Table 14.1, the acoustic facies in Figure 14.4 can be further translated into a macrobenthic community preference (see Figure 14.4). The dotted areas on this fine-scale habitat map show the presence of *L. conchilega* aggrega-tions, as seen on the imagery. Their extent correlates well with the slope areas of the sandbank environment. From this study, these are regarded as hotspots of biodiversity.

Table 14.1 Side-Scan Sonar Interpretation of Shallow Sandbank Environments

Geomorphology		Acoustic Facies		Class	Bedforms	Interpretation		Macrobenthic Community
		Reflectivity (R) Texture (T)	Pattern			Sediments		
Sandbank Troughs	R	Low	Featureless	1		(Fluid) Mud		(A) B
	T	Smooth						
	R	Medium	Featureless	2		Very-fine to fine sands, high silt-clay %		(A) B
	T	Smooth-Grainy						
Secondary relief in troughs (e.g. terraces)	R	Medium	Straight to sinuous lineations of higher reflectivity, shadow effects	3	Small to medium dunes	Fine to medium sand		C
	T	Grainy						
	R	Medium	Straight to sinuous lineations of higher reflectivity, shadow effects	4	Medium to large dunes	Medium sand with shell debris		(C) D
	T	Grainy						
	R	Medium-High	Circular to elongated patches with varying reflectivity	5	Small mound features	Fine to medium sand with silt-clay enrichment		B
	T	Grainy						
	R	High	Alternating high and low reflectivity bands	6	Ribbons	Medium to gravelly coarse sands; gravel		C D
	T	Rough			Scoured relief			
Shoreface / Sandbank Slopes	R	Low-Medium	Featureless	7		Very fine to fine sands, silt-clay enrichment		A B (C) (D)
	T	Smooth-Grainy						
	R	Medium-High	Circular to elongated patches with varying reflectivity	8	Small mound features	Fine sands with silt-clay enrichment		B
	T	Grainy						
	R	Medium	Straight to sinuous lineations of higher reflectivity, shadow effects	9	Medium dunes	Fine to medium sand		C
	T	Grainy						
	R	Medium	Straight to sinuous lineations of higher reflectivity, shadow effects	10	Large to very large dunes	Medium sand with shell debris		(A) C
	T	Grainy						

(Continued)

Table 14.1 (Continued)

Geomorphology	Acoustic Facies				Interpretation			Macrobenthic Community
		Reflectivity (R) Texture (T)	Pattern	Class	Bedforms	Sediments		
Sandbank Topzones	R	Low-Medium	Featureless	11		Fine to medium sand, loosely packed		C
	T	Grainy						
	R	Medium	Straight to sinuous lineations of higher reflectivity, shadow effects	12	Small to medium dunes	Medium sand, loosely packed		C (D)
	T	Grainy						
	R	Medium-High	Straight to sinuous lineations of higher reflectivity, shadow effects	13	Large to very large dunes	Medium sand with shell debris		C D
	T	Grainy						
	R	High	Featureless	14		Coarse sand with shell debris		D
	T	Grainy to Rough						
	R	High	Straight to sinuous lineations of higher reflectivity, shadow effects	15	Symmetrical bedforms (wave-induced)	Coarse sand with shell debris		D
	T	Grainy to Rough						

A: *Macoma balthica* community; B: *Abra alba* community; C: *Nephtys cirrosa* community; D: *Ophelia limacina* community; (): less likely.
Note: Image description and interpretation in terms of bedforms and sedimentary characteristics enables prediction of a macrobenthic community preference

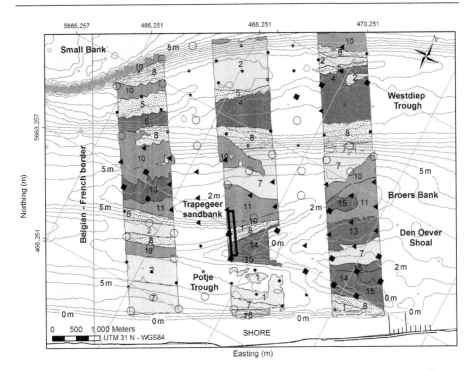

Figure 14.4 Fine-scale geomorphological map, translated into a macrobenthic community preference, based on sidescan sonar imagery of the shallow sandbank area near the Belgian–French border (Location 1, Figure 14.1). From light to dark gray, a macrobenthic community preference of *A. alba*, *N. cirrosa*, and *O. limacina* is predicted. Acoustic facies numbers are explained in Table 14.1. The dotted areas (5 and 8) show the presence of *L. conchilega* aggregations within the *A. alba* community. Their areal extent correlates well with the slope areas of this sandbank environment. Sample locations are indicated (March 2000). Open circles: *M. balthica* community; dots: *A. alba* community; triangles: *N. cirrosa* community; squares: *O. limacina* community. The rectangle on the central mosaic indicates the position of the cross-shore transect of Figure 14.3.

Acknowledgments

Research was conducted under the Belgian Science Policy project HABITAT (Contract no. MN/02/89) and the Agency of Maritime and Coastal Services-Coastal Division project HABITAT (dossier no. 99380 and 200.455; Flemish Authorities). This publication contributes also to the Belgian Science Policy project QUEST4D (Contract no. SD/NS/06B) and the EU INTERREG III B MESH project (http://www.searchmesh.net).

References

[1] S. Degraer, E. Verfaillie, W. Willems, E. Adriaens, M. Vincx, V. Van Lancker, Habitat suitability modelling as a mapping tool for macrobenthic communities: an example from the Belgian part of the North Sea, Cont. Shelf Res. 28 (3) (2008) 369–379.

[2] S. Degraer, G. Moerkerke, M. Rabaut, G. Van Hoey, I. Du Four, M. Vincx, et al., Very-high resolution side-scan sonar imagery provides critical ecological information on the marine environment: the case of biogenic *Lanice conchilega* reefs, Remote Sens. Environ. 112 (8) (2008) 3323–3328.

[3] S. Degraer, J. Wittoeck, W. Appeltans, K. Cooreman, T. Deprez, H. Hillewaert, et al., The Macrobenthos Atlas of the Belgian Part of the North Sea, Belgian Science Policy, Brussels, Belgium, 2006., ISBN 90-810081-6-1, 164, photographs, 1 cd-rom pp.

[4] M. Rabaut, M. Vincx, S. Degraer, Do *Lanice conchilega* (sandmason) aggregations classify as reefs? Quantifying habitat modifying effects, Helgol. Mar. Res. 63 (1) (2009) 37–46.

[5] E. Verfaillie, S. Degraer, K. Schelfaut, W. Willems, V. Van Lancker, A protocol for classifying ecologically relevant marine zones, Estuar. Coast. Shelf Sci. 83 (2009) 175–185.

15 Large Submarine Sand Waves and Gravel Lag Substrates on Georges Bank Off Atlantic Canada

Brian J. Todd[1], Page C. Valentine[2]

[1]Geological Survey of Canada, Dartmouth, NS, Canada, [2]US Geological Survey, Woods Hole, MA, USA

Abstract

Georges Bank is a large, shallow, continental shelf feature offshore of New England and Atlantic Canada. The bank is mantled with a veneer of glacial debris transported during the late Pleistocene from continental areas lying to the north. These sediments were reworked by marine processes during postglacial sea-level transgression and continue to be modified by the modern oceanic regime. The surficial geology of the Canadian portion of the bank is a widespread gravel lag overlain in places by well-sorted sand occurring as bedforms. The most widespread bedforms are large, mobile, asymmetrical sand waves up to 19 m in height formed through sediment transport by strong tidal-driven and possibly storm-driven currents. Well-defined curvilinear bedform crests up to 15 km long form a complex bifurcating pattern having an overall southwest–northeast strike, which is normal to the direction of the major axis of the semidiurnal tidal current ellipse. Minor fields of immobile, symmetrical sand waves are situated in bathymetric lows. Rare mobile, asymmetrical barchan dunes are lying on the gravel lag in areas of low sand supply. On Georges Bank, the management of resources and habitats requires an understanding of the distribution of substrate types, their surface dynamics and susceptibility to movement, and their associated fauna.

Key Words: Georges Bank, sand wave, barchan, gravel lag, sediment mobility, habitat

Introduction

Georges Bank is a shallow submarine bank that lies south of Nova Scotia and east of Cape Cod, Massachusetts (Figure 15.1). It bounds the seaward side of the Gulf of Maine and rises more than 300 m above the Gulf of Maine seafloor. It is approximately 280 km long and 150 km wide, and its area in depths of <200 m is 42,000 km². The bank surface gradually deepens seaward and has an average slope

Seafloor Geomorphology as Benthic Habitat. DOI: 10.1016/B978-0-12-385140-6.00015-3

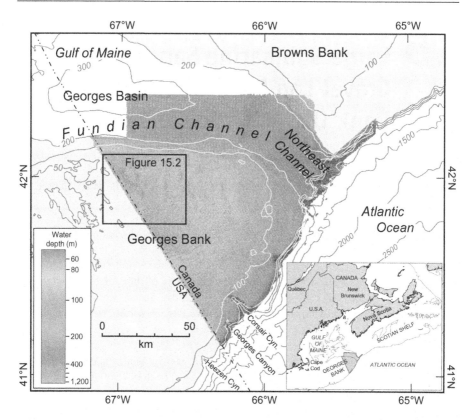

Figure 15.1 Multibeam sonar bathymetric map of the Canadian portion of Georges Bank and environs. Location of the study area is indicated by labeled box (Figure 15.2). Note that hypsometric analysis of water depth has been applied individually to this figure and subsequent figures illustrating bathymetry. Thus, the range of depths and associated colors differ for each figure. Isobaths in meters. (For interpretation of the references to color in this figure legend, the reader is referred to the web version of this book.)

of <0.05° (0.9 m/km). The 200-m isobath along the southeastern margin of Georges Bank approximates the continental shelf break. Seaward, water depths increase down the continental slope, which has an average slope of 7° (123 m/km). The international boundary between Canada and the USA transects the bank, and the eastern part of the bank (~7,500 km^2) lies in Canadian territory. Most of the Canadian portion of Georges Bank has water depths between 60 and 90 m [1–3]. The seaward margin of the bank in Canada is incised by a number of submarine canyons, the largest of which are Corsair and Georges Canyons (Figure 15.1).

Georges Bank is dominated by energetic tidal currents associated with the M$_2$ semidiurnal tidal resonance in the Bay of Fundy–Gulf of Maine system [4]. The major axis of the semidiurnal tidal current ellipse is oriented northwest–southeast across the bank, and the amplitude of this current at the seabed increases from <10 cm/s along the deeper southern flank to >100 cm/s on the relatively shallow northern edge of the bank [5]. On the shallowest portions of Georges Bank (<50 m), tidal currents are so strong that complete vertical mixing occurs throughout the year [5]. These strong currents also form large sand waves [6,7] and continuously rework the surficial sediments, thereby removing most of the clay, silt, and fine-grained sand, with coarse-grained sand remaining [8,9].

Multibeam sonar data (Figure 15.1) were collected at the mouth of Fundian Channel (in Northeast Channel) in 1999 by the Canadian Hydrographic Service using the Canadian Coast Guard Ship *Frederick G. Creed*, a Small Waterplane Area Twin Hull (SWATH) vessel. The ship was equipped with a Kongsberg EM1000 multibeam bathymetric survey system (95 kHz), with the transducer mounted in the starboard pontoon. This system produces 60 beams arrayed over an arc of up to 150°. In 1999–2000, the Canadian Hydrographic Service, in partnership with the Canadian Offshore Scallop Industry Mapping Group, collected multibeam sonar bathymetric data on the Canadian portion of Georges Bank (including the imagery shown in this chapter) using the MV *Anne S. Pierce*. This vessel was equipped with a Kongsberg EM1002 multibeam bathymetric survey system (95 kHz), with the transducer mounted on a ram extended beneath the hull. This system produces 111 beams configurable over an arc of up to 150°. Also in 2000, Clearwater Fine Foods Inc. collected multibeam sonar bathymetric data in Fundian Channel (Georges Basin) using the same system and the MV *Anne S. Pierce*.

A differential Global Positioning System was used for navigation, providing positional accuracy of ±3 m. Sound velocity profiles collected at regular intervals during multibeam data collection were used to correct the data by eliminating the effects of sound refraction caused by stratification of the water column. The bathymetric data were adjusted for tides using a tidal model that was developed for the project by the Ocean Circulation Group at the Bedford Institute of Oceanography (Charles Hannah, personal communication, 2009). Bathymetric soundings were processed using software developed by the Ocean Mapping Group at the University of New Brunswick.

The multibeam bathymetric data are presented at 5 m per pixel horizontal resolution on Georges Bank and at 10 m per pixel horizontal resolution in Fundian Channel and Northeast Channel (Figure 15.1). The shaded relief images shown in the figure were created by vertically exaggerating the topography 10 times and then artificially illuminating the relief by a virtual light source positioned 45° above the horizon at an azimuth of 315°. In the resulting images, topographic features are enhanced by strong illumination on the northwest-facing slopes and by shadows cast on the southeast-facing slopes. Thus, small topographic features are accentuated that could not be effectively shown by contours at this scale. Superimposed on the shaded relief images are colors assigned to water depth, ranging from red (shallow) to violet (deep). In order to apply the widest color range to the most frequently occurring water depths, hypsometric analysis was used to calculate the cumulative frequency of water depth. The resulting color ramp highlights subtle variations in water depth that would otherwise be obscured.

Backscatter strengths ranging from 0 to −128 decibels (dB) were logged simultaneously with the bathymetric data. To reduce the dynamic range of the recorded data, a partial correction was applied to the backscatter strength values for the varying angle of incidence by using Lambert's law for the variation with angle and assuming a flat seafloor. Backscatter strengths, with radiometric and geometric corrections applied [10], were computed with software developed by the Geological Survey of Canada using calibration values for the electronics and transducers at the time of instrument manufacture.

Some features in the multibeam data are artifacts of data collection and environmental conditions during the survey periods. The orientation of the survey track lines can, in some instances, be identified by faint parallel bands in the image. Because these artifacts are usually regular and geometric in appearance on the map, the human eye can disregard them and distinguish real topographic features and backscatter strength variations.

Geomorphic Features and Habitats

Seismic reflection profiles show that beneath the surface of Georges Bank, there is a prominent unconformity formed on late Cretaceous and Tertiary sedimentary rocks [11,12]. The surficial sediment overlying the unconformity forms a veneer of glacial debris transported to the bank during the late Pleistocene from continental areas to the north [13–15]. During the Holocene, sea level rose from a low stand 120 m below the present sea level [16,17]. Surficial sediments were reworked by marine processes during sea-level transgression and continue to be reworked under the modern oceanic regime [8,18].

Based on geophysical, geological, and photographic evidence [19], two Quaternary sediment units are mapped on Georges Bank (Figures 15.2 and 15.3). The area designated as postglacial sand and gravel (PG) is a well-sorted, generally coarse-grained sand, grading to rounded and subrounded pebble and cobble gravel (Figure 15.4A), which forms a widespread surficial lag. The lag is overlain in places by well-sorted postglacial sand (PGs) occurring mainly as bedforms (Figure 15.4B). Bedforms, the most prominent geomorphological features on the bank, are formed through sediment transport by strong tidal currents. Multibeam sonar mapping of the Canadian portion of Georges Bank has provided unprecedented views of the morphology and distribution of these bedforms. In areas where ground-truth data were sparse, identification of sediment units was based on nearby backscatter and topographic imagery that has been ground-truthed using sediment samples, photographs, and video images.

Three types of bedforms are present in our study area. In sand-rich areas, mobile sand waves composed of medium to coarse sand are wave-like geometric configurations at the water–sediment interface formed by fluid flow over an erodable granular bed (Figures 15.2, 15.3, and 15.5). In plan view, the curvilinear crests of the sand waves trend approximately southwest–northeast, normal to the major axis of the semidiurnal tidal current, and display a complex bifurcation pattern (Figure 15.5). The sand waves have wavelengths of 50–300 m and range in height from a few decimeters up to 19 m. In cross section, the features are asymmetric with gently sloping up-current (or stoss) faces and steeply dipping down-current (or lee) faces (Figure 15.5). The cross-sectional shape of the sand waves, their crest orientation

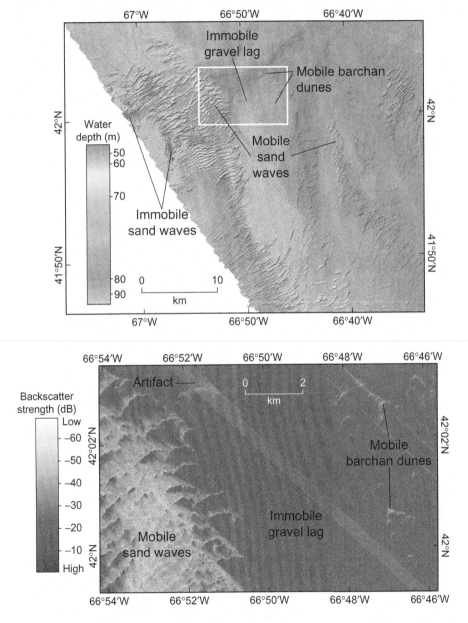

Figure 15.2 Multibeam sonar bathymetric map (upper) of the study area showing mobile and immobile sand waves and immobile gravel lag substrates. Location of the backscatter strength map (lower) is indicated by the white box. Sand has low backscatter strength (white to light green tones), while gravel lag has high backscatter strength (blue tones). Barchan dunes are evident on the backscatter strength map as discrete low backscatter strength features lying on the gravel lag. Parallel banding on the backscatter strength image is an artifact of data processing. (For interpretation of the references to color in this figure legend, the reader is referred to the web version of this book.)

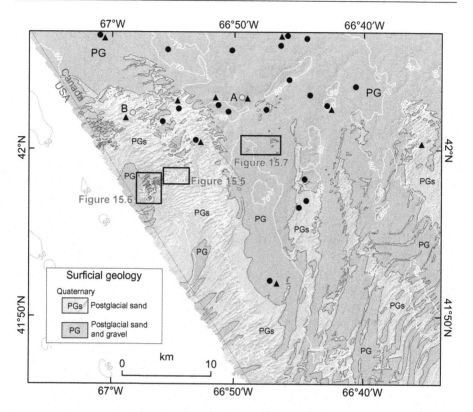

Figure 15.3 Surficial geology of the study area. A lag of postglacial sand and gravel (PG) is overlain in places by postglacial sand (PGs). Labeled blue dots (A and B) show locations of seafloor photographs in Figure 15.4. Other photograph locations are designated by black dots. Sediment sample locations indicated by black triangles. Locations of Figures 15.5–15.7 are indicated by labeled boxes. Isobaths in meters shown by labeled white lines within mapped area and by blue lines outside area. (For interpretation of the references to color in this figure legend, the reader is referred to the web version of this book.)

and distribution on the bank, and the mobility of their surfaces (Figure 15.4B) suggest that they are actively moving (presumably very slowly) and are useful for inferring the regional direction of sediment transport. Megaripples, with heights of about <2 m and wavelengths of 20–30 m, are superimposed on the up-current faces of sand waves and share the same general southwest–northeast crest orientation and complex bifurcated pattern in plan view (Figure 15.5).

A second style of sand wave occurs within relatively small areas on the bank. These bedforms have comparatively straight crests trending southwest–northeast (Figures 15.3 and 15.6), similar in orientation to crests within the sand wave fields. Although we lack direct sedimentological evidence for the composition of these features, their backscatter strength is similar to that of the sand wave fields (compare in

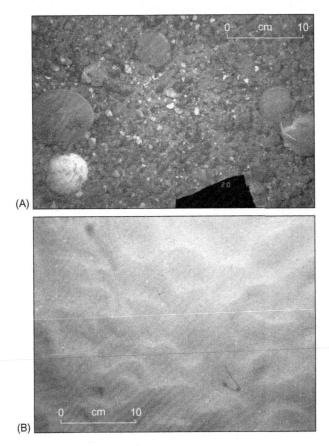

Figure 15.4 Seafloor photographs of (A) postglacial sand and gravel lag (PG) and (B) postglacial sand (PGs). Grain size analysis of sediment samples at site A indicates the PG substrate is 82% gravel, 18% sand, and 0% silt (*Hudson* 2000-047, station 105). At site B, the PGs substrate is <1% gravel, 99% sand, and <1% silt (*Bigelow* 08006, station 808077). Three-dimensional ripples in the PGs attest to frequent disturbance and reworking of the surface of the sand wave. See Figure 15.3 for locations of these samples.

Figure 15.6) and suggests that they are composed of well-sorted sand. These sand waves differ from those in the sand wave fields described above in that their simple, relatively straight crests show no bifurcation pattern. Also, in contrast to the asymmetrical profiles of mobile bedforms within the sand wave fields, these features are symmetrical in profile (Figure 15.6). We hypothesize that they are relatively immobile. Another difference between mobile and immobile sand waves is their stratigraphic position. The mobile sand waves sit on top of the regional lag gravel surface at relatively shallow water depths (<~70 m), whereas the immobile sand waves are located within bathymetric lows (>~70 m) where tidal current velocities are presumably weaker (Figure 15.6). Only the crests of the immobile sand waves

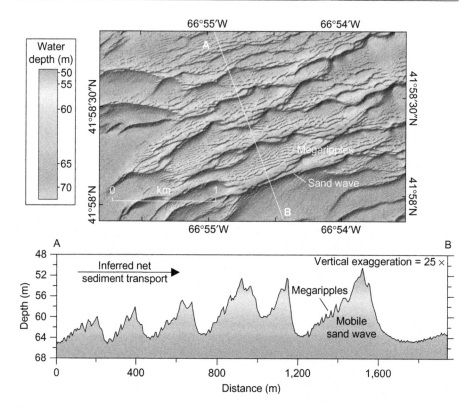

Figure 15.5 Detailed geomorphologic map of mobile sand waves and megaripples in the study area (upper) and bathymetric cross section (A–B) (lower). Inferred direction of net sediment transport is from northwest to southeast. Megaripples occur on the up-current, or stoss, faces of the sand waves. Sand waves in this image reach 14 m in height, and megaripples are <2 m in height.

reach the bathymetric level (<70 m) of the surrounding terrain. The ends of these features often terminate in distinct depressions, or moats (Figure 15.6), giving the visual impression that they are carved out of the surrounding material. They closely resemble, in their morphology and orientation to tidal current direction, sand features mapped in the Great South Channel at the western end of Georges Bank, which are immobile (but have mobile surfaces) and are separated from each other by pebble and cobble gravel lag pavements [20].

Finally, barchan dunes are present on gravel lag in sand-starved regions of the bank (Figures 15.2, 15.3, and 15.7). These dunes are composed of medium to coarse sand [21], are crescentic in planform, and are convex to the northwest, with steep lee faces facing southeastward in cross-sectional view. Their asymmetry is similar to the mobile sand waves and suggests southeastward movement. Intriguingly, the allometric relationship between barchan height and width on Georges Bank is different than the same relationship for barchan dunes measured on nearby Browns Bank

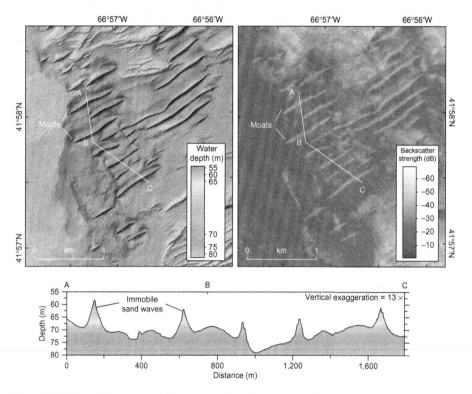

Figure 15.6 Detailed geomorphologic map of relatively immobile, straight-crested sand waves in the study area (left) and corresponding backscatter strength map (right). These features lie in a topographic depression. Bathymetric cross section (A–B–C) shown below. The sand waves reach 10 m in height with an average base width of ~90 m. Distance between crests is about 150 m. Low backscatter strength of sand waves suggests that they are sand lying on gravel lag of high backscatter strength. Note that backscatter strength of the straight-crested sand waves is similar to that of the curvilinear-crested sand waves that lie to the north and southeast.

([22]; Figure 15.1). For a given barchan dune width, their height on Georges Bank is almost twice their height on Browns Bank. This marked height difference may be a result of the much more energetic current regime on Georges Bank. Observations [23] and models [24] indicate that tidal current speeds on Georges Bank (up to 100 cm/s) satisfy the mean velocity required to erode and move sediment grains to cause barchan dune migration (~57 cm/s) [25], which possibly is accelerated under higher-velocity, storm-induced currents.

Based on multibeam bathymetric and backscatter imagery of the region and supporting ground-truth data, we have identified three major bedform regimes on Georges Bank: (1) mobile sand waves, the most widespread features, whose cross-sectional asymmetry suggests slow southeastward migration; (2) immobile sand

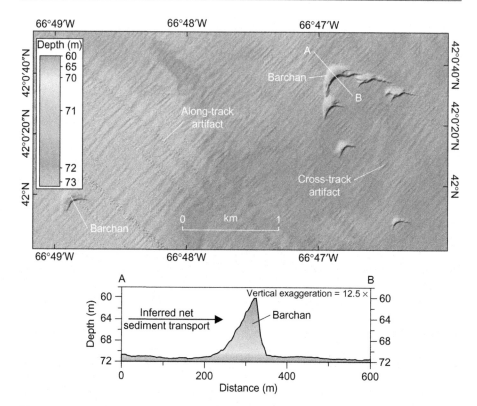

Figure 15.7 Detailed geomorphologic map of barchan dunes in the study area and bathymetric cross section (A–B) shown below. Inferred direction of net sediment transport is from northwest to southeast. The largest barchan is about 10 m in height, with a width of about 455 m from tip to tip (normal to the line of section). Along-track artifacts are bands oriented northwest–southeast (direction of survey) arising from incomplete correction of the far-field sonar beams. Cross-track artifacts appearing as ridges (~200 m long) oriented northeast–southwest arise from incomplete correction for vessel heave caused by sea surface conditions.

waves located in bathymetric lows and separated by gravel lag, whose cross-sectional symmetry suggests no migration; and (3) isolated barchan dunes located on gravel lag plains with sparse sand resources, whose cross-sectional asymmetry suggests movement to the southeast. Repetitive multibeam sonar mapping is required to detect bedform migration occurring over months or years.

Biological Communities and Surrogacy

Seabed substrates in this energetic region of Georges Bank include both mobile and immobile sand features and immobile gravel lag deposits [26]. The surfaces of both the mobile and immobile sand features are moving sand affected by strong semidiurnal tidal currents and possibly episodic storm-wave currents. Photographs of the seabed in the study area show that there is a strong relationship between substrate type (grain size and dynamics) and the associated invertebrate fauna. The dynamic surfaces of the sand features provide habitat for an invertebrate fauna of low diversity whose principal species are infaunal bivalves (e.g., razor clams *Ensis directus* and surf clams *Spisula solidissima*) and epifaunal gastropods that prey on them (e.g., moon snails *Lunatia heros*). In contrast, the static surfaces of the gravel lag deposits provide habitat for a diverse range of species types, including attached structural forms (e.g., sponges, bryozoans, tube worms, and anemones), infaunal bivalves and worms, epifaunal bivalves (e.g., sea scallops, *Placopecten magellanicus*), and arthropod scavengers. It is highly likely that these two habitat types perform different ecological functions and support different fish assemblages, and our multidisciplinary mapping results have provided the foundation for fisheries biologists to investigate this hypothesis.

Gravel lag substrates have developed as the result of erosion of sand and lie in areas of the bank that are sand-starved at present (Figure 15.3, PG; Figure 15.4A). They are more stable than habitats where sand features are actively migrating, albeit slowly (Figure 15.3, PGs; Figure 15.4B). On Georges Bank, and in similar settings, a basic understanding of substrate types based on (1) distribution, (2) surface dynamics, (3) bedform movement, and (4) associated fauna is required for managing resources and predicting habitat change through time. For example, we speculate that even a slowly moving sand feature can smother a gravel-based habitat and its associated diverse fauna in less time than it will take for a gravel habitat to recover its biological diversity as it emerges from an eroding sand veneer.

The mobile and immobile substrates we have described can be placed in a regional context. Kostylev and Hannah [27] have derived a process-driven seabed characterization scheme that is based on regional environmental attributes that include water depth, seabed frictional velocity, food resources, temperature, substrate texture, and sea surface chlorophyll concentration, among others. These attributes are reduced to two major habitat characteristics (Figure 15.8): seabed disturbance (ratio of frictional velocity to shear stress at the seabed; Figure 15.8A) and scope for growth (sum of temperature, food availability, and oxygen attributes; Figure 15.8B), which they have combined into a graphical characterization of broad-scale habitats of the continental shelf off Atlantic Canada, including Georges Bank (see figures 5 and 6 in Ref. [27]).

Their results indicate that on Georges Bank sand wave substrates represent a habitat of high disturbance and high productivity typical of bank tops and gravel lag substrates represent a habitat of low disturbance and high productivity (see figure 6 in Ref. [28]). In our study area on the bank, sand generally characterizes the southern and western parts and gravel generally typifies the northern and eastern parts (Figures 15.2 and 15.3). Both substrate types lie in a region of high productivity (Figure 15.8B) and strong currents but support very different communities because of different levels of disturbance (Figure 15.8A). We suggest that high-resolution bedform and substrate maps, such as those presented here, will aid in understanding the function of habitats and in refining regional-scale habitat classification schemes.

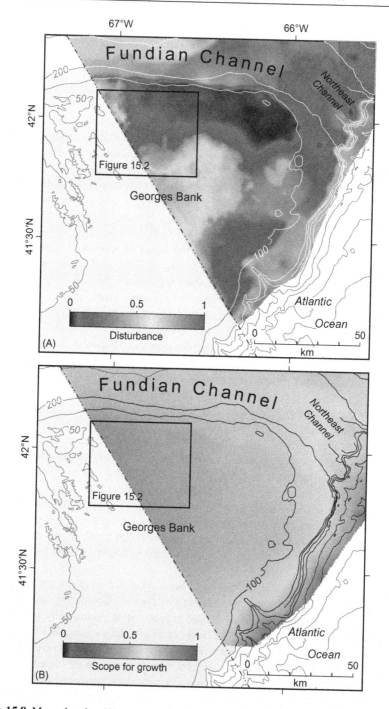

Figure 15.8 Maps showing (A) the natural disturbance axis and (B) the scope for growth axis of the habitat template model as applied to Georges Bank. Figure is modified from Kostylev and Hannah [27]. Location of the study area is indicated by labeled box (Figure 15.2).

Acknowledgments

M. Lamplugh and G. Costello of the Canadian Hydrographic Service (CHS) organized the multibeam bathymetric survey of Georges Bank in cooperation with the Canadian Offshore Scallop Association Industry Mapping Group and the Ocean Mapping Group (OMG) of the University of New Brunswick. OMG staff participated in the Georges Bank surveys. K. Paul (CHS) organized the multibeam bathymetric survey of the mouth of Fundian Channel (Northeast Channel). D. Beaver of the Geological Survey of Canada (GSC) participated in the Fundian Channel survey undertaken by Clearwater Fine Foods Inc. The Georges Bank and Fundian Channel maps are based on multibeam sonar data used in agreement with the Canadian Offshore Scallop Association Industry Mapping Group and Clearwater Fine Foods Inc. Multibeam sonar data were processed by CHS, GSC, and OMG. We thank John Hughes Clarke for the use of OMG multibeam processing software during data collection. S. Hayward (GSC) processed the backscatter strength using software developed by R.C. Courtney (GSC). We thank the officers and crews of the Canadian Coast Guard Ship (CCGS) *Frederick G. Creed* and of the MV *Anne S. Pierce* for their efforts at sea during the multibeam sonar surveys. We also thank the officers, crews, and scientific staff on a number of expeditions of the CCGS *Hudson* and the NOAA Ship *Henry B. Bigelow* to Georges Bank to collect geophysical, geological, and photographic data. Geographical Information Systems support and cartographic support was provided by S. Hayward, P. O'Regan, E. Patton, W.A. Rainey, and S. Hynes. We thank Vladimir Kostylev and Edward King (GSC), Larry Poppe and Brad Butman (US Geological Survey), Thierry Schmitt (Service Hydrographique et Océanographique de la Marine, France), and an anonymous referee for reviewing the manuscript. Vladimir Kostylev kindly provided the data used to create Figure 15.8. Any use of trade, product, or firm names in this publication is for descriptive purposes only and does not imply endorsement by the US Government. GSC Earth Science Sector contribution number is 20090413.

References

[1] Canadian Hydrographic Service, Georges Bank Eastern Portion, Chart L/C 4255, Fisheries and Oceans Canada, Ottawa, ON, Scale 1:175 000, 1990.

[2] Canadian Hydrographic Service, Georges Bank, Chart LC 8005, Fisheries and Oceans Canada, Ottawa, ON, Scale 1:300 000, 1997.

[3] P.C. Valentine, E.W. Strom, C.L. Brown, Maps Showing the Sea-Floor Topography of Eastern Georges Bank, U.S. Geological Survey, Miscellaneous Investigations Series Map I-2279-A, Scale 1:250 000, 1992.

[4] W.S. Brown, J.A. Moody, Chapter 9: Tides, in: R.H. Backus (Ed.), Georges Bank, The Massachusetts Institute of Technology Press, Cambridge, MA, 1987, pp. 100–107.

[5] B. Butman, R.C. Beardsley, Physical oceanography, in: R.H. Backus (Ed.), Georges Bank, The Massachusetts Institute of Technology Press, Cambridge, MA, 1987, pp. 88–98.

[6] G.F. Jordan, Large submarine sand waves, Science 136 (1962) 839–848.

[7] H.B. Stewart, Jr., G.F. Jordan, Underwater sand ridges on Georges Shoal, in: R.L. Miller (Ed.), Papers in Marine Geology, Macmillan, New York, NY, 1964, pp. 102–114.

[8] D.C. Twichell, B. Butman, R.S. Lewis, Chapter 4: Shallow structure, surficial geology, and the processes currently shaping the bank, in: R.H. Backus (Ed.), Georges Bank, The Massachusetts Institute of Technology Press, Cambridge, MA, 1987, pp. 31–37.

[9] B. Butman, Chapter 13: Physical processes causing surficial sediment movement, in: R.H. Backus (Ed.), Georges Bank, The Massachusetts Institute of Technology Press, Cambridge, MA, 1987, pp. 147–162.

[10] E. Hammerstad, Backscattering and Seabed Image Reflectivity. Kongsberg Maritime EM Technical Note, 2000, p. 5. Available from http://www.km.kongsberg.com/ks/web/nokbg0397.nsf/AllWeb/226C1AFA658B1343C1256D4E002EC764/$file/EM_technical_note_web_BackscatteringSeabedImageReflectivity.pdf?OpenElement.

[11] L.H. King, B. MacLean, Geology of the Scotian Shelf. Canadian Hydrographic Service, Marine Sciences Paper 7, Geological Survey of Canada, Paper 74-31, 1976, p. 31.

[12] R.S. Lewis, R.E. Sylvester, J.M. Aaron, D.C. Twichell, K.M. Scanlon, Shallow sedimentary framework and related potential geologic hazards of the Georges Bank area, in: J.M. Aaron (Ed.), Environmental Geologic Studies in the Georges Bank Area, United States Northeastern Atlantic Outer Continental Shelf, 1975–1977, U.S. Geological Survey, 1980 Open-File Report 80-240-A, pp. 5-1–5-25.

[13] F.P. Shepard, J.M. Trefethen, G.V. Cohee, Origin of Georges Bank, Bull. Geol. Soc. Am. 45 (1934) 281–302.

[14] J. Schlee, Atlantic Continental Shelf and Slope of the United States: Sediment Texture of the Northeastern Part, U.S. Geological Survey Professional Paper 529-L, 1973, p. 64.

[15] G.B.J. Fader, Geological and Geophysical Study of Georges Basin, Georges Bank, and the Northeast Channel Area of the Gulf of Maine, Geological Survey of Canada Open File 978, 1984, p. 531.

[16] K.O. Emery, L.E. Garrison, Sea levels 7,000 to 20,000 years ago, Science 157 (1967) 684–687.

[17] L.H. King, G.B.J. Fader, Wisconsinan glaciation of the continental shelf—southeast Atlantic Canada, Geol. Surv. Can. Bull. 363 (1986) 72.

[18] P.C. Valentine, E.W. Strom, R.G. Lough, C.L. Brown, Maps Showing the Sedimentary Environment of Eastern Georges Bank, U.S. Geological Survey Miscellaneous Investigations Series Map I-2279-B, scale 1:250 000, 1993.

[19] B.J. Todd, K.W. Asprey, A.S. Atkinson, R. Blasco, S. Fromm, P.R. Girouard, et al., Expedition Report: CCGS Hudson 2002-026: Gulf of Maine, Geological Survey of Canada Open File 1468, Ottawa, ON, 2003, p. 143.

[20] P.C. Valentine, T.J. Middleton, S.J. Fuller, Backscatter Intensity and Sun-Illuminated Seafloor Topography of Quadrangles 1 and 2 in the Great South Channel Region, Western Georges Bank, U.S. Geological Survey Geologic Investigations Series Map I-2698-E, Scale 1:25,000, 1 CD-ROM, 2002. Available from http://pubs.usgs.gov/imap/i2698/.

[21] B.J. Todd, B.D. Wile, D.E. Beaver, R. Murphy, M. Belliveau, J. Harding, Expedition Report: MV Hamilton Banker, Georges Bank, Gulf of Maine, Geological Survey of Canada Open File 3922, 2000, p. 57, 1999.

[22] B.J. Todd, Morphology and composition of submarine barchan dunes on the Scotian Shelf, Canadian Atlantic margin, Geomorphology 67 (2005) 487–500.

[23] J.A. Moody, B. Butman, R.C. Beardsley, W.S. Brown, P. Daifuku, J.D. Irish, et al., Atlas of tidal elevation and current observations on the northeastern American continental shelf and slope, U.S. Geol. Surv. Bull. 1611 (1984) 122.

[24] H. Xue, F. Chai, N.R. Pettigrew, A model study of the seasonal circulation in the Gulf of Maine, J. Phys. Oceanogr. 30 (2000) 1111–1135.

[25] M.C. Miller, I.N. McCave, P.D. Komar, Threshold of sediment motion under unidirectional currents, Sedimentology 24 (1977) 507–527.

[26] P.C. Valentine, B.J. Todd, V.E. Kostylev, Classification of marine sublittoral habitats, with application to the northeastern North America region, in: P.W. Barnes, J.P. Thomas (Eds.), Benthic Habitats and the Effects of Fishing, American Fisheries Society, Bethesda, MD, 2005, pp. 183–200. (Symposium 41).

[27] V.E. Kostylev, C.G. Hannah, Process-driven characterization and mapping of seabed habitats, in: B.J. Todd, H.G. Greene (Eds.), Mapping the Seafloor for Habitat Characterization, Geological Association of Canada, St. John's, NL and Labrador, 2007, pp. 171–184 (Special Paper 47).

[28] Fisheries and Oceans Canada, Framework for Classification and Characterization of Scotia–Fundy Benthic Habitats. Canadian Science Advisory Secretariat, Science Advisory Report 2005/071, 2005, 14 pp. Available from http://www.dfo-mpo.gc.ca/csas/Csas/status/2005/SAR-AS2005_071_E.pdf.

16 The Yongala's "Halo of Holes"— Systematic Bioturbation Close to a Shipwreck

Thomas C. Stieglitz[1,2,3]

[1]AIMS@JCU, Townsville, QLD, Australia, [2]School of Engineering and Physical Sciences, James Cook University, Townsville, QLD, Australia, [3]Australian Institute of Marine Science, Townsville, QLD, Australia

Abstract

Large-scale systematic bioturbation is documented in 30 m of water depth around the Yongala shipwreck in the Great Barrier Reef. The "Halo of Holes" contains over 1,200 individual depressions with diameters of up to 10 m, tightly packed in a concentric, three-quarter ring around the wreck. Two distinct zones in this halo consist of either only shallow or only deep holes. Deep holes (up to 1.5 m depth) show signs of ongoing bioturbation. Shallow holes support diverse sessile faunal assemblages of sponges, soft corals, and hard corals in an otherwise flat seafloor dominated by marine plants. The holes are most likely of biogenic origin, but the animal(s) responsible for these earthworks are unknown to date. The bioturbators are ecosystem engineers, creating a habitat for assemblages that are absent elsewhere on the adjacent seafloor. The previously undocumented halo of holes and its biota indicate that the "ecosystem wreck" extends further beyond the spatial confines of the hull of a wreck than previously considered.

Key Words: Bioturbation, shipwreck, ecosystem engineering, Yongala, Great Barrier Reef

Introduction

The wreck of the Yongala is one of Australia's most famous wreck-diving sites, attracting thousands of recreational divers each year. The 110-m long passenger vessel sunk in 1911 during a cyclone, *ca.* 25 km off the coast East of Cape Bowling Green in the central Great Barrier Reef (Figure 16.1). Both the hull and the

Seafloor Geomorphology as Benthic Habitat. DOI: 10.1016/B978-0-12-385140-6.00016-5

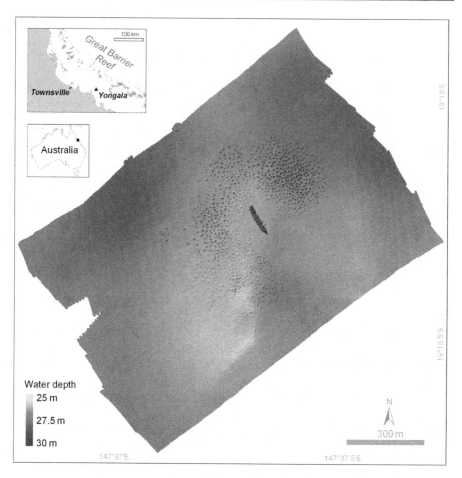

Figure 16.1 Site map and plan view of color-coded, sun-shaded, multibeam bathymetry of the seafloor around the wreck of the Yongala. The "Halo of Holes" is prominently visible around the wreck. The color bar indicates water depth with respect to the prediction datum at Stanley Reef as per Seafarer Tide Charts, Australian Hydrographic Office (Lowest Astronomical Tide). (For interpretation of the references to color in this figure legend, the reader is referred to the web version of this book.)

superstructure provide a habitat for a large range of sessile and mobile marine species, in diversity similar to coral reefs in the region [1].

The wreck lies on the inner edge of the mid-shelf in *ca.* 28 m of water depth and rises to *ca.* 15 m below the water surface. Despite regular dive tourism, the wreck site and the adjacent seafloor is, in ecological terms, in near-unimpacted conditions. The area is located in a Marine National Park ("Green Zone," a no-fishing zone) and additionally protected by the Australian Historic Shipwrecks Act (1976), which restricts dive activities and anchoring near the wreck site. Generally, the prevailing tidal current direction is southeast/northwest (up to a current velocity of 1 m/s), approximately parallel to the wreck's hull [2].

The wreck and surrounding seafloor were mapped with a Reson Seabat 8101 multibeam echosounder in November 2004. Large-scale systematic excavations, presumably of biogenic origin, were discovered in the "Halo of Holes" around the shipwreck (Figure 16.1). In September 2008, *ca.* 1,300 photographs of the seafloor were taken with an automatically triggering still camera suspended from a float, which was drifting slowly with the prevailing current. No sediment samples are available. Generally, sediments in the region are of mixed carbonate and terrestrial origin to approximately even proportions [3].

Geomorphic Features and Habitat

The multibeam survey covered an area of $1.3 \times 0.9 \, \text{km}^2$ around the wreck. Raw soundings were reduced to a bathymetry grid with a horizontal resolution of 0.5 m. The multibeam data clearly delineates a conspicuous halo of holes around the shipwreck (Figure 16.1). Individual holes are up to 10 m in diameter and up to 1.5 m deep (Figure 16.2). They are distributed in a systematic fashion, with the wreck located in the center of the halo. Elsewhere, the seafloor is flat, with no macroscopic structure other than the wreck. Water depth in the survey area ranges from 25.8 to 29.6 m (including in the holes). The wreck is located on a gentle rise, which extends to *ca.* 500 m south of the wreck. Holes are found across the full depth range of the survey.

Locations of individual holes were manually extracted from the multibeam bathymetry data, and holes were classified based on the slope of their rim as calculated from multibeam data. A hole is classed as shallow if 50% or more of its rim has a slope

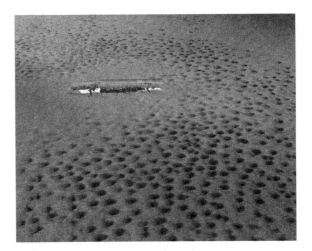

Figure 16.2 Three-dimensional perspective view of the "Halo of Holes," viewed from northeast. Deep and shallow holes are seen in the foreground and background, respectively.

between 4.5° and 14°. A hole with more than 50% of rim slope >14° is classified as deep (Figure 16.3A). Rim slope was used instead of hole depth for computational reasons; where holes are located close to each other it is difficult to define a seafloor surface around the holes, from which the hole depth can be measured, whereas slope is independent of this surface. Isolated shallow holes that were missed by this categorization were added manually based on visual inspection of the bathymetry.

In all, 1,231 holes are found within a distance of 700 m from the center of the wreck, the majority of them located in a halo ranging from 100 to 350 m from the center of the wreck (1,071 holes; 87%). Within the halo, hole-to-hole distances are frequently in the same order or less than individual hole diameter, that is, in the 5–10 m range. Outside the halo, only a small number of holes are found, generally distributed further apart from each other. Less than 4% (56) of the holes are further than 400 m away from the wreck. The halo forms a 270° ring with a gap in the southeast sector (Figure 16.3A). In the northeast and northwest sectors, holes are generally deep, whereas in the southwest sector, mostly shallow holes are found.

The seafloor was classified by neighborhood statistics into the geomorphic features Plain, Halo$_{(deep)}$, and Halo$_{(shallow)}$: a plain is a low-gradient, low-relief surface; Halo$_{(deep)}$ and Halo$_{(shallow)}$ are areas dominated by deep and shallow holes, respectively (Figure 16.3B). This deviation from standard International Hydrographic Organization (IHO) terminology, which was developed for larger spatial scales, is required to distinguish a group of holes (halo) from the IHO term "hole," which refers to a single hole. A feature may include holes from a different individual hole class, for example, Halo$_{(shallow)}$ may contain a small number of deep holes. More specifically, Halo$_{(deep)}$ contains 82% of deep holes (468) and 18% of shallow holes (104), while in Halo$_{(shallow)}$, 95% and 5% of the holes are deep and shallow, respectively. Isolated holes are considered "outliers" and are included in plains (64 holes, or 5% of all holes in the survey area).

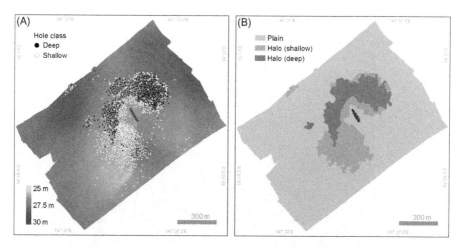

Figure 16.3 (A) Hole classes overlaid on color-coded, sun-shaded, multibeam bathymetry; (B) geomorphic feature discrimination derived from the hole distribution. (For interpretation of the references to color in this figure legend, the reader is referred to the web version of this book.)

The holes are most likely of biogenic origin, but the animal(s) responsible for these earthworks remain unknown to date. The conspicuous distribution is likely reflecting, at least to some degree, the purpose of the building of the hole. It is possible that it represents a pattern in either feeding or nesting behavior of demersal species. The bioturbators may range from small fish species to large stingrays (or may not be fishes at all). Possible explanations for the asymmetric pattern in hole depth include physical and biological processes, for example, current-driven sediment transport/infilling of holes, predator-driven or social schooling, or rotational cropping by a benthic forager. However, as the process of generation of the holes is not understood to date, any such potential explanation for the pattern remains speculation. Notwithstanding the lack of understanding of the ecological processes underlying the generation and distribution of the holes, the Yongala's halo of holes is an important, previously undocumented example for ecosystem engineering of the soft seafloor surrounding structural habitat (i.e., the wreck). Here, the role the holes play as habitat for epibenthic assemblages is documented.

Biological Communities

A total of 1,305 oblique-angle photographs of the seafloor were taken with a still camera suspended from a float, which was drifting slowly with the prevailing currents in September 2008. The camera was suspended *ca.* 1.5 m above the seafloor, and photographs were georeferenced using position data recorded with a handheld GPS mounted on the drifting float. Layback of the camera to GPS location is negligible for this experimental setup.

A simple classification of the biota and bedform (small-scale geomorphology) was applied to the image data. Sessile epibenthic assemblages were identified at a broad ecological level into the following categories: "none," "marine plants," "isolate," and "garden." In addition, and diverging from standard practice, the category "rubble" was included in this classification because the presence of rubble may represent a temporal stage in the colonization by biota of the seafloor. The implications are discussed below. The category "none" was applied where no or close to no macroscopic growth was observed (Figure 16.4). Marine plants include assemblages dominated by algae (including *Caulerpa*, *Halimeda*, and *Udotea* species) and seagrass (including *Halophila* and *Halodule* species), representing a common assemblage on the flat inter-reefal seafloor of the Great Barrier Reef [3]. The category rubble consists of loose rubble in the centimeter to decimeter size range as well as meter-scale patches of rubble covered in crustose coralline or other algae. An isolate is a singular individual or small group of individual animals (or colonies) of sponges and soft corals (including sea whip and gorgonian species). Finally, an inter-reefal garden consists of a dense and diverse assemblage of marine plants, sponges, soft corals, and the occasional hard coral (mostly *Acropora* spp.) [4]. A coverage index (low/medium/dense) was also recorded, but is not reported here as it does not provide any meaningful information in addition to the biological classification for the purpose of this study. Mobile species (fish, sea cucumbers, sea stars, sea snakes, etc.)

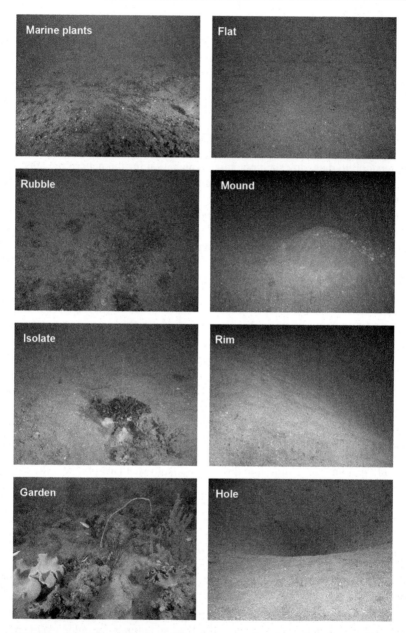

Figure 16.4 Example photographs for each category of epibenthic assemblage (left) and bedform (right). The top right image is an example for both flat seafloor as bedform as well as the category none for epibenthic assemblage. The example for rubble is a photograph of the bottom of a deep hole.

were present but were not included in the analysis. In addition, bedform was defined as "flat," "mound," "rim" (of a hole), or "hole." Only the dominant feature was recorded for each image. Mound, rim, and hole are features of biogenic origin on meter to tens-of-meter scale. Flat describes a seafloor that may include small-scale bioturbation on centimeter to decimeter scale.

Overall, category none (uncolonized seafloor) was recorded on 19% of the 1,305 images. With 62%, marine plants were the most common epibenthic assemblage; rubble was found on 9%, and isolates and gardens were rare, occurring on 6% and 4% of all images, respectively (Figure 16.5). Twenty-one percent of the images were recorded on flat seafloor, and the number of observations in the two zones of the Halo was approximately equal, with 38% (498) and 41% (537) in the $Halo_{(shallow)}$ and $Halo_{(deep)}$, respectively.

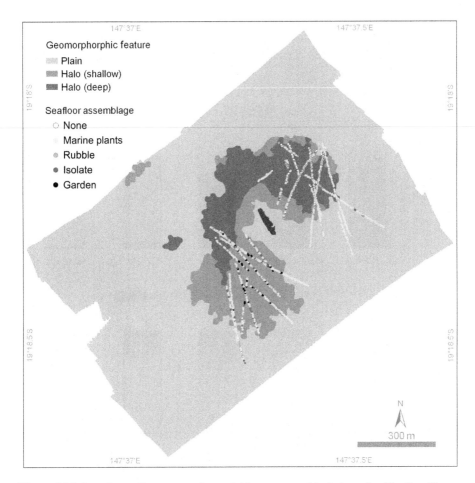

Figure 16.5 Locations of image records overlaid on geomorphic feature classification. The spatial relationship of isolates and gardens with the feature $Halo_{(shallow)}$ is prominently visible.

The surrounding plain was dominated by marine plants (90%). Categories none, rubble, and isolates were rare, with 4%, 2%, and 3%, respectively. No garden was found in this feature (0%, zero occurrence). This geomorphic feature consists almost exclusively of flat seafloor (99%).

Halo$_{(shallow)}$ was also dominated by marine plants (64% of all locations). Categories none and rubble make up 8% and 5% of the observations, respectively. At 12% and 10%, respectively, isolates and gardens occur comparatively frequently in this part of the halo. Bioturbation in form of mounds (1%), rims (7%), and holes (12%) occurs repeatedly on the otherwise flat seafloor (80%).

Halo$_{(deep)}$, unlike the other two features, shows no clear dominance of marine plants (46%) and is comparatively barren with 37% category none recorded. Rubble is significantly more frequent (16%) in this feature, whereas isolates (1%) and gardens (0.2%, one occurrence) are extremely rare. Mounds (4%), rims (16%), and holes (7%) occur frequently on the otherwise flat seafloor (73%).

It is evident from this data that sessile fauna in isolates and gardens occur almost exclusively in Halo$_{(shallow)}$ (Figures 16.5 and 16.6A). Isolates are preferentially associated with flat seafloor, whereas gardens are found dominantly in holes (Figure 16.6B).

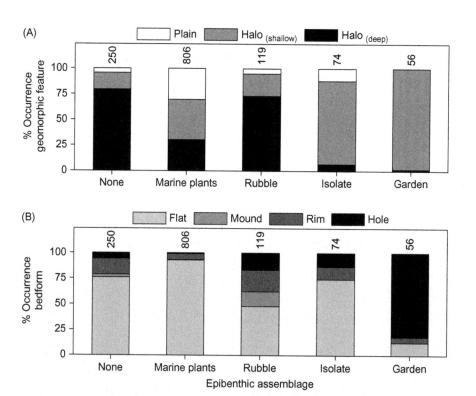

Figure 16.6 Relative distribution of (A) geomorphic features and (B) bedforms associated with each epibenthic assemblage. Total numbers of observations for each assemblage are indicated.

Categories none (no benthic cover) and rubble are chiefly associated with Halo$_{(deep)}$ (Figure 16.6A).

A simple colonization sequence with respect to bioturbation activities is suggested here to explain the observed distribution of biota in the halo. An initial hole is created and maintained for some time. The initially small and shallow hole is regularly excavated to a deep hole, the size of those observed in Halo$_{(deep)}$. During this period, colonization is unlikely because of ongoing sediment turnover or removal, and the seafloor remains bare (e.g., 37% category none in Halo$_{(deep)}$). Subsequently, rubble transported on the flat seafloor by bottom currents accumulates inside holes (e.g., 16% rubble in the same zone). The rubble accumulation (and subsequent consolidation by algae) reduces or prevents further bioturbation, eventually providing a stable substrate for the development of isolates and finally species-rich gardens in "older" holes. Cessation of excavation activities, infilling by rubble, and rim erosion reduce the depth of the holes with time. This effect, together with an estimated age of some observed organisms (sponges/corals) of up to 20 years or more (C. Battershill, personal communication), suggests that Halo$_{(shallow)}$ represents an older, "disused" section of the halo. In the other feature, Halo$_{(deep)}$, bioturbation appears to be ongoing with more than half of this feature not colonized by sessile biota (categories none and rubble together make up 53% in this feature). In addition, bioturbation mounds, usually less than a meter in diameter, are commonly observed in this feature (81% of all mounds are in Halo$_{(deep)}$). The close spatial relationship of a mound with a hole suggests that the material used for the building of the mound is excavated from the hole; however, such a removal process alone does not explain the large size of the holes in Halo$_{(deep)}$.

Where no macroscopic disturbance occurs on the plain, marine plants grow, indicating that the seafloor morphology is stable at least on the timescale of the plant growth. Some isolates, but no gardens, occur on the plain (Figure 16.6A), indicating that isolates do not develop into gardens outside the halo. In summary, the chosen epibenthic assemblage classification scheme effectively represents a temporal sequence in seabed colonization following bioturbation (none→rubble→isolate→garden).

In conclusion, the observations at the Yongala indicate the potential for large artificial reefs such as wrecks to impact on the surrounding seascape. The systematic disturbance of the seabed around the wreck provides a habitat for benthic organisms that is distinctly different in morphology from the surrounding flat seafloor. The animals responsible for the earthworks (to date unidentified) are thus allogenic ecosystem engineers (engineering by activity, not presence), transforming a habitat from one state into another [5]. Their activity ultimately results in a significant change in epibenthic assemblage composition and biomass over small spatial scales.

Similar observations of halos of holes around other shipwrecks, but also around mid-shelf shoals, indicate that such effects are not restricted to the Yongala's wreck [6]. The previously undocumented biota found in association with the halo of holes indicates that the ecosystem wreck extends further beyond the spatial confines of the hull of a wreck than previously considered.

This study provides an example for the recognized importance of small-scale habitat diversity and structure [7], and illustrates the utility of multibeam sonar applications to understanding seafloor ecology not only on regional (tens to hundreds of kilometers scales) but also on "single ecosystem scales" of tens to hundreds of meters.

Surrogacy

The relationship of epibenthic assemblage and geomorphic feature or bedform is not bijective in this study. Species-rich gardens only occur in shallow holes in the $Halo_{(shallow)}$, but neither every shallow hole nor every hole in $Halo_{(shallow)}$ is colonized by these assemblages. Beyond the use in statistical relationships for the consideration of temporally sequential colonization, no statistical analyses of the relationship between geomorphic features and epibenthic assemblages were carried out.

Acknowledgments

The multibeam data was collected with a pole-mounted sounder on *RV James Kirby*, funded by the Cooperative Research Centre for Coastal, Estuarine and Waterways Management. Data was processed with software SWATHED developed by John Hughes Clarke, University of New Brunswick, CA, and with IVS Fledermaus. Image data was collected from the AIMS vessel *RV Cape Ferguson*. Gavin Coombes, Gregg Suosaari, Ron Schroeder, and Erol Eriksson provided field support with image data collection and many thoughts on the "Halo of Holes." Thanks to Peter Doherty, Stuart Kininmonth, Rhondda Jones, Len Zell, Mike Cappo, Ian Banks, Chris Battershill, Andrew Viduka, Paul Crocombe, Richard Fitzpatrick, and others for offering thoughts on the processes occurring at the Halo of Holes. Stuart Kininmonth is thanked for an initial statistical analysis of the hole pattern and for teaching of GIS tools. This is a contribution from the Nereis Park–The Bioturbation World (www .nereispark.org). This chapter is dedicated to Prof. Dr. Paul Müller (Biogeography, University of Trier, Germany), who passed away unexpectedly in May 2010; he had a great influence on the author's work.

References

[1] H.A. Malcolm, A.J. Cheal, A.A. Thompson, Fishes of the Yongala Historic Shipwreck, CRC Reef Research Technical Report 26, CRC Reef, Townsville, 1999. http://www.reef.crc. org.au/publications/techreport/TechRep26.html.

[2] IMOS, Integrated Marine Observing System National Reference Station (NRS) Yongala Mooring, 2010. http://imos.org.au/emii_data.html.

[3] C.R. Pitcher, W. Venables, N. Ellis, I. McLeod, F. Pantus, M. Austin, et al., Great Barrier Reef Seabed Biodiversity Mapping Project: Phase 1, Report to CRC Reef, CSIRO Marine Research, Brisbane, 2002.

[4] M. Cappo, R. Kelley, Connectivity in the Great Barrier Reef World Heritage Area—an overview of pathways and processes, in: E. Wolanski, (Ed.), Oceanographic Processes

of Coral Reefs: Physical and Biological Links in the Great Barrier Reef, CRC Press, Boca Raton, FL, 2001, pp. 161–188. http://www.reef.crc.org.au/resprogram/programC/seabed/Seabedphase1rpt.htm.

[5] C.G. Jones, J.H. Lawton, M. Shackak, Organisms as ecosystem engineers, Oikos 69 (1994) 373–386.

[6] T.C. Stieglitz, E. Eriksson, G. Coombes, J. Clancy, AIMS Cruise 4323: SHOALS3 Cruise Report, Australian Institute of Marine Science, Townsville, 2006.

[7] J.E. Hewitt, S.E. Thrush, J. Halliday, C. Duffy, The importance of small-scale habitat structure for maintaining beta diversity, Ecology 86 (2005) 1619–1626.

17 Submarine De Geer Moraines in the Kvarken Archipelago, the Baltic Sea

Aarno T. Kotilainen[1], Anu M. Kaskela[1], Saara Bäck[2], Jouni Leinikki[3]

[1]Geological Survey of Finland (GTK), Espoo, Finland, [2]Ministry of Environment, Government, Finland, [3]Alleco Oy, Mekaanikonkatu, Helsinki, Finland

Abstract

The Kvarken Archipelago is located in the glaciated epicontinental basin, the Baltic Sea. The majority of the Kvarken Archipelago is very shallow (0–25 m) and shoaly, with ~7,000 islands and islets. Glacioisotatic land uplift rate is ~8.0–8.5 mm/year. The sea freezes yearly, and the annual mean temperature is +3.4°C. The salinity varies between 3‰ and 6‰. The seafloor of this archipelago consists mainly of till (~70%). The boulder-rich De Geer moraines are the most characteristic geomorphic features within the area, creating a unique, washboard-like submarine landscape. Because the area is a transition zone of salinity levels at critical levels to both marine and limnic species, the diversity of marine life is poor. The local bladder wrack, *Fucus radicans*, is characteristic of hard bottoms in the shallow areas with salinity up to 4.5‰. The Kvarken Archipelago was included on UNESCO World Heritage List in 2006 for its geological values.

Key Words: Kvarken Archipelago, Baltic Sea, geomorphic features, De Geer moraines, seabed sediments, biology, habitat mapping

Introduction

The Kvarken Archipelago is located in the European epicontinental basin, the Baltic Sea. The Baltic Sea is one of the largest brackish water bodies in the world. The Kvarken area, between Finland and Sweden, is the narrowest part of the Gulf of Bothnia. It forms a submarine sill (25 m) that separates the Bothnian Sea in the south from the Bothnian Bay in the north (Figure 17.1).

Bedrock of the Kvarken Archipelago consists mainly of Paleoproterozoic crystalline bedrock that outcrops in places. These deeply eroded remnants of ancient mountain ranges form the base of the flat landscape of the Kvarken Archipelago. Due to erosion,

Seafloor Geomorphology as Benthic Habitat. DOI: 10.1016/B978-0-12-385140-6.00017-7

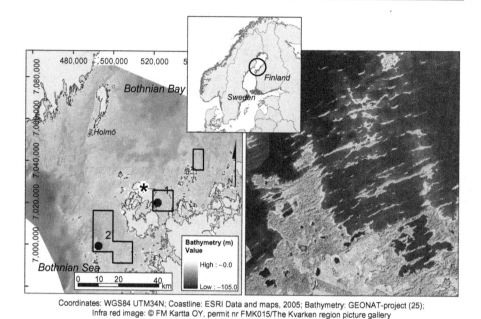

Coordinates: WGS84 UTM34N; Coastline: ESRI Data and maps, 2005; Bathymetry: GEONAT-project (25);
Infra red image: © FM Kartta OY, permit nr FMK015/The Kvarken region picture gallery

Figure 17.1 Bathymetric map of the Kvarken area, the Baltic Sea. Geological survey areas
are outlined. Black dots (1, 2) indicate detail biological study areas. Inset: the location of the
Kvarken in the Baltic Sea. The areal infrared photograph from Björköby area (the location
indicated by star) shows De Geer moraines, fladas, and glo-lakes.

no sedimentary bedrock exists in the area. The region has undergone several glaciations
during the past approximately 3 million years, which has resulted in periods of repeated
glacial erosion and sediment accumulation. During the latest glaciation, the Kvarken
Archipelago was located close to the center of the Weichselian ice sheet, which reached
a thickness of approximately 3 km during the glacial maximum [1]. The seafloor of this
region consists mainly of till. Boulder-rich De Geer moraines characterize the area [2–
5], creating a unique, washboard-like terrestrial and submarine landscape.

The majority of the Finnish side of the Kvarken Archipelago is very shallow
(0–25 m) and shoaly, with approximately 7,000 islands and islets [5]. Due to the relatively rapid rate of isostatic uplift of 8.0–8.5 mm/year [6,7], the bathymetry of the
region has changed dramatically since deglaciation. During and just after deglaciation (around 10,000 years ago), the archipelago was submerged to a water depth of
250–280 m. The ongoing land uplift leads to the continuous erosion of seafloor. It
exposes older seafloor deposits and gradually raises them above sea level.

In the Kvarken Archipelago, sea ice cover lasts between 140 and 150 days per
year [8]. The annual mean temperature is +3.4 °C [9]. Salinity decreases from 5‰
to 6‰ in the southern part to 3–4‰ in the northern part of the study area [10], and
it approaches zero at river mouths; hence, marine species live at their limits of distribution. The Kvarken Archipelago was included on UNESCO World Heritage
List in 2006 for its geological values. The major part of the Kvarken Archipelago is

included in Nature Conservation Programmes and Natura 2000 Network. The Natura 2000 Network is a joint network of bird protection areas and valuable habitats within the European Union. The main threats to the marine area are eutrophication, pollution, shipping accidents, and seafloor constructions like wind farms; however, the Kvarken Archipelago is largely unmodified by human activities.

Geomorphic Features and Habitats

In the boreal summers of 2003–2005, and in 2007, part of the Kvarken Archipelago was mapped and sampled using the Geological Survey of Finland's (GTK's) research vessels *Geola, Geomari, Kaita,* and *Kaiku* (Figure 17.1). Fine-scale seafloor geologic information was obtained using continuous single-channel seismic profiling (400–700 Hz), sub-bottom profiler (echosounder) (MeriData MD 28 kHz), multibeam echosounder (MBE) (Atlas FS 20), and sidescan sonar (Klein SA 350, 100 kHz) imaging, and various sediment sampling methods. Altogether a total length of approximately 800 km acoustic–seismic profiles were recorded, covering approximately 350 km^2 (MBE coverage only around 2 km^2). Survey lines are situated approximately 500 m apart, except in the detailed study area (Figure 17.1), where 50 m track line spacing was used. In addition, 49 long-sediment cores and 44 surface-sediment cores were recovered from the seafloor.

The broad-scale geomorphic features described here are based on the Baltic Sea scale analysis of the marine landscapes [11].

The seafloor morphology of the Kvarken area is characterized by broad-scale geomorphic features including mounds, plains, basins, and some sea valleys (holes) (Table 17.1). However, the seafloor bathymetry follows mainly the surface of the bedrock, and the relief of the area is relatively low. Moraines control the local, fine-scale morphology, and bathymetry of the area.

The classification of the broad-scale geomorphic (topographic) features shown here is based on feature names and definitions from the International Hydrographic

Table 17.1 Definitions and Areas (km^2) for Large-Scale Geomorphic Features Mapped in the Kvarken Area, the Baltic Sea

Geomorphic Feature	Definition	Surface Area (km^2)
Basins	A depression in the seafloor, more or less equi-dimensional in plain and of variable extent.	3,510
Mounds	Low elevations of shelf seafloor. Includes hills, banks, and reefs. Mounds do not consider the water depth above the feature.	4,329
Plains	Large areas where relief stays low and homogenous.	2,691
Sea valley and holes	Steep-sided small depressions or relatively shallow depressions.	1,287

Source: According to Kaskela et al. [13].

Organization [12] whenever feasible. Full-coverage multibeam data are not available for the whole Kvarken area, and presently there is multibeam data only from small areas. Therefore, a broad-scale bathymetry data set (resolution ~500 m) was used to give a general overview of the region. Bathymetry and substrate data sets were reprojected to UTM34N and rescaled to 200 × 200 m raster grids in ArcGIS. Broad geomorphic features were identified [13] combining the Bathymetric Position Index (BPI) structures [14], derived from the Benthic Terrain Modeler (BTM) analysis of bathymetry data, with the substrate data and the bathymetric zones. Combinations were done using ArcGIS Raster calculator tool by adding grid values. In the BTM analysis, the fine- and broad-scale BPI features were not separated (e.g., local basins, broad basins) due to lack of high-resolution data (Figure 17.2).

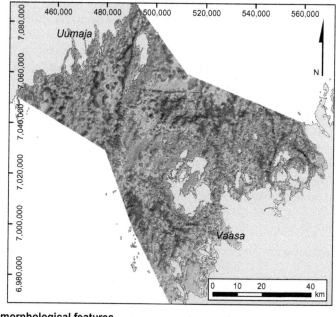

Broad geomorphological features

Coordinates: WGS84 UTM34N; Coastline: ESRI Data and maps, 2005; Feature data: BALANCE -project (11)

Figure 17.2 The broad-scale seabed topographic features of the Kvarken, the northern Baltic Sea. The seabed topographic features were defined through overlay analysis of bathymetry data set (and its derivatives) and substrate data set.

Basins: Basins are relatively small and shallow, and they were found in several areas of the Kvarken, covering altogether ~30% (3,510 km^2) of the seafloor. Coarse substrate basins (26%) are more dominant than mud basins (4%).

Mounds: Mounds cover 37% (4,329 km^2) of the Kvarken area. Large mounds are found especially in area between Raippaluoto Island and Sweden. Coarse substrate (complex) mounds (27%) are the most typical geomorphic feature type of the area. Different types of moraines and boulder reefs are included within these complex mound feature types.

Plains: Plains cover 23% (2,691 km^2) of the area. Large coarse substrate plains occur especially in the northern Kvarken and the west coast of Finland, south to the city of Vaasa.

Sea valleys: Valleys are characterized for the southwestern part of the Kvarken. Valleys locate also at the northern Kvarken, for example, to the west of the Holmö island. Valleys cover 11% (1,287 km^2) of the Kvarken area. Most likely some sea valleys are formed by the preglacial or Early Weichselian river system [15].

Fine-scale geomorphic features—moraine formations: Typical terrestrial sedimentary deposits of the area, moraine formations like De Geer moraines (called washboard moraines), hummocky moraines, large transversal "Rogen type" (or ribbed) moraines, and drumlins (Figure 17.3) occur also in the submarine area of the Kvarken Archipelago.

The boulder-rich De Geer moraine fields are the most characteristic submarine features that make the Kvarken Archipelago unique [3] in the whole world. During the surveys, submarine De Geer moraines were found at the seafloor in several areas of the Kvarken Archipelago. The formations were boulder rich, curvy, and branched (Figures 17.1 and 17.4). The mean visible length was over 100 m. Height and width of formation were around 3 and 40 m, respectively [4]. De Geer moraines occurred at

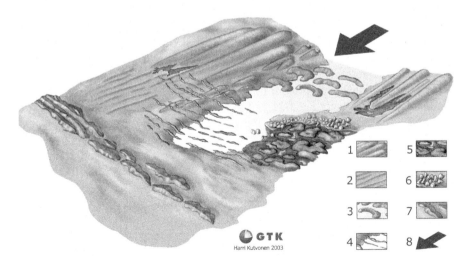

Figure 17.3 Schematic figure of the moraine formations in the northern Kvarken area:
1. drumlins, 2. flutings, 3. transversal moraines (Rogen type), 4. De Geer moraines,
5. hummocky moraines, 6. boulder-rich surface, 7. end moraines, 8. latest ice flow direction.
Source: Drawing by Harri Kutvonen 2003, GTK ©.

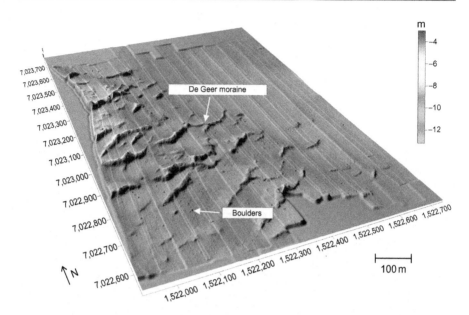

Figure 17.4 Three-dimensional perspective view of the (shallow) submarine De Geer moraines in the vicinity of the island of Raippaluoto. Bathymetry is based on multibeam echosounding data. In places the De Deer moraines are covered by postglacial soft sediments (on the right).

approximately 100 m intervals, and the orientation of formations varied [4] reflecting the orientation of the retreating ice sheet. Local (fine-scale) bathymetry varies substantially due to the heterogenous morphology of moraine formations, especially in the De Geer moraine formation fields. The combination of submarine De Geer moraines and continuous land uplift produces the range of marine and coastal habitats. For example, globally rare flada and glo-lakes are constantly formed in hollows between De Geer or Rogen moraines (Figure 17.1).

Sediments: The seafloor of the Kvarken Archipelago consists mainly of till (~70%). Bedrock outcrops exist rarely (average of 3%). Modern mud accumulation areas cover approximately 8% of the seafloor. However, the spatial distribution of mud accumulation areas is quite variable. In addition to modern mud, Late and Middle Holocene gyttja clays (muddy clays) cover approximately 7% of the seafloor. Average spatial coverages of Early Holocene sulfide clays and varved clays on the seafloor in the study area are approximately 7% and approximately 4%, respectively.

In shallow depths, soft sediments are often absent because they have been eroded by wave and ice action as well as strong currents. However, these sediments cover the seafloor in basins, which are protected from erosion. The erosion, transportation, and accumulation of the sediments on the seafloor vary spatially and temporally (e.g., land uplift), making seafloor very heterogeneous.

Biological Communities

The information on underwater biological communities was collected on offshore De Geer moraines in August 2004 (Figure 17.1, Site 1). Macroalgae and macrozoobenthos were sampled by a scuba diver in water depths of 5–8 m from one to three points on four moraines. Underwater biodiversity was documented also by underwater camera and video recordings.

The biological data of 2007 from the shallow areas by the islands was used for modeling with bathymetry and substrate to create biotope maps of the area (Figure 17.1, Site 2). In the standard diving method used in the Finnish Inventory Programme for the Underwater Marine Environment (VELMU), a scientific diver records from an estimated area of 4 m² the location, depth, and percentage coverage of substrate types, and percentage coverage and average height of all macroscopic species, vegetation, and sessile zoobenthos. A stratified sampling design is used to collect data from combined zones of depth and wave exposure types [16] for modeling. The total number of observations was 300 (Figure 17.5).

The biological communities were classified into biotopes using BalMar method [17], where biotopes are named according to the dominating species. The probabilities of the found biotopes were then mapped by using Case-Based Reasoning (CBR) modeling method.

Macrophyte vegetation: There are a few older studies on underwater vegetation of the northern Kvarken area, but no description of De Geer moraine vegetation.

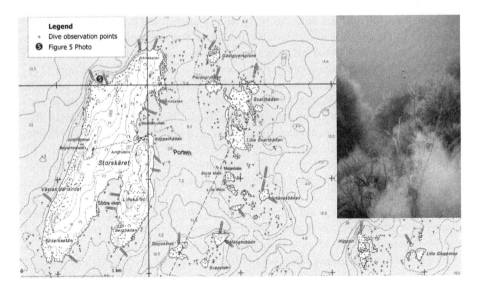

Figure 17.5 Diving transects in the study area. Red dot indicates a location of photo from a limnic phanerogam *Potamogeton filiformis*, a moss *Fontinalis* sp., and stands of *Fucus radicans* in the back. (For interpretation of the references to color in this figure legend, the reader is referred to the web version of this book).

Ehnholm [18] described algae vegetation nearby in Mickelsörarna and studied vegetation from the shores of the islands. The species number was low and the vegetation sparse. Later, Björkquist and Bergström [19] studied macroalgae vegetation in Replotfjärden Sea area, confirming that species number is also low and the rocky substrata is dominated by *Cladophora aegarophila* down to 4–5 m.

The surface of submarine De Geer moraine provides suitable substrate for macroalgae vegetation due to the stones, boulders, and gravel on which algae can attach. Sparse algae vegetation was found in all locations.

Filamentous green algae *Cladophora aegarophila* is the most dominant algae in Replotfjärden. It is growing attached to the stones in all moraines and is found even at a depth of 8 m. It was found growing as less than 1 cm dark green tufts and in some instances in very small balls. In many locations, *C. aegarophila* was covered with diatoms, giving the alga very dark color. *C. aegarophila* was found also loose lying and in some instances it formed algal mats. The species is very common in nearby Mickelsörarna area.

Another very common species is filamentous brown alga *Sphacellaria arctica*. Ehnholm [18] found the species also in Mickelsörarna. The small, tuft-forming species was growing mixed with *C. aegarophila*. Another species that is growing mixed among *Cladophora* is the red filamentous alga *Ceramium tenuicorne*. Although this species is commonly growing in shallow shores up to the Gulf of Bothnia, it was found in a few locations in Replotfjärden.

Macrozoobenthos: The benthic fauna in the northern Kvarken area is relatively scarce [20]. The Bothnian Bay, including the northern Kvarken, is of a more limnic character when compared with the rest of the Baltic Sea. The macrobenthic fauna of the Bothnian Bay is low in biomass as well as species numbers and species composition compared for instance with the high salinity area of the central Baltic. This is due to the mixing of salt water from the Bothnian Sea and fresh water runoff from the rivers. Biodiversity in the Baltic Sea is at its lowest at approximately 5 psu, which is commonly known as the boundary for physiological stress caused by salinity [21].

Common macrozoobenthic species for the Bothnian Bay are, for example, *Saduria entomon*, *Monoporeia affinis*, and *Macoma balthica*. In fact, these species alone contribute to 95% of the total macrofaunal biomass in the Bothnian Bay [22,23]. The amphipod *Corophium volutator* is a characteristic species for shallow soft bottoms in the southern parts of the Kvarken area [10,24].

In this study, the dominant species of benthic fauna was the amphipod *C. volutator*, found at a depth range between 5 and 8 m, in both hard- and soft-bottom habitats, at all sampling stations.

Other abundant species are the baltic tellin *Macoma balthica*, the amphipod *Monoporeia affinis*, and the isopod *Asellus aquaticus*. Very common is also hydroid *Cordylophora caspia* that is growing attached to stones.

Biological communities of the shores of the islands: The divers found 32 species, of which 9 were limnic, 20 marine, and 3 brackish water species. The dominant species were *Fucus radicans*, *Potamogeton perfoliatus*, and *Potamogeton pectinatus*.

In the brackish waters of the Baltic Sea, the coexistence of marine and limnic species is common. In the Kvarken area, such specialities include bladder wrack (*Fucus radicans*) and a moss species *Fontinalis antipyretica* growing along each other on the same boulder. The seabed is characterized by the mosaics of hard and soft substrata, hard- and soft-bottom species (Figure 17.5).

Surrogacy

No statistical analyses have been carried out on this data set to examine relationships between physical surrogates and benthos.

Acknowledgments

The results presented are based upon the outputs of the BSR INTERREG IIIB Neighbourhood Programme (BALANCE), INTERREG IIIA Kvarken MittSkandia Programme (GEONAT), VIMMA, and VALKO projects (VELMU) (Ministry of Environment and Ministry of Trade and Industry). We would like to thank our colleagues participating in these projects for data, support, and feedback. Anonymous reviewers provided very constructive and valuable comments on this manuscript, which are gratefully acknowledged. We would also like to thank our marine geology colleagues at GTK for their encouragement and feedback.

References

[1] J.I. Svendsen, H. Alexanderson, V.I. Astakhov, I. Demidov, J.A. Dowdeswell, S. Funder, et al., Late Quaternary ice sheet history of northern Eurasia, in: Quaternary environments of the Eurasian North (QUEEN), Quat. Sci. Rev. 23(11–13) (2004) 1229–1271.

[2] J. Nuorteva, Akustisilla luotausmenetelmillä saatu kuva merenpohjan kvartäärikerrostumista (in Finnish). Geologian tutkimuskeskus, Tutkimusraportti 82—Report of Investigation 82, 1988, 32 pp.

[3] T. Aartolahti, Glacial geomorphology in Finland, in: J. Ehlers, S. Kozarski, P. Gibbard, (Eds.), Glacial Deposita in North–East Europe, Balkema Publishers, Rotterdam, 1995, pp. 37–50.

[4] A. Reijonen, Vedenalaisten moreenimuodostumien monimuotoisuus Merenkurkun Saaristossa (in Finnish), Unpublished M.Sc. Thesis 228, Helsingin yliopisto, Geologian laitos, 2004, 102 pp.

[5] O. Breilin, A. Kotilainen, K. Nenonen, M. Räsänen, The unique moraine morphology, stratotypes and ongoing geological processes at the Kvarken Archipelago on the land uplift area in the western coast of Finland, in: A.E.K. Ojala, (Ed.), Quaternary Studies in the Northern and Arctic Regions of Finland: Proceedings of the Workshop Organized Within the Finnish National Committee for Quaternary Research (INQUA), Kilpisjärvi Biological Station, Finland, January 13–14 2005. Geological Survey of Finland, Espoo, 2005, pp. 97–111 (Special Paper 40).

[6] M. Ekman, A consistent map of the postglacial uplift of Fennoscandia, Terra Nova 8 (1996) 158–165.

[7] J. Mäkinen, V. Saaranen, Determination of post-glacial land uplift from three precise levelling in Finland, J. Geod. 72 (1998) 516–529.

[8] R. Heino, Klimatet i Vasa skärgård, in: T. Osala, (Ed.), Vaasan Saaristo, Vasa skärgård, O & G Förlaget, Vasa, 1988, pp. 66–79.

[9] V.A. Helminen, Ilmasto. Suomen kartasto,131:4–10. Maanmittaushallitus, Helsinki, 1987.

[10] P. Sevola, Vedet ja vesiluonto, 80–140. in: T. Osala, (Ed.), Vaasan Saaristo, O & G Kustannus, Vaasa, 1988, 424 pp.

[11] U. Alanen, J.H. Andersen, J. Bendtsen, U. Bergström, K. Dahl, G. Dinesen, et al., Towards marine landscapes in the Kattegat and Baltic Sea, in: Z. Al-Hamdani and J. Reker (Eds.), BALANCE Interim Report No. 10. 2007, 118 pp.

[12] IHO, Standardization of Undersea Feature Names: Guidelines Proposal form Terminology. International Hydrographic Organisation and International Oceanographic Commission, Monaco, 2001, 40 pp.

[13] A.M. Kaskela, A.T. Kotilainen, Z. Al-Hamdani, J.O. Leth, J. Reker, Seabed topographic features in a glaciated shelf sea, the Baltic Sea (Submitted).

[14] E. Lundblad, D.J. Wright, J. Miller, E.M. Larkin, R. Rinehart, T. Battista, et al., A benthic terrain classification scheme for American Samoa, Mar. Geod. 26 (2) (2006) 89–111.

[15] K. Nenonen, Pleistocene stratigraphy and reference sections in southern and western Finland, Ph.D. Thesis, Geological Survey of Finland, Kuopio, Finland, 1995, 205 pp.

[16] M. Isæus, B. Rygg, Wave exposure calculation for the Finnish coast. Oslo, Norwegian Institute for Water Research, NIVA Report, 2005, 24 pp.

[17] H. Backer, J. Leinikki, P. Oulasvirta, Baltic Marine Biotope Classification System (BMBCS)—definitions, methods and EUNIS compatibility. Alleco Oy Technical Report, 2004, 47 pp., 5 app.

[18] G. Ehnholm, Bidrag till kännedom on algfloran in Kvarken, Memoranda Soc. Fauna Flora Fennica 13 (1938) 21–24.

[19] L. Björkquist, U. Bergstöm, Hårdbottensvegetationen- en dykinventering i Västra finlands yttre skärgård, Kvarenrådets publikationer I, 1997, 48 pp.

[20] S. Ekman, Bottniska vikens lägre djurliv, in: O. Elofsson, K. Curry-Lindahl (Eds.), Natur I Ångermanland och Medelpad. Svensk Natur, 1953, pp. 180–190.

[21] P. Snoeijs, Marine and brackish waters, Acta Phytogeogr. Suec. 84 (1999) 187–212.

[22] K. Leonardsson, A. Laine, A. Andersin, Benthic conditions and macrofauna in the Gulf of Bothnia. Helsinki Commission, HELCOM 2002, Environment of the Baltic Sea area 1994–1998, Baltic Sea Environ. Proc. 82B (2002) 68–73.

[23] A.O. Laine, Distribution of soft-bottom macrofauna in the deep open Baltic Sea in relation to environmental variability, Estuar. Coast. Shelf Sci. 57 (2003) 87–97.

[24] S. Ankar, K. Leonardsson, Marin inventering av Holmöarna, 1981. Makrofaunan på mjukare bottnar, Stockholms Universitet, 1981, pp. 1–44.

18 Habitats and Benthos of an Evolving Fjord, Glacier Bay, Alaska

Guy R. Cochrane¹, Luke Trusel², Jodi Harney³, Lisa Etherington⁴

¹USGS Pacific Coastal and Marine Science Center, Santa Cruz, CA, USA, ²Clark University Graduate School of Geography, Worcester, MA, USA, ³ENTRIX, Riverview, FL, USA, ⁴Cordell Bank National Marine Sanctuary, Olema, CA, USA

Abstract

Benthic habitat mapping has been completed for Glacier Bay (in 2005) and Muir Inlet (in 2010) following multibeam sonar surveys in these areas. Though different classification schemes were used, the classes are based on geomorphologic features, substrate classes, and depth zones, which make the results comparable. There is greater diversity of habitat (and epifauna) in Glacier Bay as compared to Muir Inlet. Substrate classes ranging from sand to coarser sediments make up the majority of habitat in Glacier Bay, whereas mud is the dominant substrate in Muir Inlet. These results suggest that Muir Inlet and other inlets to the north of Glacier Bay are very efficient traps of fine glacial sediment and should be considered distinct biotopes for ecological management purposes.

Key Words: benthic habitat, glacial sedimentation, marine protected area, Glacier Bay, Muir Inlet

Introduction

Glacier Bay National Park and Preserve is located in southeastern Alaska (Figure 18.1). It is a fjord system that bifurcates into two main northern tributaries: the West Arm and Muir Inlet, also known as the East Arm. This study focuses on the main southern bay and Muir Inlet, the 41-km long, 1–4-km wide fjord. Historic deglaciation of the region is well documented [1], resulting in a dynamic estuarine environment with vigorous tidal currents and rapid sedimentation. These factors lead to seafloor instability and constantly change benthic environments. The diverse settings in the estuary generate productive food webs; recreational, commercial, and subsistence fisheries; large populations of marine mammals and seabirds; and vessel-based tourism [2].

The cool and temperate maritime climate of southeast Alaska and high-elevation mountain ranges permit widespread glaciation at a relatively low latitude setting.

Seafloor Geomorphology as Benthic Habitat. DOI: 10.1016/B978-0-12-385140-6.00018-9

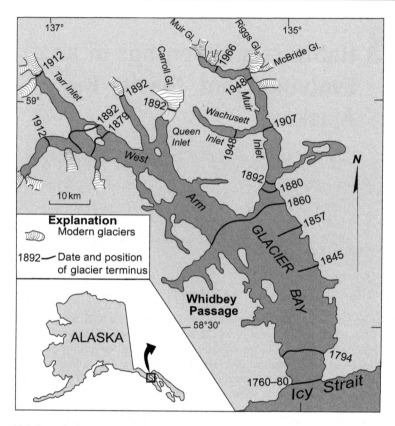

Figure 18.1 Location map of Glacier Bay National Park showing terminus positions and dates of retreat of the Little Ice Age glacier that completely filled the bay somewhat more than 200 years ago. The 1794 terminous line near the mouth of the bay is where Capt. George Vancouver and crew observed the massive glacier face during their hunt for the Northwest passage. The 1879 glacier terminous position was mapped by John Muir during his first of several visits to Glacier Bay.
Source: Modified from Ref. [1].

Abundant precipitation in the form of snow is supplied to the surrounding mountain ranges to feed many valley glaciers that flow into Glacier Bay. These temperate glaciers are some of the largest and most active in southeast Alaska, with glacial sediment fluxes being among the highest recorded worldwide [3]. Locally, sediment accumulation rates at the terminus of Muir Glacier have exceeded several tens of meters per year during its recent retreat [4]. Such high accumulation rates decrease rapidly with distance from the glacial point source [5] are mostly confined within the ice-contact basin [6], and locally act as a major physical control on the diversity and spatial distribution of fjord biota [7].

The oceanographic setting of Glacier Bay is also complex, as seen in bathymetric variability, strong tidal currents, and strong seasonal water-column stratification, which results from high rates of glacial ablation, snow melt, and precipitation [8]. Muir Inlet is a glacial fjord characterized at its southern end by a shallow basement sill topped by a morainal bank [1], steeply sloping walls (up to 77°), and multiple deep basins separated by transverse sills also topped with morainal banks. Areas of shallow seafloor have higher tidal current velocities and greater vertical mixing [8,9]. In contrast, adjacent deep-basin waters are highly stratified and have minimal benthic currents. Average root mean square (RMS) current velocities (derived from an ADCIRC tidal circulation model) for the main stem of Muir Inlet are small, ranging from 0.01 to 0.139 m/s. However, peak instantaneous velocities are approximately 50–100% higher than RMS velocity values, particularly over areas of shallow bathymetry [8]. Model results demonstrate the large tidal range within Glacier Bay, with values averaging 3.86 m in the lower Bay, increasing to an average value of 4.59 m at the head of the Muir Inlet [8].

For habitat mapping, multibeam sonar data were collected in Glacier Bay by the USGS on the R/V *Davidson* [1]; habitat derived from these data were published by Harney et al. [3]. In Muir Inlet, multibeam data were collected on the R/V M*aurice Ewing* in early June 2004, and habitat analysis has recently been completed [10]. Multibeam records in Muir Inlet were supplemented with seismic reflection profiles and several sediment cores. Published analyses from this cruise thus far have focused on tidewater-glacier dynamics and history inferred through glacial-sequence stratigraphy [11] and reconstruction of glacial discharge dynamics from rhythmic sedimentary deposits [12].

For visual ground-truth, nearly 42 h of underwater video were collected and logged real time on 52 transects, primarily in the main bay [3]. Visually observed seafloor characteristics (geomorphology, sediment texture, and biota) were digitally recorded in real time at 30-s intervals by a geologist and a biologist watching the towed video using the methodology of Anderson et al. [13].

- Primary (>50%) and secondary (>20%) substrate type (e.g., boulder/cobble, rock/sand, mud/mud)
- Substrate complexity (rugosity)
- Seafloor slope
- Benthic biomass (low, medium, or high)
- Presence of benthic organisms and demersal fish
- Small-scale seafloor features (e.g., ripples, tracks, and burrows).

Geomorphic Features and Habitats of Muir Inlet

The classification of Muir Inlet seafloor habitats [10] was done using the classes of the 2008 draft version of the Coastal and Marine Ecological Classification Standard (CMECS) hierarchy [14] as a pilot study for the National Park Service. The classification was performed using a supervised manual classification of seafloor substrate based on multibeam backscatter intensity, two derivative bathymetric properties (seafloor rugosity and slope), and knowledge from the ground-truthing sources. CMECS has a surface geology component; Table 18.1 shows substrate types for Muir Inlet

Table 18.1 Substrate Distribution for Muir Inlet

CMECS Class	CMECS Subclass	Percentage of Total Area	Area (km²)
Unconsolidated bottom	Mud	88.7	64.6
	Mixed sediments	3.3	2.4
	Cobble/gravel	0.4	0.3
Rock bottom	Boulder/rubble	1.4	1.0
	Bedrock	6.2	4.5

Table 18.2 Substrate Distribution by Depth Zone for Muir Inlet

CMECS Subclass	Deep Infralittoral: 5–30 m Water Depth		Circalittoral: 30–80 m Water Depth		Circalittoral (offshore): 80–200 m Water Depth		Mesobenthic: 200–1,000 m Water Depth	
	Percent	Area (km²)	Percent	Area (km²)	Percent	Area (km²)	Percent	Area (km²)
Mud	0.05	0.04	3.53	2.57	28.79	20.95	56.35	41.00
Mixed sediments	0	0	0.11	0.08	1.94	1.41	1.26	0.91
Cobble/gravel	0	0	0.03	0.02	0.34	0.25	0.03	0.02
Boulder/rubble	0	0	0.01	0.01	0.56	0.41	0.79	0.57
Bedrock	0.01	0.01	0.43	0.31	3.95	2.88	1.83	1.33

Table 18.3 Major Geoform Distribution for Muir Inlet

Megageoform	Mesogeoform	Percentage of Total Area	Area (km²)
Fjord	Delta	5.8	4.6
	Floor	38.7	30.7
	Moraine	15.5	12.3
	Wall	40.0	31.7

divided into CMECS classes. Table 18.2 further divides these substrate classes into CMECS depth zones. CMECS also has a geoform component to accommodate geomorphological features that may contain many geologic elements. Table 18.3 shows the CMECS geoforms found in Muir Inlet; Table 18.4 divides up the geoform areas into depth zones.

Table 18.4 Major Geoform Distribution by Depth Zone for Muir Inlet

CMECS Subclass	Deep Infralittoral: 5–30 m Water Depth		Circalittoral: 30–80 m Water Depth		Circalittoral (offshore): 80–200 m Water Depth		Mesobenthic: 200–1,000 m Water Depth	
	Percent	Area (km²)	Percent	Area (km²)	Percent	Area (km²)	Percent	Area (km²)
Delta	0	0	0.37	0.29	3.68	2.92	1.80	1.43
Floor	0	0	0	0	0.69	0.55	38.0	30.13
Moraine	0	0	1.38	1.10	10.20	8.08	3.93	3.11
Wall	0.05	0.04	2.67	2.11	22.89	18.15	14.36	11.39

Geomorphic Features and Habitats of Glacier Bay

Habitats in Glacier Bay proper (Figure 18.2, Table 18.5) were mapped prior to those of Muir Inlet [3]. The classification is based on the Greene et al. [15] system. Seafloor features between 1 m and 1 km in scale are defined as "meso/macrohabitat" types. Several interesting seafloor features, including sand waves, gullies, mud mounds, slumps, and depressions (flat basins), were identified in the acoustic data and verified using towed video. Codes for these polygon-delineated features follow the Greene et al. [15] classification system, although some codes were created specifically for this environment. In the Greene classification system, mounds and depressions are described by the same code ("m"). Because we felt it was important to distinguish bathymetric highs and lows, we employed Greene's "m" class for mounds (high features) and created a unique "v" class for large depressions and flat-floored basins. We created a unique "x" class for polygons in nearshore settings within approximately 200 m of the high-tide shoreline. These nearshore polygons were also assigned an underscored descriptor to specify the bathymetric class ("x_x" for polygons shallower than 75 m water depth; "x_y" for those in 75–200 m water depth). This distinguishes shallow, more energetic nearshore settings from shoreline margins that are steeper and deeper.

Biological Communities

Etherington et al. [8] identified three general groups of benthic habitats in Glacier Bay and Muir Inlet based on geological and physical habitat characteristics and dominant benthic associations (Figure 18.3). These benthic habitat groups include a shallow-water (depth less than 75 m) high-current sand and cobble habitat; a deep-water (depth greater than 200 m) mud habitat; and a moderate-depth (depth between

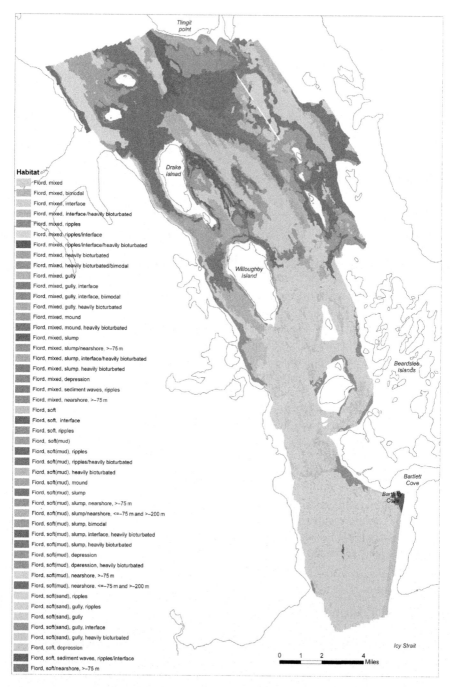

Figure 18.2 Benthic habitats in Glacier Bay, southeast Alaska [3]. Classification follows Greene et al. [15]. Full resolution available at http://pubs.usgs.gov/of/2006/1081/images/fig6.html.

Table 18.5 Total Area of Habitats in Glacier Bay, Southeast Alaska (Classification Follows Greene et al. [15])

Code	Habitat Description	Area (m²)	Area (%)
Im	Fjord, mixed	135,787,405	31
Im _b	Fjord, mixed, bimodal	48,000	0.01
Im _i	Fjord, mixed, interface	4,815,053	1
Im _i_t	Fjord, mixed, interface heavily bioturbated	195939.0409	0.05
Im _r	Fjord, mixed, ripples	81,834	0.02
Im _r_i	Fjord, mixed, ripples interface	120,000	0.03
Im _r_i_t	Fjord, mixed, ripples interface heavily bioturbated	845,745	0.21
Im _t	Fjord, mixed, heavily bioturbated	9,066,893	2
Im _t_b	Fjord, mixed, heavily bioturbated bimodal	44,582	0.01
Img	Fjord, mixed, gully	22,623,546	5.2
Img_i	Fjord, mixed, gully, interface	3,547,276	0.81
Img_i_b	Fjord, mixed, gully, interface, bimodal	1,594,900	0.37
Img_t	Fjord, mixed, gully, heavily bioturbated	3,522,000	0.81
Imm	Fjord, mixed, mound	497,025	0.11
Imm_t	Fjord, mixed, mound, heavily bioturbated	1,371,600	0.31
Ims	Fjord, mixed, slump	5,414,590	1.24
Ims/x_x	Fjord, mixed, slump nearshore bathy class $>-75\,\mathrm{m}$	563,342	0.13
Ims_i_t	Fjord, mixed, slump, interface heavily bioturbated	4,236,882	0.1
Ims_t	Fjord, mixed, slump, heavily bioturbated	17,424,569	4
Imv	Fjord, mixed, depression	833,925	0.2
Imw_r	Fjord, mixed, ripples	15,600	0.01
Imx_x	Fjord, mixed, nearshore bathy class $>-75\,\mathrm{m}$	9,593,602	2.1
Is	Fjord, soft	25,927,354	5.941
Is _i	Fjord, soft, interface	3,200	0.00071
Is _r	Fjord, soft, ripples	17,200	0.004
Isg	Fjord, soft(sand), gully	25,587,410	5.87
Isg_i	Fjord, soft(sand), gully, interface	144,400	0.03
Isg_t	Fjord, soft(sand), gully, heavily bioturbated	32,482,503	7.45
Isv	Fjord, soft, depression	614,265	0.14
Isw_r_i	Fjord, soft, ripples interface	86,399	0.012
Isx_x	Fjord, soft, nearshore bathy class $>-75\,\mathrm{m}$	91,100	0.02
Is(m)	Fjord, soft(mud)	26,688,575	6.12
Is(m) _r	Fjord, soft(mud), ripples	17,272	0.003
Is(m) _r_t	Fjord, soft(mud), ripples heavily bioturbated	25,317	0.01
Is(m) _t	Fjord, soft(mud), heavily bioturbated	3,587,580	0.82
Is(m) s/x_x	Fjord, soft(mud), nearshore bathy class $>-75\,\mathrm{m}$	5,199	0.001
Is(m)m	Fjord, soft(mud), mound	1,961,018	0.45
Is(m)s	Fjord, soft(mud), slump	65,448,806	15

(Continued)

Table 18.5 (Continued)

Code	Habitat Description	Area (m²)	Area (%)
Is(m)s/x_x	Fjord, soft(mud), slump, nearshore bathy class >−75 m	476,735	1.1
Is(m)s/x_y	Fjord, soft(mud), slump, nearshore bathy class ≤−75 m and >−200 m	1,723,051	0.4
Is(m)s_b	Fjord, soft(mud), slump, bimodal	27,329	0.01
Is(m)s_i_t	Fjord, soft(mud), slump, interface heavily bioturbated	5,410,350	1.24
Is(m)s_t	Fjord, soft(mud), slump, heavily bioturbated	6,277,790	1.44
Is(m)v	Fjord, soft(mud), depression	5,888,071	1.35
Is(m)v_t	Fjord, soft(mud), depression, heavily bioturbated	1,397,369	0.32
Is(m)x_x	Fjord, soft(mud), slump, interface nearshore bathy class >−75 m	3,650,170	0.84
Is(m)x_y	Fjord, soft(mud), nearshore bathy class ≤−75 m and >−200 m	1,744,902	0.4
Is(s) _r	Fjord, soft(sand), ripples	119,152	0.02
Is(s)g_r	Fjord, soft(sand), gully, ripples	266,359	0.06

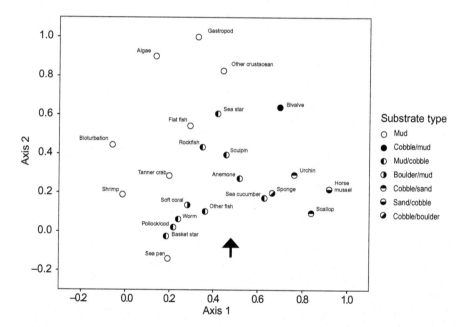

Figure 18.3 Taxa distribution within the ordination space of all video observations. The location of a taxa point in the multidimensional space denotes the centroid of transects that contained the given taxa. Species that coexist are close together in ordination space, while those that are in differing areas occupy different locations in ordination space. Color symbols represent the primary substrate type wherein the species were predominantly found. Multiple colors in the symbols indicate that the taxa most often was in the color on the left, but also was predominant in the color on the right. Bioturbation represents the presence of mounds, holes, and (or) tracks in the substrate. Note the general increase in grain size from mud on the left to larger grain sizes/harder substrates toward the right of the figure. (For interpretation of the references to color in this figure legend, the reader is referred to the web version of this book.)

75 and 200 m) cobble and mud habitat. The association of groups of taxa with these three habitat types is the result of the interaction of various physical and biological factors. One factor that could influence animal presence and abundance is the recruitment of organisms to the habitat, which would be dependent on currents influencing supply and delivery of individuals, as well as whether suitable substrate is available for settlement. Another important component of benthic habitat type could be food supply and the role of currents in delivering organic matter to the benthos. Because sedimentation rates and current speeds are high in Glacier Bay, the stability of the substrate and the amount of sediment resuspension from the seafloor are important factors that have the potential to bury organisms and clog feeding appendages. Substrate type can also influence an organism's ability to seek refuge from predation, whether the organism uses burial, hiding within cryptic or complex habitats, or escape techniques. Biological data were not incorporated into the CMECS classification of Muir Inlet [10].

References

[1] K.C. Seramur, R.D. Powell, P.R. Carlson, Evaluation of conditions along the grounding line of temperate marine glaciers: an example from Muir Inlet, Glacier Bay, Alaska, Mar. Geol. 140 (1997) 307–327.

[2] J.F. Piatt, S.M. Gende, Proceedings of the Fourth Glacier Bay Science Symposium. U.S. Geological Survey Scientific Investigations Report 2007-5047, 2007.

[3] J.N. Harney, G.R. Cochrane, L.L. Etherington, P. Dartnell, N.E. Golden, H. Chezar, Geologic characteristics of benthic habitats in Glacier Bay, southeast Alaska. U.S. Geological Survey Open-File Report 2006-1081, 2005.

[4] R.D. Powell, Grounding-line systems as second-order controls on fluctuations of tidewater termini of temperate glaciers—Glacial marine sedimentation, paleoclimatic significance, in: J.B. Anderson, G.M. Ashley (Eds.), Glacial Marine Sedimentation: Paleoclimatic Significance, Geological Society of America, Boulder, CO, 1991, pp. 75–94 (Special Paper 261).

[5] J.P.M. Syvitski, On the deposition of sediment within glacier-influenced fjords—oceanographic controls, in: R.D. Powell A. Elverhøi (Eds.), Modern glacimarine environments—Glacial and marine controls of modern lithofacies and biofacies, Mar. Geol. 85 (1989) 301–329.

[6] E.A. Cowan, R.D. Powell, Ice-proximal sediment accumulation rates in a temperate glacial fjord, southeastern Alaska, in: J.B. Anderson, G.M. Ashley (Eds.), Glacial Marine Sedimentation: Paleoclimatic Significance, Geological Society of America, Boulder, CO, 1991, pp. 61–73 (Special Paper 261).

[7] D. Carney, J.S. Oliver, C. Armstrong, Sedimentation and composition of wall communities in Alaskan fjords, Polar Biol. 22 (1999) 39–49.

[8] L. Etherington, G.R. Cochrane, J. Harney, J. Taggart, A. Mondragon, E. Andrews, et al., Glacier Bay seafloor habitat mapping and classification: first look at linkages with biological patterns, in: J.F. Piatt, S.M. Gende (Eds.), Proceedings of the Fourth Glacier Bay Science Symposium, U.S. Geological Survey, Washington, DC, 2007, pp. 71–74.

[9] D.F. Hill, S.J. Ciavola, L. Etherington, M.J. Klaar, Estimation of freshwater runoff into Glacier Bay, Alaska and incorporation into a tidal circulation model, Estuar. Coast. Shelf Sci. 82 (2009) 95–107.

[10] Luke D. Trusel, G.R. Cochrane, L.L. Etherington, R.D. Powell, L.A. Mayer, Marine benthic habitat mapping of Muir Inlet, Glacier Bay National Park and Preserve, Alaska, with an evaluation of the Coastal and Marine Ecological Classification Standard III, U.S. Geological Survey, Washington, DC, 2010. U.S. Geological Survey Scientific Investigations Map 3122.

[11] E.A. Cowan, K.C. Seramur, B.A. Willems, R.D. Powell, S.P.S. Gulick, J.M. Jaeger, Retreat sequences from the last glacial maximum and little ice age preserved in Muir Inlet, Glacier Bay National Park, southeastern Alaska, Geol. Soc. Am. Abs. Prog. 40 (2008) 226.

[12] C.L. Jackolski, E.A. Cowan, J.M. Jaeger, R.D. Powell, High-resolution glacial discharge records from deep-water tidal rhythmites in an Alaskan fjord, Eos. Trans. Am. Geophys. Union 87 (2006) 52.

[13] T.J. Anderson, G.R. Cochrane, D.A. Roberts, H. Chezar, G. Hatcher, A rapid method to characterize seabed habitats and associated macro-organisms, in: B.J. Todd, H.G. Greene (Eds.), Mapping the Seafloor for Habitat Characterization, Geological Association of Canada, Canada, 2007, pp. 71–79 (Special Paper 47. St. John's).

[14] C.J. Madden, K.L. Goodin, R. Allee, M. Finkbeiner, D.E. Bamford, Draft Coastal and Marine Ecological Classification Standard, version III, National Oceanic and Atmospheric Administration, Washington, DC, 2008.

[15] G.H. Greene, M.M. Yoklavich, R.M. Starr, V.M. O'Connell, W.W. Wakefield, D.E. Sullivan, et al., A classification scheme for deep seafloor habitats, Oceanolog. Acta 22 (1999) 663–678.

19 Geomorphic Features and Benthic Habitats of a Sub-Arctic Fjord: Gilbert Bay, Southern Labrador, Canada

Alison Copeland[1], Evan Edinger[1,2], Trevor Bell[1], Philippe LeBlanc[1], Joseph Wroblewski[3], Rodolphe Devillers[1]

[1]Geography Department, Memorial University, St. John's, NL, Canada, [2]Biology Department, Memorial University, St. John's, NL, Canada, [3]Ocean Sciences Centre, Memorial University, St. John's, NL, Canada

Abstract

Gilbert Bay is a shallow-water, low-gradient, sub-Arctic fjord in southeastern Labrador, Atlantic Canada. The bay is composed of a series of basins separated by sills that shallow toward the head. A major side bay with shallow but complex bathymetry includes important spawning and juvenile fish habitat for a genetically distinct local population of Atlantic cod (*Gadus morhua*). Six acoustically distinguishable substrate types were identified in the fjord, with two additional substrate types recognized from field observations, including areas outside multibeam sonar coverage. Ordination and Analysis of Similarity (ANOSIM) of biotic data generalized five habitat types: hard-substrate habitats developed on cobble-boulder gravel and bedrock bottoms; coralline-algae-encrusted hard-substrate habitats; soft-bottom habitats developed on mud or gravelly mud bottoms; current-swept gravel with a unique biotic assemblage; and nearshore ice-scoured gravels in waters shallower than 5 m depth. Greatest within-habitat biodiversity was found in the coralline-algae-encrusted gravel habitat.

Key Words: fjord, fiord, sill, basin, coralline algae, epifauna, infauna, sub-Arctic, Labrador, Marine Protected Area

Introduction

Fjords are geomorphic and biological systems of great interest to geomorphology, oceanography, and marine biology. Fjords typically have complex bathymetry and host highly diverse and heterogeneous habitats [1], making them challenging

Seafloor Geomorphology as Benthic Habitat. DOI: 10.1016/B978-0-12-385140-6.00019-0

environments for marine habitat mapping using acoustic remote sensing [2]. The fjords of Newfoundland and southern Labrador commonly have relatively low relief, are surrounded by forested watersheds, and exist in boreal to sub-Arctic climatic conditions, making them somewhat different from archetypal fjords of the Canadian High Arctic [3].

The Gilbert Bay Marine Protected Area (MPA) was gazetted in 2005 to protect the resident Atlantic cod (*Gadus morhua*) stock in the bay. Cod captured in Gilbert Bay display a golden-brown coloration that distinguishes them from other Atlantic cod and earned them the name "golden cod" or "bay cod" [4,5]. Gilbert Bay cod are a genetically distinct population [6] and display high site fidelity within the bay [4,7]. Fisheries and Oceans Canada commissioned habitat mapping research in Gilbert Bay to support management of the MPA.

Gilbert Bay is a 28 km long, 1–2.5 km wide fjord on the southeastern coast of Labrador, in the boreal to sub-Arctic region of eastern Canada (Figure 19.1). Gilbert Bay is a relatively complex bathymetric system, with seven basins along the length of the fjord separated by six sills [8]. The basins range in depth from 32 m near the head of the bay to 163 m at the mouth. The sills range in depth from 4 to 65 m. The mouth of Gilbert Bay is blocked by islands and is open to the sea through three entrances, all less than 350 m wide. There is a complex of three arms branching to the southwest from the main reach of the bay, at River Out and The Shinneys. These

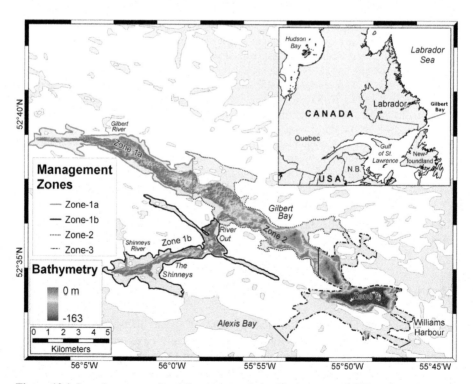

Figure 19.1 Location map and multibeam sonar-derived bathymetry of Gilbert Bay.

arms contain shallow-water basins, shoals, islands, and tidal channels. The surface of Gilbert Bay is covered by land-fast ice from late November to early May [5]. During the spring and summer, both the Gilbert River and the Shinneys River discharge significant amounts of freshwater from forested watersheds into the bay, turning the sea around the river mouths brown with tannins.

Most of Gilbert Bay is shallow. On the basis of a multibeam sonar survey [9], slightly more than half of the mapped fjord is shallower than 30 m, only 6% is deeper than 100 m, and the mean depth is 33.2 m [8] (Figure 19.1). Seabed depth and slope increase toward the mouth of the bay. Toward the fjord head, the seabed is flat floored, with less than 5° slope, whereas the margins of the outermost basin have steep slopes up to 68° [8].

Water clarity in Gilbert Bay is often quite low in the summer—a consequence of tannin-rich runoff from the rivers during spring melt and phytoplankton blooms [5]. The bay has a limited sediment supply; the only sources of mud and sand into the bay are the Gilbert River and the Shinneys River. Most modern sediment is trapped near the mouths of these rivers by shallow sills. The floor of Gilbert Bay reflects its glacial origin. Till composed of mixed sand, gravel, and boulders was deposited as moraines that now form basin sills. Eskers, which are narrow ridges of sandy gravel, are oriented parallel to the fjord coast on the mud basin floors.

Multibeam Sonar Data Acquisition and Ground-Truthing

The multibeam sonar survey of Gilbert Bay was conducted in 2002 using EM100 and EM3000 transducers. The survey covered 32 km^2 of the total 82 km^2 area of the MPA [9]. There was an extensive littoral gap in shallow areas (mostly <5 m depth) of the bay. Depth and acoustic backscatter values were gridded to 5 m spatial resolution and 0.1 m bathymetric resolution for habitat analyses. Bottom slope angle for each grid cell was calculated from the 5 m bathymetric grid [8].

The correspondence of depth, slope, and backscatter values to bottom substrate type and benthic biota was determined by extensive ground-truthing using benthic grab samples and bottom video imagery in October of 2006 and 2007 [10,11]. Paired grab samples and drop-video casts were made at 129 stations covering all parts of the bay (Figure 19.2). Position and depth of each grab sample or start and end point of drop-video transect was determined using the boat's GPS and depth sounder, with spatial accuracy better than 3 m and depth resolution of 0.1 m. Twenty-eight drop-video transects of variable length were recorded in shallow water where multibeam sonar data were unavailable.

Sediment Sample and Biota Analysis

Grab samples were photographed, subsampled in duplicate for sediment grain size and organic matter content, and then wet-sieved through 1 mm mesh to recover flora and fauna. Invertebrates and algae present in the washed sample were collected and preserved in 70% ethanol. Video imagery was reviewed, and the percentage of seabed covered by mud, sand, pebble, cobble, boulder, and coralline algae was estimated by the time that each substrate was visible during each video transect [12].

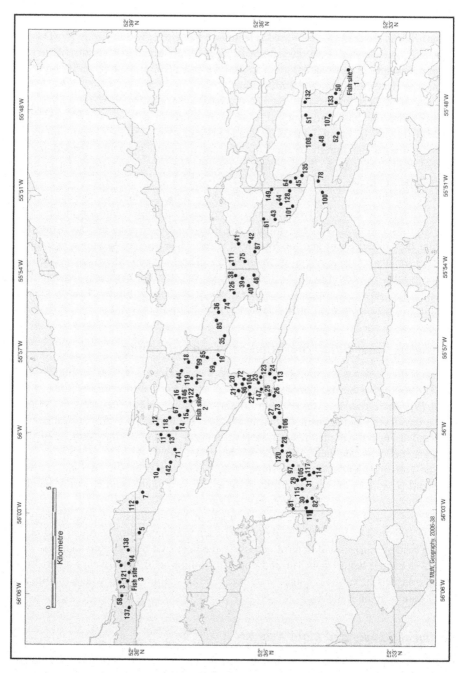

Figure 19.2 Distribution of sampling stations from 2006 field season [10]. "Fish sites" 1, 2, and 3 are the standard fish fauna survey sites [26]. Distribution of sampling stations and towed video transects from 2007 field season is indicated in Figure 19.6.

Grab-sampled biota and biota observed in video were identified to the lowest taxonomic level possible under a dissecting microscope following regional field guides [13,14]. Foraminifera were identified to skeletal type only (calcareous or agglutinated). Organisms recorded in video only were identified by comparison with specimens recovered in grab samples and by reference to published photographs and video footage. Some sponges in video observations were identified to growth form and color only, with a probable species identification indicated.

Habitat Classification of Samples and Mapping

Field samples of substrates were grouped into habitats by statistical analysis of their biota using multidimensional scaling and Analysis of Similarity (ANOSIM) conducted at the presence–absence level [15]. Substrate classes that had statistically unique biotic assemblages were classified as habitats, while substrate classes with statistically indistinguishable biota were combined into habitats [10,16].

Supervised classification of depth, slope, and backscatter in the multibeam sonar data set was employed to generate the final substrate and habitat maps [10,11]. Accuracy of the substrate maps was assessed using 20% of the ground-truthing data set reserved for validation.

Geomorphic Features, Substrates, and Habitats

Nine substrate types were identified in Gilbert Bay, of which six had distinct acoustic characteristics allowing them to be mapped using multibeam sonar.

Sills, eskers, fjord walls, and other hard-substrate features: Three substrates—muddy gravel, sandy gravel, and coralline-algae-encrusted gravel—generally occurred on top of the interbasin sills, along submerged eskers in the basins, or in shallow current-swept or wave-exposed waters. Muddy gravel substrate was characterized by pebble, cobble, and boulder gravel in a mud matrix. The gravel component was either draped by mud or exposed (Figure 19.3A). Sandy gravel substrate also contained pebble to boulder gravel, but in a sand matrix (Figure 19.3B). Pebbles, cobbles, and boulders that were more than 50% covered by branching coralline red algae were classified as coralline-algae-encrusted gravel substrate. The key features of this class were extensive algal growth and the presence of algae-forming structures, rather than algae simply encrusting the gravel clasts. This class included both algae-coated lithic clasts and rhodoliths lacking a gravel core, which were composed completely of calcium carbonate algal skeletons (Figures 19.3C and 19.4).

Coralline-algae-encrusted gravel was distributed along the margins of the bay, in the shallow upper part of the bay, and in River Out (Figures 19.5 and 19.6). The nearshore gravel substrate is present in very shallow waters of the bay, where it was mapped by continuous video transects outside the area of multibeam sonar coverage (Figure 19.6).

Two additional gravel-bottom substrate classes were defined by towed video from The Shinneys and River Out, but were not effectively sampled by grab sampler. The nearshore gravel class (Figure 19.3G) was characterized by pebbles, cobbles, and boulders, the latter in some cases over a meter in diameter, in very shallow water. Mud

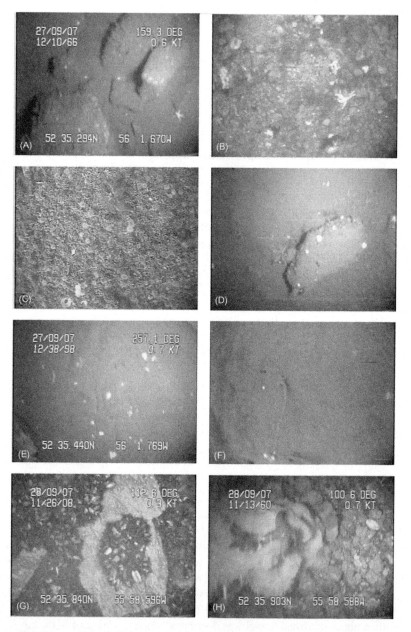

Figure 19.3 Photo plate of substrates mapped in Gilbert Bay: (A) muddy gravel; (B) sandy gravel; (C) coralline-algae-encrusted gravel; (D) gravelly muddy sand; (E) gravelly mud; (F) mud; (G) nearshore gravel; (H) current-swept gravel. Red laser points are 10 cm apart but are not clearly visible in all photos. In (C), green objects are echinoids (*Strongylocentrotus droebachiensis*), while red-brown background is rhodolith. (For interpretation of the references to color in this figure legend, the reader is referred to the web version of this book.)

Figure 19.4 Rhodoliths composed of the branching calcareous red alga *Lithothamnion glaciale* sampled from River Out.

or sand was usually absent, and coralline algae were never observed on this substrate. Finally, the current-swept gravel substrate class (Figure 19.3H) was characterized by sediment-free pebbles, cobbles, and scattered small boulders encrusted by nonbranching coralline algae. This substrate was only found in the high-current channel through Mogashu Tickle (Figure 19.6). Bedrock walls were rare in Gilbert Bay, occurring mostly around the margins of the deepest basin closest to the mouth of the fjord, and were acoustically indistinct from the muddy gravel substrate.

Basins: Finer-grained substrates were generally found in the basins between sills, except for shallow water mud habitat found in The Shinneys. Gravelly sandy mud substrate (Figure 19.3D) was defined by a matrix composed primarily of mud, with 7–41% medium (2ø) to very fine sand (4ø) by weight. This is primarily a poorly sorted, matrix-dominated substrate with scattered gravel clasts. In contrast, the gravelly mud class (Figure 19.3E) had a mostly mud (<4ø) matrix with a small amount of fine sand and isolated gravel clasts. Samples were classified as mud (Figure 19.3F) if their matrix was dominated by silt and clay with no more than 4% fine sand and they contained no gravel. Gravelly mud dominates the shallow basins near the head of the fjord, whereas mud is more common in the deeper basins toward the mouth. The only place where mud was mapped in shallow water was at the head of The Shinneys near the river mouth (Figure 19.6).

Overall, the mapped seabed of Gilbert Bay was composed of 41% gravelly mud, 38% muddy gravel, 10% each of gravelly sandy mud and mud, 8% coralline-algae-encrusted gravel, and 4% sandy gravel [10]. The area occupied by both current-swept gravel and bedrock walls was less than 1% [11]. Nearshore gravel was not included

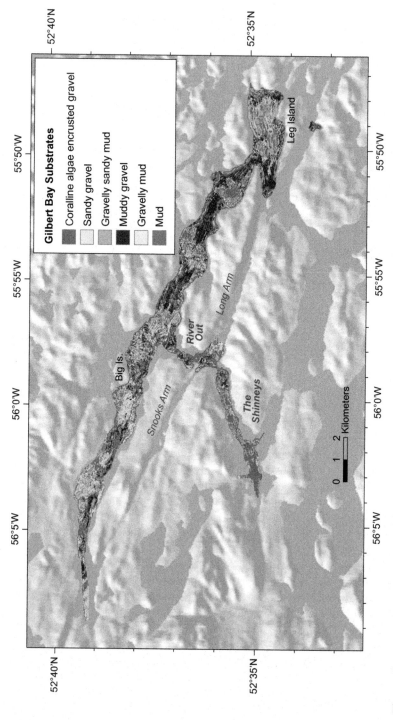

Figure 19.5 Benthic substrate map of Gilbert Bay created by supervised classification of multibeam sonar data from ground-truth samples collected in 2006. Blank patches in the map represent islands, shoals, or other gaps in the sonar coverage [10]. (For interpretation of the references to color in this figure legend, the reader is referred to the web version of this book).

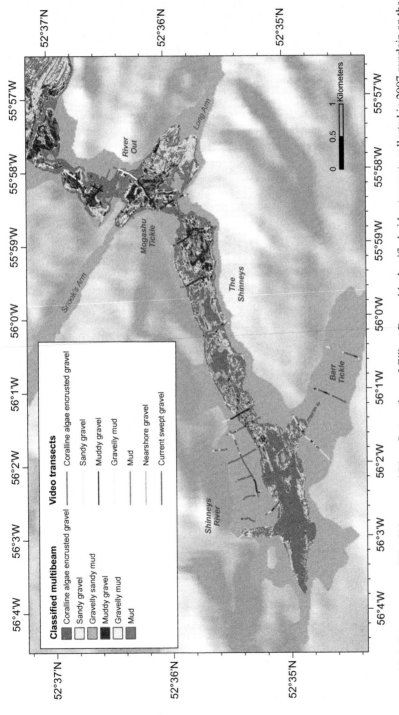

Figure 19.6 Substrate map of The Shinneys and River Out portions of Gilbert Bay, with classified video transects collected in 2007 overlain on the multibeam-derived substrate map from 2006 [11].

in the seabed cover estimate because it occupies the littoral gap between the water-depth limit of multibeam coverage and the shoreline. The total classified surface area of the bay is 111% due to classification ambiguity. Whereas 61% of the bay was uniquely classified, 22% met depth, backscatter, and slope criteria for two or three classes, and 16% remained unclassified. Classification accuracy as determined by the test data set was 69%, with 20 of 29 test points correctly classified, 5 incorrectly classified, and 4 unclassified [10].

Submerged glacio-fluvial features such as eskers and fans were identified in the upper part of the bay and at the mouth of River Out [8,10]. The shallowest parts of these features were dominated by cobble gravel and coralline algae, whereas deeper sections were draped in mud. Winnowing was particularly evident in the tidal rapids in the channel through Mogashu Tickle. With distance from Mogashu Tickle, the current-swept gravel transitioned to sandy gravel, muddy gravel, and soft-bottom substrates (Figure 19.6).

Surrogacy: From Substrate Maps to Habitat Maps

Average species richness was higher in the grab samples than in the video transects (Figure 19.7), primarily because of high diversity of polychaete worms in the

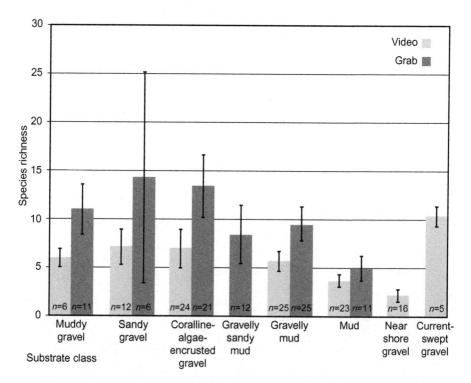

Figure 19.7 Species richness of substrates sampled in Gilbert Bay. Mean ±95% confidence interval; number of samples in each substrate type and sample type indicated on bars [11].

bottom grabs. Species richness was generally higher in the gravelly substrates than in the muddy substrates. Among the gravelly substrates, highest species richness was observed in the current-swept Mogashu Tickle gravel, followed by the coralline-algae-encrusted gravel. Lowest species richness was observed in the mud substrate.

Statistical analysis of the sampled biota using ANOSIM and Percentage Similarity (SIMPER) identified five habitats in Gilbert Bay: (Tables 19.1–19.4)

1. *Gravel bottom habitat*: Gravel bottom habitat combines muddy gravel and sandy gravel substrate classes (Figure 19.3A and B), as they were statistically indistinguishable [10]. Gravel bottom habitat was dominated by epifaunal taxa, such as *Spirorbis* sp., tube worms, bryozoans, chitons, foraminifera, and encrusting sponges. The limited areas of bedrock sampled had biota statistically equivalent to that of the other gravel bottom habitats.
2. *Soft-bottom habitat*: The mud, gravelly mud, and gravelly sandy mud substrates contained a statistically indistinguishable biotic assemblage dominated by polychaetes, particularly

Table 19.1 ANOSIM Results Table for Video Data

	Muddy Gravel	Sandy Gravel	Coralline-Algae-Encrusted Gravel	Gravelly Sandy Mud	Gravelly Mud	Mud
Muddy gravel	–					
Sandy gravel	51.5	–				
Coralline-algae-encrusted gravel	1.8	16.0	–			
Gravelly sandy mud	0.1	0.1	0.1	–		
Gravelly mud	0.1	0.1	0.1	41.7	–	
Mud	0.1	0.1	0.1	34.8	47.1	–

Numbers indicate probability of p-values, in percent (i.e., $p = 0.05$ is represented as 5.0) [10].

Table 19.2 ANOSIM Results Table for Grab Samples

	Muddy Gravel	Sandy Gravel	Coralline-Algae-Encrusted Gravel	Gravelly Sandy Mud	Gravelly Mud	Mud
Muddy gravel	–					
Sandy gravel	41.5	–				
Coralline-algae-encrusted gravel	0.1	26.7	–			
Gravelly sandy mud	5.5	4.2	0.1	–		
Gravelly mud	0.1	0.1	0.1	9.7	–	
Mud	0.2	2.4	0.1	44.8	47.5	–

Numbers indicate probability of p-values, in percent (i.e., $p = 0.05$ is represented as 5.0) [10].

Table 19.3 Characteristic Taxa of Each Substrate, Video Transect Data

	Muddy Gravel	Sandy Gravel	Coralline-Algae-Encrusted Gravel	Gravelly Sandy Mud	Gravelly Mud	Mud	Nearshore Gravel	Current-Swept Gravel	Bedrock
1	*Strongylocentrotus droebachiensis* green sea urchin (25.16)	*Leptasterias polaris* polar sea star (20.83)	*Strongylocentrotus droebachiensis* green sea urchin (23.32)	Infaunal bivalve (siphon pits) (23.48)	Burrows (32.84)	Burrows (42.94)	*Strongylocentrotus droebachiensis* green sea urchin (85)	*Henricia* sp. (13)	Bivalve shell hash (59)
2	*Halichondria panicea* sponge (20.19)	*Chlamys islandica* (live) scallop (16.41)	*Leptasterias polaris* polar sea star (22.58)	Trails (18.44)	Infaunal bivalve (siphon pits) (16.46)	Infaunal bivalve (siphon pits) (25.75)	*Mytilus edulis* Mussel (7.6)	*Crossaster papposus* spiny sunstar (13)	Branching bryozoans (27)
3	Bivalve shell hash (12.12)	*Asterias vulgaris* sea star (10.95)	*Chlamys islandica* (live) scallop (10.99)	Ophiuroidea brittle star (16.04)	Trails (7.97)	Trails (18.65)		*Leptasterias polaris* polar sea star (13)	Hydroids (7)
4	*Crossaster papposus* spiny sunstar (11.16)	*Strongylocentrotus droebachiensis* green sea urchin (10.12)	Branching coralline red algae (live) (10.46)	Burrows (15.48)	*Halichondria panicea* sponge (6.95)	Ophiuroidea brittle star (10.40)		*Strongylocentrotus droebachiensis* green sea urchin (13)	
5	*Leptasterias polaris* polar sea star (8.41)	*Halichondria panicea* sponge (9.86)	Encrusting coralline red algae (4.23)	*Halichondria panicea* sponge (11.16)	*Strongylocentrotus droebachiensis* green sea urchin (6.67)			Hydroids (8)	

6	Ophiuroidea brittle star (7.02)	Crossaster papposus spiny sunstar (8.53)	Branching coralline red algae (dead) (4.21)	Bivalve shell hash (3.99)	Ophiuroidea brittle star (6.16)	Halichondria panicea sponge (8)
7	Chlamys islandica (live) scallop (3.82)	Bivalve shell hash (5.09)	Hydroids (4.12)	Strongylocentrotus droebachiensis green sea urchin (3.68)	Polychaete tubes (4.62)	Barnacle (8)
8	Branching bryozoan (3.70)	Ophiuroidea brittle star (4.23)	Crossaster papposus spiny sunstar (4.10)		Articulated bivalve shell (dead) (3.90)	Chlamys islandica (live) scallop (8)
9		Finger sponge (4.23)	Metridium senile frilled anemone (3.98)		Branching bryozoan (3.57)	Gersemia rubiformis soft coral (4)
10			Halichondria panicea sponge (3.09)		Crossaster papposus spiny sunstar (3.36)	Encrusting coralline algae (4)

Numbers indicate the percent contribution to within-substrate faunal similarity, computed at presence–absence scale [10,11].

Table 19.4 Characteristic Taxa of Each Substrate, Grab Sample Data

	Muddy Gravel	Sandy Gravel	Coralline-Algae-Encrusted Gravel	Gravelly Sandy Mud	Gravelly Mud	Mud
1	*Spirorbis borealis* (21.06)	*Spirorbis granulatus* (39.56)	*Anomia squamula* jingle shell (33.35)	Mud polychaete tubes (16.43)	*Thyasira flexuosa* bivalve (20.41)	*Thyasira flexuosa* bivalve (59.32)
2	*Spirorbis granulatus* (8.23)	*Spirorbis borealis* (18.28)	Branching coralline red algae (19.58)	*Thyasira flexuosa* bivalve (14.58)	*Ctenodiscus crispatus* mud star (15.08)	*Ctenodiscus crispatus* mud star (29.66)
3	*Serpula* sp. (7.15)	*Serpula* sp. (8.77)	*Spirorbis granulatus* (13.66)	Unidentified polychaete (14.23)	Mud polychaete tubes (12.56)	*Periploma papyratium* bivalve paper spoon shell (11.02)
4	*Tubulipora* sp. bryozoan (5.51)	*Balanus balanus* barnacle (8.14)	Encrusting coralline red algae (5.25)	*Goniada maculata* chevron worm (10.96)	*Nucula tenuis* nut shell (10.64)	
5	*Stomachetosella sinuosa* bryozoan (4.99)	Encrusting coralline red algae (6.82)	*Tonicella rubra* chiton (5.13)	*Nuculana tenuisulcata* bivalve (7.85)	Unidentified polychaete (9.31)	
6	Muddy polychaete tube (4.75)	*Anomia squamula* (5.80) jingle shell	*Balanus balanus* barnacle (5.07)	*Nucula tenuis* nut shell (4.43)	*Goniada maculata* chevron worm (6.79)	
7	*Nucula tenuisulcata* bivalve (4.58)	*Escharella immersa* bryozoan (2.32)	*Porella* sp. bryozoan (1.67)	*Pectinaria granulata* (3.49) trumpet worm	Maldanid polychaete (4.81)	
8	*Balanus balanus* barnacle (4.15)	*Tubulipora* sp. bryozoan (2.32)	Calcareous forams (1.52)	*Anomia squamula* jingle shell (3.24)	Sandy polychaete tube (2.54)	
9	*Turritellopsis acicula* needle shell (3.53)		*Tubulipora* sp. bryozoan (2.71)	*Ophiura robusta* (2.86) brittle star	*Nuculana tenuisulcata* bivalve (2.23)	
10	Calcareous foraminifera (3.34)		*Hiatella arctica* (1.29) boring bivalve	*Turritellopsis acicula* (2.58) needle shell	*Pectinaria granulata* (1.71) trumpet worm	

Numbers indicate the percent contribution to within-substrate faunal similarity, computed at presence–absence scale [10,11].

the tube-dwelling Maldanidae family, burrowing bivalves, particularly *Thyasira flexuosa, Nucula tenuis,* and *Nuculana tenuisulcata,* mud stars (*Ctenodiscus crispatus*), and ophiuroids.

3. *Coralline-algae habitat*: The coralline-algae-encrusted gravel substrate supported the most diverse biota within the bay, including hydroids, nestling and boring bivalves, polychaete worms, chitons, sea urchins, bryozoans, and anemones.

4. *Current-swept gravel habitat*: The current-swept gravel in Mogashu Tickle hosted biota rarely sampled elsewhere in Gilbert Bay, including the soft coral *Gersemia rubiformis* and basket star *Gorgonocephalus arcticus*. Other characteristic taxa included filter-feeding epifauna such as hydroids and encrusting sponges, green sea urchins (*Strongylocentrotus droebachiensis*), the sea stars *Henricia* sp., *Crossaster papposus*, and *Leptasterias polaris*, and Iceland scallop (*Chlamys islandica*). The current-swept gravel in Mogashu Tickle could not be mapped through classification of the multibeam sonar data because it did not exhibit unique acoustic characteristics.

5. *Nearshore gravel habitat*: Nearshore gravel also generated a statistically distinct habitat due to its impoverished fauna. Generally it appeared devoid of attached biota, with the exception of blue mussels (*Mytilus edulis*), which occupied gaps among the gravel (Figure 19.3G). It is likely therefore that this habitat is scoured by land-fast ice during the winter and spring. The only other biota present were large numbers of green sea urchins (*Strongylocentrotus droebachiensis*), which, together with the ice scour, would maintain an algae-free gravel surface.

The final map of interpreted habitats of Gilbert Bay does not include current-swept gravel, nearshore gravel, or bedrock, because their distributions are quite limited, not acoustically distinct, or not surveyed by multibeam sonar (Figure 19.8). The coralline-algae-encrusted gravel habitat was found along the margins of the bay, especially in River Out and The Shinneys, and on the tops of the sills near the head of the fjord. Closer to the fjord mouth, the sills are too deep to support dense growth of coralline algae.

Discussion

Geomorphic and Biotic Contributions to Habitat

The habitats identified in Gilbert Bay were all geomorphic in origin, with the exception of the coralline-algae-encrusted gravel. The current-swept gravel habitat in Mogashu Tickle is unique within the bay, and hosts very high invertebrate abundance, including a high density of scallops [11], and species diversity, including species not sampled elsewhere in Gilbert Bay (Figure 19.7; [11]). The only biologically structured habitat, coralline-algae-encrusted gravel, was found in shallow water areas throughout the bay, but was best developed in River Out and The Shinneys. An extensive rhodolith bed was identified in River Out. The rhodolith bed differed from most other coralline-algae-encrusted gravel habitat in that there was no gravel core to the rhodoliths, and no algae-free gravel was observed in videos. MPA regulations protect the head of the bay, River Out, and The Shinneys from scallop dragging and other fishing.

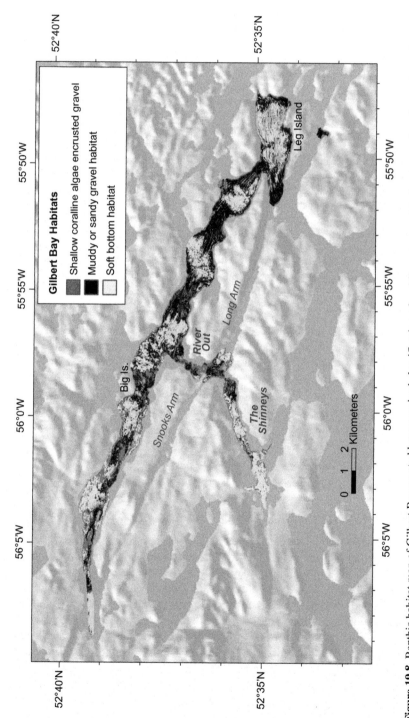

Figure 19.8 Benthic habitat map of Gilbert Bay created by supervised classification of multibeam sonar data from ground-truth samples collected in 2006. Blank patches in the map represent islands, shoals, or other gaps in the sonar coverage [10]. (For interpretation of the references to color in this figure legend, the reader is referred to the web version of this book).

The almost complete absence of large kelp (laminariales) and seagrasses may be a consequence of very high tannin concentrations in the waters of Gilbert Bay during the summer, resulting in a shallow photic zone limit of 7.6–11.5 m [5]. The high tannin concentrations in the water originate from the forested watersheds feeding it and the limited water exchange at the entrances to the bay [5]. The shallow photic zone, combined with extensive ice scour in water shallower than 5 m, likely restricts macroalgal growth, except for coralline red algae, which typically requires less light than most macroalgae [17,18]. Large macroalgae including *Laminaria*, *Saccharina*, *Agarum*, *Alaria*, *Fucus*, and *Ascophyllum* are common on hard-substrate bottoms in the Canadian High Arctic and in boreal and sub-Arctic fjords surrounding the North Atlantic [18,19].

Gilbert Bay cod habitat: River Out and The Shinneys contained extensive areas of coralline-algae-encrusted gravel and sandy gravel, with mud in shallow basins between shoals composed of coralline algae on gravel or bedrock. This area may provide feeding habitat for juvenile Gilbert Bay cod due to the high invertebrate abundance and diversity of potential prey. Benthic invertebrates made up a higher than expected component of the stomach contents of Gilbert Bay cod, particularly in The Shinneys [4].

The substrate of a known cod spawning site in the Inner Shinneys consisted of coralline-algae-encrusted boulders and silt [20]. Juvenile cod require structured habitat to avoid predation; in other Newfoundland and Labrador bays, this is typically provided by gravel and marine vegetation, including eelgrass (*Zostera marina*; [21–23]) or macroalgae [24]. In Gilbert Bay, the required structured habitat is likely provided by complex bathymetry, coarse gravel, and coralline algae. Geomorphic features, such as moraine-capped sills and eskers, contribute to habitat complexity [25], while coralline algae and gravelly mud likely provide high density of invertebrate food sources.

The age-specific nature of substrate choices have been documented in other Newfoundland cod populations [23], which is likely the case with Gilbert Bay cod also. Therefore, one single "cod habitat" can likely not be defined for Gilbert Bay, but our work enhances knowledge of the potential habitats available to fish in different parts of the bay [26].

Anthropogenic Impacts

Coralline-algae-encrusted gravel is the most likely habitat to suffer negative effects from human disturbance. This habitat is vulnerable to disturbance where scallop dragging occurs in the outer two-thirds of the main bay and where dragging depths (8–20 m) coincide with the mapped occurrence of rhodolith and coralline-algae-encrusted gravel habitat (4–15 m water depth). This is cause for concern given the documented impacts that scallop dragging has on coralline-algae-based habitats (*cf.* maerl) in other locations [27].

Gilbert Bay scallop harvesters reported fishing for scallops primarily on coralline algae, pebble, cobble, and small boulder substrates [20]. Iceland scallops are among the top three taxa contributing to similarity in samples in the sandy gravel and coralline-algae-encrusted gravel substrate classes. Scallops also appeared in the characteristic biota for the muddy gravel class, but were less important. High scallop densities were observed in Mogashu Tickle and immediately adjacent parts of the Outer Shinneys. Analysis of the scallop growth rates and size-frequency distributions in Gilbert Bay in relation to fishing effort strongly suggests that the Gilbert Bay scallop stock is overfished [28,29]. Scallop fishing as historically practiced in Gilbert Bay was incompatible with the habitat conservation objectives of the MPA. As a conservation measure, the spatial footprint of the scallop fishery in Gilbert Bay was limited to a subset of its pre-MPA.

References

[1] J.P.M. Syvitski, D.C. Burrell, J.M. Skei, Fjords: Processes and Products, Springer-Verlag, New York, NY, 1987.

[2] C.B. Cogan, T.T. Noji, Marine classification, mapping, and biodiversity analysis, in: B.J. Todd, H.G. Greene, (Eds.), Mapping the Seafloor for Habitat Characterization. Geological Association of Canada, St. John's, Newfoundland, 2007, pp. 129–140 (Special Paper 47).

[3] J.E. Dale, A.E. Aitken, R. Gilbert, M.J. Risk, Macrofauna of Canadian Arctic fjords, Mar. Geol. 85 (1989) 331–358.

[4] C.J. Morris, J.M. Green, Biological characteristics of a resident population of Atlantic cod (*Gadus morhua* L.) in southern Labrador, ICES J. Mar. Sci. 59 (2002) 666–678.

[5] K.R. Gosse, J.S. Wroblewski, Variant colourations of Atlantic cod (*Gadus morhua*) in Newfoundland and Labrador nearshore waters, ICES J. Mar. Sci. 61 (2004) 752–759.

[6] D.E. Ruzzante, J.S. Wroblewski, C.T. Taggart, R.K. Smedbol, D. Cook, S.V. Goddard, Bay-scale population structure in coastal Atlantic cod in Labrador and Newfoundland, Canada, J. Fish Biol. 56 (2000) 431–447.

[7] J.M. Green, J.S. Wroblewski, Movement patterns of Atlantic cod in Gilbert Bay, Labrador: evidence for bay residency and spawning site fidelity, J. Mar. Biol. Assoc. U.K. 80 (2000) 1077–1085.

[8] A. Copeland, T. Bell, E. Edinger, L. Hu, J. Wroblewski, Habitat Mapping in Gilbert Bay Labrador, a Marine Protected Area, Phase 1. Contract Report to Fisheries and Oceans Canada, St. John's, Newfoundland, 2006, p. 68.

[9] A. Roy, Final field report, Phase II *Matthew* Hydrographic Survey. Northern Labrador. Port Manvers, Kiglapait Harbour, and Proposed Marine Protected Area of Gilbert Bay. *GGCS Matthew* 02–050, Document No. 2602429, Canadian Hydrographic Service, St. John's, Newfoundland, 2002, p. 23.

[10] A. Copeland, E. Edinger, T. Bell, L. Hu, J. Wroblewski, R. Devillers, Habitat Mapping in Gilbert Bay Labrador, a Marine Protected Area, Phase II final report. Marine Habitat Mapping Group report 07–02, Memorial University, St. John's, Newfoundland, 2007, p. 108.

[11] A. Copeland, E. Edinger, P. LeBlanc, T. Bell, R. Devillers, J. Wroblewski, Marine Habitat Mapping in Gilbert Bay, Labrador—a Marine Protected Area, Phase III final report. Marine Habitat Mapping Group report 08–01, Memorial University, St. John's, Newfoundland, 2008, p. 71.

[12] P.B. Mortensen, L. Buhl-Mortensen, Distribution of deep-water gorgonian corals in relation to benthic habitat features in the Northeast Channel (Atlantic Canada), Mar. Biol. 144 (2004) 1223–1238.

[13] K.L. Gosner, Guide to Identification of Marine and Estuarine Invertebrates: Cape Hatteras to the Bay of Fundy, John Wiley and Sons Inc., New York, NY, 1971.

[14] K.L. Gosner, A Field Guide to the Atlantic Seashore: Invertebrates and Seaweeds of the Atlantic Coast from the Bay of Fundy to Cape Hatteras, Houghton Mifflin, Boston, MA, 1979.

[15] K.R. Clarke, R.M. Warwick, Change in Marine Communities: An Approach to Statistical Analysis and Interpretation, second ed., PRIMER-E, Plymouth, UK, 2001.

[16] V.E. Kostylev, B.J. Todd, G.B.J. Fader, R.C. Courtney, G.D.M. Cameron, R.A. Pickrill, Benthic habitat mapping on the Scotian Shelf based on multibeam bathymetry, surficial geology and sea floor photographs, Mar. Ecol. Prog. Ser. 219 (2001) 121–137.

[17] M.J. Dring, Biology of Marine Plants, second ed., Cambridge University Press, Cambridge, 1991.

[18] A. Wulff, K. Iken, L.M. Quartino, A. Al-Handal, C. Wiencke, M.N. Clayton, Biodiversity, biogeography and zonation of micro- and macroalgae in the Arctic and Antarctic, Bot. Mar. 52 (2009) 491–507.

[19] E.N. Edinger, T.M. Brown, T.J. Bell, K. Belliveau, N.R. Catto, A.I. Copeland, et al., Climate change impacts on coastal marine systems across the Canadian Arctic, in: T. Thorsnes, K. Picard (Eds.), Abstracts and Proceedings 8th GEOHAB conference. Norwegian Geological Foundation No. 2, Geological Survey of Norway, Trondheim, p. 38.

[20] C.J. Morris, J.M. Simms, T.C. Anderson, Overview of commercial fishing in Gilbert Bay, Labrador; fish harvesters local knowledge and biological observations, Can. Manuscr. Rep. Fish. Aquat. Sci. 2596 (2002). vii + p. 34.

[21] V. Gotceitas, S. Fraser, J.A. Brown, Use of eelgrass beds (*Zostera marina*) by juvenile Atlantic cod (*Gadus morhua*), Can. J. Fish. Aquat. Sci. 54 (1997) 1306–1319.

[22] R.S. Gregory, J.T. Anderson, Substrate selection and use of protective cover by juvenile Atlantic cod *Gadus morhua* in inshore waters of Newfoundland, Mar. Ecol. Prog. Ser. 146 (1997) 9–20.

[23] D. Cote, S. Moulton, P.C.B. Frampton, D.A. Scruton, R.S. McKinley, Habitat use and early winter movements by juvenile Atlantic cod in a coastal area of Newfoundland, J. Fish Biol. 64 (2004) 665–679.

[24] D.C. Schneider, M.J. Norris, R.S. Gregory, Predictive analysis of scale-dependent habitat association: Juvenile cod (*Gadus* spp.) in eastern Newfoundland, Estuar. Coast. Shelf Sci. 79 (2008) 71–78.

[25] P.J. Auster, Are deep-water corals important habitats for fishes? in: A. Freiwald, J.M. Roberts (Eds.), Cold-Water Corals and Ecosystems, Springer-Verlag, Heidelberg, 2005, pp. 747–760.

[26] J.S. Wroblewski, L.K. Kryger-Hann, D.A. Methven, R.L. Haedrich, The fish fauna of Gilbert Bay, Labrador: a marine protected area in the Canadian subarctic coastal zone, J. Mar. Biol. Assoc. U.K. 87 (2007) 575–587.

[27] J.M. Hall-Spencer, P.G. Moore, Scallop dredging has profound, long-term impacts on maerl habitats, ICES J. Mar. Sci. 57 (2000) 1407–1415.

[28] S. Liu, Growth rate of the Iceland scallop *Chlamys islandica* in Gilbert Bay, Labrador, a Marine Protected Area, M.Sc. Thesis, Environmental Science, Memorial University, St. John's, Newfoundland, 2009, p. 96.

[29] J.S. Wroblewski, T.J. Bell, A.I. Copeland, E.N. Edinger, C.Y. Feng, J.D. Saxby, et al., Toward a sustainable Iceland scallop fishery in Gilbert Bay, a marine protected area in the eastern Canada coastal zone. J. Cleaner Prod. 17 (2009) 424–430.

20 Seabed Character and Habitats of a Rocky Antarctic Coastline: Vestfold Hills, East Antarctica

Philip E. O'Brien[1], Jonathan S. Stark[2], Glenn Johnstone[2], Jodie Smith[1], Martin J. Riddle[2]

[1]Geoscience Australia, Canberra, ACT, Australia, [2]Australian Antarctic Division, Channel Highway, Kingston, TAS, Australia

Abstract

The Vestfold Hills is an area of ice-free coast in East Antarctica. The coast is a complex of small islands, embayments, and fjords. Most of the coast is rocky, but a few sandy beaches are present. Water depths vary from 0 to 200 m in the survey area. High-resolution multibeam bathymetry, underwater video, and detailed diving surveys reveal a mosaic of rock outcrops, sediment-covered basins, and transition zones similar to pediments of desert landscapes. Iceberg scouring is common, but some areas are protected by peninsulas and islands. Biological communities are controlled by substrate and light level, with rocky substrates in shallow water, where ice-free conditions persist for most of the summer, dominated by macroalgae. Areas of sediment have invertebrate fauna and some macroalgae. Rocky areas in deep water or where sea ice persists through the summer are invertebrate dominated.

Key Words: Antarctica, coast, habitat, geomorphology, multibeam bathymetry, macroalgae

Introduction

The Vestfold Hills is one of the largest ice-free areas on the East Antarctic coast, with an area of 400 km² (Figure 20.1). It lies between 68°21′S and 68°41′S and 77°49′E and 78°35′E. Its coastline is a complex of small rocky islands and fjords with water depths from 0 to 160 m in the deepest places (Long Fjord). Further offshore, depths increase rapidly to >200 m. Islands and hills on the mainland reach elevations of up to 60 m in the coastal zone, rising to 156 m inland near the edge of the ice sheet. The Vestfold Hills are an Archean basement block surrounded by younger Precambrian metamorphic rocks north and south, and westwards by the Phanerozoic Prydz Bay Basin [1]. The eastern extent of the Vestfold Hills block is unknown because it is obscured by the ice sheet. The Vestfold Hills form an elevated

Seafloor Geomorphology as Benthic Habitat. DOI: 10.1016/B978-0-12-385140-6.00020-7

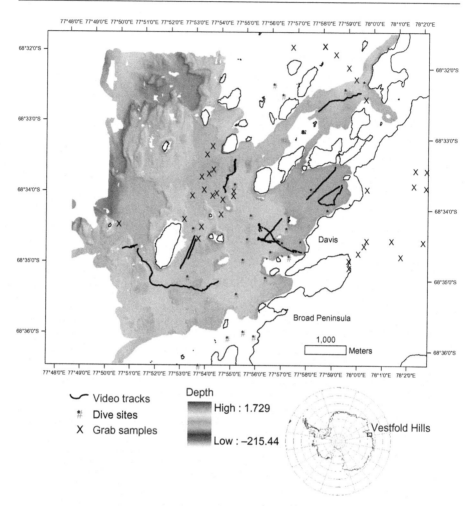

Figure 20.1 Location of the study area (inset), multibeam bathymetry, video transects, grab sample, and photoquadrat sites from the Vestfold Hills coast.

plateau surrounded by deep troughs to the south and west and probably deeper but more gently sloping bedrock to the north and east, though there is not much information on sub-ice topography in those areas.

The climate is cold-polar, with Mean Annual Maximum Temperature of −7.4°C and Mean Annual Minimum Temperature of −13°C. Snowfall is low, with a mean annual precipitation of 70.5 mm water equivalent. Mean wind speeds vary from 17 to 26 km/h, though maximum gusts of 206 km/h have been recorded [2]. Sea ice is present for most of the year. Freezing takes place intermittently from late February, and breakout typically takes place in December. Breakout occurs progressively from south to north in response to prevailing winds from the northeast and the geometry of the coast. This gradient has a major impact on marine communities. Oceanographic studies have described water masses and currents in Prydz Bay [3].

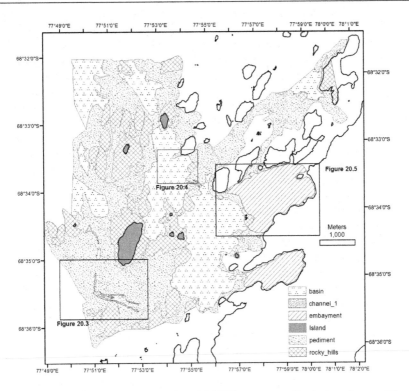

Figure 20.2 Geomorphic features of the Vestfold Hills coast near Davis, based on multibeam data and video transects. Boxes show the location of following figures.

Up to seven water masses have been identified in Prydz Bay; however, the Vestfold Hills coast is likely bathed by surface water during the most active time of the year. This water mass has potential temperatures between −1.8°C and +2.1°C and salinities of 30.6–34.2 psu [2]. Circulation is dominated by a clockwise gyre, with average velocities in the order of 0.015 m s^{-1}. The tidal range along the eastern side of Prydz Bay is around 1–2 m, but velocities measured offshore from the study area were <0.01 m s^{-1}.

The area has experienced low human impact, with the only activity taking place at the Australian station of Davis. Primary productivity is high during spring and during the ice-free period of the year, dropping to low during the months of sea-ice cover. During the field season from December 2009 to March 2010, the area was mapped and sampled by the Australian Antarctic Program, comprising multibeam and towed video by Geoscience Australia and the Deployable Geospatial Support Team of the Royal Australian Navy and sampled by Australian Antarctic Division divers. Sediment data from previous studies were also used. Bathymetry data were collected using a Konsberg EM 3002, 300 kHz, dual-head multibeam system. In total, 45 km^2 of high-resolution multibeam data were collected, with a minimum coverage of 1 sounding per 40 cm^2 (Figure 20.1). Video transects were collected using a towed video system.

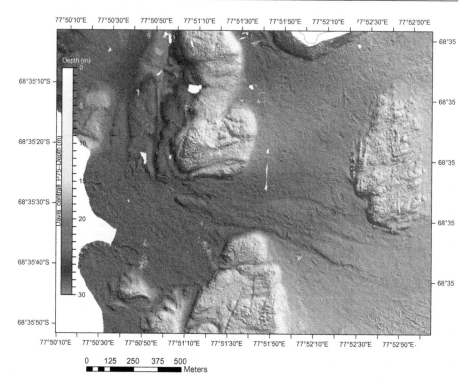

Figure 20.3 Shaded relief image of multibeam data from the Vestfold Hills (see Figure 20.2 for location). Bedrock outcrops appear as shallow areas, with joint planes and other structures visible. Bedrock banks are bordered by gently sloping pediment surfaces with iceberg scours and channels, visible in the lower left side of the map. Image is derived from a 0.75 m grid with illumination from the northeast.

The dive program visited 30 sites (Figure 20.1), acquiring 2,000 photoquadrats using a still camera mounted on a frame to ensure standard scales between photoquadrats. Some 400 push cores were collected to measure sediment grain size, chemistry, macrofauna, meiofauna, and microbial communities. A few sediment samples were also collected by shipek grab in areas not visited by the dive program. An additional set of sediment grain size data were collected by a previous study [4].

Geomorphic Features and Habitats

The geomorphology and habitats presented here are based on interpretation of the multibeam data and a preliminary assessment of the other data sets (Figure 20.2).

Rocky banks: On the multibeam data, bedrock outcrops rise above the surrounding seafloor; these are rounded banks with small- and medium-scale roughness and depths from 0 to 25 m (Figure 20.3). Joint planes and faults are commonly visible in the topography.

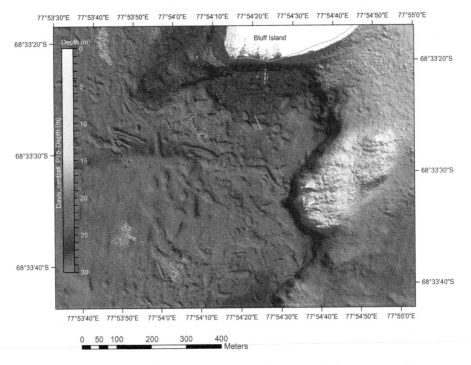

Figure 20.4 Sediment-floored depression southeast of Bluff Island (see Figure 20.2 for location). The seafloor is covered with abundant iceberg scours. Image is derived from a 0.75 m grid with illumination from the northeast.

Basins: Between islands and bedrock outcrops are basins with flat to slightly undulating floors covered with sediment. They vary from 5 to 45 m in the Davis area. They are typically floored with olive gray to green muddy fine sand to sandy mud. The mud component lacks cohesion and is probably mostly siliceous phytodetritus. Ice-keel scours are abundant and range from enclosed, rounded wallow marks 20 m across (Figure 20.4) to meandering tracks. The predominant direction of larger scours is north to south and NNW to SSE, probably reflecting oceanic circulation in Prydz Bay [3] driving icebergs, whereas the fine-scale scours tend to be oriented from NE to SW, indicating wind-driven sea ice as the cause.

Pediments: The transition areas between bedrock outcrops and sediment-floored depressions are gently sloping with thin sandy sediment cover, probably on shallow bedrock (Figure 20.3). Slopes are lower than the sides of rocky banks, and the surfaces may undulate, possibly reflecting underlying near-surface bedrock. Pebbles are common in these areas. Iceberg scours are present in some areas, but others lack apparent scours, possibly because the sediments are too coarse to retain well-defined scour topography. These areas most closely resemble pediments seen in many arid areas. A pediment is defined as "a plain of eroded bedrock (which may or may not be covered with thin veneer of alluvium) in an arid area, developed between mountain and basin areas" [5]. Though the areas off Davis are unlikely to have alluvial cover,

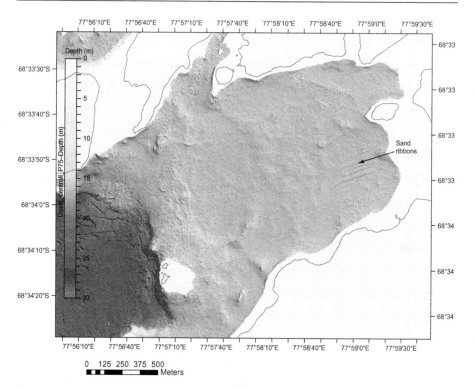

Figure 20.5 Enclosed embayment lacking iceberg scours and with sand ribbons (see Figure 20.2 for location). Image is derived from a 0.75 m grid, with illumination from the northeast. The speckled texture indicates abundant boulders on the seafloor.

their morphology, bedrock outcrops, and thin sediment quite closely resemble typical pediments.

Channels: Channels wrap around several areas of bedrock (Figure 20.3). These channels are floored with muddy sand, and videos across one channel show benthos consistently leaning south. Other, narrower channels occur in a few places that appear more like drainage features, running downslope from high points to low points at an angle to the probable prevailing current direction. In other settings, they may be relict fluvial features. However, the area has a history of glacial rebound since the Last Glacial Maximum (~20 ka), so it is unlikely the channels were exposed. Although they are clearly modified by iceberg scours, their persistence suggests recent current activity in the channels.

Enclosed embayments: Some shallow areas <10 m deep are enclosed by peninsulas or islands so that drifting ice does not scour the seafloor. These areas exhibit a relatively smooth seafloor, but with abundant large boulders (Figure 20.5). These boulders are probably relics of past glaciation [6]. The beds are muddy sand with areas of current ripples and, in the Airport Beach embayment, linear sediment ridges suggesting current activity.

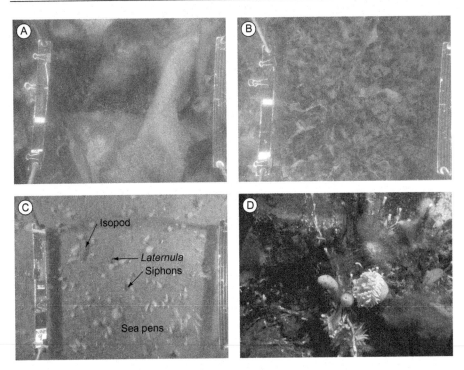

Figure 20.6 (A) *Himantothalus* is a dominant algae on hard substrate in the photic zone in areas of minimum sea-ice cover over summer. Scale bar on right is 37 cm long. (B) The algae *Iridea* is abundant in areas of low sea-ice cover and hard substrate but is also common in areas of sedimentary seafloor, where it may be drift material or growing on ice-rafted boulders. Scale bar on right is 37 cm long. (C) Typical sediment-covered seafloor offshore from the Vestfold Hills. Sediments are typically muddy sand to sandy mud with drop stones—typical benthos include burrowing bivalves (*Laternula*), sea pens, and mobile species such as Isopods and holothurians. Scale bar on right is 37 cm long. (D) Hard seafloor in an area with persistent ice cover. The community is dominated by attached invertebrates, including sponges, anemonies, ascidians, and fan worms.

Biological Communities

Qualitative assessment of the data suggests that the benthic communities in the region can be broadly classified into three "end member" groupings governed by substrate type, water depth, and the persistence of sea ice at any site:

1. In areas which are ice-free for significant parts of the year, hard substrates are dominated by algae (*Himantothalus* and *Iridea*; Figure 20.6A and B). Attached invertebrates such as ascidians, sponges, fan worms, serpulid polychaetes, and anemonies are also present, though covering a smaller proportion of the substrate than algae.
2. Sediment-covered areas are mostly inhabited by invertebrates such as sea pens and burrowing bivalves (*Laternula* sp; Figure 20.6C). Patches of algae are present in sediment-covered

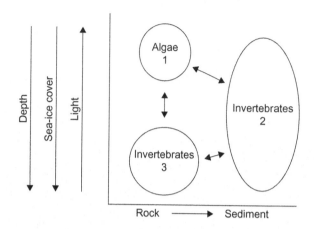

Figure 20.7 Diagram illustrating preliminary interpretation of the relationships between environmental parameters and benthic communities along the Vestfold Hills coast. Community 1 is macroalgae-dominated rocky surfaces in the photic zone. Community 2 is dominated by invertebrates that are favored by sedimentary substrates. Community 3 is an invertebrate community favored by high sea-ice cover and deeper water, probably reflecting low light levels that reduce algal growth.

areas but it is less widespread and may be displaced from rock areas by current activity. Mobile invertebrates such as holothurians, echinoderms, and nemerteans are common in sediment areas. Preliminary analysis of data has shown no detectable differences between sediment communities in different geomorphic units with sedimentary seafloors, such as basins, channels, or pediments.

3. Attached invertebrates, which become the dominant benthos on rocky substrates as the persistence of sea ice through the year increases (Figure 20.6D). In such areas, ascidians, sponges, fan worms, serpulid polychaetes, and anenomies are dominant attached forms with sea stars and mobile holothurians also prominent. Dense patches of serpulid polychaetes form "reefs" on the flanks of some fjords. Attached benthos also become more important as water depth increases.

Surrogacy

Qualitative observations of the photoquadrats and video transects suggest that the main controls on benthic communities in the Vestfold Hills region are substrate and light availability (Figure 20.7). Rocky seafloor in areas of short-lived sea-ice cover are dominated by algae. Rocky areas with persistent sea ice and in depths greater than about 30 m are invertebrate dominated. Sediment surfaces tend to be invertebrate dominated. Other factors, such as the frequency of ice-keel scouring and current activity, may yet emerge from the analysis.

Acknowledgments

Rather than single out some individuals and miss others, we would like to thank everyone who supported the work at Davis in 2009 and 2010, from those involved directly to those who operated the station. Without them, the program would not have been possible. O'Brien and Smith publish with the permission of the CEO, Geoscience Australia.

References

[1] Stagg, H.M.J. (1975) The structure and origin of Prydz Bay and Mac. Robertson Shelf, East Antarctica, Tectonophysics 114, 315–340.

[2] R.A. Nunes Vaz, G.W. Lennon., Physical oceanography of the Prydz Bay region of Antarctic waters, Deep Sea Res. 43 (5) (1994) 603–641.

[3] Bureau of Meteorology, Climate Statistics for Australian Locations. Summary Statistics DAVIS, 2010, http://www.bom.gov.au/climate/averages/tables/cw_300000.shtml.

[4] D. Franklin, The sedimentology of Holocene Prydz Bay: sedimentary patterns and processes and their implications for climate reconstruction, Ph.D. Thesis, Institute of Antarctic and Southern Ocean Studies, University of Tasmania, Hobart, 1996.

[5] D.G.A. Whitten, J.R.V. Brooks, The Penguin Dictionary of Geology, Penguin Books, Harmondsworth, 1972.

[6] E.A. Colhoun, K.W. Kiernan, A. McConnell, P.G. Quilty, D. Fink, C.V. Murray Wallace, et al., Late Pliocene age of glacial deposits at Heidemann Valley, East Antarctica: evidence for the last major glaciation in the Vestfold Hills, Antarct. Sci. 22 (2010) 53–64.

21 Fringing Reefs of the Seychelles Inner Granitic Islands, Western Indian Ocean

Jenny S. Collier, Stuart R. Humber

Department of Earth Science and Engineering, Imperial College London, London, UK

Abstract

Seabed morphology and backscatter texture were mapped within two of the Marine National Parks of the inner granitic islands of the Seychelles using high-resolution (675 kHz) sidescan sonar imaging. Individual islands are surrounded by a gently seaward-dipping marine terrace 0.5–2 km wide where water depths slowly increase to 8–10 m. Deeper water channels, with depths between 14 and 24 m, are found between these terraces. Bare sediment is the dominant substrate, comprising more than 70% of the seabed surveyed in each area. On the terraces, four biological communities were recognized, each giving a different backscatter response. Three of these communities were found to contain corals, each characterized by a different density of coral cover, species composition, and dominant colony morphology type. These reefs are more extensive along northern shorelines of individual islands. In the channels, a single biological community was found colonizing distinctive pinnacles that rise 4–6 m above the surrounding seafloor. The distribution of biological communities appears to be linked to both water depth and, on the terraces, exposure to the dominant southeast Trade Winds.

Key Words: Seychelles, reefs, sidescan sonar

Introduction

The Seychelles Plateau is a continental fragment situated just south of the equator in the western Indian Ocean (Figure 21.1). The plateau measures approximately 300×150 km, and water depths are generally less than 65 m. In the center of the plateau lie a group of islands that expose Precambrian granite. Topography is rugged, with elevations reaching a maximum of 912 m on the main Island of Mahé. These islands are generally referred to as the "inner granitic islands" to distinguish them from coralline cays that form the remaining islands of the Seychelles Republic. The region lies outside the cyclone belt and is subject to dominant southeast Trade Winds. Water temperatures range between 26°C and 30°C. The islands are surrounded by

Seafloor Geomorphology as Benthic Habitat. DOI: 10.1016/B978-0-12-385140-6.00021-9

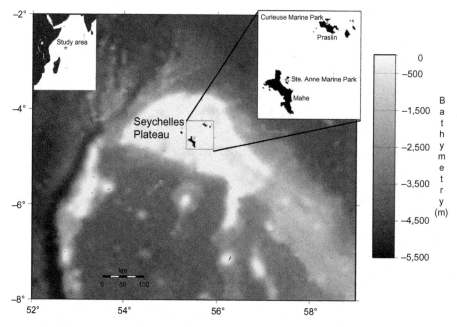

Figure 21.1 Satellite and ship-board derived bathymetry map (etopo30) showing the location of the two survey areas in the western Indian Ocean.

reefs that were badly affected by the spring 1998 Indian Ocean coral bleaching event. Detailed assessments conducted immediately following this event showed that on the reefs live hard coral cover had fallen from average values of around 50% to less than 5% in many areas [1].

The surveys presented here were conducted in autumn 1998 and 2001 within the Sainte Anne and Curieuse Marine National Parks [2,3]. Backscatter data were collected with a 675 kHz Imagenex sidescan sonar. The across- and along-track resolution was 4–7 and 60 cm, respectively. Navigation was via dGPS, with a horizontal accuracy of ±85 cm. CTD casts were taken to determine water column acoustic velocity structure for data processing. A bathymetric digital elevation model (DEM) was derived from manual picks of sidescan sonar instrument altitude, assuming a constant tow-depth. Recorded tidal height data were used to reduce these picks to mean sea level. In addition to the parallel tracks collected for the backscatter mosaics, a similar number of orthogonal tracks were used to make the DEM. The average cross-over error was 20 cm, and individual depths were averaged into 10 m^2 spatial bins before gridding [4].

Provisional identification of seabed type was provided by manta tows conducted simultaneously with the first geophysical surveys. A more detailed regional ground-truthing program was then designed with reference to the first set of sidescan sonar mosaics. A total of 23 sites were assessed by field-station biologists. At each site, the biologists randomly placed two 50-cm quadrants, recorded the physical and biological character of the substrate, took a number of underwater photographs and, where possible, collected sediment samples. The observations were recorded on a standard data sheet that was based on a coral reef rapid assessment technique [1]. A more detailed assessment was then made using a six-point index for the percentage cover of hard coral, dead standing coral, soft coral, algal turf, macroalgae, seagrass, and bare sediment within the quadrants. Organisms were identified to genus or species wherever possible. The sediment samples were subsequently washed, dried, and subjected to partical size analysis using standard techniques. Oblique-view aerial photographs were collected from a low-flying helicopter, and other conventional vertical view aerial photographs were compiled from government sources. The aerial photographs were used to aid interpretation in water depths less than 8 m.

The aim of the study was to provide primary mapping of the benthic habitats for environmental management and also to explore the potential for sidescan data to detect and quantify changes in the seabed following coral bleaching. The former objective is presented here; the latter can be found in Collier and Humber [3]. This work was conducted as part of the Royal Geographical Society's "Shoals of Capricorn" program [5].

Geomorphic Features and Habitats

The bathymetry data showed low-angled, seaward-dipping terraces with widths varying between 0.5 and 2 km to surround each of the four island coastlines surveyed (Figures 21.2–21.3). The terraces extend perpendicularly from the beach, through the intertidal zone to water depths of 8–10 m. The edge of each terrace is marked by a steeper topographic gradient, where water depths increase to more than 14 m (Figure 21.3B). The deeper water forms channels between the granite islands. The morphology of the terrace surfaces themselves is highly irregular, whilst that of the channels—apart from the coral pinnacles (see later)—is relatively featureless. Owing to the tropical location and presence of framework-building corals, there is a strong relationship between local geomorphology and biological community, as discussed later.

Biological Communities

The interpretation of seabed type was done manually using the sidescan backscatter mosaics and aerial photographs (Figure 21.2). A range of attributes such as threshold amplitudes and spatial statistics (moments and gray level co-occurrence matrices) were calculated for each backscatter mosaic to aid the delineation of boundaries between classes. On the terraces, five seabed classes were distinguished—including three containing coral and one dominated by seagrass. Here, we adopt names for these coral communities following the scheme given by Riegl and Piller [6,7]. In the deep-water channels, three seabed classes were distinguished, one of which also

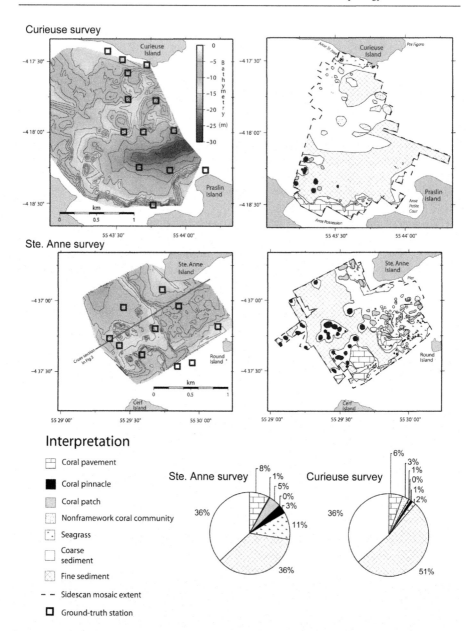

Figure 21.2 Seafloor geomorphology (left) and benthic habitats (right) for the two study areas. Bathymetry is in meters below mean sea level, contours are every 2 m, and ground-truth locations are marked with black squares. The interpretation of benthic habitats was made from bathymetry, sidescan backscatter imagery, aerial photographs, and ground-truth observations. The pie charts show the spatial coverage of each recognized seabed class. (For interpretation of the references to color in this figure legend, the reader is referred to the web version of this book.)

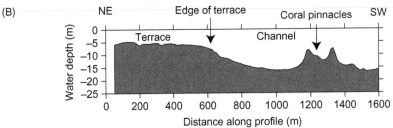

Figure 21.3 (A) Oblique aerial photograph of the south coast of Ste. Anne Island taken from a low-flying helicopter in November 1998. The image shows biological communities on the terrace that can be correlated with those seen in the sidescan mosaic. (B) Bathymetric cross section showing the geomorphology of a terrace and channel. The exact location of this section is shown in Figure 21.2.

contained corals. Textural analysis based on first-order statistics was performed on the acoustic backscatter data to test for statistical relationships with the ground-truth observations. Unsupervised cluster analysis and Mann–Whitney U-tests showed a correlation between the recorded backscatter response and biological community type that we attribute to the link between community composition and rugosity on a millimeter to tens of meters scale (see Ref. [3] for details).

Sediment: Bare sediment was recognized by its uniform backscatter texture (low standard deviation of pixels) in the sidescan mosaics. Bare sediment is estimated to cover 72% and 87% of the seabed within the Ste. Anne and Curieuse survey areas, respectively (Figure 21.2). Two sediment categories were established based on mean backscatter intensity. The boundary between these was arbitrarily drawn, and there are gradations within each class. Correlation with the sediment grain size analyses showed the high backscatter intensity class corresponds to medium-to-fine sand on the Wentworth scale (labeled "coarse sediment" in Figure 21.2), whereas the low backscatter intensity class corresponds to medium sand-to-fine sandy silt (labeled "fine sediment" in Figure 21.2). Coarse sediment was found on the terraces and fine sediment in the channels. Locally coarse sediment was also found surrounding coral pinnacles within the channels. Ripples were seen in many locations, both in the acoustic imagery and by the divers. Wavelengths and heights varied between 15–20 and 3–5 cm, respectively, with larger ripples in general being found within the coarser sediments on the terraces.

Seagrass beds: Seagrass beds were characterized by having a higher intensity and more textured response in the sidescan imagery than bare sediments, but less textured than areas containing corals. Sediments collected from the seagrass beds were found to be medium-to-coarse sand, which elsewhere when bare gave lower backscatter intensity. It seems likely therefore that the vegetation itself (leaves or roots) scattered energy back to the sonar. No sediment structures were visible in the acoustic imagery of these regions. The seagrass beds, which consist predominantly of *Syringodium* sp. and *Cymodoca* sp. with the occasional presence of *Thalassodendran* sp., are found in water depths less than 4 m and are most extensive along southern coastlines.

Coral pavements: These features have a regionally extensive and continuous form unlike the other coral-containing seabed classes that are more isolated and circular in shape. We have assigned these features the name "coral pavement" as they form hard rocky platforms with only thin veneers of loose sediment. This is quite unlike the other two coral-containing seabed classes found on the terraces, which have significant deep, soft-sediment components. The coral pavements have a significant topographic expression, with a steep bathymetric slope at their seaward edges. In several places, particularly offshore the islands of Cerf and Round, they are cut laterally by channels that link to onshore, freshwater runoff spillways. Coral pavements occupy the shallowest water depths of any of the coral communities found in the survey areas. The water depth at their perimeters is typically just 4–6 m, and they rise to within 2–3 m of the mean sea surface. Divers report the pavements to host a wide variety of encrusting, branching, massive, and submassive form corals (*Acropora* sp., *Porites* sp., and *Pocillopora* sp.) with, on average, about 30% total coral cover. Less than 10% of this coral was thought to be living at the time of the surveys. Unusually, large *Porites* sp. heads, up to 4 m in diameter, are found on the seaward edges of the pavements and on the flatter seabed just beyond their perimeters. The coral pavements preferentially fringe the northern coastlines.

Coral patches: The second coral-containing seabed type found on the terraces is referred to here as "coral patches." These have a moderate backscatter image texture and relatively sharply defined boundaries with the surrounding seafloor (Figure 21.4). The patches typically rise approximately 2 m above the adjacent seafloor and are circular to oval-shaped, with diameters of 15–80 m. The size of

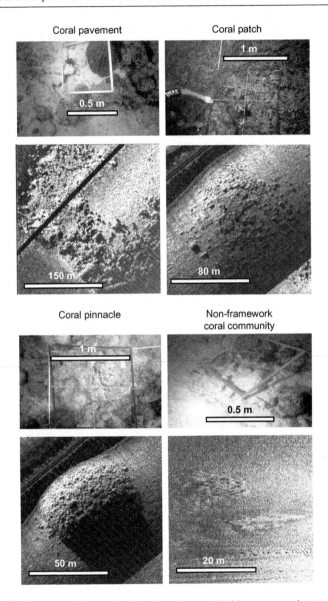

Figure 21.4 Underwater photographs (above) and example sidescan sonar images (below) for the four coral-containing community types. Low backscatter intensity is black, high intensity white.

individual patches does not appear to correlate with water depth. Within the patches, the sidescan shows rough and smooth areas to be interspersed, covering about equal proportions of the seabed. There is no detectable zoning of these textures within individual patches. Direct observation showed the smooth areas to be unconsolidated sediment and the rough areas predominantly compact, branching *Acropora* sp., with secondary, small-sized, massive-formed *Porites* sp. colonies. The *Acropora* sp. was largely dead but still standing at the time of the surveys. In places the coral cover is extremely dense, giving an appearance of a "thicket of branches."

Nonframework coral communities: This seabed class has the lowest backscatter contrast to the surrounding bare sediment, and comprises areas of coarse, grainy image texture (Figure 21.4). These regions generally follow the shape of the seafloor and do not superimpose their own morphology. For this reason, we assign the name "nonframework coral communities" to this class. In the sidescan mosaics, these textured areas have diffuse boundaries with the smoother surrounding bare sediment areas, but otherwise do not show any systematic pattern of internal texture. Direct observation showed these areas to contain a mixed assemblage of *Acropora* sp. and *Porites* sp. with generally small (10–20 cm diameter) colonies and rare encrusting forms. Hard corals were reported to cover less than 10% of the seabed within these zones, with the remaining area covered by bare sand and macroalgae. The community also included grazing invertebrates and bivalves. These communities, found on the terraces of the southern coastlines, were significantly more extensive within the Curieuse marine park than Ste. Anne.

Coral pinnacles: In the deep-water channels between the islands, a single biological community was recognized—here referred to as coral pinnacles. These are morphologically significant features occurring in 10–14 m of water and rising on average 4 m above the surrounding seafloor. They form rounded mounds, with moderately steep sides and flatter tops. Overall, they are near-circular shape in plan-view and measure 30–75 m across. In general, the pinnacles with larger diameters and those in deeper water are taller. Direct observations showed the presence of a diverse coral community of mixed, mainly massive forms, but dominantly the genus *Porites* sp. (in particular *P. lutea* and *P. solida*) together with *Goniopora* sp. and *Faviidae* sp. Individual massive colonies are typically less than 30 cm in diameter, but occasionally colonies up to 3 m in diameter were found. Most of the massive corals were alive at the time of assessment, but these were interspersed with minor small areas of dead branching *Acropora* sp. The sidescan data shows the pinnacles are characterized by a tight packing of individual coral colonies. Interspersed sandy areas are estimated to be less than 20%. There is no evidence of textural changes either down the sides or across the tops of individual pinnacles. At the base of pinnacles, there is a marked change in slope and a zone of relatively high backscatter. Sediments collected from these regions were poorly sorted with high mean grain sizes (typically >1 mm) and are characterized by large pieces of coral, presumably derived from the pinnacle itself. Individual massive coral colonies were occasionally detected in the geophysical data free standing around the bases of the pinnacles.

In summary, the areas surveyed encompassed a diverse range of shallow-water marine habitats including seagrass beds, coral pavements, coral patches, and deep coral pinnacles. There is a clear distinction between the biological communities found on the terraces compared to the deeper water channels. On the terraces themselves, more extensive reefs were found along northern coastlines. This suggests that exposure to the dominant southeast Trade Winds exerts a control on the community distribution, with the more sheltered northern coastlines of individual islands being more conducive to reef development. The pattern is complex, however, and further surveys would be needed to properly establish these dependencies.

Acknowledgments

This project was run as part of the Royal Geographical Society's *Shoals of Capricorn* program. The geophysical surveys were conducted on vessels generously supplied and manned by the Seychelles Marine Parks Authority and the Seychelles Coastguard. We thank Bill Burnett, Martin Callow, Jan Robinson, and Helga Vogt for biological ground-truthing.

References

[1] J. Turner, R. Klaus, U. Engelhardt, The reefs of the granitic islands of the Seychelles, in: D. Souter, D. Obura, O. Linden, (Eds.), Coral Reef Degradation in the Indian Ocean, CORDIO/SAREC, Marine Science Program, Stockholm University, Sweden, 2000, pp. 77–86.

[2] The Marine National Parks, The Seychelles Centre for Marine Research & Technology—Marine Parks Authority, www.scmrt-mpa.sc.

[3] J.S. Collier, S.R. Humber, Time-lapse side-scan sonar imaging of bleached coral reefs: a case study from the Seychelles, Remote Sens. Environ. 108 (2007) 339–356.

[4] P. Wessel, W.H.F. Smith, Free software helps map and display data, EOS Trans. AGU 72 (1991) 441.

[5] J. Burnett, J. Kavanagh, T. Spencer (Eds.), Shoals of Capricorn Programme Field Report 1998–2001: Marine science, training and education in the Western Indian Ocean: Royal Geographical Society (with the Institute of British Geographers), 2001.

[6] B. Riegl, W.E. Piller, Distribution and environmental control of coral assemblages in northern Safaga Bay (Red Sea, Egypt), Facies 36 (1997) 141–162.

[7] B. Riegl, W.E. Piller, Coral frameworks revisited—reefs and coral carpets in the northern Red Sea, Coral Reefs 18 (1999) 241–253.

22 Coral Reefs and Reef Islands of the Amirantes Archipelago, Western Indian Ocean

Sarah Hamylton, Tom Spencer, Annelise B. Hagan

Cambridge Coastal Research Unit, Department of Geography, University of Cambridge, Cambridge, UK

Abstract

A series of habitat maps derived from remotely sensed Compact Airborne Spectrographic Imager (CASI) data of the reefs and reef islands of the Amirantes archipelago, Seychelles, western Indian Ocean, are presented. Broad-scale (>1 km) coral reef structures and geomorphic units are described, along with a range of habitats associated with lagoons, reef-flats, and sand cays. Five different geomorphic units are outlined, comprising three different types of platform reefs, atolls, and submerged atolls. Differences between these geomorphic units (and the surficial biological communities they support) have been determined, historically, by the presence or absence of elevated reef deposits of various kinds that allow island stability and island growth to take place and, at the present time, by the variation in contemporary process environments (wave and current fields) across the Amirantes Bank.

Key Words: Coastal management, habitat

Introduction

The Amirantes archipelago lies southwest of the extensive, shallow-water Seychelles Bank (maximum recorded depth of 65 m) in the western Indian Ocean and comprises a group of carbonate islands and islets extending over a distance of approximately 152 km, from 4°52′S (African Banks) to 6°14′S (Desnoeufs) (Figure 22.1). Most of the islands are sea-level coral reef platforms with varying degrees of subaerial sand cay and coral island development. They have evolved over the last 6,000 years since regional postglacial sea level approached its present level [1]. Due to their remote nature, the islands remain largely undeveloped, with the exception of limited tourism infrastructure (i.e., single hotels) on four of the islands. The Amirantes Bank is an elongate structure, measuring approximately 180 × 35 km, deepest in its central zone (up to −70 m) with a marginal rim at water depths of 11–27 m. Approximately 95 km further south are the atolls of Alphonse and Bijoutier/St. François, which form the Alphonse Group. Of the 14 islands in the Amirantes archipelago, 13 were

Seafloor Geomorphology as Benthic Habitat. DOI: 10.1016/B978-0-12-385140-6.00022-0

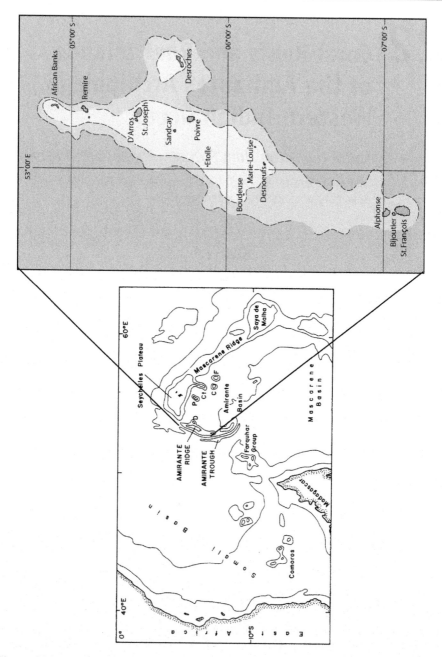

Figure. 22.1 Regional setting of the Amirantes Ridge, western Indian Ocean (left), and the islands of the Amirantes archipelago (right); approximate bathymetric contours = 1 and 4 km. *Source*: Adapted from Ref. [3].

mapped, the exception being Desroches, a shallow submerged atoll, 19–21 km in diameter, lying 16 km east of the Amirantes Bank.

The climate of the western Indian Ocean is of a humid tropical type, with mean monthly temperatures always above 20°C and rainfall totals in excess of 700 mm. Seasonal and interannual climatic variability is determined by (1) the SE Asian Monsoon and the seasonal reversal of winds associated with it; (2) monsoon-related movements of the Inter-Tropical Convergence Zone; (3) changes in the position and intensity of the South Indian Ocean subtropical high pressure; and (4) variations in ocean circulation systems and sea surface temperatures. The surface ocean circulation of the western Indian Ocean is characterized by a subtropical, anticyclonic gyre to the south (between 40°S and 15°S) and reversing monsoon gyres to the north of 10°S, with the addition of large volumes of water (up to 15 Sv in June/July) from the Pacific Ocean through the Indonesian throughflow [2].

This case study presents the results of a collaborative expedition between the Cambridge Coastal Research Unit and Khaled bin Sultan Living Oceans Foundation in January 2005. Airborne remote sensing Compact Airborne Spectrographic Imager (CASI-2) imagery was acquired over the Amirantes Bank from which both the geomorphic and biological character of the 13 reef islands could be examined (Figure 22.2). This had a spatial resolution of 1 m and a spectral resolution of 19 wavebands. Biological assessments of benthic character in this study are supported by over 1,500 ground reference records of the terrestrial and marine environments. Further detail on production of the habitat maps and individual island descriptions can be found in the *Atlas of the Amirantes* [3].

Geomorphological Features

Of the seven reef types identified in the Seychelles [1], coral reef platform, atoll, and drowned atoll are present in the Amirantes [3]. The platform reefs can be subdivided into three characteristic morphologies: those dominated by intertidal reef-flat sands (Type 1), those that overtop a rock platform (Type 2), and those that have undergone substantial sedimentary infill which support vegetated islands (Type 3).

Type 1 coral reef platform (African Banks, Remire Reef, Sand Cay, and Etoile): In Type 1 platform reefs, the entire surface has been covered by intertidal reef-flat sands. There is no subaerial cay at Remire Reef, and the exposed land area at African Banks (North Island) is extremely small relative to the total area of the platform within the breaker zone (Table 22.1). Sand Cay and Etoile are very small (0.3 km^2) and, as repeat aerial photography shows, highly mobile sedimentary landforms, surrounded by extensive subtidal sand sheets and stabilized locally by seagrass beds. They appear to be positioned on localized highs in the pre-Holocene topography of the Amirantes Bank.

Type 2 coral reef platform (Marie-Louise, Desnoeufs, and Boudeuse): In the southern Amirantes archipelago, the Type 2 platform reef islands consist of raised reefs, bedded calcareous sandstones (often phosphatized), and beachrock ridges, which imply complex histories of landform development. These islands are

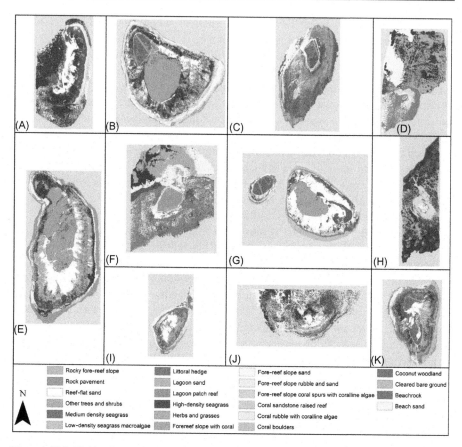

Figure. 22.2 Habitat maps of the Amirante islands: (A) African Banks, (B) Alphonse, (C) Marie-Louise, (D) Boudeuse, (E) Bijoutier/St. François, (F) Desnoeufs, (G) D'Arros/St. Joseph, (H) Etoile, (I) Remire Reef, (J) Sand Cay, and (K) Poivre.

surrounded by very narrow peripheral reefs (the area between the breaker zone and island marginal sands being only 0.08–0.24 km^2; Table 22.1) but sit on the margins of extensive and relatively shallow rock platforms, often incised with characteristic patterns of narrow, shallow, anastomosing channels. The subaerial islands thus occupy less than 10% of the total area classified from CASI imagery at Marie-Louise and Desnoeufs, and less than 1% at Boudeuse.

Type 3 coral reef platform (D'Arros and Poivre): At each of the two Type 3 platform reef islands, infilling of the platform surface has allowed the development of subaerial islands that exceed 2 km^2 in total area. The percentage cover of the area of the reef platform inside the breaker zone by islands is comparable to the Type 2 reefs, but habitat areas seaward of the breaker zone are much less extensive. Thus, the islands at D'Arros and Poivre account for a much greater proportion (38% and 13%, respectively) of the total area classified at these two locations (Table 22.1).

Table 22.1 Morphological Statistics for the Atolls and Reef Platforms of the Amirantes

Atolls	Total Areas/km²					Peripheral Reef Area as Proportion of Total Reef Platform Area	Lagoon Area as a Proportion of Total Reef Platform Area
	Overall Classified[1]	Reef Platform[2]	Peripheral Reef[3]	Land[4]	Lagoon		
St. Joseph	31.54	17.62	10.43	1.72	5.48	59.19	31.1
Alphonse	23.44	13.61	7.42	1.63	4.55	54.52	33.43
Bijoutier/St. François	71.33	37.99	19.48	0.46	18.05	51.28	47.51
Reef Platforms						Land area as proportion of total reef platform area	Land area as proportion of total classified area
African Banks (Type 1)	20.66	8.05	7.99	0.06		0.75	0.29
Remire Reef (Type 1)	19.30	11.61	11.61	0.00		0.00	0.00
Boudeuse (Type 2)	9.00	0.11	0.08	0.03		24.1	0.29
Marie-Louise (Type 2)	7.89	0.94	0.20	0.74		78.59	9.36
Desnoeufs (Type 2)	5.93	0.72	0.24	0.48		67.06	8.14
D'Arros (Type 3)	5.48	3.26	1.16	2.10		64.42	38.82
Poivre (Type 3)	20.24	14.67	12.01	2.66		18.13	13.14

Areas are delineated as follows:
[1] Overall classified: Total area classified from CASI imagery and shown on island habitat maps.
[2] Reef platform: Total area inside the breaker zone at each island, including area of any subaerial islands.
[3] Peripheral reef: Area between the breaker zone and island marginal sediments and rocks.
[4] Land: Area covered by terrestrial habitat categories.

Atolls (St. Joseph, Desroches, Alphonse, and Bijoutier/St. François): The atolls are small by global standards [1]. They are characterized by wide reef-flats, typically occupying 50–60% of the reef platform inside the breaker zone (Table 22.1), shallow lagoons, and poor lagoon–ocean exchange. At St. François, the largest of the atolls, extensive sand sheets on the windward coast infill the lagoon. This process is more advanced at St. Joseph where the lagoon occupies only around 30% of the reef platform inside the breaker zone (Table 22.1) and has a maximum depth of approximately 6 m.

Submerged atoll (Desroches): Although this was not included in the CASI analysis, it was classified for comparison from a Landsat TM image. The total area classified at Desroches was 189 km², making it more than twice as large as the largest sea-level atoll, Bijoutier/St. François, in the Alphonse Group. The atoll rim takes the form of a 1–3 km wide submerged reef platform, occupying 82 km² (43%) of the total classified area. The platform is found at depths of 4–7 m on its eastern and southern sides and at less than 3 m water depth on its northern edge. On the western margin, a narrow rim, with water depths of 4–8 m, is backed on its lagoonward side by a shelf at 15–18 m. The lagoon floor lies at between 23 and 27 m, much deeper than the shallow lagoons seen at the three sea-level atolls (6 m), and the lagoon, at 57% of the total classified area, occupies a far greater area than at the other atolls (where the coverage is 17–25%).

Biological Features

The spatial distribution of habitat types by island location (Figure 22.3) shows that islands on the western margin of the Amirantes Bank are characterized by a restricted range of terrestrial and littoral habitats, whereas those on the eastern side of the Bank show a greater range of habitats, particularly in subaerial environments. This may be because of their greater exposure to incident waves driven by the southeast Trade Winds, which encourage reef growth and subsequent island formation.

Seagrass was the most well-represented benthic cover type (13–84% cover), encompassing low-, medium-, and high-density communities. The dominant seagrass species were *Thalassodendron ciliatum* and *Thalassia hemprichii*. Fore-reef slope material, reef-flat sand, and lagoon sand were also abundant (Figure 22.3). Seagrass was abundant because the extensive reef-flats meet the habitat requirements of *Thalassodendron* and *Thalassia*. These include a marine environment, adequate rooting substrate, sufficient immersion in seawater, and illumination to maintain growth [4]. Like reef-flat sand and lagoon sand, seagrass beds adopt a lateral distribution of wide leaves that appear horizontally extensive when viewed from above, which is likely an adaptation for capturing light. Airborne remote sensing therefore lends itself well to mapping these coverages. Conversely, some of the habitats, such as beach sand and the fore-reef categories, were comparatively limited in the spatial extent of what was recorded because of their more vertical distribution, which could not be viewed as efficiently from above.

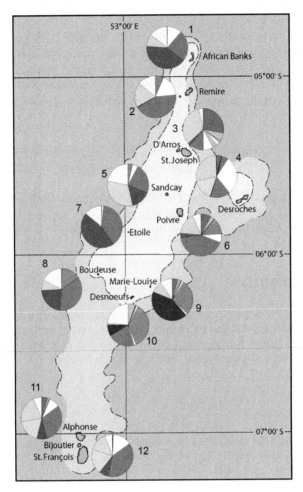

Figure. 22.3 Breakdown of habitat coverages of the reef islands of the Amirantes archipelago and surrounding reefs. Colours employed in the pie charts relate to the colour scheme employed in the legend of Figure 22.2.

Surrogacy

Collectively, the maps illustrate considerable inter-island habitat diversity across the Amirantes Bank. This variation reflects the presence or absence of elevated reef deposits of various kinds, related to past sea-level and environmental histories on the Bank. In conjunction with this, more recent biophysical controls include the degree to which regionally variable wind and surface current fields determine windward versus leeward contrasts across individual islands. At the island scale, the habitat structure of seagrass patches on Etoile Cay has been related to incident wave power using a spatially explicit analysis of variance model [5], and the linearity of reef-flat patches of seagrass around the rim of Alphonse Atoll has been related to both the absolute amount of incident wave energy and the transfer of this energy over spur and groove morphological features on the upper fore-reef using a spatial multivariate regression model [6]. The combined influence of adjacent groove depths and incident wave power (modelled using linear wave theory) explained 81% of the variation in seagrass patch linearity across the overall Atoll habitat map. The estimated regression function read:

$$\text{Patch linearity} = -0.234 + 2.584 \times 10^{-3*} \text{ wind factor}$$
$$+3.93 \times 10^{-3*} \text{ groove depth} \tag{22.1}$$

The t-test values were 9.337 and 12.429 for the mean wind force and groove depth, respectively (561 degrees of freedom; $p < 0.001$), suggesting it to be highly likely that the estimated coefficients are different from zero.

Acknowledgments

The following organizations are thanked for funding and logistical support in the field: Khaled bin Sultan Living Oceans Foundation, Seychelles Centre for Marine Research and Technology—MPA, Island Development Company, Seychelles and Great Plains Seychelles. Dr. Chris Banks is thanked for assistance with image data processing.

References

[1] D.R. Stoddart, Coral reefs of the Seychelles and adjacent regions, in: D.R. Stoddart (Ed.), Biogeography and Ecology of the Seychelles Islands, W. Junk, The Hague, the Netherlands, 1984, pp. 63–81.
[2] F. Schott, J.P. McCreary, The monsoon circulation of the Indian Ocean, Progr. Oceanogr. 51 (2001) 1–123.
[3] T. Spencer, A.B. Hagan, S.M. Hamylton, P. Renaud, Atlas of the Amirantes, University of Cambridge, UK, 2009.
[4] M.A. Hemminga, C.M. Duarte., Seagrass Ecology, Cambridge University Press, Cambridge, UK, 2000.
[5] S. Hamylton, T. Spencer, Classification of seagrass habitat structure as a response to wave exposure at Etoile Cay, Seychelles, EARSeL Proceedings 6 (2007) 82–94.
[6] S. Hamylton, T. Spencer, Geomorphological modelling of tropical marine landscapes: optical remote sensing, patches and spatial statistics, Continent. Shelf Res. 31 (2011) 151–161.

23 Hyperspectral Remote Sensing of the Geomorphic Features and Habitats of the Al Wajh Bank Reef System, Saudi Arabia, Red Sea

Sarah Hamylton

Cambridge Coastal Research Unit, Department of Geography, University of Cambridge, Cambridge, UK

Abstract

Results of a simultaneous field and airborne hyperspectral remote-sensing campaign to assess geomorphological and biological characteristics of lagoonal reef communities inside the Al Wajh Bank, Red Sea, are presented. For a study area encompassing fine-scale geomorphic features, such as a reef ridge network and reef patches, the percentage composition of four different reef community components (live coral, dead coral, macroalgae, and carbonate sand) is quantified from a combination of phototransects collected *in situ* and linear spectral unmixing of remotely sensed image data. The distribution of live coral, a key carbonate producer, is modeled as a function of water depth, incident wave power, and suspended sediment concentration using spatial regression techniques. A spatially lagged autoregressive component is included to account for endogenous effects acting through a neighborhood context (e.g., coral larval recruitment and the availability of suitable antecedent topography for settlement). Incorporation of this component yields a model that explains a substantial amount of variation in the distribution of live coral cover ($R^2 = 0.76$).

Key Words: spectral unmixing, spatial regression, bathymetry, wave power, suspended sediment concentration

Introduction

The Al Wajh Bank is situated along the Red Sea coastline of the Kingdom of Saudi Arabia. The Bank surrounds a shallow shelf platform of variable depth (range 30–60 m), and the water depth outside the Bank drops steeply to approximately

Seafloor Geomorphology as Benthic Habitat. DOI: 10.1016/B978-0-12-385140-6.00023-2

100 m [1]. Surface water temperatures of the Red Sea range between 22°C in the north and 36°C in the south in the summer [2]. Evaporation exceeds precipitation by about 2 m/year [1], making this one of the hottest and saltiest (generally 35 psu) bodies of seawater in the world. However, localized hypersaline conditions reaching as high as 113 psu have been reported in lagoons where limited water flux occurs [3]. The mountain ranges that flank the Red Sea reduce the main wind systems to a north–south aligned channel that blows along its length. Intraseasonal variability in winds and associated water circulation is driven by atmospheric pressure and wind stress associated with two distinct northeasterly and southwesterly monsoons [4]. Because much of the Red Sea is extremely hot and arid in nature, there is a marked constancy in a number of climatic parameters, including storms, rainfall, and tidal fluctuations (mean range 60 cm) [5]. The Bank is situated in an area relatively free from anthropogenic impact due to its remote nature; however, regional fish stocks have been heavily depleted.

This case study presents the findings of a collaborative expedition between the Cambridge Coastal Research Unit and the Khaled bin Sultan Living Oceans Foundation in May 2008. Airborne AISA Eagle hyperspectral imagery was acquired over the Wajh Bank, from which both the geomorphic and biological character of the Bank could be examined. The AISA Eagle instrument measured 128 contiguous spectral bands from 400 to 994 nm at a spectral and spatial resolution of 5 nm and 1 m respectively. A detailed assessment was carried out of geomorphic features (dimension *ca.* 100 m) inside a localized study area of the Bank using a combination of phototransect information collected in the field and linear spectral unmixing techniques. The study area covered approximately 20 km² (1.5 km wide by 13 km long) and was located along the northern coast of the inner Wajh Barrier (see inset box in Figure 23.1). This area encompassed considerable variability in terms of environmental gradients (e.g., areas were exposed and sheltered to wave energy, shallow- and deep-water depths, in- and offshore) and provided an appropriate site to develop a model that established relationships between physical environmental characteristics and benthic coverage. The composition of coral community assemblages appears to be structured by a combination of endogenous factors such as coral competition, coral community fecundity, and provision of suitable settlement sites for larvae, and exogenous factors such as water depth, exposure to incident wave power, and water clarity [6].

Geomorphic Features and Habitats

The barrier reef system is comprised of a continuous line of reefs stretching for *ca.* 100 km and separated by several narrow (<200 m width) channels [5]. The outer edge of the Bank lies *ca.* 26 km offshore, running parallel to the shoreline for approximately 50 km, before changing course landward to enclose the reef system around a central lagoon (Figure 23.1). Inside the Bank, the depth of the lagoon shelf gets progressively shallower toward the coast, an alluvial sandy plain.

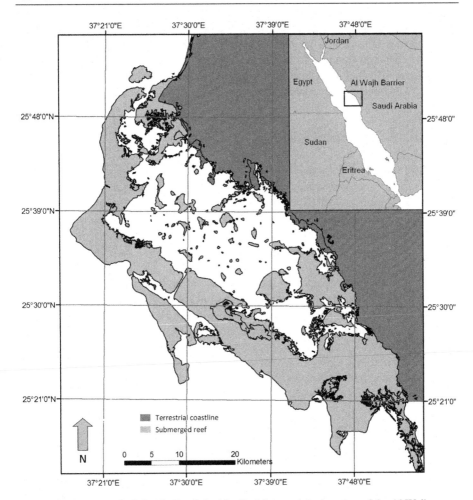

Figure 23.1 The Al Wajh Bank, Saudi Arabia, Red Sea, and the location of the Al Wajh Bank in the Red Sea (inset). The area marked in red to the north of the Bank is the study area where fine-scale community assessments and a surrogacy analysis were carried out. (For interpretation of the references to color in this figure legend, the reader is referred to the web version of this book.)

The shallow shelf inside the barrier supports a range of islands and associated reef formations, including platform or patch reefs, submerged reef ridge networks, and cay reefs. Within the study area, three geomorphic features were apparent where spectral unmixing and biological assessments were carried out.

Shallow platform: On the inner rim of the Bank to the north of the study area, this was separated by a deep channel (one of two sites of hydrological flux in and out of the lagoon area) from the lagoon to the south.

Reef ridge network: A series of linear, narrow, elevated reef structures with steep sides crossing the center of the study area perpendicular to the coastline.

Shallow plateau (ranging from 4 to 30 m water depth): On the coastal side of the inner lagoon. This platform supports numerous shallow reef landforms, such as coral patches, platforms, and cay reefs.

Biological Communities

Assessments of benthic composition are supported by underwater phototransect data and the results of a linear mixture model developed on the basis of spectra collected *in situ* using a Ramses TRIOS sensor. The coverage of four benthic community types (live coral, dead coral, macroalgae, and sand) was estimated using a staged approach in which linear spectral unmixing was applied to a subset of wavebands identified from the hyperspectral remote-sensing data using a multiple discriminant function analysis. Spectra collected in the field were used to estimate the proportion of each benthic cover type by decomposing the reflectance of each pixel into a linear equation comprised of the reflectance spectrum of each individual coverage and solving for the proportion of coverages. Full details of this procedure are described elsewhere [7].

Twelve phototransects were recorded at different locations within the study area. These included coral patches on the shallow plateau to the south of the study area and the reef ridge network across the center of the study area (Figure 23.2). Two 20 × 2 m phototransects, a shallow (5 m depth) and a deep (10 m depth), were sampled at each site in both sheltered and exposed locations within the study area. Overall, the relative

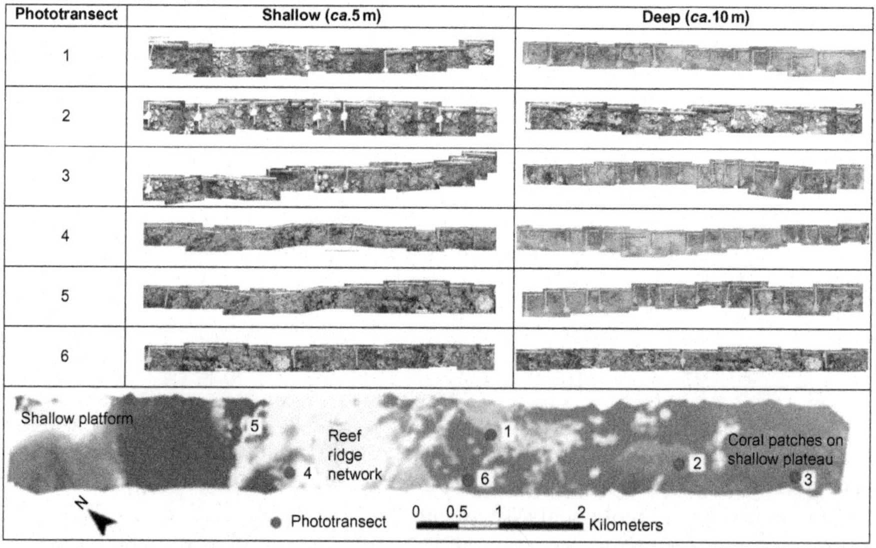

Figure 23.2 Phototransects used for validating benthic estimations derived from the spectral unmixing algorithm, one shallow and one deep transect per site. Locations plotted on the RGB image composite of the study area.

coverage of different benthic substrates was estimated from 120 photographs. Where corals were present, their status (live or dead) and growth form were identified. Coral cover ranged from 30% to 90%, and communities showed distinct characteristics depending on their location and local environmental conditions (Table 23.1).

A high correlation was found between live coral cover recorded in the field and that estimated by the linear unmixing analysis ($R^2 = 0.88$). From the unmixed benthic coverages, a high cover of live coral was found to be associated with the three morphological units, which were to the north of the study area around the shallow platform, covering a dominant proportion of the ridge network and in conjunction with the patches in the south of the study site. Mean live coral cover across the whole study areas was 35.1% (standard deviation 17.6). Dead coral occupied the shallower areas of these reef structures, with high coverage along the tops of several prominent ridges among the network across the center of the study site and the exposed platform to the northeast. Mean dead coral cover was 49.4% (standard deviation 22.1). Mean macroalgal cover was 51.8% (standard deviation 37.8), with a macroalgal mat covering a large area of the lagoon floor, particularly in flat areas toward the south of the study area. There was limited coverage of macroalgae in areas where reefs were well developed, that is, around the ridge network in the center of the study site and in association with reef patches. Carbonate sand occurred in association with the live reefs and appeared to be at particularly high proportions in areas directly adjacent to coral, that is, in rings around the reef patches on the shallow shelf to the south of the study area. Mean carbonate sand cover was 36.6% (standard deviation 36.2).

Table 23.1 Summary of the Key Characteristics of Coral Communities in Different Local Environmental Conditions of the Al Wajh Bank

Location	Community Description	Environment
Outer Reef Ridge Network (*Sites 4 and 5 shallow*). Reef crest/upper reef slope: Shallow sites of well developed reefs subject to high wave exposure (5 m).	Stout taxa of low growth forms, for example, digitate and submassive.	Rich coral assemblages were limited to outer reef crest and flat. Physically structured by wind-driven waves during low tides. High water clarity.
Shallow Plateau and Ridge Network (Deep). Sites 2 and 3, sites 4 and 5 (deep). Mid-Bank shallow submerged lagoonal patch reefs of low exposure and high turbidity.	Mixed assemblages of digitate, encrusting, branching, tabular, and massive corals.	On the sides and tops of submerged patch reefs, and in lagoons behind barrier reefs. Low wave energy. Resuspension of fine sandy/silty sediments can reduce water clarity.
Inner Reef Ridge Network. Sites 1 and 6. Shallow (5 m) and deep (>10 m) sites of moderate wave exposure.	Branching, tabulate *Acropora*, massive, and mushroom corals.	Wide depth range of moderate wave exposure.

Site numbers relate to those labeled in Figure 23.2.

Surrogacy

The distribution of live coral cover inside the Al Wajh lagoon was modeled using a multivariate regression model that encompassed water depth, exposure to incident waves, and the concentration of suspended sediment in the water column (Figure 23.3). Two models were developed: an ordinary least squares regression model and a spatially lagged autoregressive model. For spatial regression, cases referred to geographical locations and information contained at neighboring data cases was used in the model. To derive information on the independent variables, bathymetry was estimated from the AISA Eagle hyperspectral remote-sensing dataset using ratio transform techniques [8]; wave exposure was estimated from fetch lengths and the local wind field using GIS-based methods [9]; and the concentration of suspended sediment was estimated by correlating water samples collected *in situ* to the derivative reflectance of the remotely sensed imagery in an infrared band of wavelength 650 nm [10].

The transition from an initial ordinary least squares model to the spatial model was accompanied by a marked growth in explanatory power ($R^2 = 0.26$ to $R^2 = 0.76$). The theoretical implication that follows is that neighborhood context interactions

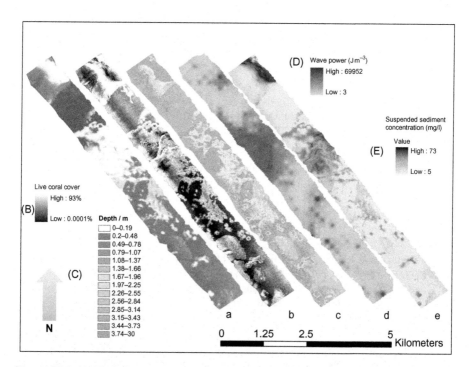

Figure 23.3 (A) Three band QuickBird imagery of the study area; (B) grayscale unmixed image output depicting the abundance of coral, white areas indicate areas of high coral cover; (C–E) spatial distribution of the modeled values for the three independent variables: (C) bathymetry, (D) wave power, and (E) suspended sediment concentration.

(arising from processes which themselves are spatially autocorrelated) play an important role in determining live coral cover at Al Wajh. Such spatially dependent influences could include dynamics associated with coral recruitment (spawning and larval dispersal), as well as physical factors such as the availability of suitable substrate for settlement of early life stage corals and elevated topographical remnants of antecedent populations. These physical factors may be particularly influential along this Red Sea coastline where present living reefs, both on the barrier and inside the lagoon, are thought to have developed over the past 6,000 years during Holocene sea-level rise on top of topographic highs formed by earlier relict reef structures [5,6].

Acknowledgments

This work would not have been possible without the generous support of the Khaled bin Sultan Living Oceans Foundation. Dr. Tom Spencer at the University of Cambridge is thanked for guidance on the manuscript. Dr. Rainer Reuter, University of Oldenburg, is thanked for assistance with fieldwork.

References

[1] S.E. Morcos, Physical and chemical oceanography of the Red Sea, Oceanography and Marine Biology Annual Review 8 (1970) 73–202.

[2] F.J. Edwards, Climate and oceanography, in: A.J. Edwards, S.M. Head (Eds.), Red Sea, Pergammon Press, Oxford, 1987, pp. 45–69.

[3] A.H. Meshal, Hydrography of a hypersaline coastal lagoon the Red Sea, Estuarine, Coastal and Shelf Science 24 (1987) 167–175.

[4] S.P. Murray, W. Johns, Direct observations of seasonal exchange through the Bab el Mandab Strait, Geophysical Research Letters 24 (1997) 2557–2560.

[5] C.R.C. Sheppard, A. Price, C. Roberts, Marine Ecology of the Arabian Region: Patterns and Processes in Extreme Tropical Environments, Academic Press, London, 1992.

[6] De Vantier, L. (2000) Final report on the study on coastal marine habitats and biological inventories in the northern part of the Red Sea coast in the Kingdom of Saudi Arabia. Unpublished report to the National Commission for Wildlife Conservation and Development, Riyadh, Saudi Arabia, Japan International Cooperation Agency.

[7] S. Hamylton, Estimating the coverage of coral reef benthic communities from airborne hyperspectral remote sensing data: multiple discriminant function analysis and linear spectral unmixing, International Journal of Remote Sensing (2011). in press

[8] R.P. Stumpf, K. Holderied, Determination of water depth with high-resolution satellite imagery over variable bottom types, Limnology and Oceanography 48 (2003) 547–556.

[9] J. Ekebom, P. Laihonen, T. Suominen, A GIS-based step-wise procedure for assessing physical exposure in fragmented archipelagos, Estuarine, Coastal and Shelf Science 57 (2003) 887–898.

[10] Z. Chen, P.J. Curran, J.D. Hansom, Derivative reflectance spectroscopy to estimate suspended sediment concentration, Remote Sensing of Environment 40 (1992) 67–77.

24 Mesophotic Coral Reefs of the Puerto Rico Shelf

Roy A. Armstrong[1], Hanumant Singh[2]

[1]Department of Marine Sciences, University of Puerto Rico, Mayaguez, Puerto Rico, [2]Woods Hole Oceanographic Institution, Woods Hole, MA, USA

Abstract

Mesophotic coral ecosystems (MCEs) in the Puerto Rico Shelf can be divided into two broad categories, low-gradient platforms, and high-gradient slopes. The distribution of MCEs is determined by a combination of suitable hard substrates and physical factors such as temperature, light availability, and low sedimentation. Water turbidity, by limiting the penetration of light in the water column, is a determinant factor in the amount and maximum depth of living coral cover. The insular shelf and slope areas between 30 and 100 m in Puerto Rico have an area of approximately 3,900 km² (or 46% of the total area between 0 and 100 m), representing potential habitat for these deeper reef communities. The benthic imaging capabilities of the Seabed autonomous underwater vehicle have been used since 2002 to map and characterize MCEs throughout the Puerto Rican insular shelf. Some of these reefs are structurally complex with high coral cover and abundant fish and invertebrate fauna. The geomorphology and benthic community structure of two MCE areas are described.

Key Words: Puerto Rico, mesophotic reefs, AUV

Introduction

Mesophotic coral ecosystems (MCEs) can be defined as light-dependent coral, algal, and sponge communities that occur in the deepest half of the photic zone (starting at 30 m and extending to over 150 m) in tropical and subtropical regions [1]. The distribution of MCEs in the US Caribbean and elsewhere, including ecologically relevant parameters such as percent living coral cover, reef rugosity, incidence of bleaching, and species diversity, are largely unknown. MCEs are known habitats of commercially important fish species and could serve as refugia for coral reef organisms during times of global climate change.

MCE habitats in Puerto Rico and the US Virgin Islands (USVI) can be divided into two broad categories: low-gradient platforms composed of insular shelves and banks, and high-gradient slopes. Geomorphology can exert a fundamental control on

Seafloor Geomorphology as Benthic Habitat. DOI: 10.1016/B978-0-12-385140-6.00024-4

the occurrence and distribution of MCEs by providing favorable hard substrates for colonization and by directing the downslope transport of sediment [2]. In addition to geomorphology, physical factors such as temperature, light availability, and low sedimentation are determinant factors in the distribution of these reefs.

Two low-gradient platforms, the Bajo de Cico bank and the Hind Bank Marine Conservation District (MCD), are described in this case study (Figure 24.1). These well-developed but morphologically different MCEs occur in clear oligotrophic waters with high water circulation. At the MCD and Vieques Island, on the eastern Puerto Rican Shelf, a mean light attenuation coefficient of $0.035\,m^{-1}$ results in the euphotic zone (1% of surface values) reaching depths of approximately 131 m. In Bajo de Cico, off western Puerto Rico, a mean attenuation coefficient of $0.042\,m^{-1}$ results in a euphotic zone of 109 m [3]. This difference is due to seasonal intrusions of more turbid riverine waters from western Puerto Rico. Other relevant oceanographic features include a mixed-layer depth reaching a maximum of 100 m in January–March (dry season) and a minimum of 25 m in September–October (wet season). Mixed-layer temperature and salinity follow the same seasonal pattern, with temperatures ranging from 26°C to 30°C and salinities from 34 to 36.3 ps [4].

Both of these MCEs, as well as other surveyed Puerto Rican Shelf mesophotic reefs, appear to be mostly undisturbed by anthropogenic factors due to their greater depths and considerable distance from land sources of runoff and pollution. The low levels of coral mortality observed in these relatively pristine reefs are largely due to coral bleaching and disease.

Since the insular shelf and slope areas between 30 and 100 m in Puerto Rico and the USVI have an area of approximately $3,900\,km^2$, or 46% of the total area between 0 and 100 m (Figure 24.1), it is impractical to rely solely on diving to adequately survey these potential reef habitats. Deep insular shelf MCEs covers large areas east of Puerto Rico between the islands of St. Thomas and Vieques. Since 2002, we have used the imaging capabilities of the Seabed autonomous underwater vehicle (AUV) to map and characterize MCEs throughout the Puerto Rico Shelf [5–10]. The Seabed AUV, designed for high-resolution underwater optical and acoustic imaging, is a stable platform that provides high-resolution color imagery of benthic environments. More information on the Seabed AUV components, sensors, control systems, and navigation can be found in Ref. [5].

Geomorphic Features and Habitats

In Puerto Rico, MCEs achieve their greatest development on low-gradient platforms, where relic reefs and terraces provide favorable hard substrates for colonization. The area of potential MCE habitat in high-gradient slopes is minimal when compared to low-gradient platforms. High-gradient slope MCEs off southwestern Puerto Rico have been described in Ref. [5,10,11]. We describe here two representative and ecologically important low-gradient platform mesophotic reefs, Bajo de Cico and the Hind Bank MCD.

Two distinct types of low-gradient platform MCEs are found in Puerto Rico. One type is characterized by a structurally complex, high-rugosity coral reef dominated

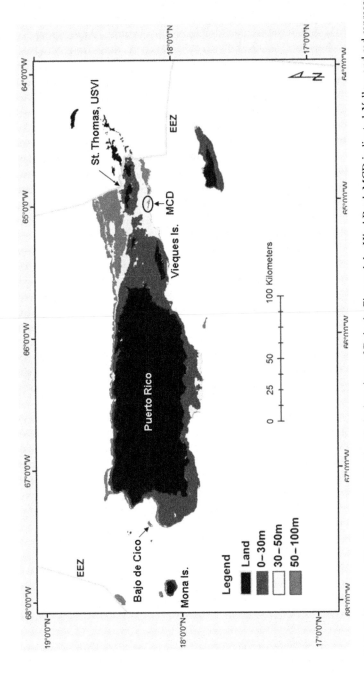

Figure 24.1 Bathymetry of the Puerto Rico Shelf with the locations of Bajo de Cico and the Hind Bank MCD indicated. Yellow and red represent the upper mesophotic (30–50 m) and the lower mesophotic (50–100 m) zones, respectively. EEZ = Exclusive Economic Zone. (For interpretation of the references to color in this figure legend, the reader is referred to the web version of this book.)

by a flattened morphotype of *Montastrea annularis* complex. These formations are common on the insular shelf east of Puerto Rico at depths of 30–45 m. The MCD is representative of this category. The other type of MCE formation is associated with extensive algal rhodolith deposits and dominated by benthic algae, sponges, and corals of the genus *Agaricia*. These reefs are typical of oceanic islands and isolated banks in the Mona Passage, west of Puerto Rico at depths of about 50–100 m or more. Bajo de Cico is an example of this MCE type.

Bajo de Cico Bank

Bajo de Cico is an isolated bank with a relatively flat top and a surface area of approximately 11 km² located 25 km west of Puerto Rico. A narrow ridge in the southwest corner reaches depths of 25 m, while the rest of the bank is a flat homogeneous platform that dips gently from 40 to 90 m depth before dropping abruptly to over 200 m depth (Figure 24.2). The reef top is highly rugose, with many rock outcrops and sand channels [12]. A low-relief MCE begins at 45 m and extends to depths of nearly 100 m, as documented by the AUV imagery.

Hind Bank MCD

The Hind Bank MCD is a 41 km² protected area located at the edge of the Puerto Rico Shelf, approximately 12 km south of St. Thomas, USVI. The MCD is a low-gradient platform habitat with a depth range from 30 to 62 m (average depth 42 m) that dips gently southward before reaching the shelf edge (Figure 24.3). The geomorphology of this MCE consists of a series of coral ridges 100 m across separated from each other by sandy grooves ranging from 50 to 300 m wide [2]. The primary benthic habitats include well-developed coral reefs, colonized hard bottom, algal plain, and sand flats [7].

Biological Communities

Benthic characterizations of the study areas were based on Seabed AUV phototransects using a high-resolution (1280 × 1024, 12-bit) camera in a submersible housing and a 150 W-s strobe for illumination. From an altitude of 3.5 m the area of each image was nearly 4 m². Four 1-km long digital phototransects were obtained at the Hind Bank MCD and provided data on benthic species composition and abundance at depths between 32 and 54 m. In Bajo de Cico, three AUV transects covered a total distance of about 5.5 km at depths between 55 and 100 m. More details on AUV data acquisition and analysis for the MCD and Bajo de Cico study areas can be found in Refs. [7] and [3], respectively.

Bajo de Cico Bank

In Bajo de Cico, extensive rhodoliths present along low topographic relief terraces are colonized by sponges of the genus *Agelas* and *Aplysina*, turf, fleshy and calcareous macroalgae, and corals at depths of 45–100 m (Figure 24.4). The maximum

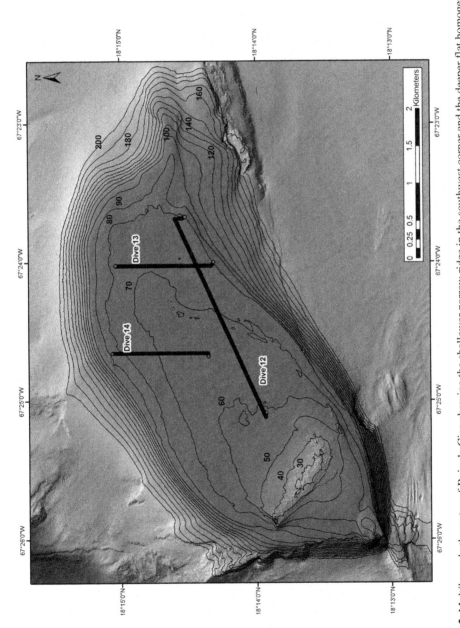

Figure 24.2 Multibeam bathymetry of Bajo de Cico showing the shallower narrow ridge in the southwest corner and the deeper flat homogeneous platform with the location of the three AUV transects. Grid size of the bathymetry is 5 m.

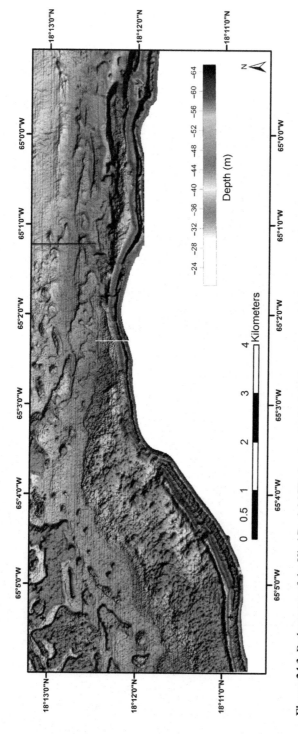

Figure 24.3 Bathymetry of the Hind Bank MCD showing the high bank reef zones (red) on the western end and the deeper algae and pavement zones to the east. (For interpretation of the references to color in this figure legend, the reader is referred to the web version of this book.)

Figure 24.4 Low-relief mesophotic reef at a depth of 50 m in Bajo de Cico showing high live cover, mostly by the macroalgae *Lobophora variegata* and sponges. Several flattened colonies of the genus *Agaricia* can be seen in the foreground.
Source: Photo by J. Sabater.

average coral cover, mostly *Agaricia lamarcki* and *A. grahamae*, was 16% at 70–80 m depth. Sponge species richness was high, with 49 species present [2]. *Lobophora variegata* was the most prominent component of the benthic algae assemblage with a mean cover of 42.0% [12]. Dominance by macroalgae, a rich and abundant sponge fauna, and relatively low coral cover, composed mostly of *Agaricia* sp., characterize the mesophotic habitat at Bajo de Cico. The relatively low coral cover is likely due to the low-relief antecedent topography and the predominantly unconsolidated substrate, composed of rhodoliths and coarse sand, which is largely unsuitable for coral settlement and growth.

Hind Bank MCD

At the western end of the MCD, well-developed coral reefs were found at depths of 33–42 m. The average living coral cover was 43%, with maximum cover of 70% found in some areas at 40 m depth [7]. Within the scleractinian cover composition, the image data analysis revealed dominance by *M. annularis* complex at all transects (Figure 24.5). *Porites astreoides* and *Mycetophyllia* sp. were found at all depths between 33 and 47 m. The *M. annularis* complex and *Agaricia* sp. both grew as plate-like formations, representing an adaptation to reduced sunlight at deeper depths [13]. The eastern MCD consists of small coral reef patches separated by hardground areas along with algal mats, sand, and sparse seagrasses. The most dominant sponges were the giant barrel (*Xestospongia muta*) and rope sponges of

Figure 24.5 Mesophotic reef at the western end of the MCD showing dominance by
M. annularis complex. Image obtained by the AUV from an altitude of 3.5 m at a depth
of 43 m. Scale bar = 1 m.

the genus *Aplysina*. Also found in abundance was an encrusting sponge of the
genus *Cliona*. A total of 17 species of sponges could be identified from the AUV
images.

The mesophotic reefs on the eastern Puerto Rico Shelf consist of structurally complex, extensive high coral cover habitats dominated by the *M. annularis* complex. This
agrees with reports of Black Jack Reef, on the south coast of Vieques Island [9], where
M. annularis complex accounted for 76% of the total cover by live coral at depths
between 36 and 40 m. On the other hand, the mesophotic reefs off western Puerto Rico
consist of extensive algal rhodolith deposits dominated by benthic algae and a rich
sponge community. Coral cover, mostly by flattened species of the genus *Agaricia*,
accounts for only up to 16% of the live benthic cover. These reefs are typical of oceanic islands and isolated banks in the Mona Passage up to depths of approximately
100 m.

The AUV images also showed that MCEs in Puerto Rico appear to be unaffected
by terrestrial sources of pollution and human impacts. Incidence of coral bleaching
and disease are reduced when compared to the shallower (<30 m) reefs. The potential of MCEs as a source of larvae for the recovery of shallow coral reef communities, which have experienced declines in coral cover of up to 80% in the last
30 years [14], remains to be ascertained.

Surrogacy

Due to the logistical complexity of working in the mesophotic zone, there is little information on the factors that influence MCE community structure and their relative importance. Mesophotic reefs, due to their depth range, are largely sheltered from the direct physical damage from storm surge. Although some reports suggest that the lower limit of hermatypic corals is not imposed by temperature gradients [15], an instance of a large, cryptic mortality at the MCD has been attributed to intrusions of cold, turbid waters associated with the thermocline [16].

Geomorphology, depth, and water turbidity (and its surrogate, distance from land) are factors affecting the community structure of MCEs in Puerto Rico [3]. Rivero-Calle [3] compared five known MCE areas around the island and found that coral cover, macroalgae cover, and total live cover tend to increase with distance from land and decrease with water turbidity. These relationships were considerably stronger in the lower mesophotic ranges (50–100 m) than the upper mesophotic ranges (30–50 m). The effect of depth is indirect and related to the amount of incident light, since light is attenuated exponentially with depth. Light is the limiting factor for the maximum depth distribution of zooxanthellate corals [14].

Acknowledgments

The Seabed AUV development and use was made possible by the Bernard M. Gordon Center for Subsurface Sensing and Imaging Systems (Gordon-CenSSIS), under the Engineering Research Centers Program of the National Science Foundation (Award Number EEC-9986821). We thank Sara Rivero-Calle for analyzing the Bajo de Cico AUV images, William Hernandez for Figure 24.1, and Jorge Sabater for Figures 24.2–24.4.

References

[1] L.M. Hinderstein, J.C.A. Marr, F.A. Martinez, M.J. Dowgiallo, K.A. Puglise, R.L. Pyle, et al., Theme section on Mesophotic coral ecosystems: characterization, ecology, and management, Coral Reefs 29 (2010) 247–251.

[2] S.D. Locker, R.A. Armstrong, T.A. Battista, J.J. Rooney, C. Sherman, D.G. Zawada, Geomorphology of mesophotic coral ecosystems, Coral Reefs 29 (2010) 329–345.

[3] S. Rivero-Calle, Ecological Aspects of Sponges in Mesophotic Coral Ecosystems, Master's Thesis, University of Puerto Rico at Mayaguez, 2010, 85 pp.

[4] J. Capella, D.E. Alston, A. Cabarcas-Nuñez, H. Quintero-Fonseca, R. Cortes-Maldonado, Oceanographic considerations for offshore aquaculture on the Puerto Rico-U.S. Virgin Island platform, in: C.J. Bridger, B.A. Costa-Pierce, (Eds.), Open Ocean Aquaculture: From Research to Commercial Reality, The World Aquaculture Society, Baton Rouge, LA, 2003, pp. 247–261.

[5] H. Singh, R.A. Armstrong, R. Eustice, C. Roman, O. Pizarro, J. Torres, Imaging Coral I: imaging coral habitats with the SeaBED AUV, Subsurf. Sens. Technol. Appl. 5 (1) (2004) 25–42.

[6] R.A. Armstrong, H. Singh, Remote Sensing of Deep Coral Reefs in Puerto Rico and the U.S. Virgin Islands Using the Seabed Autonomous Underwater Vehicle, in: SPIE Europe Remote Sensing Conference Proceedings #6360-10, Stockholm, Sweden, 2006, September 11–14.

[7] R.A. Armstrong, H. Singh, J. Torres, R. Nemeth, A. Can, C. Roman, et al., Characterizing the deep insular shelf coral reef habitat of the Hind Bank Marine Conservation District (US Virgin Islands) using the Seabed Autonomous Underwater Vehicle, Continent. Shelf Res. 26 (2006) 194–205.

[8] R.A. Armstrong, Deep zooxanthellate coral reefs of the Puerto Rico—U.S. Virgin Islands insular platform, Coral Reefs 26 (4) (2007) 945.

[9] S. Rivero-Calle, R.A. Armstrong, F.J. Soto-Santiago, Biological and physical characteristics of a mesophotic coral reef: Black Jack reef, Vieques, Puerto Rico, in: Proceedings of the 11th International Coral Reef Symposium, Ft. Lauderdale, FL, July 2008, 2009, pp. 567–571.

[10] R.A. Armstrong, H. Singh, S. Rivero, F. Gilbes, Monitoring Coral Reefs in optically-deep waters, in: Proceedings, 11th International Coral Reef Symposium, Ft. Lauderdale, FL, July 2008, 2009, pp. 593–597.

[11] C. Sherman, M. Nemeth, H. Ruíz, I. Bejarano, R. Appeldoorn, F. Pagán, et al., Geomorphology and benthic cover of mesophotic coral ecosystems of the upper insular slope of southwest Puerto Rico, Coral Reefs 29 (2010) 347–360.

[12] J.R. García-Sais, R. Castro, J. Sabater-Clavell, M. Carlo, R. Esteves, Characterization of benthic habitats and associated reef communities at Bajo de Cico Seamount, Mona Passage, Puerto Rico. Final Report submitted to the Caribbean Fishery Management Council, San Juan, Puerto Rico, 2007, pp. 91.

[13] R.R. Grauss, I.E. Macintyre, Variation in growth forms of the reef coral *Montastrea annularis*: a quantitative evaluation of growth response to light distribution using computer simulation, in: K. Rutzler, I.G. Macintyre, (Eds.), The Atlantic Barrier Reef Ecosystem at Carrie Bow Cay, Belize I. Structure and Communities, Smithsonian Institution Press, Washington, DC, 1982, pp. 441–464. Smithsonian Contributions in Marine Science 12.

[14] T.A. Gardner, I.M. Cote, J.A. Gill, A. Grant, A.R. Watkinson, Long term region-wide declines in Caribbean corals, Science 301 (2003) 958–960.

[15] S.E. Kahng, J.R. Garcia-Sais, H.L. Spalding, E. Brokovich, D. Wagner, E. Weil, et al., Community ecology of mesophotic coral reef ecosystems, Coral Reefs 29 (2010) 255–275.

[16] T.B. Smith, J. Blondeau, R.S. Nemeth, S.J. Pittman, J.M. Calnan, E. Kadison, et al., Benthic structure and cryptic mortality in a Caribbean mesophotic coral reef bank system, the Hind Bank Marine Conservation District, U.S. Virgin Islands, Coral Reefs 29 (2010) 289–308.

25 Geomorphic Features and Infauna Diversity of a Subtropical Mid-Ocean Carbonate Shelf: Lord Howe Island, Southwest Pacific Ocean

Brendan P. Brooke[1], Matthew A. McArthur[1],
Colin D. Woodroffe[2], Michelle Linklater[2],
Scott L. Nichol[1], Tara J. Anderson[1],
Richard Mleczko[1], Stephen Sagar[1]

[1]Marine and Coastal Environment Group, Geoscience Australia, Canberra, ACT, Australia, [2]School of Earth and Environmental Science, University of Wollongong, NSW, Australia

Abstract

Lord Howe Island in the southwest Pacific Ocean is surrounded by a shallow (20–120 m) subtropical carbonate shelf 24 km wide and 36 km long. On the mid shelf, a relict coral reef (165 km^2) extends around the island in water depths of 30–40 m. The relict reef comprises sand sheet, macroalgae, and hardground habitats. Inshore of the relict reef a sandy basin (mean depth 45 m) has thick sand deposits. Offshore of the relict reef is a relatively flat outer shelf (mean depth 60 m) with bedrock exposures and sandy habitat. Infauna species abundance and richness were similar for sediment samples collected on the outer shelf and relict reef, while samples from the basin had significantly lower infauna abundance and richness. The irregular shelf morphology appears to determine the distribution and character of sandy substrates and local oceanographic conditions, which in turn influence the distribution of different types of infauna communities.

Key Words: marine geomorphology, benthic biodiversity, marine sediment Holocene coral reef, multibeam sonar

Introduction

Lord Howe Island (31°33′S, 159°04′E) sits approximately 580 km east of Australia in the southwest Pacific Ocean (Figure 25.1). This remote island is the subaerial remnant of a hot-spot volcano that erupted between 6.6 and 7.2 Ma and comprises a

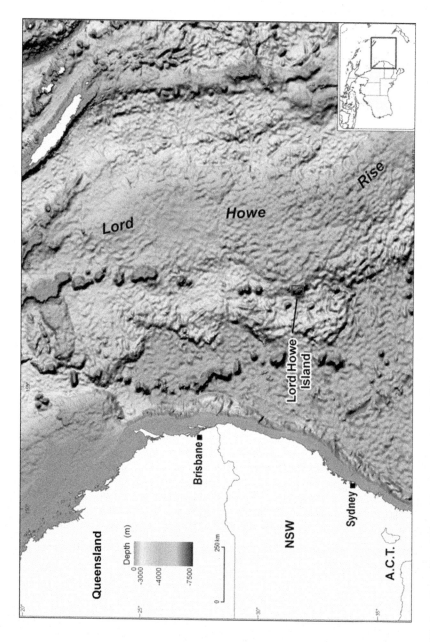

Figure 25.1 Location of Lord Howe Island, approximately 580 km east of the southeast coast of Australia, on the western margin of the Lord Howe Rise.

mid-ocean basaltic sequence [1]. Erosion of these basalts has dominated island evolution since the cessation of volcanism, creating a rhomboidal-shaped shelf around Lord Howe Island, 24 km wide and 36 km from north to south. As part of a large collaborative program of marine biodiversity research (CERF Marine Biodiversity Hub: www.marinehub.org), the shelf around Lord Howe Island was selected as an area representative of a shallow (20–120 m) mid-ocean carbonate shelf environment. In this chapter, we describe and examine the geomorphic features and their importance to the pattern of infauna diversity on this carbonate shelf.

Lord Howe Island sits at the boundary of tropical and temperate water masses, with the mean annual sea surface temperature (SST) around the island varying between 18°C and 23°C [2]. In summer, the poleward flowing, warm-water East Australian Current (EAC) extends to around 34°S, bathing the Lord Howe Island shelf in warm waters. In winter, the boundary of the EAC shifts to 30°S and the island comes under the influence of the cooler Tasman Front [3,4]. The tidal range at Lord Howe Island is 1.5 m at springs and 0.8 m at neaps, with a mean significant wave height of 2.5 m from the southwest [5]. The island has a highly energetic wind regime (mean speed ~30 km/h), predominantly from the east in summer and southwest in winter.

Lord Howe Island was included on the World Heritage List in 1982 based on the high degree of preservation of its volcanic island and coral reef environments. The surrounding waters, which have been declared a State Marine Park since 1999 and a Commonwealth Marine Park since 2000 [6,7], contain a high degree of endemic taxa, including 10% of the marine invertebrates [8]. Subtropical coralline algae dominate the shelf sediments, with coral reefs restricted to the shallow margin of Lord Howe Island [2,9,10]. However, a much more extensive relict reef feature occurs in the middle of the shelf, rising from water depths of around 50 m to a mean depth of approximately 34 m [11,12].

Descriptions of the geomorphology and infauna of the Lord Howe Island shelf are based on the analysis of detailed bathymetric data (8 m grid) and sediment samples collected in 2008 aboard the *RV Southern Surveyor* (voyage SS06). Seabed topographic data were collected using a Kongsberg Simrad EM 300 multibeam echosounder. Coverage of the shallow inshore waters, where the vessel could not be operated, was provided by Laser Airborne Depth Sounder (LADS) derived bathymetry for a swath across the lagoon and the archipelago of islands to the northeast, and Quickbird satellite image derived bathymetry for the very shallow (<15 m depth) nearshore zone around the island. For areas outside the extent of these data, lower resolution singlebeam echosounder data and digitized Admiralty Chart data were employed (Figure 25.2) [13]. Bottom sediments and infauna were examined using a Smith-McIntyre grab sampler, while seabed type and sediment thickness were mapped using a Topas PS18 parametric 1.5 kHz acoustic sub-bottom profiler [13].

A brief overview of broad-scale habitats on the Lord Howe Island shelf is based on the new shelf topographic and sediment data and descriptions of benthic substrates and biota identified in underwater video footage collected by the Australian Institute

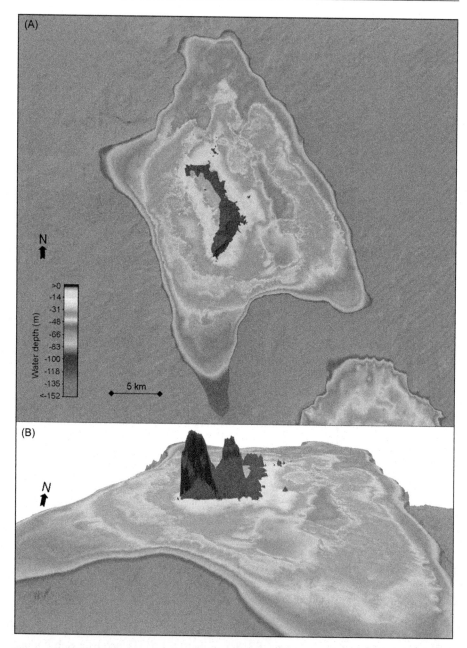

Figure 25.2 DRMs of the Lord Howe Island shelf based on multibeam sonar bathymetry and a range of supplementary bathymetry data. (A) The shelf around Lord Howe Island showing the extensive relict reef feature on the middle shelf (red-gray). (B) Three-dimensional view of the southern and eastern areas of the shelf (viewed from the S) showing the deeper sections of the basin (green-yellow) inboard of the relict reef (red-gray) (VE = 5). (For interpretation of the references to color in this figure legend, the reader is referred to the web version of this book.)

of Marine Science (AIMS) in 2004 [14] and the New South Wales Government's Department of Environment, Climate Change and Water (NSW DECCW) in 2009 [7].

Geomorphic Features and Habitats

A range of geomorphic features were delineated using a digital relief model (DRM) of the seabed derived from the combined bathymetry data (Figures 25.2 and 25.3; Table 25.1). Features were mapped by onscreen digitizing of polygons in ArcGIS, with feature boundaries defined by distinct breaks in slope, changes in seabed morphological parameters derived from the DRM, and multibeam backscatter data supplemented with sediment information derived from the grab samples and sub-bottom profiles [15]. The geomorphic terms used to describe the fine-scale reef and shelf features are drawn from Woodroffe [16] and Woodroffe et al. [11].

Outer shelf: The outer shelf is defined as the area between the break in slope on the shelf edge and the break in slope and change in roughness at the outer margin of the relict reef. It covers an area of $212 \, km^2$ and has a mean water depth of 60 m (Table 25.1; Figure 25.3). The major break in slope at the shelf edge occurs in water depths between 90 and 125 m, where the outer shelf connects with the steeply sloping flanks of the volcano. Generally, the outer shelf is relatively flat; however, on the northern part of the shelf there are mounds and ridges with up to 10 m of relief. Terraces (~5 m relief) are evident in water depths of 57–80 m at various locations, and there are occasional shallow (a few meters) depressions (10–100 m wide).

Relict reef: This distinct seabed feature is a relict barrier reef that last grew in the early to middle Holocene [11] (Figure 25.3; Table 25.1). The relict reef rises from a water depth of around 50 m and has a rough upper surface that sits in depths of 30–40 m. This feature is characterized by extensive areas of hardground and sand deposits, mostly less than around 1 m thick but up to 4 m thick in depression in the relict reef surface based on sub-bottom profiles [13]. Overall, the relict reef ranges in width from 0.5 to 0.8 km. It covers an area of approximately $165 \, km^2$ (~20 times the size of the modern reef), which is approximately 32% of the shelf area. The relict reef lies close to the northwest corner of Lord Howe Island, but on the eastern side of the island the reef's steep inner flank is more than 6 km offshore (Figures 25.2 and 25.3). There are prominent spur and groove features on the southern, western, and eastern margins of the relict reef, and three former reef passages dissect it in the north, northeast, and east (Figure 25.3).

Basin: A sandy basin forms most of the seabed between the relict reef and inner shelf and is interpreted as a palaeolagoon that originally formed as the Holocene barrier reef grew [15] (Figure 25.3). The basin covers some $81 \, km^2$ and has a mean water depth of 45 m, which is up to 20 m deeper than the adjacent relict reef (Figures 25.2; Table 25.1). Sub-bottom profiles over the basin areas show a number of reflectors that indicate this is a long-term depositional zone, with up to 20 m of sandy sediment [13]. Basins sediments comprised sand with little to no gravel, unlike samples from the outer shelf and relic reef (Figure 25.3).

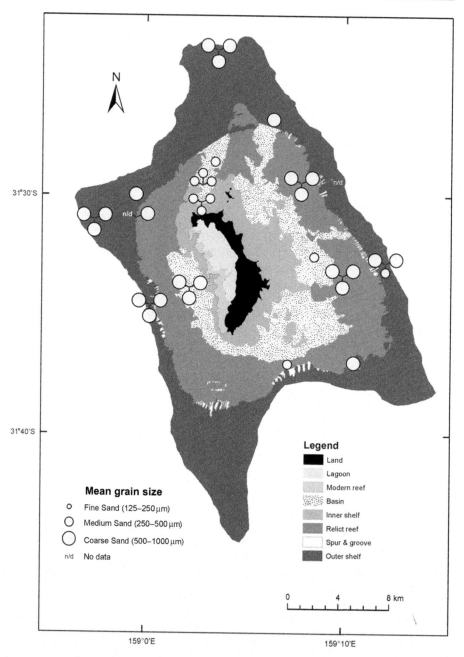

Figure 25.3 Map of the broad-scale geomorphic features of the shelf around Lord Howe Island. Spur and groove structures on the relict reef are also indicated. The location of sediment and infauna samples collected on the shelf are shown (center point of symbols), with two or three samples collected at most sites. The size of the sample symbols indicates mean grain size.

Table 25.1 Physical Characteristics of the Major Geomorphic Features on the Lord Howe Island Shelf

Major Geomorphic Features	Area (km^2)	Depth Range (m)	Mean Depth (m)	Characteristic Substrate
Outer shelf	212	143–30	61	Sandy sediment (sand flats and dunes), areas of bedrock
Relict reef	165	53–24	34	Hardground and sand patches
Basin	81	67–21	48	Thick sand deposits (several to 10 m thick)
Inner shelf	49	55–2	35	Patch reefs and bedrock
Fringing reef	9	32–0	11	Modern coral reef
Lagoon	6	18–2	2	Sand and patch reefs

Inner shelf: This zone of relatively rough seabed surrounds the island in water depths of predominantly 20–30 m and covers an area of 49 km^2 (Figures 25.2 and 25.3; Table 25.1). This feature includes low-relief coral patch reefs and pinnacles several meters high that rise to within a few meters of the surface. Boulders and gravelly seabed are common, as are patches of sand. There are several small basaltic islands and bedrock reefs, especially around the Admiralty Islands (northeast of Lord Howe Island).

Modern reef and lagoon: A coral reef (9 km^2) and sandy lagoon (6 km^2) fringe the western side of the island [2] (Figure 25.3) but are not examined in this chapter.

Habitats

The NSW Marine Parks Authority is currently reviewing the zoning of the Lord Howe Island Marine Parks [7] (http://www.mpa.nsw.gov.au/lhimp.html), which includes the development of a habitat map of the shelf based on a range of survey data, including those described in this chapter.

The outer shelf is characterized by large patches of sandy seabed, bedrock (basalt) exposures, and rubble (Figure 25.4E, F). Sandy areas examined are often reported to be bare of benthos, while brown algae and filter feeders (Gorgonians, crinoids) are recorded on hard substrates, especially toward the outer edge of the shelf [14]. The relict reef appears to comprise large patches of hardground and sand, with more sand toward the inner half of the reef structure [7] (Figure 25.4B–D). Dense algal communities (encrusting/coralline red algae; brown and green macroalgae) and areas with little discernible epifauna characterize the relict reef hardground sites that have been examined [7,14]. Other biota reported on the relict reef include patches of coral,

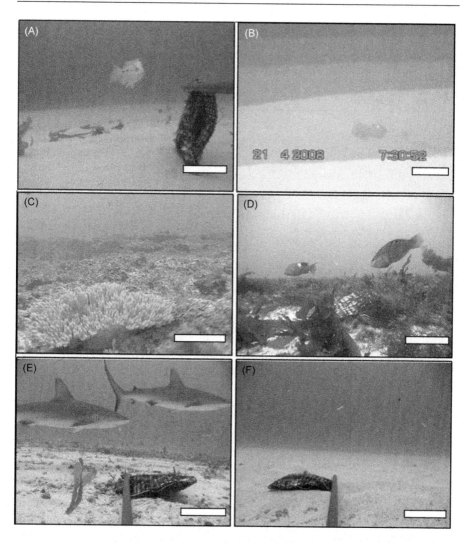

Figure 25.4 Benthic habitats of the Lord Howe Island shelf (screen shots from towed video and baited video sites). All scale bars indicate 20 cm in the foreground of the images, (A) sandy seabed of the basin, with black-spot pigfish (*Bodianus unimaculatus*); (B) seabed sand dunes on the relict reef, 30–50 cm high; (C) low-lying reef with soft coral (Family: Alcyoniidae); (D) low-lying reef covered in dense green macroalgae and sand, with Luculentus wrasse (*Pseudolabrus luculentus*) and the golden-spot pigfish (*Bodianus perditio*); (E) outer-shelf gravelly sands (foreground) adjacent to edge of relict reef (background), with Galapagos sharks (*Carcharhinus galapagensis*); (F) coarse sand on the outer shelf. Images A, C–F courtesy of M. Cappo and P. Speare from AIMS 2004 LHI survey, © the Australian Institute of Marine Science.

solitary corals, and sponges. Little discernable epifauna has been reported at sandy sites on the relict reef (Figure 25.4B). Basin areas have either a flat seafloor with no epifauna or patches of seabed dunes (Figure 25.4A) [7,14]. On the inner shelf, extensive coral communities are common. Low-relief reefs form linear features, especially on the northeast and eastern side of the island. Patches of basalt, green macroalgae, and a mix of coral, coralline algae, and green macroalgae are common on the inner shelf [7].

Biological Communities: Infauna

The shelf around Lord Howe Island supports a diverse infaunal assemblage, with 163 operational taxonomic units (OTUs) recorded from a total of 2,139 infaunal organisms. However, this assemblage was numerically dominated by five species (*Maeridae* sp., *Kalliapseudes* sp., *Copepoda* sp., *Asellota* sp., *Grammaridae* sp.) that accounted for 55% of the total infauna recorded, while most species recorded were rare (≤ 2 individuals recorded, 47% of species). Several new species are currently being described by specialist taxonomists. The three geomorphic features sampled (basin, relic reef, and outer shelf) support different infaunal assemblages, with numbers of total infauna increasing offshore, while species richness was significantly lower in the basin than in either the relic reef or outer shelf (Figure 25.5). High infaunal abundances on the outer shelf and relic-reef habitats were driven mostly by high numbers of *Maeridae* sp. Conversely, basin sediments supported significantly less individuals with species characterized by *Copepoda* sp. and *Lovenia elongatus* and the absence of other dominant species.

Surrogacy

The preliminary observations of geomorphology and infauna on the Lord Howe Island shelf show a clear link between the spatial distribution of infauna communities and the outer shelf, relict reef, and sandy basin features (Figure 25.5). Sediment grain size and percent gravel generally increase offshore and also influence infaunal species distributions (Figure 25.3). For example, most numerically dominant taxa (e.g., *Maeridae* sp., *Kalliapseudes* sp., *Asellota* sp., *Grammaridae* sp.) are strongly correlated with the proportion of gravel, with sediment devoid of gravel also lacking these species. Sediment thickness and seabed stability are also likely to be important factors in predicting infaunal assemblages across the shelf. For example, the outer shelf and relict reef have mostly thin deposits of sediment that are likely more regularly mobilized by wave-generated currents than deposits in the less exposed and relatively deep basin areas. Conversely, basin sediments are significantly thicker and the basin is an area of long-term sediment accumulation [13]. Infaunal species that feed on the microbes (e.g., *Copepoda* sp.) and nutrients (e.g., *Lovenia elongatus*) found on the surface of sediments were strongly associated with

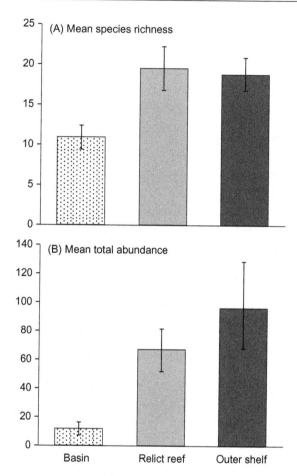

Figure 25.5 Infauna diversity on the Lord Howe Island shelf. (A) Mean infauna species richness (±1 standard error) for samples collected on three major seabed geomorphic features; (B) mean number of infauna individuals per sample (±1 standard error) for samples collected on three major geomorphic features.

the more protected basin environment. Inversely, the high proportion of suspension feeding taxa (e.g., *Kalliapseudes* sp. and *Spiochaetopterus* sp.) on the outer relic reef may denote the need for strong oceanic and wave-generated currents to deliver their food. A detailed analysis of the covariance of infauna species with fine-scale physical parameters (grain size, shelf morphometric, and seabed exposure parameters) is currently being undertaken as part of the CERF Marine Biodiversity Hub's surrogacy research program.

Shelves and banks of similar morphology to the Lord Howe Island shelf are found in other regions of the Australian marine estate, such as Norfolk Island and the Queensland Plateau; and at numerous locations in a range of temperate and sub-tropical settings, for example the Bahamas, Channel Islands, Canary Islands, and Hawaiian Islands. However, the distinctive setting of the Lord Howe Island shelf at the interface of tropical and temperate carbonate provinces, and its high degree of endemism, make this a unique site of global significance.

Acknowledgments

This work is an outcome of collaboration between Geoscience Australia and the University of Wollongong, and has been partly funded through the Commonwealth Environment Research Facilities (CERF) program, an Australian Government initiative supporting world-class, public-good research. The CERF Marine Biodiversity Hub is a collaborative partnership between the University of Tasmania, CSIRO Wealth from Oceans Flagship, Geoscience Australia, Australian Institute of Marine Science, and Museum Victoria. We are grateful to the Marine National Facility for time on RV *Southern Surveyor*, and to the captain and crew, plus Ian Atkinson, Andrew Hislop, Gareth Crooke, and Jack Pittar (Geoscience Australia) for field support. Multibeam data were acquired and processed by Cameron Buchanan and Michele Spinoccia (Geoscience Australia). Thanks to Peter Speare and Mike Cappo (AIMS) for providing underwater images of the LHI shelf. Professors Nic Bax and Dawn Wright, and an anonymous reviewer, provided constructive reviews of the draft manuscript. This chapter is published with permission of the Chief Executive Officer, Geoscience Australia.

References

[1] I. McDougal, B.J.J. Embleton, D.B. Stone, Origin and evolution of Lord Howe Island, southwest Pacific Ocean, J. Geol. Soc. Aust. 28 (1981) 115–176.

[2] J. Veron, T.J. Done, Corals and coral communities of Lord Howe Island, Aust. J. Mar. Freshwater Res. 30 (1979) 203–236.

[3] P. Marchesiello, J.H. Middleton, Modeling the East Australian current in the western Tasman Sea, J. Phys. Oceanogr. 30 (2000) 2956–2971.

[4] B.R. Stanton, An oceanographic survey of the Tasman Front, N.Z. J. Mar. Freshwater Res. 15 (1981) 289–297.

[5] M.E. Dickson, Shore platform development around Lord Howe Island, southwest Pacific, Geomorphology 76 (2006) 295–315.

[6] Marine Parks Authority, User's guide to the zoning plan: Lord Howe Island Marine Park. New South Wales Marine Parks Authority, Sydney, Australia, 2004. <www.mpa.nsw. gov.au>

[7] Marine Parks Authority, Natural values of Lord Howe Island Marine Park. New South Wales Marine Parks Authority, Sydney, Australia, 2010. <www.mpa.nsw.gov.au>

[8] W.F. Ponder, I. Loch, P. Berents, An assessment of the marine invertebrate fauna of the Lord Howe Island shelf. Report prepared for Environment Australia. Australian Museum, Sydney, 2000.

[9] V.J. Harriott, P.L. Harrison, S.A. Banks, The coral communities of Lord Howe Island, Mar. Freshwater Res. 46 (1995) 457–465.

[10] D.M. Kennedy, C.D. Woodroffe, B.G. Jones, M.E. Dickson, C.V.G. Phipps, Carbonate sedimentation on subtropical shelves around Lord Howe Island and Balls Pyramid, Southwest Pacific, Mar. Geol. 188 (2002) 333–349.

[11] C.D. Woodroffe, B.P. Brooke, M. Linklater, D.M. Kennedy, B.G. Jones, C. Buchanan, et al., Response of coral reefs to climate change: expansion and demise of the southern-most Pacific coral reef, Geophys. Res. Lett. 37 (2010) L15602, 10.1029/2010GL044067.

[12] C.D. Woodroffe, M.E. Dickson, B.P. Brooke, D.M. Kennedy, Episodes of reef growth at Lord Howe Island, the southernmost reef in the southwest Pacific, Global Planet. Change 49 (2005) 222–237.

[13] B.P. Brooke, C.D. Woodroffe, M. Linklater, M.A. McArthur, S.L. Nichol, B.G. Jones, et al., Geomorphology of the Lord Howe Island shelf and submarine volcano. SS06-2008 Post-Survey Report. Geoscience Australia Record 2010/26. Geoscience Australia, Canberra, Australia, 2010.

[14] P. Speare, M. Cappo, M. Rees, J. Brownlie, W. Oxley, Deeper Water Fish and Benthic Surveys in the Lord Howe Island Marine Park (Commonwealth Waters): February 2004, Australian Institute of Marine Science, Townsville, Australia, 2004.

[15] M. Linklater, An assessment of the geomorphology and benthic environments of the Lord Howe Island Shelf, Southwest Pacific Ocean, and implications for quaternary Sea level, Unpublished Environmental Science Honours Thesis, School of Earth and Environmental Sciences, University of Wollongong, Australia, 2009.

[16] C.D. Woodroffe, Coasts: Form, Process and Evolution, Cambridge University Press, Cambridge, UK, 2002.

26 Geomorphology of Reef Fish Spawning Aggregations in Belize and the Cayman Islands (Caribbean)

William D. Heyman[1], Shinichi Kobara[2]

[1]Department of Geography, Texas A&M University, College Station, TX, USA, [2]Department of Oceanography, Texas A&M University, College Station, TX, USA

Abstract

Commercially important reef fish such as grouper and snapper use specific geomorphologic locations as spawning grounds throughout the wider Caribbean. In the Cayman Islands and Belize, fish choose shelf-edge reef promontories for spawning, and share these areas with other species. These multispecies spawning aggregation sites are generally located near the inflection points of convex-shaped reefs, in 20–40 m water depth, adjacent to sharp shelf edges where water depth drops to several hundred meters. Reef geomorphology may be the key determinant for the selection of reef fish breeding habitat. This paper presents the three-dimensional structure of four representative spawning aggregation sites in the Cayman Islands and Belize. We are evaluating the hypothesis that similar geomorphologic patterns exist at other spawning sites in the Caribbean.

Key Words: spawning aggregation, conservation, grouper, snapper, Caribbean, Belize, Cayman Islands, coral reef

Introduction

Caye Glory and Gladden Spit in Belize, and Grand Cayman East (GCE) and Little Cayman West (LCW) in the Cayman Islands, have a common geomorphologic signature and all harbor important breeding grounds for commercially important reef fishes (Figure 26.1). Most commercially important reef fishes, including many species of grouper and snapper, travel relatively long distances over the course of days or weeks to aggregate and spawn at very specific times and in very specific places. Locations of known sites were compiled using published literature and new field surveys and indicate commonality in the underlying geomorphology and species seasonality of reef fish spawning aggregation (FSA) sites throughout the wider Caribbean.

Seafloor Geomorphology as Benthic Habitat. DOI: 10.1016/B978-0-12-385140-6.00026-8

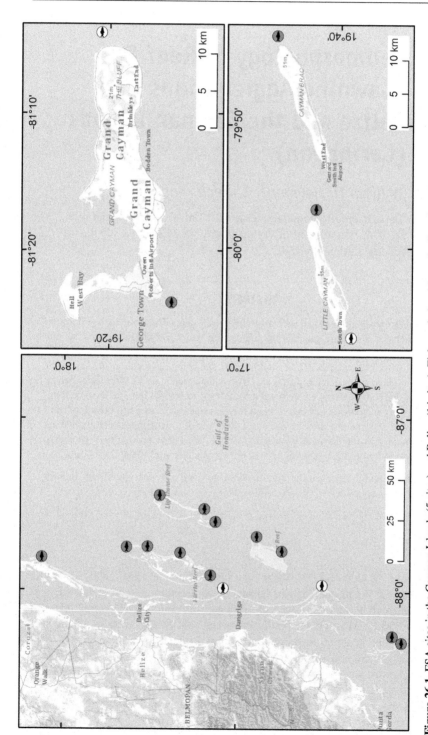

Figure 26.1 FSA sites in the Cayman Islands (5 sites) and Belize (14 sites). Fish symbols without shading show the focal sites in this study: Grand Cayman East (GCE), Little Cayman West (LCW), Cay Glory or Emily (EML), and Gladden Spit (GS). Background map source: World Topo Map by ESRI, Redlands, CA.

GCE and LCW are about $6.5\,km^2$ and $3.4\,km^2$, respectively. They both include slowly sloping shelves that extend from the shore for 0.7 km to 30 and 45 m depth, respectively and then drop abruptly into depths of around 370 and 480 m (Figure 26.2). The Cayman Islands (Grand Cayman, Little Cayman, and Cayman Brac) are low-lying carbonate islands with narrow shelves (<1 km) and fringing reefs [1,2]. The Cayman Islands rest on a major submarine ridge, 6 km north of the Cayman Trench, which plunges to depths of 6,000 m. Tucker et al. [3] described five historical Nassau grouper aggregation sites in the Cayman Islands. All were located at the shelf edges of the fringing reefs of the three islands, and two of these (GCE and LCW) are focal points for this study.

Caye Glory (locally called Emily) and Gladden Spit are about 3 and $7\,km^2$, respectively, and are both located along the main spine of the Belize Barrier Reef (Figure 26.1). The geomorphology of these sites is similar to those described above—reef promontories that jut offshore to abrupt and steep shelf edges in 35 m water depth, but that drop to deep waters that ultimately plummet into the Cayman Trench, reaching 4,000 m within 10 km of the barrier reef (Figures 26.1 and 26.2). The main difference is that the Belize sites are along an offshore barrier reef, while the Cayman sites form a fringing reef along the mainland shelf. Emily was known as the most productive grouper fishing site in the country [4,5]. Gladden Spit has served as a regionally important snapper fishery since the 1920s [6].

All of the four study sites were heavily fished multispecies FSA sites. They are all presently protected for at least some portion of the year through their inclusion in marine reserves or seasonally protected areas. In spite of some damage still being caused from line fishing and anchoring, they all now remain in relatively natural states.

The oceanography at all four of these reef promontories is highly complex but little studied. The promontory sites share similar morphologies and thus likely also share physical oceanographic conditions. All four sites are subject to the interaction between far-field ocean currents moving through deep trench waters and the sharply bending reef that juts seaward with nearly vertical walls. These conditions create complex flow dynamics that are different than adjacent areas. Gladden Spit has shown twice the current speed and thrice the variability of adjacent nonpromontory sites [7]. The dynamics of this site are under further study but early results indicate that the promontory may create eddies that retain larvae near these sites after spawning [8].

Geomorphic Features and Habitats

Each of the four study sites are centered on reef promontories that are formed at the intersections between a gradually sloping shelf (5–10° grade), a sharply sloping wall (>45° grade) that plummets into deep water (i.e., >250 m), and a convex bend in the reef (in this case, ranging approximately 70–120°). Geomorphic features were described using terminology from the International Hydrographic Organization [9] and supplemented with standard terminology developed for coral reefs of the Caribbean [10].

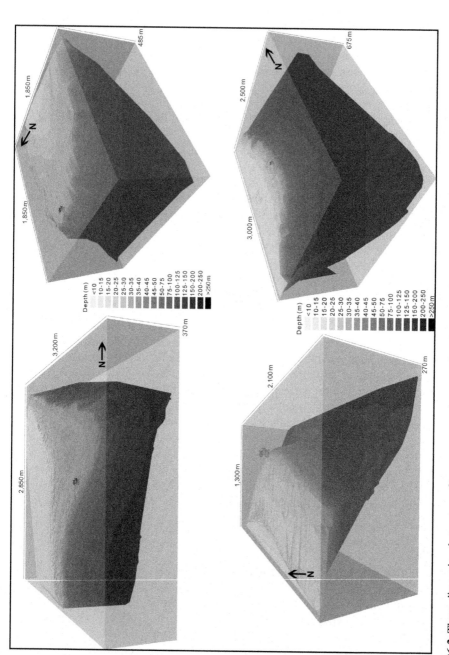

Figure 26.2 Three-dimensional structure of two representative spawning sites in the Cayman Islands and two in Belize. GCE (upper left) and LCW (upper right), Emily (Cay Glory) (bottom left), and Gladden Spit (bottom right).

In the Cayman Islands, benthic habitats extend from the shore to deep water through gradually sloping (3–5° slope) areas of sand, some interspersed with patch reefs in 5–15 m water depth, out to more developed fore-reef communities in 15–20 m water depth. Between 0.04 and 0.7 km from shore (i.e., nearer to the shelf edge) channels extend perpendicular to the shoreline and gradually converge to spur and groove coral reef structure with broad (10-15m) ridges. The ridges are largely calcium carbonate built by deposition of reef-building corals and have medium relief (3–5 m). The benthic community on the ridges is dominated by hard and soft corals, sponges, and fleshy and calcareous algae, typically with 30–40% live coral cover interspersed with channels filled with sand [11,12].

At Gladden Spit in Belize, benthic habitat can be broken into six zones going seaward from the reef crest: reef crest, inner fore-reef hardground, inner fore-reef sand dominated, outer fore-reef spur and groove, fore-reef slope spur and groove, and the steep fore-reef slope. The shallow, inner fore-reef hardground is nearly flat and extends for about 200 m from 30 m seaward of the reef crest and about 3 m water depth. The hard benthos has less than 1 m relief and is dominated by low, rounded coral heads of *Diploria* sp. and *Montastrea anularis*, and interspersed with gorgonians. The hardground gives way to sand-dominated inner fore-reef approximately 200 m from the reef crest in about 10 m water depth and gradually drops to about 15 m water depth approximately 400 m from the reef crest. The reef structure close to the spawning site has low-relief (1 m) coral and gorgonian cover on a gently sloping forereef with 5–10% coral cover. Poorly developed spur and groove morphology begins 400 m north of the promontory and continues to the north [7].

All sites were mapped with a singlebeam acoustic sonar system. Data were collected with a Lowrance® echosounder and AirMar® transducer and used to create bathymetric maps around all known FSA sites using triangulated irregular network (TIN) interpolation in ArcGIS (following techniques in Refs. [13] and [14]) (Figure 26.2). The distances between transect lines vary between 15 and 250 m, though most are between 30 and 50 m. This variation was purposeful. Less data were collected on flat areas, more at shelf edges. Coordinates of known Nassau grouper FSA sites in the Cayman Islands were obtained from the Cayman Islands Department of Environment. Benthic habitat data were only collected with sparse diver surveys that lead to generalized descriptions of the sites. Instead, our purpose was to look at the relationship between gross geomorphology and the presence of multispecies reef fish FSAs.

Biological Communities

Benthic communities were described qualitatively from diver observations in the earlier section. The main focus of this study was the relationship between gross geomorphology and the location of FSA sites. Coordinates of FSA sites in Belize were obtained from the Belize Spawning Aggregations Working Committee, which conducts regular underwater visual surveys of FSA sites throughout the country [15]. The positions were recorded using a handheld GPS from a boat following scuba divers. The center point of the FSA sites was overlayed on bathymetry data (Figure 26.2).

In the Cayman Islands, Kobara and Heyman [16] observed that all five Nassau grouper FSA sites occurred near (<200 m) from shelf-edge dropoffs into deep water, at reef promontories, in 25–45 m water depth. Sites were not preferentially windward or associated with reef channels. These five reef promontories also served as FSA sites for other species.

In Belize, similar to the Cayman Islands, all Nassau grouper FSA sites occurred at shelf edges near the tips of reef promontories immediately adjacent to deep waters [15]. Several thousands of fishes from various species aggregate and spawn en masse at this and other reef promontory locations (Figure 26.3). Moreover, these 12 sites are shared with other grouper and snapper spawning aggregations and are therefore considered multispecies FSAs [15].

Twelve FSA sites in Belize and four in the Cayman Islands are documented multispecies spawning sites. The other sites may be multispecies sites but there has not been sufficient study at the predicted times. Nine species were documented to spawn at Caye Glory ([15]; Table 26.1). Seventeen species of reef fish spawn at Gladden Spit, with Cubera snapper, mutton snapper, and dog snapper most prominent in abundance, biomass, and egg production (Table 26.1). The locations of the FSA of each species were mapped in more detail but all were located within a 6 ha area near the reef promontory tip [7].

Most of the 36 documented and verified FSA sites in the Caribbean were found near shelf edges (29 sites or 81%) and dropoffs (23 sites or 64%) (Figure 26.4) [17]. Yet a comprehensive analysis of the geomorphology of all known spawning locations in the wider Caribbean had not been available before the work of Kobara [17]. Findings from that study illustrate that most known FSA sites occur in 25–45 m water depth along shelf edges of convex-shaped reef promontories jutting into deep water (Kobara et al., in preparation). Kobara and Heyman [15] demonstrated that the location of FSA sites could be predicted based on gross geomorphology. Understanding geomorphology of FSAs might provide a fishery-independent way to locate potential FSA sites in other locations.

Surrogacy

Different than many other case studies in this atlas, this study addresses surrogacy from the standpoint of transient use of the area by spawning fishes, rather than by measures of benthic composition. Analyses of reef geomorphology may serve to help guide marine spatial planning efforts, particularly as proxies for critical life habitat of key species. This case demonstrates that benthic cover may be less important than gross geomorphology as a surrogate for the location of multispecies spawning aggregation sites.

There have been at least 84 reef FSAs found in the Caribbean [18] (Figure 26.4), most of them heavily overfished [19]. Recent studies suggest that geomorphology of marine environments may dictate the location of critical life habitat for a variety of marine species [20].

Figure 26.3 Spawning event of Cubera snapper at Gladden Spit, Belize. Several thousands of fishes from various species aggregate and spawn en masse at this and other reef promontory locations.
Source: Photo Courtesy of Douglas David Seifert.

Table 26.1 Spawning Aggregation Species at Four Representative Sites in the Cayman Islands and Belize

Site	Spawning Aggregation Species		
Grand Cayman East (GCE)	*Epinephelus striatus* *L. cyanopterus*	*Mycteroperca tigris*	*Lutjanus analis*
Little Cayman West (LCW)	*Epinephelus striatus* *M. venenosa* *Caranx latus* *C. bartholomaei*	*Mycteroperca tigris* *Lutjanus analis* *C. ruber* *Decapterus macarellus*	*M. bonaci* *L. jocu* *C. lugbris* *Kyphosus incisor*
Emily (EML) Caye Glory	*Epinephelus striatus* *Lutjanus jocu* *Lactophrys triqueter*	*Mycteroperca bonaci* *Calamus bajonado* *L. trigonus*	*M. venenosa* *C. calamus* *Haemulon album*
Gladden Spit (GS)	*Epinephelus striatus* *M. venenosa* *L. jocu* *Caranx hippos* *C. bartholomaei* *Scomberomorus cavalla* *Haemulon album* *Calamus bajonado*	*Mycteroperca bonaci* *Lutjanus analis* *Ocyurus chrysurus* *C. ruber* *Trachinotus falcatus* *Chaetodipterus faber* *Canthidermis sufflamen* *Lactophrys trigonus*	*M. tigris* *L. cyanopterus* *Seriola dumerili* *C. latus* *Decapterus macarellus* *Lachnolaimus maximus* *Xanthichthys ringens* *L. triqueter*

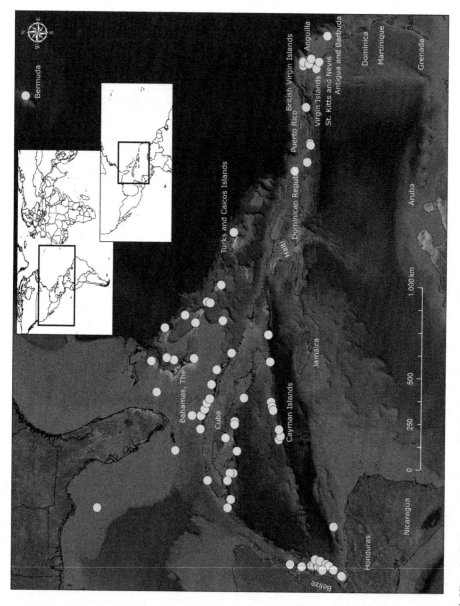

Figure 26.4 Historically known transient reef FSA sites in the wider Caribbean.

References

[1] P.G. Bush, Grand Cayman, British West Indies, in: B. Kjerfve (Ed.), CARICOMP—Caribbean Coral Reef, Seagrass and Mangrove Sites, UNESCO, Paris, 1998, pp. 347. Coastal region and small island papers 3.

[2] P.L. Colin, D.Y. Shapiro, D. Weiler, Aspects of the reproduction of two groupers, *Epinephelus guttatus* and *E. striatus* in the West Indies, Bull. Mar. Sci. 40 (1987) 220–230.

[3] J.W. Tucker, P.G. Bush, S.T. Slaybaugh, Reproductive patterns of Cayman Islands Nassau Grouper (*Epinephelus striatus*) populations, Bull. Mar. Sci. 52 (1993) 961–969.

[4] A.K. Craig, The grouper fishery of Cay Glory, British Honduras, Ann. Assoc. Am. Geogr. 59 (1969) 252–263.

[5] J. Carter, D. Perrine, A spawning aggregation of dog snapper, *Lutjanus jocu* (Pisces, Lutjanidae) in Belize, Central America, Bull. Mar. Sci. 55 (1994) 228–234.

[6] W.D. Heyman, B. Kjerfve, R.T. Graham, K.L. Rhodes, L. Garbutt, Spawning aggregations of Cubera snapper, *Lutjanus cyanopterus* (Cuvier), on the Belize Barrier reef over a six-year period, J. Fish Biol. 67 (2005) 83–101.

[7] W.D. Heyman, B. Kjerfve, Characterization of transient multi-species reef fish spawning aggregations at Gladden spit, Belize, Bull. Mar. Sci. 83 (2008) 531–551.

[8] T. Ezer, W.D. Heyman, C. Houser, B. Kjerfve, Numerical simulations and observations of high-frequency flow variability at a reef fish spawning aggregation site in the Caribbean. Ocean Dynamics, 60 (5) (2010) 1307–1318.

[9] IHO, Standardization of Undersea Feature Names: Guidelines Proposal form Terminology, fourth edition, International Hydrographic Organisation and International Oceanographic Commission Bathymetric Publication No. 6.

[10] P.J. Mumby, A.R. Harborne, Development of a systematic classification scheme of marine habitats to facilitate regional management and mapping of Caribbean coral reefs, Biol. Conserv. 88 (2) (1999) 155–163.

[11] P. Blanchon, Architectural variation in submerged shelf-edge reefs: the hurricane-control hypothesis, in: Proceedings of the 8th International Coral Reefs Symposium, Panama, 1 (1997) 547–554.

[12] L. Whaylen, C.V. Pattengill-Semmens, B.X. Semmens, P.G. Bush, M.R. Boardman, Observations of a Nassau grouper, *Epinephelus striatus*, spawning aggregation site in Little Cayman, Cayman Islands, including multi-species spawning information, Environ. Biol. Fish. 70 (2004) 305–313.

[13] J.B. Ecochard, W.D. Heyman, N. Requena, E. Cuevas, F.B. Biasi, Case Study: Mapping Half Moon Caye's Reef Using the Adaptive Bathymetric System, The Nature Conservancy, Arlington, VA, 2003. www.reefresilience.org.

[14] W.D. Heyman, J.L.B. Ecochard, F.B. Biasi, Low-cost bathymetric mapping for tropical marine conservation: a focus on reef fish spawning aggregation sites, Mar. Geodes. 30 (2007) 37–50.

[15] S. Kobara, W.D. Heyman, Sea bottom geomorphology of multi-species spawning aggregation sites in Belize, Mar. Ecol. Prog. Ser. 405 (2010) 243–254.

[16] Kobara, S., and Heyman W.D. Geomorphometric patterns of Nassau grouper (*Epinephelus striatus*) spawning aggregation sites in the Cayman Islands. Mar. Geodes. 31 (2008) 231–245.

[17] SCRFA Global Spawning Aggregations Database. http://www.scrfa.org/database/ (accessed 10.01.2010).

[18] S. Kobara, Regional analysis of seafloor characteristics at reef fish spawning aggregation sites in the Caribbean. Doctoral Dissertation. Texas A&M University, 2009.

[19] Y. Sadovy de Mitcheson, A. Cornish, M. Domeir, P.L. Colin, M. Russell, K.C.A. Lindeman, Global baseline for spawning aggregations of reef fishes, Conserv. Biol. 22 (5) (2008) 1233–1244.

[20] D. Wright, W.D. Heyman, Marine and coastal GIS for geomorphology, habitat mapping and marine reserves, Mar. Geodes. 31 (4) (2008) 1–8.

27 Submerged Reefs and Aeolian Dunes as Inherited Habitats, Point Cloates, Carnarvon Shelf, Western Australia

Scott L. Nichol[1], Tara J. Anderson[1], Chris Battershill[2], Brendan P. Brooke[1]

[1]Marine and Coastal Environment Group, Geoscience Australia, Canberra, ACT, Australia, [2]Australian Institute of Marine Science, Townsville, QLD, Australia

Abstract

The Carnarvon shelf at Point Cloates, Western Australia, is characterized by a series of prominent ridges and hundreds of mounds that provide hardground habitat for coral and sponge gardens. The largest ridge is 20m high, extends 15km alongshore in 60m water depth, and is interpreted as a drowned fringing reef. To landward, smaller ridges up to 1.5km long and 16m high are aligned to the north–northeast and are interpreted as relict aeolian dunes. Mounds are less than 5m high and may also have a subaerial origin. In contrast, the surrounding seafloor is sandy, with relatively low densities of epibenthic organisms. The dune ridges are estimated to be Late Pleistocene in age and their preservation is attributed to cementation of calcareous sands to form aeolianite, prior to the postglacial marine transgression. On the outer shelf, sponges grow on isolated low-profile ridges at ~85 and 105m depth and are also interpreted as partially preserved relict shorelines.

Key Words: continental shelf, submerged shoreline, epibenthic, bathymetry, multibeam sonar, Ningaloo Reef

Introduction

The Carnarvon shelf is the northern sector of the Dirk Hartog Shelf that extends 280km along the central western margin of Australia [1] (Figure 27.1). Distinctive features of this subtropical carbonate shelf, also known as the Carnarvon Ramp [2], are the marked decrease in shelf width northwards, from 33km at Cape Cuvier to 7km at North West Cape, and an indistinct shelf edge. The Carnarvon sector of the shelf incorporates Ningaloo Reef, which forms a 270-km long fringing reef located 0.2–5km from the coastline. At Point Cloates, the shelf seaward of Ningaloo Reef is

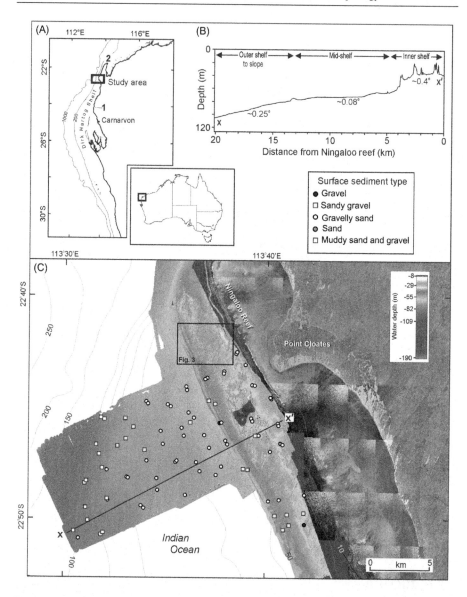

Figure 27.1 (A) Location of study area and localities mentioned in text; 1: Cape Cuvier, 2: North West Cape; (B) representative profile of the shelf to upper slope at Point Cloates; (C) multibeam sonar bathymetric map of the continental shelf at Point Cloates based on a 3-m grid, showing sediment type from grab samples.
Aerial photo source: Landgate, Western Australian Land Information Authority.

approximately 20 km wide and divides into a low-gradient inner shelf (0.4°), nearly flat mid-shelf (0.08°), and slightly steeper outer shelf (0.25°). The shelf edge is not well defined, with the outer-shelf grading to the continental slope (1.3°) beyond approximately 150 m water depth. The regional oceanography is influenced by the warm, southward-flowing Leeuwin Current that impinges onto the shelf and the northward-flowing Ningaloo Current that forms a counterclockwise circulation across the shelf at Point Cloates [3]. Swell waves arrive predominantly from the south to southwest, with a northeast component in summer. The region is at the southern limit of cyclone impact and on average experiences a cyclonic storm once a decade [4]. Tides on the open coast are microtidal, ranging from 0.6 (mean neap range) to 1.8 m (mean spring range).

Ningaloo Reef is within a World Heritage-listed area, recognized for the near-pristine condition of the ecosystem and its "outstanding natural heritage significance" [5]. The region has negligible human impact from land-based activities but is increasingly the focus for tourism. In August–September 2008, the shelf seaward of Ningaloo Reef at Point Cloates was mapped and sampled using the Australian Institute of Marine Science research vessel *Solander*. Bathymetric and backscatter data were collected using a Kongsberg EM3002(D) 300-kHz multibeam sonar system over a survey area of 281 km^2, together with underwater video footage at 42 representative sites and 89 sediment grab samples at 39 stations [6] (Figure 27.1).

Oceanographic measurements from two instrument moorings deployed in 32 and 54 m water depth offshore from Point Cloates recorded significant wave heights that ranged from 0.5 to 4.5 m over a 34-day period [6]. Tidal currents flooded to the south and ebbed to the north. Nontidal currents at the deeper water mooring maintained a south to west–southwest flow direction, consistent with the Leeuwin Current. Maximum near-bed velocities ranged from 0.27 to 0.29 m/s but were maintained for only 2–3 h. The threshold velocity for sand transport (~0.2 m/s) was achieved for only approximately 1% of the deployment time.

Geomorphic Features and Habitats

Geomorphic features were mapped in ArcGIS based on a bathymetric grid and a slope map at 3 m spatial resolution. Maps and perspective views of geomorphic features were generated in IVS Fledermaus version 7.2. Features were mapped by on-screen digitizing of polygons, with feature boundaries defined by breaks in slope and/or a change in seabed roughness as indicated by the multibeam bathymetry and backscatter (e.g., edge of a bedform field). At this scale of mapping, the standard nomenclature for undersea feature names [7] does not readily apply, so appropriate terms from the literature are adopted. Six types of geomorphic feature are mapped for the Point Cloates shelf, with an additional category to represent spatially complex areas of two feature types, as follows:

1. *Ridge*: The combined area of ridges on the Point Cloates shelf is approximately 10 km^2, with most ridges located on the inner shelf in water depths of 20–60 m (Figure 27.2; Table 27.1). The largest ridge is a shore-parallel feature that extends 15 km alongshore in 60 m water depth and is up to 20 m high and 200 m wide (Figure 27.3). On the basis of its continuity and form, this ridge is interpreted as a drowned fringing reef that likely formed during the Late Pleistocene to early Holocene. As such, it is the predecessor to the modern Ningaloo Reef. Landward of this ridge, a series of smaller ridges up to 16 m high are uniformly aligned to the north–northeast, with some converging at their landward end. These

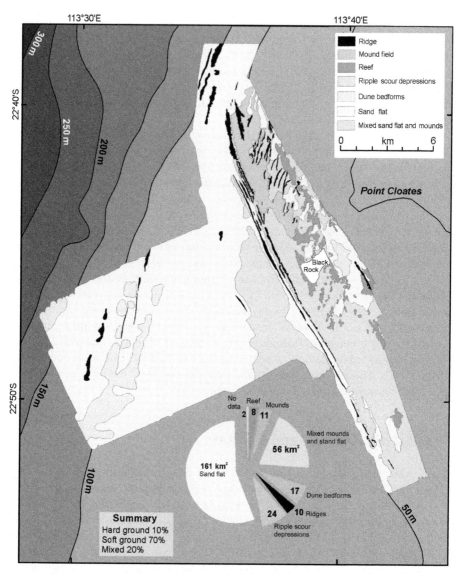

Figure 27.2 Geomorphic features of the Carnarvon shelf interpreted from multibeam bathymetry.

smaller ridges are interpreted as relict aeolian dunes, based on their shape and orientation, which is consistent with the onshore parabolic dune field at Point Cloates. Preservation of these dunes on the shelf is attributed to lithification of dune sands to form aeolianite, prior to the postglacial marine transgression. Ridges are also mapped on the outer shelf at Point Cloates, where they form low-profile (<3 m high) features that are 200 m to 4 km long with a combined area of approximately 3 km². These ridges follow a general parabathic trend in 85–105 m water depth and may also be relict reef or shoreline features.

Table 27.1 Areas of Geomorphic Features Mapped on the Carnarvon Shelf

Geomorphic Feature	Area (km^2)	Percentage of Mapped Area
Ridge	9.46	3.2
Mound	10.74	3.7
Reef	7.96	2.8
Sand flat	160.24	55.6
Dune bedforms	17.24	6.0
Rippled scour depression	24.17	8.4
Mixed mound and sand flat	56.36	19.6
Black rock reef shoal (not mapped)	1.96	0.7
Total	288.13	100

2. *Mound*: Landward of the large shore-parallel ridge, a dense field of low mounds covers an area of approximately 11 km^2 on the inner shelf (Figure 27.3; Table 27.1). The mounds are 2–5 m high and tens of meters in diameter, with isolated mounds scattered in no particular pattern. Some mounds are interconnected and broadly aligned in a shore-parallel direction. Others are oriented in the same north–northeast direction to the relict dune ridges. Together, these mounds are interpreted as relict coastal deposits, with the interconnected shore-parallel mounds preserving evidence of former shorelines at lower sea level and the shore-oblique mounds likely related to the aeolian dune system.

3. *Reef*: The combined area of reef on the inner shelf at Point Cloates is approximately 8 km^2, excluding an unmapped area of 2 km^2 that surrounds Black Rock, which rises to sea level (Figure 27.2; Table 27.1). Reefs are distinguished from ridges on the basis that they are irregular in outline and generally wider, ranging up to 500 m across. Most reefs rise to within 15 m below sea level from basal water depths of 30–35 m and are characterized by relatively smooth surfaces. Some of the larger reefs (up to 1 km^2) are shore-parallel and abut the landward ends of relict dune ridges, which sit lower than the reefs. This suggests that the reefs existed when the dunes were active during lower sea levels and formed a topographic barrier to further dune extension. On this basis, it is possible that these areas of irregular, smooth reef are remnants of Last Interglacial reef that occur beneath the Holocene reef and lagoon of northern Ningaloo Reef and as outcrop onshore [8].

4. *Sand flat*: Sand flat forms the most extensive geomorphic feature in the mapped area, covering an area of approximately 160 km^2 (Figure 27.2; Table 27.1). The mid- and outer-shelf are dominated by sand flat (52% of mapped area), where it forms a gently sloping surface that steepens toward the shelf break. On the inner shelf, sand flats occur as localized areas between reefs, with the largest area (7.6 km^2) located seaward of a tidal passage into Ningaloo lagoon. Sediment samples from the inner- and mid-shelf sand flat mostly comprise gravelly sand with some sandy gravel, whereas outer-shelf samples are mostly muddy sand and gravel.

5. *Dune bedforms*: The only subaqueous dune bedforms mapped at the scale of the 3-m grid used here are sand lobes and sand waves, with a combined area of approximately 17 km^2 (Figure 27.2; Table 27.1). Sand lobes only occur on the inner shelf in a small area (1.6 km^2) seaward of the tidal passage into Ningaloo lagoon, where they form low (~1 m) lobes that are oriented to the northwest and are 200–800 m long. Sand waves occur on the outer shelf in water depths of 75–170 m. The most extensive sand wave fields are in the 75–95 m depth

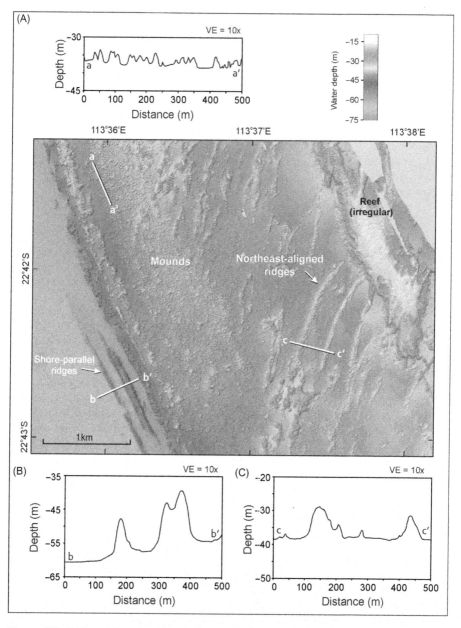

Figure 27.3 Bathymetric map of the inner-shelf offshore from Point Cloates, with representative profiles across: (A) mounds; (B) shore-parallel ridges; and (C) northeast-oriented ridges.

interval, where they cover approximately $14\,km^2$ and include straight-crested, symmetrical bedforms with wavelengths of 60–70 m and heights up to 1.2 m.

6. *Rippled scour depressions*: On the mid-shelf, an area of $24\,km^2$ seaward of the large shore-parallel ridge is interpreted as a rippled scour depression feature (Figure 27.2; Table 27.1).

Seabed depressions are up to 0.5 m deep and are of varied extent (10^3–10^5 m^2) with surface sediments of gravelly coarse sand. The areas between depressions are sandy and level with the nearby flat seabed. Based on the concentration of coarse sediments within the depressions, they are interpreted as the product of current scour, possibly associated with the northward-flowing Ningaloo current.

7. *Mixed mound and sand flat*: An area covering approximately 56 km^2 on the inner shelf is characterized by mixed mounds and sand flat, with the mounds in lower density and number than in the northern part of the mapped area (Figure 27.2). The mounds in this area have similar dimensions to those in the north, with some smaller interconnected mounds forming a shore-parallel trend. However, this pattern becomes less distinct with distance south.

8. *Sediments*: Sediment samples were collected from sand flats and sand waves on the inner-, mid-, and outer-shelf, plus the rippled scour depression area on the mid-shelf. Samples were analyzed for percentage mud, sand, gravel, and calcium carbonate content at Geoscience Australia using standard laboratory techniques [6]. Of the 89 samples collected, 57 are classified as gravelly sand and 18 as sandy gravel, with the remaining samples comprising gravel ($n = 2$), sand ($n = 1$), and mixed mud, sand and gravel ($n = 11$) (Figure 27.1). Mean grain size of the sand and mud fractions ranges from 87 (very fine sand) to 995 μm (coarse sand). Sorting ranges from moderate ($n = 23$) to poor ($n = 44$) and very poor ($n = 22$). Bulk calcium carbonate content ranges from 81% to 100%, with calcareous material including fragments of coralline algae, benthic forams, bryozoans, and other skeletal material. Muddy sands and gravels are limited to the outer shelf in water depths of 90–100 m; across the inner- and mid-shelf there is no clear spatial pattern in sediment texture.

Biological Communities

Assessments of shelf benthos at Point Cloates are based on underwater video and photography data collected along 500-m transects at 42 locations. Percent cover of substrata (rock, boulders, cobbles, rubble, gravel, sand, and mud), benthic categories (hard corals, sponges, octocorals, rhodoliths, red algae), and key taxa (e.g., Kebab sponge, *Caulospongia* sp.) were recorded for the Frame of View (FoV) at 15 s intervals along each transect. Sponges were classified into three size classes: small (<15 cm in height), medium (15–49 cm), and large (>50 cm), and further differentiated into growth form (globular, lamellate, and branching). Relief was defined as either a soft-sediment "bedform" such as hummocks, ripples, or sand waves, or by the vertical "relief" of hard substratum, in which relief classes ranged from flat (0 m), low (<1 m), moderate (1–3 m), to high relief (>3 m), or rock walls (high relief with >80° incline) following established protocols [9].

Habitat complexity, epibenthic diversity, and percent cover decreased with distance offshore (Figure 27.4). The inner-shelf supported the most complex habitats, characterized by highly rugose ridges and mounds that supported the highest epibenthic diversity and percent cover of the study area (89% combined cover) (Figure 27.5). Epifauna (49% of total cover) in these habitats comprised mixed sessile invertebrates (20%), hard corals (20%), and sponges (9%), while epiflora was mostly coralline paint (27%), with some green and red algae (~8 and ~5%, respectively) (Figure 27.6). On high-relief features such as large ridges, the cover of hard corals was as high as 70%. In contrast, the lower relief sandy habitats adjacent to ridges and mounds supported

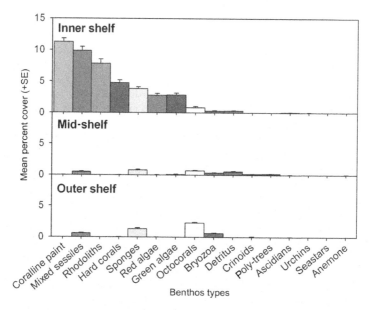

Figure 27.4 Abundance of epifauna and epiflora on the Carnarvon shelf at Point Cloates, measured as percent cover of the seabed as observed for 500-m long tow video transects.

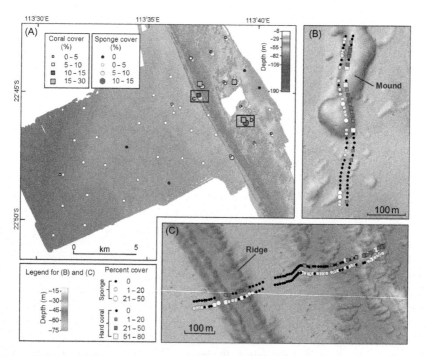

Figure 27.5 Distribution of sponges and hard corals offshore from Point Cloates, shown as: (A) mean percent cover per video transect; (B) and (C) mean percent cover for each 15-s video frame along video transects that include a 5-m high mound and a 20-m high ridge, respectively.

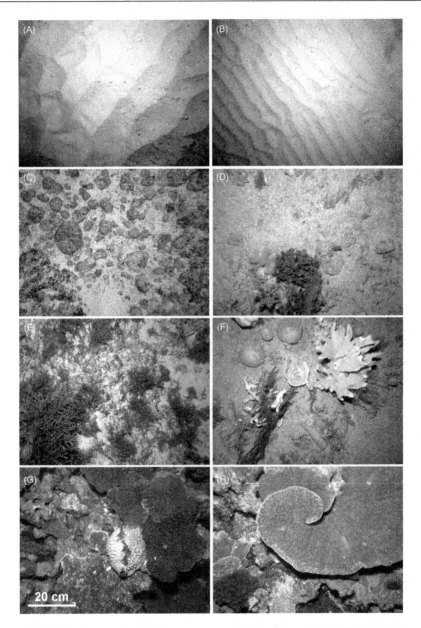

Figure 27.6 Still photographs of habitat types on the Point Cloates shelf. (A) Mid-shelf sand ripples; (B) inner-shelf sand ripples; (C) inner-shelf rhodolith bed over coarse sand; (D) biogenic rubble from mid-shelf, with a diverse assembly of sessile invertebrates including foliose coral, bryozoans, hydroids, and soft corals; (E) inner-shelf reef with a cover of filamentous red algae and sessile invertebrates; (F) low-lying biogenic reef from outer shelf, with a diverse assembly of sessile invertebrates including sponges, gorgonians, and crinoids; (G) high-relief reef covered in foliose corals and sponges, inner shelf; (H) high-relief reef covered in foliose corals, coralline algae, and bryozoans, seaward margin of Ningaloo fringing reef. Scale bar is approximate and applies to all photos.

fewer epifauna and epiflora (57% combined cover). Here, epiflora dominated (35.5% of total cover) with rhodoliths (16%), coralline paint (13%), and low amounts of red and green algae (~4 and ~3%, respectively). Epifauna were also common (22% of total cover), comprising mostly mixed sessile invertebrates (14%), sponges (~5%), and low numbers of hard corals (2.5%).

The mid-shelf was characterized by mostly continuous sediments of either sand waves or sand rippled depressions, with only localized deposits of low-lying biogenic rubble. The area of rippled scour depressions supported only a few isolated epifauna and epiflora (4.4% combined cover). Epifauna (2.5% of total cover) included rare occurrences of sponges, octocorals, sessile invertebrates, bryozoa, and associated crinoids, along with sparsely distributed polychaete trees (Eunicids). Epiflora were also rare (<2% of total cover) with negligible amounts of detritus, rhodoliths, coralline paint, and red and green algae.

Sand waves on the mid-shelf supported few isolated epifauna and epiflora (~3% combined cover). Again, epiflora (<2% of total cover) comprised negligible amounts of coralline paint, rhodoliths, detritus, and red algae. Most of these floras were observed on rubble located in troughs of sand waves. Epifauna were equally rare (<2% combined cover) and comprised sessile invertebrates, sponges, octocorals, and bryozoa.

The outer shelf was characterized by mainly soft sediments, with fewer bedforms observed than on the mid-shelf and more low-lying biogenic rubble and reef. Epibenthos were scarce on the outer shelf and are therefore not quantified.

Flat-relief habitats were observed in video in all three shelf zones and were characterized by gravel (62% of habitat type), sand (25%), rock outcrop (9%), and cobbles (4%). In combination, these flat habitats were mostly bare with less than 5% epibenthos cover. On flat sedimentary areas, observed taxa included octocorals (2.5%), sponges (1%), and bryozoa (<1%). However, discrete outcrops of flat rock supported higher levels of epibenthos (41% cover), that included octocorals (19%), small sponges (14%), and green algae (8%). In some sand flat areas, evidence for successional extension of sponge communities was observed. In particular, localized scouring of the bed around a sponge garden was noted to provide opportunity for recruitment via asexually generated propagules. Evidence to support this observation was found in benthic sled samples from these areas that included numerous sponge buds and fragments of sponges [6]. Overall, sponges observed on the Point Cloates shelf were classed as small (61%), with few medium-sized (29%) and less large-sized (10%) sponges.

Surrogacy

At the time of writing, no statistical analyses have been carried out on this data set to examine fine-scale relationships between physical surrogates (sediment, seabed morphometrics, backscatter, exposure) and benthos, but are planned as part of the ongoing research undertaken by the Marine Biodiversity Hub. Nonetheless, the observations presented here show that a strong control on the spatial distribution of key benthic habitats and communities is produced by the occurrence of relict reefs and coastal landforms. These features appear to create the substrate and local oceanographic conditions that link to the ecological processes that determine the distribution of benthic biota on the Carnarvon Shelf.

Acknowledgments

This work has been partly funded through the Commonwealth Environment Research Facilities (CERF) program, an Australian Government initiative supporting public-good research. The CERF Marine Biodiversity Hub is a collaborative partnership between the University of Tasmania, CSIRO Wealth from Oceans Flagship, Geoscience Australia, Australian Institute of Marine Science, and Museum Victoria. Multibeam data was acquired and processed by Cameron Buchanan, Justy Siwabessy, and Mike Sexton (Geoscience Australia). We thank the Master and crew of RV *Solander*, plus Ian Atkinson and Stephen Hodgkin (Geoscience Australia) for survey support. Nic Bax and two anonymous reviewers are thanked for their constructive reviews of the manuscript. This chapter is published with permission of the Chief Executive Officer, Geoscience Australia.

References

[1] M.A. Carrigy, R.W. Fairbridge, Recent sedimentation, physiography and structure of the continental shelves of Western Australia, J. Roy. Soc. West. Aust. 38 (1954) 65–95.

[2] N.P. James, L.B. Collins, Y. Bone, P. Hallock, Subtropical carbonates in a temperate realm: modern sediments on the southwest Australian shelf, J. Sediment. Res. 69 (1999) 1297–1321.

[3] M. Woo, C. Pattiaratchi, W. Schroeder, Dynamics of the Ningaloo current off point cloates, Western Australia, Mar. Freshwater Res. 57 (2006) 291–301.

[4] Bureau of Meteorology, North West Cape Project. www.cawcr.gov.au/bmrc/wefor/research/nw_cape_project.htm, 2010 (last accessed 14.04.10).

[5] UNSECO, Ningaloo Coast. http://whc.unesco.org/en/tentativelists/5379/, 2010 (last accessed 20.07.10).

[6] B. Brooke, S. Nichol, M. Hughes, M. McArthur, T. Anderson, R. Przeslawski, et al., Carnarvon Shelf Survey Post-Survey Report. Geoscience Australia Record 2009/02, Canberra, 2009, 90 pp.

[7] International Hydrographic Organization, Standardization of Undersea Feature Names, International Hydrographic Bureau, Monaco, 2001. Bathymetric Publication No. 6, 12 pp

[8] L.B. Collins, Z.R. Zhu, K.-H. Wyrwoll, A. Eisenhauer, Late Quaternary structure and development of the northern Ningaloo Reef, Australia, Sediment. Geol. 159 (2003) 81–94.

[9] T.J. Anderson, G.R. Cochrane, D.A. Roberts, H. Chezar, G. Hatcher, A rapid method to characterize seabed habitats and associated macro-organisms, in: B.J. Todd, H.G. Greene, (Eds.), Mapping the Seafloor for Habitat Characterization, Geological Association of Canada, Canada, 2008, pp. 71–79 (Special Paper 47) .

28 Seafloor Morphology and Coral Habitat Variability in a Volcanic Environment: Kaloko–Honokohau National Historical Park, Hawaii, USA

Ann E. Gibbs, Susan A. Cochran

US Geological Survey, Santa Cruz, CA, USA

Abstract

Multiple Holocene volcanic flows coalesce within Kaloko-Honokohau National Historical Park (KAHO) on the island of Hawaii to create a complex offshore morphology. The volcanic-dominated morphology includes flat to gently sloping volcanic benches, boulder fields, cliffs and ledges, pinnacles, ridges, arches, and steep shelf escarpments. Each of these environments provide distinct habitat zones for coral species, ranging from isolated heads of *Porites lobata* and *Pocillopora meandrina* to dense thickets of *Porites compressa*. In contrast to coral habitat elsewhere in the Hawaiian Islands, where coral typically populates relict carbonate platforms, coral cover in KAHO is typically only a thin veneer of live coral and rubble on exposed volcanic pavement. In only a few locations does coral or accreted carbonate reef obscure the underlying volcanic surface.

Key Words: coral habitat variability, Hawaii, Kaloko-Honokohau National Park, benthic habitats

Introduction

Kaloko–Honokohau National Historical Park (KAHO) is one of three National Park lands along the leeward, west, or Kona, coast of the island of Hawaii, USA. The park includes 596 acres (2.4 km^2) of submerged lands and marine resources within its official boundaries [1] (Figure 28.1). The offshore region of KAHO, part of the insular shelf of the island of Hawaii, comprises a volcanic embayment that extends nearly 3.5 km alongshore and varies in width between 120 and 875 m from the shoreline to the 40 m isobath, the limit of the high-resolution bathymetry [2] (Figure 28.2).

Seafloor Geomorphology as Benthic Habitat. DOI: 10.1016/B978-0-12-385140-6.00028-1

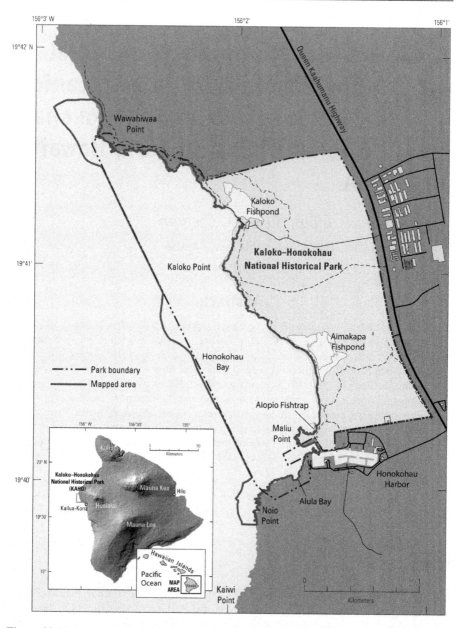

Figure 28.1 Location map showing the boundary of Kaloko–Honokohau National Historical Park, the mapped area, and surrounding geographical locations.

Marine resources located within KAHO include a diverse coral-reef community and habitat for many marine fauna, including green sea turtles and a variety of fish and invertebrates. Many submerged archeological, cultural, and popular recreational resources are also located within the park [1,3].

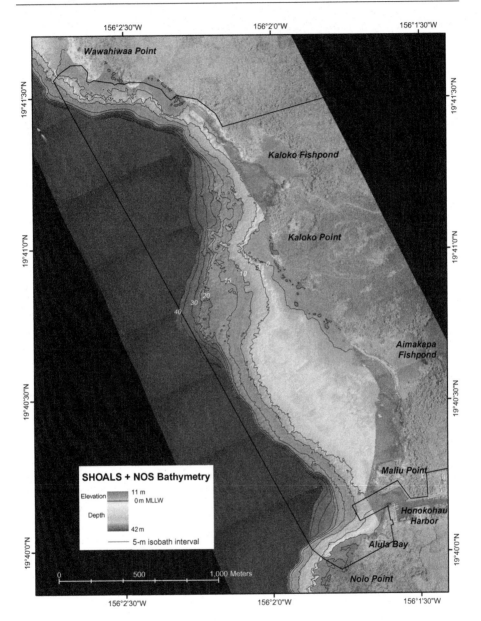

Figure 28.2 Bathymetric map of Kaloko–Honokohau.

KAHO is located adjacent to a moderately well-developed area of the Kona coast. The park is bordered on the south by the Honokohau small-boat harbor, which is the primary fueling, maintenance, and berthing facility on the central Kona coast, and on the north by a luxury residential/resort and golf course development currently (2011) under construction. A county landfill and wastewater treatment facility that discharges treated effluent are located within a mile upslope of the harbor and the park. Across the highway from the east boundary are a rock quarry and a light industrial and business district [1]. Future development proposed for lands adjacent to the east boundary of the park include 5000 homes, a hospital, wastewater facilities, and commercial light-industrial facilities.

Although typically considered a low-wave-energy coast, the Kona coast can experience large wave events associated with seasonally varying wave sources. The Kona coast is semiprotected by the islands of Maui, Molokai, and Lanai from all but the most westerly of north Pacific swell. Wave energies that affect the Kona coast, and KAHO in particular, are primarily from westerly north Pacific swell, Southern swell, Kona storm waves, waves associated with the passage of low-pressure systems and typhoons (hurricanes), and locally generated wind waves [4–7]. Tides along this section of coast are mixed, semidiurnal with a tidal range of 0.64 m [8]. The mean sea-level trend at Hilo, Hawaii (the only continuously recording tide gauge on the island) is 3.27±0.35 mm/year from 1927 to 2006 [9].

Details of the nearshore oceanography around KAHO are not well known and measurements of tides, waves, currents, and water column properties such as temperature, salinity, and turbidity of the marine waters in KAHO are limited. Recent studies [10,11] show that, in general, currents and circulation are primarily controlled by tides and winds and with distinct differences between the northern and southern parts of Honokohau Bay. Mean flow directions are principally alongshore (N–S) in the north and cross-shore (E–W) in the south. Near-surface velocities are on the order of 0.5 m/s or less with lower velocities (<0.2 m/s) near the seafloor. Current direction and velocities vary seasonally and in response to high-swell and wind events. Temperatures and salinity also vary seasonally and with daily tides, ranging from 23.0°C to 27.7°C and 22.91–35.03 psu in north to 23.0–27.5°C and 29.52–35.01 psu in the south. Lower values reflect intermittent submarine-groundwater discharge (SGD). SGD is widespread in KAHO and may influence temperature, salinity, and nutrient concentrations to water depths up to 5 m below the surface and extending up to 1,000 m offshore [12]. Near-bottom turbidity is generally low but increases during periods of high wave activity. Wave properties measured at the two locations [10] indicate that, in general, southern Honokohau Bay is more exposed to large, long-period North Pacific swell than the northern bay, which is more protected from these waves by Wawahiwaa Point, as reflected in the smaller significant wave heights and shorter dominant wave periods. Similarly, more frequent southerly mean wave directions were measured at the northern site, suggesting a greater influence of southerly swell at that location.

Throughout the Hawaiian Islands, modern coral growth typically occurs as a thin veneer, with little vertical accretion except in locations protected from substantial wave energy [5,6,13,14]. In contrast to coral-reef environments elsewhere in Hawaii, where modern coral communities typically rest unconformably on Pleistocene fossil reef [14–16] or early Holocene reef [14,15], coral communities along the Kona coast of Hawaii grow directly atop volcanic substrate. The combination of geologically young lava flows in KAHO (1.5–10 ka; [17]), and persistent, isostatic subsidence of the island associated with volcanic loading, has precluded a substantial accretion of framework reef along most of the Kona coast.

Coral communities in KAHO generally follow the physiographic zonation of the Kona coast as described in Refs. [5,6] where a shallow boulder zone transitions to a deeper reef bench zone, a reef slope zone, and finally a largely dead coral-rubble zone below depths of about 30 m. The well-defined zonation of the coral community structure is likely the result of variation of wave stress on the coral from breaking waves [5,6]. Our results [2] show variability in this model across KAHO, reflecting differences in wave stresses and possibly other controls (e.g., SGD, recreational activities, and harbor development).

Geomorphic Features and Habitats

In 2003, a GIS-based, benthic-habitat classification map was created for the park using color aerial photography, Scanning Hydrographic Operational Airborne Lidar Survey (SHOALS) bathymetric data, georeferenced underwater video, and still photography [2,18].

Benthic habitats were mapped using standards and classification schemes similar to those established for mapping coral reefs in the USA and its territories on the basis of their seafloor geomorphology, geographic zonation, and biological cover [19–22]. Features were interpreted and boundaries digitized using georectified aerial photography and ground-truthed using underwater video imagery and photographs.

A minimum mapping unit of 100 m² was used, and some modifications to the classification scheme were made where necessary to improve identification of benthic habitats and geologic substrates within the study area. Over 1,185 polygons, covering more than 2,479 km², were digitally delineated and classified with four attributes as defined in [21]: the *major structure* or underlying substrate, the *dominant structure*, the *major biologic cover* found on the substrate, and the *percent cover* of the major biologic cover (Table 28.1).

Seafloor geomorphology was mapped using a combination of aerial photography and high-resolution, lidar-derived bathymetry. Based on observed differences in seafloor morphology (e.g., slope, structure, complexity) and water depth, eleven unique habitat zones were delineated within the study area. The habitat zones include: shoreline/intertidal, shallow cliff, shallow bench (narrow and broad), intermediate bench, shelf break, shelf escarpment, pinnacles/ridges, deep bench, deep cliff, and deep slope (Figure 28.3; Table 28.2). These habitat zones and their associated benthic communities are briefly described in the following sections and more completely in [2].

Shoreline/intertidal: The shoreline/intertidal habitat zone comprises 3.2% of the study area. It is constrained either by water depth (between about −1 and +1 m MLLW), a wet/dry line, or deeper black color of the rocks as seen on the aerial photograph. Between Honokohau Harbor and Noio Point, the shoreline is delineated by a sheer cliff that drops 2–3 m to the seafloor, and no significant intertidal habitat zone is present.

Table 28.1 List of Classification Attributes

Major Structure	Dominant Structure	Major Biologic Cover (Macroscopic)	Percentage Cover
Unconsolidated	Mud	Unknown	Unknown
sediment	Sand	Uncolonized	10−<50
Reef and	Aggregate reef	Macroalgae	50−<90
hardbottom	Spur and groove	Seagrass	90–100
	Individual patch reef	Coralline algae	
	Aggregated patch reef	Coral	
	Volcanic pavement with 10–50% rocks/boulders	Emergent vegetation	
	Volcanic pavement	Octocoral	
	Volcanic pavement with >50% Rocks/Boulders		
	Volcanic pavement with sand channels		
	Reef rubble		
Other	Unknown		
	Land		
	Artificial		
	Artificial/historical		
Unknown	Unknown		

Shallow bench (narrow, broad): Comprising 32% of the study area, the shallow bench habitat zone is the largest in the study area. It is a nearly flat, shallow, pahoehoe-type basalt platform that extends seaward from the shoreline to water depths of less than 6 m throughout the study area. This habitat zone has been subdivided into two categories, narrow (<50 m wide; 20% of the total habitat zone) and broad (up to 700 m wide; 80% of the total habitat zone) benches. The two are primarily distinguished by their morphological relationship and continuity with the adjacent subaerial volcanic flows. The narrow shallow bench is low lying (slope <~4°) and less than 5 m below sea level. It is commonly wave washed and fronted by a low cliff, 1–3 m high, along most of its reach. The broad shallow bench is also a shallow (<~6 m), low-lying (average slope <3°) volcanic platform fronted by a 2–3 m high cliff, but has a more undulating surface characterized by many low ridges and troughs (<1 m high). Large waves commonly shoal and break across this broad, shallow platform.

Shallow cliff: The shallow cliff habitat zone comprises less than 1% of the study area and fronts both the narrow and the broad shallow benches. The cliff is a near-vertical wall between 1 and 3 m high, although in places multiple steps separate the shallow bench from the intermediate bench below.

Intermediate bench: The intermediate bench habitat zone comprises 21% of the study area and is the third largest habitat zone. Where the shallow cliff is present, the intermediate bench begins at its base and slopes gently (<12°) seaward to water

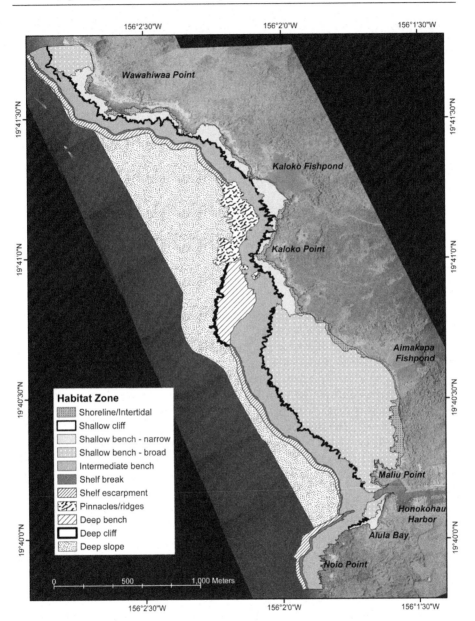

Figure 28.3 Delineation of the 11 habitat zones described in the text, overlaid on shaded relief bathymetry. Hatched lines indicate areas lacking bathymetric data.

depths of around 20 m. South of the harbor, this zone directly abuts the 23 m high cliff that marks the shoreline. The width of the bench varies between 25 and 250 m, is typically approximately 75 m wide, is narrowest in the northern part of the study area, and is widest off the harbor mouth. The seaward boundary of this habitat zone

Table 28.2 Total Area and Relative Percentage of Total Area of the 11 Habitat Zones

Habitat Zone	Area (m²)	Percentage Total Area
Shallow cliff	7,047	0.3
Deep cliff	7,456	0.3
Shelf break	37,134	1.5
Pinnacles/ridges	78,109	3.2
Shoreline/intertidal	78,431	3.2
Deep bench	83,365	3.4
Shelf escarpment	115,106	4.7
Intermediate bench	512,431	20.8
Deep slope	751,210	30.5
Shallow bench—combined	793,362	32.2
Shallow bench—narrow	157,640	6.4
Shallow bench—broad	635,722	25.7
Total	2,463,731	100

is generally the shelf break, except in the northern part of Honokohau Bay, where it abuts the pinnacles/ridges and deep-bench habitat zones.

Shelf break: The shelf break habitat zone comprises 1.5% of the total study area and marks the transition between the gently sloping intermediate bench and the steep shelf escarpment below. This habitat spans depths from 9 to 31 m (average 15–20 m), and average slope values are between 12° and 20°. The shelf break habitat zone does not occur adjacent to pinnacle/ridges and deep cliff habitat zones.

Shelf escarpment: The shelf escarpment habitat zone comprises 4.7% of the study area and extends the length of the study area, except near pinnacles/ridges and deep cliff habitat zones. The inner boundary of this habitat zone is the seaward edge of the shelf break in water depths of between about 15 and 20 m. It continues to depths exceeding the limit of the lidar bathymetry data (>40 m). Limited fathometer transects show that the offshore slope flattens in water depths between 60 and 80 m, marking the end of the shelf escarpment zone and the beginning of the deep slope. Slopes on the shelf escarpment typically range between about 20° and 45°.

Pinnacles/ridges: The pinnacles/ridges habitat zone comprises 3.2% of the study area and is dominantly a region of basalt pinnacles, ridges, and arches in the northern part of Honokohau Bay, between Kaloko Fishpond and Kaloko Point. This habitat zone extends approximately 50–150 m across the shelf, in water depths of 9–25 m, before transitioning as an irregular, steep escarpment to the deep sloping shelf habitat below. The pinnacles/ridges habitat zone marks a well-defined transition from the generally smooth, gently sloping surface of the intermediate bench to an irregular, undulating, and mounded surface, where the pinnacles rise more than 5 m from the adjacent seafloor to within 5 m of the surface. The ridges and intervening channels are typically oriented in a shore-normal direction.

Deep bench: The deep bench habitat zone comprises 3.4% of the study area and is restricted to a limited area offshore of Kaloko Point in the central portion of the park. It extends as much as 250 m across the shelf between the intermediate bench habitat zone on the east and the deep cliff habitat zone on the west. It is bounded on the north by the pinnacles/ridges habitat zone and on the south by the intermediate bench and the shelf escarpment habitat zones. This habitat zone is a mostly smooth, pahoehoe-type volcanic pavement with large areas of rubble and sand. It is slightly (1–3 m), but distinctly, lower than the adjacent intermediate bench and slopes gently seaward (~2°) across the shelf between about 12 and 22 m of water depth. It ends along the crenulated edge of the approximately 3–7 m high, near-vertical deep cliff in approximately 18–22 m of water.

Deep cliff: The deep cliff habitat zone comprises less than 1% of the study area and bounds the seaward edge of the deep bench habitat zone. The deep cliff is a near-vertical wall, typically 3–7 m high, and its upper edge ranges in water depths between 15 and 30 m.

Deep slope: The deep slope habitat zone comprises 30.5% of the total area and is the second largest habitat zone mapped within the study area. The deep slope fronts the entire study area and is predominantly a gently sloping sand sheet. This habitat zone begins at the base of the shelf escarpment, deep cliff, and pinnacles/ridges habitat zones described earlier, in water depths of between 60 and 80 m, where there is a distinct change in slope and transition from rubble and rock-covered or exposed volcanic pavement to predominantly sand.

Biological Communities

Assessment of the benthic habitat and biological communities (dominant structure/ substrate, the type and percentage of major biologic cover, and the relative distribution of principal coral species) were determined using both qualitative analysis of photographic and video imagery and quantitative results of Rapid Assessment Transect (RAT) surveys. RAT surveys were conducted at 17 random sites within the study area to quantify the distribution of coral, invertebrates, algae, and fish within the park [2,23].

Descriptions below indicate the dominant structure and benthic cover (major biologic cover and percent cover), and principal coral species for each habitat zone. Relative distribution of dominant structure classes and percent sand and coral in each habitat zone are shown in Figures 28.4 and 28.5, respectively.

Shoreline/intertidal: The dominant structure in the shoreline/intertidal habitat zone is primarily volcanic pavement, scattered volcanic rocks, boulders, or sand. The pavement surface is commonly covered by a carpet of undifferentiated species of macroalgae.

Shallow bench: The dominant structure on the narrow bench varies from uncolonized volcanic pavement and sand to colonized (10–<50%) volcanic pavement, 10–<50% rocks and boulders, and >50% boulders. Where present, coral cover in this habitat zone is predominantly *Pocillopora meandrina*, with secondary *Porites lobata* and low-lying encrusting coral species (Figure 28.6A). The dominant structure and benthic cover of the broad-bench habitat zone is much more variable. Near

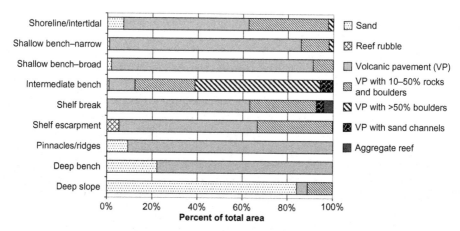

Figure 28.4 Percent of "dominant structure" classes in each habitat zone, except shallow and deep cliffs.

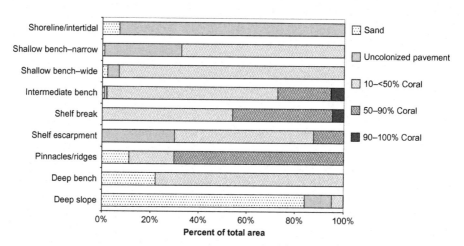

Figure 28.5 Percent of sand and coral on hardbottom in each habitat zone, except shallow and deep cliffs.

the shoreline (water depth <2 m) the habitat is predominantly volcanic pavement colonized with carpets of the lavender-colored octocoral *Anthelia edmonsoni* (Figure 28.6B). Coral cover increases seaward with water depth and surface complexity, and scattered small colonies of primarily *P. lobata* (10–<50%) colonize the flat volcanic pavement, with locally higher percentages of coral cover on the lee side of the ridges and within cracks in the volcanic surface. Toward the seaward edge of the shallow platform, the volcanic surface is covered with scattered rocks and boulders (typically <50%) colonized with relatively larger, lobate colonies of *P. lobata* and *P. meandrina* (still <50% coral cover). The volcanic pavement in this area is also commonly bored by urchins (*Echinometra mathaei*).

Figure 28.6 Representative photos of the shallow- and intermediate-bench habitat zones. (A) Seaward edge of the narrow bench colonized with scattered *Pocillopora meandrina* and *Porites lobata* on a volcanic pavement; (B) a blanket of the lavender-colored octocoral *Anthelia edmonsoni* colonizing volcanic pavement in the shallowest regions of the broad bench; (C) poorly colonized volcanic pavement with scattered rocks and rubble on the intermediate bench; and (D) *Porites* species colonizing boulders south of Kaloko Fishpond.

Shallow cliff: The dominant structure and benthic cover of the shallow cliff habitat zone is uncolonized volcanic pavement, although isolated colonies of *P. lobata* and *P. meandrina* are present locally on the cliff face (<10%).

Intermediate bench: This habitat zone claims the highest variety of dominant substrate structure, percent coral cover, and overall coral diversity among all the habitat zones within the study area (Figures 28.4 and 28.5). It is characterized by the dominance of poorly (10–<50%) to moderately (50–<90%) colonized boulders and scattered rocks and rubble on volcanic pavement (Figure 28.6C). In the relatively protected embayment of southern Honokohau Bay, the highest percent coral cover (90–100%) and the only area of vertically accreted carbonate reef (thickness unknown) in the study area are within this habitat zone. Dominant coral species include encrusting and lobate forms of *P. lobata*, undifferentiated *Porites* species (Figure 28.6D), and, in southern Honokohau Bay, thickets of *Porites compressa*.

Shelf break: The shelf break habitat zone marks a dramatic transition in benthic habitat relative to adjacent habitat zones. For example, fewer boulders and rocks

Figure 28.7 Representative photos of the shelf break, shelf escarpment, pinnacle/ridges, and deep-slope habitat zones. (A) Moderately high coral cover on shelf break with *Porites lobata*, *Porites compressa*, and *Anthelia edmonsoni*; (B) scattered coral and rubble on shelf escarpment; (C) moderately high coral cover and platey corals on vertical edge of a ridge; and (D) angular boulders and sand at base of shelf escarpment colonized by small, lobate colonies of *Porites lobata*.

exist on the shelf break compared to the intermediate bench, and coral cover is distinctly higher compared to the shelf escarpment (Figures 28.5 and 28.6). A wide range of habitat classifications are included within this habitat zone, with almost no uncolonized volcanic pavement, but also the highest relative percentage of aggregate reef in the study area (Figures 28.5 and 28.7A).

Shelf escarpment: The dominant structure and benthic cover of the shelf escarpment habitat zone is primarily uncolonized to poorly (10–<50%) colonized volcanic pavement that is essentially monotypic, patchy thickets of live and skeletal *P. compressa*, and rubble (Figure 28.7B). The soft octocoral *Anthelia edmonsoni* is commonly associated with the *P. compressa*. The extreme slope of the escarpment causes this habitat to be highly unstable and vulnerable to damage from boulders, cobbles, and larger coral heads tumbling down the slope, as well as damage from higher wave energy conditions associated with severe, but episodic, storms. Small rocks and rubble are also common within this habitat, while large boulders are rare. In only a few areas does coral cover exceed

50%. Near the shelf break and in areas where the slope is somewhat lower, *P. lobata* and *P. meandrina* also colonize this habitat.

Pinnacles/ridges: The complex structure, vertical relief, and relatively protected position of this habitat zone within northern Honokohau Bay results in a wide range of coral cover and relative diversity. Relative topographic highs have low (10–<50%) to moderately high (50–<90%) coral cover of predominantly *P. meandrina* and *Porites* sp., while the intervening lows are typically uncolonized pavement with a thin cover of sand and rubble. Where tops of the ridges and pinnacles reach close to the water surface, coral cover is typically low (10–<50%) to moderate (50–<90%) and predominantly *P. meandrina* and small and encrusting *P. lobata*. Along the edges and walls of these shallow features, and in general as water depth increases, percent coral cover and relative diversity increases to moderately high (50–<90%) and high (90–100%) (Figure 28.7C).

Deep bench: The dominant structure and benthic cover of the deep bench habitat zone is mostly volcanic pavement with low (10–<50%) but variable coral cover of primarily small colonies of lobate *P. lobata* and *P. meandrina*, and patches of uncolonized pavement, sand, and rubble. Locally there are areas of higher coral cover and larger heads of *P. lobata*. A relatively large sand sheet abuts the intermediate bench along the south boundary of this habitat zone.

Deep cliff: The dominant structure and benthic cover of the deep cliff habitat zone is uncolonized volcanic pavement.

Deep slope: The dominant structure and benthic cover of the deep-slope habitat zone is predominantly uncolonized sand with 10–<50% rocks and boulders and volcanic pavement. Small colonies of lobate *P. lobata* were observed on some of the boulders (<10% overall) (Figure 28.7D). Sea cucumbers and garden eels were also observed to inhabit this habitat zone.

Surrogacy

No statistical analyses have been carried out on this data set to examine relationships between physical surrogates and benthos.

Acknowledgments

This project was funded by the US Geological Survey (Pacific Coral Reef Project, Coastal and Marine Geology Program) and by the National Park Service (Kaloko-Honokohau NHP). Michael Field (USGS) and Sallie Beavers (NPS) were instrumental in bringing the project to fruition. Josh Logan and Eric Grossman (USGS) were invaluable in the field and provided many insightful conversations during the course of the study. Will Smith and colleagues from the University of Hawaii Coral Reef Assessment and Monitoring Program (CRAMP) team performed the third-party field-check observations and provided the accuracy assessment calculations.

References

[1]   K. DeVerse, Appendix A: Kaloko-Honokohau National Historical Park resource over-
      view, in: L. HaySmith, F.L. Klasner, S.H. Stephens, G.H. Dicus (Eds.), Pacific Island
      Network vital signs monitoring plan. Natural Resource Report NPS/PACN/NRR—
      2006/003 National Park Service, Fort Collins, Colorado, 2006, http://science.nature.nps.
      gov/im/units/pacn/monitoring/plan/PACN_MP_AppendixA_KAHO.pdf.
[2]   A.E. Gibbs, S.A. Cochran, J.B. Logan, E.E. Grossman, Benthic habitats and offshore
      geological resources of Kaloko-Honokohau National Historical Park, Hawaii. U.S.
      Geological Survey Scientific Investigations Report 2006–5256, <http://pubs.usgs.gov/
      sir/2006/5256/> 2007, 62 pp.
[3]   J.D. Parrish, G.C. Smith, J.E. Norris, Resources of the marine waters of Kaloko-
      Honokohau National Historical Park. Cooperative National Park Resource Studies Unit,
      Technical Report 74, <http://manoa.hawaii.edu/hpicesu/techr/074.pdf> 1990, 115 pp.
[4]   R.M. Moberly, T. Chamberlain, Hawaiian Beach Systems, Hawaii Institute of
      Geophysics, University of Hawaii, 2007, Honolulu, 1964, 95 p.
[5]   S.J. Dollar, Wave stress and coral community structure in Hawaii, Coral Reefs 1 (1982)
      71–81.
[6]   S.J. Dollar, G.W. Tribble, Recurrent storm disturbance and recovery; a long-term study
      of coral communities in Hawaii, Coral Reefs 12 (1993) 223–233.
[7]   S. Vitousek, M.M. Barbee, C.H. Fletcher, B.M. Richmond, A.S. Genz, Puʻukohola-
      Heiau national historic site and Kaloko-Honokohau Historical Park, Big Island of
      Hawaii; Coastal Hazard Report. NPS Geologic Resources Division, 2009, http://www.
      soest.hawaii.edu/coasts/nps/index.php.
[8]   National Oceanic and Atmospheric Administration National Centers for Coastal Ocean
      Science, Tidal station locations and ranges, 2005, http://tidesandcurrents.noaa.gov/
      tides11/tab2wc3.html#166.
[9]   National Oceanic and Atmospheric Administration National Ocean Service, Sea Level
      Trends; Sea Levels Online: <http://tidesandcurrents.noaa.gov/sltrends/sltrends.shtml>,
      2010.
[10] C.D. Storlazzi, M.K. Presto, Coastal circulation and water column properties along
      Kaloko-Honokōhau National Historical Park, Hawaii. Part I: Measurements of waves,
      currents, temperature, salinity and turbidity: April–October 2004. U.S. Geological
      Survey Open-File Report 2005-1161, <http://pubs.usgs.gov/of/2005/1161/> 2005,
      30 pp.
[11] M.K. Presto, C.D. Storlazzi, J.B. Logan, E.E. Grossman, Submarine groundwater dis-
      charge and seasonal trends along the coast of Kaloko-Honokōhau National Historic
      Park, Hawaii, Part I: Time-series measurements of currents, waves and water proper-
      ties: November 2005–July 2006. U.S. Geological Survey Open-File Report 2007-1310,
      <http://pubs.usgs.gov/of/2007/1310/> 2007, 39 pp.
[12] E.E. Grossman, J.B. Logan, M.K. Presto, C.D. Storlazzi, Submarine groundwater
      discharge and fate along the coast of Kaloko-Honokohau National Historical Park,
      Island of Hawaii, Part 3: Spatial and temporal patterns in nearshore waters and coastal
      groundwater plumes, December 2003–April 2006. U.S. Geological Survey Scientific
      Investigations Report 2010-5081, <http://pubs.usgs.gov/sir/2010/5081/> 2010, 76 pp.
[13] R.W. Grigg, Holocene coral reef accretion in Hawaii: a function of wave exposure and
      sea level history, Coral Reefs 17 (1998) 263–272.

[14] C.H. Fletcher, C. Bochicchio, C.L. Conger, M. Engels, E.J. Feirstein, N. Frazer, et al., Geology of Hawaii Reefs, in: B.M. Riegl, R.E. Dodge, (Eds.), Coral Reefs of the U.S.A., Springer, Berlin, 2008, pp. 435–488.

[15] M.S. Engels, C.H. Fletcher, M.E. Field, C.D. Storlazzi, E.E. Grossman, J.J. Rooney, et al., Holocene reef accretion: southwest Molokai, Hawaii USA, J. Sediment. Res. 74 (2) (2004) 255–269.

[16] J. Rooney, C. Fletcher, E. Grossman, M. Engels, M. Field, El Niño influence on Holocene reef accretion in Hawaii, Pac. Sci. 58 (2) (2004) 305–324.

[17] E.E. Wolfe, J. Morris, Geologic map of the Island of Hawaii: U.S. Geological Survey Miscellaneous Investigations Series I-2524-A, <http://pubs.usgs.gov/of/2007/1089/HawIsland_zone5_2007.pdf> 1996, 18 pp., 3 sheets.

[18] A.E. Gibbs, S.A. Cochran, An integrated approach to benthic habitat mapping and GIS: an example from the Hawaiian Islands, in: Xiaojun Yang, (Ed.), Remote Sensing and Spatial Information Technologies for Coastal Ecosystem Assessment and Management: Principles and Applications, Springer, Germany, 2009, pp. 211–231.

[19] M.S. Kendall, M.E. Monaco, K.R. Buja, J.D. Christensen, C.R. Kruer, M. Finkbeiner, et al., Methods used to map the Benthic Habitats of Puerto Rico and the U.S. Virgin Islands, U.S. National Oceanic and Atmospheric Administration. http://ccma.nos.noaa.gov/products/biogeography/benthic/index.

[20] M.S. Coyne, T.A. Battista, M. Anderson, J. Waddell, W. Smith, P. Jokiel, et al., Benthic habitats of the main Hawaiian Islands. NOAA Technical Memorandum NOS NCCOS CCMA 152, 2003, http://ccma.nos.noaa.gov/products/biogeography/hawaii_cd/.

[21] NOAA National Centers for Coastal Ocean Science, Shallow-water benthic habitats of American Samoa, Guam, and the Commonwealth of the Northern Mariana Islands. NOAA Technical Memorandum NOS NCCOS 8, 2005, http://ccma.nos.noaa.gov/ecosystems/coralreef/us_pac_mapping.aspx.

[22] S.O. Rohman, M.E. Monaco, Mapping Southern Florida's shallow-water coral ecosystems; an implementation plan. NOAA Technical Memorandum NOS NCCOS 19, 2005, 39 pp. http://ccma.nos.noaa.gov/publications/biogeography/FloridaTm19.pdf.

[23] K.S. Rodgers, P.L. Jokiel, E.K. Brown, Rapid assessment of Kaloko-Honokohau and Pu'uhonua o Honaunau, West Hawaii. Hawaii Coral Reef Assessment and Monitoring Program (CRAMP), Hawaii Institute of Marine Biology, 2004, 56 pp.

29 Habitats and Benthos at Hydrographers Passage, Great Barrier Reef, Australia

Robin J. Beaman[1], Thomas Bridge[2], Terry Done[3],
Jody M. Webster[4], Stefan Williams[5], Oscar Pizarro[5]

[1]School of Earth and Environmental Sciences, James Cook University, Cairns, QLD, Australia, [2]School of Earth and Environmental Sciences, James Cook University, Townsville, QLD, Australia, [3]Australian Institute of Marine Sciences, Townsville, QLD, Australia, [4]School of Geosciences, the University of Sydney, Sydney, NSW, Australia, [5]Australian Centre for Field Robotics, the University of Sydney, Sydney, NSW, Australia

Abstract

Hydrographers Passage lies on the shelf edge of the central Great Barrier Reef in northeastern Australia. The survey location is approximately $800 \, km^2$ in area and ranges in depth from 14 to 300 m. The mapped geomorphic features ($>1 \, km$) include a broader-scale slope, terrace, and platform, with smaller reefs and dune features. Sediment grabs from two cross-shelf transects show a generally similar composition of poorly-sorted, muddy sand, and iron stained carbonate gravel. In contrast, dune sediments are a well-sorted sand with no iron staining. Images from two autonomous underwater vehicle (AUV) transects provide the data for a hierarchical clustering of substrate types into five substrate groups: sand, gravel, rubble, sediment-covered limestone, and reef. The AUV imagery reveals a clear distinction between the benthos associated with hard substrate, and soft substrate habitats at a finer-scale than can be shown within the broader-scale geomorphic features. Maximum entropy modeling is used to generate a habitat preference map for azooxanthellate, filter-feeding octocorals.

Key Words: mesophotic, coral reefs, substrate preferences, autonomous underwater vehicle, vertical zonation, community composition

Introduction

Hydrographers Passage lies on the continental margin of the central Great Barrier Reef (GBR) in northeastern Australia (Figure 29.1). The GBR margin represents the world's largest extant tropical siliciclastic/carbonate depositional system, which extends for about 2,300 km in length [1]. This system is composed of shoreline, shelf, shelf edge and slope, and basin elements. The Hydrographers Passage site is currently the

Seafloor Geomorphology as Benthic Habitat. DOI: 10.1016/B978-0-12-385140-6.00029-3

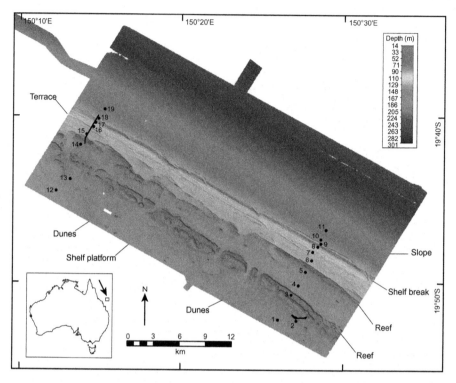

Figure 29.1 Multibeam sonar bathymetric map of Hydrographers Passage at a grid pixel resolution of 25 m. Black lines show the AUV transects across the seabed. Black dots indicate the sediment grab locations. Inset shows the site location on the northeastern Australia outer shelf.

best-studied section of the GBR shelf edge and upper slope, comprising geomorphic features such as modern coral reefs, drowned or submerged reefs, a shelf platform draped with extensive dune fields, seaward terraces, and a smooth upper slope. Depths range from about 70 m on the shelf platform, rising to 14 m over the shoal reefs, dropping across terrace features to the shelf break at about 100 m, and then to 300 m at the seaward limit of the study area. The site is considered to be in pristine condition and is relatively undisturbed.

The area is swept by the East Australia current, with water temperatures ranging from about 25°C at the sea surface to a near-seabed temperature of about 20°C. Strong tidal currents flow through narrow gaps in the reef matrix and result in upwelling of nutrients for delivery into the shelf edge system of the southern GBR [2]. Hydrographers Passage is named for the narrow shipping lane through the shoal reefs, which is frequently used by merchant vessels accessing coal-loading facilities on the adjacent Australian mainland. In September–October 2007, the area was mapped and sampled using the Australian research vessel *Southern Surveyor* [3]. Bathymetry data were acquired with a Simrad EM300 30-kHz multibeam sonar to map an area 800 km². Ground-truthing utilized sediment grabs and images collected by an autonomous underwater vehicle (AUV) from the Australian Centre for Field Robotics along two cross-shelf transects.

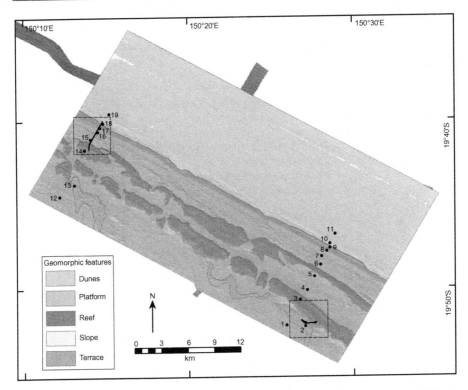

Figure 29.2 Geomorphic features from Hydrographers Passage, as listed in Table 29.1. Black lines show the location of AUV transects across the seabed. Black dots indicate the sediment grab locations. Dashed boxes delineate close-up views of the AUV transects colored by substrate groups in Figure 29.3.

Geomorphic Features and Habitats

Distinct geomorphic features are clearly observed within the bathymetric map, which has a grid pixel size of 25 m (Figure 29.1). The classification of the geomorphic features described here use undersea features names as recommended by the IHO [4] and also used in the literature [5,6]. The boundaries of geomorphic features were first identified in ArcGIS using a slope analysis on the bathymetry grid and then hand-drawn as vector lines in ArcMap (Figure 29.2). Fledermaus three-dimensional visualization software was also used to assist during the geomorphic boundary delineation. Table 29.1 records the definitions and surface area of the mapped features at Hydrographers Passage.

The smallest mapped geomorphic features are reefs at about 1 km in size but ranging up to 12 km in length. Dune fields are about 9 km in size, while broader-scale features are the underlying shelf platform, terrace, and slope features at 40 km in size across the full width of the survey area. Superimposed upon these mapped geomorphic features are finer-scale sand waves within the dune areas, numerous coral pinnacles on the reefs, and down-cutting channels across the shelf platform

Table 29.1 Definitions and Surface Area for the Geomorphic Features Mapped Within the Hydrographers Passage Survey Area

Geomorphic Feature	Definition	Surface Area (km^2)
Slope	Located seaward from the shelf edge to the upper edge of a continental rise or the point where there is a general reduction in slope	399.36
Terrace	Relatively flat horizontal or gently inclined surface, which is bounded by a steeper ascending slope on one side and by a steeper descending slope on the opposite side	106.14
Platform	Low-gradient, low-relief surface of extensive horizontal dimensions	189.50
Reef	Rock lying at or near the sea surface that may constitute a hazard to surface navigation	86.13
Dunes	Mounds or ridges of drifted unconsolidated sand	18.25

and terrace with dimensions of tens of meters. The following geomorphic feature descriptions also include sediment descriptions taken from the 19 grabs obtained along the two transects across the geomorphic features.

Slope: The mapped portion of the upper continental slope is 399 km^2, or about half of the total survey area. Depths range from 300 m at the deeper limit to about 130 m at the base of the steep scarp marking the shelf break or the seaward limit of the GBR shelf. The surface is relatively smooth and has a gentle gradient of about 1° toward the northeast. There are no indications of submarine canyons within its surface. Sediments are poorly sorted, muddy, coarse sand, and gravel (grabs 10, 11, 18, and 19). The coarse fraction is generally >30% *Halimeda* algae, 30% benthic forams, 10% bryozoa, 10% mollusk, and the remainder grainstone fragments. Most grains are worn and broken, with iron staining.

Terrace: Between the reefs and the shelf break is a wide terrace feature with a gentle slope of <1° in gradient and about 106 km^2 in area. The seaward extent is marked by the steep step which is the shelf break, starting at about 100 m and dropping to 130 m at the upper limit of the continental slope. The landward extent abuts the bases of the numerous submerged reefs lying parallel to the shelf edge at about 90 m. Sediments are moderately to poorly sorted muddy sand to fine gravel (grabs 6, 7, 8, 9, 15, 16, and 17). The coarse fraction is 30% benthic forams, 20% *Halimeda*, 20% mollusk, 10% bryozoans, 5% echinoderms, and the remainder coral fragments. Most grains show signs of iron staining.

Platform: The platform is the main GBR outer shelf upon which the reefs and dune fields are located. The gradient is near flat or <0.5° and has a mapped area of 189 km^2 that extends toward the modern patch reefs found on the middle to outer shelf (not shown here) and seaward, where it forms the bed of the inter-reefal channels lying between the submerged reefs. Depths range from about 70 m along the southwest edge of the survey area to about 80 m within the deepest channels feeding into the adjacent terrace. Sediments range from poorly sorted coarse sand and rubble to moderately sorted medium to coarse sand (grabs 1, 4, 5, and 12). The coarse fraction comprise 30% benthic forams, 15% *Halimeda*, 10% bryozoa, 5% mollusk fragments, and also encrusted grainstone pieces. Fragments are either freshly broken or appear worn.

Reef: The most distinct geomorphic feature at Hydrographers Passage are the twin parallel lines of barrier reefs up to 1,500 m wide, which extend across the width of the survey area. Their combined area is about 86 km². At several locations on the inner line of reefs, depths rise 40–50 m above the surrounding flat platform to about 14 m, but the tops generally lie between 20 and 36 m. The outer line of reefs is more subdued in relief, rising about 20 m above the surrounding platform, with tops generally lying between 50 and 60 m. These features are clear evidence of the submerged reefs that occupy the GBR shelf edge [5,6]. Sediments are poorly sorted sand to coarse gravel (grabs 3 and 14). The coarse fraction comprise 60% benthic forams, 15% *Halimeda*, 10% mollusk fragments, and 5% bryozoa. Most grains appear unstained.

Dunes: The dune fields found on the landward side of the reefs are a response to the strong tidal streams through the reef matrix passages. Their combined area is 18 km², a conservative estimate because individual sand wave crests are also observed right across the platform, as distinct from the three main fields of dunes mapped here. Dune crests are linear to curvilinear, and the lee and stoss sides are symmetrical in shape corresponding to their tidal influence. Trough to crest heights range from 2 to 4 m, in depths of 65–70 m. Sediments are well-sorted medium to coarse sand (grabs 2 and 13). The coarse fraction was too fine to determine composition percentages but comprise broken benthic forams, *Halimeda*, and mollusk fragments.

Biological Communities

Assessment of benthos at Hydrographers Passage utilized the still images obtained by the AUV at two cross-shelf transects: (A) in the northwest about 3.6 km long, from a deep reef across the terrace to the upper slope; and (B) in the southeast about 2.7 km long, from a shallow reef across the platform and over a dune field (Figure 29.3). The AUV maintained a constant height of 2 m above the seabed while acquiring stereo image pairs at 1-s intervals, with field-of-view dimensions of 1.5 × 1 m. The dense spatial overlap of imagery was necessary for the accurate navigation of the AUV [7], but the high number of images (>7,000 image pairs) also proved challenging to analyze in order to quantify the benthic distribution and assemblage composition along each transect.

A method was developed to identify macrofauna trends based on the hypothesis that substrate is an important factor in the distribution of benthos [8]. Thus, a measure of the substrate distribution along the AUV transects leads to a better understanding of the associated benthos. This method used every tenth image, providing a quadrat for examination at approximately 4.5 m intervals. The data collected used a grading system based on the relative abundance of five substrate types observed in the images: sand, gravel, rubble, sediment-covered limestone, and rough limestone (Table 29.2). The scoring method followed [8] where 1 = <5% cover of any given substrate type, 2 = 5–10%, 3 = 11–30%, 4 = 30–80%, and 5 = >80% in each image. To identify discrete substrate groups that could be mapped in ArcGIS, a hierarchical

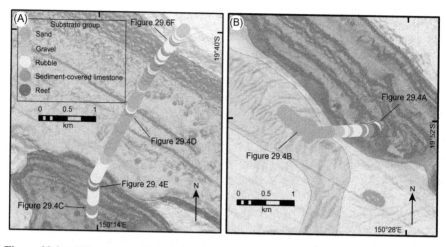

Figure 29.3 AUV transects colored by substrate group, with the underlying geomorphic features colored as per Figure 29.2. The positions of the seabed images shown in Figure 29.4 are indicated against the AUV transects: (A) northwest transect; (B) southeast transect. See Figure 29.2 for transect locations within the survey area.

Table 29.2 Descriptions of the Substrate Types Classified from the AUV Imagery

Substrate Types	Description
Sand	Unable to distinguish individual grains in images; grain size <2 mm
Gravel	Larger than sand but smaller than rubble; grain size ~3–30 mm
Rubble	Clasts >30-mm grain size but not firmly attached to the substrate
Sediment-covered limestone	Substrate covered with soft sediment but appears hard underneath
Rough limestone	Limestone protrudes above the surrounding seafloor; no soft sediment

cluster analysis was conducted on the raw data to identify the proportion of substrate types within each cluster. The five substrate groups were thus named after the dominant substrate type in each cluster.

Northwest transect: Figure 29.3A shows the substrate groups as colored points along the transect. Starting at a depth of 57 m on the top of a submerged reef, the substrate is dominated by rubble, sediment-covered limestone, and reef. The associated benthos are a diverse assemblage of mesophotic reef taxa. Above depths of 70 m, the most abundant taxa are photosynthetic and often zooxanthellate (Figure 29.4C and E). These include various types of algae: fleshy green algae (*Caulerpa*), crustose coralline algae (*Mesophyllum* and *Lithothamnion*), foliose coralline algae (*Halimeda*), and cyanobacteria (probably *Lyngbia*). The most abundant macrofauna are zooxanthellate octocorals (particularly of the family Xeniidae) and the sponge *Carteriospongia*, which contains a symbiotic cyanobacteria. The shallower submerged reef also has a relatively sparse but diverse community of zooxanthellate scleractinian corals, such as

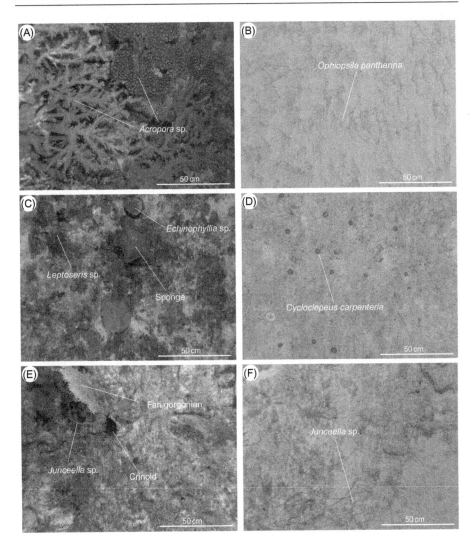

Figure 29.4 AUV images taken on hard (left side) and soft (right side) substrata at Hydrographers Passage. See Figure 29.3 for the locations of the images.

Leptoseris, Echinophyllia, Montipora, Acropora, Goniopora, and *Seriatopora.* Below 70 m, the photosynthetic taxa become less abundant and are replaced by azooxanthellate, filter-feeding taxa, which are predominantly sponges and octocorals.

At the base of the reef in about 86 m, the rubble substrate gives way to mostly sand and minor gravel across the relatively flat, 2-km wide terrace feature. The only abundant taxon in this area is the large benthic foram *Cycloclepeus carpenteri* (Figure 29.4D). At minor intervals along the terrace, sediment-covered limestone and reef substrate appear, which correspond to finer scale pinnacles and minor scarps. Here, azooxanthellate, filter-feeding octocorals appear again. In the vicinity of the shelf break at about 100 m, rubble and sediment-covered limestone then dominate, and rubble continues as the main substrate group below the shelf break to 120 m. In contrast to the shallower reef, the biota in this zone is dominated by azooxanthellate filter-feeding taxa, particularly octocorals, for example, *Annella* and *Chironepthya*, and members of the families Isididae and Ellisellidae (Figure 29.4F). Other azooxanthellate filter-feeding taxa also occur, such as sponges and crinoids. Near the limit of the transect at about 130 m on the upper slope, gravel begins to dominate the substrate and macrobenthos are relatively sparse.

Southeast transect: Figure 29.3B shows the substrate groups along the transect, which commences in a depth of 17 m on an inner submerged reef. For much of the reef feature at depths less than 40 m, sediment-covered limestone and reef substrate is found, with minor rubble substrate (Figure 29.4A). The shallow areas on the top of the reefs are dominated by photosynthetic taxa, with some sections having almost 100% cover of zooxanthellate scleractinia coral, including *Acropora* and *Pocillopora* spp. Other shallow areas contain abundant communities of zooxanthellate octocorals (particularly *Cespitularia*), while others contain sparse macrofauna. However, even the areas with sparse macrofauna contain no algal growth apart from crustose coralline algae. Below about 40 m depth and still on the high-gradient reef, rubble becomes the dominant substrate. Biota in this region resembles the community at the start of the northwest transect, which is dominated by algae and cyanobacteria, interspersed with sparse zooxanthellate scleractinia and octocorals.

At 55 m, near where the reef base merges with the surrounding relatively flat platform, gravel becomes the main substrate and the large macrofauna abruptly disappears with the transition into the soft sediment. The transect continues across the platform until a depth of about 63 m, where it crosses into a large dune field with sand as the dominant substrate. A remarkable discovery found that the dunes were the preferred habitat for countless brittlestar *Ophiopsila pantherina* (Figure 29.4B). They appeared to aggregate on the leeward (during flood tide) sides of the dunes, and were possibly filter-feeding on the near-seabed nutrients upwelling due to the strong tidal currents entering through the submerged reef matrix [9]. At an incredible density of over 418 individuals/m^2, the brittlestars form an impressive "wall of arms" capturing plankton from this inflowing shelf water.

Surrogacy

The broad-scale geomorphic features shown in Figure 29.2 represent a generally close approximation to the distribution of the predominant benthic assemblages, in association with their substrate group (Figure 29.3). However, the AUV transects reveal a finer scale detail that cannot be resolved within the geomorphic features, such as the change in mesophotic taxa below about 70 m depth from a photosynthetic and often zooxanthellate community to a predominantly azooxanthellate, filter-feeding octocoral community on the same reef feature. The AUV imagery also reveals a distinction between the benthos associated with hard substrates and soft substrates at a finer scale than can be shown within the broader scale geomorphic features. For example, the terrace geomorphic feature is mostly sand and gravel substrate with sparse benthos, but the smaller reef substrate pinnacles and scarps also found there are sites for dense patches of octocorals. The results of the sediment grabs were generally poor at predicting substrate and benthos trends, except in the hydrodynamically active dune fields. The sand (and brittlestars) collected from the dunes were a good indicator of the sand substrate and aggregated brittlestars revealed by the AUV imagery.

The community presence data from the AUV, in combination with the available environmental maps as surrogates, were used to generate environmental suitability maps for selected benthos. We utilized the open-source software Maxent, which estimates the target distribution by finding the distribution of maximum entropy (i.e., closest to uniform), subject to the constraint that the expected value of each environmental variable under this estimated distribution matches its empirical average [10]. Maxent is ideally used for single species distributions, however, we selected the occurrences of the deeper azooxanthellate, filter-feeding octocorals from the northeast AUV transect to generate a list of presence records for this community. The input environmental datasets were 5-m pixel grids derived from the shallow (<200 m) multibeam data: depth, backscatter, slope, rugosity, aspect, and Benthic Terrain Modeler zones [11]. Figure 29.5 shows a resulting habitat preference map for the octocorals, which indicate suitable locations between 70 and 110 m, from the deeper submerged reef and across

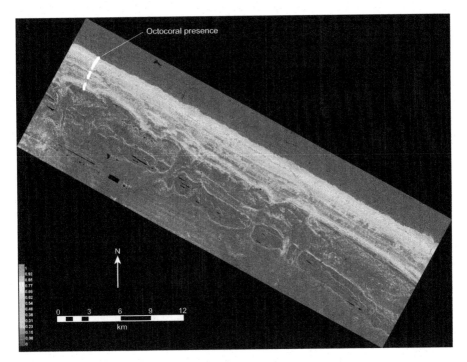

Figure 29.5 A habitat preference map for the deep azooxanthellate, filter-feeding octocorals, based upon the presence records from the northwest AUV transect, shown as white dots, and the available environmental maps. Red and yellow pixels on the map indicate more suitable locations for the octocorals.

the terrace feature, but with higher preferences for the smaller pinnacles and scarps found on the terrace.

Acknowledgments

We thank the captain and crew of the RV *Southern Surveyor* for their outstanding work on the cruise. Robin Beaman acknowledges a Queensland Smart Futures Fellowship for salary support. This work is supported by the ARC Centre of Excellence program, funded by the Australian Research Council (ARC) and the New South Wales State Government, the Integrated Marine Observing System (IMOS) through the Department of Innovation, Industry, Science and Research (DIISR) National Collaborative Research Infrastructure Scheme, the Australian Marine National Facility, and *National Geographic*.

References

[1] D. Hopley, S.G. Smithers, K.E. Parnell, The Geomorphology of the Great Barrier Reef: Development, Diversity and Change, Cambridge University Press, Cambridge, UK, 2007.

[2] J.H. Middleton, P. Coutis, D.A. Griffin, A. Macks, A. McTaggart, M.A. Merrifield, et al., Circulation and water mass characteristics of the Southern Great Barrier Reef, Aust. J. Mar. Freshwater Res. 45 (1994) 1–18.

[3] J.M. Webster, R.J. Beaman, T. Bridge, P.J. Davies, M. Byrne, S. Williams, et al., From corals to canyons: the Great Barrier Reef margin, EOS 89 (2008) 217–218.

[4] IHO, Standardization of Undersea Feature Names: Guidelines, Proposal Form, Terminology, International Hydrographic Bureau/Intergovernmental Oceanographic Commission, Monaco, 2001.

[5] P.T. Harris, P.J. Davies, Submerged reefs and terraces on the shelf edge of the Great Barrier Reef, Australia: morphology, occurrence and implications for reef evolution, Coral Reefs 8 (1989) 87–98.

[6] R.J. Beaman, J.M. Webster, R.A.J. Wust, New evidence for drowned shelf edge reefs in the Great Barrier Reef, Australia, Mar. Geol. 247 (2008) 17–34.

[7] S.B. Williams, O. Pizarro, J.M. Webster, R.J. Beaman, I. Mahon, M. Johnson-Roberson, et al., Autonomous underwater vehicle-assisted surveying of drowned reefs on the shelf edge of the Great Barrier Reef, Australia, J. Field Rob. 27 (2010) 675–697.

[8] T.C.L. Bridge, T.J. Done, R.J. Beaman, A. Friedman, S.B. Williams, O. Pizarro, et al., Topography, substratum and benthic macrofaunal relationships on a tropical mesophotic shelf margin, Central Great Barrier Reef, Australia. Coral Reefs. 30 (2011) 143–153.

[9] M. Byrne, Flashing Stars Light up the Reef's Shelf. ECOS Magazine, CSIRO Publishing, Collingwood, Australia, 2009. August–September, 28–29.

[10] S.J. Phillips, R.P. Anderson, R.E. Schapire, Maximum entropy modeling of species geographic distributions, Ecol. Modell. 190 (2006) 231–259.

[11] M.D. Erdey-Heydorn, An ArcGIS seabed characterization toolbox developed for investigating benthic habitats, Mar. Geod. 31 (2008) 318–358.

30 Two Shelf-Edge Marine Protected Areas in the Eastern Gulf of Mexico

Rebecca J. Allee[1], Andrew W. David[2], David F. Naar[3]

[1]National Oceanic and Atmospheric Administration, Gulf Coast Services Center, Stennis Space Center, MS, USA, [2]National Oceanic and Atmospheric Administration, National Marine Fisheries Service, Panama City, FL, USA, [3]College of Marine Science, University of South Florida, St. Petersburg, FL, USA

Abstract

The Madison-Swanson Marine Protected Area is located off the Florida coast in the Gulf of Mexico at the margin of the continental shelf and slope in 60–140 m of water. Prominent within Madison-Swanson is a limestone ridge thought to be the remnant of a 14,000 + year old coral reef. The Pulley Ridge Habitat Area of Particular Concern is a 100 + km-long series of N–S trending, drowned, barrier islands on the southwest Florida Shelf. It appears to be formed on top of an ancient coastal barrier island or strand line during a period when sea levels were ~65–80 m lower and is believed to be the deepest hermatypic coral reef on the continental shelf of the US. The depth ranges from ~55 m near the eastern edge of Pulley Ridge down to about 115 m on the western edge. Fisheries studies of Madison-Swanson indicate that *Mycteroperca* spp. (grouper) and *Lutjanus campechanus* (red snapper) are associated with hard bottom features, with spawning aggregations of *M. microlepis* (gag) and/or *M. phenax* (scamp) confirmed at several sites. Canonical Correspondence Analyses (CCA) conducted on fish and habitat data from Madison-Swanson indicated gag are most closely associated with relict coral reefs and to a lesser degree with greater depth and higher relief, while red snapper had a higher correlation with soft corals. Remotely Operated Vehicle (ROV) observations were used to associate fish species with habitat types for Pulley Ridge. The loose rubble found in Pulley Ridge was conducive for *Epinephelus morio* (red grouper) to excavate pits in the sediment, and they were the most abundant large grouper species in that area.

Key Words: drowned barrier islands, hermatypic corals, mounds, paleoreef, paleoshoreline, ridges, pinnacles, grouper, *Mycteroperca*, spawning aggregation

Introduction

The Madison-Swanson Marine Protected Area (MPA) and Pulley Ridge Habitat Area of Particular Concern (HAPC), two protected areas off the Florida shelf in

Seafloor Geomorphology as Benthic Habitat. DOI: 10.1016/B978-0-12-385140-6.00030-X

the eastern Gulf of Mexico (Figure 30.1), were established by the Gulf of Mexico Fishery Management Council and the US National Oceanic and Atmospheric Administration to protect benthos-associated organisms. However, the types of organisms and the reasons for the protection are quite different [1,2]. Madison-Swanson is used by an economically valuable reef fish species, the gag grouper, as a spawning ground [3,4]. Pulley Ridge contains the deepest hermatypic scleractinian coral colonies in the continental USA [5,6]. Madison-Swanson was closed to most fishing activity in 2000 to improve reproductive output and subsequently stock size of the grouper, which in turn was hoped to produce financial benefits to the fishery. Pulley Ridge was closed to most fishing activity in 2005 to maintain the biodiversity associated with the coral formations.

Madison-Swanson is a 394 km^2 area located ~60 km southwest of Cape San Blas, Florida, at the margin of the continental shelf and slope in 60–140 m of water and is a site of spawning aggregations of gag (*Mycteroperca microlepis*) and other reef fish species [4]. High-resolution seismic stratigraphy [7] and 300-kHz multibeam data [8] show that the Madison-Swanson MPA is a drowned river delta that is estimated to have formed between 58,000 and 28,000 years ago [9]. Prominent within Madison-Swanson is a continuous ~13 km long curved ridge ~6 m tall and ~80 m wide in ~80 m of water depth (Figure 30.2) interpreted by Gardner et al. [9] to be a barrier island that formed contemporaneously with the delta, but then was preserved with the onset of rapid sea-level rise ~14,000 years ago. An upwelling current flows perpendicularly to the long axis of the feature and has undermined the sand at its base, creating a trench along the offshore face. This current has undercut the rock structure to a sufficient degree that numerous large boulders have calved off the offshore face and now lie in the trench. This boulder field, and the ridge itself, form the type of highly rugose habitat preferred by gag. Pulley Ridge (Figure 30.3) is a 100+ km-long series of N–S trending, drowned barrier islands on the southwest Florida shelf ~250 km west of Cape Sable, Florida.

Pulley Ridge appears to be formed on top of an ancient coastal barrier island or strand line dating back approximately 14,000 years before present when sea level was ~65–80 m lower [5]. Presently, Pulley Ridge periodically underlies the Loop Current, which feeds into the Gulf Stream western boundary current. The Loop Current brings warm, nutrient-rich waters that are clear enough to allow 1–2% of sunlight to reach the ~65 m water depth [5]. Biological assessments of these protected areas were carried out in generally similar fashion: initial mapping with multibeam sonar followed by targeted observations with remote still and video cameras.

Geomorphic Features and Habitats

The formation of the Florida carbonate platform began when North America and Africa rifted apart during the opening of the Atlantic Ocean. This occurred approximately during the late Triassic to early Jurassic (~200 Ma) ([10] and references therein). This rifting caused substantial faulting and subsidence of the original continental crust, and the formation of shallow seas, coral reefs, and the deposition of

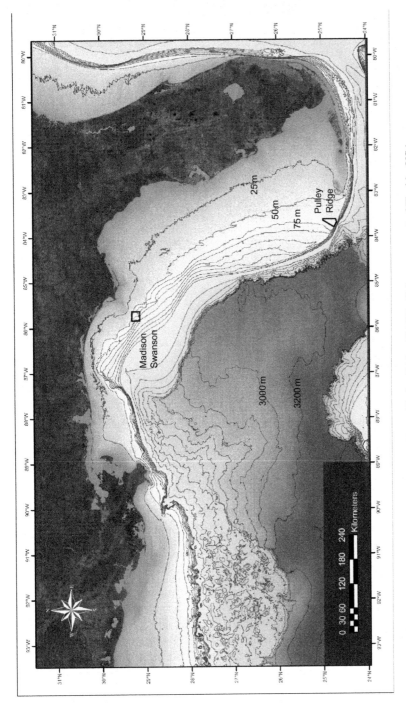

Figure 30.1 The locations of the Madison-Swanson MPA and Pulley Ridge HAPC off the west coast of Florida, USA.

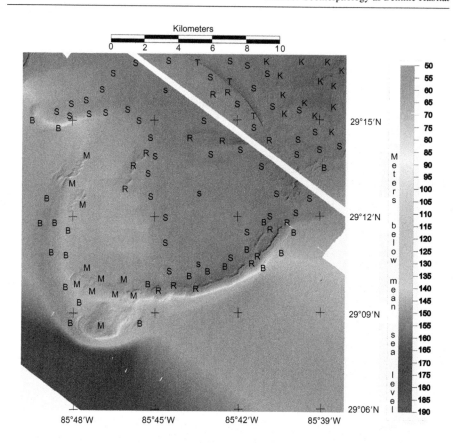

Figure 30.2 Color-shaded multibeam bathymetry (mostly 100 kHz, except for 300 kHz in NE corner separated by diagonal gap). Illumination is from the SE at an elevation of 60° with 10:1 vertical exaggeration. Following and adding to a previously published interpretation [9], the following geomorphology is identified: "R" denotes ridges, "S" denotes sedimentary back ridge area, "M" denotes mounds, "T" denotes trough, "B" denotes bedforms of variable dimensions, and "K" denotes karst-like geomorphology. The smooth (deeper) areas that exist outside the denoted area are likely to be of sedimentary origin. (For interpretation of the references to color in this figure legend, the reader is referred to the web version of this book.)

relatively flat layers of carbonate sediment, approximately 1–5 km thick, primarily composed of limestone over the original crystalline basement forming the majority of the Florida carbonate platform. The depositional history, however, was complicated during this long interval because of the repeated rise and fall of sea level during glacial events and terrestrial (siliclastic) sediment transport from the north. The most recent rise in sea level occurred after the last deglaciation (~14 Ka), and evidence of this rapid sea-level transgression has been observed in the preservation of several paleoshorelines at the approximate depths of 65, 70, and 80 m in the Pulley Ridge area [5,6] and Madison-Swanson area [9], which correlate well with similar depths near the Marquesas Keys along the southern edge of the Florida platform [11].

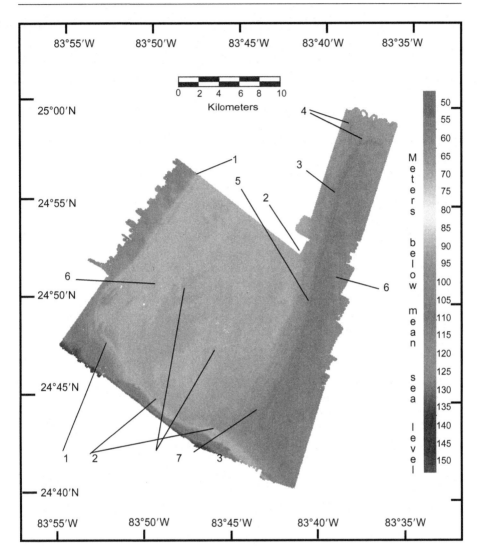

Figure 30.3 Plan view of 300 kHz multibeam bathymetry over Pulley Ridge area. There are two primary drowned barrier island ridges (Pulley Ridge West and East) at a depth of ~80 and 65 m (the westernmost orange strip next to green and the easternmost red color identified by numbers 1 and 3). Number 2 represents an intermediate paleoshoreline of about 70 m. Numbers 4, 5, 6, and 7 represent recurved spits in the north, crescent shoreline typical of barrier islands, cuspate foreland (relict inlets and ridges), and extensive cross-cutting N–S and E–W preserved bedform structures. (For interpretation of the references to color in this figure legend, the reader is referred to the web version of this book.)

Initial data collection was aboard the R/V Suncoaster operated by the Florida Institute of Oceanography. The sonar head was fixed on a secure pole mount, with metal brackets and metal stays to ensure maximum rigidity. Seafloor patch tests were conducted before and after the cruise, using the KSI acquisition and post-processing software. An Applanix POS/MV (Version 3.0) was used to measure ship attitude (heading, roll, pitch, and heave), position, and velocity. Additional post-processing has been applied to the data using CARIS HIPS software. A real-time surface sound velocity sensor was used to measure variations of the surface waters, as well as routine CTD casts.

Mounds: The Mounds is a relatively small area of exposed relict coral in the central portion of the Madison-Swanson MPA (Figure 30.2). There are no dramatic bathymetric changes in the vicinity of this feature, and the vertical relief of the carbonate outcrops can be as high as 3 m. The exposed margin of the feature is ephemeral, as currents and large storm events bury and uncover rock with the surrounding unconsolidated small grain sand, clay, and silt. Despite its small size, the Mounds is very biologically rich with large numbers of reef fishes and invertebrates.

Pinnacles: The Pinnacles of Madison-Swanson is a relict coral reef with relief approaching 10 m. Depth increases drastically beyond the offshore face, while the inshore face transitions smoothly into a sandy substrate. This area has the high rugosity associated with living coral reefs and supports high abundance of octocorals, sponges, nonhermatypic corals, and other encrusting invertebrates.

Pockmarks: EM 3000 multibeam sonar has detected very circular but low-relief (<1 m) pits or depressions that range from 5 to 15 m in diameter in both areas and most everywhere in between during transits over smooth sediments in water depths of about ~80 m ($\pm$~20 m). This abundance of pockmarks or circular depressions has been noted and documented by others using both sidescan sonar and multibeam data. Video data [12] document that these pits are created by red grouper.

Ridge—Madison-Swanson: This prominent feature within Madison-Swanson is a continuous, ~13 km long curved ridge ~6 m tall and ~80 m wide in ~80 m of water depth (Figure 30.2). There are also shorter and smaller ridges, some of which have been classified as small, partially buried calcareous pinnacles; however, the majority of the major ridge appears to be an aggregate reef (D. Palandro, personal communication) and is interpreted to have formed on top of the remnants of drowned barrier islands [9,11]. The Ridge is an area of high relief, as much as 10 m in some areas.

Ridge—Pulley Ridge: The southern edge of Pulley Ridge marks the very edge of the southern part of the Florida carbonate platform, where the depth drops down to 150 m in a short distance (Figure 30.3). The Ridge itself reflects drowned barrier island morphology at depths of ~80, ~70, ~65 m. Smaller ridge-like features are interpreted to be recurved spits, as seen on present-day barrier islands and cuspate forelands (relict inlets, ridges, and/or marsh/lagoon areas). The extensive cross-cutting N–S and E–W preserved bedform structures in Pulley Ridge were most likely formed and then modified by the rapid fluctuations in sea level described earlier.

Sediments: Madison-Swanson MPA consists of drowned, karst, hard bottom, rocky outcrops several kilometers in length, and variable thickness of fine-grained and apparently mobile coarse-grained sediments. Pulley Ridge hosts linear coral reefs that reside on drowned barrier reefs but with less sediment over the limestone basement.

Trough: Lying between the curved ridges of Madison-Swanson in about 60 m water, extending at least 10 km, is a ~8 m deep trough (Figure 30.2). The ~2 km gap between the two curved ridges appears to be covered by the sediment of the drowned delta rising 40 m above the limestone basement [9].

Figure 30.4 Coral habitats and associated fauna found in Pulley Ridge and Madison-Swanson. (A) *Agaricia* sp. coral in the Pulley Ridge HAPC. The fish to the left is a rock beauty (*Holacanthus tricolor*). (B) A sand tilefish (*Malacanthus plumieri*) mound in the Pulley Ridge HAPC. The mounds are created when the fish pile up loose rubble and normally have a single entrance at the base. The small fish above the mound is a yellowtail reef fish (*Chromis enchrysurus*). (C) A gag grouper (*Mycteroperca microlepis*) in the Madison-Swanson MPA. The red dots are from a laser measuring system on the camera array and are 10 cm apart. (D) A black grouper (*Mycteroperca bonaci*) in the Pulley Ridge HAPC. (E) A dense stand of octocorals in the Pinnacles stratum in the Madison-Swanson MPA.

Biological Communities

The principal biological interest in Madison-Swanson was a large and mobile apex predator, while in Pulley Ridge it was a sessile invertebrate (Figure 30.4). The mobility, or lack thereof, of these biological targets shaped the biological assessment strategies. In Madison-Swanson, the initial effort was complete multibeam bathymetric and backscatter mapping. These maps were used to stratify the entire reserve into seven regions. The bathymetry was used to segregate the high-relief areas, while the backscatter data were used to differentiate low-relief areas with different sediment types. Remotely Operated Vehicle (ROV) video transects were surveyed to confirm the habitat types within each strata. ROVs were a poor choice for the fishery assessment, however. The lights, noise, and movement of ROVs alter fish behavior, attracting some species and repelling others. The gag grouper is not as inquisitive as other grouper species and tend to flee upon the approach of an ROV. Therefore, stationary camera arrays, using only ambient light, were found to be the most effective tool to evaluate gag grouper populations. These were deployed in a stratified random fashion to produce statistically robust estimates of grouper abundance and distribution. Gag also migrate seasonally between winter spawning grounds on the outer continental shelf and mid-shelf foraging areas during the spring, summer, and fall. For this reason, all sampling had temporal stability with annual surveys beginning within a 2-week window in late January and concluding by mid-March.

The primary goal of the Pulley Ridge HAPC was to define the extent, health, and abundance of scleractinian corals within and adjacent to the protected area; the secondary goal was to identify and quantify the fish assemblages present. Unlike Madison-Swanson, the biological targets on Pulley Ridge, stony corals, are immobile and thus the survey design was able to exploit the mobility of ROVs without concern for target avoidance. Seasonality was similarly not a concern, other than a general avoidance of periods with high hurricane or winter storm activity that would interfere with vessel operations. The persistent nature of the corals also allowed for another modification of the survey design used to the north at Madison-Swanson; mapping and biological assessments were conducted concurrently rather than consecutively. After one portion of the HAPC was mapped, ROV survey transects were selected and executed. The ROV results contributed to the selection of subsequent areas to be mapped. Once the mapping data from the new area was processed, further ROV transects were planned and completed.

A total of 544 stationary camera array drops were conducted in the Madison-Swanson MPA between 2001 and 2009 (Figure 30.5). The stationary camera array utilized four orthogonally spaced cameras in a circular aluminum frame. Each camera had a 75° field of view, resulting in a 15° gap in coverage between adjacent cameras and a total coverage of 300°. At each station the cameras were deployed for 30 min. A 20 min segment of the recorded video was later analyzed, and all species identified to the lowest possible taxonomic level. Thirty-three ROV dives were conducted between 2007 and 2009 in and adjacent to the Pulley Ridge HAPC (Figure 30.6). Starting and ending positions for each dive were preselected, and the support vessel moved along the transect at speeds between 1 and 3 km/h. A downweight was used to stabilize the ROV umbilical against the prevailing currents, while a 30-m "leash" below the weight allowed a significant range to investigate targets to either side of the planned transect. The survey protocol involved continuous video recording and downward-looking digital still photographs taken every 2 min. The video was used to identify all vertebrate and major invertebrate species as well as the gross habitat types. The still images were used to determine percent cover of species, species groups, and benthos types.

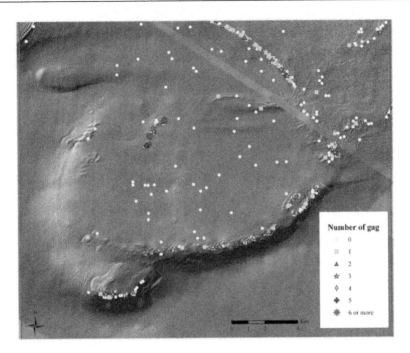

Figure 30.5 Stationary camera array deployment locations in the Madison-Swanson MPA. The Ridge is in the north, the Pinnacles is the high-relief feature in the lower half of the MPA, and the Mounds are the small feature west of the center of the area. Varying numbers of gag seen on each deployment are represented by different symbols and colors. (For interpretation of the references to color in this figure legend, the reader is referred to the web version of this book.)

A distinct difference in habitat types existed inside and outside the HAPC. Sites to the north of the HAPC, as well as some dives to the west of the HAPC, primarily supported a heterotrophic octocoral-dominated community lacking the reefal accumulation which is characteristic inside the HAPC. Habitats within the HAPC, as well as some dives to the west of the HAPC, were consistently similar, but drastically different from the other sites. The habitat in this area was characterized as rock rubble with varying coverage of algae, coralline algae, hermatypic corals, solitary and encrusting sponges, octocorals, and antipatharians. Sand tilefish (*Malacanthus plumieri*) mounds and red grouper (*Epinephelus morio*) pits were common in this area.

The Pulley Ridge HAPC was characterized as having anywhere from 70% to 100% rock rubble covered in varying degrees, with encrusting organisms providing some vertical relief. Coral species observed included *Agaricia undata*, *Agaricia lamarcki*, *Montastraea cavernosa*, and *Leptoseris cucullata*. We observed *Agaricia* spp. only on rock rubble habitat inside the HAPC and a single observation on a single dive to the west of the HAPC. In the areas it exists, *Agaricia* can be quite abundant; we had several ROV transect with over 100 colonies sighted. *Agaricia* was found in depths between 61 and 89 m, with plates of coral ranging in diameter between 5 and 30 cm.

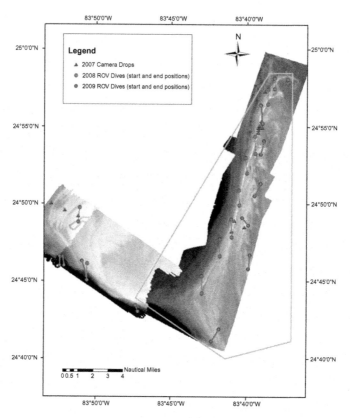

Figure 30.6 Multibeam bathymetry of the Pulley Ridge HAPC. ROV dive transects are represented by connected lines with different survey years shown in different colors. The HAPC boundary is indicated with the green line. (For interpretation of the references to color in this figure legend, the reader is referred to the web version of this book.)

One interesting observation was the abundance of dead *Agaricia*. It was common to see plates of coral that contained significant areas of dead tissue; however, no stressed or bleached corals were seen. Corals appeared to be either healthy or dead.

Overall, 106 species of fishes have been identified within Madison-Swanson and 73 within Pulley Ridge. These values must be evaluated in the context of the total effort and methods employed. Nearly 550 camera deployments were made in Madison-Swanson compared to 33 ROV dives in Pulley Ridge. Also, it was previously noted that the camera array is less disturbing to fish than the ROV. It is highly likely a greater number of species are present in Pulley Ridge than in Madison-Swanson. In Madison-Swanson, gag were present at 45% of all sites surveyed. Other economically valuable species were seen at similar levels: scamp at 65%, red snapper (*Lutjanus campechanus*) at 48%, and red grouper at 38%. In Pulley Ridge, the rock rubble habitat (the only habitat found in the HAPC) had the highest fish diversity with 51 different species. Some fish species were observed in all habitat types: tattlers (*Serranus phoebe*), reef butterflyfish

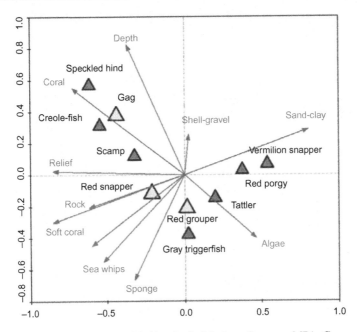

Figure 30.7 CCA of fish species and habitat in the Madison-Swanson MPA. Gag are most closely associated with relict coral reefs and to a lesser degree with greater depth and higher relief.

(*Chaetodon sedentarius*), and yellowtail reef fish (*Chromis enchrysurus*). Some species were observed exclusively over rock rubble: cherubfish (*Centropyge argi*), blue chromis (*Chromis cyaneus*), yellowtail damselfish (*Microspathodon chrysurus*), bicolor damsel-fish (*Chromis partitus*), yellowhead wrasse (*Halichoeres garnoti*), parrotfish (*Sparisoma* spp.), blue tang (*Acanthurus coeruleus*), and yellowhead jawfish (*Opistognathus aurifrons*).

Surrogacy

A series of Canonical Correspondence Analyses (CCA) were conducted on fish and habitat data from Madison-Swanson. CCA of fish species and habitat indicated gag are most closely associated with relict coral reefs and to a lesser degree with greater depth and higher relief (Figure 30.7); red snapper had a higher correlation with soft corals. The analysis of habitat and strata revealed the Pinnacles was associated with increasing depth and coral density, the Mounds was associated with increasing depth, and the Ridge was associated with greater sponge, sea whip, and soft coral density. Species and strata associations indicated gag most correlated with the Pinnacles and red snapper with the Ridge.

A total of 131 grouper were observed on the ROV dives in Pulley Ridge in 2009; 51 grouper were observed over pavement, 40 over moderate relief outcrops, 28 over

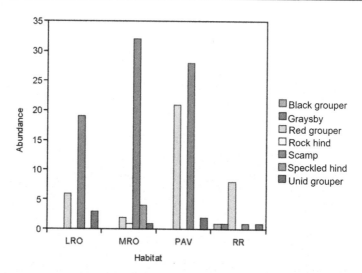

Figure 30.8 Grouper abundances by habitat type in the Pulley Ridge HAPC. LRO: low relief outcrops, MRO: moderate relief outcrop, PAV: pavement, and RR: rock rubble. No grouper were observed over sand. LRO, MRO, and PAV habitats were found outside the HAPC. RR was the only habitat observed in the HAPC. Unid grouper: unidentifiable grouper (not able to be identified down to species).

low-relief outcrops, and 10 over rock rubble (Figure 30.8). Even though grouper abundances were not high on rock rubble, diversity of grouper species was the highest on this habitat type along with moderate relief outcrops. The loose rubble in the HAPC was conducive for red grouper to excavate pits in the sediment, and they were the most abundant grouper species in that area; however, scamp was the most frequently observed grouper outside the HAPC. Rock hind (*Epinephelus adscensionis*) and speckled hind (*Epinephelus drummondhayi*) were only found on moderate relief outcrops. Black grouper (*Mycteroperca bonaci*) were only observed on rock rubble.

Acknowledgments

This project was supported in part by grants from MARFIN (2001) and NOAA's Coral Reef Conservation Program (2002–2009). Dr. Christopher Gledhill of NOAA Fisheries—Pascagoula provided the canonical correspondence analyses. Stacey Harter of NOAA Fisheries—Panama City analyzed the Pulley Ridge biological data. The NOAA ships *Gordon Gunter*, R/V *Caretta*, R/V *Gandy*, the NASA ship M/V *Freedom Star*, and the Florida Institute of Oceanography ships R/V *Suncoaster* and R/V *Weatherbird* provided the research platforms for these efforts. ROV services were provided by NURC-UNCW and NURC-UConn, and particular recognition is due to Lance Horn and Glenn Taylor of NURC-UNCW. Brian Donahue has played a leading role in all aspects of the USF EM 3,000 multibeam mobilization, acquisition, data post-processing, figure preparation, and data dissemination. More than two dozen individuals from the NOAA Fisheries Laboratories in Panama City, FL and Pascagoula, MS, and the University of South Florida in St. Petersburg, FL, assisted with the collection, processing, and analysis of data. A special thank you is given to David Palandro for his consultation on classifying these habitats.

References

[1] Gulf of Mexico Fishery Management Council (GMFMC), Regulatory Amendment to the Reef Fish Fishery Management Plan to set 1999 Gag/Black Grouper Management Measures (revised), Gulf of Mexico Fishery Management Council, Tampa, FL, 1999, 84 p.

[2] Gulf of Mexico Fishery Management Council (GMFMC), Essential Fish Habitat Amendment 3. Addressing essential fish habitat requirements, habitat areas of particular concern, and the adverse effects of fishing on fishery management plans of the Gulf of Mexico, March 2005.

[3] F.C. Coleman, C.C. Koenig, L.A. Collins, Reproductive styles of shallow water groupers (Pisces: Serranidae) in the eastern Gulf of Mexico and the consequences of fishing spawning aggregations, Environ. Biol. Fish. 47 (1996) 129–141.

[4] C.C. Koenig, F.C. Coleman, L.A. Collins, Y. Sadovy, P.L. Colin, Reproduction in gag (*Mycteroperca microlepis*) (Pisces: Serranidae) in the eastern Gulf of Mexico and the consequences of fishing spawning aggregations, in: F. Arraguin-Sánchez, J.L. Munro, M.C. Balgos, D. Pauly, (Eds.), Biology, Fisheries and Culture of Tropical Groupers and Snappers, ICLARM Conf. Proc. 48 (1996) 307–323.

[5] B.D. Jarrett, A.C. Hine, R.B. Halley, D.F. Naar, S.D. Locker, A.C. Neumann, et al., Strange bedfellows—a deep-water hermatypic coral reef superimposed on a drowned barrier island; southern Pulley Ridge, SW Florida platform margin, Mar. Geol. 214 (2005) 295–307.

[6] A.C. Hine, R.B. Halley, S.D. Locker, B.D. Jarrett, W.C. Jaap, D.J. Mallinson, et al., Coral reefs, present and past, on the West Florida shelf and platform margin, in: B. Riegl, R.E. Dodge, (Eds.), Coral Reefs of the USA, Springer, Dordrecht, 2008, pp. 127–174.

[7] H.A. McKeown, P.J. Bart, J.B. Anderson, High-resolution stratigraphy of a sandy, Ramp-Type Margin—Apalachicola, Florida, USA. Late Quaternary Stratigraphic Evolution of the Northern Gulf of Mexico Margin SEPM Special Publication No. 79, Copyright © 2004. SEPM (Society for Sedimentary Geology), ISBN 1-56576-088-3, 2004 pp. 25–41.

[8] J.V. Gardner, P. Dartnell, K.J. Sulak, Multibeam mapping of the West Florida Shelf, Gulf of Mexico: USGS Open-File Report 02-005. http://geopubs.wr.usgs.gov/open-file/of02-005/ Digital Data: http://geopubs.wr.usgs.gov/open-file/of02-005/site/data.html, http://geopubs.wr.usgs.gov/open-file/of01-448/—Also see http://geopubs.wr.usgs.gov/open-file/of01-448/ for details related to metadata and acquisition of data from cruise, 2002.

[9] J.V. Gardner, P. Dartnell, L.A. Mayer, J.E. Hughes Clarke, B.R. Calder, G. Duffy, Shelf-edge deltas and drowned barrier-island complexes on the northwest Florida outer continental shelf, Geomorphology 64 (2005) 133–166.

[10] A.F. Randazzo, The sedimentary platform of Florida: mesozoic to cenozoic, in: A.F. Randazzo, D.F. Jones (Eds.), The Geology of Florida, University Press of Florida, Gainesville, FL, 1997, pp. 39–56, Chapter 4.

[11] S.D. Locker, R.A. Armstrong, T.A. Battista, J.J. Rooney, C. Sherman, D.G. Zawada, Geomorphology of mesophotic coral ecosystems: current perspectives on morphology, distribution, and mapping strategies, Coral Reefs (2010). Published Online: DOI 10.1007/s00338-010-0613-6

[12] K.M. Scanlon, F.C. Coleman, C.C. Koenig, C. Christopher, D.C. Twichell, R.B. Halley, Landscape modification and sediment mixing by red grouper and other fish on the West Florida Shelf. Anonymous Abstracts with Programs—Geological Society of America, 36(5) (2004) 302.

31 Nontropical Carbonate Shelf Sedimentation. The Archipelago Pontino (Central Italy) Case History

Eleonora Martorelli[1], Silvana D'Angelo[2], Andrea Fiorentino[2], Francesco L. Chiocci[1]

[1]Department of Earth Sciences, University of Rome "La Sapienza", Roma, Italy, [2]Department for Soil Defense, Geological Survey of Italy—ISPRA, via Curtatone, Rome, Italy

Abstract

The Tyrrhenian Sea is a young back arc basin with a continental shelf characterized by siliciclastic sediments, delivered mainly by rivers. However, areas of carbonate sediments are also found on the shelf, as is characteristic of mid-high latitudes. The Archipelago Pontino in the eastern central Tyrrhenian Sea consists of five islands formed mainly of volcanic rocks of Plio-Pleistocene age. The low siliciclastic sediment input in this area can be attributed to the lack of significant river runoff, because there are no major rivers on the islands, which are located about 30 km away from the mainland. The relatively high-energy hydrodynamic regime around the islands also prevents the deposition of muddy sediments. The biogenic sediment fraction comprises mainly coralline algae, foraminifers, bryozoans, and undifferentiated bioclasts, and to a lesser extent bivalves, crinoids, echinids, gastropods, ostracods, pteropods, serpulids, and sponge spicules. A variety of depositional units that characterize particular marine habitats have been identified in the area. The most significant habitats include those of *Posidonia oceanica*, rhodoliths (unattached red algae), and "coralligène" (coralline algal build-ups).

Key Words: Tyrrhenian Sea, nontropical carbonate shelf, habitat mapping, coralline build-ups, rhodoliths

Introduction

The Tyrrhenian Sea is in the central part of the Mediterranean Sea between the mainland of Italy and the islands of Corsica, Sardinia, and Sicily (Figure 31.1). The sea is located in a geodynamically active area, with extensive tectonics and a young volcanism related to the opening of a back arc basin that started during Late Miocene to

Seafloor Geomorphology as Benthic Habitat. DOI: 10.1016/B978-0-12-385140-6.00031-1

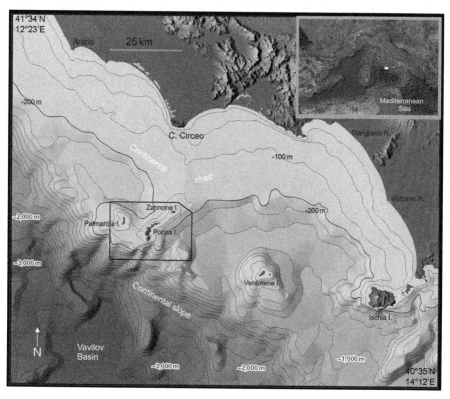

Figure 31.1 Location map and bathymetry of the Archipelago Pontino.

Pliocene times [1]. The continental shelf surrounding the basin was mainly built during the Plio-Pleistocene, after the rifting phase. As a result of its recent development, the morphology of the shelf is still evolving. The width of the continental shelf on the eastern margin of the Tyrrhenian Sea varies from 80 km (northward) to less than 1 km (southward).

The Tyrrhenian Sea is characterized by a microtidal regime, which has only a secondary influence on the coastal dynamics. Average surface salinity is about 37.5–38.5 psu; surface temperatures range from a minimum of 13°C to a maximum of 25°C. A seasonal thermocline is present at depths between 15 and 40 m, caused by summer heating of the surface waters that isolates the warmer and less dense upper layer; in winter, mixing of the surface waters causes the thermocline to progressively deepen until it disappears. Below 100–200 m water depth, the water temperature is constantly around 13–14°C.

The Italian continental shelf is predominantly characterized by terrigenous, siliciclastic sediments delivered mainly by rivers. However, areas of carbonate sediments are found on the shelf in discrete areas surrounding islands such as the Archipelago Pontino. The archipelago consists of five islands located off the coast of central Italy between 40°N and 41°N and 12°E and 13°E. It extends along the outer margin of the shelf near the continental slope, some 30 km south of the Circeo promontory (Figure 31.1).

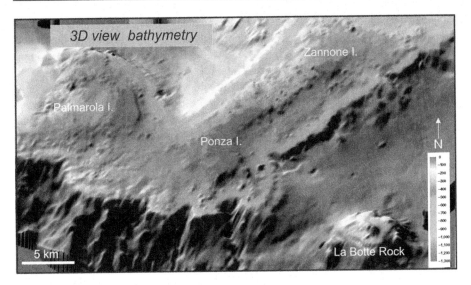

Figure 31.2 Shaded relief of the Western Pontine Islands.

The islands of Ponza, Palmarola, and Zannone (the Western Pontine Islands) are formed almost completely by volcanic rocks of Plio-Pleistocene age. The submerged areas around these islands, down to approximately 160 m water depth, are of particular interest to the study of environmental geology in Italy as they are different to most of the Italian continental margin. Although much of the shelf sediment is predominantly detritic and of fluvial origin, the area around the islands is characterized mainly by recent bioclastic carbonate deposits.

In terms of the naturalness of the study area, the Archipelago Pontino is in pristine environmental condition. This is because of their distance from the mainland and the absence of large human settlements. Moreover, human impacts are prevented by the inclusion of the archipelago within a marine protected area, as an extension of the "Parco Nazionale del Circeo."

Geomorphic Features and Habitats

Shelf: The shelf area around Ponza, Palmarola, and Zannone is less than 1.5 km wide to the south of Ponza and approximately 8 km in the channel between Ponza and Palmarola. The shelf break is clearly defined at depths varying from 95 m (N–NNE of Palmarola) to 160–165 m (toward the south), where it has a fingered termination cut by gullies and canyon heads (Figure 31.2).

Wind and wave data from the Western Pontine Islands (collected by Italian Air Force and Navy, National Wave Measurement Network) indicate that the wave trend is mainly from the west for both minor and major events, according to the higher fetch of the western sector (*ca.* 150 nm) relative to that in the east (*ca.* 50 nm). Surface currents flow mainly to the NW along the Pontine shelf area [2].

Sediments: The low siliciclastic sediment input in this area can be attributed to the lack of a significant river runoff on the islands and their distance from the mainland; the nearest rivers are more than 60 km away. The overall sediment starvation of the shelf is further enhanced by relatively high-energy hydrodynamics that prevent the deposition of muddy sediments. This setting, together with the widespread occurrence of hard substrate on the shelf, support the development of carbonate assemblages dominated by benthic populations typical of coastal detritic and coralligenous biocoenoses (respectively "Détritique Côtier" and "Coralligène," *sensu* Peres and Picard [3]). Peres and Picard specifically classified the Mediterranean by subdividing the seafloor into plains (e.g., mesolittoral, infralittoral, circalittoral) characterized by different environmental conditions and biocoenoses.

Nontropical carbonate sedimentation is characteristic of mid-high latitudes; a synthesis of the world studies on this subject was published by Nelson [4,5], including examples from southwest and southern Australia, New Zealand, British Isles, west of Scotland, and northeast Atlantic Ocean. Concerning the Mediterranean, the main studies have focused on the Balearic Islands [6,7], the Alboran Banks [8], the Algerian shelf [9,10], and the Moroccan shelf [11].

Nelson [4] introduced the term "nontropical" to emphasize that carbonate deposits can be found in areas other than tropical environments. Different classifications of biogenic assemblages were reported by Lees and Buller [12] and Lees [13] based on temperature and salinity gradients, and by Carannante et al. [14] and James [15] based mainly on the composition of the assemblages; the last two authors pointed out that temperature is not the only parameter controlling nontropical assemblages, as carbonate sedimentation is also influenced by factors such as nutrient supply, turbidity, and depth of deposition. Pomar [16] defined specific depth zones of assemblage distribution depending on light availability: euphotic (down to 30 m water depth), oligophotic (between 30 and 100 m), and aphotic (deeper than 100 m). In the Archipelago Pontino, the low turbidity of waters allows for a deeper penetration of the light, which induces a downward shift of the bathymetric boundaries between such zones.

Biological Communities

The carbonate fraction is formed mainly by coralline algae, foraminifers, bryozoans, and undifferentiated bioclasts, with secondary echinoderms, mollusks, ostracods, pteropods, serpulids, and sponge spicules that vary in abundance according to the water depth.

Benthic forams are homogenously distributed across the area, except in shallow-water settings; the abundance of planktonic forams increases with distance offshore. Bryozoans are very common, with a peak in abundance between 160 and 180 m. However, the component that most characterizes the Archipelago sediments is represented by rhodoliths, unattached structures of red coralline algae that are common in the western Mediterranean.

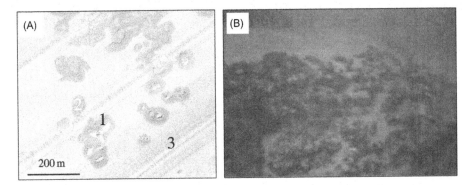

Figure 31.3 Coralline buildups: (A) sidescan sonar mosaic of a seafloor colonized by coralline buildups (black: high backscatter, white: low backscatter); 1: coralline buildup; 2: sediments produced by coralline buildups dismantling; 3: uncolonized seafloor. (B) ROV image of coralline buildup encrusting hard substrate.

Sediment–Organism Interaction

Surveys of the Pontine Islands area were carried out during the Italian geological mapping project (CARG) using sidescan sonar, seafloor sampling (grab sampling, dredging and direct rock sampling by scuba diving), remotely operated vehicles (ROVs), and textural–compositional analysis of seafloor sediments. A variety of depositional units that characterize specific marine habitats have been identified as a result of these surveys.

Coralline buildups: Apart from the widespread occurrence of *Posidonia oceanica* (L.) Delile meadows, which are dealt with in [17] and were found around the three islands in water depth up to 40–45 m by Ardizzone and Belluscio [18], another significant habitat is represented by coralline build-ups that crop out on the outer shelf (Figure 31.3). Coralline buildups extensively colonize the hard substrate to form wide and nearly continuous structures (sometimes resembling the shape of tropical reefs). The buildups extend laterally from approximately 1 to 100 m and, in deeper waters, are often completely or partially buried by fine-grained sediments.

The "coralligène" assemblage includes corals, gorgonias, and sea sponges and represents one of the most diverse ecosystems in the Mediterranean Sea. It also forms a significant natural habitat and ecosystem for fish nursery grounds.

Bioclastic sand and gravel made up of well-preserved bryozoans and coralline algae occur only in association with the "coralligène" structures and represent the dismantling facies of bioherms (Figure 31.3A). This association is demonstrated by the absence of bioclastic sands/gravels in areas that are not colonized by coralline buildups, where palimpsest deposits are found instead.

Rhodoliths: One of the most unusual habitats on the seafloor around the Pontine Islands is characterized by rhodoliths. These unattached red algae structures form an irregular coarse-grained deposit that is so widespread that it effectively forms a continuous cover on the seafloor of the Archipelago. The rhodoliths are most abundant

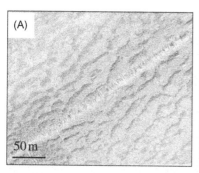 

Figure 31.4 Rhodoliths: (A) sidescan sonar image of a rhodoliths bottom affected by intense reworking by bottom currents (black: high backscatter, white: low backscatter); (B) photo of bioclastic sand containing living rhodoliths (pralines).

at water depths between 40 and 100 m (and were generally alive at the time of sampling), especially on saddle areas between Ponza–Palmarola and Ponza–Zannone, where their growth is favored by the light conditions and bottom currents. Different morphotypes are closely related to differences in hydrodynamics, sedimentation rates, and substrate stability [19], such that their distribution can be used as a tool to interpret the environments in which they are found. Pralines (subspherical, globular lumps) are more abundant where hydrodynamics are more intense (e.g., north of Palmarola, on the southern side of Palmarola–Ponza channel, and in the Ponza–Zannone channel). Boxworks (larger irregular concretions) are more common where hydrodynamics are lower (e.g., NE of Ponza). This bioclastic sediment (Figure 31.4B) is made of sands with gravel, containing mainly coralline algae, biogenic fragments, and benthic foraminifera. Coralline algae, both maerl (branching concretions) and praline facies, dominate the gravel fraction; a subordinate sediment fraction includes terrigenous clasts. This sediment is interrelated to either the coastal detritic seafloor (*sensu* Peres and Picard [3]) or to the rhodalgal lithofacies (*sensu* Carannante [14]). These types of seafloor form a uniform facies with high backscatter in sidescan sonar records, but become heterogenous on areas of the seabed that are swept by bottom currents (Figure 31.4A).

Bioclastic sediment also occurs mainly between 50 and 80 m water depth as patches or ribbons with a distinct pattern of low (finer grained sediments, mainly gravelly sand) and high backscatter (coarser grained sediment; e.g., sand with gravel) that follow the bathymetric contours. The distance between the ribbons ranges from a few meters to 20–40 m. Ribbons in the Western Pontine occur on the saddle areas and are interpreted as longitudinal bedforms as they trend subparallel to modern shelf currents.

Surrogacy

No statistical analyses have been carried out on this data set to examine relationships between physical surrogates and benthos.

Acknowledgments

We thank the Hydrographical Navy Institute of Italy for providing bathymetric data for the studied area. We are grateful to all reviewers who helped improve our chapter.

References

[1] L. Jolivet, C. Faccenna, C. Piromallo, From mantle to crust: stretching the mediterranean, Earth Planet. Sci. Lett. 285 (1–2) (2009) 198–209, Elsevier.

[2] V. Artale, M. Astraldi, G. Buffoni, G.P. Gasparini, Seasonal variability of gyre-scale circulation in the northern Tyrrhenian Sea, J. Geophy. Res. Oceans 99 (C7) (1994) 14.127–14.138.

[3] J.M. Peres, J. Picard, Nouveau Manuel de Bionomie Benthique de la Mer Mediterranée, Rec. Trav. St. Mar. 31 (47) (1964) 1–137.

[4] C.S. Nelson, An introductory perspective on non-tropical shelf carbonates, Sediment. Geol. 60 (Special Issue) (1988) 3–12.

[5] C.S. Nelson, Non-tropical shelf carbonates—modern and ancient, Sediment. Geol. (Special Issue) (1988) 60.

[6] J.J. Fornos, W.M. Ahr, Temperate carbonates on a modern, low-energy, isolated ramp: the Balearic platform, Spagna, J. Sediment. Res. 67 (2) (1997) 364–373.

[7] M. Canals, E. Ballesteros, Production of carbonate particles by phytobenthic communities on the Mallorca-Menorca Shelf, northwestern Mediterranean Sea, Deep-Sea Res. 44 (1996) 611–629.

[8] J.D. Milliman, Y. Weiler, D.J. Stanley, Morphology and carbonate sedimentation on shallow banks in the Alboran Sea, in: D.J. Stanley (Ed.), The Mediterranean Sea: A Natural Sedimentation Laboratory, Dowden, Hutchinson & Ross, Stroudsburg, PA 1972, pp. 241–259.

[9] J. Caulet, Recent biogenic calcareous sediments on the Algerian continental shelf, in: D.J. Stanley (Ed.), The Mediterranean Sea: A Natural Sedimentation Laboratory, Dowden, Hutchinson & Ross, Stroudsburg, PA 1972, pp. 261–277.

[10] L. Leclaire, La sedimentation holocène sur le versant méridional du Bassin Algéro-Baléares (Précontinent algérien), Mém. Mus. (Série C – Sc. De la Terre) 24 (1972) 1–391.

[11] C.P. Summerhayes, phosphate deposits on the Northwest African continental shelf and slope, Unpublished Ph.D. Thesis, University of London, 1970.

[12] A. Lees, A.T. Buller, Modern temperate-water and warm-water shelf carbonate sediments contrasted, Mar. Geol. 13 (1972) 1767–1773.

[13] A. Lees, Possible influence of salinity and temperature on modern shelf carbonate sedimentation, Mar. Geol. 19 (1975) 159–198.

[14] G. Carannante, M. Esteban, J.D. Milliman, L. Simone, Carbonate lithofacies as paleolatitude indicators/problems and limitations, Sediment. Geol. 60 (1998) 333–346.

[15] N.P. James, The cool-water carbonate depositional realm, in: N.P. James, J.A.D. Clarke, (Eds.), Cool-Water Carbonates, SEPM. Special Publication, Tulsa, Oklahoma, 1997, 56, pp. 1–22.

[16] L. Pomar, Types of carbonate platforms: a genetic approach, Basin Res. 13 (2001) 313–334.

[17] S.D'Angelo, A. Fiorentino, Phanerogam meadows: a characteristic habitat of the Mediterranean shelf. Examples from the Tyrrhenian Sea. This volume, pp.159–167.

[18] G.D. Ardizzone, A. Belluscio, Le praterie di Posidonia oceanica delle coste laziali, in: Il mare del Lazio, Università degli studi di Roma "La Sapienza", Regione Lazio assessorato opere e reti di servizi e mobilità, Roma, 1996, pp. 194–217.

[19] D. Basso, Deep rhodolith distribution in the Pontian Islands, Italy: a model for the paleoecology of a temperate sea, Palaeogeogr. Palaeoclimatol. Palaeoecol. 137 (1998) 173–187.

32 Habitats of the Cap de Creus Continental Shelf and Cap de Creus Canyon, Northwestern Mediterranean

Claudio Lo Iacono[1], Covadonga Orejas[2],
Andrea Gori[3], Josep Maria Gili[3], Susana Requena[3],
Pere Puig[3], Marta Ribó[3]

[1]Unidad de Tecnología Marina, Consejo Superior de Investigaciones Científicas (UTM-CSIC), Barcelona, Spain, [2]Instituto Español de Oceanografía (IEO), Santander, Spain, [3]Instituto de Ciencias del Mar, Consejo Superior de Investigaciones Científicas (ICM-CSIC), Barcelona, Spain

Abstract

The Cap de Creus continental shelf and Cap de Creus canyon are located in the southernmost sector of the Gulf of Lions, in the northwestern Mediterranean. The Cap de Creus continental shelf contains sandy and muddy sediments and an abrupt morphology, with rocky outcrops, relict bioherms, erosive features, and planar bedforms. The Cap de Creus canyon breaches the shelf at a depth of 110 m and denotes a marked difference in the morphology between the northern and the southern flank, reflecting a different depositional regime. The most common substrates correspond to coarse and medium sands (28%) and silty sediments (40%). The most common megabenthic assemblages of the shelf correspond to the communities of "offshore detritic" (31.95%) and "coastal terrigenous muds" (36.99%), mostly dominated by sea pens, alcyonaceans, and ceriantharians. The northern flank of the Cap de Creus canyon is predominantly depositional, whereas the southern flank is erosional. Rocky outcrops provide the substratum for cold-water coral (CWC) communities' development, in which the white coral *Madrepora oculata* is the most abundant species.

Key Words: Swath bathymetry, Continental Shelf, Submarine Canyon, Geomorphology, Habitat mapping, Northwestern Mediterranean

Introduction

Continental shelves and submarine canyons represent peculiar environments that are geologically heterogenous on a small spatial scale (form hundreds to tens of meters) and therefore play a central role in increasing the ecosystem biodiversity [1–4].

Seafloor Geomorphology as Benthic Habitat. DOI: 10.1016/B978-0-12-385140-6.00032-3

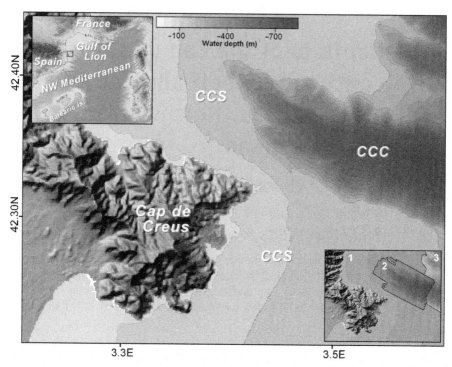

Figure 32.1 Bathymetric map of Cap de Creus continental shelf (CCS) and Cap de Creus canyon (CCC) based on a 10 m resolution grid. Top-left inset: location map showing the northwestern Mediterranean and the survey area, outlined in red. Bottom-right inset: coverage of the multibeam surveys on which this work is based (1, light green: Spanish Fishery General Secretary-ESPACE Project; 2, dark green: Fugro N.V., AOA Geophysics, University of Barcelona; 3, yellow: INDEMARES Project). (For interpretation of the references to color in this figure legend, the reader is referred to the web version of this book.)

The study area of this work includes the Cap de Creus continental shelf and the Cap de Creus canyon (depth range 40–800 m), located in the Gulf of Lions, where a complex network of submarine valleys cut the continental shelf and slope regions [5] (Figures 32.1 and 32.2). The total surface of the area is around 611 km^2, 74.7% of which is covered by the continental shelf and a small portion of the upper slope (456 km^2), whereas the remaining 25.3% is covered by the submarine canyon (155 km^2). The continental shelf in this region displays a rough morphology and its width is dramatically reduced between the Cap de Creus promontory and the canyon (Figure 32.1). Shelf sediments are mainly composed of muds and change to coarser fractions along the outer shelf, reflecting an intensification of bottom currents flowing from the North [6]. Numerous studies of sediment transport and accumulation processes have been recently conducted in the submarine canyons from the Gulf of Lions, providing a unique temporal and spatial perspective of their functioning [7,8]. These studies recognized the importance of major storms and dense shelf-water cascading

Figure 32.2 Three-dimensional perspective model of the study area: rf: relict features; pb: planar bedforms; es: erosional scour; sr: suboutcropping rocks; us: upper slope; CCC: Cap de Creus canyon; gf: giant furrows; tc: tributary channel. Forked arrows: bottom current paths of the area based on Ref. [6] and inferred from the orientation of the observed morphologic features.

events in exporting shelf waters and particles toward deep-sea regions [7,8]. The preferential direction of the coastal currents, the narrowing of the shelf and the coastal topographic constrains result in most of the sediment transport occurring through the Cap de Creus canyon, where observed sediment fluxes associated with a approximately 1 m/s near-bottom current, mainly flowing across the southern canyon flank, are two orders of magnitude higher than in the eastern and central submarine canyons [9].

In this study, we present a characterization of the main morphological and depositional features and related habitats of the Cap de Creus continental shelf and canyon head, in the northwestern Mediterranean (42.21°–42.44°N, 3.12°–3.50°E), integrating swath bathymetry, video images, and sediment samples. No statistical analyses have been carried out to analyze relationships between abiotic surrogates and benthos.

Geomorphic Features and Habitats

The Cap de Creus continental shelf has been mapped during a number of oceanographic cruises conducted since the 1990s by the Spanish National Research Council (CSIC) [10]. High-resolution swath-bathymetry data of the continental shelf presented in this work were kindly provided by the Spanish Fishery General Secretary ("Secretaria General de Pesca") and acquired in the frame of the ESPACE Project (Estudio de la Plataforma Continental Española) (Figure 32.1). Additional bathymetric data were acquired in specific sectors of the shelf as a part of the projects DEEPCORAL and INDEMARES [11]. Multibeam data from the Cap de Creus Canyon has been acquired during two surveys conducted in 2004 by Fugro N.V., AOA Geophysics, and the University of Barcelona, and in 2010 by the Instituto de Ciencias del Mar (ICM-CSIC) in the frame of the INDEMARES Project (Figure 32.1). The combination of the different swath mapping data sets allowed the production of 10 m cell size high-resolution bathymetric images of the area for depths ranging from 20 to 850 m (Figures 32.1 and 32.2). Twenty-two video surveys were conducted from 2005 to 2007 using ROVs and the manned submersible JAGO (IFM—GEOMAR) in a depth range of 50–400 m. Megabenthic communities in the area were observed in video records, and in some cases, organisms were collected. Video analysis was performed with the Final Cut software (Apple Inc.). Three classes (soft sediment, rocky boulders, and massive hardrock outcrops) have been defined in order to characterize the different substrates in the video transects as well as the correspondence between organisms and substrate type [4].

Several sediment samples have been collected in the area since the 1970s [6,12]. Analysis of the main characteristics of sediments (composition and grain size) and of their associated benthic assemblages resulted in a high-resolution habitat mapping of the shelf and of the canyon [11]. The totality of data have been merged using Quantum-GIS software, an open source Geographic Information System (GIS), and ArcInfo mapping software, in order to produce maps and carry out spatial analysis operations.

The Cap de Creus continental shelf: The continental shelf mapped around the Cap de Creus promontory covers a surface of around 400 km², displaying a width ranging from a minimum of 2.7 km near the cape to a maximum of 12 km along the regions south and north of the cape. The average slope of the shelf is 1°, with maximum values of 2.5°, in correspondence with the easternmost sector of the cape. The inner shelf area shows reduced extensions and very steep rocky outcrops along the coastal belt for depths of up to 60 m. The shelf edge lies at an average depth of 120–130 m and is N–S oriented, except for sector where it is breached by the southern flank of the Cap de Creus canyon, showing a NW–SE trend (Figure 32.3A).

The shelf alternates between smooth areas, where sandy and muddy sediments prevail, to rough rocky areas (7.16%), mainly present along the coast and in the outer shelf for depths between 95 and 130 m (Figure 32.3A). Rocky outcrops, elevated by about 6 m above the general seafloor, trend parallel to the coastline at depths of 90–100 m (Figures 32.2 and 32.3A). These outcrops correspond to circular highs located north of the Cape and concentrated along a 17 km long and 500 m wide NW–SE trending area and to 4.5 km long linear ridges located south of the Cape along a 600 m wide NW–SW trending area ("rf" in Figure 32.2). These features correspond to relict coastal sedimentary features and/or relict bioherms whose formation is related to the last sea-level rise stage [10]. North of the Cap de Creus promontory, between the described circular highs and the southern flank of the canyon, juxtaposing planar sedimentary beds occur for a depth range of 100–120 m, giving rise to sinuous steps up to 3 m high ("pb" in Figure 32.2). The area, covering a surface of around 12 km³, is characterized by very coarse sediments, which have been described as detritic communities in section "Macrobenthic Communities" (Figure 32.3B).

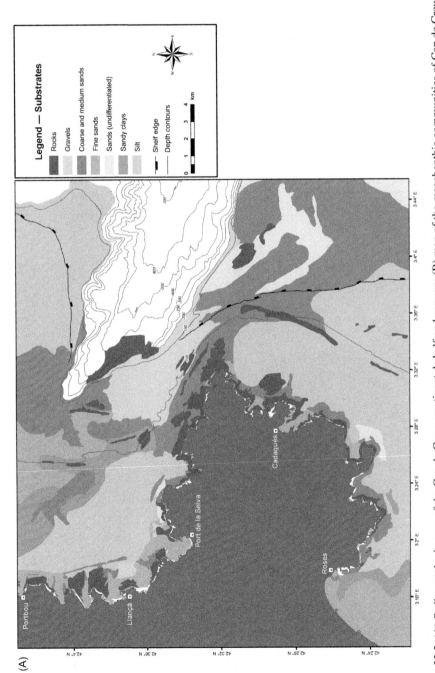

Figure 32.3 (A) Sedimentologic map of the Cap de Creus continental shelf and canyon; (B) map of the megabenthic communities of Cap de Creus continental shelf and canyon.

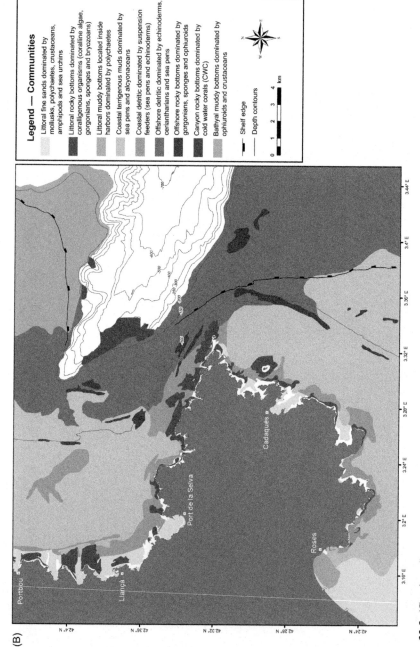

Figure 32.3 (Continued)

Table 32.1 Areal Extent (%, km^2) of Substrates and Communities of the Cap de Creus Continental Shelf and Canyon

Substrate	Surface Area (%)	Surface Area (km^2)
Rock	7.16	28.11
Gravels	4.79	18.82
Coarse and medium sands	27.72	108.66
Sands (undifferentiated)	4.96	19.45
Fine sands	10.33	40.49
Sandy clays	4.86	19.02
Silt	40.18	157.59
Total	100	392.14

Community	Surface Area (%)	Surface Area (km^2)
Coastal detritic	6.53	25.51
Coastal terrigenous muds	36.99	145.33
Coralligen	3.16	12.43
Fine sands	5.69	22.21
Offshore detritic	31.95	125.28
Offshore rocks	1.81	7.13
Cold-water corals	1.19	4.68
Bathyal muds	12.12	47.47
Port communities	0.56	2.10
Total	100	392.14

South of the Cap de Creus canyon, the shelf region is marked by a rough morphology for depths deeper than 120 m. A NW–SE-oriented 10 m deep elliptical hole (scour) with the major axis 5 km long and the minor axis 1 km long, occurs at a depth of 130 m and reflects a strongly erosive regime controlled by bottom current flowing from the north across the shelf ("es" in Figure 32.2). An outcropping and suboutcropping rocky area occurs between 120 and 135 m, rising from the general seafloor for 10 m on average and covering up to 34 km^2 ("sr" in Figure 32.2). North the Cap de Creus canyon, the shelf gently slopes to the upper slope region to depths of 160 m, with an average gradient of 2°. In this area, NE–SW-oriented linear bedforms occur for depths from 125 to 135 m. For deeper depths the seafloor is breached by a tributary channel of the Cap de Creus canyon ("tc" in Figure 32.2).

Silty sediments constitute the dominant sediment fraction on the Cap de Creus shelf (40.18%) (Table 32.1). Fine sediments accumulated in the Cap de Creus shelf close to the coast, as well as in deeper areas [13] (Figure 32.3A). From the inner shelf to a depth of up to 90 m, sediments are mainly composed of silt, mud, and fine sands (60.3%) (Table 32.1; Figure 32.3A). From the depth of 100 m, along the canyon head and along the outer shelf south the Cap de Creus canyon, sediments are composed by coarse and medium sands (27.7%) or detritic bioclastic gravels (4.79%) (Table 32.1; Figure 32.3A). The occurrence of such coarse sediments at these depths is related to the bottom current acceleration documented along the outer shelf regions of the area

[14] (Figure 32.2). Finally, silty sediments characterize the northern portion of the shelf from depths from 100 m to the upper slope region, partially sinking to the Cap de Creus canyon through its northern flank [14,15] (Figure 32.3A).

The Cap de Creus canyon: The Cap de Creus canyon is the southwesternmost submarine canyon of the Gulf of Lions, before the constriction of the Cap de Creus promontory (Figure 32.1). The canyon incises the shelf edge at depths of 110 m for more than 50 m and progressively widens for maximum amplitude of up to 6 km and depths of more than 650 m (Figure 32.2). The thalweg displays a deeply incised V-shaped morphology, 2.5° steep on average, descending to a depth of 615 m. From this depth, the thalweg amplitude increases, widening from 2 to 3.5 km. The flanks of the canyon present a different morphological configuration, indicating different depositional settings and hydrodynamic regimes. The southern flank is mainly characterized by broad areas with rocky outcrops, steep and overhanging walls, and terraces indicating an ongoing, predominantly erosive regime [16] (Figure 32.2). Giant furrows, tens of kilometers long, occur for depths between 150 and 1,400 m (out of the limits of the area investigated in this work) ("gf" in Figure 32.2) [15]. These erosive morphological features are generated by dense-water cascading and represent the preferential routes for sediment transport along the southern canyon flank [8,16,17]. The northern flank displays a smooth morphology, with rounded gullies and scars, suggesting a depositional regime for this sector (Figure 32.2). The observed contrast in flank aspect is mainly ascribed to the strong bottom currents and associated high suspended sediment loads flowing down the canyon preferentially along the southern flank [17] (Figure 32.2).

Sediments on the canyon head are composed of coarse sands and shell debris until a depth of 400 m. At deeper depths, the canyon axis is composed of soft consolidated muds mixed to interlayered shell debris [14]. Along the southern flank, where sediment accumulation rates are very low [6], thin layers of gravels and very coarse sands mixed to fine fractions occur along the upper canyon rim until a distance of 15 km from the canyon head. The northern flank of the canyon is mainly composed of clays, with sediment accumulation rates of up to 1.5 mm/year, probably due to advection via nepheloid layer transport controlled by regional southward currents [6].

Megabenthic Communities

The geological configuration of the continental shelf and of the canyon of Cap de Creus (Figure 32.3A) gives rise to different kinds of megabenthic communities (Figure 32.3B). Considering the shelf and the canyon together, more than 1,200 species have been recorded to date (Table 32.2). Benthic suspension feeders are the dominant functional group. Description of the megabenthic communities is based on Pérès and Picard (1964) [18] and Desbruyères et al. (1972–1973) [12]. Original French names after Pérès and Picard (1964) are in cursive within brackets.

The Cap de Creus continental shelf: The coastal rocky bottoms of the Cap de Creus continental shelf are dominated by the community of coralligenous

Table 32.2 Number of Benthic Species Recorded in the
Cap de Creus Continental Shelf and Canyon

Phylum	Class	Order	Species Number
Porifera			21
Cnidaria	Scyphozoa		6
	Hydrozoa		101
	Anthozoa (Hexacorallia)		40
	Anthozoa (Octocorallia)		25
Ctenophora			1
Entoprocta			1
Mollusca	Bivalvia		38
	Cephalopoda		16
	Gasteropoda		78
	Polyplacophora		4
	Scaphopoda		2
Annelida			173
Echiura			1
Sipuncula			1
Arthropoda (Crustacea)	Malacostraca	Amphipoda	100
		Decapoda	128
		Isopoda	4
		Mysidacea	12
		Euphausiacea	11
		Cumacea	10
		Tanaidacea	3
		Stomatopoda	2
	Copepoda		133
	Cirripedia		5
	Ostracoda		4
Bryozoa			87
Brachiopoda			2
Chaetognatha			5
Echinodermata	Crinoidea		2
	Holothuridea		1
	Echinoidea		9
	Asteroidea		6
	Ofiuroidea		9
Chordata (Cephalochordata)			1
Chordata (Tunicata)	Ascidiacea		31
Chordata (Vertebrata)	Chondrichthyes		15
	Osteichthyes		159
	Reptilia		3
	Aves		15
	Mammalia		6
Total			1,271

Source: References [11,18]; this work.

Figure 32.4 Megabenthic organisms of some of the biocenoses inhabiting the continental shelf and canyon of Cap de Creus. (A) Crinoids field of the genus *Leptometra*, which are characteristic of the community of coastal and offshore detritics; (B) community of coastal terrigenous muds, dominated by sea pens; (C, D) community of offshore detritic with exemplars of *Echinus melo* and *Cidaris* (C) characteristic in areas with presence of boulders, and of *Cerianthus membranaceus* (D), a frequent species in those communities in the soft-bottom patches; (E, F) offshore rocky community, mainly composed by gorgonias from the genus *Eunicella* (E), and sponges from genus *Axinella* (F); (G–I) community of white corals, two of the CWC species present in the canyon with low densities: *Dendrophyllia cornigera* (G) and *Lophelia pertusa* (H); the most abundant species is *Madrepora oculata* (I), forming, as in the picture, dense aggregations in rocky boulders. Bar scales: A–F: 10 cm, G and H: 5 cm, I: 20 cm.
Source: Images: A–F: Gavin Newman; G–I: JAGO team, IFM-GEOMAR. Communities have been named after Pérès and Picard (1964).

(*biocoenose coralligène*) (3.16%) (Table 32.1; Figure 32.3B), which is mainly characterized by coralline algae and suspension feeders species (gorgonians, sponges, and bryozoans). Sediments of the coastal belt and of the inner shelf are dominated by the community of fine sands (*biocoenose des sables fins et bien calibres*) (5.69%) and the community of costal detritic (*biocoenose des fonds détritiques côtières*) (6.53%) (Table 32.1; Figure 32.3B). The community of fine sands is dominated by the mollusk *Nucula sulcata*, polychaetes such as *Lumbrinereis gracilis* and *L. latreilli*, crustaceans such as the amphipod *Ampelisca diadema*, and sea urchins such as *Echinus melo*. The community of costal detritic is characterized by the presence of filter feeders as crinoids (Figure 32.4A), taking advantage of the strong hydrodynamic conditions of this area. The muddy areas of the continental shelf (36.99%) are dominated by the community of coastal terrigenous muds (*biocoenose des vases terrigenes côtières*) (Figure 32.3B). This biocenosis is characterized by high densities of sea pens from the genus *Pteroides* and *Pennatula* (Figure 32.4B), and of the alcyonarian *Alcyonium palmatum*.

Along the outer shelf, species diversity as well as organism abundance decrease compared to the coastal areas. The community of the offshore detritic (*biocoenose des fonds détritiques du large*) dominates the outer shelf (31.95%) (Table 32.1; Figure 32.3B) and is characterized by patchily distributed dominant taxa: echinoderms (mainly sea urchins such as *Echinus melo* and *Cidaris cidaris*), ceriantharians (*Cerianthus membranaceus*), and isolated sea pens (Figure 32.4B–D); crinoids (from the genus *Leptometra*) have been found in this community (Figure 32.4A). The rocks outcropping along the outer shelf are colonized by the offshore rocky community (*biocoenose de la roche du large*) (1.81%), which is mainly composed by cnidarian species (mainly gorgonians from the genus *Eunicella*), sponges (genus *Axinella*), and ophiurids (genus *Ophioderma*) (Figure 32.4E–F).

The Cap de Creus canyon: The communities of the Cap de Creus canyon described in this chapter are located from 100 to 400 m depth, which to date correspond to the surveyed depth range. Sediments present on the southern flank of the canyon are patchily distributed and correspond to gravels mixed to fine sediments, with the dominant benthic communities corresponding to the offshore detritic (*biocoenose des fonds détritiques du large*), characterized by sea pens, alcyonaceans, or ceriantharians. The northern flank of the canyon is mainly characterized by soft sediments, and the corresponding communities have not yet been described. The most characteristic and studied community in the Cap de Creus canyon is the one growing on the rocky substrates (Figure 32.3B). The rocky areas of the canyon (different-sized boulders and vertical walls) shelter the highly diverse benthic community of white corals (*biocoenose des coraux blancs*). Four cold-water coral (CWC) species are the most conspicuous in the Cap de Creus canyon: *Madrepora oculata, Lophelia pertusa, Dendrophyllia cornigera*, and *Desmophyllum cristagalli* (Figure 32.4G–I). *M. oculata* (Figure 32.4I), always associated with hard substrate, is the most abundant species and is the coral framework maker of the Cap de Creus canyon. The populations of this species have been quantitatively analyzed and densities of around 11 colonies/m^{-2} have been documented in some locations [4]. The populations of *M. oculata* are present in different scenarios, occurring in some cases in large patches with different size colonies. This indicates that populations are healthy, as new recruits are present simultaneously with old individuals. *M. oculata* is also the dominant species in other Mediterranean CWC communities [19,20]; this dominance differs from the east Atlantic Ocean, where *L. pertusa* is the most abundant species [21]. The reason for this difference is still not clear. However, it is likely that *M. oculata* has a wider temperature tolerance than other CWC and is able to survive in the relatively warm Mediterranean deep waters (around 13°C at 200–300 m depth in the Cap de Creus canyon) [8,17]. *L. pertusa, D. cornigera*, and *D. cristagalli* do not form real populations in the canyon and appeared mostly as isolated colonies. The CWC community presents a high diversity of associated species, mainly dominated by crinoids, brachiopods, sponges, bivalves, polychaetes, hydroids, and bryozoans, among others.

Further surveys in the Cap de Creus canyon will accomplish a complete and exhaustive mapping of the habitats present in the area.

Acknowledgments

This research was supported by the European projects INDEMARES (Life + NAT/E/000732), HERMES (Goce-CT-2005-511234-I), HERMIONE (Contract number: 226354), INTERREG ("Pirineus Mediterrànis: La muntanya que uneix"), and the Spanish national projects DEEPCORAL (CTM2005-07756-C02-02/MAR) from the Spanish Ministry of Science and Technology and SHAKE (CGL2011-30005-C02-02) of the Spanish Ministry of Science and Innovation. Some bathymetric data has been kindly provided by the Spanish Fishery General Secretary ("Secretaria General de Pesca"). We are grateful to Dan Orange (AOA Geophysics Inc.) for having provided the bathymetric data set of the Cap de Creus canyon. We thank the captain and the crew of the R/V *Garcia del Cid* for their kind assistance during the surveys, and the UTM technician staff. We are grateful to Carlos Dominguez (ICM-CSIC) for his assistance in photo editing.

References

[1] K.J. Sinka, W. Boshoffb, T. Samaaic, P.G. Timm, S.E. Kerwath, Observations of the habitats and biodiversity of the submarine canyons at Sodwana Bay, S. Afr. J. Sci. 102 (2006).

[2] C. Lo Iacono, E. Gràcia, S. Diez, G. Bozzano, X. Moreno, J.J. Dañobeitia, et al., Seafloor characterization and backscatter variability of the Almería Margin (Alboran Sea, SW Mediterranean) based on high-resolution acoustic data, Mar. Geol. 250 (2008) 1–18.

[3] C. Lo Iacono, J. Guillén, P. Puig, M. Ribó, M. Ballesteros, A. Palanques, et al., Large-scale bedforms along a tideless outer shelf setting in the western Mediterranean, Continent. Shelf Res. 30(17) (2010) 1802–1813.

[4] C. Orejas, A. Gori, C. Lo Iacono, P. Puig, J.M. Gili, M.R.T. Dale, Cold-water corals in the Cap de Creus canyon, northwestern Mediterranean: spatial distribution, density and anthropogenic impact, Mar. Ecol. Prog. Ser. 397 (2009) 37–51.

[5] S. Berné, B. Loubrieu, the CALMAR ship board party, Canyons and recent sedimentary processes on the Western Gulf of Lions margin, C.R. Acad. Sci., Ser. IIa: Sci. Terre Planets 328 (1999) 471–477.

[6] A.L. DeGeest, B.L. Mullenbach, P. Puig, C.A. Nittrouer, T.M. Drexler, X. Durrieu de Madron, et al., Sediment accumulation in the western Gulf of Lions, France: the role of Cap de Creus Canyon in linking shelf and slope sediment dispersal systems, Continent. Shelf Res. 28 (2008) 2031–2047.

[7] X. Durrieu de Madron, P. Wiberg, P. Puig, Sediment dynamics in the Gulf of Lions: the impact of extreme events. Introduction special issue, Continent. Shelf Res. 28 (2008) 1867–1876.

[8] M. Canals, P. Puig, X. Durrieu de Madron, S. Heussner, A. Palanques, J. Fabrés, Flushing submarine canyons, Nature 444 (2006) 354–357.

[9] A. Palanques, X. Durrieu de Madron, P. Puig, J. Fabres, J. Guillén, A. Calafat, et al., Suspended sediment fluxes and transport processes in the Gulf of Lions submarine canyons: the role of storms and dense water cascading, Mar. Geol. 234 (2006) 43–61.

[10] G. Ercilla, J.L. Diaz, B. Alonso, M. Farrán, Late Pleistocene-Holocene sedimentary evolution of the northern Catalonia continental shelf (northwestern Mediterranean Sea), Continent. Shelf Res. 15 (1995) 1435–1451.

[11] C. Orejas, J.M. Gili, Caracterización física y ecológica de la franja costera, plataforma continental y cañón submarino de Cap de Creus. Technical Report, Fundación Biodiversidad, 2009, 103 pp.

[12] D. Desbruyères, A. Guille, J. Ramos, Bionomie benthique du plateau continental de la côte catalane espagnole, Vie et Milieu 23 (1972–1973) 335–363.

[13] P. Arnau, C. Liquete, M. Canals, River mouth plume events and their dispersal in the Northwestern Mediterranean Sea, Oceanography 17 (2004) 23–31.

[14] A. DeGeest, Cap de Creus Canyon: A Link between Shelf and Slope Sediment Dispersal Systems in the Western Gulf Lions, France, Masters thesis, Texas A & M University, 2005, 88 pp.

[15] C. Ulses, C. Estournel, J. Bonnin, X. Durrieu de Madron, P. Marsaleix, Impact of storms and dense water cascading on shelf-slope exchanges in the Gulf of Lion (NW Mediterranean), J. Geophys. Res. 113 (2008).

[16] G. Lastras, M. Canals, R. Urgeles, D. Amblas, M. Ivanov, L. Droz, et al., A walk down the Cap de Creus canyon, Northwestern Mediterranean Sea: recent processes inferred from morphology and sediment bedforms, Mar. Geol. 246 (2007) 176–192.

[17] P. Puig, A. Palanques, D.L. Orange, G. Lastras, M. Canals, Dense shelf water cascades and sedimentary furrow formation in the Cap de Creus Canyon, Northwestern Mediterranean Sea, Continent. Shelf Res. 28 (2008) 2017–2030.

[18] J.M. Pérès, J. Picard, Nouveau Manuel de bionomie benthonique de la Mer Mediterranee, Recueil des Travaux de la Station Marine d'Endoume 31 (47) (1964) 1–137.

[19] M.A. Taviani, A. Freiwald, H. Zibrowius, Deep coral growth in the Mediterranean Sea: an overview, in: A. Freiwald, J.M. Roberts, (Eds.), Cold-Water Corals and Ecosystems, Springer-Verlag, Berlin, 2005, pp. 137–156.

[20] A. Freiwald, L. Beuck, A. Rüggeberg, M. Taviani, D. Hebbeln, The white coral community in the Central Mediterranean Sea revealed by ROV surveys, Oceanography 22 (2009) 58–74.

[21] J.H. Fösa, P.B. Mortensen, D.M. Furevik, The deep-sea coral Lophelia pertusa in Norwegian waters: distribution and fishery impacts, Hydrobiología 471 (2002) 1–12.

33 Rock Ridges in the Central English Channel

Roger A. Coggan, Markus Diesing

Centre for Environment, Fisheries and Aquaculture Science (Cefas), Lowestoft, UK

Abstract

The English Channel is a tide-dominated continental shelf sea situated between France and England. The central region of the Channel is a low-depositional environment, and previous to this study the seabed was generally perceived to comprise mostly lag gravel, with a few isolated rock outcrops. Our geophysical and biological analysis revealed an extensive system of rock ridges located 30 km south of the Isle of Wight in water depths ranging between 40 and 80 m below Chart Datum. The feature extends 100 km in an east–west direction and 15 km in a north–south direction, covering *ca.* 1,100 km² of seafloor. The rock habitat supports a substantial coverage of fauna including sponges, bryozoans, hydroids and anemones. Three major geomorphic feature types were identified, namely flats, rock ridges and a palaeovalley. Surrogacy can be high for taxa that exploit niche habitats but is generally low when considering habitats and biotopes assigned according to the European Nature Information System (EUNIS) habitat classification scheme.

Key Words: habitat, biotope, benthos, multibeam, bathymetry, continental shelf, English Channel

Introduction

The English Channel is a funnel-shaped, ENE–WSW-trending, shallow shelf sea between France and England. Hydrodynamically, it is a tide-dominated environment but is also influenced by long swell waves approaching from the open Atlantic Ocean. The Channel is situated at the boundary zone between Lusitanean and Boreal biogeographical provinces [1]. The human impact on the benthic environment is assessed to be predominantly high to very high [2], including shipping, selective extraction (demersal fishing and aggregates), and obstruction (cables, wrecks), among others [3].

An extensive system of exposed rock ridges lies approximately 30 km south of the Isle of Wight (Figure 33.1) in depths between 40 and 80 m below Chart Datum. It covers about 1,100 km² of seafloor, extending 100 km east-to-west and 15 km north-to-south. The site is currently being considered as a Special Area of Conservation under the European Union's Habitats Directive.

Seafloor Geomorphology as Benthic Habitat. DOI: 10.1016/B978-0-12-385140-6.00033-5

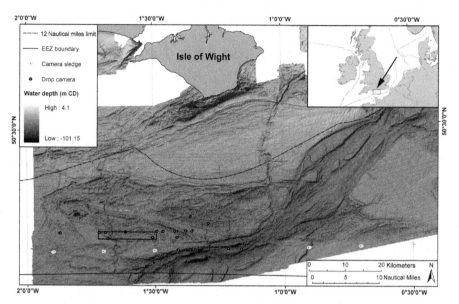

Figure 33.1 Singlebeam echosounder bathymetric map of the seabed south of the Isle of Wight based on a 75 m grid. Camera sampling locations are indicated. Black rectangle indicates area surveyed with multibeam as shown in Figure 33.2. Artificial illumination is from the northwest. Inset showing location of study site indicated by arrow.
Source: Bathymetric data © British Crown & SeaZone, 2007. Lic. No. 042007.005. Not to be used for navigation.

We made a broad-scale analysis of the ridge system based on singlebeam bathymetry data, originally collected for navigational charting purposes between 1978 and 2003, and a finer scale analysis based on our own surveys using multi-beam sonar and underwater cameras. The multibeam survey comprised a series of east–west transects with some cross lines and one significant area of full coverage. Information on benthic epifauna and habitats was gained from video transects of the seabed collected at 28 camera sampling stations (Figure 33.1) using a drop camera or towed sledge moving at a constant, slow speed (0.5–0.75 knots; 0.25–0.4 m s^{-1}), for about 300 m. Analysis of the video segmented each record into different habitat classes based on the five substrate types used in the European Nature Information System (EUNIS) habitat classification [4], namely rock (including outcrops and boulders) and coarse sediment, sand, mud and mixed sediments, with a minimum segment duration set at 2 min observation time. Typically this resulted in one or two video segments (i.e., broad habitats) per site, though the actual maximum was four. For each video segment, taxa were identified to the lowest practical taxonomic level, recording relative abundance. Each segment was then assigned a EUNIS bio-tope class, up to EUNIS Level 5, according to the observed substrate type and faunal assemblage. Full details on data acquisition, processing, analysis and interpretation can be found in [5–7].

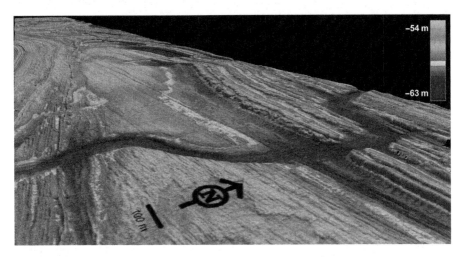

Figure 33.2 Three-dimensional representation of the seabed in the area surveyed with multibeam (see Figure 33.1 for location). Vertical exaggeration is 5×. Bedrock ridges, flat bedrock with a thin cover of mobile sediment and a meandering palaeovalley tributary are visible. See text for more details.

Geomorphic Features and Habitats

Figure 33.1 shows the processed singlebeam bathymetry gridded into 75 m × 75 m bins. The seabed is gently sloping toward the south, with an average gradient of roughly 1:1000 (0.057°). The bathymetric map shows distinct seabed textures, which can be directly related to bedrock geology [8]. Large areas are characterized by a flat seabed displaying a smooth texture. These areas correlate well with massively bedded, homogeneous Upper Cretaceous chalk in the subsurface. The chalk bedrock is covered by gravel and sandy gravel, forming a thin veneer of coarse-grained lag deposit at the seabed. Conversely, the seabed in the center of Figure 33.1 is much rougher, displaying a series of closely spaced ridges up to 5 m high and several kilometers long. These are produced by the much more varied and partly cyclic lithologies of the underlying Jurassic (predominantly interbedded mudstones, shales and thin limestones of the Kimmeridge Clay Formation), Lower Cretaceous (predominantly sandstones, shales, mudstones and siltstones of the Wealden Group) and Palaeogene rocks (predominantly sands, clays and mudstones of the London Clay Formation). All the aforementioned seabed types are dissected by a large-scale erosive structure, the so-called Northern Palaeovalley [9], which forms a roughly 100 km long structure running from northeast to southwest. A network of inner channels can also be recognized from the bathymetric data.

The full-coverage multibeam survey covered 16 km × 2 km of seabed in 47–65 m depth of water (see Figure 33.1 for location) and provided finer scale information and detail on the seabed morphology and habitats. Figure 33.2 shows bedrock ridges of the Kimmeridge Clay Formation up to 5 m high. Steeper (up to 25°) irregular

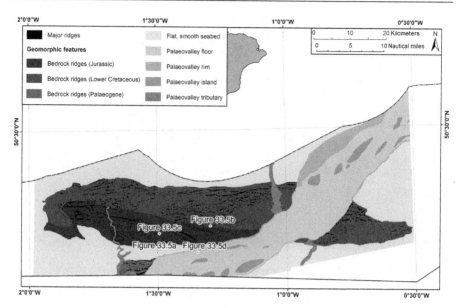

Figure 33.3 Map of geomorphic features. Locations of ground-truth sites detailed in Figure 33.5 are indicated.

scarps face toward the south, while less steep (less than 10°) bedding planes dip toward the north. The Jurassic bedrock is folded and faulted, as evidenced by bends and offsets. The Kimmeridge Clay is composed of a complex series of sedimentary rhythms [10] and here lies only 60 km southeast of the type location at Kimmeridge Bay in Dorset. The bedrock ridges develop due to the fact that the beds are slightly tilted and the varying lithologies display a differential resistance to erosion. The ridges are separated by troughs, some of which contain sediment. Expanses of relatively flat-lying bedrock are visible south of the ridges in Figure 33.2. These areas are at least partly covered with mobile sediment, indicated by the presence of large subaqueous dunes (*sensu* [11]). A meandering palaeovalley tributary is cutting through the bedrock from north-to-south.

Based on these results, we mapped three major geomorphic feature types (Figure 33.3): (i) flat, smooth seabed, (ii) bedrock ridges and (iii) the palaeovalley. Areas dominated by bedrock ridges were further subdivided based on bedrock age, as there was a distinct correlation between bedrock age and the seabed texture [7,8]. Major ridges were extracted through the application of terrain analysis of the singlebeam data set using Benthic Terrain Modeler software [12] and are highlighted in Figure 33.3. The bedrock ridges classes might contain flat bedrock covered by thin sediment, which is only identifiable from the higher resolution multibeam data. The palaeovalley was further subdivided into floor, rims, islands and tributaries. Table 33.1 shows the area covered by each of the major seabed types in Figure 33.3.

The geomorphic features map (Figure 33.3) was subsequently translated into a EUNIS habitat map (Figure 33.4) by incorporating additional environmental data on energy status (magnitude of tidal currents), biological zone (infralittoral, circalittoral,

Table 33.1 Areal Extent of the Different Geomorphic Features Mapped

Geomorphic Feature	Area (km²)
Flat, smooth seabed	1127
Palaeovalley (total)	974
Bedrock ridges (total)	1129
Jurassic	349
Lower Cretaceous	752
Palaeogene	29

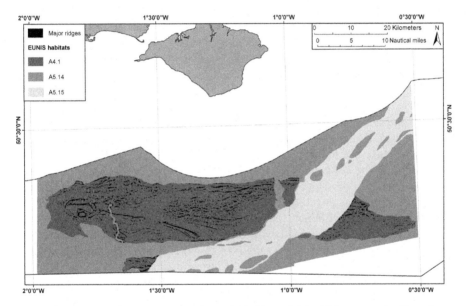

Figure 33.4 Map of EUNIS Level 3 and Level 4 habitats. A4.1—high energy circalittoral rock, A5.14—circalittoral coarse sediment, A5.15—deep circalittoral coarse sediment.

etc.) and substrate type (rock, coarse sediments, sand, mud and mixed sediments), which underpin the EUNIS classification hierarchy. This hierarchy relies solely on physical properties down to Level 3 for rock and Level 4 for sediment habitats; hence the modeled EUNIS map (Figure 33.4) shows the classes A4.1 "high energy circalittoral rock", A5.14 "circalittoral coarse sediment" and A5.15 "deep circalittoral coarse sediment."

Biological Communities

Biological communities were compared across the three main geomorphic features, namely flat seabed, bedrock ridges and the palaeovalley, considering both the taxa recorded and the biotopes identified during the video analysis. Sixty-two taxa were identified to various levels of taxonomic precision and the analysis presented here is based on their relative frequency of occurrence (i.e., how many video segments they occurred in). Table 33.2 presents summary univariate indices for epifaunal taxa and biotopes. Table 33.3 gives an index of the relative frequency of occurrence of taxa, aggregated to class level. Table 33.4 shows how each of the 13 EUNIS biotope classes was distributed among the geomorphic features.

The flat, smooth seabed was characterized by coarse and mixed sediments (Figure 33.5A) and supported a sparse benthic epifauna in which anemones, crabs and starfish were the most common taxa recorded (Table 33.3). Five biotopes were recognized (Table 33.4), with A5.141 (*Pomatoceros* and *Balanus* on coarse sediment) and A5.445 (brittlestar beds on mixed sediment) each occurring at three of the seven stations sampled. Atypically, a rock biotope (A4.134, *Flustra* and colonial ascidians) occurred at two locations as the local sediment included many cobbles and had become consolidated. The flat, smooth seabed showed the lowest richness and diversity of fauna (taxa) and biotopes among the three main geomorphic features (Table 33.2).

Within the system of bedrock ridges, the video records typically showed a series of ridges up to 5 m high with mobile sediment lying between them (Figures 33.5B and C). This gave rise to an alternating pattern of two or three biotopes as the camera passed from the sediment-filled troughs, up the steeper irregular scarp faces and down the less steep dip slopes. Hence, five rock and four sediment biotopes were recorded from stations targeting the rock ridges (Table 33.4). Scarp and dip slopes could support slightly different faunal communities, the steeper scarp slopes featuring taxa frequently associated with faster moving water, such as the hydroid *Tubularia indivisa* that characterizes the A4.112 biotope, while the more gentle dip slopes featured taxa such as the bryozoan *Flustra foliacea*, sponges, hydroids and ascidians, which characterize the A4.131 and A4.134 biotopes.

At deeper stations, sponges became more prevalent, encrusting forms giving way to cushion and erect forms, such as *Polymastia boletiformis* and ultimately massive forms such as *Pachmatisma johnstonia* (Figure 33.5D, photo 1), giving rise to the record of biotope A4.12 (sponge communities on deep circalittoral rock).

Within the bedrock ridges, sponges were the most frequently observed taxa (Table 33.4) and the most diverse, with 11 species and four morphological forms (e.g., "arborescent") recorded. The foliose bryozoan *Flustra foliacea* was also common and widespread. Ascidians were common, but their small body size generally precluded precise identification. Lithophilic taxa such as the anemones, hydroids and soft corals (exclusively *Alcyonium digitatum*) were also prevalent. Among the mobile epifauna, the starfish *Henricia* sp. and the sun-star *Crossaster papposus* were most frequent, being recorded in 15 and 9 video segments respectively. Fauna and biotope richness and diversity was higher in the bedrock ridges compared to the flat, smooth seabed, but similar to that for the palaeovalley (Table 33.2).

Four rock and four sediment biotopes were recorded in the palaeovalley (Table 33.4), the rim of which descended approximately 20 m in this area, sometimes very steeply but at other times over a distance of ≈250 m (Figure 33.5C). Parts of

Table 33.2 Univariate Indices Relating to Taxa and Biotopes Recorded from Video Analysis (Using Margalev's Richness and Shannon Diversity)

Index	Geomorphic Feature		
	Flat	Ridge	Palaeovalley
Stations	7	13	8
Segments	11	29	18
Taxon number (S)	26	47	43
Taxon frequency (N)	71	309	199
Taxon richness	5.9	8.0	7.9
Taxon diversity	3.0	3.4	3.4
Biotopes	5	9	8
Biotope richness	1.7	2.3	2.4
Biotope diversity	1.5	1.9	1.8

Note: The source data for taxa is not numerical abundance, but the number of video segments in which the taxa occurred.

Table 33.3 Relative Abundance of Taxa (Aggregated to the Class Level) within the Three Geomorphic Features

Taxon	Flat	Ridge	Palaeovalley
PHYLUM, Class (Common Name)			
ANNELIDA			
Polychaeta (bristle worms)	0	9	1
BRYOZOA			
Gymnolaemata (foliose bryozoans)	2	33	16
CHORDATA			
Ascidiacea (sea squirts)	0	18	5
CNIDARIA			
Hexacorallia (anemones)	15	37	12
Hydrozoa (hydroids)	2	37	17
Octocorallia (soft corals)	0	15	1
CRUSTACEA			
Eumalacostraca (crabs)	6	0	1
Maxillopoda (barnacles)	1	0	
ECHINODERMATA			
Asteroidea (starfish)	11	25	22
Holothuroidea (sea cucumbers)		0	
Ophiuroidea (brittlestars)	3	1	1
MOLLUSCA			
Gastropoda (snails)	0	0	1
Pelecypoda (bivalves)	3	1	1
PORIFERA			
Demospongiae (sponges)	1	105	81

Note: Calculated as $O*(S/N)$, where O is the class frequency after aggregation (i.e., if there are five different anemones in one video segment, then n for Hexacorallia = 5, not 1), S = the number of stations in which the class occurred and N = the number of stations sampled (as in Table 33.2). Zeros indicate positive values < 1; blanks indicate absence.

Table 33.4 Percent Frequency Occurrence of Biotopes Classified to a Maximum of EUNIS Level 5 (with N Indicating Actual Frequency of Observation)

EUNIS	Description	Flat	Ridge	Palaeovalley	N
A4.111	*Balanus* and *Tubularia* on extremely tide-swept circalittoral rock	–	–	100	1
A4.112	*Tubularia* on strongly tide-swept circalittoral rock	–	100	–	1
A4.12	Sponge communities on deep circalittoral rock	–	29	71	7
A4.131	Bryozoan turf and erect sponges on strongly tide-swept circalittoral rock	–	75	25	4
A4.134	*Flustra* and colonial ascidians on strongly tide-swept circalittoral rock	13	60	27	15
A4.214	Faunal and algal crusts on moderately tide-swept circalittoral rock	–	100	–	3
A5.14	Circalittoral coarse sediment	29	71	–	7
A5.141	*Pomatoceros* with barnacles and bryozoan crusts on unstable circalittoral cobbles and pebbles	38	50	13	8
A5.15	Deep circalittoral coarse sediment	–	20	80	5
A5.27	Deep circalittoral sand	–	100	–	1
A5.444	*Flustra* and *Hydrallmania* on tide-swept circalittoral mixed sediment	100	–	–	1
A5.445	*Ophiothrix* and/or *Ophiocomina* brittlestar beds on sublittoral mixed sediment	75	–	25	4
A5.45	Deep circalittoral mixed sediments	–	–	100	1

the rim and floor comprised coarse, consolidated cobble, colonized by *Flustra foliacea*, encrusting sponges and ascidians, hence the rock biotopes A4.131 and A4.134 were recorded here as well as on the bedrock ridges. However, the deep sponge community (A4.12) was more common here and especially characteristic of the steeper parts of the rim featuring boulder slopes and broken bedrock (Figure 33.5D, photos 1 and 2). The floor of the palaeovalley was typically a thin layer of coarse or mixed sediment overlaying bedrock. In unconsolidated areas, where the substrate was still mobile, there were few epifauna. One brittlestar bed was observed (A5.445), as was one area where the pebbles and cobbles were encrusted with the keel-worm *Pomatoceros* and barnacles (A5.141). The palaeovalley as whole had similar levels of richness and diversity to the bedrock ridges (Table 33.2).

Surrogacy

The extent to which the geomorphic features can be used as surrogates for the occurrence of taxa and biotopes can be judged from the information in Tables 33.3 and 33.4. The issue relates to the specificity and fidelity of the taxon group or biotope in

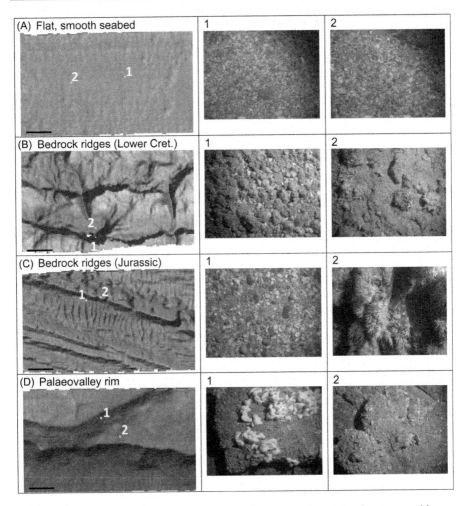

Figure 33.5 Multibeam sonar and associated seabed photographs illustrating geomorphic feature classes identified in the study area. Scalebar in lower left corner of the multibeam images is 50 m. Field-of-view of the seabed photographs is approximately 80 cm × 80 cm. The locations of sites A–D are indicated in Figure 33.3.

question: complete specificity means it always occurs in association with a feature; complete fidelity means it only occurs in association with that feature. For example, sponges were largely absent from the flat, smooth seabed areas, as they are typically confined to rock habitats that occur nearly exclusively in the bedrock ridges and palaeovalley. Apart from sponges there is little evidence of a potential for surrogacy when taxa are grouped at the class level, as the diversity of organisms within each class are usually able to exploit a wide range of habitats. It is likely that surrogacy would be stronger when taxa are considered at the species level, as species tend to evolve to exploit particular niche habitats. Some evidence of this is seen in Table 33.3, where the class Octocorallia was represented by a single species,

Alcyonium digitatum, which is nearly exclusive to steep tide-swept rock faces; thus, class Octocorallia appears almost exclusively in the bedrock ridge geomorphic feature.

Scale issues arise when considering surrogacy for biotope classes. Local conditions of substrate type and environment determine the specific biotope that will develop at any location, yet broad-scale mapping at the scale of geomorphic features necessarily highlights the dominant physical elements of those features; hence the assignment of a rock habitat class, A4.1, to the area of bedrock ridges (Figure 33.4). However, this should not imply that the area is exclusively rock or that sediment biotopes do not occur within that area on a finer spatial scale. Surrogacy now hinges on the closeness of match between the geomorphic feature and the habitat or biotope definition. The observation of near equal numbers of rock and sediment habitats in both the bedrock ridge and palaeovalley features suggests that geomorphic features are too broad scale to be used as surrogates for EUNIS habitat or biotope classes.

References

[1] W.P. Dinter, Biogeography of the OSPAR Maritime Area, Bundesamt für Naturschutz, Bonn, 2001.

[2] B.S. Halpern, S. Walbridge, K.A. Selkoe, C.V. Kappel, F. Micheli, C. D'Agrosa, et al., A global map of human impact on marine ecosystems, Science 319 (2008) 948–952.

[3] P.D. Eastwood, C.M. Mills, J.N. Aldridge, C.A. Houghton, S.I. Rogers, Human activities in UK offshore waters: an assessment of direct, physical pressure on the seabed, ICES J. Mar. Sci. 64 (2007) 453–463.

[4] European Environment Agency, EUNIS Habitat Classification, version 2006 11, European Environment Agency, Copenhagen, 2006. (internet application: <http://eunis.eea.europa.eu/habitats.jsp >)

[5] R. Coggan, M. Diesing, The seabed habitats of the central English Channel: A generation on from Holme and Cabioch, how do their interpretations match-up to modern mapping techniques? Continent. Shelf Res. 31, Supplement 1 (2011)S132-S150.

[6] R. Coggan, M. Diesing, K. Vanstaen, Mapping Annex I Reefs in the central English Channel: evidence to support the selection of candidate SACs, Science Series Technical Report 145, 2009, Cefas, Lowestoft.

[7] M. Diesing, R. Coggan, K. Vanstaen, Widespread rocky reef occurrence in the central English Channel and the implications for predictive habitat mapping, Estuar. Coast. Shelf Sci. 83 (2009) 647–658.

[8] J.S. Collier, S. Gupta, G. Potter, A. Palmer-Felgate, Using bathymetry to identify basin inversion structures on the English Channel shelf, Geology 34 (2006) 1001–1004.

[9] J.W.C. James, B. Pearce, R.A. Coggan, A. Morando, Open shelf valley system, Northern Palaeovalley, English Channel, U.K. (this volume).

[10] R. Hamblin, A. Crosby, P. Balson, S. Jones, S. Chadwick, I. Penn, et al., United Kingdom offshore regional report: the geology of the English Channel. HMSO for the British Geological Survey, London, 1992.

[11] G.M. Ashley, Classification of large-scale subaqueous bedforms: a new look at an old problem, J. Sediment. Petrol. 60 (1990) 160–172.

[12] E.R. Lundblad, D.J. Wright, J. Miller, E.M. Larkin, R. Rinehart, D.F. Naar, et al., A benthic terrain classification scheme for American Samoa, Mar. Geodes. 29 (2006) 89–111.

34 Characterization of Shallow Inshore Coastal Reefs on the Tasman Peninsula, Southeastern Tasmania, Australia

Vanessa Lucieer[1], Neville Barrett[1], Nicole Hill[1], Scott L. Nichol[2]

[1]Institute for Marine and Antarctic Studies, University of Tasmania, Hobart, Tasmania, Australia, [2]Geoscience Australia, Canberra, ACT, Australia

Abstract

Multibeam data of coastal bedrock reefs and associated sediments on the southeastern coast of Tasmania, Australia, were recently collected as part of the Commonwealth Environmental Research Facilities Marine Biodiversity Hub. The main aim of the surveys was to test a range of physical environmental parameters of nearshore reefs in southeast Tasmania as surrogates to predict patterns of benthic biodiversity. Two data sets were collected: (1) high-resolution bathymetry and seabed acoustic reflectance (backscatter) from a multibeam system and (2) high-resolution spatially rectified stereo still photographs of the seafloor collected by an autonomous underwater vehicle. In this analysis, the degrees of change of the slope of the seabed over a distance of 6 m and the seabed morphology are examined as surrogates for classifying seabed substrate. The distribution of the biological communities over different seabed morphologies is examined as a first step in the process of linking biological data to seabed substrate.

Key Words: Habitat mapping, coastal waters, multibeam, autonomous underwater vehicle, reef

Introduction

The island of Tasmania has a remarkably long coastline for its size. It has a highly indented shoreline with large estuaries, harbors and embayments, and many offshore islands including those in the Furneaux Group, King, Maria, Bruny, and Macquarie islands [1]. The State of Tasmania has more coastline per unit land area than any other Australian state and comprises 8.2% of the Australian coastline (6,472 km Tasmania—11,000,000 km Australia) [2]. Southeast Tasmania is characterized by a

Seafloor Geomorphology as Benthic Habitat. DOI: 10.1016/B978-0-12-385140-6.00034-7

particularly complex coastline with an abundance of islands, peninsulas, and estuaries. The southeast coast experiences variable oceanic conditions, primarily as a result of weather systems that move along the east coast of Australia [3]. South-facing shores are exposed to constant and often extremely large swells originating from gales in the southern ocean, while east-facing shores are exposed to less frequent but occasionally large swells derived from easterly weather patterns in the Tasman Sea. Despite this exposure, the convoluted nature of this coastline also provides a substantial component of sheltered waters and associated habitats. The marine habitats of the South East Coast of Tasmania display a high degree of naturalness, with low-level human activity impacting the marine ecosystems.

In 1999, the Tasmanian Aquaculture and Fisheries Institute (TAFI) initiated its seabed habitat mapping project, SeaMap Tasmania (www.utas.edu.au/tafi/seamap/). The aim of this project was to improve the understanding and management practices of the marine environment in estuarine and coastal areas of the state. The goal was to map the inshore marine habitats and consolidate all of this information into a single marine database that would provide the state with a comprehensive baseline of seabed information. This mapping initiative has supported coastal research and planning of Marine Protected Area development, environmental impact modeling and assessment, and spatial management of the state's fisheries. It also has underpinned the assessment of localized coastal developments and research, including risk assessment (tsunami modeling, sea-level rise, and oil-spill response). The comprehensive database was one of the first of its kind in Australia and now consists of over 5,000 km^2 of seabed habitat information at a scale of 1:25,000 from the coastline to 40 m water depth. This starting point provided a perfect platform to further develop mapping techniques capable of a finer-scale resolution as part of the Commonwealth Environmental Research Facilities (CERF) Marine Biodiversity Hub project (www.marinehub.org).

The inshore area of the Tasman Peninsula was one of several areas identified to be mapped and sampled as part of the CERF Marine Biodiversity Hub project (www.marinehub.org). The main aim of the survey was to acquire data to enable a range of physical environmental parameters of nearshore reefs (10–70 m) in southeast Tasmania to be tested as surrogates to predict patterns of benthic biodiversity. Two data sets were collected: (1) high-resolution bathymetry and seabed acoustic reflectance (backscatter) from a multibeam system, and (2) high-resolution spatially rectified stereo still photographs of the seafloor collected by an autonomous underwater vehicle (AUV).

A Simrad EM3002(D) 300 kHz multibeam sonar was used to survey an area of 117 km^2 offshore from the Tasman Peninsula, between High Yellow Bluff and Cape Hauy (Figure 34.1). Motion referencing and navigation data were collected using an Applanix position and orientation system coupled with a C-Nav GPS system. EM3002 data were acquired using Kongsberg's Seabed Information System (SIS) software. The data were processed using Caris Hips and Sips software to generate both bathymetric and backscatter grids at a spatial resolution of 2 m. The AUV *Sirius*, operated by the Integrated Marine Observing System's (IMOS) Autonomous Underwater Vehicle facility at the University of Sydney (www.acfr.usyd.edu.au/research/projects/subsea/

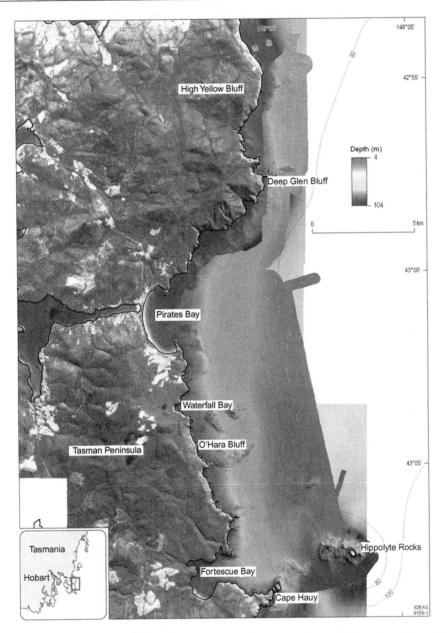

Figure 34.1 Multibeam sonar bathymetric map of southeastern Tasmania based on a 2-m grid (inset, location map of the Tasmanian study sites).
Courtesy of Geoscience Australia [4].

auvSIRIUS.shtml), was used to collect 12 transects of fine-scale imagery over a range of depths and substrates mapped by the multibeam. Substrata, geomorphology, and prevalence of key taxa were classified from the digital still images.

Data derived from AUV imagery were compared with the derived spatial products from the multibeam sonar data to test the effectiveness of multibeam products as surrogates for identifying seabed habitats and their associated biological communities. Data presented here on the southeastern coast of Tasmania (Figure 34.1) were collected as part of a collaboration between the TAFI (University of Tasmania) and Geoscience Australia (GA) [4].

Geomorphic Features and Habitats

The Tasman Peninsula survey area incorporates a series of highly fractured reefs located directly offshore of headlands and coastal bluffs, and isolated reefs located on the midcontinental shelf. The total mapped area of reefs was 14.4 km^2 with the remaining survey area (102.6 km^2) characterized by flat sandy seabed that extends into three coastal embayment's at Pirates Bay, Waterfall Bay, and Fortescue Bay (Figure 34.1). The inshore reefs range in area from 0.15–2.7 km^2 and are located in water depths of 10–70 m, extending up to 3 km onto the continental shelf. In plain view and in profile these reefs are very irregular, with localized peaks up to 20 m high, likely formed in dolerite. In contrast, the shoreline reefs (closer to the coast) are typically low profile and in places have a stepped profile and are possibly formed on outcrops of sedimentary rocks (siltstone or sandstone) (Figure 34.2A). The largest area of continuous reef on the Tasman Peninsula shelf surrounds the Hippolyte Rocks, located offshore from Cape Hauy. Formed in granite of Devonian age, the Hippolyte Rocks reef covers 2.13 km^2 and rises from 90 m water depth at its base to ~6 m at its shallowest point. A distinctive feature of this reef is a series of three peaks with near-vertical sides that are 20–30 m high (Figure 34.2B).

Some special features of significance on the reef habitats on the southeast coast include areas of high aesthetic value to the diving community, including sponge gardens, *Macrocystis* giant kelp forests, and marine cave systems. Large sections of sedimentary coast have rock types providing a mixture of patchy broken and low-profile reef extending for 1–2 km offshore throughout the 20–45 m depth range and reef communities (>40 m and up to 80 m depth) are found around the Hippolyte Rocks where this hard granite structure has resisted erosion in deeper continental shelf waters. Some of these areas are subject to strong currents with associated rich sponge communities.

A variety of physical characteristics of the seabed were derived from the multibeam bathymetry (e.g., slope, aspect, curvature) and backscatter (hardness and roughness). These were initially used to classify substratum habitats and will also be used to examine the relationship between the physical structure of the seabed and the benthic communities derived from the AUV imagery.

The Benthic Terrain Modeller (BTM), an ArcGIS Spatial Analyst extension [5], was used to classify the multibeam bathymetric data set into substratum habitat classes. The BTM application relies on the concept of bathymetric position index (BPI), a measure

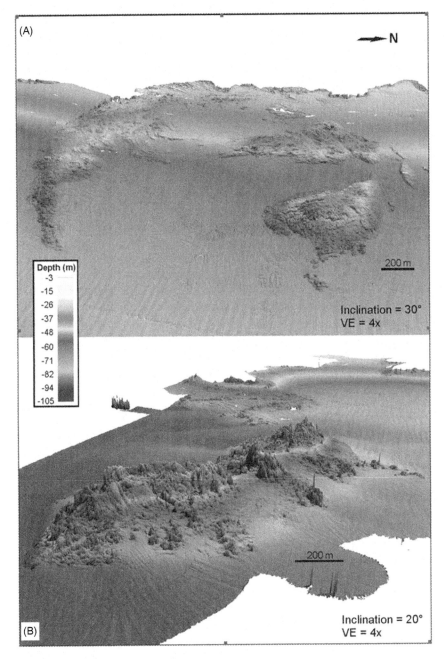

Figure 34.2 Perspective views of reefs along the Tasman Peninsula, showing: (A) inshore reefs at O'Hara Bluff and (B) high-relief reef surrounding the Hippolyte Rocks offshore from Cape Hauy. Scale bars are indicative for the position on the image at which they are shown.

of the elevation of a point on the seabed relative to its surroundings. Positive BPI values denote regions that are higher than the surrounding area, i.e., crests. Conversely, negative values characterize depressions, while values near zero show either flat areas (where the slope is near zero) or areas of constant slope (where the slope is significantly greater than zero). The BPI was applied at two scales: a fine-scale search window of 2 (2-m grid × 2 fine-scale BPI) identified features at a scale of 4 m; a broadscale search window of 10 (2-m grid × 10 broadscale BPI) identified features at a scale of 20 m. This process was used to classify substratum habitats. The four habitat classes that were separated using the BTM data were shallow unconsolidated sediments (<40 m), shallow reef (<40 m), deep reef (>40 m), and deep unconsolidated sediments (>40 m) (Figure 34.3).

Over the entire survey region, substrata was dominated by flat homogeneous soft sediments that comprised 91% of the seabed, with the remaining 9% comprising a range of low- and medium-profiled reef habitats and boulders. This reflects the relatively limited offshore extension of reef systems in this region, despite the coastline being predominantly rock. Again, reflecting the predominance of sediment, seabed relief was generally flat, with over 93% of the area being classified in the 0–5° of slope category, 4% of the area in the 5–10° of slope category, and the remaining 3% greater than 10°; these higher slopes are generally associated with reef. Within the 0–40 m depth range, 58% of the habitat consisted of unconsolidated habitat and 42% consolidated substrate; within the >40 m depth strata, 80% consisted of unconsolidated substrate and 20% of consolidated habitat. The predominant aspect of seabed habitats on the Tasman peninsula was 30% westerly, 18% southwesterly, 18% northwesterly, and 12% to the south.

Biological Communities

The majority of the coastline of the Tasman Peninsula is subject to moderate to high wave energy. This was reflected in the reef communities surveyed by SeaMap Tasmania [6], with typical exposed coast species such as bull kelp (*Durvillaea*), cray weed (*Phyllospora*), kelp (*Ecklonia*), red algae, and sponges dominating. In shallow waters on the east-facing coast, *Durvillaea* rarely extended below 5 m (due also to light availability), whereas on the higher-energy south-facing coast near the Hippolyte, *Durvillaea* extended below 5 m and in places below 10 m. Sponge communities were abundant along this coast due to the extensive area of reef below the euphotic zone (~33 m), particularly in areas of high water movement, such as the ends of the capes, headlands, and island groups. Small pockets of mixed fucoid algae are found on reefs in the more sheltered embayments at Fortescue Bay and Pirates Bay, with areas of patchy sea grass identified on the sediments. *Macrocystis* forests occur throughout this region on reefs at depths of 5–25 m and are particularly abundant in bays that provide reef habitat at suitable depths and moderate exposures, such as Fortescue Bay and Lagoon Bay.

In 2008–2009, 12 AUV transects were completed over reef systems on the Tasman Peninsula. Fine-scale imagery of the seabed over representative but discrete spatial areas (deployment location for O'Hara Bluff is shown in Figure 34.3) was collected using a USBL (ultra short base line) system attached to the AUV *Sirius* system to accurately record the geographic position of acquired imagery relative to the multibeam

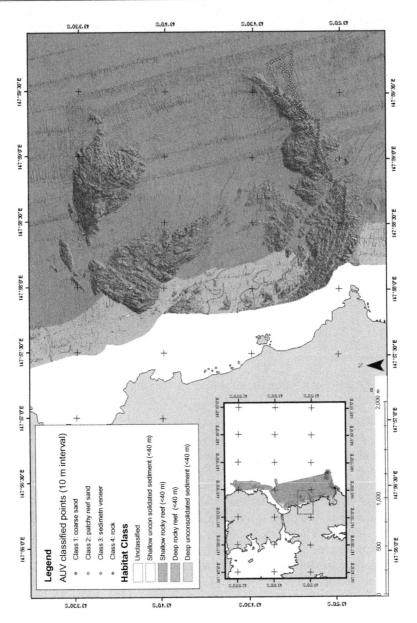

Figure 34.3 Predicted substratum map generated from a 2-m bathymetric grid applying slope and aspect at both fine scale BPI (4 m) and broadscale BPI (20 m) indexes of the O'Hara Bluff study area, southeastern Tasmania.

maps. Reefs ranged in depth from 20 to 95 m and a total of 126,731 overlapping stereo image pairs were taken. Each image captured ~2.1 m² (1.6 m × 1.3 m) of seafloor and its benthic biota. In initial analyses, 1,246 nonoverlapping images (spaced ~40 m along the transect path) were scored using digital imaging software. Images

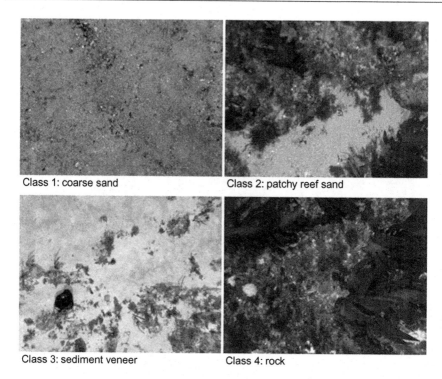

Class 1: coarse sand   Class 2: patchy reef sand

Class 3: sediment veneer   Class 4: rock

Figure 34.4 Examples of substrate classes scored at O'Hara Reef using the AUV images.

were scored for their dominant substrate (mud, sand, coarse sand, gravel, cobble, boulders, rock, patchy reef/sand, sediment veneer, or screw shell), rugosity (flat/none, low, moderate, high), apparent slope (flat, low-moderate, high-vertical), and sponge structure (none, low profile, medium profile, high profile) (Figure 34.4). Using 50 random points in each image, the benthos was identified to the lowest possible taxonomic unit. For most sessile invertebrates, this was morphotype; representative algae were identified to species, otherwise to functional groups; and mobile invertebrates (less frequently observed) were assigned to broad categories (e.g., starfish, sea urchin, mollusks). Sponge morphotypes were additionally assigned morphological categories such as arborescent, encrusting, cup, and fan. Cnidarians were also assigned to functional categories such as soft corals, sea whips, solitary hard corals, and anemones.

The major substrates captured in images along the AUV transect were coarse sand (33%), patchy reef/sand (27%), reef (20%), and sediment veneer on hard substrate (15%). A majority of the images showed a relatively flat seafloor (63%), while 28% covered a low to moderate slope and 8% covered a steep slope. Sponges were a dominant organism in all transects, with 30% of images recording low-profile sponge cover, 36% moderate-profile sponge cover, and 10% high-profile sponge cover. There was, however, a distinct zonation of the biological communities recorded on hard substrate features. Brown and red algae dominated communities to a depth of 40m, with virtually no algae recorded past 50m (Table 34.1). *Ecklonia radiata* was the dominant

Table 34.1 Percent Composition by Depth of Biological and Physical Cover on Reef Systems (Hard Substrate) and Adjacent Sediments (Sand Substrate) on the Eastern Tasman Peninsula

| | | Algae | | | | Invertebrates | | | | | | | | | | | | Substrate | | |
Sub-strate	Depth (m)	Chloro-phyta	Phaeo-phyceae	Rhodo-phyta	Drift Algae	Ascidiacea	Anthozoa	Biological Matrix	Bryozoa	Crino-idea	Echino-idea	Hydrozoa	Ophiuroidea	Porifera	Screw Shell	Unknown Biology	Biological Rubble	Rock Substrate	Sand	Images Scored
Hard	20–30	7.3	42.8	46.3	0.0	0.0	0.0	0.8	0.9	0.0	0.5	0.0	0.0	0.5	0.0	0.0	0.0	0.9	0.0	13
Hard	30–40	1.8	48.5	20.5	1.0	0.0	0.5	18.1	4.7	0.7	0.0	0.1	0.0	1.6	0.0	1.8	0.3	0.4	0.0	111
Hard	40–50	0.4	6.2	10.1	0.3	0.0	3.9	59.6	1.4	0.9	0.0	0.3	0.0	10.6	1.5	3.6	0.8	0.2	0.0	220
Hard	50–60	0.0	0.0	2.4	0.3	0.1	4.4	64.0	1.1	1.3	0.0	0.8	0.0	16.7	2.3	5.6	0.3	0.6	0.0	236
Hard	60–70	0.0	0.0	0.6	0.3	0.3	7.2	64.9	1.4	0.4	0.0	0.9	0.0	16.8	1.3	4.7	0.9	0.3	0.0	157
Hard	70–80	0.0	0.0	0.4	0.0	0.1	5.8	55.0	0.9	0.7	0.0	1.2	0.3	14.2	2.0	6.3	11.9	1.0	0.0	71
Hard	80–90	0.0	0.0	1.5	0.0	0.0	8.3	62.1	0.4	0.2	0.0	0.0	0.0	12.7	3.1	4.6	5.8	0.8	0.0	18
Sand	30–40	0.0	4.4	2.0	5.1	0.0	0.0	0.4	0.2	0.0	0.0	0.0	0.0	0.0	11.6	0.0	12.8	0.0	63.4	18
Sand	40–50	0.0	0.0	0.0	0.1	0.0	0.0	0.1	0.0	0.0	0.0	0.0	0.0	0.1	2.7	0.1	0.7	0.0	96.2	49
Sand	50–60	0.0	0.1	0.0	0.0	0.0	0.0	0.7	0.0	0.0	0.0	0.0	0.0	0.4	4.8	0.2	1.3	0.1	92.2	66
Sand	60–70	0.0	0.0	0.0	0.0	0.0	0.0	0.9	0.2	0.0	0.0	0.0	0.0	0.3	2.5	0.2	3.2	0.0	92.6	60
Sand	70–80	0.0	0.0	0.0	0.0	0.2	0.2	5.8	0.2	0.0	0.0	0.0	0.0	1.0	10.8	1.2	11.4	0.0	69.1	103
Sand	80–90	0.0	0.0	0.0	0.0	0.0	0.0	1.3	0.0	0.0	0.0	0.0	0.0	0.1	9.4	0.1	2.8	0.1	86.6	107
Sand	90–100	0.0	0.0	0.0	0.0	0.0	0.0	5.4	0.0	0.0	0.0	0.0	0.0	0.1	23.2	0.4	9.6	0.0	61.2	19

Data are the percentage of scored points assigned to each category based on photographic imagery derived from an autonomous underwater vehicle, with 50 random points scored per image.

brown algae recorded, and red foliose algae formed a significant component of the visible algal cover in the deeper sections of the algal zone. Beyond 40 m depth, reefs were primarily dominated by invertebrates. A particular common feature in these images was the dominance of what we term "biological matrix" that consisted of hydroids, bryozoans, and other small invertebrates interspersed with sediment. This matrix was impossible to differentiate further into individual components at the resolution of the imagery used here (Figure 34.4). The proportion of points scored as biological matrix was relatively consistent across the deeper depth bands from 40 to 90 m (~60% of the overall cover; Table 34.1). Presumably, the presence of this fine biogenic layer is at least partly due to the attenuation of wave energy with depth. Sponges formed the next most conspicuous grouping, consisting of ~15% of all points scored below the algal zone (Table 34.1). Sponges were also the most morphotypically diverse group, with 105 morphotypes recognized within this survey area. Arborescent (21%) and encrusting sponges (18%) were the most common morphotypes observed. Anthozoans, bryozoans, hydrozoans, and crinoids were also recorded in images, with anthozoans being the most significant component, becoming more abundant with increasing depth and contributing 8% of the total cover by 90 m. Echinoderms, mollusks, and ascidians were recorded infrequently (Table 34.1). On sandy substrates little biota was scored, although "biological rubble," a term we use here for biological fragments (such as hard bryozoans) that have been dislodged from reef areas, comprised up to 13% cover in some depth bands. The invasive New Zealand screw shell was found at all depths, was primarily associated with sandy substrates, and while displaying no consistent trend with depth, was a dominant feature between 90 and 100 m, where it formed 23% of the cover (Table 34.1).

Surrogacy

We investigated the correlation of seabed morphologies (specific seabed shapes as opposed to degrees of slope mentioned earlier) with substrate classes derived from the AUV imagery (Figure 34.5). Seabed morphology (seabed shape, not differentiating habitat type) was calculated from the 2-m resolution bathymetric grid using the Landserf tool [7,8] to identify six different morphometric classes. The AUV-derived coarse sand class was highly correlated with flat areas (plains), patchy reef/sand with the passes (mixed ridge and channel features), sediment veneer with the plains, and rock with the ridge and channel features. For the O'Hara reef system, 75% of the area was classified as plain, 19% as ridge, 5% as channels, 0.2% as pit, 0.4% as peak, and 0.4% as passes. The shape of the seabed can be an important indicative surrogate for substrate. Our next step in this process is to examine the relationship between the substrates we have now defined and their correlated multibeam derivatives, to see to what extent we can use these to predict our mapped patterns of benthic community structure.

The combination of AUV-derived imagery and high-resolution multibeam data at matching fine scales has provided a significant improvement to our understanding of benthic habitats and biological assemblages in southeastern Tasmania. The

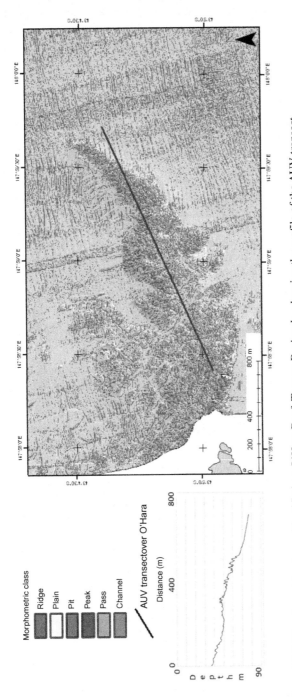

Figure 34.5 Morphometric classes identified on O'Hara Reef, Tasman Peninsula, showing the profile of the AUV transect.

combination of high-resolution and spatially referenced imagery makes the AUV an invaluable tool for quantifying seabed habitats and biota at fine scales in a repeatable manner. The new multiscaled biophysical data provides the first baseline information on seabed habitats and biota for these reef and sediment habitats.

Acknowledgments

We would like to thank J. Hulls, S. Williams, O. Pizaro, M. Jacuba, D. Mercer, and G. Powell for conducting the AUV surveys, L. Myer and J. Dowdney for the scoring of AUV images, and the R.V. *Challenger* crew (M. Francis and J. Gibson) and shipboard party, including I. Atkinson and C. Buchanon, for collection of multibeam data. Advice for AUV transect design was provided by T. Anderson and P. Dunstan. This paper was prepared with the support of the Commonwealth Environment Research Facilities (CERF) program, an Australian Government initiative supporting world-class, public good research. The CERF Marine Biodiversity Hub is a collaborative partnership between the University of Tasmania, CSIRO Wealth from Oceans Flagship, Geoscience Australia, Australian Institute of Marine Science, and Museum Victoria. The authors would also like to acknowledge the support of the IMOS Australian Autonomous Underwater Vehicle Facility and the Tasmanian Aquaculture and Fisheries Institute, University of Tasmania.

References

[1] C. Sharples, Indicative Mapping of Tasmanian Coastal Vulnerability to Climate Change and Sea-Level Rise: Explanatory Report, third ed., Consultant Report to Department of Primary Industries & Water, Tasmania, 2006, 173 pp., plus accompanying electronic (GIS) maps. Report available at http://www.coastalvulnerability.info

[2] R. Mount, Tasmania's Coastline and Coastal Waters, Information Paper, State of the Environment Unit, Resource Planning and Development Commission, Hobart, Tasmania, 2001, 3 pp.

[3] M.A. Hemer, K. McInnes, J.A. Church, J. O'Grady, J.R. Hunter, 2008, Variability and Trends in the Australian Wave Climate and Consequent Coastal Vulnerability, report to the Department of Climate Change, Canberra, Australia.

[4] S.L. Nichol, T.J. Anderson, M. McArthur, N. Barrett, A. Heap, J.P.W. Siwabessey, et al., Southeast Tasmania Temperate Reef Survey: Post Survey Report, Geoscience Australia Record 2009/43, Geoscience Australia, Canberra, 2009, 73 pp. Available online: http://www.ga.gov.au/resources/publications/marine-publications.jsp

[5] E. Lundblad, D.J. Wright, J. Miller, E.M. Larkin, R. Rinehart, T. Battista, et al., A benthic terrain classification scheme for American Samoa, Mar. Geod. 29 (2006) 89–111.

[6] N. Barrett, J.C. Sanderson, M. Lawler, V. Halley, A. Jordan, Habitat Mapping of Inshore Marine Habitats in South Eastern Tasmania for Marine Protected Area Planning and Marine Management, Technical Report Series 7, Tasmanian Aquaculture and Fisheries Institute, Tasmania, 2001, 75 pp., ISSN 1441-8487.

[7] P. Fisher, J. Wood, What is a mountain? Or the Englishman who went up a Boolean geographical concept but realised it was fuzzy, Geography 83 (1998) 247–256.

[8] J. Wood, Landserf Version 2.2. www.landserf.org, 2005 (accessed 30.06.10).

35 Rocky Reef and Sedimentary Habitats Within the Continental Shelf of the Southeastern Bay of Biscay

Ibon Galparsoro, Ángel Borja, J. Germán Rodríguez, Iñigo Muxika, Marta Pascual, Irati Legorburu

AZTI-Tecnalia, Marine Research Division, Herrera Kaia, Portualdea s/n, Pasaia, Spain

Abstract

Southeastern Bay of Biscay (Basque coast) seafloor characterization and benthic habitat mapping was carried out integrating data from multibeam echosounder, topographic and bathymetric LiDAR, video, and sediment and biological sampling ranging from the intertidal zone up to 100 m depth over 1,096 km². The area shows high geomorphologic diversity from which rocky reefs, sedimentary habitats, and mixed rock and sediment seascapes are dominant. Rocky bottoms are dominant along the shore and they reach the outer part of the continental shelf; meanwhile, sandbanks are distributed from beaches and river mouths down to muddy depths. Marine habitats along the Basque coast are related to geomorphology and hydrography. The analysis of biological and environmental data shows that wave energy, in the near-bottom, and sedimentary characteristics are the main environmental factors explaining the composition and spatial distribution of sedimentary benthic communities. Within the continental shelf, habitat suitability modeling has been applied, focusing on biological resource management. Moreover, different approaches have been carried out, especially in areas of process-driven habitat mapping, for ecosystem-based management, goods and services valuation, and marine spatial planning.

Key Words: Bay of Biscay, geomorphology, geodiversity, seascape, multibeam echosounder, LiDAR, benthos, habitat mapping, habitat modeling, Marine Spatial Planning

Introduction

The Basque continental shelf is located in the southeastern part of the Bay of Biscay (Figure 35.1). It is very narrow, ranging from 7 to 20 km [1], and comprises the total length of the coastline of *ca.* 150 km. Structural features dominate the morphology of the continental shelf, where horsts and anticlines, found generally in Cretaceous

Seafloor Geomorphology as Benthic Habitat. DOI: 10.1016/B978-0-12-385140-6.00035-9

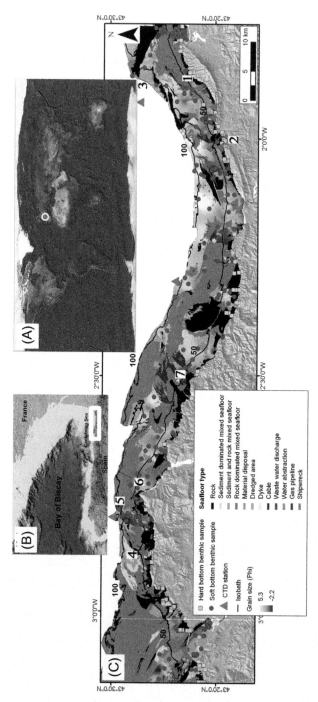

Figure 35.1 Study area location at (A) world scale, (B) within the Bay of Biscay, and (C) morphosedimentary map and available biological and oceanographical data sample location. The numbers represent the location of the geomorphologic habitats described in the text and in next figures.

Source: Modified from [9].

rocks, form areas starved of soft Neogene sediment. Faults and synclines filled with Tertiary material underlie sandy depressions [2]. Coastal rock lithologies are mainly sandstone, calcareous sandstone, limestone, clay, limonite, marl, and marly limestone [3]. Offshore, the external section of the continental shelf is a sedimentary Neogene and Pleistocene prism, developed by progradation [4].

In terms of oceanographical characteristics, waves from the northwest direction (swell) are dominant over the region [5], and the recorded periods range from 5 to 22 s, with the most frequent being between 8 and 12 s [6]. The tidal wave is semidiurnal, but despite the importance of tidally induced surface water fluctuations, the contribution of the tides to the generation of currents is somewhat modest (except within the estuaries) [7].

Marine habitats along the Basque coast are related to geomorphology and hydrography. Sandbanks are distributed from beaches and river mouths (in terms of sediment supply, the Basque Country is drained by 12 main rivers, which discharge $1.57 \; 10^6 \, \text{t} \cdot \text{yr}^{-1}$ of suspended material [8]) down to muddy depths. Rocky bottoms are dominant along the shore, reaching the outer part of the continental shelf.

In terms of naturalness, after recent recovery of the most polluted aquatic systems, fishing seems to be the main pressure on offshore marine habitats, but not enough scientific research has been applied to this particular subject [9].

In 2005, a seafloor mapping program commenced with the aim of benthic habitat mapping and seafloor characterization of the Basque continental shelf. This investigation integrates different remote sensing and *in situ* sampling techniques to cover a continuum from land to circalittoral marine environments and covers a total area of 1,096 km². Among them, multibeam echosounder (MBES) (operating up to 100 m water depth, producing 1 m horizontal resolution grid bathymetry and backscatter), topographic LiDAR (terrestrial land to mid-intertidal zone), bathymetric LiDAR (up to 20 m water depth, producing 2 m horizontal resolution grid) [10], and aerial photography [11,12] techniques have been used. *In situ* subtidal samples correspond to biological benthic data, which include 413 grabs from soft bottom (period 2003–2008), and 405 samples from rocky seafloor taken by divers (period 1992–2009). Oceanographic data were obtained from 21 CTD stations (sampled since 1998 at each season of the year), within a monitoring network [13]. Moreover, data from three offshore oceanographic buoys (from January 2007 to March 2009) and six littoral oceano-meteorological stations (from 2001 to 2009) are available.

Geomorphic Features and Habitats

The Basque continental shelf shows high geomorphological diversity. Among the different seascapes the following were identified [9]: (i) rocky seafloor (14% of the total); (ii) sedimentary seafloor (35% of the surface, where the sedimentary features cover *ca.* 4% of the sedimentary seafloor); (iii) mixed rock and sediment seafloor (49% of the total); and (iv) areas of dredged material disposal (2% of the total) and other peculiar structures such as dredging marks and waste water disposal sites.

Morphological Features

Rocky reefs: The shallow rocky reef seascape is present as a continuous belt of rock adjacent to the coastline, interrupted only by sandbanks off the major estuary mouths, which correspond to sedimentary infill of the major paleochannels (Figures 35.1 and 35.2). It is characterized by its roughness and steep slope (~10%) in water depths above 35–40 m. Below this water depth, the slope decreases to 2–3%. This slope change corresponds to one of the most marked terraces identified. The slope of the rocky seafloor depends on the orientation of the strata in relation to the coastline. Where rock strata are perpendicular to the coastline the seafloor has a low slope, whereas in areas where rock strata are parallel to the coastline, the rocky reef shows higher slope and reaches 30–40 m water depth close to the baseline of the coastal cliffs. The shallow rocky seafloor shows numerous incisions that correspond to paleoriver channels. This seascape faces the major amount of wave energy and the presence of large sublittoral rock blocks are related to coastal cliff erosion.

At deeper water depth, the rocky substratum is flatter and smoother, with the presence of sand patches between rocky strata. This pattern produces a mixture of sedimentary and rocky seafloor, which is the dominant seafloor type in the continental shelf. Apart from these, some rock outcrops are present along the continental shelf where some of them rise *ca.* 30 m above the seafloor (Figure 35.3).

Sedimentary seafloor: Ten major sandbanks have been identified, which represent the extension of the present estuaries. The mean grain size corresponds to fine sand (mean median 2.1 Phi). The mean composition is 75% sand, 18% mud, and 4% gravel, with 35% $CaCO_3$ content, 4.1% organic matter content, and a redox potential of +161 mV.

Various sedimentary bedforms are present in the area (Figure 35.1). At shallow-water depth, wave-induced morphologies such as rhythmic surf zone sandbars and troughs, and sandbank morphologies associated to the wave closure depth, could be identified. At deeper water depth, Infralittoral Prograding Wedge (Figure 35.4) and sorted bedforms have been identified (Figure 35.5) [9]. Sorted bedforms are present as elongate slightly depressed (up to 0.5 m in depth, relative to the surrounding upper shoreface) features, which lie perpendicular to the isobaths. Most of the sorted bedforms have developed just outside the fair-weather surf zone water depth (20–25 m), down to water depths of 90–100 m. The largest ones are around 1,650 m in width and 4,400 m in length. Sorted bedform areas are easily identifiable in MBES records, both by the bathymetric depression and by the higher backscatter response than the surrounding sediment (Figure 35.5). Sedimentological samples located in sorted bedform areas showed a mean grain size of 1.25 Phi, with 80.3% sand, 11.7% gravel and shell debris, and 7.7% mud, with 3.2% organic matter and

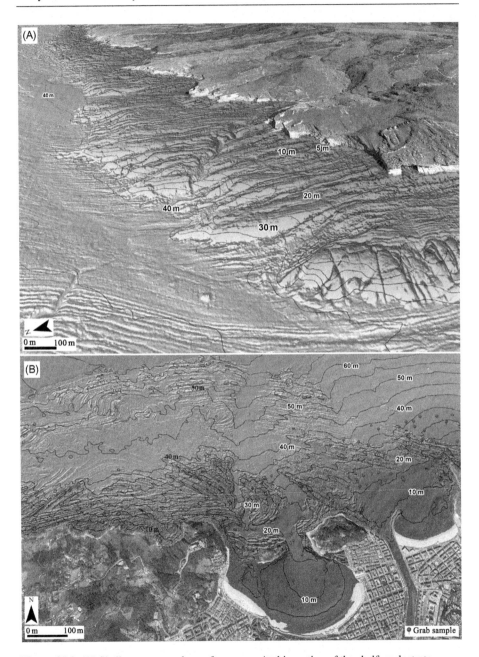

Figure 35.2 (A) Shallow-water rocky reef seascape. At this section of the shelf, rock strata are approximately oriented to the North. Rock morphology change can be identified at 40 m depth (see text for explanation). A paleoriver channel can also be identified (see Figure 35.1 for the seascapes location, as number 1). (B) Seascape showing shallow-water rocky reef, a paleochannel, and bay. Both seascapes are generated by the integration of topographic LiDAR (Digital Elevation Model at 1 vertical exaggeration), bathymetric LiDAR, and MBES information (vertical exaggeration 4×). (See Figure 35.1 for the seascapes location, as number 1 and 2).

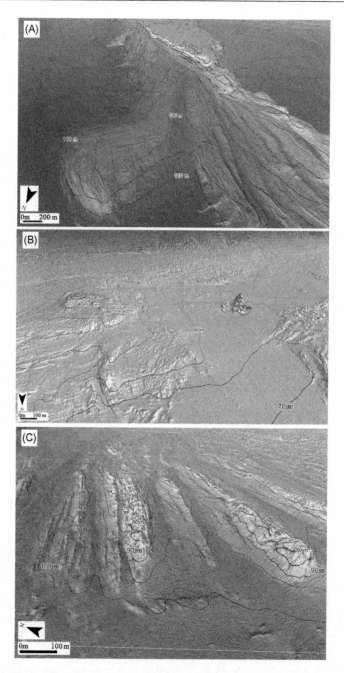

Figure 35.3 Three different rock outcrop seascapes. Digital Elevation Model vertical exaggeration 10×. (See Figure 35.1 for the seascape location, Figure 35.3A is number 3; Figure 35.3B is number 4, and Figure 35.3C is number 5).

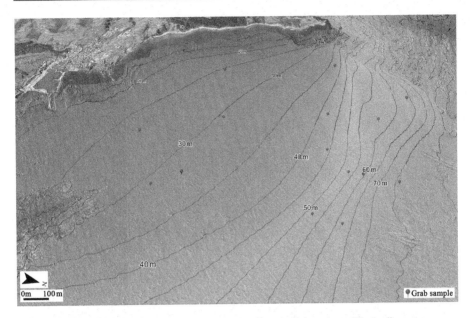

Figure 35.4 Sedimentary seafloor seascape and grab sample locations. The sedimentary seafloor morphology changes at 30–35 m water depth, which corresponds to wave closure depth. The steeper slope observed at the 45–55 m range corresponds to a Infralittoral Prograding Wedge front. In the lower left of the picture, some dredge disposal marks can be identified (See Figure 35.1 for the seascape location, as number 6).

38.8% $CaCO_3$, which is higher than the surrounding material. Video records indicate that the boundaries are sharp between the surrounding fine sand and coarse sand inside the sorted bedform.

Biological Communities

Hard-bottom communities: Rocky seafloor benthic characterization was based on survey tracks from 0 to 50 m water depth (Figure 35.1). In the 405 samples, 1,147 macrobenthic invertebrate and 215 macroalgae species have been identified.

Regarding macroalgae, the zonation in the intertidal and subtidal zones is determined by tides and exposure to wave action. However, other physical (i.e., topography, substratum nature, sedimentation level), chemical (i.e., salinity, pollution), and biological factors have also been identified as being important [13]. Gorostiaga et al. [14], applying multivariate analysis, observed that subtidal macroalgae changed gradually, with no discrete communities being clearly distinguishable. These gradual changes in flora were related mainly to increasing sedimentation levels. According to Díez et al. [15], the vegetation of the western Basque coast is fairly homogenous although somewhat more diverse than that of the eastern coast, due to a greater variation in environmental variables such as water quality, sedimentation,

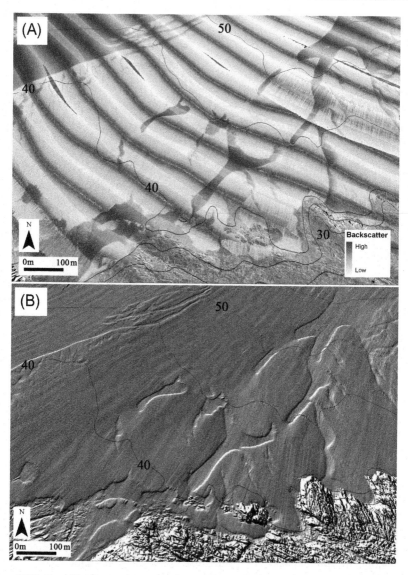

Figure 35.5 Sorted bedform seascape. (A) Non-calibrated multibeam backscatter response, with darker colors representing areas of coarse or gravely sand; (B) shaded relief model extracted from 1 m horizontal resolution Digital Elevation Model (with a vertical exaggeration of 10×). Sorted bedforms can be identified by the depression they produce (see Figure 35.1 for the seascape location, as number 7). (For interpretation of the references to color in this figure legend, the reader is referred to the web version of this book.)

and wave exposure levels. Discrete communities have been recognized, on the basis of the dominant species, to facilitate the description of vegetation. The exposed or extremely exposed infralittoral rock is dominated by *Gelidium corneum* (Figure 35.6), whilst the more sheltered zones are dominated by *Cystoseira baccata*

Figure 35.6 Algae dominated (*Gelidium corneum*) shallow water, high-exposure rocky reef.

[16–18]. In rocky areas affected by sedimentation, *G. corneum* tends to disappear. With moderate sediment increase *C. baccata* and *Codium decorticatum* can dominate, whilst in highly sedimented zones *Zanardinia typus* and *Aglaothamnion cordatum* are more abundant [15].

In terms of hard-substratum macrofaunal distribution, there is also a gradual change of species, which could be directly associated to the same environmental features that drive the changes in vegetation. Alternatively, it could be due to other physical parameters, or related to the macroalgae species present at various depths.

The polychaete species *Lysidice ninetta* characterizes the rocky bottom between intertidal zone depths and about 15 m and decreases in frequency with depth. Other characteristic species at those depths are the gastropod *Bittium reticulatum*, the bivalves *Hiatella arctica* and *Musculus costulatus,* and, at more than 10 m water depth, the polychaetes *Platynereis dumerilii* and *Spirobranchus polytrema* [15].

The gastropod *B. reticulatum* is one of the most frequent species at approximately 10–15 m water depth, and the most frequent species at about 15–25 m water depth. Other characteristic species at 15–25 m water depth are the crustacean *Verruca stroemia*, the mollusks *H. arctica* and *Nassarius reticulatus*, and the polychaete *L. ninetta* [15].

At more than 25 m water depth, the crustacean *V. stroemia* is the most characteristic macrofaunal species, appearing in all sampling sites. The mollusks *B. reticulatum*, *H. arctica*, and *Ocenebra erinaceus* are also notable [15] (see Figures 1 and 2 of Supplementary Material for hard-substrate habitats images).

Soft-bottom communities: The assessment of soft-bottom benthos is based upon 413 grabs, in which 1,202 species were identified. A BIO-ENV analysis of PRIMER was carried out to relate the sedimentological and oceanographical conditions to species distribution. Most of the variability on species composition was explained by the sediment composition and sediment resuspension produced by wave action. This result was used for habitat classification and their spatial delimitation by environmental information layer combination in a GIS environment. Moreover, the pan-European EUNIS habitat classification [19] was used as the base classification for management and conservation purposes, but it was adapted to the specific characteristics of the Basque continental shelf habitats [20]. Some of the most significant soft-bottom habitats are described later.

The exposed coarse sand and gravel sand, present in highly dynamic areas, supporting natural disturbance produced by currents and wave action, is characterized by robust infaunal polychaetes such as *Polygordius appendiculatus*, *Protodorvillea kefersteini*, and Terebellidae, the oligochaete *Grania*, bivalves such as *Tellina (Moerella) donacina*, and holothurians such as *Leptosynapta inhaerens*.

In slightly gravel-mixed sands in areas of open exposed coast, the *Glycera lapidum* polychaete is present. This species is rarely considered characteristic of a community, but in this case its dominance could be related to the exclusion of other species. This habitat could be affected by continuous or periodical disturbance of the sediment due to wave action, which impedes the establishment of more stable communities. Other *taxa* include spionid polychaetes such as *Spio martinensis* and *Spiophanes bombyx*, *Nephtys* spp. and, in some areas, the bivalve *Spisula elliptica*. Due to the variability of sediment regime in this habitat, seasonal or spatial variations of this community may occur.

In coarse sands of the open exposed coast, in water depths >15–20 m, the habitat is characterized by the presence of nematodes, likewise the annelid *P. appendiculatus*, nemerteans, *Pisione remota*, oligochaetes of genus *Grania*, *Sphaerosyllis bulbosa*, *G. lapidum*, and *P. kefersteini*.

The infralittoral habitat consisting of fine sand, in shallow-water depth, is characterized by the presence of the sea urchin *Echinocardium cordatum*, the bivalve *Mactra stultorum*, the polychaetes *Magelona johnstoni*, *S. bombyx*, *Mediomastus fragilis*, *Owenia fusiformis*, and *Paradoneis armata*, the amphipods *Siphonoecetes kroyeranus* and *Hippomedon denticulatus* and the nemertean *Tubulanus polymorphus*.

The infralittoral habitat consisting of muddy sand contains a variety of polychaetes (*M. johnstoni*, *Magelona filiformis*, *O. fusiformis*, *P. armata*, genus *Scolaricia*, *Prionospio (Prionospio) steenstrupi*, *Myriochele danielsseni*, *Chaetozone gibber*), bivalves (*Tellina fabula*), gastropods (*N. reticulatus*), and amphipods (*S. kroyeranus*, *Urothoe pulchella*).

At deeper water depths (>27 m) this habitat is more stable than shallower ones, and consequently shows higher species diversity. Clean fine sands, with mud content less than 5%, are characterized by the presence of copepods as well as *Echinocyamus pusillus*, *S. bombyx*, *Abra alba*, *Lumbrineris cingulata*, *Abra prismatica*, *P. (Prionospio) steenstrupi*, nemerteans, *M. filiformis*, *C. gibber*, *Ampelisca brevicornis*, and *T. polymorphus*.

The circalittoral habitat with muddy sand and fine contents between 5% and 20% is present at depths deeper than 27 m and shows rich infaunal communities comprised of *E. cordatum*, *M. stultorum*, *M. johnstoni*, *S. bombyx*, *M. fragilis*, *O. fusiformis*, and *S. kroyeranus*.

The infralittoral region with muddy sand and fine contents higher than 20%, and at depths deeper than 27 m is present in sheltered bays or inlets. The dominant species include *Nephtys cirrosa*, *H. denticulatus* and other amphipods of genus *Hippomedon*, *E. cordatum*, *Urothoe brevicornis*, and *Dispio uncinata*. It may also contain the species *Gastrosaccus sanctus*, *Bathyporeia elegans*, and *Scolelepis bonnieri*.

The circalittoral sandy mud habitat, with fine contents higher than 20% at water depths >27 m, is present in deep bays, inlets, and in deep water where the wave energy action is low. A community of *L. cingulata, Thyasira flexuosa, Tellina compressa, S. bombyx, C. gibber, Ampharete finmarchica, Prionospio fallax, Aponuphis bilineata, Spiophanes kroyeri, M. filiformis*, nemerteans, *Chone filicaudata, Ampelisca tenuicornis, M. danielsseni*, and *A. brevicornis* is present.

The circalittoral region consisting of fine mud, at water depths >27 m, is characterized by the presence of the polychaete of genus *Monticellina* and the species *Galathowenia, Magelona minuta, Monticellina dorsobranchialis, T. flexuosa, S. kroyeri, Abyssoninoe hibernica, Chaetozone setosa, A. tenuicornis, A. finmarchica, Paradiopatra calliopae, Maldane glebifex, Prionospio (Prionospio) ehlersi, Terebellides stroemii, P. fallax, A. alba*, and *Euclymene* (see Figures 1 and 3 of Supplementary Material for soft-substrate habitats images).

Surrogacy

Habitat modeling approaches have been mainly focused on commercially exploited species, and the derived results have been used for fisheries management and coastal zone management purposes. Borja et al. [21] provided information on the standing stock of the goose barnacle (*Pollicipes pollicipes*), and of the significant positive correlation between biomass, coverage, and density, and environmental factors such as wave height and energy derived from waves received at the coast. In the case of *G. corneum*, a significant positive correlation between biomass and irradiance and a negative correlation with wave energy was also found [13].

Galparsoro et al. [22] considered the identification of seafloor morphological characteristics, together with wave energy conditions, that determine the presence of European lobster (*Homarus gammarus*), and predicted suitable habitats over the Basque continental shelf with a Boyce index of 0.98 ± 0.06, showing the quality of the model [23].

The production and access to high-quality, high-resolution data on the seabed, together with the increasing interest in integrative studies and ecosystem-based management of the marine environment, has triggered a general interest in habitat mapping and modeling. Hence, different approaches are being carried out, especially in areas of integrative analysis of biological, physical, and oceanographical processes. The results could serve for a better understanding of the habitats, and with the objective of ecosystem-based management of marine waters, the integration of marine uses, goods and services valuation of coastal ecosystems, and, finally, Marine Spatial Planning of the Basque continental shelf.

Acknowledgments

This project was supported by the Department of Environment, Land Use, Agriculture and Fisheries of the Basque Government. Part of the survey work was funded by MESH Atlantic project (Interreg Atlantic Area Transnational Programme of the European Regional Development

Fund). Irati Legorburu and Marta Pascual were funded by the Fundación Centros Tecnológicos grants. Professor Michael Collins (School of Ocean and Earth Science, University of Southampton (UK) and AZTI-Tecnalia (Spain)) kindly advised us on some details of the manuscript. This paper is contribution number 554 from AZTI-Tecnalia (Marine Research Division).

References

[1] A. Uriarte, Sediment Dynamics on the Inner Continental Shelf of the Basque Country (N. Spain). PhD Thesis. University of Southampton, 1998, pp. 302.

[2] A. Pascual, A. Cearreta, J. Rodríguez-Lázaro, A. Uriarte, Geology and Palaeoceanography. Elsevier Oceanography Series, in: Á. Borja, M. Collins (Eds.), Oceanography and Marine Environment of the Basque Country, vol. 70, Elsevier, pp. 53–73. Chapter 3.

[3] EVE, Mapa geológico del País Vasco. Mapa, memoria y bases de datos, Ente Vasco de la Energía (2003), ISBN: 84-8129-054-8.

[4] G. Boillot, L. Montadert, M. Lemoine, B. Biju-Duval, (Eds)., Les marges continentales actuelles et fossiles autour de la France. Edited by Masson, E. Paris, 1984, ISBN 2225801347. 342 pp.

[5] A. Uriarte, M. Collins, A. Cearreta, J. Bald, G. Evans, Sediment supply, transport and deposition: contemporary and Late Quaternary evolution, in: Á. Borja, M. Collins, (Eds.), Elsevier Oceanography Series. In Oceanography and Marine Environment of the Basque Country, vol. 70, Elsevier, pp. 97–131. Chapter 5.

[6] P. Castaing, Le transfert à l'océan des suspensions estuariennes. Cas de la Gironde. PhD Thesis. Université de Bordeaux, 1981, pp. 277.

[7] A. Fontán, M. González, N. Wells, M. Collins, J. Mader, L. Ferrer, G. Esnaola, A. Uriarte, Tidal and wind-induced circulation within the Southeastern limit of the Bay of Biscay: Pasaia Bay, Basque Coast, Continent. Shelf Res. 29 (2009) 998–1007.

[8] L. Ferrer, A. Fontán, J. Mader, G. Chust, M. González, V. Valencia, A. Uriarte, M.B. Collins, Low-salinity plumes in the oceanic region of the Basque Country, Continent. Shelf Res. 29 (2009) 970–984.

[9] A. Borja, J. Bald, J. Franco, J. Larreta, I. Muxika, M. Revilla, J.G. Rodríguez, O. Solaun, A. Uriarte, V. Valencia, Using multiple ecosystem components, in assessing ecological status in Spanish (Basque Country) Atlantic marine waters, Mar. Pollut. Bull. 59 (2009) 54–64.

[10] I. Galparsoro, Á. Borja, I. Legorburu, C. Hernández, G. Chust, P. Liria, A. Uriarte, Morphological characteristics of the Basque continental shelf (Bay of Biscay, northern Spain); their implications for Integrated Coastal Zone Management, Geomorphology 118 (2010) 314–329.

[11] G. Chust, I. Galparsoro, A. Borja, J. Franco, B. Beltrán, A. Uriarte, Detección de cambios recientes en la costa vasca mediante ortofotografía, Lurralde 30 (2007) 59–72.

[12] G. Chust, I. Galparsoro, Á. Borja, J. Franco, A. Uriarte, Coastal and estuarine habitat mapping, using LIDAR height and intensity and multi-spectral imagery, Estuar. Coast. Shelf Sci. 78 (2008) 633–643.

[13] Á. Borja, F. Aguirrezabalaga, J. Martínez, J.C. Sola, L. García-Arberas, J.M. Gorostiaga, Benthic communities, biogeography and resources management, in: Á. Borja, M. Collins (Eds.), Oceanography and Marine Environment of the Basque Country, vol. 70, Elsevier, pp. 455–492.

[14] J.M. Gorostiaga, A. Santolaria, A. Secilla, I. Díez., Sublittoral benthic vegetation of the eastern Basque coast (N. Spain): structure and environmental factors, Bot. Mar. 41 (1998) 455–465.

[15] I. Díez, A. Santolaria, J.M. Gorostiaga, The relationship of environmental factors to the structure and distribution of subtidal seaweed vegetation of the western Basque coast (N Spain), Estuar. Coast. Shelf Sci. 56 (2003) 1041–1054.

[16] J.M. Limia, J.M. Gorostiaga, Flora marina bentónica sublitoral del tramo de costa comprendido entre Pta. Covarón y Pta. Muskes (Vizcaya, N.E. España). Actas del VI Simposio Nacional de Botánica Criptogámica: 1987, pp. 81–88.

[17] Á. Borja, V. Valencia, L. García, A. Arresti, Las comunidades bentónicas intermareales y submareales de San Sebastián - Pasajes (Guipúzcoa, norte de España). Actas del IV Coloquio Internacional de Oceanografía del Golfo de Vizcaya. 1995, pp. 165–181.

[18] I. Díez, A. Santolaria, A. Secilla, J.M. Gorostiaga, Eds., Comunidades fitobentónicas submareales de la zona exterior de la Reserva de la Biosfera de Urdaibai. Consideraciones sobre su estado ecológico. Investigación aplicada a la reserva de la Biosfera de Urdaibai. Edited by Gobierno Vasco, Vitoria, 2000, pp. 151–157.

[19] C.E. Davies, D. Moss, M.O. Hill, EUNIS Habitat Classification revised. Report to the European Topic Centre on Nature Protection and Biodiversity; European Environment Agency. October 2004, pp. 310.

[20] I. Galparsoro, G. Rodríguez, Á. Borja, I. Muxika, Elaboración de mapas de hábitats y caracterización de fondos marinos de la plataforma continental vasca. Informe inédito elaborado por AZTI-Tecnalia para el Dirección de Biodiversidad; Viceconsejería de Medio Ambiente; Departamento de Medio Ambiente, Planificación Territorial, Agricultura y Pesca del Gobierno Vasco, 2009, pp. 74.

[21] A. Borja, P. Liria, I. Muxika, J. Bald, Relationships between wave exposure and biomass of the goose barnacle (*Pollicipes pollicipes*, Gmelin, 1790) in the Gaztelugatxe Marine Reserve (Basque Country, northern Spain), ICES J. Mar. Sci. 63 (2006) 626–636.

[22] I. Galparsoro, Á. Borja, J. Bald, P. Liria, G. Chust, Predicting suitable habitat for the European lobster (*Homarus gammarus*), on the Basque continental shelf (Bay of Biscay), using Ecological-Niche Factor Analysis, Ecol. Model. 220 (2009) 556–567.

[23] M.S. Boyce, P.R. Vernier, S.E. Nielsen, F.K.A. Schmiegelow, Evaluating resource selection functions, Ecol. Model. 157 (2002) 281–300.

Supplementary Material

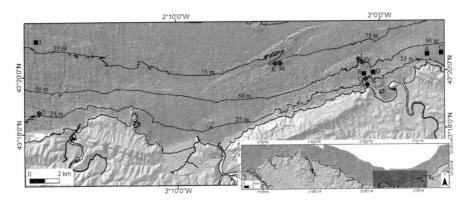

Figure 1 Underwater image positions within the Basque shelf. Points corresponds to images shown in Figure 2 of supplementary material; meanwhile, rectangles corresponds to images shown in Figure 3 of supplementary material.

Figure 2 Examples of rocky habitats within the Basque shelf. Data obtained from a multibeam echosounder (left column) and an *in situ* seafloor image taken by an underwater video camera (right column), in the points highlighted in the echosounder image. (A) A rock reef at 13 m water depth; (B) the area is dominated by red algae (in this case, probably *Mesophyllum lichenoides*). (C) Rock reef at 21 m water depth; (D) brown (*Dictyota dichotoma*) and red (*Halopteris filicina*) algae dominated the community. (E) Sediment affected rocky seafloor at 28 m water depth; (F) *Brown and red macroalgae*. (G) Rock outcrop at 45 m water depth and seafloor image location; (H) different species of encrusting algae and sponges could be identified. (I) Rock outcrop at 50 m water depth; (J) encrusting red algae, the cirripeda *Verruca stroemia* and some other species can be identified. (K) Rock outcrop at 70 m water depth; (L) some epifauna and encrusting fauna can be identified. (M) Rock outcrop at 75 m water depth; (N) different species of epifauna can be identified.

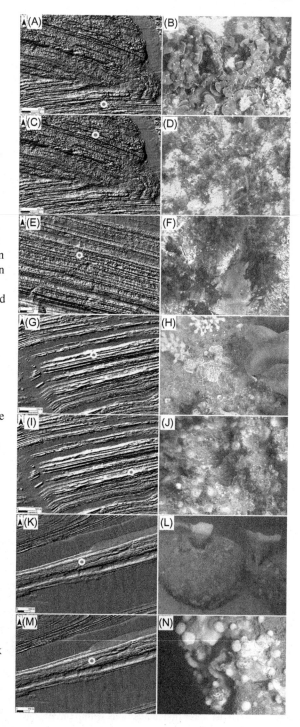

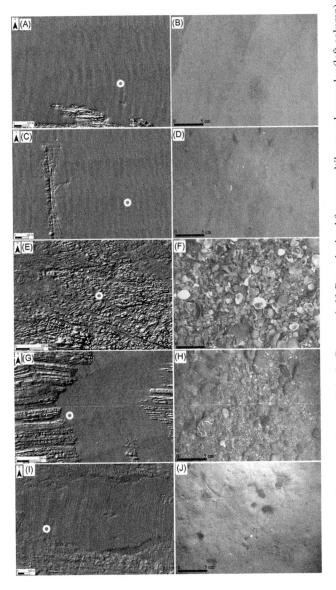

Figure 3 Some examples of sedimentary habitats within the Basque shelf. Data obtained from a multibeam echosounder (left column) and an *in situ* seafloor image taken by a submarine video camera (right column), in the points highlighted in the echosounder image: (A) sandy bottom at 37 m water depth; (B) the ripple marks indicate wave action in the seafloor (grain size corresponds to medium sand with 98% of sand content). (C) Sandy bottom at 40 m water depth next to an urban wastewater outfall; (D) the presence of tolerant species to organic enrichment (mainly *Nassarius* spp.), indicate the influence of the outfall (the grain size corresponds to fine sand). (E) Gravel hatch between rock strata at 42 m water depth; (F) a tanatocenosis dominated by mollusks, shows the highly dynamic area. (G) Wave-induced seabed features at 50 m water depth; (H) some epifauna and encrusting fauna can be distinguished in the pebbles. (I) Sedimentary seafloor at 80 m water depth; (J) some tracks of echinoderms and mollusks, together with holes of annelids and/or crustaceans can be identified. The grain size of the area corresponds to fine sand (75% sand and 25% of mud content).

36 Rock Reefs of British Columbia, Canada: Inshore Rockfish Habitats

K. Lynne Yamanaka[1], Kim Picard[2], Kim W. Conway[2], Robert Flemming[1]

[1]Pacific Biological Station, Nanaimo, BC, Canada, [2]Pacific Geoscience Centre, Sydney, BC, Canada

Abstract

Juan Perez Sound in northern British Columbia and Northumberland Channel in the south are geographically remote and distinct in their bedrock geologic composition, structure, and geomorphology. Within the 70-km^2 study areas (0–300 m in water depth), inshore rockfish (*Sebastes* spp.) and spotted ratfish (*Hydrolagus colliei*) habitats are delineated from a multiband raster (multibeam bathymetry, six derived terrain layers, and backscatter strength) using point locations of these observed species to extrapolate class probability layers. The spotted ratfish class probability layer, representing noninshore rockfish habitat, is then subtracted from the inshore rockfish layer to derive an exclusive inshore rockfish habitat probability layer. This layer is consistent with bedrock areas identified within the study sites. Correspondence analysis of visual substrata and fish species data confirm that inshore rockfish habitat is well described by bedrock and cobble substrata with their corresponding dominant rockfish species assemblage and spotted ratfish (noninshore rockfish) habitat is described by gravel and mud substrata with their corresponding dominant round fish and flat fish species groups.

Key Words: Juan Perez Sound, Haida Gwaii, Northumberland Channel, Strait of Georgia, British Columbia, Canada, rockfish, inshore rockfish habitat, fish habitat, habitat classification

Introduction: Geographic and Geological Setting of British Columbia, Canada

The western margin of Canada is a tectonically complex region where many geological processes have contributed to a coastal and nearshore zone mainly dominated by bedrock geology. The coastal and submarine landscape has developed as a consequence of regional accretionary tectonics, resulting in the development of a broad zone of deformation and mountain building forming a margin parallel, insular belt of mountains apparent as two main island land masses, Haida Gwaii and Vancouver Island, while deep sedimentary basins have repeatedly formed on the adjacent continental shelf due to the same tectonic processes (Figure 36.1). Lithology defines

Seafloor Geomorphology as Benthic Habitat. DOI: 10.1016/B978-0-12-385140-6.00036-0

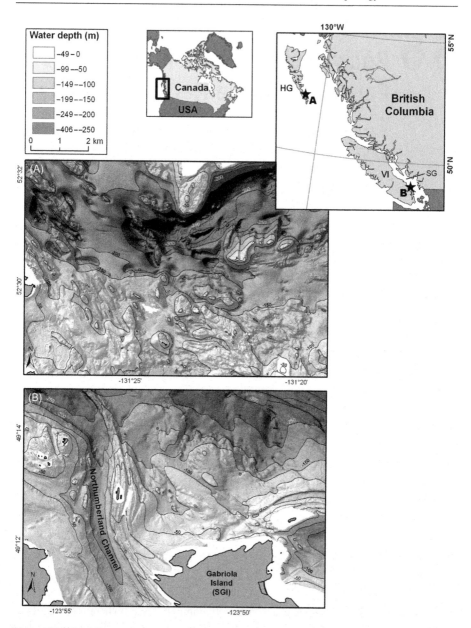

Figure 36.1 Multibeam sonar bathymetric maps for the northern study area. (A) Juan Perez Sound, Haida Gwaii archipelago, and the southern study area; (B) Northumberland Channel entrance, Southern Gulf Islands archipelago. Insets: Location map showing western British Columbia (box) in Canada map (left) and the location of the study areas (stars) in British Columbia (right). HG, Haida Gwaii; VI, Vancouver Island; SG, Strait of Georgia; SGI, Southern Gulf Islands.

much of the character of bedrock exposures, and the geomorphologies of the land-forms also reflect geologic age, present and past tectonic stress regimes, and deformation style of the rock units.

The recent glacial and sea-level history of the region has resulted in a paucity of coastal sediments and a sediment-starved continental shelf, leaving many bedrock areas exposed and unburied by more recent sediments. Outcrops and bedrock terrain within shelf and inshore areas (0–300 m in water depth) are also exposed, in part, due to the removal of large volumes of coastal and shelf sediments to the deep sea during the last ice advance 25–15 ka BP. Abundant glacial landforms along the margin include many fjords and coastal channels, as well as shelf troughs or sea valleys and moraines. Many areas of exposed bedrock remain unsedimented since deglaciation, due to the dynamic oceanographic regime where strong tidal currents are locally enhanced by a complex bedrock landscape of knolls, channels, and narrows.

The benthic habitat provided by bedrock is often complex and functions as habitat at varying scales from micro to macro [1,2]. The slope and the roughness or rugosity of bedrock outcrops are important considerations in the assessment of bedrock habitats. These rocky habitats are normally associated with epifauna because hard substrate is utilized by sessile invertebrate and algal species as they colonize rock surfaces, and they may themselves then become habitat for other organisms, as has been demonstrated for corals, sponges, and kelp beds. Bedrock habitats, including offshore banks, have been studied in California [3], Oregon, and areas of Washington, while studies of Alaska rock reef benthic habitats are under way ([4] and references therein).

Rockfish are valuable and even iconic figures for the rock reefs where they reside. About 65 species of rockfish occur along the west coast of North America, occupying various habitats from the intertidal down to the abyss [5]. The inshore rockfish group includes five demersal species (*Sebastes* spp.) that stratify by depth and aggregate over rock habitats from the sea surface to 300 m in depth. Yelloweye rockfish (*Sebastes ruberrimus*) and quillback rockfish (*S. maliger*) are highly prized food fish and are important to commercial, recreational, and subsistence fisheries. Fisheries for inshore rockfish are conducted with hook-and-line gear types, which are not associated with significant seafloor disturbances [6]. These rockfish display extreme longevity—115 years for yelloweye rockfish and 95 years for quillback rockfish—and are slow to mature and experience sporadic good recruitment. This naturally low stock productivity, coupled with increasing fishing pressure, has led to conservation concerns for inshore rockfish and the recent designation of rockfish conservation areas (RCAs) or no-take zones as a spatial fishery management tool [7].

During the initial planning for the implementation of the RCAs, rock reef areas between 0 and 200 m in depth were identified using a combination of fishery catch data and seafloor complexity (second derivative of slope) using available bathymetry from nautical charts. With the recent acquisition of multibeam bathymetry for some areas of the coast, we take a more detailed view, using new methods, to examine rock reef occurrences and assess their potential importance and function as rockfish habitat. Two representative areas, both 70 km² in sea surface area, were selected for detailed study of rock reef characteristics and habitat associations. One area in northern British Columbia (BC) waters, located within the Haida Gwaii archipelago, is Juan Perez Sound, a fjord where multiple exposed bedrock knolls are apparent on the seabed. In southern BC, Northumberland Channel, an area on the western edge of Strait of Georgia, adjacent to the southern Gulf Islands archipelago was also selected for study. Here, folded and faulted sedimentary bedrock form rugged and extensive underwater ridges. These areas are geographically remote from one another and are distinct in their bedrock geologic composition, structure, and geomorphology.

Geomorphic Features and Habitats

Northern Study Area: Juan Perez Sound

Juan Perez Sound indents the eastern side of Moresby Island, a mountainous island formed of layered carbonate, volcanic, and clastic sedimentary rocks of Mesozoic age [8] (Figure 36.1A). A deep (up to 365 m) mid fjord basin is bordered by an eastern sill marking the terminus of a glacier that occupied and excavated the sound between 15 ka and 12 ka BP [9]. Ice streamlined, elongate mounds or drumlins are apparent in the southwestern portion of the study area (Figure 36.2A). Following the last glaciation, relative sea level fell up to 150 m below present and areas more shallow than this depth still contain remnants of a drowned terrestrial landscape, dominated by river systems, alluvial fans, and delta plains [9–11]. Terraces associated with this low sea-level stand are visible in the south central portion of the study area. The deep central basin is infilled with post-transgressive muds except where a few bedrock outcrops project above the basin floor (Figure 36.2A). Current scour is apparent where tidal currents have created moats or depression adjacent to these bedrock knolls.

Southern Study Area: Northumberland Channel

Bedrock in the southern Gulf Islands area, both onshore and offshore, is composed of thick sequences of Mesozoic (Cretaceous) clastic sedimentary rocks including shale, sandstone and conglomerate. Sequences are frequently interbedded with sharp contacts between these rock types, and where erosion of these sections has occurred, emergent beds of conglomerate and sandstones may project from the more erodible shales [12–14]. This differential erosion adds to the complexity of the bedrock surface and increases rugosity. The eroded layers create linear depressions following the bedding planes, which provide a depositional center where reworked sediments accumulate. Bedrock ridges and intervening depressions are frequently traceable from onshore into the offshore (Figure 36.2B). At a scale of hundreds of meters to kilometers, broad syn- and anticlinal folds characterize the sedimentary rocks in the study area [12,14]. These elongate bedrock features frequently have steep to moderate slopes where exposure to tidal currents is variable. During glaciation, tills were deposited over much of the region [15] and are apparent in multibeam data as undulatory, broad plateaus (Figure 36.2B). In the Gabriola Island area, remnants of glacial deposits are found [16], and along the eastern portion of the study area, ice streamlined glacial sediments are observed. Unlike Juan Perez Sound, only very limited relative sea-level lowering accompanied deglaciation in the Strait of Georgia [17];

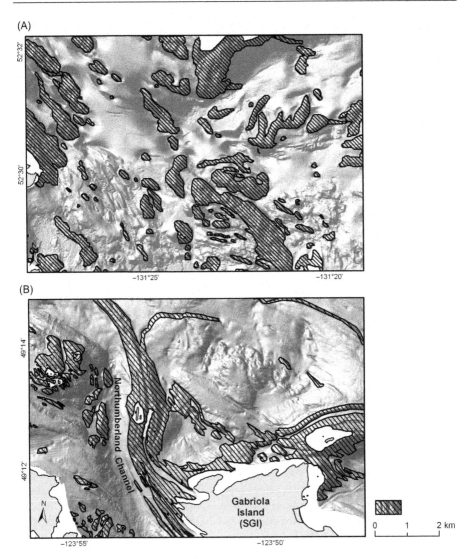

Figure 36.2 Distribution of bedrock for the northern study area. (A) Juan Perez Sound, Haida Gwaii archipelago, and southern study area; (B) Northumberland Channel entrance, Southern Gulf Island archipelago.

therefore, associated low stand sedimentary units are not found. In deeper water, in the northern portion of the study area, a drape of recent mud mantles the older units.

Habitat Identification

Potential inshore rockfish habitat was delineated by performing a supervised classification of multibeam bathymetry and derived terrain layers, as well as a backscatter strength layer [18,19]. Rather than attempting to obtain some number of distinct classes, the goal of this process was to develop a layer describing the probability of observing a rockfish in each cell, in effect classifying the area into one of two classes: either rockfish present or rockfish absent. Point observations of inshore rockfish species: yelloweye, quillback, copper (*S. caurinus*), and tiger (*S. nigrocinctus*) from *in situ* submersible and remotely operated vehicle (ROV) video were used as training sample sites. A statistical relationship between the point observations and the remotely sensed seafloor characteristics is estimated, allowing a classification to be extrapolated [20].

Multibeam bathymetry (Canadian Hydrographic Service) and backscatter data were available in several data sets of varying extent and resolution. These raster data sets were normalized, mosaiced at 5-m resolution, and clipped to each study area. Three types of terrain layers were derived from the bathymetry: (1) bathymetric position index (BPI) at four different scales to identify benthic features ranging from very fine (5–25 m) to broad (125–250 m) [21–23]; (2) percent slope (ESRI ArcView); and (3) surface area or rugosity [24,25]. These six layers were combined along with depth and backscatter data into a multiband raster using the ESRI ArcGIS "Composite Bands Tool" for the supervised classification process.

The point observations for inshore rockfish species to be used as training sample sites were converted to a 5-m grid coincident with the terrain data. A given 5 m × 5 m grid cell in which any number of point observations occur is coded as present. This grid was then used with the "Create Signatures Tool" to select the areas from the terrain data from which multivariate statistics for the presence class were calculated. These statistics are then used to develop a single class probability layer for the entire study area. The resulting class probability layer represents the probability of observing inshore rockfish species over the given combination of terrain variables.

This class probability layer was further refined to develop an exclusive probability layer. The training sample process was repeated using point observations of spotted ratfish (*Hydrolagus colliei*), a species that was observed most commonly over mud, to develop a class probability layer for spotted ratfish habitat or nonrockfish habitat. This layer was subtracted from the probability layer for inshore rockfish, resulting in a layer where high positive values represent exclusive inshore rockfish habitat and high negative values represent exclusive spotted ratfish (nonrockfish) habitat (Figure 36.3). These exclusive inshore rockfish habitat areas are associated with bedrock areas shown in Figure 36.2.

Biological Communities

A range of seafloor habitats, from flat mud to steep complex bedrock, and associated benthic fish assemblages, are assessed from recorded *in situ* observations and video data. Benthic fish are identified to species in association with a primary (50% cover) substratum category (i.e., bedrock, cobbles, gravel, and mud). For Juan Perez Sound, ~2,200 paired fish and substrata observations occurred in water depths ranging from 30 to 270 m. These observations were recorded over 11 transects conducted in May 2005 during a research cruise using the manned submersible *Aquarius* (Nuytco

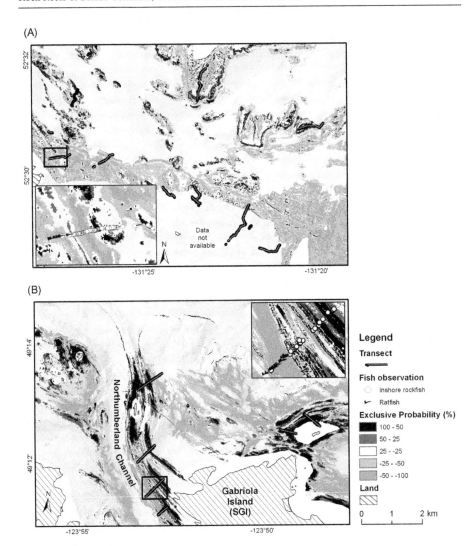

Figure 36.3 Inshore rockfish (*Sebastes* spp.) habitat shown as high positive values and spotted ratfish (*Hydrolagus colliei*) habitat shown as high negative values on an exclusive probability of occurrence map for the northern study area. (A) Juan Perez Sound and southern study area; (B) Northumberland Channel entrance. The training sample points are shown as discrete data points (see legend).

Research Ltd.). Similarly, for Northumberland Channel, over 750 paired observations were recorded from 20 to 120 m water depth using a Phantom HD 2 + 2 (Deep Ocean Engineering) ROV. Six ROV transects were conducted during research cruises in February and November 2009.

In the study areas, 27 fish species are observed over the five substrata categories (Table 36.1). A greater number of fish species, and higher fish densities, are

Table 36.1 Fish Species Identified by Substrata and Study Area: Juan Perez Sound and Northumberland Channel

Fish Species	Total Number	Juan Perez Sound % Frequency by Substratum				Total Number	Northumberland Channel % Frequency by Substratum			
		Mud	Bedrock	Cobble	Gravel		Mud	Bedrock	Cobble	Gravel
Spiny dogfish (*Squalus acanthias*)[a]	0	–	–	–	–	10	20	0	80	0
Spotted ratfish (*Hydrolagus colliei*)[a]	313	6	24	12	58	72	3	1	38	58
Pacific cod (*Gadus macrocephalus*)[a]	11	0	0	18	82	22	0	5	95	0
Pacific tomcod (*Microgadus proximus*)[a]	21	0	0	0	100	0	–	–	–	–
Walleye pollock (*Theragra chalcogramma*)[a]	79	0	0	0	100	5	0	0	80	20
Eelpouts (*Zoarcidae*)[a]	120	2	0	5	93	17	12	0	53	35
Sturgeon poacher (*Agonus acipenserinus*)[a]	15	0	13	7	80	0	–	–	–	–
Silvergray rockfish (*Sebastes brevispinis*)[b]	52	0	77	2	21	0	–	–	–	–
Darkblotched rockfish (*S. crameri*)[b]	14	0	7	21	71	0	–	–	–	–
Greenstriped rockfish (*S. elongatus*)[b]	36	0	14	19	67	65	34	17	45	5
Puget Sound rockfish (*S. emphaeus*)[b]	23	0	100	0	0	0	–	–	–	–
Yellowtail rockfish (*S. flavidus*)[b]	47	6	87	4	2	0	–	–	–	–
Canary rockfish (*S. pinniger*)[b]	18	0	78	17	6	0	–	–	–	–

Species										
Redstripe rockfish (*S. proriger*)[b]	34	85	12	0	3	0	–	–	–	–
Pygmy rockfish (*S. wilsoni*)[b]	31	74	26	0	0	0	–	–	–	–
Sharpchin rockfish (*S. zacentrus*)[b]	145	51	39	0	10	0	–	–	–	–
Copper rockfish (*S. caurinus*)[b,c]	0	–	–	–	–	13	85	8	0	8
Quillback rockfish (*S. maliger*)[b,c]	574	87	5	3	5	150	81	3	12	4
Tiger rockfish (*S. nigrocinctus*)[b,c]	14	100	0	0	0	20	100	0	0	0
Yelloweye rockfish (*S. ruberrimus*)[b,c]	53	91	2	2	6	30	63	7	30	0
Kelp greenling (*Hexagrammos decagrammus*)[d]	36	92	0	8	0	33	76	0	18	6
Lingcod (*Ophiodon elongatus*)[d]	51	47	10	25	18	55	31	9	47	13
Rex sole (*Glyptocephalus zachirus*)[e]	12	0	0	0	100	0	–	–	–	–
Pacific halibut (*Hippoglossus stenolepis*)[e]	11	9	18	0	73	0	–	–	–	–
Rock sole (*Lepidopsetta bilineata*)[e]	0	–	–	–	–	12	25	0	67	8
Dover sole (*Microstomus pacificus*)[e]	60	3	7	5	85	17	18	0	71	12
English sole (*Parophrys vetulus*)[e]	19	0	0	26	74	7	14	0	86	0

Fish listed by common name, scientific name, total number, and percent frequency by substratum category. Species groups denoted by superscript alphabets.

[a] Round fish.
[b] Rockfish.
[c] Inshore rockfish.
[d] Hexagrammids.
[e] Flat fish.

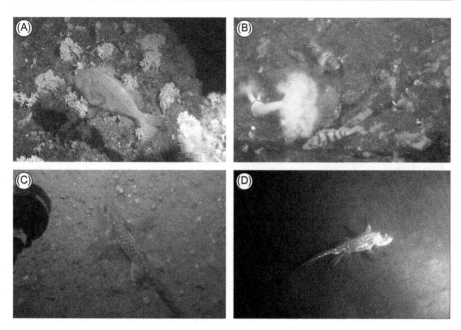

Figure 36.4 Images of substrata categories and their associated fish species. (A) Bedrock with yelloweye rockfish (*Sebastes ruberrimus*); (B) cobble with quillback rockfish (*S. maliger*) and tiger rockfish (*S. nigrocinctus*); (C) gravel with spotted ratfish (*Hydrolagus colliei*); and (D) mud with spotted ratfish. Rock reef substrata (inshore rockfish habitat) are composed of bedrock and cobble, in contrast to gravel and mud substrata (spotted ratfish habitat).

observed in Juan Perez Sound than in Northumberland Channel. This may be, in part, due to the wider range of habitats and depths covered in the northern study area and/or a seasonal effect for some species. Various fisheries have also taken place in the two study areas, perhaps more intensely in the south than the north due to its proximity to population centers. Quillback rockfish, a targeted species in fisheries, is the most commonly encountered fish in both study areas, followed by spotted ratfish, a nonfished species.

Rock reef habitats: Bedrock and cobble substrata are dominated by rockfish (*Sebastes* spp.) (Table 36.1, Figure 36.4). Of the inshore rockfish species, tiger rock-fish are solely observed on bedrock substratum. Copper and quillback rockfish are observed >80% of the time on bedrock substratum, and yelloweye rockfish are observed >90% of the time on bedrock substratum in the northern Juan Perez Sound study site and ranged over various substrata in the southern Northumberland Channel study site, where they are observed over bedrock only 63% of the time (Figure 36.4 A and B). Kelp greenling (*Hexagrammos decagrammus*) are observed 92 and 76% of the time over bedrock substratum in the north and south, respectively. Lingcod (*Ophiodon elongatus*) are the most cosmopolitan species, with observations over a wide range of habitats; they tend to rock reef habitats in the north and nonrock reef habitats in the south.

Nonrock reef habitats: Gravel and mud substrata are dominated by round fish and flat fish (Table 36.1, Figure 36.4). Spotted ratfish are observed 58% of the time over mud and 6 and 38% over gravel in the north and south, respectively (Figure 36.4C and D). Other round fish and flat fish species occurred over gravel and mud habitats in over 80–100% of the observations, with the exception of Pacific halibut (*Hippoglossus stenolepis*) in the north and rock sole (*Lepidopsetta bilineata*) in the south. These two species of flat fish occurred about 75% of the time over this habitat. Two species of rockfish, darkblotched (*S. crameri*) and greenstriped (*S. elongatus*), did occur over the mud substratum 71 and 67% of the time, respectively, unlike the majority of the rockfish species group.

Surrogacy

Correspondence analysis is used to statistically analyze and graphically display the relationships among substrata categories (rows) and among fish species (columns) [18,19,26]. In both study areas, inshore rockfish species are situated in a cluster away from the origin (center of the graph) in the bedrock subspace (Figure 36.5). This represents a strong association among these inshore rockfish species and between these species and the bedrock category of substrata. In contrast to the inshore rockfish species group, spotted ratfish is situated in the mud subspace but also at a distance from the origin, representing an inverse correspondence with inshore rockfish and association with the mud category of substrata (Figure 36.5). Rock reef habitats are well represented by the bedrock substratum and, to a lesser extent, the cobble substratum, and show high correspondence with the rockfish species group, particularly the inshore rockfish and kelp greenling. The gravel and mud substrata represent nonrock reef habitats and show associations with the round fish, especially spotted ratfish, and the flat fish species group.

Lingcod are the most unassociated species with respect to substrata, as depicted by their nearness to the graph origin (Figure 36.5). The gravel substratum in Juan Perez Sound is undifferentiated and positioned near the origin, largely due to its relatively low frequency of occurrence (Figure 36.5A). Greenstriped rockfish in the Northumberland Channel study site fall into the cobble–gravel subspace and span both the rock reef and nonrock reef areas (Figure 36.5B).

Bedrock substrata, regardless of its geologic origin in Juan Perez Sound or the Northumberland Channel, provides habitat for inshore rockfish. Correspondence analysis establishes repeatable relationships between bedrock and inshore rockfish species, as well as between mud and spotted ratfish. These links are also derived through multivariate statistics on remotely sensed terrain data and fish species presence data. The resulting Exclusive Probability layer shows high positive values representing exclusive inshore rockfish habitat over those areas known to be bedrock, and high negative values representing exclusive spotted ratfish habitat over areas of nonbedrock. These new methods provide a more detailed and robust analysis of inshore rockfish habitat in these two distinct areas of coastal British Columbia and can be applied in other areas where appropriate data sets exist.

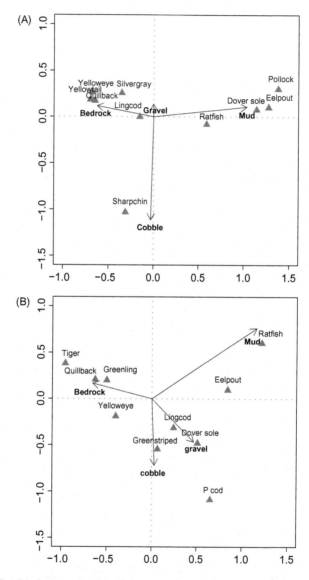

Figure 36.5 Graphical representation of the correspondence analyses of substrata categories (rows) and the 10 most abundant fish species (columns) for Juan Perez Sound (A) and Northumberland Channel (B). Inshore rockfish species (quillback, tiger, and yelloweye) are closely associated species that occur over bedrock substratum, in contrast to spotted ratfish (ratfish), which are disassociated from inshore rockfish and occur over the mud substratum.

Acknowledgments

We would like to acknowledge Chris Grandin and Lisa Lacko for 2006 draft work on benthic habitat classification in Juan Perez Sound. We would also like to thank the Canadian Hydrographic Service for the collection of multibeam sonar data, as well as Dr. J. Vaughn Barrie and Dr. Gary H. Greene for the discussions on habitat mapping.

References

[1] H.G. Greene, M.M. Yoklavich, R.M. Starr, V.M. O'Connell, W.W. Wakefield, D.E. Sullivan, et al., A classification scheme for deep seafloor habitats, Oceanolog. Acta 22 (6) (1999) 663–678.

[2] H.G. Greene, J.J. Bizzarro, V.M. O'Connell, C.K. Brylinsky, Construction of digital potential marine benthic habitat maps using a coded classification scheme and its applications, in: B.J. Todd, H.G. Greene, (Eds.), Mapping the Seafloor for Habitat Characterization, Geological Association of Canada, St. John's, NL, 2007, pp. 141–155. (Special Paper 47).

[3] M.M. Yoklavich, V.M. O'Connell, Twenty years of research on demersal communities using the Delta submersible in the northeast Pacific, in: J.R. Reynolds H.G. Greene (Eds.), Marine Habitat Mapping Technology for Alaska, vol. 10, Alaska Sea Grant College Program, University of Alaska Fairbanks, Fairbanks, AK, 2008, pp. 143–155.

[4] J.R. Reynolds, H.G. Greene, (Eds.), Marine Habitat Mapping Technology for Alaska, Alaska Sea Grant College Program, University of Alaska Fairbanks, Fairbanks, AK, 2008

[5] M.S. Love, M. Yoklavich, L. Thorsteinson, The Rockfishes of the Northeast Pacific, University of California Press, Berkeley and Los Angeles, CA, 2002. 404 p.

[6] R. Chuenpagdee, L.E. Morgan, S.M. Maxwell, A.N. Elliot, D. Pauly, Shifting gears: assessing collateral impacts of fishing methods in US waters, Front. Ecol. Environ. 1 (10) (2003) 517–524.

[7] K.L. Yamanaka, G. Logan, Developing British Columbia's inshore rockfish conservation strategy, Mar. Coast. Fish. Dyn. Manage. Ecosyst. Sci. 2 (2010) 28–46.

[8] R.I. Thompson, J.W. Haggert, P.D. Lewis, Late Triassic through early Tertiary evolution of the Queen Charlotte Basin, with a perspective on hydrocarbon potential, in: Evolution and Hydrocarbon Potential of the Queen Charlotte Basin, British Columbia, Geological Survey of Canada, St. John's, NL, Paper 90-10, 1991, pp. 3–29.

[9] H.W. Josenhans, D.W. Fedje, K.W. Conway, J.V. Barrie, Post glacial sea levels on the western Canadian continental shelf: evidence for rapid change, extensive subaerial exposure and early human habitation, Mar. Geol. 125 (1995) 73–94.

[10] H.W. Josenhans, D.W. Fedje, R. Pienitz, J.R. Southon, Early humans and rapidly changing Holocene sea levels in the Queen Charlotte Islands—Hecate Strait, British Columbia, Canada, Science 277 (1997) 71–74.

[11] D.W. Fedje, H.W. Josenhans, Drowned forest and archaeology on the continental shelf of British Columbia, Geology 28 (2000) 99–102.

[12] P.S. Mustard, The upper Cretaceous Nanaimo Group, Georgia Basin, in: J.W.H. Monger, (Ed.), Geology and Geological Hazards of the Vancouver Region, Southwestern British Columbia, Geological Survey of Canada, St. John's, NL, 1994, pp. 27 Bulletin 481.

[13] P.S. Mustard, G.E. Rouse, Stratigraphy and evolution of Tertiary Georgia Basin and sub-jacent Upper Cretaceous sedimentary rocks, southwestern British Columbia and north-western Washington State, in: J.W.H. Monger, (Ed.), Geology and Geological Hazards of the Vancouver Region, Southwestern British Columbia, Geological Survey of Canada, St. John's, NL, 1994, pp. 97. Bulletin 481.

[14] T.D.J. England, R.M. Bustin, Architecture of the Georgia Basin, southwestern British Columbia, Bull. Can. Petrol. Geol. 46 (1998) 288.

[15] J.V. Barrie, K.W. Conway., Rapid sea-level change and coastal evolution on the Pacific margin of Canada, Sediment. Geol. 150 (2002) 171–183.

[16] K. Picard, Surficial Geology and Shaded Seafloor Relief, Nanaimo, British Columbia, Geological Survey of Canada, St. John's, NL, 2010. Map 2118A, scale 1:50 000.

[17] T.S. James, I. Hutchinson, J.V. Barrie, K.C. Conway, D. Mathews, Relative sea-level change in the northern Strait of Georgia, British Columbia, Géogr. Phys. Quat. 59 (2005) 113–127.

[18] Buhl-Morensen, P., Dolan, M. and Buhl-Mortensen, L. 2009. Prediction of benthic bio-topes on a Norwegian offshore bank using a combination of multivariate analysis and GIS classification. ICES J. Mar. Sci. 66, 2026-2032, doi:10.1093/icesjms/fsp200.

[19] M.F.J. Dolan, P. Buhl-Mortensen, T. Thorsnes, L. Buhl-Mortensen, V.K. Bellec, R. Boe, Developing seabed nature-type maps offshore Norway: initial results from the MAREANO programme, Norw. J. Geol. 89 (2009) 17–28.

[20] C.N. Rooper, M. Zimmermann, A bottom-up methodology for integrating underwater video and acoustic mapping for seafloor substrate classification, Cont. Shelf Res. 27 (7) (2007) 947–957.

[21] A.D. Weiss, Topographic position and landform analysis, ESRI International User Conference (Poster), San Diego, CA, 2001.

[22] P. Iampietro, R. Kvitek, Quantitative seafloor habitat classification using GIS ter-rain analysis: biological significance of rugosity and topographic position index (TPI) in assessment of fish and invertebrate habitat on the Monterey Peninsula, CA, USA, GeoHab 2002 Poster, Moss Landing Marine Laboratories, CA, 2002.

[23] C.E. Whitmire, Integration of High-Resolution Multibeam Sonar Imagery with Observational Data from Submersibles to Classify and Map Benthic Habitats at Heceta Bank, Oregon, Master's Thesis, 2003.

[24] ESRI, ArcGIS 9.3 Desktop Help, 2008.

[25] J.S. Jenness, Surface area and ratio for ArcGIS. <http://www.jennessent.com/arcgis/sur-face_area.htm> 2004 (accessed 16.04.10).

[26] O. Nenadic, M. Greenacre, Correspondence analysis in R, with two- and three-dimen-sional graphics: the ca package, J. Stat. Software 20 (3) (2007) 13.

37 Seabed Habitats of the Southern Irish Sea

Karen A. Robinson[1], Andrew S.Y. Mackie[2],
Charles Lindenbaum[1], Teresa Darbyshire[2],
Katrien J.J. van Landeghem[3], William G. Sanderson[1]

[1]Countryside Council for Wales, Maes y Ffynnon, Ffordd Penrhos, Bangor, Gwynedd, Wales, UK, [2]Amgueddfa Cymru—National Museum Wales, Cathays Park, Cardiff, Wales, UK, [3]School of Ocean Sciences, Bangor University, Menai Bridge, Anglesey, Wales, UK

Abstract

The southern Irish Sea is a shallow sea (with maximum depths of 160 m) that lies between Wales and the Republic of Ireland in the NE Atlantic region, covering an area of ~33,000 km². The area includes many broadscale geomorphic features including plains, banks, ridges, reefs, valleys, shoals, bioherms, and sediment wave fields. Biological diversity has been recorded across many of these features during studies such as the BIOMÔR [A.S.Y. Mackie, P.G. Oliver, E.I.S. Rees, BIOMÔR Rep. 1 (1995) 263], SWISS [A.S.Y. Mackie, E.I.S. Rees, J.G. Wilson, Marine Biodiversity in Ireland and Adjacent Waters, Ulster Museum, Belfast, 2002], and HABMAP [K.A. Robinson, T. Darbyshire, K. Van Landeghem, C. Lindenbaum, F. McBreen, S. Creavan, et al., BIOMÔR Rep. 5 (2009) 148] projects. Of particular interest is the occurrence of a horse-mussel (*Modiolus modiolus*) bioherm that forms part of a Special Area of Conservation (SAC) off the Welsh coast. The bioherm is inhabited by rich macroinfaunal and epifaunal assemblages and has a characteristic acoustic signature due to its wave-like structure. The species diversity and affinities of the fauna associated with eight geomorphic features and five broad sediment categories are investigated using multivariate techniques. Gravelly sediments in the plains, valleys, banks, and *Modiolus* bioherm support the highest species richness. Conversely, sandbanks and shoals have low-diversity assemblages of animals adapted to mobile sands.

Key Words: Irish Sea, habitats, bioherm, *Modiolus modiolus*, diversity, multivariate analyses

Introduction: The Southern Irish Sea Area

The southern Irish Sea is a shallow sea lying on the northwest European Continental Shelf between Wales and the Republic of Ireland, covering an area of 33,000 km² (Figure 37.1) and reaching depths of 160 m. The sea contains a channel running along its center roughly oriented north–south and gradually shelves on both sides.

Seafloor Geomorphology as Benthic Habitat. DOI: 10.1016/B978-0-12-385140-6.00037-2

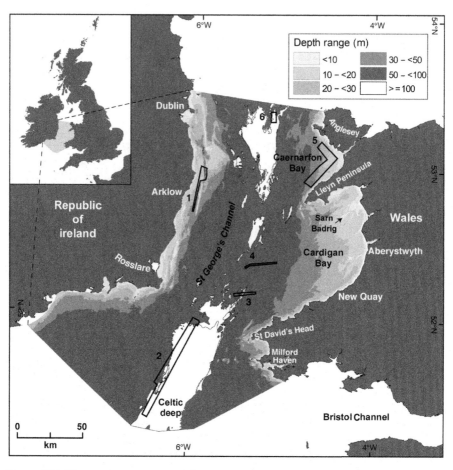

Figure 37.1 The southern Irish Sea study area, showing survey locations (Areas 1–6) investigated during the HABMAP project.

Three primary sources of water movement exist through the southern Irish Sea—tidal flows, weather-related and density-driven movements [1]. Of these, tidal flows are the most significant, with residual currents flowing in a northward direction through the area. A gradient of increasing salinity exists from north to south [1], and seasonal fronts are known to form to the north and south of the study area [2]. However, tidal mixing is sufficient across much of the region to ensure that the water column remains well mixed throughout the year.

The morphology and sediments of the southern Irish Sea are products of historic glacigenic processes and subsequent exposure to waves, storms, and tidal flows [3–5]. The Irish Sea Ice Stream poured into the Irish Sea several times during the Quaternary period [6,7], and multiple generations of glacial incisions have been documented [8,9]. The last glaciation and postglacial thawing is represented by extensive areas of lag deposits, mainly reworked glacial and postglacial sediments in a complex mosaic of various sediment types and bedforms [4,5,10]. For example, the Sarns off the west coast of Wales mark the boundaries of glacial flows [11], while drumlins, ribbed moraines, eskers, and iceberg scour marks to the north and west of Anglesey clearly document ice advance and ultimate retreat [12,13].

Previous studies conducted in the area have shown the present-day seabed environment to be highly variable and to contain a diverse range of both physical conditions and biological communities [14–16]. The diversity of these physical environments is complemented by a richness of benthic habitats, as demonstrated during the Habitat Mapping for Conservation and Management of the Southern Irish Sea (HABMAP) surveys [16] (survey areas shown in Figure 37.1), and has contributed to the designation of several Special Areas of Conservation around the Welsh and Irish Coast under the EU Habitats Directive.

The Irish Sea is one of the busiest maritime regions of the United Kingdom and Ireland, and has great economic value to both countries in terms of its fisheries, aggregate resources, oil and gas reserves, and offshore power generation. Main pressures from anthropogenic activities come from fishing, dredging, shipping, and power generation, and the sea has been classified as a high impact area [17]. Climate change impacts are also becoming evident from rising sea levels and temperature, the latter bringing about a gradual change in the biology of seabed communities in the eastern Irish Sea, and contributing to the spread of non-native species [18].

Main Geomorphic Features of the Southern Irish Sea

A number of studies and data sources were used to compile a summary of the geomorphic features in the southern Irish Sea, including the British Geological Survey seabed sediment and geology maps [4,5,19,20] and data collected during the HABMAP [16,21], Irish Sea Marine Aggregate Initiative [22,23], BIOMÔR [14,15], Training and Research in the Irish Sea (TRIS), Strategic Environmental Assessment 6 [24,25], Countryside Council for Wales Marine Monitoring [26–28], and Welsh Sandbanks [29] surveys (Figure 37.2).

Plains: Sediment plains form the main bed type in the southern Irish Sea, covering $\sim$21,630 km^2. The sediments associated with these large, low-gradient plains are generally coarse, ranging from gravels to sands. Finer sediments can be found in deeper waters toward the Celtic Deep and in sheltered inshore regions along the Welsh and Irish coasts [30–34].

Sediment waves: The seafloor sediments in many areas of the Irish Sea form a mobile layer that is actively involved in the present hydraulic regime. Waves (composed of both sand and gravel) are therefore commonly encountered along sediment transport paths in response to tidal currents and dependent on sediment supply [33,34]. These cover an area of $\sim$7,250 km^2 and are ubiquitous in the Irish Sea [22,24].

Sediment wave fields were identified during the HABMAP survey [16], notably off Arklow, Ireland (Figure 37.1: HABMAP Area 1) and along the St. George's Channel (midpart of study area). Two locations were surveyed in the St. George's Channel (Figure 37.1: Areas 3 and 4), both of which were found to have similar sediment

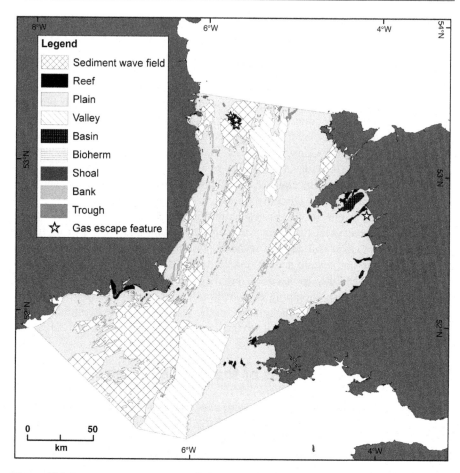

Figure 37.2 Broadscale geomorphic features found within the southern Irish Sea.

types, benthic communities, and bedforms. They were characterized by mostly coarse or mixed sediment plains, with occasional waves. Of key interest was a number of straight-crested features with adjacent deep scour marks (Figure 37.3C). Further multidisciplinary investigation of these features showed their surfaces to be relatively coarse, mobile sand. These unusual bedforms were first recorded in the Irish Sea several decades ago [10,23,35,36] and are found elsewhere on the NW European Shelf, including off the north coast of Ireland (unpublished data, BGS). Similar high and symmetrical sedimentary forms are also documented from the North American shelf seas [37,38].

Banks: A number of large sand and gravel banks are present in the southern Irish Sea, approximating to 130 km^2 of seabed. The largest of these occur along and parallel to the east coast of Ireland and are oriented roughly north–south at a small angle to the peak tidal current direction. Sandbanks also occur off the Welsh coast, though they are smaller than their Irish counterparts.

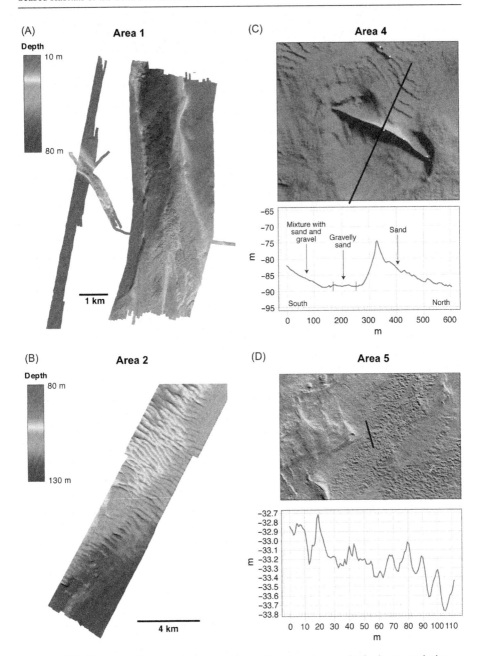

Figure 37.3 Features of interest detected using multibeam echosounder bathymetry during the HABMAP project. (A) Trough and adjacent gravel sediments and waves in the Arklow Bank location (Area 1); (B) sediment waves in the Celtic Deep location (Area 2); (C) ridge feature in the St. George's Channel (Area 4), with cross section showing ridge profile and sediment types along illustrated line; and (D) *Modiolus modiolus* reef (bioherm) in Caernarfon Bay (Area 5), with cross section showing reef undulations along illustrated line.

Shoals: Several small shoals lie off the Welsh coast, composed of shallow coarse sediments, covering a total area of about 6km^2. Most notable are Bastram Shoal and Devil's Ridge, just south of the Lleyn Peninsula.

Valleys: Two main valley features occur within the area, covering $\sim 3,510 \text{km}^2$. The Celtic Deep lies at the southern edge of the study area and comprises a wide muddy expanse of sediment in water deeper than 100 m. The valley gradually grades into shallower coarse sediment plains and wave fields (Figure 37.3B: Area 2). A second large valley begins at the northern edge of the study area and is similarly composed of muddy sediments in deep waters, surrounded by shallower coarse sediment plains.

Troughs: A number of steep-sided, glacially incised troughs occur in the Irish Sea. Only a few small ones occur within our study area, covering $\sim 47 \text{km}^2$ of seabed. Examples include a trough off the Irish Coast at Arklow (Figure 37.3A: Area 1), which reaches a depth of around 80 m, and off the west coast of Anglesey, where troughs extend to water depths of over 100 m. Sediments within one of the west Anglesey troughs were sampled during the HABMAP project (Figure 37.1: Area 6) and found to be composed of boulder clay and mixed sediments.

Basins: One basin of particular note is in Tremadog Bay in the northernmost part of Cardigan Bay. It lies between the Lleyn Peninsula and Sarn Badrig and is known locally as "muddy hollow." It covers $\sim 130 \text{km}^2$ and has a maximum depth of 40 m. A similarly muddy, but much smaller, basin (16km^2) known as "The Gutter" occurs in shallower waters southwest of Aberystwyth, and several other lesser areas occur further south.

Bioherm: Mussel reefs (and occasional *Sabellaria* reefs) are the main type of sub-tidal biotic structures found in the southern Irish Sea, covering an approximate area of 10km^2. Of particular note is a well-studied horse-mussel (*Modiolus modiolus*) bioherm found in shallow waters (25–35 m depth) off the north coast of the Lleyn Peninsula (Figure 37.1: Area 5, south; Figure 37.2). *Modiolus modiolus* beds are known to occur in the Irish Sea [26–28,39–41], often forming distinct reef-like structures that can be detected using sidescan sonar or multibeam echosounder [41,42]. The bioherm lying off the Lleyn Peninsula was surveyed during the HABMAP project [16] (Figures 37.3D and 37.4) and covers an area of 4km^2 in relatively tide-swept waters on mixed or coarse sediments.

The Caernarfon Bay *Modiolus* bioherm is an important component of the Pen Llŷn ar Sarnau Special Area of Conservation (SAC), designated under the EC Habitats Directive (Council Directive 92/43 EEC). The main reef occurs in an area of high tidal flow between 25 and 35 m below chart datum, and forms a series of undulating structures that differ substantially from the surrounding lag gravel bedforms and their associated sand ripples. *The Modiolus* reef appears on sidescan sonar and multibeam records as dark (reflections) and white (shadows), forming a characteristic "mottled" image (Figure 37.3D). The repetitive wave-like morphology of the reef, shown in a multibeam echosounder (MBES) bathymetry cross section (Figure 37.3D), differs significantly in wavelength and amplitude from adjacent gravel waves [28]. Records from boomer profiles show the *Modiolus* bedform is built up to as much as 1 m over the underlying gravelly substratum [28].

Reefs: Rocky reefs are common in shallow waters off the Welsh and Irish coasts, amounting to around 380km^2. Home to a diverse range of animal and plant

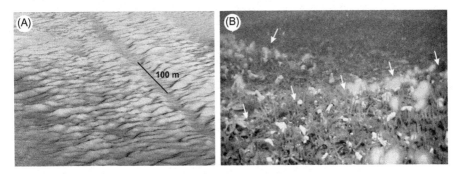

Figure 37.4 Panoramic views of the Caernarfon Bay *Modiolus modiolus* reef showing undulations in the structure as detected by (A) multibeam echosounder bathymetry and (B) underwater photography (arrows indicate orientation of ridges).
Source: Image supplied by Paul Kay, Marine Wildlife Photo Agency.

communities, they are designated features of several SACs in Wales. Important rocky reef areas are off the coast of Pembrokeshire (southwest Wales) and around the Lleyn Peninsula and Anglesey (north Wales). Of particular note are the Cardigan Bay Sarns (boulder, rock, and cobble reefs formed during the last ice age), including Sarn Badrig (shown in Figure 37.1).

Gas escape features: Various bedforms have been linked to methane escape in different parts of the Irish Sea. Pockmarks of varying sizes, ranging from <20 m across to 250 m by 150 m and 14 m deep, occur in the northwestern Irish Sea mud valley [43]. Twenty-three elongate seep mounds have been observed to be aligned along the underlying Codling Fault Zone in the sedimentary Kish Bank Basin region off Dublin (typical sizes: >250 m long, 80 m wide, and 5–10 m high). These are associated with carbonate-cemented sandstone and have been interpreted as methane-derived authigenic structures [25]. Holden's reef in the shallow (10 m) northeastern part of Cardigan Bay has also been identified as a methane-derived structure.

Biological Communities

Biological data have been collected throughout the study area over a number of years and during a range of projects including BIOMÔR, SWISS, and HABMAP [14–16]. The data used in the new macrofaunal analyses presented here were originally analyzed and interpreted separately [14,16,26,29,44]. The data set comprised 165 quantitative grab-sampled stations and 1,137 macrofaunal taxa; colonial epifaunal species were excluded. However, the multivariate analyses (Figure 37.5) were conducted on presence–absence data owing to the different sampling regime of the Irish sandbank study (Figure 37.6) [44].

Cluster analysis performed on the data set (not shown) revealed five broad clusters, which respectively comprised groups A–E, F–G, H–K, L, and N–M as depicted in the nonmetric multidimensional scaling (MDS) plots (Figure 37.5). Examination

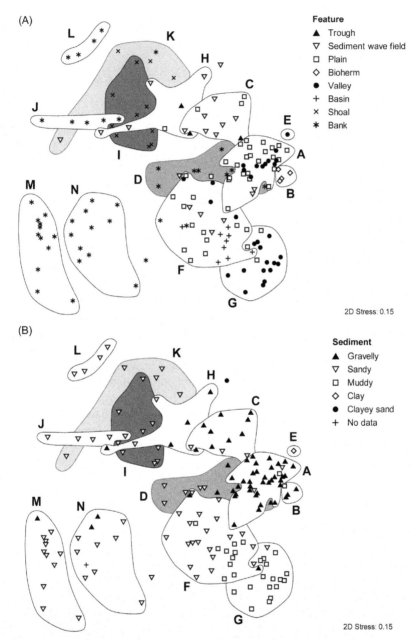

Figure 37.5 Benthic macroinfaunal relationships in the Irish Sea area (groups delineated by a presence–absence cluster analysis). (A) Nonmetric MDS plot with stations coded by geomorphic feature and (B) nonmetric MDS plot with stations coded by broad sediment category. Groups D, I, and J colored to aid interpretation. (For interpretation of the references to color in this figure legend, the reader is referred to the web version of this book.)

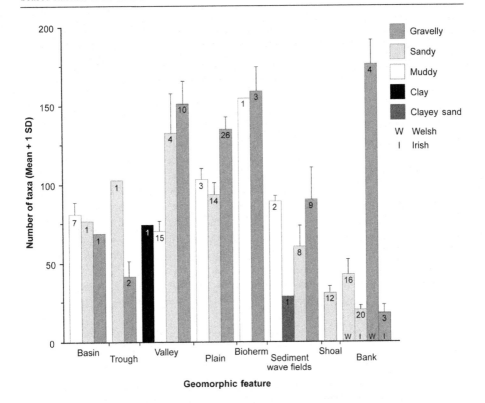

Figure 37.6 Species richness of macroinfauna by broad sediment category for each geomorphic feature. Species richness calculated as number of taxa per $0.2\,m^2$ ($2 \times 0.1\,m^2$ Van Veen Grab samples; 0.5-mm mesh sieve) station, except for Irish (I) bank stations that are per $0.5\,m^2$ ($5 \times 0.1\,m^2$ Day grab samples; 1.0-mm mesh sieve).

of the species composition of the 14 groups, together with species richness (Figure 37.6) of their constituent stations, indicated a general trend of change from widespread species-rich stable gravelly plains and valleys (A), through moderately rich gravelly sediment fields (C) in HABMAP Areas 3 and 4 (Figure 37.1) and Area 1 (H), to impoverished faunal assemblages associated with the more mobile sands of Welsh shoals and sandbanks (I–L). In addition, it was noticeable that the larger Irish banks (M–N) supported much less diverse faunas compared to the Welsh banks, confirming the findings from an earlier comparison [44].

The second obvious trend was for a more gradual decrease in species richness from the rich gravelly plains (A), through the predominantly inshore sandy and muddy sediments of the plains, sediment wave fields, and basins of Cardigan and Caernarfon Bays (F), to the muddy valley stations in the Celtic Deep (G). The faunal characterizations within each main geomorphic feature are summarized below in relation to the habitat (biotope) classification manual for Britain and Ireland [45].

Plains: The species-rich gravelly plain stations of Group A (Figure 37.5) were widespread throughout St. George's Channel and in Cardigan and Caernarfon Bays

(Figures 37.1 and 37.2). They were generally classified [14,16] as "*Mediomastus fragilis, Lumbrineris* spp., and venerid bivalves in circalittoral coarse sand or gravel," although elements of other biotopes appeared (e.g., from *Modiolus* assemblages). The large bivalve *Glycymeris glycymeris* was often present in offshore areas. Sandy plains (F, in part) in the two large Welsh Bays supported a fauna intermediate between "*Lagis koreni* and *Phaxas pellucidus* in circalittoral sandy mud" and "*Fabulina fabula* and *Magelona mirabilis* with venerid bivalves and amphipods in infralittoral compacted fine muddy sand." The offshore plains east of the Celtic Deep were more similar to "*Abra prismatica, Bathyporeia elegans*, and polychaetes in circalittoral fine sand."

Sediment waves: Gravelly wave field stations from HABMAP Areas 3 and 4 in St. George's Channel predominated in Group C (Figure 37.5). The fauna, including that of the unusual straight-crested ridge features (Figure 37.3C), was described as "*Hesionura elongata* and *Protodorvillea kefersteinia* in offshore coarse sand" [16]. Sandy and gravelly waves (H) in Area 1 off the Irish coast resembled the "*Moerella* spp. with venerid bivalves in infralittoral gravelly sand" biotope. Sandy wave fields in Caernarfon Bay (F, in part) were attributed to "*Lagis koreni* and *Phaxas pellucidus* in circalittoral sandy mud." The sediment wave stations were moderately rich in species (Figure 37.6).

Banks: Group D stations were largely associated with the shallow sandy and gravelly sediments of two low banks off New Quay in Cardigan Bay [29]. Faunal characterization was variable, with a mixture of species indicative of stable gravelly (e.g., tubiculous terebellimorph polychaetes) and mobile sand (e.g., small interstitial polychaete *Hesionura elongata*) sediments. The gravelly stations supported a very rich fauna (Figure 37.6). The other banks were predominantly sandy and formed separate faunal groupings (Figure 37.5) indicative of less stable sediments. Group J was wholly representative of Turbot Bank, south of Milford Haven. *Nephtys cirrosa, Eurydice truncata*, and *Gastrosaccus spinifer* were the most abundant species in fine to medium sands. This low-diversity (11–34 taxa) assemblage was attributed to "infralittoral mobile clean sand with sparse fauna." Group L comprised the majority of the medium sand stations at Bais Bank, northwest of St. David's Head. Nemertea and *Protodriloides chaetifer* dominated a very sparse fauna of only 7–13 taxa.

Another species-poor sandbank was Blackwater Bank (M), northeast of Rosslare, Ireland. This fine sandbank was characterized [44] as approaching the "infralittoral mobile clean sand with sparse fauna" or "*Nephtys cirrosa* and *Bathyporeia* spp. in infralittoral sand" biotope definitions [45]. The other Irish sandbank, Kish Bank (N), southeast of Dublin, had a related fauna. However, several biotopes were identified at the medium- to very fine-sand sediment stations, including "*Glycera lapidum* in impoverished infralittoral mobile gravel and sand," "*Abra prismatica, Bathyporeia elegans*, and polychaetes in circalittoral fine sand," and "*Abra alba* and *Nucula nitidosa* in circalittoral muddy sand or slightly mixed sediment." Species richness was higher (14–42 taxa) than on Blackwater Bank (10–15 taxa).

The different sieve size (1.0-mm mesh) used in the Irish sandbank survey [44] could potentially explain the lower species representations compared to the other studies (Figure 37.6), though this would be countered by the use of five, rather than two, grab samples. Experience from North Sea sand wave areas (Mackie, personal observation) indicates an average 20.6% reduction in taxa when using a 1.0-mm sieve rather than a 0.5-mm sieve. Given that the large linear Irish banks had less than 50% of the taxa found at the Welsh sandbanks, it is likely that the lower richness of the former is a real characteristic rather than an artifact.

Shoals: The nine stations of Group I were mainly associated with two medium sand shoals south of the Lleyn Peninsula (Figure 37.1), Bastram Shoal and Devil's Ridge. The more exposed Bastram Shoal can be attributed to "*Hesionura elongata* and *Microphthalmus similis* with other interstitial polychaetes in infralittoral mobile coarse sand." The nearby Devil's Ridge stations were characterized [29] as "*Nephtys cirrosa* and *Bathyporeia* spp. in infralittoral sand." The medium- and coarse-sand stations of the Tripods, to the north of the Lleyn, formed the core of Group K. The fauna of this shoal was indicative of a less rich variant of "*Hesionura elongata* and *Microphthalmus similis* with other interstitial polychaetes in infralittoral mobile coarse sand." Species richness of the shoal stations was low (Figure 37.6).

Valleys: The valley stations were distributed between four cluster groups (A, E, F, and G) according to sediment category. Species richness was highest in the gravels and sands (Figure 37.6). Gravelly sediments had the same fauna as the gravelly plains (A). The fauna of the sandy sediments north of the Celtic Deep (HABMAP Area 2; Figure 37.1) was allied to that of the sandy plains and sediment wave fields (F) and attributed to "*Echinocyamus pusillus*, *Ophelia borealis*, and *Abra prismatica* in circalittoral fine sand" [16]. The Celtic Deep fauna (G) was characterized as the general "Offshore circalittoral mud" biotope because no precise match was possible. Species included the polychaetes *Levinsenia gracilis*, *Mediomastus fragilis*, and *Nephtys incisa*.

Group E, found in the valley to the west of Anglesey (HABMAP Area 6), had a unique boulder clay inhabiting fauna dominated by a small spioniform polychaete of the genus *Uncispio*. The only previously known species of this genus was described in California, USA.

Troughs: The two trough stations in HABMAP Area 1 off Arkow (Figure 37.1) formed part of faunal cluster H (*Moerella* spp. with venerid bivalves in infralittoral gravelly sand), though species richness was lower than for the sediment wave field stations (Figure 37.6). One sandy station in Area 6 to the west of Anglesey had more affinity with the sediment wave stations of Group C.

Basins: The fauna of the inshore muddy basins of Cardigan Bay formed a distinct subgrouping within Group F (Figure 37.5). The species compositions [14] showed affinities with the "*Amphiura filiformis*, *Mysella bidentata*, and *Abra nitida* in circalittoral sandy mud" and "*Melinna palmata* with *Magelona* spp. and *Thyasira* spp. in infralittoral sandy mud" biotopes. Species richness was moderately high (Figure 37.6).

Bioherm: *Modiolus modiolus* communities are known to be rich in species [26,46,47]. They have a limited distribution in UK coastal waters but are known from several parts of the Irish Sea [47]. The mussel can form persistent dense beds that build up as biogenic reefs/bioherms through the accumulation of shell and fecal deposits [42,47], which are in turn colonized by epifaunal species. Patches of muddy sand and shells accumulate within the bioherm, and some dead *Modiolus* shells become aligned by the current and imbricate.

The Caernarfon Bay *Modiolus* reef (B) has recently been studied using a multidisciplinary approach [16,26,27]. Dive surveys, video footage, and sediment profile images have shown a dense epifaunal presence, typically including the soft coral *Alcyonium digitatum*, serpulid polychaete *Pomatoceros* sp., and brittlestar *Ophiothrix fragilis*—as well as various hydroids and bryozoans (Figure 37.4B). The predominantly mixed gravelly sediment stations support a very rich macrofauna (Figure 37.6). The infauna and epifauna associated with the reef's ridges and troughs differs, with the troughs having a slightly poorer species composition [26,27].

Reefs: None were sampled in the studies included in this review.

Gas escape features: No known gas-associated locations were sampled in any of the surveys. However, the remarkable discovery of a new sea spider (Pycnogonida) belonging to the genus *Sericosura* [16,48] may be indicative of an undiscovered seep on shallow stony ground (20 m) in the northern part of HABMAP Area 1 (Figures 37.1 and 37.3A). The genus is usually found associated with deeper hydrothermal vents or seep areas associated with oceanic ridges at depths of 850–2,600 m. The fauna in this location was identified as the serpulid polychaete "*Pomatoceros* (*lamarcki*) with barnacles and bryozoan crusts on unstable circalittoral cobbles and pebbles." An abundant hydroid epifauna was additionally characteristic of "*Sertularia* (*argentea*) and *Hydrallmania falcata* on tide-swept sublittoral sand with cobbles or pebbles." A total of 165 infaunal and epifaunal taxa were recorded from the dredge sample.

Surrogacy

Relationships between benthic macroinfauna and seabed environmental variables were investigated in many studies [14–16,49,50] using the BIO-ENV routine in PRIMER [51,52]. Canonical correspondence analysis (CCA) using CANOCO [53,54] was also employed. These studies reported strong relationships with various sediment parameters and depth. For example, in a reexamination [50] of the southeastern Irish Sea benthos [14], the best Harmonic Rank Correlation ($\rho_w = 0.71$) was obtained from three combined variables—gravel, depth, and silt. These three variables were identified in CCA as the most important of the six that cumulatively explained a significant proportion (33.82%) of the variance in the species data. Different combinations of sediment variables have the potential to be good predictors of different macroinfaunal assemblages in the southern Irish Sea area [21].

References

[1] K.F. Bowden, Physical and dynamical oceanography of the Irish Sea, in: The North-West European Shelf Seas: The Sea Bed and Sea in Motion. Physical and Chemical Oceanography and Physical Resources, Elsevier Oceanography Series 24B, 1980, pp. 391–414. Edited by F.T. Banner, M.B. Collins and K.S. Massie. Elsevier, New York.

[2] D.W. Connor, P.M. Gilliland, N. Golding, P. Robinson, D. Todd, E. Verling, UKSeaMap: The Mapping of Seabed and Water Column Features of UK Seas, Joint Nature Conservation Committee, Peterborough, 2006. ISBN 86107 590 1, 80 pp. + Annexes.

[3] R.A. Chadwick, D.W. Holliday, Deep crustal structure and Carboniferous basin development within the lapetus convergence zone, northern England, J. Geol. Soc. Lond. 148 (1991) 41–53.

[4] D.R. Tappin, R.A. Chadwick, A.A. Jackson, R.T.R. Wingfield, N.J.P. Smith, United Kingdom Offshore Regional Report: The Geology of Cardigan Bay and the Bristol Channel, HMSO for the British Geological Survey, London, 1994, 107 pp.

[5] D.I. Jackson, A.A. Jackson, D.J.A. Evans, R.T.R. Wingfield, R.P. Barnes, M.J. Arthur, United Kingdom Offshore Regional Report: The Geology of the Irish Sea, HMSO for the British Geological Survey, London, 1995, 123 pp.

[6] W.B. Wright, The Quaternary Ice Age, Macmillan, London, 1937. 478 pp.

[7] G.F. Mitchell, The Pleistocene history of the Irish Sea: second approximation, Sci. Proc. R. Dubl. Soc. 4 (1972) 181–199.

[8] R.T.R. Wingfield, Glacial incisions indicating Middle and Upper Pleistocene ice limits off Britain, Terra Nova 1 (1989) 538–548.

[9] R. Wingfield, The origin of major incisions within the Pleistocene deposits of the North Sea, Mar. Geol. 91 (1990) 31–52.

[10] M.R. Dobson, W.E. Evans, K.H. James, The sediment on the floor of the Southern Irish Sea, Mar. Geol. 11 (1971) 27–69.

[11] M.A.I. Hession, Quaternary geology of the South Irish Sea, Ph.D. Thesis, University of Wales, Aberystwyth, 1988.

[12] V. Blyth-Skyrme, C. Lindenbaum, E. Verling, K. Van Landeghem, K. Robinson, A. Mackie, et al., Broad-Scale Biotope Mapping of Potential Reefs in the Irish Sea (North-West of Anglesey), JNCC Report 423, JNCC, Peterborough, 2008, ISSN 0963-8091.

[13] K.J.J. Van Landeghem, A.J. Wheeler, N.C. Mitchell, Seafloor evidence for palaeo-ice streaming and calving of the grounded Irish Sea Ice Stream: implications for the interpretation of its final deglaciation phase, Boreas 38 (2009) 119–131.

[14] A.S.Y. Mackie, P.G. Oliver, E.I.S. Rees, Benthic biodiversity in the southern Irish Sea. Studies in marine biodiversity and systematics from the National Museum of Wales, BIOMÔR Rep. 1 (1995) 263.

[15] A.S.Y. Mackie, E.I.S. Rees, J.G. Wilson, The South-west Irish Sea Survey (SWISS) of benthic biodiversity, in: J.D. Nunn (Ed.), Marine Biodiversity in Ireland and Adjacent Waters, 8, MAGNI publication, Ulster Museum, Belfast, 2002, pp. 166–170.

[16] K.A. Robinson, T. Darbyshire, K. Van Landeghem, C. Lindenbaum, F. McBreen, S. Creavan, et al., Habitat Mapping for Conservation and Management of the Southern Irish Sea (HABMAP) I: seabed surveys. Studies in Marine Biodiversity and Systematics from the National Museum of Wales, BIOMÔR Rep. 5 (1) (2009) 148, Appendices 74 pp.

[17] B.S. Halpern, S. Walbridge, K.A. Selkoe, C.V. Kappel, F. Michelis, C. D'Agrosa, et al., A global map of human impact on marine ecosystems, Science 319 (2008) 948–952.

[18] Charting Progress 2: The State of UK Seas, Published by the Department for Environment Food and Rural Affairs, on behalf of the UK Marine Monitoring and Assessment Strategy Community, ©Crown copyright 2010. http://chartingprogress.defra.gov.uk/, 2010.

[19] J.W.C. James, R.T.R. Wingfield, Cardigan Bay (Sheet 52°N–06°W), Seabed Sediments (1:250,000 Offshore Map Series), published by the British Geological Survey, UK.

[20] J.W.C. James, R.T.R. Wingfield, Anglesey (Sheet 53°N–06°W), Seabed Sediments (1:250,000 Offshore Map Series), published by the British Geological Survey, UK.

[21] F. McBreen, J.G. Wilson, A.S.Y. Mackie, C. Nic Aonghusa, Seabed mapping in the southern Irish Sea: predicting benthic biological communities based on sediment characteristics, in: J. Davenport, G. Burnell, T. Cross, M. Emmerson, R. McAllen, R. Ramsay, et al. (Eds.), Challenges to Marine Ecosystems: Proceedings of the 41st European Marine Biology Symposium. Hydrobiologia 606 (2008) 93–103.

[22] M. Kozachenko, R. Fletcher, G. Sutton, X. Monteys, K. Van Landeghem, A. Wheeler, et al., A geological appraisal of marine aggregate resources in the southern Irish Sea, Technical Report Produced for the Irish Sea Marine Aggregates Initiative (IMAGIN) Project, University College Cork, Ireland, 2008, 390 pp., ISBN 978-0-9556109-2-9.

[23] P.F. Croker, M. Kozachenko, A.J. Wheeler, Gas-related seabed structures in the western Irish Sea (IRL-SEA6), Technical Report Produced for Strategic Environmental Assessment—SEA6, 120 pp. www.offshore-sea.org.uk/consultations/SEA_6/SEA6_Gas_CMRC.pdf, 2005.

[24] R. Holmes, D.R. Tappin, DTI strategic environmental assessment area 6, Irish Sea, seabed and surficial geology and processes, British Geological Survey Commissioned Report, CR/05/057, 2005.

[25] A. Judd, P. Croker, L. Tizzard, C. Voisey, Extensive methane-derived authigenic carbonates in the Irish Sea, Geo-Mar. Lett. 27 (2007) 259–267.

[26] E.I.S. Rees, W.G. Sanderson, A.S.Y. Mackie, R.H.F. Holt, Small-scale variation within a *Modiolus modiolus* (Mollusca: Bivalvia) reef in the Irish Sea. III. Crevice, sediment infauna and epifauna from targeted cores, J. Mar. Biol. Assoc. UK 88 (1) (2008) 151–156.

[27] W.G. Sanderson, R.H.F. Holt, L. Kay, K. Ramsay, J. Perrins, A.J. McMath, et al., Small-scale variation within a *Modiolus modiolus* (Mollusca: Bivalvia) reef in the Irish Sea. II. Epifauna recorded by divers and cameras, J. Mar. Biol. Assoc. UK 88 (1) (2008) 143–149.

[28] C. Lindenbaum, J.D. Bennell, E.I.S. Rees, D. McClean, W. Cook, A.J. Wheeler, et al., Small-scale variation within a *Modiolus modiolus* (Mollusca: Bivalvia) reef in the Irish Sea. I. Seabed mapping and reef morphology, J. Mar. Biol. Assoc. UK 88 (1) (2008) 133–141.

[29] T. Darbyshire, A.S.Y. Mackie, S.J. May, D. Rostron, A macrofaunal survey of Welsh sandbanks, Countryside Council for Wales CCW Report 539, 2002, 113 pp.

[30] R.H. Belderson, Holocene sedimentation in the western half of the Irish Sea, Mar. Geol. 2 (1964) 147–163.

[31] R. McQullin, J.E. Wright, B. Owens, T.R. Lister, Recent geological investigations in the Irish Sea, Nature 222 (1969) 365–366.

[32] H.M. Pantin, C.D.R. Evans, The quaternary history of the central and southwestern Celtic Sea, Mar. Geol. 57 (1984) 259–293.

[33] R.H. Belderson, R.D. Pingree, D.K. Griffiths, Low sea-level tidal origin of Celtic Sea sand banks—evidence from numerical modelling of M_2 tidal streams, Mar. Geol. 73 (1982) 99–108.

[34] R.H. Belderson, M.A. Johnson, N.H. Kenyon, Bedforms, in: A.H. Stride, (Ed.), Offshore Tidal Sands: Processes and Deposits, Chapman & Hall, London, 1982

[35] J.G. Harvey, Large sand waves in the Irish Sea, Mar. Geol. 4 (1966) 49–55.

[36] R.T.R. Wingfield, Giant sand waves and relict periglacial features on the sea bed west of Anglesey, Proc. Geol. Assoc. 98 (1987) 400–404.

[37] P.C. Valentine, S.J. Fuller, L.A. Scully, Sea Floor Image Maps Showing Topography, Sun-Illuminated Topography, Backscatter Intensity, Ruggedness, Slope, and the Distribution of Boulder Ridges and Bedrock Outcrops in the Stellwagen Bank National Marine Sanctuary Region Off Boston, Massachusetts, U.S. Geological Survey in cooperation with the National Oceanic and Atmospheric Administration.

[38] J.V. Barrie, K.W. Conway, K. Picard, H.G. Greene, Large-scale sedimentary bedforms and sediment dynamics on a glaciated tectonic continental shelf: examples from the Pacific margin of Canada, Cont. Shelf Res. 29 (2009) 796–806.

[39] D.G. Erwin, B.E. Picton, D.W. Connor, C.M. Howson, P. Gilleece, M.J. Bogues, Inshore Marine Life of Northern Ireland, Department of Environment Northern Ireland/Ulster Museum: HMSO, Belfast, 1990.

[40] R. Brown, Strangford Lough: The Wildlife of an Irish Sea Lough, Institute of Irish Studies, Queen's University, Belfast, 1990. 228 pp.

[41] R. Nunny, A sidescan sonar survey of Strangford Lough, in: M. Service, (Ed.), The Impact of Commercial Trawling on the Benthos of Strangford Lough, Industrial Science Division TI/3160/90, Belfast, 1990, pp. 1–7.

[42] D.J. Wildish, G.B.J. Fader, P. Lawton, A.J. MacDonald, The acoustic detection and characterization of sublittoral bivalve reefs in the Bay of Fundy, Cont. Shelf Res. 18 (1998) 105–113.

[43] F. Yuan, J.D. Bennell, A.M. Davis, Acoustic and physical characteristics of gassy sediments in the western Irish Sea, Cont. Shelf Res. 12 (10) (1992) 1121–1134.

[44] C. Roche, D.O. Lyons, J. Fariñas Franco, B. O'Connor, Benthic surveys of sandbanks in the Irish Sea, Irish Wildlife Manuals 29 (2007) 48 pp.

[45] D.W. Connor, J.H. Allen, N. Golding, L.K. Howell, L.M. Lieberknecht, K.O. Northen, et al., The Marine Habitat Classification for Britain and Ireland Version 04.05, JNCC, Peterborough, ISBN 1-861-07561-8 (internet version). www.jncc.gov.uk/MarineHabitatClassification, 2004.

[46] G. Thorson, Life in the Sea, McGraw-Hill, London, 1971.

[47] T.J. Holt, E.I.S. Rees, S.J. Hawkins, R. Seed, Biogenic reefs. An overview of dynamic and sensitivity characteristics for conservation management of marine SACs. UK Marine SACs Project for SNH, DOE (NI), CCW, EN, JNCC & SAMS, IX, 1998, 170 pp.

[48] R.N. Bamber, Two new species of *Sericosura* Fry & Hedgpeth, 1969 (Arthropoda: Pycnogonida: Ammotheidae), and a reassessment of the genus, Zootaxa 2140 (2009) 56–68.

[49] A.S.Y. Mackie, C. Parmiter, L.K.Y. Tong, Distribution and diversity of Polychaeta in the southern Irish Sea, in: D.J. Reish (Ed.), Proceedings of the Fifth International Polychaete Conference, Qingdao, China, 1995. Bull. Mar. Sci. 60 (2) (1997) 467–481.

[50] A.S.Y. Mackie, Macrofaunal assemblages and their sedimentary habitats: working toward a better understanding, Irish Sea Forum Semin. Rep. 32 (2004) 30–38.

[51] K.R. Clarke, R.N. Gorley, PRIMERv5: User Manual/Tutorial, PRIMER-E Ltd, Plymouth, 2006.

[52] K.R. Clarke, R.M. Warwick, Change in Marine Communities: An Approach to Statistical Analysis and Interpretation, PRIMER-E LTD, Plymouth, 2001.

[53] C.J.F. ter Braak, Canonical correspondence analysis: a new eigenvector technique for multivariate direct gradient analysis, Ecology 67 (1986) 1167–1179.

[54] C.J.F. ter Braak, CANOCO—A FORTRAN Program for Canonical Community Ordination, Microcomputer Power, Ithaca, NY, 1987–1992. LWA-88-02: 95 pp.

38 Habitats and Demersal Fish Communities in the Vicinity of Albatross Bank, Gulf of Alaska

Jennifer R. Reynolds[1], Sean C. Rooney[1], Jonathan Heifetz[2], H. Gary Greene[3], Brenda L. Norcross[1], S. Kalei Shotwell[2]

[1]University of Alaska Fairbanks, School of Fisheries and Ocean Sciences, Fairbanks, Alaska, [2]National Oceanic and Atmospheric Administration, National Marine Fisheries Service, Alaska Fisheries Science Center, Auke Bay Laboratories, Juneau, Alaska, [3]Tombolo Habitat Institute, Eastsound, Washington

Abstract

The outer shelf and upper slope in the vicinity of Albatross Bank, Gulf of Alaska, have been shaped by glaciation, recent sedimentation, and mass wasting. A series of flat sedimentary bedrock banks in about 50–100 m water depth are variably covered by glacial deposits and modern sediment. Sediment redistribution and pavement formation reflect the energy of bottom currents. Three sites, Snakehead fishing ground, 8-Fathom Pinnacle, and 49-Fathom Bank, were surveyed using a multibeam echosounder mapping system to 800 m and by manned submersible diving between 15 and 360 m. Habitat maps produced from these surveys emphasize seafloor physiography and geological substrate. Twenty-three fish species were observed, dominated by rockfishes (*Sebastes* spp. and *Sebastolobus* spp., 69% of fishes observed). Macroinvertebrate distribution was also characterized. At a scale of tens to hundreds of meters, seven fish communities associated with different depth zones and substrate characteristics were identified.

Key Words: shelf, bank, pinnacle, moraine, glacial, till, pavement, macroinvertebrate, fish community, rockfish

Introduction

Albatross Bank is located in the Gulf of Alaska's outer continental shelf near Kodiak Island (Figure 38.1). Geologically, it sits in a broad forearc above the Alaska subduction zone, 50–100 km north of the trench and 250 km south of the Alaska arc

Seafloor Geomorphology as Benthic Habitat. DOI: 10.1016/B978-0-12-385140-6.00038-4

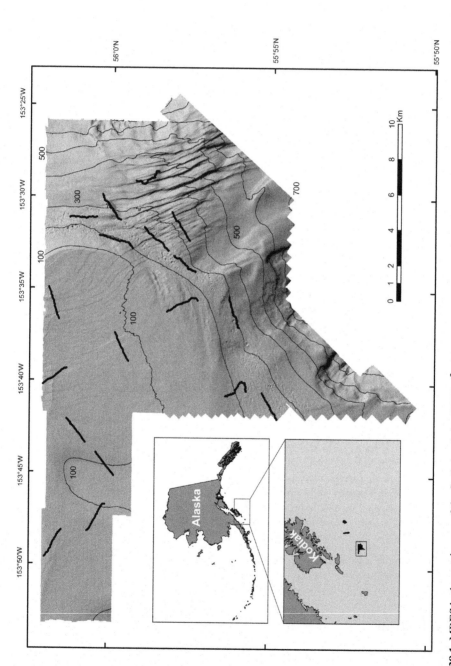

Figure 38.1 MBES bathymetric map of the Snakehead site (235 km²). Inset location map shows the Snakehead site in relation to the other study sites, to Kodiak Island, and to the Alaska mainland. The 100 kHz bathymetry data are gridded at 10 m. Contour interval is 100 m. Solid lines indicate 18 *Delta* submersible dive tracks.

volcanoes. Earthquakes and vertical tectonic processes from active plate collision form the general morphology of the region. Bedrock at the outer shelf is composed of Miocene and younger sedimentary strata, mostly flat lying, and hosts seeps that vent methane-bearing fluids [1,2]. The area is cut by shelf-parallel active faults and is subjected to uplift and subsidence on the order of meters during subduction-related seismic cycles [3].

The substrate has been shaped by glaciation, recent sedimentation, and mass wasting [2]. The seafloor of the outer shelf is principally composed of flat banks of sedimentary bedrock that shoal at about 50–100 m water depth. Glaciers and ice streams have eroded the banks and carved transverse troughs through them, and glacial sediments of generally coarse-grained materials blanket the banks, troughs, and upper slope. Modern sediment transport is reworking the glacial materials. The net result is an outer shelf with variable fine sediment that has buried some areas and left bedrock and glacial deposits exposed in others.

Albatross Bank is influenced by strong ocean currents. The Alaskan Stream flows along the shelf break and continental slope in a generally southwest direction [4]. Through interaction with currents and local bathymetric features, including the transverse troughs, slope waters are transported onto the banks [5]. These nutrient-rich waters support many fish and invertebrate species, resulting in Albatross Bank being one of the most important fishing grounds in the Gulf of Alaska [6]. Because of fishing activity, the seafloor and associated biota in the study area are not pristine. Bottom trawl gear is fished primarily in water depth less than 500 m, and longline gear in water depth to 1,000 m.

A multibeam echosounder (MBES) survey was conducted in 2003, using a hull-mounted Reson SeaBat 8111 sonar system (100 kHz).[1] The bathymetry data were gridded at 10 m, and backscatter resolution varied between 2 and 10 m. Using the methods of Greene et al. [7–9], an initial series of interpretive habitat maps was developed based on the MBES bathymetry and backscatter data and on geological information from the literature. The habitat maps were used to select areas for visual surveys with the *Delta* submersible [10]. The *Delta* is a small (4.75 m) submersible with a 365 m depth limit, and accommodates one scientist and a pilot. Twenty-two dives were conducted in June and July 2005, and a single dive in August 1999. Video data from these dives were used to ground-truth the habitat maps and update them as needed.

Geomorphic Features and Habitats

Albatross Bank is divided by transverse troughs into Southern, Middle, and Northern Albatross Bank. The study area is composed of three sites in the vicinity of the shelf edge at Southern Albatross Bank and the trough dividing it from Middle Albatross Bank (Figures 38.1–38.2). Seafloor images show that the bedrock is a sandy siltstone with subhorizontal bedding. It is exposed at a shelf-edge pinnacle and at a small

---

[1] Reference to trade names does not imply endorsement by the National Marine Fisheries Service, U.S. National Oceanic and Atmospheric Administration (NOAA).

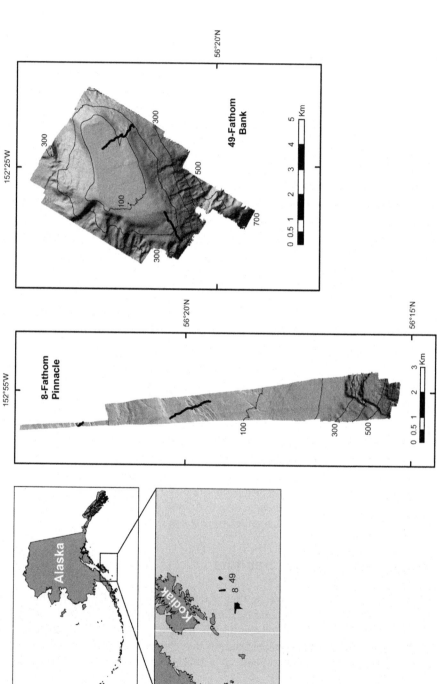

Figure 38.2 MBES bathymetric maps of the 8-Fathom Pinnacle site (17 km²) and 49-Fathom Bank site (57 km²). Location map shows the 8-Fathom Pinnacle and 49-Fathom Bank sites in relation to the other study site, to Kodiak Island, and to the Alaska mainland. The 100 kHz bathymetry data are gridded at 10 m. Contour interval is 100 m. Solid lines indicate *Delta* submersible dive tracks, with two dives at each site.

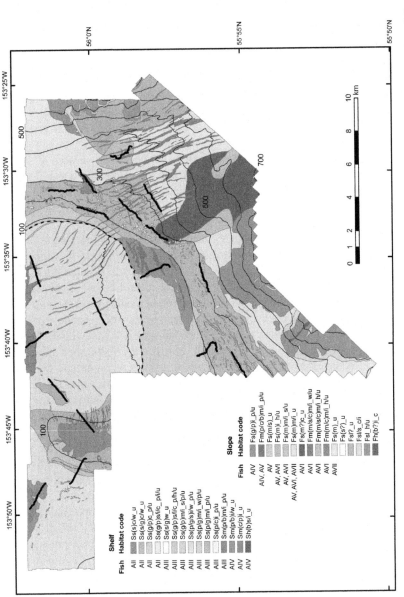

Figure 38.3 Habitat map of the Snakehead site. Physiographic setting and substrate characteristics are represented using habitat codes following the classification scheme of Greene et al. [7–9]. Habitat types are briefly described in Table 38.1. The fish communities AI, AII, etc. associated with each habitat, shown in the figure legend, are used to order the legend. However, note that this is not a map of fish community distribution because it does not include the depth ranges of the communities. Bathymetric contour interval is 100 m. Dashed line is the shelf edge.

bank on the upper slope, where it is underlain by a conglomerate. Generally speaking, however, bedrock exposures are limited and most of the substrate is unsorted glacial till, primarily subrounded and rounded clasts up to boulder size.

Snakehead fishing ground: This site lies between the shelf edge and a small trough that indents the main bank (Figure 38.1). It covers 235 km² and includes both shallow bank and upper continental slope in 60–810 m depth. The shelf edge lies at about 110 m.

In general, the shelf surface is irregular on the scale of meters, with glacial till up to boulder size in higher areas and current-driven accumulations of fine sediment or winnowed pavements in low areas (Figure 38.3; Tables 38.1 and 38.2). The pavements are dominated by pebble and gravel, with variable amounts of cobbles and boulders. Sediment waves and scour around boulders also indicate locally strong bottom currents.

At the northern edge of the site are high-standing, flat bedrock surfaces of sandstone or siltstone, as shallow as 60 m water depth. These are mantled by a thin deposit of glacial till dominated by pebbles and cobbles, with patches of unsorted sediment up to boulder size and gravel–pebble sediment waves. The steeper edges of these bedrock areas have abundant large boulders at their bases.

On the east side of the site, the bathymetric map of the shelf and upper slope shows dozens of low ridges oriented north–northwest, interpreted as a field of winnowed gravel/sand waves. However, these ridges are also covered by till. Because deposition of the till has not obliterated the preexisting topography, it must be only a thin veneer and was most likely deposited as ice-rafted debris or from a floating ice sheet.

Fine sediment cover exists locally. On the west side of the site, a shallow canyon cuts into the shelf. The floor of the canyon is filled by sand and sandy mud, sloping up to sand and gravel waves on its eastern wall. Bathymetry of the shelf edge and uppermost slope indicates areas smoothed by recent sedimentation around the shelf break. The fine sediment is sandy mud at the shelf edge and upper slope, but grades into mud with depth. The fine sediment cover is locally variable, but increases with depth, reaching 100% mud in deeper areas of the study site. It also increases along the slope from northeast to southwest.

On the south side of the bank, older recessional moraines emerge from the sediment cover as arcuate, uneven ridges. The substrate on these moraines is similar to that of the shelf, with poorly sorted till of sand to boulder size, pavements dominated by gravel and pebble, and loose cobbles and boulders locally up to 85%. Farther downslope, the upper slope is indented by numerous iceberg keel marks and depressions from ice grounding or ice melt. This surface is also covered by till, probably ice-rafted debris, and the substrate is similar to the shelf but with greater amounts of fine sediment cover. Nevertheless, the sediment around some boulders in this region exhibits current scour. In heavily sedimented areas, the underlying till emerges as patches of cobbles and boulders. Below the zone of ice gouging, the slope is incised by canyon headwalls, slumps, and landslides.

8-Fathom Pinnacle: The pinnacle is a rocky outcrop that rises from a 46 m bank at the shelf edge of Southern Albatross Bank. The MBES map shows a narrow strip covering a 17 km² portion of the pinnacle, bank, and upper slope in 20–715 m water depth, though the pinnacle is known to rise to 15 m (8 fathoms) (Figure 38.2).

Table 38.1 List of the Habitat Codes Used for the Snakehead Site in Figure 38.3

Habitat Code	Habitat Description	Area (m²)
Site 1: Snakehead fishing ground		
Ss(p/g/s)i/w_p/u	Pavement of pebble–gravel–sand, with boulder–cobbles locally to 80%. Patches of sand and gravel waves, and mud. Glacial till.	90,078,399
Ss(s/g)c/w_u	Sand and gravel waves and stringers in canyon.	7,727,942
Ss(s)c/w_u	Sand waves and stringers in canyon.	5,635,042
Sm(c/p)i_u	Unsorted rocks, pebble to boulder size, dominated by cobbles and pebbles. Glacial till (draping bedrock?).	3,237,027
Ss(p/g)m/i_w/p/u	Narrow ridges with pavement of variable pebble–gravel–sand–cobble, with boulders scattered and in local concentrations up to 70%. Glacial till draping underlying winnowed gravel/sand waves (?).	2,163,153
Ss(g/p)c_p/u	Pavement dominated by gravel–pebble, on canyon wall.	1,863,704
Sm(p/b)i/w_u	Patchy, including boulders on pebble-dominated surface, gravel–pebble pavement and waves; boulder piles. Glacial till.	1,559,112
Ss(p/g)m/i_p/u	Pebble–gravel pavement with gravel patches, glacial till on mounds and ridges.	1,250,519
Ss(g/p)s/i/c_p/h/u	Gravel–pebble pavement surface with boulder/cobble areas, on slump deposit on canyon wall. Glacial till.	864,433
Ss(s/g)w_u	Sand and gravel waves and stringers.	316,447
Sh(b)s/i_u	Boulders on scarp. Glacial till (draping edge of bedrock?).	156,920
Ss(g/p)s/i/c_ p/i/u	Gravel–pebble pavement with boulder/cobble areas, on slump scarp (interface) on canyon wall. Glacial till.	75,664
Ss(g/p)m/i_s/p/u	Iceberg keel scours with gravel–pebble–cobble pavement and patches of gravel and mud. Glacial till.	49,098
Sm(p/b)m/i_p/u	Mounds with pebble pavement, loose gravel and gravel waves, and boulder–cobble rock piles. Glacial till.	36,766
Ss(p/c)i_p/u	Pebble–cobble pavement, glacial till.	7,993
Fs(m)_u	Unconsolidated mud.	48,521,302
Fs(m)i_h/u	Unconsolidated mud on hummocky glacial till (up to boulder size), with iceberg keel and grounding depressions.	37,758,447
Fsl_h/u	Landslide covered in unconsolidated sediment.	23,603,747
Fs(m/s)_u	Unconsolidated sandy mud.	22,505,784
Fs(m?)c_u	Unconsolidated sediment (mud?) in canyon.	17,464,524
Fsl?_u	Unstable seafloor (landslide?).	12,508,246
Fm(m/s/c)m/i_h/u	Mounds with unconsolidated sandy mud and cobbles on hummocky glacial till (up to boulder size).	6,719,785
Fm(m/s/c)m/i_w/u	Narrow ridges with unconsolidated sandy mud and cobbles on glacial till (up to boulder size), draping underlying gravel/sand waves (?).	6,714,034
Fs(g/p)i_p/u	Recessional moraine with pavement dominated by gravel–pebble.	6,701,282

(*Continued*)

Table 38.1 (Continued)

Habitat Code	Habitat Description	Area (m²)
Fsl/s_c/i	Landslide scarp (interface) in consolidated sediment.	3,304,576
Fs(m)m/i_s/u	Unconsolidated mud in iceberg keel scours in glacial till (up to boulder size).	2,336,918
Fs(m)m/i_u	Unconsolidated mud in ice grounding or melt depressions in glacial till (up to boulder size).	1,834,703
Fm(p/c/b)m/i_p/u	Recessional moraine mounds/ridges with pavements of pebble–cobble–boulder, and loose cobbles and boulders up to 85%, on continental slope	1,140,855
Fm(m/c)m/i_h/u	Mounds with unconsolidated mud on glacial till (up to boulder size), hummocky, possible moraine.	746,466
Fs(s?)_u	Unconsolidated sediment (sand?).	98,423
Fh(b?)i_c	Isolated boulder (dropstone?).	28,496

Shelf and slope habitats are grouped separately, and are listed in order of decreasing areal extent. For each habitat type, a brief description is provided. The first character of the codes indicates S = continental shelf or F = continental slope (flank). The second character indicates s = soft substrate, m = mixed hard and soft substrate, or h = hard substrate. The next characters, in parentheses, indicate sediment grain size where m = mud, s = sand, g = gravel, p = pebble, c = cobble, and b = boulder. For further details of the habitat codes see [7–9]. Area (m²) is the total area covered by the habitat type within this site.

Table 38.2 List of the Habitat Codes Used for the 8-Fathom Pinnacle and 49-Fathom Bank in Figure 38.4

Habitat Code	Habitat Description	Area (m²)
Site 2: 8-Fathom Pinnacle		
Sm(g/p)e_f/c/u	Outcrop surface of sandy siltstone, subhorizontal bedding, orthogonal fractures. Patches of thin, unconsolidated sediment.	6,077,109
Shm/e_f/c	Outcrop mounds of sandy siltstone with fractures. Minor patches of loose sediment.	894,311
Smm/i_d/c/u	Recessional moraine with partial unconsolidated sediment cover.	501,993
Ss_u	Unconsolidated sediment.	313,722
Ss(s?)_u	Sand (?) stringers.	146,107
Shs/e_c	Scarp in outcrop of sandy siltstone.	60,022
Fs_u	Unconsolidated sediment.	4,447,342
Fm(m?)l_h/c/u	Landslide deposit including conglomerate bedrock (mud matrix, clasts up to boulder size), partly covered by unconsolidated mud.	2,210,835
Fs_h/u	Hummocky unconsolidated sediment, possible old landslide.	1,162,750
Fm(m?)l/s/e_c/u	Landslide scarp exposing conglomerate bedrock (mud matrix, clasts up to boulder size), partly covered by unconsolidated mud.	565,658

(Continued)

Table 38.2 (Continued)

Habitat Code	Habitat Description	Area (m²)
Fm(m?)s/e_c/u	Bedding scarp in siltstone, with unconsolidated mud (?).	88,741
Site 3: 49-Fathom Bank		
Sm(g/p)t/e_f/c/u	Siltstone outcrop with fractures and patches of unconsolidated sediment.	7,019,429
Sht/e/s_c	Fault scarp in siltstone.	81,648
Shm/t/e_f/c	Outcrop mounds of fractured siltstone. Minor patches of loose sediment.	13,687
Fm(m/s/g)e_p/c/u	Patchy sandy mud and pavements of gravel/sand/ pebble, on outcrop of consolidated sediment (mud matrix, clasts up to boulder size).	10,474,281
Fm(g/m)l/s_p/h/c/u	Landslide with flat sandy mud with gravel-rich pavement, cobble/boulder patches and large blocks of consolidated sediment.	7,699,701
Fm(m/s?)e_h/c/u	Unconsolidated sandy mud on hummocky outcrops of consolidated sediment (mud matrix, clasts up to boulder size).	4,541,832
Fm(m/s?)g/e_c/u	Gullied slope, consolidated sediment (mud matrix, clasts up to boulder size) with unconsolidated sediment (mud?).	1,821,965
Fsl/s_i/u	Landslide scarp (interface).	604,920
Fs(s?)_u	Unconsolidated sediment (sand?).	338,197
Fs(g/m)s_p/u	Pavement of gravel/sand/pebble or pebble/gravel/ sand over unconsolidated sandy mud, at fault scarp.	97,579
Fm(m/s?)m/e_c/u	Mound of consolidated sediment locally covered with unconsolidated sediment (sandy mud?).	47,054
Fs(g/m)m_s/p/u	Pavement of gravel/sand/pebble or pebble/gravel/ sand over unconsolidated sandy mud, in current scour depression around mounds.	33,831
Fs(g/m)m_p/u	Mounds (outcrops?) covered by unconsolidated sandy mud and pavement of gravel/sand/pebble.	7,950

Shelf and slope habitats are grouped separately, and are listed in order of decreasing areal extent. For each habitat type, a brief description is provided. The first character of the codes indicates S = continental shelf or F = continental slope (flank). The second character indicates s = soft substrate, m = mixed hard and soft substrate, or h = hard substrate. The next characters, in parentheses, indicate sediment grain size where m = mud, s = sand, g = gravel, p = pebble, c = cobble, and b = boulder. For further details of the habitat codes see [7–9]. Area (m²) is the total area covered by the habitat type within each site.

The 8-Fathom Pinnacle has both bedrock exposures and glacial deposits. The upper part has an outcrop surface of fine-grain sedimentary bedrock, likely sandy siltstone, locally covered by unconsolidated sand and sandy mud (Figure 38.4). The bathymetry exhibits a series of northeast-striking ridges; both these ridges and the intervening flat seafloor are of the same bedrock. This siltstone erodes into rounded cobbles and boulders. In several locations, it appears to be cemented and exhibits a hard material with a pitted surface. At the summit of the pinnacle, this hard material

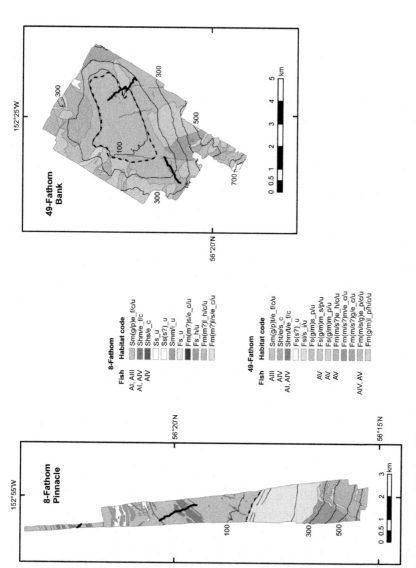

Figure 38.4 Habitat maps of the 8-Fathom Pinnacle and 49-Fathom Bank study sites. Physiographic setting and substrate characteristics are represented using habitat codes following the classification scheme of Greene et al. [7–9]. Habitat types are briefly described in Table 38.2. Corresponding fish community codes AI, AII, etc., also shown in the figure legends, are used to order the legends. However, note that these are not maps of fish community distribution because they do not include the depth ranges of the communities. Bathymetric contour interval is 100 m. Dashed line is the shelf edge.

probably acts as a cap rock that preserves the pinnacle from erosion. Beyond the base of the pinnacle, the seafloor at 60–80 m water depth has east–northeast-striking ridges interpreted as recessional moraines. No visual ground-truth data from dives are available from this terrain. Between 80 m and the shelf break at 110–125 m, the substrate is probably bedrock exposures of sandy siltstone locally covered by patches of unconsolidated sediment.

Below the shelf break, unconsolidated sediment cover increases rapidly with depth. Bedding scarps on the seafloor indicate that the siltstone bedrock continues to at least 160 m depth, but it is only locally exposed. A second break in slope occurs at 230–240 m depth. Below that point, mass wasting dominates the slope surface, where landslides have exposed a lumpy, irregular bedrock surface similar to the outcrops of conglomerate below the shelf break of the 49-Fathom Bank.

49-Fathom Bank: This submerged isolated bank rises from approximately 400 m water depth on the upper continental slope (Figure 38.2). It is located at the mouth of the transverse trough that divides Southern and Middle Albatross Bank. The MBES map covers 57 km^2 in 84–805 m water depth, including part of the upper slope. The summit of this bank is 7.2 km^2 (Figure 38.4).

In contrast to the other two sites, which sit at the edge of the continental shelf, at 49-Fathom Bank glacial till has not been found. The surface of the bank is composed of sandy siltstone similar to that on the 8-Fathom Pinnacle. The siltstone has a loose, patchy cover of pebbles, cobbles, and boulders. Small mounds on the bank are erosional remnants of the same bedrock. Below the shelf break at 90–110 m, the exposed bedrock is consolidated sediment of moderately sorted clasts up to boulder size in a matrix of mud, and has the appearance of a conglomerate rather than a glacial deposit. The substrate on the slope varies between exposures of this consolidated conglomerate, pavements of gravel–sand–pebble (in order of dominant clasts), and unconsolidated sandy mud. The slopes have extensive landslide scarps, landslide deposits, and gullied surfaces.

Biological Communities

Assessment of biological communities is based on analysis of submersible observations of fishes and macroinvertebrates [10]. The focus was on the abundance and community composition patterns of fishes relative to substrate types and water depth. Rockfishes (*Sebastes* spp. and *Sebastolobus* spp.) were the most abundant group, accounting for 69% of the fishes observed. The four main nonrockfish taxa were ronquils (Bathymasteridae), Dover sole (*Microstomus pacificus*), poachers (Agonidae), and sculpins (Cottidae). Multivariate analysis was used to examine associations among fishes and seafloor habitat, including structure-forming macroinvertebrates. For this analysis, depth was binned in 50 m intervals. The *Delta* dive transects, and hence the fish communities identified in this analysis, were mapped to a maximum depth of 365 m.

Seven different fish communities and habitat associations have been defined. They are primarily correlated with depth and, to a lesser extent, with substrate grain

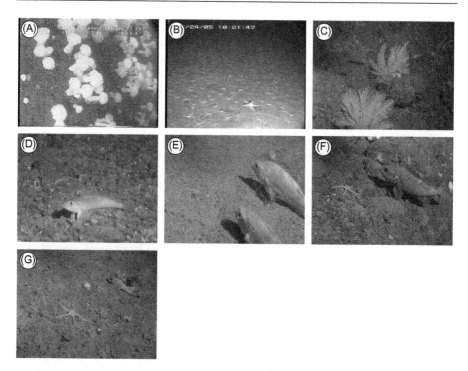

Figure 38.5 Images of seven habitats on Albatross Bank associated with fish communities AI-AVII. (A). Bedrock outcrop covered by an aggregation of sea anemones (*Metridium farcimen*), habitat Shm/e_f/c, fish community AI; (B) sandy mud substrate covered by large numbers of brittlestars (*Ophiura sarsii*), habitat Ss(s)c/w_u, fish community AII; (C) northern rockfish (*Sebastes polyspinis*) adjacent to gorgonian corals growing on boulders, habitat Ss(p/g/s)i/w_p/u, fish community AIII; (D) ronquils (bathymasteridae) lying atop a mix of sand, gravel, and pebbles, habitat Sm(c/p)i_u, fish community AIV; (E) two Pacific ocean perch (*Sebastes alutus*) in close proximity to sandy mud seafloor, habitat Fs(m/s)_u, fish community AV; (F) rougheye rockfish (*Sebastes aleutianus*) and ophiuroids resting atop muddy sand and pebble bottom, habitat Fm(m/s/c)m/i_h/u, fish community AVI; (G) shortspine thornyhead (*Sebastolobus alascanus*) with ophiuroids and burrowing sea anemones on a sandy mud substrate, habitat Fs(m)m/i_u, fish community AVII.
Source: All images are from *Delta* submersible dives in 2005. Images A–B are from videoframe grabs; images C–G are from a hand-held digital still camera.

size and seafloor physiography (Figure 38.5). The communities are designated AI, AII, and so on, in which A refers to Albatross Bank, and are numbered in order of increasing depth range. The fish communities AIV, AV, and AVI share similar substrates, but cover different (overlapping) depth ranges. Details are in [10].

Fish Community AI: This community was the shallowest, at depths of 15–36 m on 8-Fathom Pinnacle. It had the second highest density of fish due to high numbers of juvenile rockfish associated with sea anemones (*Metridium farcimen*). This community was associated with fractured outcrop of sandy siltstone locally covered by gravel, pebbles, and shell hash. The substrate did not provide refugia such as cracks or crevices for fishes, but several biogenic sources contributed to the overall habitat complexity including patches of red algae (rhodophyta), brown algae (*Agarum clathratum* and *Laminaria* spp.), and dense aggregations of large sea anemones.

Fish Community AII: This community was at depths of 85–100 m at the Snakehead site, in the head of the sediment-filled shelf canyon. Fish density and diversity were lowest in this community. The sediment consisted primarily of sandy mud. Sand and gravel waves (<20 cm amplitude) added some habitat complexity to this low-relief habitat. Brittle stars (*Ophiura sarsii*) covered up to 20% of the seafloor.

Fish Community AIII: This community was at depths of 70–165 m on the current-swept surface of banks at all three sites. The AIII community had the highest density and diversity of fishes. Two main types of substrate were associated with it. At the Snakehead site the exposed substrate was a gravel–pebble pavement with occasional cobbles and boulders, and local concentrations of sand and gravel waves. On the summit of 49-Fathom Bank and the bank surrounding the 8-Fathom Pinnacle, the substrate consisted of fractured sandy siltstone, locally thinly covered with unconsolidated sediment. Here the fish were associated with cobbles and boulders eroded from the siltstone, which formed mounds up to 5 m tall. Coverage of the cobbles and boulders by small sessile invertebrates (e.g., bryozoans, hydroids, and tunicates) and large sessile invertebrates (crinoids, sponges, and corals) added to the habitat complexity. While the total area of the seafloor covered by crionoids, sponges, and corals was low (<1%), they were locally abundant and covered up to 20% of the seafloor. Corals were diverse and included sea fans (Order Gorgonacea) 20–50 cm tall, sea whips (Order Pennatulacea), which reached over a meter in height, and hydrocorals (Order Stylasterina), which reached 20 cm tall.

Fish Community AIV: This community was on the shelf and upper slope of all sites, at 65–200 m. The AIV community had a relatively low density and diversity of fishes. On the Snakehead shelf, the community was located over shallow, till-covered bedrock and the boulder-covered slopes at the edges of the bedrock areas. At the summit of 49-Fathom Bank and the bank surrounding the 8-Fathom Pinnacle, the substrate for this community was bedrock mounds of fractured sandy siltstone.

On the upper slope of all three sites, the AIV community was associated with pavements dominated by gravel and pebbles, with patches of cobbles and boulders. At the Snakehead site, these slope pavements were developed on recessional moraines. On the flanks of the 49-Fathom Bank, the pavements were developed on a moderately sorted, consolidated conglomerate with clasts up to boulder size. These pavements were interrupted by areas of sandy mud. Invertebrates were common, primarily bryozoans and hydroids. Large invertebrates (e.g., corals) were rare and smaller than those in higher relief substrates.

Fish Community AV: This community was on the upper slope at depths of 103–260 m at the Snakehead site and 49-Fathom Bank. The AV community had one of the highest densities of fishes, largely due to a single taxon, Pacific ocean perch (*Sebastes alutus*). On the Snakehead site the substrate was recessional moraines, with sandy mud or mud, occasional cobbles and boulders, and locally developed sand–gravel–pebble pavement. On the upper slope of the 49-Fathom Bank the substrate was a partly exposed bedrock of moderately sorted, consolidated conglomerate with clasts up to boulder size. This bedrock was overlain by areas of sandy mud, sometimes with pavements of gravel–sand or gravel–pebble.

Most invertebrates associated with this community were small and included burrowing sea anemones and zoanthids. Invertebrate taxa varied greatly in abundance, and percent cover ranged between 0–40% of the seafloor. Crinoids were the only common large invertebrate and formed locally dense aggregations on pebble substrates.

Fish Community AVI: This community was at 205–360 m on the Snakehead site. It had low density but high diversity of fish taxa. The substrate was sandy mud and mud with gravel, overlying locally exposed glacial deposits. Scattered cobbles and boulders occurred throughout the habitat, covering up to 40% of the seafloor. The seafloor was hummocky on a 10 m scale. Ridges were subject to strong currents, and many of the boulders were recessed in scour depressions. The AVI community was also found in the head of a sedimented slope canyon in the southeast portion of the Snakehead site. Invertebrates included hydroids, burrowing sea anemones, zoanthids, and other small emergent epifauna, which typically covered 20–30% of the substrate. Sea whips up to 1 m high were the only large invertebrates.

Fish Community AVII: This community was encountered at depths from 325 to 350 m on the Snakehead site. It had both low diversity and low density. The substrate was mud, with small isolated patches of cobbles or boulders. Invertebrates were mostly small burrowing sea anemones, zoanthids, and several taxa of ophiuroids. The percent cover of invertebrates was 20–30%. Shortspine thornyhead (*Sebastolobus alascanus*) and Dover sole were the most common fishes and were associated with habitat dominated by mud. However, while hard substrate made up less than 1% of their sampled habitats, the patches of hard substrate had some of the highest fish densities encountered on Albatross Bank.

Surrogacy

This study employed statistical methods to study associations between fish communities and benthic habitat, and hence identified habitat types that may be useful as surrogates in a search for potential habitat of these fish communities. However, the study did not examine the predictive capability of specific surrogates.

Acknowledgments

Funding for this research was provided by the Alaska Fisheries Science Center's Auke Bay Laboratories through the Cooperative Institute for Arctic Research, and by the Rasmuson Fisheries Research Center at the University of Alaska Fairbanks. We thank Dr. Dana Hanselman, the crew of the R/V *Velero IV*, and Delta Oceanographics for their help during fieldwork for this project.

References

[1] R. von Huene, M.R. Hampton, M.A. Fisher, D.J. Varchol, G.R. Cochrane, Near-surface geologic structures, Kodiak Shelf, Alaska. 1:500,000, Map MF-1200, U.S. Geological Survey, Miscellaneous Field Studies, 1980.

[2] M.A. Hampton, Geology of the Kodiak Shelf, Alaska: environmental considerations for resource development, Continent. Shelf Res. 1 (1983) 253–281.

[3] G. Carver, J. Sauber, W. Lettis, R. Witter, B. Whitney, Active faults on northeastern Kodiak Island, Alaska, in: J.T. Freymueller, P.J. Haeussler, R.L. Wesson, G. Ekström (Eds.), Active Tectonics and Seismic Potential of Alaska, 2008, pp. 167–184. American Geophysical Union Geophysical Monograph 179.

[4] T. Weingartner, Physical and Geological Oceanography: coastal boundaries and coastal and ocean circulation, in: P. Mundy (Ed.), The Gulf of Alaska: Biology and Oceanography, Alaska Sea Grant College Program AK-SG-05-01CD, University of Alaska Fairbanks, Fairbanks, 2005, pp. 35–48.

[5] P.J. Stabeno, N.A. Bond, A.J. Hermann, N.B. Kachel, C.W. Mordy, J.E. Overland, Meteorology and oceanography of the northern Gulf of Alaska, Continent. Shelf Res. 24 (2004) 859–897.

[6] C. Coon, Retrospective analysis on the effects of bottom fishing in the Gulf of Alaska and Aleutian Islands. MSc Thesis, University of Alaska Fairbanks, Fairbanks, Alaska, USA, 2006, 228 pp.

[7] H.G. Greene, M.M. Yoklavich, R.M. Starr, V.M. O'Connell, W.W. Wakefield, D.E. Sullivan, et al., A classification scheme for deep seafloor habitats, Oceanol. Acta 22 (1999) 663–678.

[8] H.G. Greene, J.J. Bizzarro, J.E. Tilden, H.L. Lopez, M.D. Erdey, The benefits and pitfalls of geographic information systems in marine benthic habitat mapping, in: D.J. Wright, A.J. Scholz (Eds.), Place Matters, Oregon State University Press, Portland, 2005, pp. 34–46.

[9] H.G. Greene, J.J. Bizzarro, V.M. O'Connell, C.K. Brylinsky, Construction of digital potential marine benthic habitat maps using a coded classification scheme and its application, in: B.J. Todd, H.G. Greene (Eds.), Mapping of the Seafloor for Habitat Characterization, 2007, pp. 141–155. Geol. Assoc. Can. Spec. Paper 47.

[10] S.C. Rooney, Habitat Analysis of Major Fishing Grounds on the Continental Shelf off Kodiak, Alaska. MSc Thesis, University of Alaska Fairbanks, Fairbanks, Alaska, USA, 2008, 107 pp.

39 Seabed Habitat of a Glaciated Shelf, German Bank, Atlantic Canada

Brian J. Todd[1], Vladimir E. Kostylev[1], Stephen J. Smith[2]

[1]Geological Survey of Canada, Dartmouth, NS, Canada,
[2]Fisheries and Oceans Canada, Dartmouth, NS, Canada

Abstract

An area of 5,320 km^2 in water depths of 30–250 m has been mapped on German Bank on the southern Scotian Shelf in Atlantic Canada. The Scotian Shelf is a formerly glaciated continental margin characterized by a topographically rugged inner shelf. Bedrock is exposed at the seafloor on much of German Bank. Till occurs as a widespread sediment blanket and was deposited beneath or at the margins of the ice sheet directly onto bedrock during the Wisconsinan glaciation. Ice-distal glaciomarine silt overlies the older till and is primarily confined to small basins on the bank. Limited accumulations of postglacial sediments, composed of well-sorted sand grading to rounded and subrounded gravel, are also present. Analysis of seafloor photographs revealed broad-scale gradients in benthic fauna composition dependent on water depth and covarying chemical and physical variables such as oxygen saturation and temperature variability. Statistical analyses of benthic megafauna and commercial species (lobster, scallops, crab) show that the distribution of benthic assemblages is associated with sediment type or broad-scale (kilometers to tens of kilometers) geomorphological features.

Key Words: German Bank, Canada, Atlantic Ocean, Nova Scotia, Gulf of Maine, glaciated shelf

Introduction

German Bank is located off southern Nova Scotia on the Scotian Shelf in the eastern Gulf of Maine (Figure 39.1) and is the offshore extension of the southern Nova Scotia landmass. Much of German Bank is exposed bedrock comprising Cambro–Ordovician metasedimentary rocks intruded by Late Devonian–Carboniferous granitoid plutons [1]. Bedrock has been modified by glacial erosion and is separated by a rugged erosional surface from the discontinuous overlying Quaternary sediments [2].

Seafloor Geomorphology as Benthic Habitat. DOI: 10.1016/B978-0-12-385140-6.00039-6

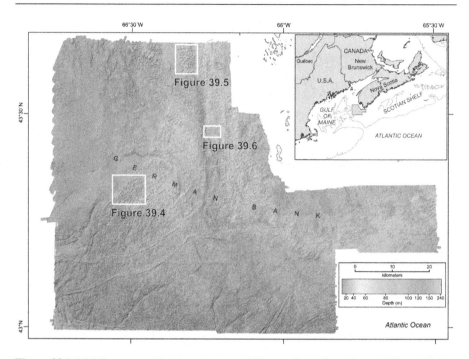

Figure 39.1 Multibeam sonar bathymetric map of German Bank. Locations of Figures 39.4–39.6 are indicated by labeled white boxes. Artificial sun illumination is from the northwest, and the vertical exaggeration is 15×. Note that hypsometric analysis of water depth has been applied to this figure and subsequent figures illustrating bathymetry. Thus, the range of depths and associated colors are not common but are unique to each figure. Blue line on inset map of Atlantic Canada denotes the 200 m isobath. (For interpretation of the references to color in this figure legend, the reader is referred to the web version of this book.)

German Bank lies within the photic zone, with water depths on the shallowest parts shoaling to 30 m. The southern and western margins of the bank are approximately demarcated by the 100 m isobath. Consequently, German Bank has an irregular outline, extending roughly 90 km seaward from land at its greatest extent. The channel to the south of German Bank and the basin to the west both attain depths greater than 200 m.

Seasonal circulation on German Bank is tidally dominated with a persistent westward and north-westward flow toward the Gulf of Maine [3,4]. Tidal current speeds reach 70 cm/s over eastern parts of German Bank. The northwest flow of Scotian Shelf water across German Bank contributes to the broad-scale counter-clockwise ocean circulation within the Gulf of Maine. The average yearly bottom-water temperature on German Bank increases from 4°C in the east to 8°C in the west, varying seasonally from less than 2°C in the deeper part of the bank (>100 m) to almost 10°C in the shallow inshore zone [5]. Average bottom salinity is 32‰ in shallow eastern waters, increasing to 34‰ in the deeper part of the bank to the west [5]. Based on the summer water density difference between surface and 30 m depth, water masses on the bank are well mixed, with stratification slightly higher in the eastern near-shore part of the study area. Spring phytoplankton bloom production reaches 6 µg/ml (estimated from SeaWifs, G. Harrison, personal communication, 2004). The bottom waters of German Bank are highly saturated with oxygen, with July average satura-tion 100% to the east near the Nova Scotia shore, decreasing to 60% saturation to the southwest [6] in deeper waters.

The German Bank benthic environment has experienced, in places, very high impacts from human activities [7,8]. The region supports both commercial sea scallop (*Placopecten magellanicus*) [9] and lobster (*Homarus americanus*) fisher-ies [10]. An improved understanding of the benthic environment is required to aid in striking a temporal and spatial balance between these two colocated fisheries. Understanding the different physical habitats (both geological and oceanographic) of the bank and their associated benthic assemblages is necessary information to yield improved fisheries management practice. To this end, multibeam sonar surveys were conducted over 5,320 km^2 of German Bank from 1997 to 2003, augmented by the collection of 2,133 km of geophysical profiles, 86 sediment samples, 1,134 seafloor photographs, and 97 video transects [11]. All these data were utilized in the interpre-tation of German Bank surficial geology and benthic habitat.

Substrate Geology and Geomorphic Features

If a strict interpretation of the definition of a bank were applied [12], German Bank would not qualify because it is not an *isolated* elevation of the seafloor. Rather, it is an area of the Canadian Atlantic continental shelf, bathymetrically contiguous with the adjacent landmass, that has been designated a bank since at least 1812 [13]. Regionally, at horizontal distances of tens of kilometers (small scale), German Bank can be characterized as bathymetrically smooth glaciated continental shelf, with a regional gradient of less than 1° to the southwest. At distances of tens to hundreds of meters (large scale), geomorphic features are evident in outcropping bedrock and within the overlying glacial and postglacial sediments.

Bedrock: Metasedimentary and igneous rocks outcrop over the central portion of the map area (designated as IM on Figure 39.2). Seabed relief on bedrock reaches 40 m in places (over a horizontal distance of ~500 m) but is generally less than 6 m (Figure 39.3). At small scale, the igneous bedrock exhibits two jointing directions (~40° and 190°); these zones of inherent weaknesses have been exploited by erosion, lending a distinctive cross-hatch pattern to the bedrock topography (Figure 39.4). At large scale, areas mapped as bedrock outcrop are in fact a complex of substrates including exposed rock, accumulations of shell hash, sand, and gravel within inter-stices in the bedrock (Figure 39.7A–D).

Glacial sediment: Till is widespread on German Bank and was deposited directly on bedrock beneath and at the margins of the ice sheet during the late Wisconsinan substage of the Pleistocene epoch [14]. The till (designated as T on Figure 39.2) is unsorted, unstratified, and unconsolidated glacial drift consisting of a mixture of mud, sand, and gravel and displays a characteristic chaotic to transparent seismic-reflection

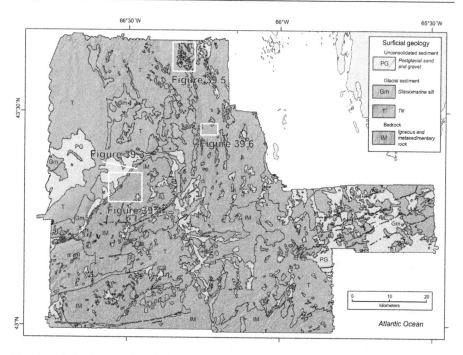

Figure 39.2 Surficial geology map of German Bank. Locations of Figures 39.4–39.6 are indicated by labeled white boxes. Location of seismic profile in Figure 39.3 is indicated by white line. (For interpretation of the references to color in this figure legend, the reader is referred to the web version of this book.)

configuration (Figure 39.3). Glacial landforms of till are widespread on German Bank and include drumlins and moraines (see later). They were subjected to modification during sea-level transgression but were not buried after glacial retreat because of the low volume of postglacial sediment transported offshore.

Drumlins were formed by deposition beneath the active ice sheet on German Bank and are streamlined, oval-shaped mounds of glacial deposits with long axes oriented northwest–southeast, aligned parallel to the direction of glacial ice flow (Figure 39.5). The drumlins are typically 500–800 m in length and 100–300 m in width, with elevations 10–15 m above the surrounding seafloor. Seismic-reflection profiles reveal that the mounds are constructed of acoustically incoherent sediment, interpreted as till, with intermound areas deeply draped by stratified sediment (>10 m thick in places) characterized by continuous coherent reflections and interpreted as glaciomarine sand and silt.

De Geer moraines on German Bank form fields of numerous parallel ridges of cobbles and boulders; each ridge displays a simple or curved line in planform (Figure 39.6) and strike approximately west-southwest–east-northeast. Individual moraine crests can be traced horizontally from short segments of 2 km up to almost 10 km. De Geer moraines are formed at or close to the grounding lines of

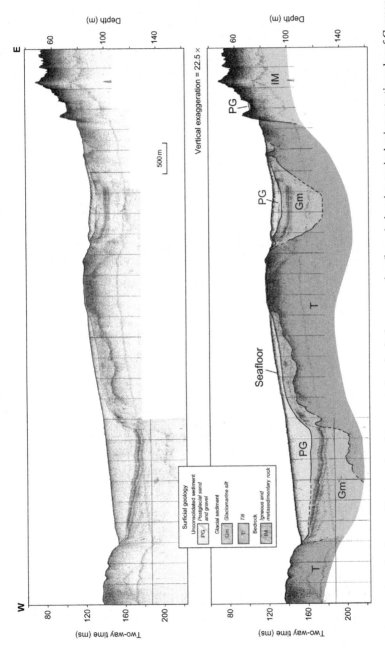

Figure 39.3 Seismic-reflection profile (upper) and interpreted geological cross section (lower) showing typical seismostratigraphy of German Bank. See Figure 39.2 for location of profile (CCGS *Hudson* 76016, Day 173, 0356–0450).

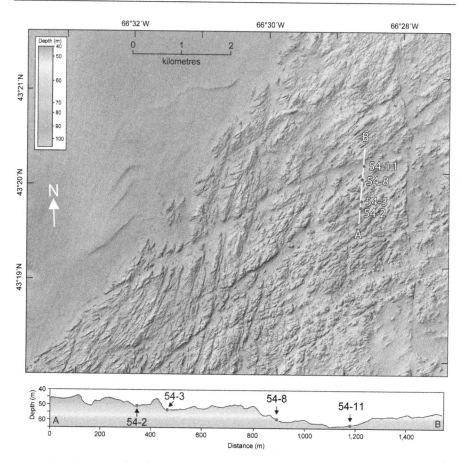

Figure 39.4 Rugged German Bank bedrock topography to the southeast juxtaposed
with smooth glacial and postglacial sediment topography to the northwest. Artificial sun
illumination is from the northwest, and the vertical exaggeration is 15×. Labelled red dots
indicate locations of seafloor photographs in Figure 39.7. Position of topographic
cross-section AB is indicated by white line. Vertical exaggeration of the cross section is 5.8×.
(For interpretation of the references to color in this figure legend, the reader is referred to the
web version of this book.)

water-terminating glaciers [15] and are the most ubiquitous glacial landforms on
German Bank. The moraines are generally symmetrical in cross section and the
horizontal distance between moraines, normal to strike, varies from 30 to 200 m.
Moraine height varies from 1.5 to 8 m, and width varies from 40 to 130 m. Cobbles
and boulders were observed and recovered from De Geer moraines [16]. In places,
De Geer moraines are subdued in relief or are absent, resulting from partial to
complete burial by sediment (sandy gravel) transported from glacial fronts and/or
reworked during sea-level transgression of German Bank. This sediment blanket

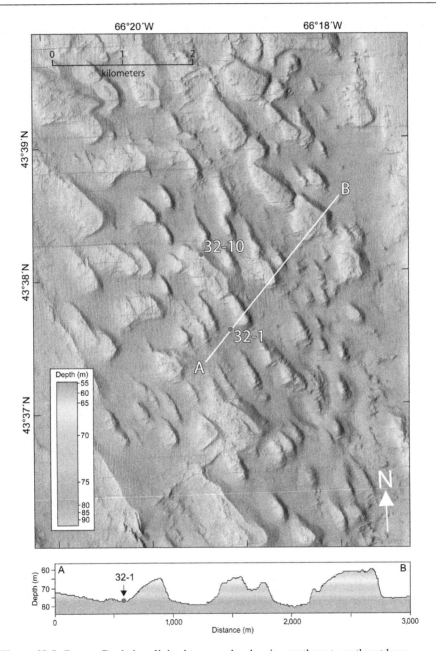

Figure 39.5 German Bank drumlinized topography showing northwest–southeast long axis strike. Labeled red dots indicate locations of seafloor photographs in Figure 39.7. Artificial sun illumination is from the west, and the vertical exaggeration is 15×. Position of topographic cross-section AB is indicated by white line. Vertical exaggeration of the cross section is 13.8×. (For interpretation of the references to color in this figure legend, the reader is referred to the web version of this book.)

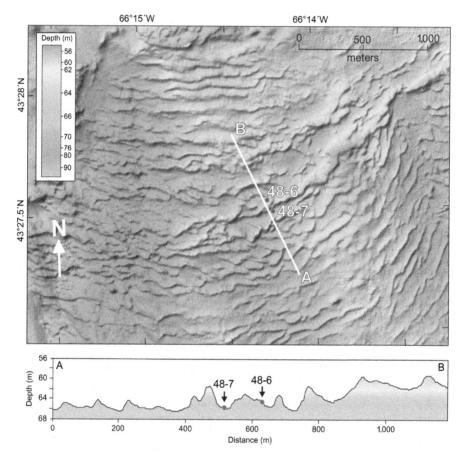

Figure 39.6 German Bank De Geer moraine swarm showing approximate west-southwest–east-northeast strike of moraine crests. Artificial sun illumination is from the north, and the vertical exaggeration is 15×. Labeled red dots indicate locations of seafloor photographs in Figure 39.7. Position of topographic cross-section AB is indicated by white line. Vertical exaggeration of the cross section is 6.3×. (For interpretation of the references to color in this figure legend, the reader is referred to the web version of this book.)

is generally thin (a few meters) but can reach an appreciable thickness (~10m) in places.

Ice-distal glaciomarine silt (Gm on Figure 39.2) is widespread in basins on German Bank. This sediment is poorly sorted clayey and sandy silt with some gravel. At the map scale of Figure 39.2, there are only two mapped seafloor exposures of glaciomarine silt in the extreme east of the study area. However, the glaciomarine silt is commonly observed in the subsurface in seismic profiles (Figure 39.3). This unit exhibits a characteristic parallel to subparallel seismic-reflection configuration; the reflectors conformably drape the irregular top of the underlying till and bedrock. In places, the glaciomarine silt is interbedded with the till.

Unconsolidated sediment: Postglacial sediments, composed mainly of sand with minor gravel (PG on Figure 39.2), are derived from current and wave reworking of glaciogenic material. In the shallow, eastern region of German Bank, postglacial sediments occur in elongated mounded deposits reaching hundreds of meters in length and 17 m in thickness, oriented southeast–northwest, aligned with the current flow direction. In deeper water of western German Bank, postglacial sediments are deposited mainly in a basin-fill external morphology (Figure 39.3) up to 20 m thick in places.

Biological Communities

High-resolution seafloor imagery was obtained on German Bank using Campod, an instrumented tripod that includes forward- and downward-looking video cameras and a downward-looking 35 mm still camera [17]. The Campod provides a very large-scale (tens of centimeters) view of the seafloor. Physical habitat characteristics and the type of benthic fauna on German Bank were interpreted from 1,134 seafloor photographs. All visible species of megabenthos were identified to the highest possible taxonomic resolution.

The benthic habitat observations presented here are based on seabed photographs. These images are at a much larger scale (i.e., smaller area) than the surficial geological units (bedrock, till, glaciomarine silt, and postglacial sand and gravel) mapped in Figure 39.2 and discussed in the previous section. Because the surficial geological mapping is undertaken at a standard scale of 1:50,000, generalization must necessarily occur, with the result that a given mapped geological unit may contain a number of substrate types (e.g., the map unit *till* may contain substrates such as patches of sand and muddy sediment). Because of this marked scale difference between the geological and biological analyses, we discuss biological communities identified from photographs in relation to the local substrates they occupy; we then discuss their relation to the bank-wide distribution of geological units.

Fauna on hard substrates (exposed bedrock and gravel, Figure 39.7A and D) includes a variety of encrusting and erect sponges, sea stars (*Asterias* sp., *Crossaster papposus*, *Solaster endeca*, and *Hippasteria phrygiana*), tunicates (*Boltenia ovifera*), brachiopods (*Terebratulina* sp.), and soft corals (*Gersemia* sp.). Hard substrates are commonly overgrown with dense mattes of hydrozoa and bryozoa. Fish (flounder, hake, skate, and cod) are commonly observed in this topographically complex habitat. Calcareous polychaete tubes, likely of *Filograna implexa*, and sponges are characteristic benthic fauna on gravel. On the patches of sand and shell hash infilling crevasses in bedrock, fauna is scarce, with spider crab (*Hyas aranaeus*), urchins (*Strongylocentrotus* sp.), and infrequent occurrence of flat fish (Figure 39.7B and C).

The seabed texture on drumlin-forming *till* is an immobile pebble and cobble gravel. Abundant epifauna with low diversity, such as anemones, sea stars (*Asterias* sp.), and erect and encrusting breadcrumb sponges (*Halichondria panicea*) is common (Figure 39.7E and F). The seabed texture between the drumlins is coarse-grained

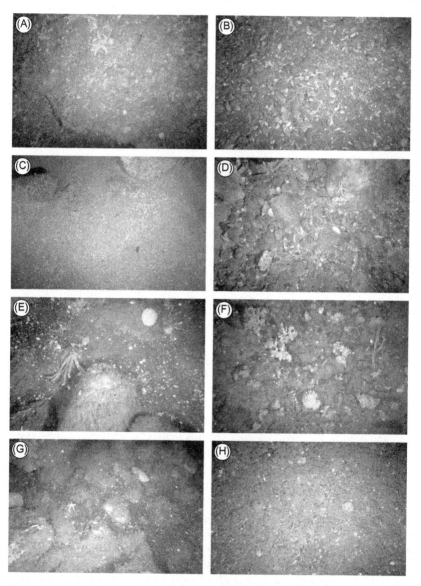

Figure 39.7 Seafloor photographs on German Bank. The approximate area of each photograph is 0.96 m². All photographs are from expedition *Hudson* 2003-054 [16] and are referred to as station number followed by photograph number. Images (A) through (D) are located on the area mapped as bedrock (Figures 39.2 and 39.4). (A) Exposed rock, station 54-2, 51 m; (B) accumulation of shell hash, station 54-3, 54 m; (C) sand, station 54-8, 57 m; (D) gravel, station 54-11, 64 m; (E) boulders on top of drumlin (mapped as till), station 32-10, 62 m (Figure 39.5); (F) sandy gravel between drumlins (mapped as glaciomarine silt), station 32-1, 78 m (Figure 39.5). Images (G) and (H) are located on the area mapped as till (Figure 39.6). (G) Cobbles and boulders on De Geer moraine, station 48-6, 63 m; (H) sand between De Geer moraines, station 48-7, 66 m.

sand where groundfish, Jonah crab (*Cancer* sp.), and sea scallops (*Placopecten magellanicus*) are common.

Cobbles and boulders within the moraines are covered with the encrusting red algae *Lithothamnium* sp. and densely colonized by fauna typical of southern Nova Scotia nearshore environments (Figure 39.7G). The frilled anemone *Metridium senile* is very abundant, along with the stalked tunicate *Boltenia ovifera*. Prolific fauna includes northern horse mussel (*Modiolus modiolus*), as well as encrusting, erect sponges (*Haliclona oculata*), mound-shaped sponges, and echinoderms (*Asterias* sp., *Henricia* sp., and *Strongylocentrotus* sp.). On the sandy gravel between moraines, fauna is scarce (Figure 39.7H), with spider crab (*Hyas aranaeus*), urchins (*Strongylocentrotus* sp.), flat fish, and sea scallops (*Placopecten magellanicus*) present.

In deeper parts of German Bank, at 100–140 m water depth, muddy sediments host a low abundance and diversity of megafauna. Benthic megafauna include the dominant burrowing anemone *Cerianthus borealis* and shrimps *Pandalus* sp., with less common Jonah crabs (*Cancer* sp.) and hermit crabs (*Pagurus* sp.). Some hake (*Urophycis* sp.), monkfish (*Lophius* sp.), and several species of flat fish are found above and on the seabed.

Surrogacy

Statistical analysis of the megafaunal community data was undertaken to establish the relationship between the distribution of benthic fauna and environmental variables. A total of 127 taxa and biogenic features were described from the analysis of seabed photographs collected at 70 stations. The frequency of occurrence of each species was calculated for each station based on their presence/absence in individual photographs. Physical habitat at the 70 stations was described in terms of relative abundance of gravel classes (boulders, cobbles, pebbles, and granules), sand, and mud, as well as topographic complexity (rugosity, slope, and benthic position index at two scales) computed from multibeam sonar bathymetric data [11]. Seafloor cover by sediment of different grain size classes, as well as shell hash and bedrock, was assessed semiquantitatively on a scale from 1 (some cover) to 5 (full cover) on each photograph and averaged for a station. Each station was also assigned a geological unit (bedrock, till, glaciomarine silt, and postglacial sand and gravel) based on the surficial geological interpretation (Figure 39.2).

Environmental variables used in the analysis include RMS (Root Mean Square) tidal current speed, average annual bottom temperature, variability in bottom temperature, bottom-water salinity, density difference or stratification, spring average chlorophyll-a concentration, and oxygen saturation [3–6,18].

Statistical analysis of the relationship between biological factors of megafaunal community data as assessed from large-scale seabed photographs and environmental variables (BioEnv) revealed that the single variable that best explains the distribution of bottom fauna is summer oxygen saturation at the seabed. The best combination of variables related to benthic community composition is water depth, oxygen saturation, seabed cover by cobbles and boulders, and seabed cover by sand. However, Analysis of Similarity (ANOSIM) between the small-scale mapped surficial geology units (bedrock, till, glaciomarine silt, and postglacial sand and gravel; see Figure 39.2) and the benthic community structure showed no difference in observed fauna between different surficial geology units.

Bycatch data listing catches of commercial and noncommercial species of fish and invertebrates collected by observers onboard commercial scallop fishing vessels [19] were also analyzed. Because of the general homogeneity of oceanographic processes on German Bank, no distinct broad-scale gradients in benthic fauna composition were deduced from analysis of the bycatch data. The best single variable explaining the spatial pattern in the BioEnv analysis of this data set was the interannual variability in water temperature. Three additional variables that contributed to the pattern were water stratification, seabed slope, and spring chlorophyll-a concentration. ANOSIM analysis of the data shows that the four mapped surficial geology units (bedrock, till, glaciomarine silt, and postglacial sand and gravel; see Figure 39.2) have a significant relationship with the distribution of bycatch fauna.

We presented analyses of two distinctly different types of data: one set of data derived from seabed photographs and the second set from the observer bycatch data. The type of information, biases, and the scale of sampling of these data sets differ, and these differences are reflected in the statistical results. Despite their differences, in both the seabed photograph data set and in the bycatch data set, the seabed substrate plays an important role in determining faunal composition, albeit at different spatial scales.

In this study of German Bank, substrate type had a significant effect on faunal composition determined from seabed photographs, while geological units did not. In the bycatch data set, where substrate type could not be determined directly, geological units had a significant effect on the fauna. This apparent contradiction in the results of the two analyses is explained by the scale difference of the sampled areas. In the case of the bycatch data, the sample area is large and more representative of the geological units than of local substrate. For commercial species (lobsters, crabs, and scallops), the two scales of geological units and local substrate can be linked: the spatial variability in the distribution in bycatch data was statistically explained by the broadly defined surficial geology units, and an assemblage characterized by a high abundance of these species was commonly observed on seabed photographs from sandy flat valleys between moraine ridges (postglacial sand and gravel) [20]. The dynamic nature of these habitat–animal relationships should be taken into account by fishery managers because, in the case of scallops, commercial fishing can mask these relationships over time. For example, as originally abundant scallop beds are fished, animal densities will more rapidly decrease in preferred habitat [21].

Acknowledgments

We thank Kim Conway and Kim Picard (Geological Survey of Canada), Jennifer R. Reynolds (University of Alaska), and Katrien J.J. Van Landeghem (University of Liverpool) for reviewing the manuscript. This is Earth Science Sector contribution number 20100038.

References

[1] G. Pe-Piper, L.F. Jansa, Pre-Mesozoic basement rocks offshore Nova Scotia, Canada: new constraints on the origin and Paleozoic accretionary history of the Meguma Terrane, Geol. Soc. Am. Bull. 111 (1999) 1773–1791.

[2] G. Drapeau, L.H. King, Surficial geology of the Yarmouth—German Bank map area. Canadian Hydrographic Service, Marine Sciences Paper 2, Geological Survey of Canada Paper 72-24, 1972, 6 pp., scale 1:300 000.

[3] P.C. Smith, The mean and seasonal circulation off southwest Nova Scotia, J. Phys. Oceanogr. 13 (1983) 1034–1054.

[4] D.R. Lynch, J.T.C. Ip, C.E. Naimie, F.E. Werner, Comprehensive coastal circulation model with application to the Gulf of Maine, Cont. Shelf Res. 16 (1996) 875–906.

[5] C.G. Hannah, J.A. Shore, J.W. Loder, C.E. Naimie, Seasonal circulation on the western and central Scotian Shelf, J. Phys. Oceanogr. 31 (2001) 591–615.

[6] Fisheries and Oceans Canada, BioChem: database of biological and chemical oceanographic data, Version 8, 2005. http://www.medssdmm.dfompo.gc.ca/biochem/Biochem_e.htm, 2006.

[7] B.S. Halpern, S. Walbridge, K. Selkoe, C.V. Kappel, F. Micheli, C. D'Agrosa, et al., A global map of human impact on marine ecosystems, Science 319 (2008) 948–952.

[8] Fisheries and Oceans Canada, Presentation and review of the benthic mapping project in scallop fishing area 29, Southwest Nova Scotia, 16 February 2006. Fisheries and Oceans Canada, Canadian Science Advisory Secretariat Proceedings Series 2006/047, 2006, 42 pp. http://www.dfo-mpo.gc.ca/CSAS/Csas/Proceedings/2006/PRO2006_047_E.pdf.

[9] S.J. Smith, M.J. Lundy, A brief history of scallop fishing in Scallop Fishing Area 29 and an evaluation of a fishery in 2002. Canadian Science Advisory Secretariat Research Document 2002/079, 2002, 23 pp., http://www.dfo-mpo.gc.ca/csas/Csas/DocREC/2002/RES2002_079e.pdf.

[10] Fisheries and Oceans Canada, Framework assessment for lobster (*Homarus americanus*) in Lobster Fishing Area (LFA) 34. Canadian Science Advisory Secretariat Advisory Report 2006/024, 2006, 17 pp. http://www.dfo-mpo.gc.ca/csas/Csas/status/2006/SAR-AS2006_024_E.pdf.

[11] B.J. Todd, Surficial geology and sun-illuminated seafloor topography, German Bank, Scotian Shelf, offshore Nova Scotia, Geological Survey of Canada Map 2148A, 2009, scale 1:50 000.

[12] International Hydrographic Organization, Standardization of undersea feature names. Bathymetric Publication No. 6, International Hydrographic Bureau, Monaco, 2008, 30 pp. http://www.gebco.net/data_and_products/undersea_feature_names/documents/b_6.pdf.

[13] N. Holland, A Chart of the Coast of New England from the South Shoal to Cape Sable Including Georges Bank from Holland's Actual Surveys, John Norman, Boston, MA, 1812.

[14] B.J. Todd, P.C. Valentine, O. Longva, J. Shaw, Glacial landforms on German Bank, Scotian Shelf: evidence for Late Wisconsinan ice-sheet dynamics and implications for the formation of De Geer moraines, Boreas 36 (2007) 148–169.

[15] D.I. Benn, D.J.A. Evans, Glaciers and Glaciation, Arnold, London, 1998.

[16] B.J. Todd, W.A. Rainey, P.R. Girouard, P.C. Valentine, C.B. Chapman, G.D. Middleton, et al., Expedition report CCGS *Hudson* 2003-054: German Bank, Gulf of Maine. Geological Survey of Canada Open File report 4728, 2004, 129 pp.

[17] D.C. Gordon, Jr., D.L. McKeown, G. Steeves, W.P. Vass, K. Bentham, M. Chin-Yee, Canadian imaging and sampling technology for studying benthic habitat and biological

communities, in: B.J. Todd, H.G. Greene (Eds.), Mapping the Seafloor for Habitat Characterization, Geological Association of Canada, St. Johns, Newfoundland, 2007, pp. 29–37 (Special Paper 47).

[18] Todd, B.J., Kostylev, V.E. Surficial geology and benthic habitat of the German Bank seabed, Scotian Shelf, Canada. Continent. Shelf Res. DOI: 10.1016/j.csr.2010.07.008, 2010.

[19] S.J. Smith, S. Rowe, M.J. Lundy, J. Tremblay, C. Frail, Scallop fishing area 29: stock status and update for 2007. Fisheries and Oceans Canada, Canadian Science Advisory Secretariat Research Document 2007/029, 2007, 71 pp. http://www.dfo-mpo.gc.ca/CSAS/Csas/DocREC/2007/RES2007_029_e.pdf.

[20] M.J. Tremblay, S.J. Smith, B.J. Todd, P.M. Clement, D.L. McKeown, Associations of lobsters (*Homarus americanus*) off southwestern Nova Scotia with bottom type from images and geophysical maps, ICES J. Mar. Sci. 66 (2009) 2060–2067.

[21] S.J. Smith, J. Black, B.J. Todd, V.E. Kostylev, M.J. Lundy, The impact of commercial fishing on the determination of habitat associations for sea scallops (*Placopecten magellanicus*, Gmelin), ICES J. Mar. Sci. 66 (2009) 2043–2051.

40 Identifying Potential Habitats from Multibeam Echosounder Imagery to Estimate Abundance of Groundfish: A Case Study at Heceta Bank, OR, USA

Julia E.R. Getsiv-Clemons[1], W. Waldo Wakefield[1], Curt E. Whitmire[1], Ian J. Stewart[2]

[1]NOAA Fisheries, Fishery Resource Analysis and Monitoring Division, Northwest Fisheries Science Center, Newport, OR, USA
[2]NOAA Fisheries, Fishery Resource Analysis and Monitoring Division, Northwest Fisheries Science Center, Seattle, Washington, USA

Abstract

Heceta Bank is one of the largest banks off the western coast of North America, extending 55 km from north to south and rising above the continental shelf to 67 m water depth. Due to heterogeneous substrate of varying relief, the bank supports a diverse assemblage of demersal fishes and is an important fishing ground off the coast of Oregon, USA. The top of the bank is comprised of boulders and cobbles eroded from outcrops of sedimentary rocks, while layers of finer grain size material cover the lower relief flanks. Using observations of fish–substrate associations from 19 remotely operated vehicle dives, we identified 57 distinct substrate types and 9 habitat types. In a separate exercise, we overlaid observations of substrate types onto high-resolution multibeam imagery to delineate four potential habitat classes. Using this map of potential habitats, we estimated bankwide abundance and variance values for a select group of resident fishes.

Key Words: California Current Large Marine Ecosystem, continental margin, bank, multibeam sonar imagery, habitat, demersal fish, ROV

Introduction

Over the past 20 years, a growing number of US West Coast fisheries biologists, marine ecologists, and marine geologists have conducted collaborative studies to quantify fish and associated invertebrate populations in the context of their marine habitats [1–6]. An expanding number of these regional research programs along the

Seafloor Geomorphology as Benthic Habitat. DOI: 10.1016/B978-0-12-385140-6.00040-2

US West Coast have formed the basis for a coast-wide network of sites where sea-floor mapping and direct observation are supporting ongoing habitat-based ground-fish research. Technologies, methods, and classification schemes for mapping marine habitats have evolved and advanced over this period and will continue to do so [7,8]. Work at one site in the eastern North Pacific off Oregon, Heceta Bank, is illustrative of these collaborative studies and is highlighted here as a case study (Figure 40.1).

Primary investigations were initiated on Heceta Bank in 1987 by a team of researchers at Oregon State University when very little information was known about the bank's lithology, structure, and biological communities [9]. The research team established six reference sites that were revisited each subsequent year (Figure 40.2). Over the course of four years (1987–1990), submersible dives were conducted on the bank to study the associations of fishes and invertebrates with habitats [10–12].

In 1998, a high-resolution multibeam echosounder survey that included most of Heceta Bank was conducted using a Simrad EM300 (30 kHz) system (725 km^2; Figures 40.1–40.2). This survey was embedded in a more extensive survey of por-tions of the northern California and Oregon margin [13] and was augmented by two additional surveys in 2002 (C. Goldfinger and R. Embley, unpublished data) and 2004 (R. Embley and S. Merle, unpublished data). For this study, Heceta Bank (hereafter "bank") is defined to the north, south, and east by the extent of the EM300 survey, and to the west by the 400 m depth contour. These data were gridded to a resolution of 5 m on the top of the bank, corresponding to the shallowest depths of this feature, and to 10 m along the flanks of the bank, where depths extend from 175 to 400 m.

Video transects and directed sampling were conducted with an advanced scien-tific remotely operated vehicle (ROV), the Remotely Operated Platform for Ocean Sciences (*ROPOS*) Canadian Scientific Submersible Facility, in 2000 and 2001 during 19 dives that revisited five of the six reference sites and also explored new areas of the bank (Figure 40.2) [5]. Eighty-eight hours of video from the 19 ROV dives have been analyzed for fish and invertebrate communities and habitat types. A major goal of this study was to create the most detailed and distinctive habitat classes that could serve as proxies for fish presence. These video data were combined with the multibeam data to construct a habitat map, which could be used to extrapolate bank-wide abundance of several species of groundfish observed during ROV dives.

Geomorphic Features and Habitats

The outer continental shelf off Oregon is marked by a series of rocky banks, includ-ing (from north to south) Nehalem, Daisy, the Stonewall/Heceta Complex, and Coquille [17]. The Stonewall/Heceta Bank Complex is one of the largest rocky banks off the west coast of North America, extending 90 km in a north–south direction. This bank feature more than doubles the width of the continental shelf off central Oregon. Heceta Bank, located at the southern and most offshore portion of the complex (Figure 40.1), was named after Bruno de Heceta (Hezeta) y Dudagoitia (1743–1807), a Spanish Basque explorer of the northwestern coast of the US in 1775 [18,19]. It extends 55 km north to south, 18 km east to west, and rises from the continental shelf

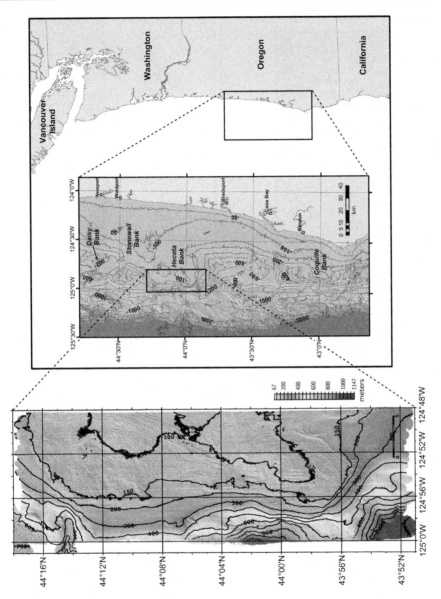

Figure 40.1 Location of Heceta Bank relative to the northwestern coast of the US and state of Oregon. The left panel shows the portion of Heceta Bank that was surveyed with a high-resolution Simrad EM300 multibeam echosounder (30 kHz) system [13] in 1998 (data gridded at 5 m in the shallowest depths of the bank to 10 m along the continental shelf edge). Bathymetric depths range from 67 m at the top of the bank to 1160 m at the southwestern edge of the survey. The bathymetry map in the middle shows the continental margin off central Oregon, highlighting Heceta and other prominent banks. The black box in the middle panel represents the extent of the multibeam survey. (For interpretation of the references to color in this figure legend, the reader is referred to the web version of this book.)

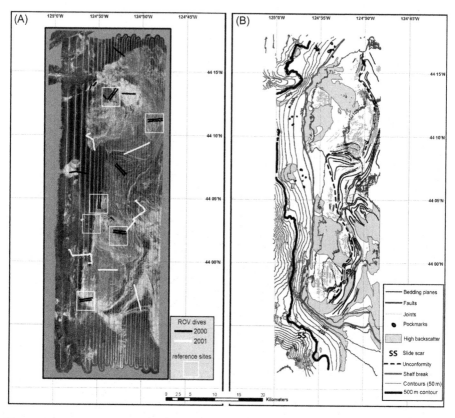

Figure 40.2 (A) EM300 high-resolution acoustic reflectivity (backscatter) imagery at Heceta Bank, collected with the multibeam echosounder Simrad (30 kHz) system in 1998 [13]. Grid cell size is 10 m. The tracks show the ROV video transects for 2000 in green and 2001 in yellow. Magenta boxes mark the six reference sites established in 1987 [9]. (B) Detailed geologic interpretation, adapted from Torres et al. [14], of Heceta Bank from bathymetric and backscatter imagery, dated core samples and seismic reflection profiles [15,16]. Locations of pockmarks and areas of high acoustic backscatter are also shown. (For interpretation of the references to color in this figure legend, the reader is referred to the web version of this book.)

to 67 m water depth (Figure 40.1). Heceta Bank is composed of late Miocene and Pliocene sedimentary rocks that rise above surrounding Pleistocene sediment and are likely to be fault controlled [20]. Its diversity of habitats and relief hosts a wide array of fishes and invertebrates, including numerous species of rockfish (genus *Sebastes*), many of which are of commercial importance.

The oceanographic habitat of Heceta Bank is a temperate eastern boundary current upwelling system within the northern portion of the California Current Large Marine Ecosystem. Oregon lies within a region of highly productive seasonal upwelling, where episodic northerly winds during April through September drive currents southward and provide a supply of cold nutrient-rich but oxygen-poor water. During the remainder of the year, southerly winds result in downwelling conditions and northward currents [21]. The currents and water masses are also influenced by a combination of freshwater sources from major river systems (e.g., Columbia River) and seafloor and coastal topographic features such as submarine canyons and rocky banks (e.g., Astoria Canyon and Stonewall/Heceta Complex) and headlands (e.g., Cape Blanco) [22–24]. The interaction of the bank topography with flow results in more favorable conditions for primary productivity that in turn enhances production at higher trophic levels [25,26]. The high levels of productivity in this region of the Oregon margin have also been associated with recent hypoxic events, especially inshore of the bank [27]. In a recent study, Juan-Jordá et al. [28] identified Heceta Bank as one of five oceanographic habitats in the Northern California Current where high chlorophyll-a concentrations were associated with increased abundances of commercially important groundfishes.

The high-resolution bathymetric and backscatter survey of Heceta Bank and associated ROV dives that are at the focus of the current study also formed the basis for a new and detailed geological interpretation of the bank, and a description and interpretation of methane seeps that are a common feature of the outer continental shelf and upper slope in this region [14]. Torres et al. [14] composite map of the bank includes the location of bedding planes, faults, and pockmarks with associated seeps (Figure 40.2). The seafloor morphology and texture of Heceta Bank is primarily the result of subaerial and wave-based erosion during eustatic changes in sea level. The imagery from the multibeam echosounder survey reveals striking outcroppings of differentially eroded, jointed and folded strata located partly in zones of high backscatter (Figures 40.1 and 40.2). The outer edge of the bank is clearly defined by a change to low backscatter muds of the upper slope and by wave-cut scarps and benches formed during lowered sea level. The more resistant strata on the top of the bank appear as distinct topographic ridges. The southwestern flank of the bank shows a history of slides, as evidenced by slide scars [29]. The seeps mapped by Torres et al. [14] conformed to four structural types: pockmarks in areas of unconsolidated sediment seaward of the western flank of the bank; seeps discharging out of rock outcrops within the interior of the bank; seeps aligned along underlying structures; and areas of unconsolidated sediment with large carbonate deposits (measured at ~4 km^2 in one case).

Substrate classification: From the ROV videos of Heceta Bank, we identified a total of nine substrate types varying by sediment type and relief (Table 40.1). Two consolidated sediment types, rock ridge (R) and flat rock (F), were differentiated by a presence or absence of relief. Seven sediment types, boulder (B, >25.5 cm grain size), cobble (C, 25.5–6.4 cm), pebble (P, 6.4–0.4 cm), sand (S, 2–0.06 mm), mud (M, < 0.06 mm), unconsolidated (U, fine-grained sediment of undetermined grain size), and carbonate rock (A, sedimentary rock with carbonate deposits generated from seeps) were differentiated by grain size and lithology. Because substrate types often occur in mixture, a two-letter code [10,11] was used to describe the substrate within the ROV video's field-of-view (FOV). The first letter described the primary (>50% FOV) observed substrate type, while the second letter described the secondary (>20% FOV) type. For example, the code "BC" would indicate a FOV comprised of greater than 50% boulders with at least 20% cobbles. If boulders occupied >70% of the FOV, the code was "BB." Each uniform stretch of substrate, or

Table 40.1 Habitat Terms Used in Text and Related Scales, Descriptions, and Data Sources

Habitat Term	Spatial Scale Term[*]	Spatial Scale Dimensions	Description	Data Source (s)
Substrate type	Macro	1–10 m	Types include: rock ridge (R), flat rock (F), boulder (B), cobble (C), pebble (P), sand (S), mud (M), unconsolidated (U), carbonates (A)	ROV
Substrate patch	Macro	1–10 m	Two-letter code: primary (>50%) and secondary (>20%); 57 patch types encountered	ROV
Potential habitat	Meso	10 m–1 km	Four classes: steep ridge (SR), low-relief ridge and mixed substrate (LRRM), cobble and boulder (CB), unconsolidated sediment (U)	ROV/ MB/[30]
Habitat	Meso	10 m–1 km	Nine habitats were identified from a cluster analysis of substrate patches and associated fishes	ROV/ dendrogram

H.G. Greene, M.M. Yoklavich, R.M. Starr, V.M. O'Connell, W.W. Wakefield, D.E. Sullivan, J.E. McRea Jr., G.M. Cailliet, A classification scheme for deep seafloor habitats, Oceanol. Acta 22 (1999) 663–678.
*Denotes reference to the Greene et al. classification scheme for deep seafloor habitats. "MB" stands for Simrad EM300 multibeam echosounder survey data and associated imagery [13].

"patch," that persisted for at least 30 s of video time (a minimum distance of ~8 m) was assigned a two-letter substrate code (Table 40.1). During ROV dives conducted in 2000 and 2001, 57 substrate patch types were identified, out of a possible 81 two-letter substrate combinations.

Potential habitat classification: In order to estimate bank-wide abundances of demersal fishes, we first needed to delineate potential habitats, or areas of distinct textural characteristics, in the multibeam imagery. Although many of these potential habitats manifest themselves as characteristic bedforms in the bathymetry imagery, or areas of discrete acoustic reflectivity in the backscatter imagery, much of the habitat variability was only evident in the *in situ* video from the ROV. By overlaying the ROV dive tracks and symbolizing the observed substrate types on the multibeam imagery in ArcGIS (Geographical Information System) software (Environmental Systems Research Institute, Inc., Redlands, California, USA), we were able to visually digitize the textural boundaries apparent in both the backscatter and bathymetry imagery that represented potential habitat boundaries (Figure 40.3). In addition to the observational data, a previous study by Whitmire et al. ([30]; Figure 40.3D) provided another classification of potential habitats based primarily on quantitative interpretation of the multibeam pixel data. The Whitmire et al. [30] map revealed two classes (cobble/boulder and mud/sand) based on textural differences in the backscatter data, and one class (ridge) based on relief. While this study identified two

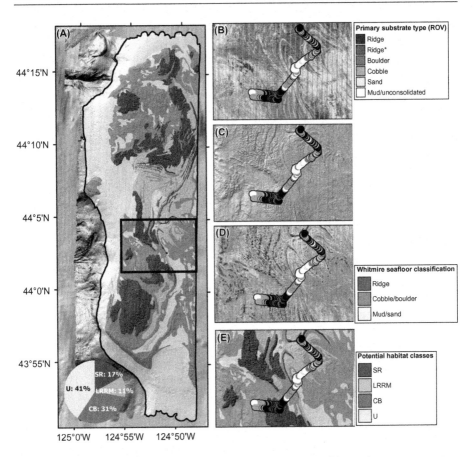

Figure 40.3 Map for Heceta Bank, Oregon, of the four potential habitats, shown as semitransparent over the multibeam bathymetry data (A and E). Potential habitats include steep ridge (SR, green), low-relief ridge and mixed substrate (LRRM, pink), cobble and boulder (CB, orange), and unconsolidated sediment (U, blue), with the relative proportion of each type displayed in the pie chart (A). The black box in panel (A) shows the region pictured in (B–E). An ROV transect is displayed in panels (B–E) and is color coded by primary observed substrate type (*ridge as secondary type). Other panels include: (B) EM300 backscatter imagery, (C) EM300 sun-illuminated bathymetry (illumination is from the northwest), (D) potential habitats identified from quantitative classification of multibeam pixel data in a previous study [30], and (E) potential habitats identified in this study. (For interpretation of the references to color in this figure legend, the reader is referred to the web version of this book.)

potential habitat classes similar to the sedimentary ones in the Whitmire et al. [30] study, it identified two other classes with distinct characteristics of both relief and texture. Consequently, this study revealed a total of four potential habitat classes: steep ridge (SR), low-relief ridge and mixed substrate (LRRM), cobble and boulder (CB), and unconsolidated sediment (U), listed in descending order of relief and consolidation (Figure 40.3).

Steep ridge: From the final map of potential habitats, the steep ridge class (SR) accounted for 10,151 ha, or 17%, of the entire bank's area (Table 40.2). Steep ridge habitat was found mostly in the northern and southern ends of the bank, marked the bank's northern and southern topographic highs, and transitioned into low-relief ridge/mixed habitat for the central portion of the bank. By comparison, the ROV sampled 6.3 ha of steep ridge, accounting for 33% of the survey effort (Table 40.2).

Low-relief ridge and mixed substrate: LRRM was the most diverse of all potential habitats and comprised all the areas of the bank that did not fit into the more strictly defined steep ridge, cobble and boulder, or unconsolidated categories. Although the LRRM class comprised the smallest portion of the bank (11%), it was associated with the most diverse group of substrate types, contained the largest number of substrate patches, and occupied the majority of the central portion of the bank. The greater part of this potential habitat was found where the ridge expression was more eroded, with frequent transitions occurring along transects from one substrate type to another. This mixed environment was also located to the east of both the northern and southern steep ridge areas and comprised 11% of the bank with 6,354 ha. The sampling effort by the ROV was comparable with 8% of the survey effort and 1.6 ha total (Table 40.2).

Cobble and boulder: The cobble and boulder potential habitat class (CB) was found primarily in the northern and southern regions of the bank and frequently surrounded the steep ridge, accounting for 31% of the mapped area (18,617 ha). This potential habitat tended to transition from a larger clast-size boulder substrate to mixed cobble/boulder and cobble/pebble along the edge of the bank. The cobble and boulder potential habitat enclosed much of the steep ridge habitat at intermediate depths (80–200 m) and was associated with a diverse group of substrate types. In addition, there was an isolated area of high backscatter intensity on the seaward flank of the bank (260–400 m). One ROV dive revealed the area to be cobble and boulder substrate surrounded by mud and separated from other similar substrate types on top of the bank. This potential habitat class was sampled more than any other class by the ROV, with 39% of the survey effort accounting for 7.4 ha (Table 40.2).

Table 40.2 Distribution of the Potential Habitat Types Identified on Heceta Bank (Figure 40.3). Relative to the Distribution of Survey Effort by the ROV Within Potential Habitats

Habitat	Habitat Map		ROV Survey	
	Area (ha)	Portion of Bank (%)	Area (ha)	Portion of Survey Effort (%)
Steep ridge	10,151	17	6.3	33
Low-relief ridge and mixed	6,354	11	1.6	8
Cobble and boulder	18,617	31	7.4	39
Unconsolidated sediment	24,035	41	3.2	19
Total	59,157	100	18.5	100

Unconsolidated sediment: Unconsolidated sediment (U), which included sand, mud, and mixed soft sediment, comprised the largest area of the bank (41% and 24,035 ha) (Table 40.2) and represented the deepest potential habitat, extending down to a depth of 400 m. This class surrounded the bank at deeper depths (150–400 m) on the northern, southern, and seaward sides and extended seaward onto the upper continental slope.

Test for sampling bias: An ideal sampling design for delineating all habitats within the survey area would have included the collection of random, independent, and internally consistent observations of the entire geographic area for which we are interested in delineated habitat boundaries. For example, the ROV transects did not follow a random sampling design because dive locations were established with different goals in mind, such as conducting a fixed-site survey to compare contemporary and historical research, ground-truthing the newly acquired multibeam survey, and geological exploration by the interdisciplinary team of scientists.

We conducted simulated random sampling to explore the actual sampling bias of the ROV. The simulation indicated that ROV sampling was highly biased toward hard substrate, which was not surprising as the ROV research was focused on revisiting previously sampled rocky areas of the bank. Of the four potential habitats, this simulation demonstrates that SR was the most oversampled, followed by CB and LRRM, while U was undersampled (Figure 40.4).

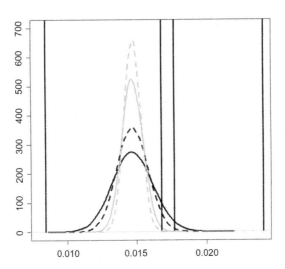

Figure 40.4 Frequency distribution of random sampling of the bank relative to the actual sampling of potential habitats. A comparison of the distribution of potential habitats from simulated random sampling (green—SR, orange—CB, pink—LRRM, blue—U) to habitat actually encountered on ROV dives (black vertical lines) shows SR was the most oversampled, followed by LRRM and CB, and that UC was undersampled on the bank. (For interpretation of the references to color in this figure legend, the reader is referred to the web version of this book.)

Biological Communities

The following description of biological communities is limited to assemblages of demersal fishes and will not include discussion of epibenthic invertebrates. The megafaunal invertebrate distribution and abundance on Heceta Bank was developed as a separate analysis, both published and in preparation [5,31–33]. All fish observations from the ROV video were classified to the lowest possible taxonomic unit, and total fish length was estimated to the nearest centimeters. For fish that could only be identified to a taxonomic group higher than species, a generalized code was used (e.g., "FF" for unidentified flatfish). In cases where the fish observed was one of two possible species, we created new codes to accommodate this. One example is Puget Sound rockfish (*Sebastes emphaeus*) and pygmy rockfish (*S. wilsoni*), which tended to school together. Some individuals could be identified to species, but not every single individual in a large school; thus, we would use the code "PRC" for pygmy–Puget complex.

The ROV video provided the opportunity to examine substrate associations of fishes in relation to previous work on Heceta Bank [32]. The most abundant fishes on Heceta Bank included a number of species that attain smaller sizes as adults and also a category of unidentified small/juvenile rockfishes (*Sebastes* spp.) that ranged in size from 5 to 10 cm. The latter category was the most common taxa, representing 22.6% of all observed fish (Table 40.3). Pygmy and Puget Sound rockfishes, and a pygmy–Puget complex (size = 5–25 cm), comprised 41% of all fish counted. Rockfishes were the most common group of fishes accounting for 95.0% of all fishes with a total of 17 species. In addition to the unidentified small-sized rockfishes, sharpchin, *S. zacentrus*, rosethorn, *S. helvomaculatus*, yellowtail, *S. flavidus*, greenstriped, *S. elongatus*, and redstripe, *S. proriger*, rockfishes were common (Table 40.3). Within the group of rockfishes that were counted, 9.7% were not identified beyond the level of genus. Other common fish taxa included unidentified sculpins (Cottidae) and Dover sole (*Microstomus pacificus*) (Table 40.3).

The following description of fish communities associated with potential habitat classes is limited to 11 taxa (highlighted in Table 40.3) based on the following criteria: (1) commercially important groundfish species with an emphasis on rockfishes, (2) taxa representing the range of substrate types observed, (3) taxa showing a variety in their selectivity and specificity of substrate, (4) rank order of abundance (ROA), and (5) species designated by management bodies as "overfished" [34].

Steep ridge: The fish assemblage associated with the steep ridge potential habitat were dominated by commercially important rockfishes, including, yellowtail rockfish, yelloweye rockfish (*S. ruberrimus*), and canary rockfish (*S. pinniger*), as well as unidentified small/juvenile and pygmy/Puget Sound rockfish complexes (Figure 40.5A–C). On Heceta Bank, Tissot et al. [32] observed a similar assemblage associated with their shallow rock ridge and large boulder habitat class. That research suggested that the shallow rock ridge and large boulder habitat was an important nursery area for young-of-the-year rockfishes because of its proximity to individuals in the postlarval settlement stage, prey availability, and shelter from predators, including the availability of structure-forming invertebrates.

Table 40.3 Rank Order of Occurrence of Fishes from ROV Transects on Heceta Bank, Oregon (Based on Raw Counts) for the Top 100 Ranked Taxa (Out of a Total of 208 Fish Taxa), Primarily Identified at Species Level

Rank	Common Name	Scientific Name	Counts	% of Total
1*	Small/juvenile rockfish, unident.	*Sebastes* spp.	10,937	22.6
2*	Pygmy rockfish	*Sebastes wilsoni*	8246	17.0
3*	Sharpchin rockfish	*Sebastes zacentrus*	6077	12.5
4	Puget Sound rockfish	*Sebastes emphaeus*	6005	12.4
5	Pygmy–Puget Sound rockfish	*Sebastes wilsoni/emphaeus*	5635	11.6
6	Rockfish, unident.	*Sebastes* spp.	4467	9.2
7*	Rosethorn rockfish	*Sebastes helvomaculatus*	2188	4.5
8*	Yellowtail rockfish	*Sebastes flavidus*	1229	2.5
9*	Greenstriped rockfish	*Sebastes elongatus*	594	1.2
10	Sculpin, unident	Cottidae	375	0.8
11*	Redstripe rockfish	*Sebastes proriger*	345	0.7
12*	Dover sole	*Microstomus pacificus*	340	0.7
13	Fish, unident	Osteichthyses	242	0.5
14	Ronquil, unident	Bathymasteridae	216	0.4
15	Poacher, unident	Agonidae	150	0.3
16*	Lingcod	*Ophiodon elongatus*	147	0.3
17*	Threadfin sculpin	*Icelinus filamentosus*	143	0.3
18	Hagfish, unident	*Eptatretus* spp.	135	0.3
19	Canary rockfish	*Sebastes pinniger*	132	0.3
20	Spotted ratfish	*Hydrolagus colliei*	120	0.2
21	Flatfish, unident	Pleuronectiformes	106	0.2
22*	Yelloweye rockfish	*Sebastes ruberrimus*	104	0.2
23	Kelp greenling	*Hexagrammos decagrammus*	94	0.2
24	Harlequin rockfish	*Sebastes variegatus*	79	0.2
25	Icelinus sculpin, unident	*Icelinus* sp.	61	0.1
26	Shortspine thornyhead	*Sebastolobus alascanus*	50	0.1
27	Rex sole	*Glyptocephalus zachirus*	41	0.1
28	Blenny-like fish, unident	Blennioidei	32	0.1
29	Harlequin-redstripe rockfish	*Sebastes variegatus/ proriger*	27	0.1
30	Eelpout, unident	Zoarcidae	24	0.0
31	Bigfin eelpout	*Lycodes cortezianus*	24	0.0
32	Greenspotted rockfish	*Sebastes chlorostictus*	20	0.0
33	Widow rockfish	*Sebastes entomelas*	17	0.0
34	Prickleback, unident	Stichaeidae	15	0.0
35	Pacific halibut	*Hippoglossus stenolepis*	12	0.0
36	Longnose skate	*Raja rhina*	11	0.0
37	Skate unident.	Rajidae	9	0.0
37	Slender sole	*Lyopsetta exilis*	9	0.0

Continued

Table 40.3 (Continued)

Rank	Common Name	Scientific Name	Counts	% of Total
38	Blackeye goby	*Coryphopterus nicholsi*	8	0.0
39	Tiger rockfish	*Sebastes nigrocinctus*	6	0.0
40	Big skate	*Raja binoculata*	4	0.0
40	Sturgeon poacher	*Agonus acipenserinus*	4	0.0
41	Spotted cusk-eel	*Chilara taylori*	2	0.0
41	Darkblotched rockfish	*Sebastes crameri*	2	0.0
41	Redbanded rockfish	*Sebastes babcocki*	2	0.0
42	Flag rockfish	*Sebastes rubrivinctus*	1	0.0
42	Pacific hake	*Merluccius productus*	1	0.0
42	Petrale sole	*Eopsetta jordani*	1	0.0
42	Sablefish	*Anoplopoma fimbira*	1	0.0
42	Shortbelly rockfish	*Sebastes jordani*	1	0.0
42	Shark, unident.	Chondrichthyes	1	0.0
42	Starry skate	*Raja stellulata*	1	0.0
		Total	48,493	100.0

*Signifies the 11 species of groundfishes that were selected for analysis of habitat associations and estimation of bank-wide abundances.

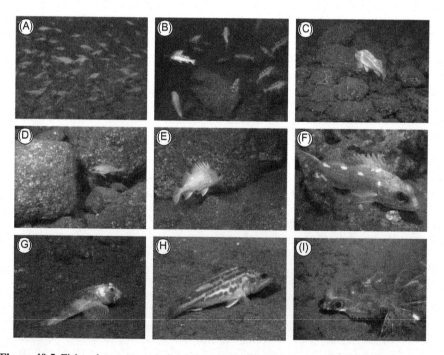

Figure 40.5 Fish–substrate associations observed in ROV video: (A) pygmy and small/juvenile rockfishes over Ridge substrate; (B–D) yellowtail, yelloweye, and redstripe rockfishes over boulder substrate; (E) greenspotted rockfish over boulder–unconsolidated substrate; (F) rosethorn rockfish on carbonate–mud substrate; (G) threadfin sculpin on unconsolidated–cobble substrate; (H and I) greenstriped rockfish and Dover sole on Mud substrate.

Low-relief ridge and mixed substrate: Given the extreme variety of substrate types contained within the LRRM potential habitat, the fishes that occupied this class represent a similarly diverse assemblage, including small/juvenile, pygmy, rosethorn, and greenstriped rockfishes, and Dover sole (Figure 40.5A, F, H, and I).

Cobble and boulder: Small/juvenile, pygmy/Puget Sound, sharpchin, redstripe, rosethorn, yellowtail, and greenspotted (*S. chlorostictus*) rockfishes were the dominant species in the cobble and boulder potential habitat (Figure 40.5A, B, D–F). For this class, a nonmetric multidimensional scaling (NMDS) analysis showed associations with sharpchin, small/juvenile, rosethorn, and pygmy/Puget Sound rockfishes [35]. This class overlaps in depth and fish assemblages, with the mid-depth boulder–cobble and deep cobble habitats identified by Tissot et al. [32]. In both studies, the pygmy/Puget Sound rockfish complex and sharpchin rockfish were particularly abundant.

Unconsolidated sediment: Greenstriped rockfish and Dover sole were the most common of the nine key species in this potential habitat. In addition, rosethorn, sharpchin, and greenspotted rockfishes occurred where boulder, cobble, and pebble added complexity to substrate of primarily mud (Figure 40.5E, F, H, and I). Noncommercial species observed in this class included sculpins (e.g., *Icelinus* sp., Figure 40.5G), unidentified ronquils and poachers, and hagfish (*Eptatretus* sp.). This class corresponded to Tissot et al.'s [32] deep mud slope habitat where Dover sole, rex sole, shortspine thornyhead, and greenstriped rockfish were abundant.

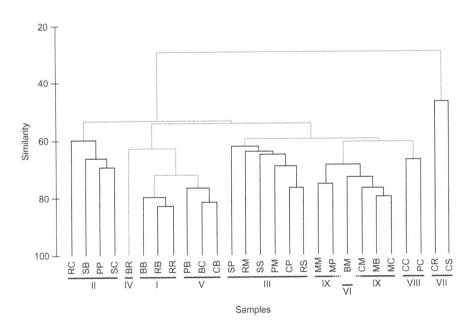

Figure 40.6 Dendrogram showing a hierarchical clustering of Bray–Curtis similarities run on a fourth root transformation of habitat specific abundances for fishes inhabiting Heceta Bank, Oregon. The fourth root transformation was applied to the abundances to de-emphasize abundant species. Nine pooled habitats are identified by numerals I–IX.

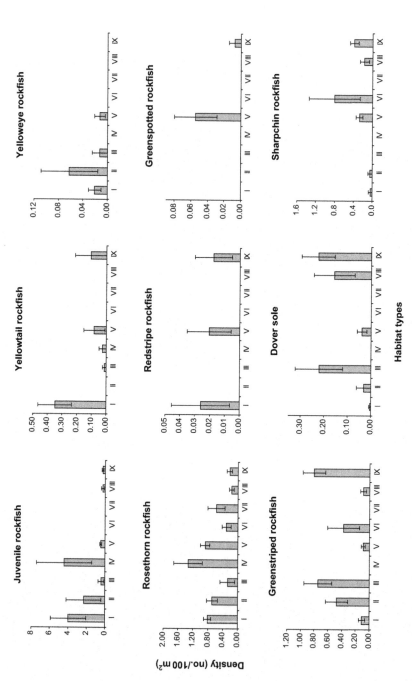

Figure 40.7 Densities (mean number/100m² ± 1 SE) of nine key fish taxa relative to the nine habitats generated from the dendrogram output of a cluster analysis (see Figure 40.6). Habitat types are arranged from left to right by decreasing relief and grain size and include: ridge/boulder (I), ridge/cobble (II), ridge/unconsolidated (III), boulder/ridge (IV), boulder/cobble (V), boulder/mud (VI), cobble/ridge (VII), cobble (VIII), and unconsolidated (IX).

Surrogacy

A major goal of the series of interrelated studies on Heceta Bank was to identify fish associations with habitat [5,12,35]. To this end, we examined similarities of substrate specific abundances using a hierarchical cluster analysis (Primer software, PRIMER-E Ltd., Lutton, Ivybridge, UK). Of the 57 types of substrate patches observed on the bank, fish were associated with 53 of them. From the dendrogram output of the cluster analysis, we identified nine habitat types (Figure 40.6). Eight of nine of these habitat types showed Bray–Curtis similarities of at least 60. The distributions of select taxa within these nine habitat types are shown in Figure 40.7. The cluster analysis also resulted in some habitat classes that were not readily discernible in the multibeam imagery, even using the methods presented earlier in the paper for identifying potential habitats.

Another goal of this study was to estimate bank-wide abundances for a group of eleven species/taxa of groundfishes (highlighted in Table 40.3). In addition, we examined the question of whether or not increased resolution of habitat information could improve the precision of population estimates. First we selected three levels of habitat specificity on the bank: (1) entire bank (to 400 m); (2) consolidated (e.g., R)

Table 40.4 Bank-Wide Abundances, Shown in Descending Order, for the 11 Taxa of Groundfishes That Were Selected for Analysis of Habitat Associations

Taxon	Entire Bank		Consolidated versus Unconsolidated		Potential Habitats	
	Abundance	SD	Abundance	SD	Abundance	SD
Pygmy rockfish	8508	3,357	6145	2,408	5538	2,253
Small/juvenile rockfish	4391	1,684	3182	1,208	3013	1,177
Rosethorn rockfish	2052	271	1540	193	1501	195
Sharpchin rockfish	1071	317	1110	363	1181	375
Greenstriped rockfish	991	168	1171	220	1197	226
Yellowtail rockfish	630	252	455	181	367	136
Dover sole	386	87	501	142	513	143
Yelloweye rockfish	149	98	107	71	120	81
Threadfin sculpin	125	41	123	55	127	56
Lingcod	110	33	105	42	106	43
Redstripe rockfish	77	47	56	34	50	27

Note: Abundance (× 1,000) and standard deviation (sd × 1,000) are reported for each of three levels of habitat specificity.

versus unconsolidated (e.g., M, S) substrate types; and (3) four potential habitats (SR, LRRM, CB, and U). We then calculated the density and variance of fish species for each habitat type (Table 40.4). Finally, we expanded fish abundance for the select group of groundfish species. In general, species that demonstrated an affinity for soft substrate (e.g., Dover sole) showed an increase in abundance with greater habitat stratification, illustrating that the one-habitat initial estimate was biased low due to nonrandom sampling; in contrast, species who prefer more consolidated substrate (e.g., redstripe rockfish) showed a decrease in abundance with increasing habitat stratification. Sharpchin showed minimal change in abundance with increasing habitat specificity.

Acknowledgments

We would especially like to thank the following colleagues who contributed to this study: Bill Barss, Bob Embley, Chris Goldfinger, Gordon Hendler, Mark Hixon, Douglas Markle, Susan Merle, Noelani Puniwai, Chris Romsos, David Stein, Brian Tissot, Mary Yoklavich, and Keri York. John Harms, Allan Hicks, and Mark Zimmermann provided constructive reviews of the paper. This portion of the Heceta Bank project was funded by the West Coast and Polar Programs Undersea Research Center of NOAA's National Undersea Research Program, the Northwest and Southwest Fisheries Science Centers, and NOAA's Pacific Marine Environmental Laboratory. Several individuals involved in various aspects of this research are funded by NOAA through the Cooperative Institute for Marine Resources Studies at Oregon State University. We would like to thank the professional personnel who operated the *ROPOS* ROV and the R/V *Ronald H. Brown*.

References

[1] V.M.O'Connell, W.W. Wakefield, Workshop proceedings: applications of side-scan sonar and laser line systems in fisheries research. Alaska Department of Fish and Game, Special Publication 9, Juneau, Alaska, 1994.

[2] M.M. Yoklavich, H.G. Greene, G.M. Cailliet, D.E. Sullivan, R.N. Lea, M.S. Love, Habitat associations of deep-water rockfishes in a submarine canyon: an example of a natural refuge, Fish. Bull. 98 (2000) 625–641.

[3] J.R. Reynolds, R.C. Highsmith, B. Konar, C.G. Wheat, D. Doudna, Fisheries and fisheries habitat investigations using undersea technology. Marine Technology Society/IEEE Oceans 2001, Conference Proceedings, Paper MTS 0-933957-29-7, 2001.

[4] N.M. Nasby-Lucas, R.W. Embley, M.A. Hixon, S.G. Merle, B.N. Tissot, D.J. Wright, Integration of submersible transect data and high-resolution multibeam sonar imagery for a habitat-based groundfish assessment of Heceta Bank, Oregon, Fish. Bull. 100 (2002) 739–751.

[5] W.W. Wakefield, C.E. Whitmire, J.E.R. Clemons, B.N. Tissot, Fish habitat studies: combining high-resolution geological and biological data, in: P.W. Barnes, J.P. Thomas, (Eds.), Benthic Habitats and the Effects of Fishing, vol. 41, American Fisheries Society, Bethesda, MD, 2005, pp. 119–138.

[6] V.M. O'Connell, C.K. Brylinsky, H.G. Greene, The use of geophysical survey data in fisheries management: a case history from southeastern Alaska, in: B.J. Todd,

H.G. Greene, (Eds.), Mapping the Seafloor for Habitat Characterization, Geological Association of Canada, St. John's, Newfoundland, 2007, pp. 319–328. (Special Paper 47)

[7] B.J. Todd, H.G. Greene, (Eds.), Mapping the Seafloor for Habitat, Characterization, Geological Association of Canada, St. John's, Newfoundland, 2007 (Special Paper 47)

[8] J.R. Reynolds, H.G. Greene, (Eds.), Marine Habitat Mapping Technology for Alaska, Alaska Sea Grant College Program, University of Alaska Fairbanks, Fairbanks, AK, 2008

[9] W.G. Pearcy, D.L. Stein, M.A. Hixon, E.K. Pikitch, W.H. Barss, R.M. Starr, Submersible observations of deep-reef fishes of Heceta Bank. Oregon, Fish. Bull. 87 (1989) 955–965.

[10] M.A. Hixon, B.N. Tissot, W.G. Pearcy, Fish assemblages of rocky banks of the Pacific Northwest (Heceta, Coquille, and Daisy Banks). U.S. Minerals Management Service, OCS Study 91-0052, Final Report, Camarillo, California, 1991.

[11] M.A. Hixon, B.N. Tissot, Fish assemblages of rocky banks of the Pacific Northwest. U.S. Minerals Management Service, Final report supplement, OCS Study 91-0025, Camarillo, California, 1992.

[12] B.N. Tissot, M.A. Hixon, D.L. Stein, Habitat-based submersible assessment of macro-invertebrate and groundfish assemblages at Heceta Bank, Oregon, from 1988 to 1990, J. Exp. Mar. Biol. Ecol. 352 (2007) 50–64.

[13] MBARI (Monterey Bay Aquarium Research Institute). "MBARI, Northern California and Oregon Margin Multibeam Survey." MBARI, Digital Data Series Number 5, CD ROM, Moss Landing, California, 2001.

[14] M.E. Torres, R.W. Embley, S.G. Merle, A.M. Tréhu, R.W. Collier, E. Suess, et al., Methane sources feeding cold seeps on the shelf and upper continental slope off central Oregon, USA, Geochem. Geophys. Geosyst. 10 (2009) Q11003, 10.1029/2009GC002518.

[15] G.E. Muehlberg, Structure and stratigraphy of Quaternary strata of the Heceta Bank, central Oregon shelf, Masters Thesis, Oregon State University, Corvallis, Oregon, 1971.

[16] L.C. McNeill, C. Goldfinger, L.V.D. Kulm, R.S. Yeats, Tectonics of the Neogene Cascadia forearc basin: investigations of a deformed late Miocene unconformity, Geol. Soc. Am. Bull 112 (2000) 1209–1224.

[17] C.G. Romsos, C. Goldfinger, R. Robison, R.L. Milstein, J.D. Chaytor, W.W. Wakefield, Development of a regional seafloor surficial geologic habitat map for the continental margins of Oregon and Washington, USA, in: B.J. Todd, H.G. Greene, (Eds.), Mapping the Seafloor for Habitat Characterization, Geological Association of Canada, St. John's, Newfoundland, 2007, pp. 219–243. (Special Paper 47).

[18] H.K. Beals, For Honor and Country, The Diary of Bruno de Hezeta, Oregon Historical Society Press, Portland, OR, 1985.

[19] D. Hayes, Historical Atlas of the Pacific Northwest: Maps of Exploration and Discovery, Sasquatch Books, Seattle, Washington, DC, 1999.

[20] L.D. Kulm, G.A. Fowler, Oregon continental margin structure and stratigraphy: a test of the imbricate thrust model, in: C.A. Burk, C.L. Drake, (Eds.), The Geology of Continental Margins, Springer-Verlag, New York, NY, 1974, pp. 261–283.

[21] A. Huyer, Coastal upwelling in the California Current System, Prog. Oceanogr. 12 (1983) 259–284.

[22] B.M. Hickey, Patterns and processes of circulation over the shelf and slope, in: M.R. Landry, B.M. Hickey, (Eds.), Coastal Oceanography of Washington and Oregon, Elsevier, Amsterdam, The Netherlands, 1989, pp. 41–116.

[23] R.M. Castelao, J.A. Barth, Coastal ocean response to summer upwelling favorable winds in a region of alongshore bottom topography variations off Oregon, J. Geophys. Res. 110 (2005) C10S04, 10.1029/2004JC002409.

[24] A. Huyer, J.H. Fleischbein, J. Keister, Two coastal upwelling domains in the northern
      California Current System, J. Mar. Res. 63 (2005) 901–929.
[25] H.P. Batchelder, J.A. Barth, P.M. Kosro, P.T. Strub, R.D. Brodeur, W.T. Peterson, et al.,
      The GLOBEC Northeast Pacific California Current System Program, Oceanography 15
      (2002) 36–47.
[26] J.A. Barth, P.A. Wheeler, Introduction to special section: coastal advances in shelf trans-
      port, J. Geophys. Res. 110 (2005) C10S01, 10.1029/2005JC003124.
[27] B.A. Grantham, F. Chan, K.J. Nielsen, D.S. Fox, J.A. Barth, A. Huyer, et al., Upwelling-
      driven nearshore hypoxia signals ecosystem and oceanographic changes in the northeast
      Pacific, Nature 429 (2004) 749–754.
[28] M.J. Juan-Jordá, J.A. Barth, M.E. Clarke, W.W. Wakefield, Groundfish species asso-
      ciations with distinct oceanographic habitats off the Pacific northwest coast, Fish.
      Oceanogr. 8 (2009) 1–19.
[29] C. Goldfinger, L.D. Kulm, L.C. McNeill, P. Watts, Super-scale failure of the southern
      Oregon Cascadia margin, Pure Appl. Geophys. 157 (2000) 1189–1226.
[30] C.E. Whitmire, R.W. Embley, W.W. Wakefield, S.G. Merle, B.N. Tissot, Characterizing
      benthic habitats at Heceta Bank, Oregon, in: B.J. Todd, H.G. Greene, (Eds.), Mapping
      the Seafloor for Habitat Characterization, Geological Association of Canada, St. John's,
      Newfoundland, 2007, pp. 111–126. (Special Paper 47)
[31] N.P.F. Puniwai, Spatial and temporal distribution of the crinoid *Florometra serra-
      tissma* on the Oregon continental shelf. Masters Thesis, Washington State University
      Vancouver, Vancouver, Washington, 2002.
[32] B.N. Tissot, W.W. Wakefield, M.A. Hixon, J.E.R. Clemons, Twenty years of fish-habitat
      studies on Heceta Bank, Oregon, in: J.R. Reynolds, H.G. Greene, (Eds.), Marine Habitat
      Mapping Technology for Alaska, Alaska Sea Grant College Program, University of
      Alaska Fairbanks, Fairbanks, AK, 2008 10.4027/mhmta.2008.15.
[33] B.N. Tissot, W.W. Wakefield, N.P.F. Puniwai, M.M. Yoklavich, J.E.R. Clemons, J. Pirtle,
      Ecological associations between structure-forming invertebrates and rockfishes on
      Heceta Bank, Oregon with special reference to corals (in preparation).
[34] S.D. Miller, M.E. Clarke, J.D. Hastie, O.S. Hamel, Pacific coast groundfish fisheries,
      in: Our Living Oceans: Report on the Status of U.S. Living Marine Resources, 6th ed.,
      (NMFS) U.S. Dept. Commer., NOAA Tech. Memo. NMFS-F/SPO-80, 2009.
[35] W.W. Wakefield, J.E.R. Clemons, B.N. Tissot, C.E. Whitmire, R.W. Embley, Habitat
      associations in demersal fishes on a deep-water rocky bank off Oregon (in preparation).

41 Open Shelf Valley System, Northern Palaeovalley, English Channel, UK

J.W. Ceri James[1], Bryony Pearce[2],
Roger A. Coggan[3], Angela Morando[1]

[1]British Geological Survey, Keyworth, Nottingham, UK,
[2]Marine Ecological Surveys Ltd, Bath, UK, [3]Cefas, Lowestoft, UK

Abstract

The Northern Palaeovalley is an open shelf valley system in the English Channel. The valley is >100 km long and varies in width from 8 to 20 km. Its floor lies at depths of 45–90 m, with margins up to 30 m high. Net sediment transport is west to east, with rock and coarse sediment in the valley floor in the west and a gradual eastward increase in sand cover and volume, culminating in two large linear sandbanks up to 28 km long along the northeast margin of the valley. The Palaeovalley supports a rich diversity of infaunal and epifaunal communities. Considerable overlap exists in the infaunal communities associated with the different geomorphic features, while the epifaunal communities exhibit far greater variability in response to the environmental gradients. Interstitial polychaetes dominate the macrobenthic communities across the area reflecting the dominance of mobile sediment deposits.

Key Words: Habitat mapping, English Channel, palaeovalley, sandbank

Introduction

The Northern Palaeovalley is an open shelf valley system in the English Channel (Figure 41.1). It is a distinctive erosional feature within the northern half of the eastern English Channel with a particularly well-defined margin on its northern side and an upper break of slope at a depth of 25–30 m with the valley floor generally at depths of 55–75 m (Figure 41.2). The valley segment described here is about 100 km long and varies in width from 8 to <20 km but extends further west for a further 60 km. It is a tide-dominated system with maximum tidal current velocities >1 m/s. The benthic environment is impacted by a number of features including shipping, fishing, cables, and wrecks, with marine aggregate extraction conducted adjacent to the Palaeovalley. The level of human impact is categorized as predominantly high to very high [1].

Seafloor Geomorphology as Benthic Habitat. DOI: 10.1016/B978-0-12-385140-6.00041-4

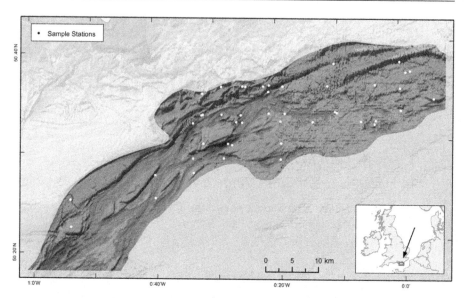

Figure 41.1 Singlebeam echosounder bathymetric map of Northern Palaeovalley (unmasked area) based on 75–> 250-m grid, including seabed sample stations. Shade source 315° NW. Singlebeam echosounder data ©British Crown & Sea Zone Solutions Ltd. 2008. All rights reserved. Data License 052008.012.

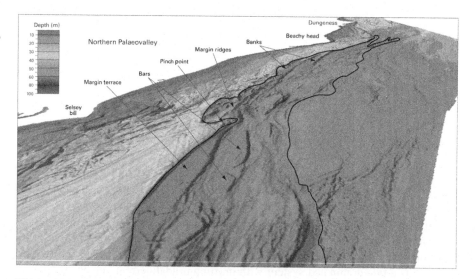

Figure 41.2 Three-dimensional perspective view of the Northern Palaeovalley and surrounding seabed looking to the east, with examples of geomorphic features indicated. Shade source 315° NW.
Singlebeam echosounder data ©British Crown & Sea Zone Solutions Ltd. 2008. All rights reserved. Data License 052008.012.

Interpretation has been based on a grid of singlebeam echosounder data collected for the United Kingdom Hydrographic Office (UKHO) to produce a regional sea-bed model (Figure 41.1). This is complemented by a more detailed survey along some widely spaced narrow corridors and lines using multibeam, sidescan sonar, and boomer sub-bottom profiler. The seabed has been sampled at 42 stations along these latter survey corridors using a 0.1 m² Hamon grab, a 2-m beam trawl, and an under-water camera (video and still imagery), either singly or in combination [2,3].

Geomorphic Features and Habitats

The resolution of the regional seabed model varies between the western and eastern halves, reflecting different survey strategies used by the UKHO. The data allowed grid-ding at 75 m × 75 m in the west, but only 250 m × 250 m in the east (Figure 41.1). The shallower eastern half of the Palaeovalley is relatively wide, from 16 to 20 km, compared to the deeper western half, which varies in width from 8 to 16 km. There is a distinct narrowing of the valley between these two halves at an 8-km wide pinch point that marks the occurrence of a steeply dipping monocline trending NW–SE across the Palaeovalley (Figure 41.2). The monocline dips to the northeast and forms the boundary between relatively soft Tertiary rocks to the east and more resistant Upper Cretaceous Chalk to the west [4]; hence the wider Palaeovalley to the east in less resis-tant Tertiary rocks and the funneling of the valley through the Chalk cored pinch point at the monocline. Underlying rock structures have also controlled the line of the north-ern margin of the Palaeovalley in its eastern half, with the southern limb of a relatively steeply dipping E–W trending anticline forming a natural barrier to further northward migration of the Palaeovalley.

Five principal geomorphic features (Figure 41.3) have been identified and associ-ated with the Palaeovalley system based on the seabed model and sub-bottom pro-filer data.

Palaeovalley floor: This is the most extensive geomorphic feature and extends along the whole length of the Palaeovalley. The poor resolution of the model in the east means that detailed features are lost, but there are indications of bars and small channels up to 5 m in elevation within the floor. Further west, the floor is character-ized by well-developed channels, some of which are incised in the narrows between Palaeovalley bars. The Palaeovalley has acted as a conduit for the eastward migration of sand associated with net easterly sediment transport [5]. The floor is sandy gravel which is relatively well sorted (Figure 41.4A) but becomes progressively sandier to the east and eventually becomes dominantly gravelly sand with megaripples and some isolated sand waves. However, much of the floor, particularly in the west, is interpreted as rock with a thin layer of sediment.

Palaeovalley bars: These are positive features with relatively steep sides gener-ally 10 m high, some up to 20 m high. Those in the west, underlain by horizontally bedded Chalk, are the best developed, have a relatively flat surface at a depth of around 55 m, and are elongated parallel to the Palaeovalley axis. These can be up to 10 km long and from 0.8 to ~2 km wide. There are two bars immediately east

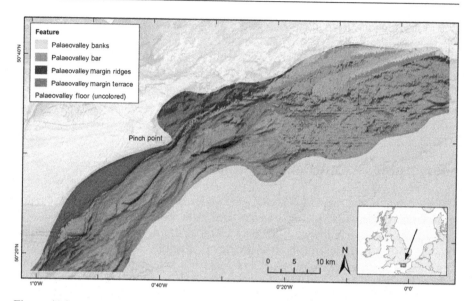

Figure 41.3 Distribution of identified geomorphic features overlain on singlebeam echosounder bathymetric map of Northern Palaeovalley (unmasked area). Shade source 315° NW.
Singlebeam echosounder data ©British Crown & Sea Zone Solutions Ltd. 2008. All rights reserved. Data License 052008.012

of the monocline pinch point where the Palaeovalley is underlain by Tertiary rocks. The surfaces of these Tertiary cored bars are slightly undulating, although they are at similar depths to the Chalk cored bars, at around 55 m, their sides are at a shallower angle. Some bars can be up to 20 m high, probably as a result of stronger and more erosive currents flowing through the pinch point and the relative softness of the Tertiary rocks. The limited seismic profiler evidence suggests that these bars are cored with rock, not formed of sediment; any sediment cover is a thin veneer, predominantly sandy gravel (Figure 41.4B).

Palaeovalley margin terrace: This feature runs along the northern margin of the Palaeovalley for about 25 km west of the monocline pinch point. At its widest point, its extent is about 4.5 km and thins out at its northern and southern limits. From the back of the terrace, there is a gentle decline of its surface from around a depth of 40–50 m. This gentle decline and relatively smooth surface results from the terrace being underlain by horizontally bedded Upper Cretaceous Chalk. The steep margin at the back of the terrace can be up to 15 m high, with its crest at a depth of 25–30 m. The steep slope at the front of the terrace is also up to 15 m high and reaches down to a depth of 65 m in the Palaeovalley floor. The surface of the terrace is marked by several meandering channels up to 300 m wide and incised to depths of up to 4.5 m. These channels flow over the front of the terrace and feed into the main Palaeovalley. Few of these channels extend over the slope at the back of the terrace. The form of this terrace and the rock cored bars within the main channel suggest that

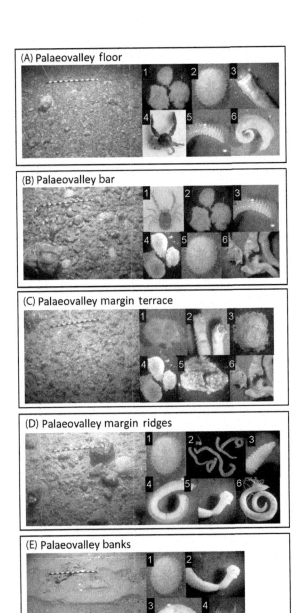

Figure 41.4 Typical sediment deposits and fauna associated with the five major geomorphic features identified within the Northern Palaeovalley. (A) Palaeovalley floor: (1) *Balanus crenatus*, (2) *Echinocyamus pusilus*, (3) *Lagis koreni*, (4) *Pisidia longicornis*, (5) *Laonice bahusiensis*, and (6) *Scalibregma inflatum*. (B) Palaeovalley bar: (1) *Galathea intermedia*, (2) *Balanus crenatus*, (3) *Laonice bahusiensis*, (4) *Epizoanthus couchii*, (5) *Echinocyamus pusilus*, and (6) *Pomatoceros triqueter*. (C) Palaeovalley margin terrace: (1) *Dendrodoa grossularia*, (2) *Sabellaria spinulosa*, (3) *Polycarpa fibrosa*, (4) *Epizoanthus couchii*, (5) Styelidae, and (6) *Pomatoceros triqueter*. (D) Palaeovalley margin ridges: (1) *Echinocyamus pusilus*, (2) *Aonides paucibranchiata*, (3) *Lumbrineris gracilis*, (4) *Notomastus latericeus*, (5) *Glycera lapidum*, and (6) *Caulleriella alata*. (E) Palaeovalley banks: (1) *Echinocyamus pusilus*, (2) *Glycera lapidum*, (3) *Ampelisca spinipes*, and (4) *Poecilochaetus serpens*.

the Palaeovalley has undergone at least two major phases of incision, with a significant hiatus between the two phases. The limited evidence suggests that the terrace is underlain by rock with a thin veneer of gravelly sediment (Figure 41.4C); however, there are linear trains of sand trapped against the back wall of the terrace and at the foot of the terrace front slope.

Palaeovalley margin ridges: These ridges lie within the northern margin of the Palaeovalley immediately east of the pinch point. The area of ridges extends for about 20 km along the margin and is about 6 km wide. The ridges include a series of scarps and dip slopes in Tertiary rocks, although some ridges may be sediment cored. They vary from <10 to 20 m in height at depths of 30–55 m. They appear to be the result of enhanced erosion in front of the pinch point. Significant accumulations of sediment may occur within the troughs between rock-formed scarps. The southern margin of these ridges with the Palaeovalley floor has a well-developed train of sand waves and megaripples up to 2 km wide over a distance of ~12 km.

Palaeovalley banks: The Palaeovalley banks are two large parallel sandbanks aligned WSW–ENE along the northern margin of the Palaeovalley. The outer bank is well developed, with a single virtually straight crest line over 16 km long that is attached at its eastern end to the coastal platform. The crest lies at a depth of 30–35 m through much of its length, gradually descending to ~45 m as it peters out westward into the Palaeovalley floor. Throughout its length, it has a relatively consistent width of around 2–3 km. It has a slightly rounded cross profile at its western end but gradually becomes more asymmetrical towards the east, with its steeper south-facing slope varying in height from 20 to 25 m. It has well-developed megaripples from <0.5 to ~1.0 m high on both stoss and lee slopes indicating that the surface sand is mobile, at least within this depth range (Figure 41.4E).

The inner bank is less well developed and extends for about 28 km. As with the outer bank, its eastern end is attached to the coastal platform, with its crest at a depth of around 30 m. The crest is relatively straight for about 10 km to the west and maintains a depth from 30 to 35 m and a width of around 2 km. Its south-facing slope varies in height from 10 to 20 m, with its north side no more than 12 m high and diminishing to the east as the bank nears the coastal platform. As it descends to the west, the inner bank almost loses its classic bank cross profile in water depths of 45–50 m, and for a length of about 4 km seems to reverse its asymmetry, with a steeper north-facing slope in parts up to 4 m in height. Hereabouts the rock-based slope margin of the coastal platform, which is up to 20 m high and lies behind the inner bank, changes its alignment from ESE to WSW and becomes parallel to the inner bank. From this point and for the rest of its 14 km length, the inner bank becomes attached to the coastal margin with sand banked against it. It maintains a consistent crest depth of around 45 m and also maintains a virtually straight south-facing slope with a height of 15–20 m. This area of the bank has east-facing sand waves up to 4 m high, with abundant megaripples up to 1 m high.

A number of factors appear to have influenced the formation of these banks. The orientation and form of the margin of the coastal platform behind the banks appears to be controlled by folding of the underlying Tertiary and Chalk bedrock into a relatively steep anticline. The margin has eroded along this anticlinal axis which has turned at this point from WSW to ESE causing the Chalk coastal platform to extend out from Beachy Head (Figure 41.2). Net sand transport in this part of the English

Channel is from west to east. This has been a long-term process: sea level reached its current elevation over 5,000 years ago, with sand being transported eastward along the Northern Palaeovalley margin and being trapped against the Chalk cored extension of the coastal platform. It would appear that both banks are a form of banner bank attached to the immobile coastal Chalk platform.

Biological Communities

The biological communities associated with the Northern Palaeovalley were investigated through the analysis of 42 Hamon grab ($0.1\,m^2$) samples in conjunction with associated seabed images. Macrofauna ($>1\,mm$) from the grab samples were identified to the highest possible taxonomic resolution and counted. Environmental data (Table 41.1) for each sampling station were used to investigate the relationship between the physical environment and the associated biological communities using multivariate statistical routines available in the PRIMER–Version 6 software [6].

The Palaeovalley as a whole is characterized by comparatively rich macrofaunal communities comprising infaunal and epifaunal species in varying proportions. There is significant overlap between the infaunal communities associated with the five geomorphic features in this area, as all are dominated by interstitial polychaetes, many of which are tolerant of a wide range of sediment classes. There is, however, some subtle variation in both the diversity and abundance of the macroinfauna (Table 41.2) in response to environmental gradients including sediment composition and current speed (Table 41.3). It should be noted that the observed differences may be confounded by inconsistencies in sampling effort, as the data used here is necessarily opportunistic and not sourced from a targeted survey designed to discriminate communities among different geomorphic features.

The floor of the Palaeovalley accounts for the largest portion of the area and comprises well-sorted deposits ranging from gravelly sands in the northeast to sandy gravels in the southwest. The macrofaunal communities are broadly similar along the extent of the Palaeovalley floor, with a mixture of interstitial fauna including the polychaetes *Scalibregma inflatum* and *Laonice bahusiensis* and the pea urchin *Echinocyamus pusilus*. The high level of sorting and instability of the Palaeovalley floor appears to have prevented the development of a diverse epifaunal community, although the barnacle *Balanus crenatus* is relatively abundant.

The Palaeovalley bars are topped with a thin layer of sediment, which has a granulometric composition and faunal associations similar to those observed in the floor of the Palaeovalley. A slightly more diverse epifaunal community is present on the bar features, which includes the creeping anemone *Epizoanthus couchii* and the tubiculous polychaete *Pomatoceros triqueter*. The barnacle *Balanus crenatus*, which is abundant along the length of the Palaeovalley floor, is also present on the bar features although in lower numbers.

The northern edge of the Palaeovalley is bounded by three very different geomorphic features with the margin terrace and ridges to the west and two large sandbanks to the east. The terrace and ridges are thought to be structured by rock formations, but a thin veneer of mixed sediment overlays these two features and hence the

Table 41.1 Summary of the Environmental Variables Used to Investigate Potential Surrogates for Macrofaunal Composition within the Northern Palaeovalley

Variable	Data Type	Details
Geomorphic feature	Categorical	As described in Figure 41.3; Palaeovalley floor = 1, Palaeovalley banks = 2, Palaeovalley bar = 3, Palaeovalley margin ridges = 4, Palaeovalley margin terrace = 5
Current velocity	Modeled numerical	Maximum amplitude of the depth averaged mean spring tidal current (m/s)
Latitude	Numerical	Recorded position of sample (latitude) *Note*: Longitude was not used in the analysis because of the high correlation between latitude and longitude
Percent retained on sieves: 63 mm, 45 mm, 31.5 mm, 22.4 mm, 16 mm, 11.2 mm, 8 mm, 5.6 mm, 4 mm, 2.8 mm, 2 mm, 1.4 mm, 1 mm, 0.71 mm, 0.5 mm, 0.35 mm, 0.25 mm, 0.18 mm, 0.13 mm, 0.09 mm, 0.06 mm, PAN	Numerical	% of sample (by weight) recorded from the Hamon grab samples
Mean	Numerical	Mean (arithmetic average) particle size (phi) recorded from Hamon grab samples [7]
Sorting	Numerical	Sorting (standard deviation) of the grain sizes (phi) recorded from Hamon grab samples [7]
Skewness	Numerical	Skewness (asymmetry) of the sediment composition (phi) recorded from Hamon grab samples [7]
Kurtosis	Numerical	Kurtosis (peakedness) of the sediment composition (phi) recorded from Hamon grab samples [7]

communities have many species in common with the Palaeovalley floor and bars. However, the sediment overlying the margin terrace contains a greater proportion of coarse particles, making them more stable, which has allowed the development of a more diverse and abundant macrofaunal community (Table 41.2). The fauna associated with the margin terrace is dominated by epilithic ascidians (sea squirts) including *Dendrodoa grossularia*, *Polycarpa fibrosa*, and Styelidae. Other species contributing to this rich epifaunal community include the tubiculous polychaete *Sabellaria spinulosa* and the creeping anemone *Epizoanthus couchii*. All of these species utilize the coarse sediment fraction as an attachment surface.

The sediments overlying the margin ridges are sandier than those overlying the terrace and so support a community that is dominated by infaunal species, very

Table 41.2 The Average Number of Species (*S*), Number of Individuals (*N*), Shannon Weiner's Diversity (*H'*), and Taxonomic Distinctness (Delta*) Recorded in 0.1 m² Hamon Grab Samples Taken from within the Five Geomorphic Features of the Northern Palaeovalley

	S	*N*	*H'* (log e)	Delta*
Palaeovalley floor (*n* = 28)	39.18	104.07	2.99	75.73
Palaeovalley bar (*n* = 4)	39.50	89.25	3.12	79.03
Palaeovalley margin ridges (*n* = 4)	48.50	116.00	3.29	76.89
Palaeovalley banks (*n* = 5)	26.00	52.00	2.79	76.54
Palaeovalley margin terrace (*n* = 1)	117.00	536.00	3.33	77.01

Table 41.3 Summary of BIO-ENV Analyses Carried Out to Investigate the Environmental Surrogates That Best Correlate with the Biological Communities Found in the Northern Palaeovalley

No. of Variables	Correlation	Variables
5	0.293	Current speed, latitude, 63 mm, 1 mm, 0.35 mm
	0.285	Current speed, 63 mm, 2.8 mm, 1 mm, 0.35 mm
	0.285	Current speed, latitude, 1 mm, 0.35 mm, sorting
3	0.282	Latitude, 1 mm, 0.35 mm
	0.276	Current speed, 1 mm, 0.35 mm
	0.264	Current speed, 63 mm, 0.35 mm
1	0.228	Current speed
	0.209	Latitude
	0.205	0.35 mm

similar in composition to the fauna sampled from the Palaeovalley banks. Both of these geomorphic features support fauna typical of sand-dominated environments, including the polychaetes *Glycera lapidum*, *Aonides paucibranchiata*, *Lumbrineris gracilis*, and *Poecilochaetus serpens*. The tubiculous amphipod *Ampelisca spinipes* and the pea urchin *Echinocyamus pusilus* are also present. The main difference between the communities supported by the banks and ridges is in their respective abundances. The banks support fewer species and individuals than the ridges (Table 41.2), which is most likely a reflection of the mobility of these sediments.

Surrogacy

ANOSIM tests were employed to investigate the differences observed in the macrofauna collected at each of the geomorphic features. No significant differences were detected between the macrofaunal assemblages associated with individual geomorphic features or between combinations of features, indicating that the features themselves do not make good surrogates for macrofaunal community composition, although this result may, in part, be a reflection of the low number of samples taken in some features. Further targeted sampling of the geomorphic features identified here would be required to test their surrogacy potential more conclusively.

The Bio-Env routine within PRIMER was used to investigate other potential environmental surrogates using data summarized in Table 41.1. The correlations between the biological communities and the environments in which they are found are weak, with a maximum correlation of 0.293 identified with a combination of five environmental variables (Table 41.3). Current speed features heavily amongst the environmental variables that best explain the patterns observed in the faunal communities as does the proportion of medium sand (0.35 mm). This conforms well with the observed gradient between mobile sandy sediments, which are too unstable to support a significant epifaunal community, and the more stable and coarser sediments upon which epifauna have become well established. However, the correlation is still weak, indicating a very low potential to predict the distribution of distinct sediment-dwelling macrofaunal communities based on any combination of these environmental variables.

It is notable that all geomorphic features within the Palaeovalley are dominated by sediment habitats, and there is clearly insufficient environmental gradient—e.g., in substrate type, depth, or hydrodynamic regime—to cause notable differences among their benthic communities. Such differences will certainly emerge when the Palaeovalley is considered in the wider spatial context of the central and eastern English Channel that includes cobble, boulder, and rock substrates.

Acknowledgments

This chapter is based on research funded by the UK Marine Aggregate Levy Sustainability Fund (MALSF) and commissioned by the Marine Environment Protection Fund (MEPF). Ceri James and Angela Morando publish with the approval of the Director, British Geological Survey.

References

[1] B.S. Halpern, S. Walbridge, K.A. Selkoe, C.V. Kappel, F. Micheli, C. D'Agrosa, et al., A global map of human impact on marine ecosystems, Science 319 (2008) 948–952.

[2] J.W.C. James, R.A. Coggan, V.J. Blyth-Skyrme, A. Morando, S.N.R. Birchenough, E. Bee, et al., The Eastern English Channel Marine Habitat Map, Cefas Science Series Technical Report 139, Cefas, Lowestoft, 2007.

[3] J.W.C. James, B. Pearce, R.A. Coggan, S. Arnott, R.W. Clark, J.F.M. Plim, et al., The South Coast Regional Environmental Characterisation, British Geological Survey Open Report 09/51, Nottingham, 2010.

[4] R.J.O. Hamblin, A. Crosby, P.S. Balson, S.M. Jones, R.A. Chadwick, I.E. Penn, et al., United Kingdom Offshore Regional Geology Report: The Geology of the English Channel, HMSO for the British Geological Survey, London, 1992.

[5] N.T.L. Grochowski, M.B. Collins, S.R. Boxall, J.C. Salomon, Sediment transport predictions for the English Channel, using numerical models, J. Geol. Soc. 150 (1993) 683–695.

[6] K.R. Clarke, R.N. Gorley, PRIMER v6: User Manual/Tutorial, Primer-E-Ltd., Plymouth Marine Laboratory, Plymouth, UK, 2006.

[7] R.L. Folk, W.C. Ward, Brazos river bar: a study of significance of grain size parameters. J. Sediment. Petrol. 27 (1957) 3–26.

42 Benthos Supported by the Tunnel-Valleys of the Southern North Sea

Bryony Pearce[1], David R. Tappin[2], Dayton Dove[3], Jennifer Pinnion[1]

[1]Marine Ecological Surveys Limited, Bath, UK, [2]British Geological Survey, Keyworth, Nottingham, UK, [3]British Geological Survey, Edinburgh, UK

Abstract

The tunnel-valleys of the southern North Sea are arcuate and linear seabed depressions. The origin of these features has been the subject of much discussion, but they are generally considered to have been formed by subglacial erosion and sediment backfill beneath the outer margins of a receding ice sheet. We present here a study of two tunnel-valleys, the Silver Pit and the Sole Pit. Extensive areas of *Sabellaria spinulosa* reefs have been identified on the western flanks of the Silver Pit that extend down to the valley floor, representing a resource of significant conservation interest. The eastern flanks were found to support a diverse faunal assemblage with widespread hydroid and bryozoans turfs and abundant ascidians. The seabed of the Sole Pit is characterized by much sandier deposits than that of the Silver Pit, and it supports an abundant bivalve community, dominated by *Abra alba*. This bivalve assemblage also contains *Coarcuta obliquata*, which has only been recorded once before in the UK.

Key Words: Tunnel-valley, benthos, habitat, seabed mapping, *Sabellaria spinulosa*, North Sea, multibeam

Introduction

Tunnel-valleys are prominent features of the southern North Sea, where they form actuate and linear seabed depressions with a limited sediment infill (Figure 42.1). In the outer Humber area, the valleys attain water depths up to 100 m and radiate outward from the mapped palaeo-ice margin of the Devensian (Weischelian) glaciations [1]. The location and orientation of the valleys indicates that their origin is associated with the ice sheet, although the precise process of formation is uncertain [2,3]. It is likely that the valleys were formed by either subglacial fluvial activity or ice-damn breakouts (Jökulhlaups).

Seafloor Geomorphology as Benthic Habitat. DOI: 10.1016/B978-0-12-385140-6.00042-6

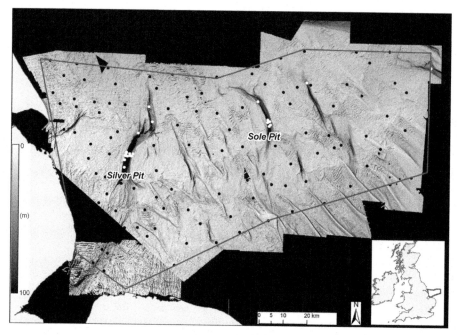

Figure 42.1 Single beam echo sounder (SBES) bathymetric map of the Humber Regional Environmental Characterization (REC) study area gridded at 100 m resolution. Stations sampled for biological analysis are also shown (those highlighted in white were taken from within the tunnel-valley features). Shade source 315°—NW.
SBES data © British Crown & Sea Zone Solutions Ltd., 2008. All rights reserved. Data License 052008.012.

Recent research in the southern North Sea has focused on buried equivalents of the Weischelian features that are located farther north and are of Elsterian age [3]. These studies indicate that such valleys are the result of steady-state subglacial fluvial erosion and subsequent sediment backfill beneath the outer tens of kilometers of a northward receding ice-sheet margin [2,3]. The absence of sediment infill in the Southern North Sea tunnel-valleys is not well understood, but may be due to either erosion during postglacial sea-level rise or an original lack of infill. Modern tidal forces in the region of the tunnel-valleys are locally strong, with mean spring tidal current velocities between 0.5 and 1.5 m/s (ABP Marine Environmental Research et al., 2008). These current speeds are certainly of sufficient magnitude to prohibit future sediment deposition [4,5].

A detailed study of two of these tunnel-valleys, the Silver Pit and the Sole Pit, was carried out as part of the Humber Regional Environmental Characterisation (REC) project funded through the Marine Aggregate Levy Sustainability Fund [6]. The study area covered 11,000 km^2 and was located offshore from the Humber Estuary on the east coast of England. Over 3,000 line kilometers of geophysical data, including multibeam echo soundings, sidescan sonar, and boomer seismic, were acquired across the study area. An irregular survey grid reflecting the orientation of tidal currents and major seabed features was constructed, with a series of corridors running

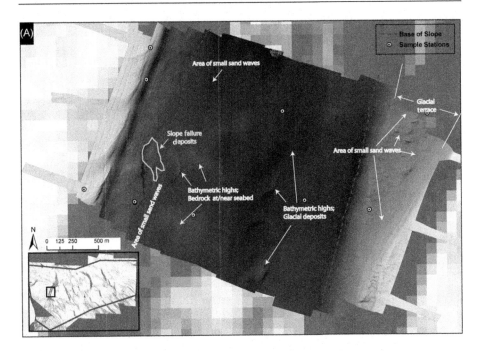

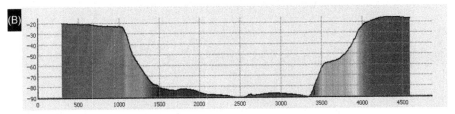

Figure 42.2 (A) Bathymetric map of the Silver Pit focused study area. Multibeam echo sounder (MBES) data (1 m resolution) overlaying slope map of gridded single beam echo sounder (SBES) data (100 m resolution). Stations sampled for biological analysis of the focused study site are also shown as are the boundaries between the valley floor and slope edges (delineated by red dashed lines) and bedform annotations. (B) Depth profile of the Silver Pit. (A) © British Crown & Sea Zone Solutions Ltd., 2008. All rights reserved. Data License 052008.012. (For interpretation of the references to color in this figure legend, the reader is referred to the web version of this book.)

approximately NW–SE and two tie lines running EW. Geophysical data were acquired on a more closely spaced and regular grid (100 m line spacing), across sections of the two tunnel valleys giving rise to near complete coverage multibeam and sidescan sonar data. Multibeam echo soundings were collected using a Kongsberg Simrad EM 710 mounted on the keel of the survey vessel. This instrument is a $1° \times 1°$ system operating at 95 kHz frequency, with 256 beams and a swath footprint of approximately three times the water depth. The resulting bathymetric data collected across the two detailed study areas are presented in Figures 42.2 and 42.3.

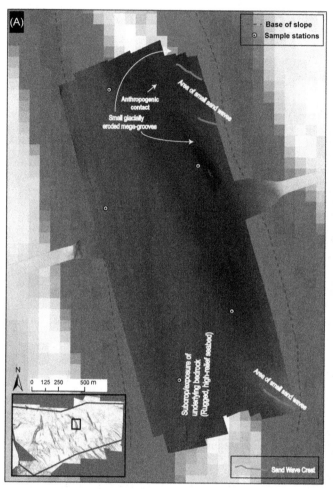

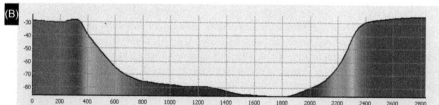

Figure 42.3 (A) Bathymetric map of the Sole Pit focused study area. Multibeam echo sounder (MBES) data (1 m resolution) overlaying slope map of gridded single beam echo sounder (SBES) data (100 m resolution). Stations sampled for biological analysis of the focused study site are also shown as are the boundaries between the valley floor and slope edges (delineated by red dashed lines) and bedform annotations. (B) Depth profile of the Sole Pit. (A) © British Crown & Sea Zone Solutions Ltd., 2008. All rights reserved. Data License 052008.012. (For interpretation of the references to color in this figure legend, the reader is referred to the web version of this book.)

Following the comprehensive interpretation of the geophysical data, a number of biological sampling stations were selected to investigate the ecological assemblages associated with the seabed features identified. The benthic macrofauna were compared with fauna identified across the wider Humber REC study area as a means of assessing the ecological significance and surrogacy potential of the tunnel-valleys.

Geomorphic Features and Habitats

Silver Pit—It is the larger of the two surveyed valleys. It is approximately 50 km long and up to 5 km wide, with maximum water depths reaching almost 100 m (Figure 42.2). The detailed survey area is located in the center of the valley, with an approximate dimension of 3.5 km × 2 km. It incorporates the shallow, relatively flat seabed, through which the valleys were cut, the valley flanks, and the valley floor. Within the detailed survey area, the valley attains a maximum depth of approximately 90 m. Typically the valley floor is gently undulating.

Within the valley, three Quaternary formations are exposed at the seafloor, where they subcrop the thin Holocene sediment cover. These formations (and their lithologies) are, in ascending order, Sand Hole (laminated clay), Egmond Ground (muddy sand), and Bolders Bank (boulder clay). On the valley floor, a thin Holocene sediment cover overlies Cretaceous Chalk.

The western flank of the valley is relatively uniform, with a slope failure in the southwest corner. The displaced sediment rests on the valley floor (Figure 42.2). The eastern flank is more complex, and the otherwise continuous slope is interrupted by a number of terraces (Figure 42.2). These terraces are most likely the result of multiple phases of subglacial erosion. The seabed on top of this terrace is irregular, with localized bathymetric highs formed of coarse material which either represents exposed Quaternary sediment or glacial lag deposits. Where sands predominate on the terrace, megaripples are also common.

Megaripples are also common in the western valley floor where finer sands occur. There are three small rounded bathymetric highs of uncertain origin. They may be glacial relicts or simply localized areas where the Chalk is more resistant to erosion. Along the middle of the valley, there are a series of linear bathymetric highs that are generally parallel to the orientation of the valley. These are probably glacial relicts that formed as moraines and are also associated with exposures of coarser-grained sediments.

Sole Pit—It is approximately 30 km long and up to 3 km wide. The detailed survey area has the approximate dimensions of 3.5 km × 1.5 km, incorporating the shallow, relatively flat surrounding seabed through which the valley was cut, the slopes (flanks) of the valley, and the valley floor, which has a maximum depth of approximately 90 m (Figure 42.3).

The valley incises four Quaternary formations that underlie thin or absent Holocene sediments on the flanks. These formations (and their lithologies) are, in ascending order, Winterton Shoal (interbedded sand and clay), Yarmouth Roads (sand interbedded with silty clay), Bolders Bank (boulder clay), and Botney Cut (laminated clay). Holocene sediments, where present on the valley floor, are thin and are underlain by the Botney Cut formation in the northeast of the valley, and by Triassic and Jurassic Lias Group formations in the west and south. The Lias Group comprises both sandstone and limestone.

Both valley flanks have continuous undisturbed slopes, although in the east there are small areas of megaripples (Figure 42.3). The morphology of the valley floor in the west and south is characterized by the curvilinear strike of the underlying rock. Individual lineations appear to be associated with occurrences of smaller hummocky features that typically constitute coarser sediments than the surrounding area. It is unclear whether these hummocky features are outcrops of rock or glacial deposits that may have nucleated along the bedding planes of the underlying rock.

In the extreme southeast, there is a large sand wave feature upon which there are smaller parasitic megaripples. The profile of the sand wave, and the smaller sand waves in the north, indicate predominantly north-trending bottom currents. In the east of the valley floor, there are two longitudinal deeps that indicate localized glacial erosion.

Biological Communities

The biological communities of the outer Humber study area were characterized by analyzing 135 mini Hamon grab ($0.1 \, m^2$) samples. These include 10 samples from the focused study area in the Silver Pit and a further six from the focused study area in the Sole Pit. Macrofauna ($>1 \, mm$) were extracted from the grab samples and identified to the highest possible taxonomic resolution, usually to species level, and counted. The quantitative faunal data recorded from these samples were supplemented with observations from seabed video and digital stills acquired at the same stations. Subsamples were taken from each of the grab samples and analyzed to give full particle size distribution based on the Wentworth scale. The nature of the substrate was also recorded from seabed images and video footage.

Benthic species abundance data recorded from the detailed survey areas, as well as from the wider Humber REC study area, were analyzed using the PRIMER (v6) multivariate statistical package [7]. The similarity percentages (SIMPER) routine was used to characterize the biological communities associated with the Silver Pit and the Sole Pit. This routine breaks down the average Bray–Curtis similarity (and dissimilarity) between pairs of samples into percentage contributions made by each species. In this way it is possible to identify the species that contribute most to the similarity of samples belonging to any given community as well as those which contribute to the separation of two communities. In this case, samples were grouped according to the geomorphic feature from which they were sampled in order to investigate differences in species compositions associated with these features.

A multivariate analysis of similarity (ANOSIM) test was used to investigate differences between the macrofauna associated with the tunnel-valleys and macrofauna recorded across the wider Humber study area. The ANOSIM test is analogous to a univariate analysis of variance (ANOVA) test based on permutations, or randomization, of the data, making minimal assumptions about the distribution of that data. This allows for a robust test of the hypothesis that no difference in species composition exists between samples taken from within the tunnel-valleys and samples taken across the wider study area.

Finally, the BIO-ENV routine was used to investigate the surrogacy potential of these geomorphic features using environmental data summarized in Table 42.1. The BIO-ENV routine searches across all possible combinations of environmental variables to identify strong correlations with the multivariate patterns identified in the biological data.

Silver Pit—It can be split into three broad morphological categories: the western slopes, the tunnel floor, and the eastern slopes. Beyond the valley itself, the margin sediments are typically sandy gravels, although there is a larger mud component on the eastern margins. The sandy gravel deposits extend down the slopes of the valley, becoming increasingly sandy toward the base of the slope. The floor of the tunnel-valley is predominantly gravelly sand, although there are occurrences of slightly gravelly sand in the west of the valley floor as well as muddy sandy gravel associated with the linear glacial relicts in the middle of the valley floor (Figure 42.4).

Western slopes—The biological communities associated with the western slopes of the Silver Pit are numerically dominated by the Ross worm *Sabellaria spinulosa*. An irregular textured signature was visible in much of the sidescan sonar data collected across this area (Figure 42.5i), which has been reported to be indicative of the biogenic aggregations built by this polychaete [8–10]. The presence of this textured signature across the slope suggests that this area may support an extensive reef complex which is likely to qualify for protection under European legislation. The majority of seabed images collected from the western slopes of the Silver Pit also show significant *S. spinulosa* aggregations (Figure 42.5i) typically in association with the pink shrimp *Pandalus montagui*. This finding confirms earlier suggestions that the *S. spinulosa* reefs in this area support a commercially important pink shrimp population [11].

The edible mussel *Mytilus edulis* was also found to be associated with the *S. spinulosa* reefs of the western slopes, present at similar densities in some locations. Succession between Sabellariid reef and mussel beds has been widely reported in the literature [12–15]. Because there is a strong overlap in the environmental requirements of these two species, it seems likely that there could be an ongoing cyclical succession in response to minor fluctuations in environmental conditions.

The erect bryozoan *Flustra foliacea* was also found to be characteristic of this *S. spinulosa* reef habitat, as were a range of interstitial polychaetes including *Lumbrineris gracilis*, *Mediomastus fragilis*, and *Pholoe inornata*.

Valley floor—It was also found to support *S. spinulosa* aggregations although the densities recorded from the grab samples were significantly lower than those recorded on the western slopes. Examination of the seabed video and still images reveals a less topographically distinct carpet-like reef (Figure 42.5ii), indicating that tube growth is being limited, possibly by the fast water currents running through the valley. The carpet reef identified across the floor of the tunnel-valley is

Table 42.1 Summary of the Environmental Variables Used to Investigate Potential Surrogates for Benthic Macrofauna in the Outer Humber Region

Variable	Data Type	Details
Geomorphic feature	Categorical	As described in Figures 1.3; Sole Pit (Floor) = 5, Silver Pit (eastern slope) = 4, Silver Pit (floor) = 3, Silver Pit (western slope) = 2, other tunnel-valleys = 1, wider Humber area = 0
% Mud	Numerical	% of sample (by weight) recorded from the Hamon grab samples.
% Sand	Numerical	% of sample (by weight) recorded from the Hamon grab samples. *Note*: % Gravel was removed from this analysis as there was a very strong correlation with % sand. % Sand should therefore be treated as a proxy for % gravel.
Sorting	Numerical	Sorting (standard deviation) of the grain sizes (phi) recorded from Hamon grab samples [29]
Skewness	Numerical	Skewness (asymmetry) of the sediment composition (phi) recorded from Hamon grab samples [29]
Kurtosis	Numerical	Kutosis (peakedness) of the sediment composition (phi) recorded from Hamon grab samples [29]
Wave strength	Modeled numerical	Annual mean wave power (kW/m) per horizontal meter of wave crest calculated using the energy period calculation (TE) © Crown. All rights reserved 2008.
Tidal current velocity	Modeled numerical	Maximum amplitude of the depth averaged mean spring tidal current (m/s)
Near bed temperature	Modeled numerical	Month averaged near bed temperature, resolution 1/9° calculated using the Atlantic Margin Model (www.myocean.eu.org)
Seafloor rugosity	Modeled numerical	Rugosity (ratio of surface area to planar area) of the seafloor calculated from single beam echo sounder bathymetric data on a 25–250 m grid using the Benthic Terrain Model (BTM) in ArcGIS V9.3. © British Crown & Sea Zone Solutions Ltd., 2008. All rights reserved. Data License 052008.012.
Slope	Modeled numerical	Slope (degrees) calculated using Spatial Analyst in ArcGIS V9.3 from single beam echo sounder bathymetric data on a 25–250 m grid © British Crown & Sea Zone Solutions Ltd., 2008. All rights reserved. Data License 052008.012.

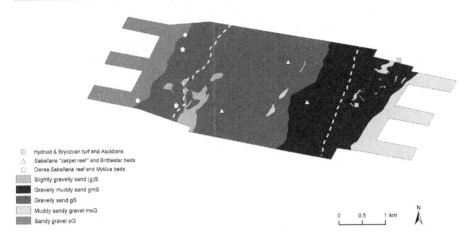

Hydroid & Bryozoan turf and Ascidians
Sabellaria "carpet reef" and Brittlestar beds
Dense *Sabellaria* reef and *Mytilus* beds
Slightly gravelly sand (g)S
Gravelly muddy sand gmS
Gravelly sand gS
Muddy sandy gravel msG
Sandy gravel sG

0   0.5   1 km   N

Figure 42.4 Sediment map of the Silver Pit focused study site derived from interpretations of geophysical data and sediment samples. Also shown are the boundaries between the valley floor and slope edges (delineated by gray dashed lines) and the locations at which the three biological communities were sampled.

also inhabited by brittlestars such as *Ophiothrix fragilis* and *Ophiura albida*, which were conspicuous in the seabed imagery (Figure 42.5ii). The nature of the association between *S. spinulosa* and the brittlestars is not clear, but both species are filter feeders and it has been postulated that dense aggregations of brittlestars could inhibit growth and recruitment of *S. spinulosa* reef through their active removal of larvae from the water column and competition for food [16].

Eastern slopes—*S. spinulosa* reefs are notably absent from the eastern slopes of the inner Silver Pit, with sporadic records of only a few individuals. The absence of significant aggregations indicates that the environmental conditions of the eastern slopes are less suited to this species than in the rest of the valley, and it is likely that this is in some part influenced by the sediment composition. Sabellariid polychaetes are dependent on a good supply of sand, which they use to construct their tubes [13,17,18]. The eastern flanks are muddier than the rest of the tunnel-valley and this alone may make the area unsuitable for the development of *S. spinulosa* reefs. Significant amounts of fine sediments may also foul the ciliary feeding apparatus of the worm, as has been reported for the congener *Sabellaria kaiparensis* [19].

The eastern slopes support a comparatively diverse and numerically abundant faunal assemblage (Table 42.2). There is an abundance of epifauna here including hydroid turfs and sea squirts (Figure 42.5iii), and an abundant infaunal community including the bamboo worm *Euclymene* and the amphipod *Urothoe elegans*. The overall diversity of the eastern slopes of the Silver Pit was found to be higher than both the western slopes and the valley floor despite there being only two samples taken from this area.

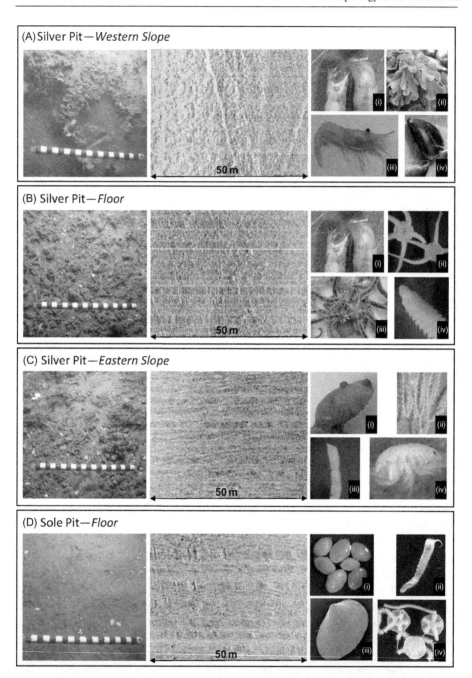

Figure 42.5 Typical sediment deposits, sidescan sonar signatures, and fauna associated with Silver Pit and Sole Pit. (i) Silver Pit—western slope: (A) *S. spinulosa*; (B) *Flustra foliacea*; (C) *Pandalus montagui*; (D) *Mytilus edulis*. (ii) Silver Pit—floor: (A) *S. spinulosa*; (B) *Ophiura albida*; (C) *Ophiothrix fragilis*; (D) *Lumbrineris gracilis*. (iii) Silver Pit—eastern slope: (A) ASCIDIACEA; (B) *Sertularia cupressina*; (C) *Euclymene*; (D) *Urothoe elegans*. (iv) Sole Pit—floor: (A) *A. alba*; (B) *A. gracilis*; (C) *C. obliquata*; (D) *A. filiformis*.

Table 42.2 The Average Number of Species (S), Number of Individuals (N), Shannon Weiner's Diversity (H'), and Taxonomic Distinctness (Delta*) Recorded in 0.1 m^2 Hamon Grab Samples Taken Within and Outside the Tunnel-Valley Features

	S	N	H' (Log$_e$)	Delta*
Silver Pit—western slope ($n = 5$)	171	517	2.55	90.55
Silver Pit—floor ($n = 3$)	118	478	2.97	86.72
Silver Pit—eastern slope ($n = 2$)	149	405	3.88	93.62
Sole Pit—floor ($n = 6$)	164	1125	1.74	93.80
All tunnel-valley stations ($n = 25$)	425	661	3.34	91.81
None tunnel-valley stations ($n = 110$)	636	317	3.83	90.61

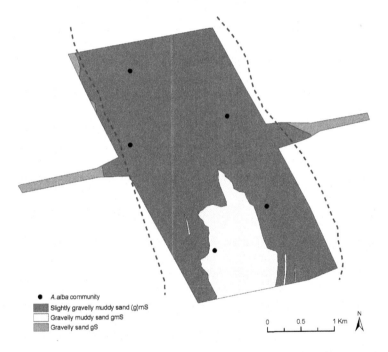

- A.alba community
- Slightly gravelly muddy sand (g)mS
- Gravelly muddy sand gmS
- Gravelly sand gS

0   0.5   1 Km   N

Figure 42.6 Sediment map of the Sole Pit focused study site derived from interpretations of geophysical data and sediment samples. Also shown are the boundaries between the valley floor and slope edges (delineated by gray dashed lines) and the locations at which the *A. alba* community was sampled.

Sole Pit—the sediments of the Sole Pit are sandier than those of the Silver Pit, with the margins and upper slopes comprised of gravelly sands and the lower slopes and valley floor comprised of slightly gravelly muddy sands (Figure 42.6). The biological communities associated with this tunnel-valley feature are also markedly different from those of the inner Silver Pit (Figure 42.5iv), although it should be noted that only six samples were taken from this feature and all were located in the valley floor. The most abundant species recorded from the Sole Pit was the bivalve mollusk,

Abra alba, with over 4,000 individuals recorded from six grab samples. This bivalve is generally found in association with muddy fine sand and is reported to be most abundant at depths around 20 m [20]. Whilst it is unusual to record such high abundances of *A. alba* at these depths (70–90 m), it is likely that this is a reflection of the sampling effort rather than an expansion in the depth range of this species.

Other abundant fauna identified in association with the floor of the Sole Pit included the brittlestar *Amphiura filiformis* and its coinhabitant the bivalve *Mysella bidentata* [21]. *A. filiformis* is a burrowing brittlestar that favors muddy environments, where its body, and all but the tips of its arms, remain below the surface [22]. A range of interstitial polychaetes including *Anobothrus gracilis* were also present, but perhaps the most conspicuous characteristic of this habitat was the absence of any epifaunal species.

Surrogacy

In order to investigate the relationship between the tunnel-valleys and the benthos across the outer Humber study area, the biological composition of samples collected from within valley features were compared with the biological composition recorded across the wider area (Figure 42.1). An ANOSIM test found no significant difference between the faunal communities within the valleys and those found in the wider environs ($R = 0.034$, 19.1% significance). However, this is likely to be influenced in part by the wide variety of habitats and faunal communities found across the outer Humber region [6].

A number of species were found to be unique to the tunnel-valleys (Table 42.3). The most notable is the small bivalve mollusk, *Coracuta obliquata*, which has only been recorded once in UK waters in the last 100 years [23]. This bivalve appears to have very specific environmental requirements, having been recorded only in samples taken from the floor of the Sole Pit. Holmes *et al.* [23] observed *C. obliquata* in sediments ranging from sand to sandy gravel, which corresponds well with our observations in the Sole Pit where this mollusk was observed in gravelly sand. A BIO-ENV analysis carried out on *C. obliquata* abundance, and the full range of environmental variables listed in Table 42.1, yielded a maximum correlation of 0.341 with geomorphic feature (Table 42.4A). This reflects the localized distribution of this species within the Sole Pit although this alone could not be used as a surrogate for the distribution of *C. obliquata*. The distribution of this species is likely to be determined by a combination of environmental factors, including some not recorded as part of this study, and also by biological interactions. Other species in the same family of mollusks (Montacutidae) are known to have commensal relationships with other invertebrates [20,21,24]. It is possible that the same is true of *C. obliquata* although no specific associations have yet been identified.

A BIO-ENV analysis carried out on the *S. spinulosa* abundance, and the full range of environmental variables listed in Table 42.1, revealed that various aspects of the hydrodynamic regime and sediment composition correlated with *S. spinulosa* abundance in this area (Table 42.4B). However, given that the highest correlation

Table 42.3 Summary of Species Found to Be Unique to the Inner Silver Pit ($n = 10$), the Sole Pit ($n = 5$), and the Wider Study Region in the Outer Humber ($n = 110$)

Outer Humber Region (259)		Silver Pit (33)		Sole Pit (12)	
Species	n	Species	n	Species	n
Bathyporeia elegans	1.59	*Ophiactis balli*	0.47	*Coracuta obliquata*	3.33
Nymphon brevirostre	1.52	*Socarnes erythrophthalmus*	0.13	*Malmgreniella andreapolis*	1.33
Alvania semistriata	0.34	*Vitreolina philippi*	0.13	*Hyala vitrea*	0.67
Fabulina fabula	0.27			*Retusa umbilicata*	0.50
Monocorophium acherusicum	0.25			*Echiurus echiurus*	0.17
Pomatoceros triqueter	0.21			*Myriochele danielsseni*	0.17
Aonides oxycephala	0.19			*Ebalia nux*	0.17
Polydora cornuta	0.18			*Palio dubia*	0.17
Alcyonidium mytili	0.18			*Devonia perrieri*	0.17
Glycera oxycephala	0.17			*Saxicavella Jeffreysi*	0.17
Urothoe poseidonis	0.17				
Branchiostoma lanceolatum	0.16				
Pisione remota	0.15				
Ischyrocerus anguipes	0.15				
Nicolea venustula	0.13				
Perioculodes longimanus	0.13				
Microporella ciliata	0.11				
Prionospio banyulensis	0.10				
Pseudopotamilla reniformis	0.10				
Elminius modestus	0.10				
Clausinella fasciata	0.10				

Only species with an average abundance greater than 0.1 per grab are shown. The total number of unique taxa is given in brackets.

identified was 0.341, these factors alone could not be used as a surrogate of *S. spinulosa* distribution.

The tunnel-valleys of the southern North Sea have not been found to be effective surrogates for any benthic species or assemblages. However, these features support a variety of marine life, including species and habitats that may be of conservation interest. For example, the *S. spinulosa* reef complex found in association with the Silver Pit has previously been noted in environmental impact assessments of adjacent aggregate extraction areas [8,25,26] indicating that it is a relatively persistent feature of the seabed. An area of *S. spinulosa* reef to the south of the Silver Pit has been included as a conservation feature of the Inner Dowsing, Race Bank, and North Ridge possible Special Area of Conservation (pSAC), which is currently being

Table 42.4 Summary of BIO-ENV Analyses Carried to Investigate the Environmental Surrogates Which Best Correlate with the Distribution Patterns of (A) *C. obliquata* and (B) *S. spinulosa* in the Outer Humber Region

(A) C. obliquata

Number of Variables	Correlation	Variables
1	0.341	Geomorphic feature
2	0.307	Geomorphic feature, seafloor rugosity
1	0.230	Wave strength
2	0.230	Geomorphic feature, wave strength
1	0.230	Near Bed Temperature

(B) S. spinulosa

Number of Variables	Correlation	Variables
2	0.379	Sorting, tidal current velocity
3	0.363	Sorting, wave strength, tidal current velocity
2	0.350	Sorting, wave strength
4	0.339	% Sand, sorting, wave strength, tidal current velocity
4	0.334	Sorting, wave strength, near bed temperature, tidal current velocity

assessed for inclusion into the UK's network of Marine Protected Areas [27]. The areas of reef identified in this focused study of Silver Pit fall outside the proposed pSAC boundary, but it is possible that this may be incorporated into the area protected on the basis of the evidence presented here.

The physical attributes of the tunnel-valleys makes them unsuitable for most anthropogenic activities which occur in the marine environment and hence it seems likely that these features will become an important component of the UK's network of Marine Protected Areas (part of the Natura 2000 network) in years to come.

References

[1] S.J. Carr, R. Holmes, J.J.M.V.D. Meer, J. Rose, The Last Glacial Maximum in the North Sea Basin: micromorphological evidence of extensive glaciation, J. Quaternary Sci. 21 (2006) 131–153.

[2] M. Huuse, H. Lykke-Andersen, Overdeepened Quaternary valleys in the eastern Danish North Sea: morphology and origin, Quaternary Sci. Rev. 19 (2000) 1233–1253.

[3] D. Praeg, Seismic imaging of mid-Pleistocene tunnel-valleys in the North Sea Basin— high resolution from low frequencies, J. Appl. Geophys. 53 (4) (2003) 273–298.

[4] R.D. Pingree, D.K. Griffiths, Sand transport paths around the British Isles resulting from M2 and M4 tidal interactions, J. Mar. Biol. Assoc. 59 (1979) 497–513.

[5] R. Proctor, J.T. Holt, P.S. Balson, Sediment deposition in offshore deeps of the western North Sea: questions for models, Estuar. Coast. Shelf Sci. 53 (2001) 553–567.

[6] D.R. Tappin, B. Pearce, S. Fitch, D. Dove, B. Geary, J. Hill, et al., The Humber Regional Environmental Characterisation. British Geological Survey Open Report OR/10/54, 2011.

[7] K.R. Clarke, R.N. Gorley, PRIMER v6: User Manual/Tutorial, PRIMER-E, Plymouth, Devon, UK, 2006.

[8] Marine Ecological Surveys Ltd, Area 480: Benthic Biological Survey. Analysis of Survey Data for 2008 and Comparison with 2002 Survey Data. Report prepared for Hanson Aggregates Marine Ltd, Southampton, UK, 2009, 166 pp.

[9] Marine Ecological Surveys Ltd, Hastings Shingle Bank Area 366–370: Benthic Monitoring Report. Report for the Resource Management Association (RMA), Southampton, UK, 2006, 86 pp.

[10] Marine Ecological Surveys Ltd, Thanet Offshore Windfarm, Benthic and Intertidal Resource Survey. Report prepared for Haskoning UK Ltd, Peterborough, UK, 2005, 86 pp.

[11] P.J. Warren, Feeding and migration patterns of the pink shrimp (Pandalus montagui) in the Wash. Laboratory Leaflet No 8, 46. Ministry of Agriculture, Fisheries and Food (MAFF), 1973.

[12] Cunningham, P.N., Hawkins, S.J., Jones, H.D., Burrows, M.T. 1994. The biogeography and ecology of Sabellaria alveolata, Nature Conservancy Council CSD Report 535.

[13] S.M.L. Pohler, The Sabellariid worm colonies of Suva Lagoon, Fiji, South Pac. J. Nat. Hist. 22 (2004) 36–42.

[14] K. Reise, A. Shubert, Macrobenthic turnover in the subtidal Wadden Sea: the Norderaue revisited after 60 years, Helgolander Meereshunter 41 (1987) 69–82.

[15] W. Riessen, K. Reise, Macrobenthos of the subtidal Wadden Sea: revisited after 55 years, Helgolander Meereshunter 35 (1982) 409–423.

[16] C.L. George, R.M. Warwick, Annual macrofauna production in a hard-bottom reef community, J. Mar. Biol. Assoc. UK 65 (1985) 713–735.

[17] C. Chen, C.F. Dai, Subtidal Sabellariid reefs in Hualien, eastern Taiwan, Coral Reefs 28 (2009) 275.

[18] D. Kirtley, A review and taxonomic revision of the family Sabellaridae (Annelida: Polychaeta), Sabecon Press, Vero Beach, FL, 1994.

[19] H.W. Wells, Sabellaria reef masses in Delaware Bay, Chesapeake Sci. 11 (1970) 258–260.

[20] N. Tebble, British bivalve seashells. A handbook for identification, British Museum (Natural History), London, 1966. 212 pp.

[21] A.B. Josefson, Resource limitation in marine soft sediments—differential effects of food and space in the association between the brittle-star Amphiura filiformis and the bivalve Mysella bidentata?, Hydrobiologia 375/376 (1998) 297–305.

[22] D. Nichols, J. Cooke, D. Whitely, The Oxford Book of Invertebrates, Oxford University Press, Oxford, 1971. 218 pp.

[23] A.M. Holmes, J. Gallichani, H. Wood, Coracuta obliquata n. gen (Chaster, 1897) (Bivalvia: Monacutidae)—first British record for 100 years, J. Conchol. 39 (2) (2006) 151–157.

[24] B. Morton, P.V. Scott, The Hong Kong Galeommatacea (Mollusca: Bivalvia) and their hosts, with descriptions of new species, Asian Mar. Biol. 6 (1989) 129–160.

[25] Emu Environmental Ltd, Area 480 Sabellaria spinulosa survey. Report prepared for Hanson Aggregates Marine Ltd, Southampton, UK, 2005.

[26] Marine Ecological Surveys Ltd, Marine Aggregate Extraction Application Area 106 East (480): Environmental Statement. Report prepared by Marine Ecological Surveys Ltd (MESL) for Hanson Aggregates Marine Ltd, Southampton, UK, 2003, 199 pp.

[27] Joint Nature Conservation Committee and Natural England, Inner Dowsing, Race Bank and North Ridge Selection Assessment Document Version 5.0. Report prepared on behalf of the Department of Environment, Food and Rural Affairs (Defra), 2010, 34 pp.

[28] ABP Marine Environmental Research, the Met Office, Garrad Hassan and Proudman Oceanographic Laboratory, Atlas of UK Marine Renewable Resources; Technical report R/3387/5 prepared for the Department of Trade and Industry, UK, 2004.

[29] R.L. Folk, W.C. Ward, Brazos River bar: a study in the significance of grain size parameters, Journal of Sedimentology and Petrology 27 (1957) 3–26.

43 Benthic Habitats and Benthic Communities in Southeastern Baltic Sea, Russian Sector

Elena Ezhova[1], Dmitry Dorokhov[1], Vadim Sivkov[1], Vladimir Zhamoida[2], Daria Ryabchuk[2], Olga Kocheshkova[1]

[1]P.P. Shirshov Institute of Oceanology, Atlantic Branch (ABIORAS), Kaliningrad, Russia, [2]A.P. Karpinsky Russian Geological Research Institute (VSEGEI), St. Petersburg, Russia

Abstract

This study concerns benthic habitats of the southeastern Baltic Sea, including part of the Gdansk Basin and adjacent shallow water areas and lagoons. The study area covers 11,633 km^2 and varies in water depth from 0 to 110 m. The study area represents a diversity of geomorphic features, from the relatively deep plain of the Gdansk Deep (sea depth 80–110 m) to its gentle slope to shallow water erosion plateau and coastal lagoon plains. Smaller geomorphic features such as banks, sandy ridges and waves, valleys, and so on are found in the study area. The bottom sediment varies from silty-clayey mud within the Gdansk Deep to coarse-grained sediments (sand, pebble, gravel, and boulders) in adjacent shallow water areas. Two broad groups of benthic habitats are soft-sediment bottoms (89% of surface area) and hard substrate (11%). Benthic faunal assemblages on hard substrata vary in terms of species diversity and abundance, but are dominated by sessile suspension feeders (blue mussels, barnacles, hydroids, bryozoans), whereas soft-bottom assemblages are dominated by selective and nonselective deposit feeders (bivalves, polychaetes). Biodiversity and biomass reach maximum values on hard substrates located between 10 and 25 m water depth; benthos have been absent from depths >83 m (the Gdansk Deep) in recent years due to oxygen depletion.

Key Words: Southeastern Baltic, Gdansk Deep, Curonian–Sambian plateau, *Macoma balthica*, *Mytilus edulis*

Introduction

The southeastern Baltic Sea (SEB) is a part of an intracontinental shelf basin of the Atlantic Ocean, located within a large depression of the Baltic Shield and Russian plate of the East European Platform. The study area comprises the Russian sector of the SEB, covering part of the Gdansk Basin between 54°30′ and 55°30′N. The recent bottom relief

Seafloor Geomorphology as Benthic Habitat. DOI: 10.1016/B978-0-12-385140-6.00043-8

of the SEB was formed after the retreat of the Weichselian ice sheet, *ca.* 14,000 years BP. The Baltic Sea underwent several stages of development, from a freshwater ice-lake to semi-enclosed brackish-water sea. Such development was a result of climatic change, gradual melting of the ice sheet, eustatic sea-level rise, and glacio-isostatic uplift of the Baltic Shield [1–7]. In general, relief is smooth, but with large positive (Curonian–Sambian Plateau) and negative (Gdansk Deep, Curonian, and Vistula lagoons) geomorphic structures. Broadscale geomorphic features of the area are lagoon plain, shallow water area, gentle slope, and relatively deep water area of the Gdansk Deep (sea depth 80–110 m). The dissolved oxygen content of bottom water in the Gdansk Deep is close to zero. Salinity, nutrient concentration, and oxygen condition in water layers below the halocline are influenced by periodical North Sea inflows. Ecological condition is far from pristine. Moderate nutrient loading, intensive shipping, limited fisheries, the presence of exotic species, oil-drilling platform with underwater pipelines, and three underwater cables all affect the environmental condition of the study area.

Methods

The study of bottom relief and bottom sediment distribution was carried out in 2003–2007 by A.P. Karpinsky Russian Geological Research Institute (VSEGEI) and Atlantic Branch of P.P. Shirshov Institute of Oceanology (ABIORAS) using the Russian research vessel *Professor Shtockman* and motorboats in the shallow waters. Bathymetric data were collected using a singlebeam Simrad EA400 echosounder. Seabed types were investigated using a CM-2 C-MAX Ltd. sidescanning sonar, and bottom sediments were sampled using grab and gravity corer. Benthic samples were collected in 2000–2007 at a total of 249 stations: 2 by dredging, 183 using an *Ocean* benthic grab (sample area of 0.25 m^2), and 64 taken by divers.

Digital maps were derived from 6,390 digitized data points gridded at 500 m using values extracted from 1:25,000 to 1:100,000 scale Russian nautical charts. The Natural Neighbor Interpolation method was used to create a bathymetric model. The main advantage of this method is the absence of high distortions of input depth values. All GIS-layers and final maps were presented in the projected coordinate system WGS_1984_UTM_Zone_34 N.

Geomorphic Features and Habitats

Bathymetry across the survey area is characterized by increasing depth toward the northwest and a smooth gentle slope of ~0.1° (Figures 43.1 and 43.2). Broadscale geographic features include the *Gdansk Deep*, the *Curonian–Sambian Plateau*, and the *Curonian and Vistula lagoons* (Figure 43.1). *The Gdansk Deep* is a large seafloor depression with a maximum depth of 110 m, elongated in a submeridional (N–S) direction. It is flanked to the east by shallow water depths of 30–35 m. The *Curonian–Sambian Plateau* is located in the northeastern part on the survey area and is joined to the Curonian Spit (Figure 43.1). Erosional terraced relief is developed within the Curonian–Sambian Plateau. A steep stepped slope separates the Plateau from the

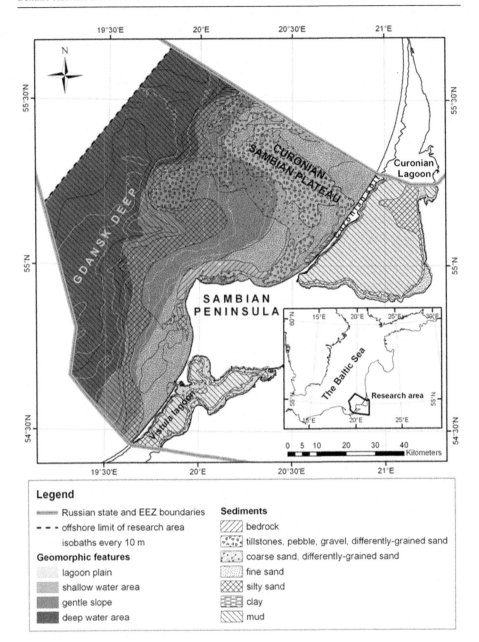

Figure 43.1 Sediment map of the Southeastern Baltic Sea, Russian EEZ.

Gdansk Deep. *The Curonian and Vistula lagoons* are shallow coastal lagoons separated by narrow spits from the sea. The Lagoons extend along the coast from the southwest to the northeast. The maximum depth of the Curonian Lagoon is 6 m and that of the Vistula Lagoon is 5 m (Figure 43.1).

In order to further subdivide the study area into general geomorphic units, the seabed slope map (Figure 43.2) was used to delineate four geomorphic features: lagoon plain, shallow water area, gentle slope, and deep water area.

Lagoon plain (LP) is represented by the Baltic Sea's two largest lagoons, the Vistula Lagoon (490 km^2) and Curonian Lagoon (1,170 km^2). Both lagoons are very shallow and floored with soft sediments. Silty-clayey mud prevails in the central part of the basins and sand occurs along the lagoon coasts.

Shallow water (SW) area extends offshore from the coast to a sea depth of 20 m in the south to 50 m in the north, over an area of 2,350 km^2. The shallow area, to depths of 20–25 m, is within the zone of active wave influence, and here the seabed is characterized by bars and furrows. Medium and fine sands prevail in the south, sand of different grain size, pebble, and gravel in the north, while bedrock and till-stones are present adjacent to the Sambian Peninsula [8]. Medium and fine sands form extended fields joined to the Curonian and Vistula spits and are observed within some parts of coastal areas of the Sambian Peninsula. These sediments are represented by depositional bottom relief (sand bodies). Erosion of coastal sedimentary rocks (siltstones, argillite, limestones) produces tillstones and pebbles, supplemented by debris of crystalline rocks delivered from Scandinavia. Surface sediments composed of sand with gravel and pebbles occur along the borders of the Curonian–Sambian Plateau and at some locations to the north of Sambian Peninsula. Glacially derived boulders and cobbles occur in isolated patches. In some places, elongated fields of drifting sands with ripple marks are observed [9].

Gentle slope (GS) borders the SW area and reaches 30–90 m water depth, and covers an area of 2,380 km^2. Gdansk Deep slope is characterized by a smooth inflection and in some places by hilly surfaces inherited from buried glacial moraine relief. Muddy sediments dominate in the south, fine and silty sand in the central part, and ill-sorted sand, pebbles, and gravel in the north.

Deep water (DW) area is the SE part of the Gdansk Deep, covering 2,820 km^2. Mud covers the surface of the Gdansk Deep's deepest part. Mud is characterized by poor sorting, high moisture, low density, periodic H_2S contamination, high content of organic matter (3–5% C_{org}), and increased methane concentration [10]. The presence of methane is manifested by pockmarks (gas craters) on the seabed and acoustic anomalies created by the concentration of gas bubbles within bottom sediments [11,12].

Combining all categories of bottom sediment, from silty-clayey mud in deepest area to sand, pebble, and gravel in the shallow zone, the study area can be described as comprising *soft bottoms* (89% of the area) and *hard substrata beds* (11%). Hard substrates, from the perspective of macrobenthos such as sessile filter-feeding animals, include bedrock, tillstones, tillstone-pebbles deposits, gravel and pebbles beds, and gravel and pebble fields with coarse ill-sorted sands in between. This habitat type occurs mostly in the shallow water area (35%) along the northern coast of the Sambian Peninsula and the Curonian Spite, and partially (18%) in the gentle slope area (Figure 43.1). The deep water area, lagoon plains, most of the slope area, and 99% of the shallow water area are represented by sands, silt, and mud sediments, collectively, regarded as soft-bottom habitat. Infaunal and epifaunal species, mostly deposit feeders, could develop communities here.

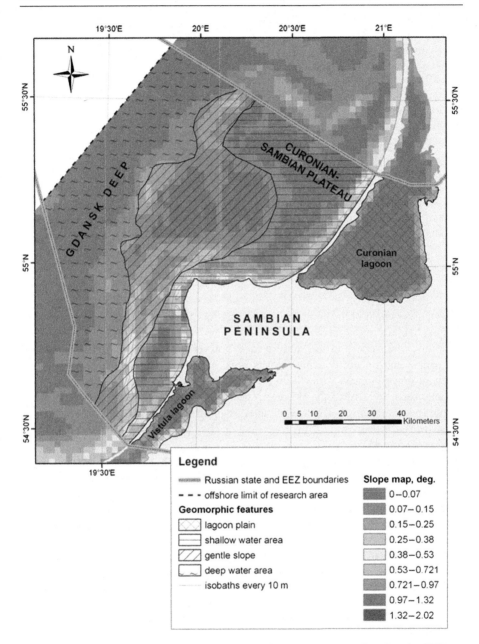

Figure 43.2 Large-scale geomorphic features of the Southeastern Baltic Sea, Russian EEZ.

Biological Communities

The description of benthos in this study is based on the results of 247 quantitative samples collected at 110 stations and two qualitative dredge stations. Material

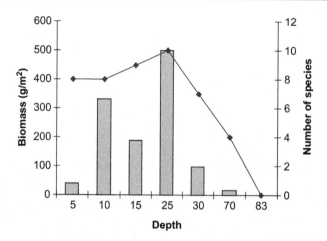

Figure 43.3 Species richness and average biomass of infauna versus water depth in the Southeastern Baltic Sea, Russian EEZ.

allowed assessment of the community structure for all main geomorphic features and different habitats in the survey area.

Benthic biodiversity in the study area is not very high. Benthic macroalgae are very sparse and do not dominate in any area. Macrozoobenthos is more abundant, and the average biomass varies from 0.75 to 3,077 g/m^2 depending on depth, bottom characteristics, and hydrology. However, species diversity is low, with only 52 macrobenthic species recorded. The average number of species was lower in the most shallow (0–5 m) and deepest (>70 m) zones and reached a peak at intermediate depths of 10–25 m (Figure 43.3).

Four different macrobenthic communities, with predominance of the bivalves *Macoma balthica* (Linnaeus, 1758), *Mya arenaria* Linnaeus, 1758, or *Mytilus edulis* Linnaeus, 1758, and spionid polychaete worms were distinguished in the marine survey area and in both lagoons. Two different main biological communities occur in each, corresponding with the bottom relief, type of sediments, and hydrology in defined zones, as follows.

Lagoon plain (LP) is characterized by soft sediments (sand, silt, mud) and inhabited by uniform soft-bottom communities. Chironomidae and Oligochaeta community occupies >70% of bottoms in the Curonian Lagoon, and polychaete worm *Marenzelleria neglecta* Sikorski and Bick, 2004 occupies 90% of bottoms in the Vistula Lagoon. Communities dominated by invasive species occupy sparse hard substrata in both lagoons. The Ponto-Caspian bivalve *Dreissena polymorpha* (Pallas, 1771) occurs in the Curonian Lagoon, and the New Zealand snail *Potamopyrgus antipodarum* (J.E. Gray, 1843) occurs in the Vistula Lagoon. Bottom sediments of the Vistula Lagoon are bioturbated up to 20–30 cm by burrows of exotic North American spionid worm *Marenzelleria neglecta* (Figure 43.4A). Lagoon plain habitats demonstrate high levels of benthic productivity, and total biomass averages 20–90 g/m^2 but reaches hundreds of grams per square meter locally.

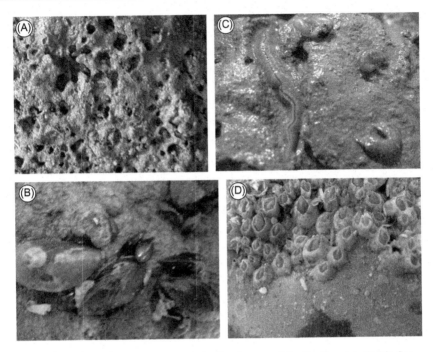

Figure 43.4 Photographs taken on hard and soft substrata of the Southeastern Baltic Sea. (A) Openings of *Marenzelleria neglecta* burrowed in silt. (B) Filter feeders on stone: bivalve *Mytilus edulis*. (C) Polychaete worm *Nereis diversicolor* on silt sand. (D) Bryozoans and barnacles *Balanus improvisus* on stone.

Shallow water (SW) area is the most diverse area and is one of the most important habitats for benthic communities. The highest level of species diversity and productivity occurs on bedrock, tillstones, pebble, and gravel. Biocenosis with a predominance of blue mussel *Mytilus edulis* (Figure 43.4B) occupies such hard substrates. *M. edulis* contributes 98–99% of total benthic biomass here. Barnacle *Balanus improvisus* Darwin, 1854, polychaetes *Hediste diversicolor* (O.F. Müller, 1776) and *Fabricia stellaris* (Müller, 1774), snail *Theodoxus fluviatilis* (Linnaeus, 1758), and amphipods are typical for the blue mussel community. Biomass reaches a maximum of 6,136.4 g/m² (average 1,065 g/m²) and a peak taxonomic diversity of 30 species.

Other parts of the shallow water area contain wide areas of fine sand, inhabited by a biocenosis of bivalve *Macoma balthica*. The share of predominant species varies from 50 to 99%, depending on depth and hydrology. Common species in *M. balthica* biocenosis are bivalves *Mya arenaria*, *Cerastoderma edule* (Linnaeus, 1758), polychaetes *Hediste diversicolor* (Figure 43.4C), *Marenzelleria neglecta*, and *Pygospio elegans* Claparède, 1863, oligochaetes, and the snail *Hydrobia neglecta* Muus, 1963. Average benthic biomass is between 70 and 220 g/m². Different species dominate locally. In areas where the near-bottom layer is enriched by a fine fraction of suspended matter, suspension feeders such as the bivalve *M. arenaria* (15–25 m depth) and worms *P. elegans* and *M. neglecta* (0–10 m) are dominant. In the area of pebbles, gravel, and tillstones mixed with coarse sand, the bivalve *M. balthica* dominates the biomass, but sessile filter feeders, such as *M. edulis*, *B. improvisus*, and bryozoans (Figure 43.4D), are also important contributors. The total benthic biomass here can reach 180–340 g/m^2.

Gentle slope (GS): *M. balthica* community is distributed over the whole area of gentle slope, regardless of sediment type. Community structure is similar to that of the SW. Biomass of total benthos is high at depths down to 70 m (up to 300 g/m^2). In the deeper areas, however, the community becomes species poor, with as few as two to five species, and the biomass decreases to only a few grams per square meter.

Deep water (DW) area: Gdansk Deep is characterized by a very poor macrobenthos. Soft sediments with a high content of organic matter, and the recurrent phenomena of oxygen depletion and hydrogen sulfide contamination, form an unfavorable environment for the sustainable development of zoobenthos. The extremely poor benthos consists of rare examples of *M. balthica*, oligochaetes, and polychaete worms present between 90 and 110 m depth when dissolved oxygen content is above zero. Benthic biomass can reach 3–6 g/m^2 in these periods. Over the last 3 years, macrobenthos has been absent below 83 m in the Gdansk Deep.

Surrogacy

No statistical analyses have been carried out on this data set to examine relationships between physical characteristics and benthos.

References

[1] E. Nilsson, On Late-Quaternary History of Southern Sweden and the Baltic Basin, vol. 4, Baltica, Vilnius, 1970. pp. 11–31.

[2] V.K. Gudelis, E.M. Emelyanov (Eds.), Geology of the Baltic Sea, Mokslas, Vilnius, 1976 (in Russian).

[3] H. Krog, The Quaternary history of the Baltic, Denmark, in: V. Gudelis, (Ed.), The Quaternary History of the Baltic, Uppsala, 1979, pp. 207–217.

[4] S. Björck, A review of the history of the Baltic Sea, 13.0–8.0 ka BP, Quat. Int. 27 (1995) 19–40.

[5] A.I. Blazhchishin, Paleogeography and Evolution of Late-Quaternary Sedimentation in the Baltic Sea, Yantarny skaz, Kaliningrad, 1998. 160 p. (In Russian).

[6] J.B. Jensen, O. Bennike, A. Witkowski, W. Lemke, A. Kuijpers, Early Holocene history of the south-western Baltic Sea, Boreas 28 (1999) 437–453.

[7] E.M. Emelyanov (Ed.), Geology of the Gdansk Basin, Baltic Sea, Yantarny skaz, Kaliningrad, 2002.

[8] VSEGEI, Atlas f Geological and Environmental Geological Maps of the Russian Area of the Baltic Sea, VSEGEI, St. Petersburg, 2010, 78 p.

[9] V.A. Zhamoida, D.V. Ryabchuk, Y.P. Kropatchev, D. Kurennoy, V.L. Boldyrev, V.V. Sivkov, Recent sedimentation processes in the coastal zone of the Curonian Spit

(Kaliningrad region, Baltic Sea), Z. Dtsch. Gesellschaft Geowissenschaften 160 (2009) 143–157.

[10] E.M. Emelyanov, Conclusion: main features of the sedimentogeneses processes, in: E.M. Emelyanov, K.M. Vypykh (Ed.), Sedimentation Processes in the Gdansk Basin (the Baltic Sea), P.P. Shirshov Institute of Oceanology of Soviet Union Academy of Sciences, Moscow, 1986, pp. 248–259 (in Russian).

[11] A.I. Blazhchishin, A.N. Yephimov, Geology and minerals of the Baltic Sea shelf, in: M.N. Alekseev, (Ed.), Geology and Minerals of Russian Shelves, GEOS, Moscow, 2002, pp. 425.

[12] N.V. Pimenov, M.O. Ulyanova, T.A. Kanapatsky, E.F. Veslopolova, P.A. Sigalevich, V.V. Sivkov, Microbially mediated methane and sulfur cycling in pockmark sediments of the Gdansk Basin, Baltic Sea // Geo-Mar Lett. 2010. V. 30(3–4). P. 439–448.

44 Inland Tidal Sea of the Northeastern Pacific

J. Vaughn Barrie[1], H. Gary Greene[2], Kim W. Conway[1], Kim Picard[1]

[1]Geological Survey of Canada, Pacific, Institute of Ocean Sciences, Sydney, British Columbia, Canada, [2]Tombolo, Eastsound, WA, USA

Abstract

Along the Pacific coast of northwestern USA and western Canada, large inland marine straits, sounds, and fjords are found with interconnecting narrow channels through island archipelagos to the open ocean. The Salish Sea is one of the world's largest inland seas, encompassing 400 islands, 7,500 km of coastline, and water depths to 650 m. The inland sea developed through Pleistocene glacial processes and the resultant geography provides for a meso- to macrotidal environment. Typical features found are shallow banks, deltas, fjords, glacial troughs, bedrock (ridges and banks), subaqueous dune fields, and sponge reefs. The past and present physical processes have created a variety of habitats, such as steep, near-vertical rock walls, and stacked boulders, which offer habitat for juvenile and adult rockfish (*Sebastes* spp.), subaqueous dunes that shelter sand lances (*Ammodytes hexapterus*), mud flats that provide habitat for a variety of shellfish and birds, and raised glacial banks that allow for the formation of siliceous (glass) sponge reefs.

Key Words: Inland sea, habitats, subaqueous dunes, sponge reefs, Fraser Delta, Salish Sea, British Columbia, Washington State

Introduction

Inland seas are landlocked seas that are only connected to the ocean through narrow channels. They usually contain many islands, channels, sounds, and straits. Inland sea was first defined for the Seto Inland Sea, separating the three main islands of Japan and connected to the Pacific Ocean through narrow channels. Other examples of inland seas include the Baltic Sea and Khor Al Daid, the inland sea of the western Qatar. The Salish Sea of the North American Pacific coast is defined by two large straits (Straits of Georgia and Juan de Fuca) and one sound (Puget Sound) surrounded by the British Columbia mainland, Washington State, and Vancouver Island (Figure 44.1). It is one of the world's largest inland seas, encompassing 400 islands, 7,500 km of coastline, water depths to 650 m in the coastal fjords, and an overall

Seafloor Geomorphology as Benthic Habitat. DOI: 10.1016/B978-0-12-385140-6.00044-X

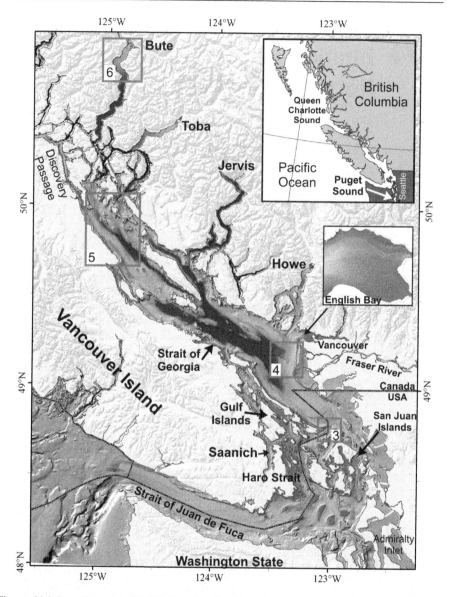

Figure 44.1 Location map of Salish Sea, showing the extent of multibeam bathymetric swath mapping coverage and locations of Figures 44.3–44.6.

marine area of 17,000 km². The largest water body, the Strait of Georgia, stretches for approximately 220 km in a NW–SE direction between Vancouver Island and the mainland; the overall width varies from 25 to 55 km. The average depth of the Strait of Georgia is 155 m and the deepest point is 420 m [1]. Large areas of the Salish Sea reach depths of between 100 and 250 m, although several shallow banks exist down the axis of the straits. The inland sea connects with the open sea in the south, first through the Gulf Islands and San Juan Islands, and then through the Strait of Juan de Fuca (Figure 44.1). Bottom topography in the Gulf/San Juan Islands area is complex but mostly shallower than 100 m, except for a few narrow, deep channels. In the north, the Strait of Georgia connects to the open shelf of Queen Charlotte Sound through four narrow channels with sill depths of 90 m or less (Figure 44.1). The Salish Sea is dominated by estuarine and tidal flow with a net outflow of low salinity water toward the Strait of Juan de Fuca in the upper layer and a net northward inflow of high salinity water in the lower part of the water column that reaches the Strait of Georgia in late summer [1].

The Salish Sea is a fore arc basin, Georgia Basin, developed with subsidence that began in the late Cretaceous (85 million years ago). The basin consists of a series of structural depressions, overdeepened by Tertiary fluvial erosion and Quaternary glaciation and partially infilled by glacial and postglacial sediments [2]. Most of the Salish Sea is presently sediment starved, with sediment capture within the coastal fjords and inlets. The one exception is in the southern Strait of Georgia, where sedimentation from the Fraser River dominates the surficial geology with Holocene sediment thicknesses varying from 0 on Pleistocene banks to greater than 300 m within the basin [3]. Present-day sedimentation rates vary from up to 10 cm/near the river mouth to less than 3 cm/in the distal parts of the prodelta [4,5].

The spectacular landscape of the Salish Sea is surrounded by the largest coastal population and development growth in Canada and an area of significant growth within the USA. This has resulted in significant development, including sewage outfalls, ferry terminals, port facilities, airports, electrical and communication cable corridors, and fishery. In the less-populated areas, coastal forestry and aquaculture occupy many coastal inlets and fjords. Consequently, the naturalness has been significantly reduced, particularly in the southern Salish Sea near the large urban areas of Vancouver, Seattle, and Victoria.

Over 10,000 km^2 of multibeam and backscatter data collected in the Salish Sea is used to define the key geomorphic features and associated habitats. Mapping has been undertaken at a scale of 1:50,000 within the areas of multibeam mapping, supported by extensive network of sediment grabs and subbottom profile survey data, all collected over the past 20 years. Detailed potential habitat maps have been produced for the transboundary (Canada/USA) region of the central Salish Sea region [6]. This chapter summarizes the primary geomorphic features and identifies known associated habitat characteristics for the entire Inland Sea. There is a limited biological database for the Salish Sea, so only those biological features associated with geomorphic features that have been studied and published are presented.

Geomorphic Features and Habitats

Bedrock—Deposition of the Fraser River sediment is not ubiquitous in the central and southern Strait of Georgia. As a result, parts of the basin have limited or no sedimentation, and bedrock ridges and banks are exposed (Figure 44.2). For example, on

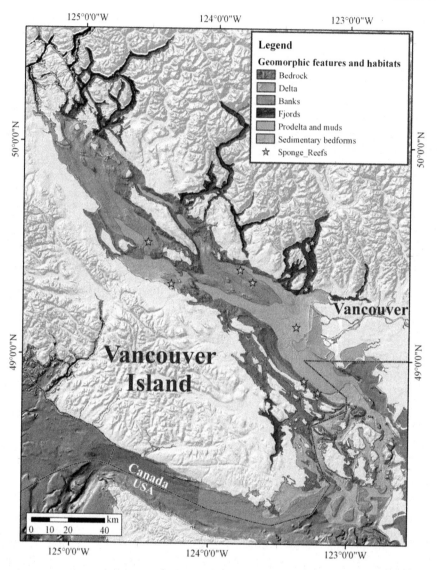

Figure 44.2 Distribution of geomorphic features and habitats within the Salish Sea overlain on shaded bathymetric relief.

the western side of the basin opposite the delta and into interconnecting channels, bedrock is exposed (Figure 44.3). The basin is moderately deformed along a series of mainly northwest-trending faults and folds with northeast to east dipping beds [7]. All the interconnecting channels through to the Pacific Ocean wholly, or in part, consist of exposed bedrock.

Fraser Delta—The Fraser River is the largest natural source of sediment in the Strait of Georgia and the largest single geomorphic feature of the Salish Sea

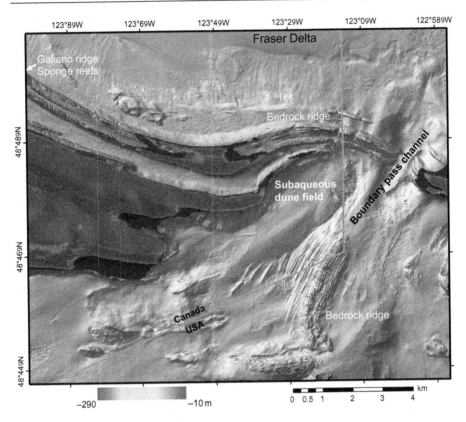

Figure 44.3 Multibeam bathymetry shaded relief image of the bedrock ridges at the entrance to Boundary Pass, the main channel connecting the Strait of Georgia and the Strait of Juan de Fuca. Note the very extensive subaqueous dune field just south of the restricted channel entrance. Sun angle is from 315° at a 45° azimuth and the vertical exaggeration is 5×. Color scale bar represents water depth. Figure location is shown in Figure 44.1 (For interpretation of the references to color in this figure legend, the reader is referred to the web version of this book.).

(Figures 44.2 and 44.4). Sediment from the river is dispersed throughout the Strait through a variety of transport pathways [8]. Sedimentation from the Fraser River dominates the surficial geology of the southern Strait of Georgia. It is also the site of a significant anthropogenic modification (population of greater Vancouver is 2.1 million) including ferry terminals, port facilities, airport, dredging, sewage disposal, fishery, and a critical electrical transmission, and communication cable corridor to Vancouver Island [9].

Banks—Originally the Salish Sea would have been filled with sediment during the nonglacial to glacial transition at the beginning of the Fraser Glaciation. Ice moving south from the Coast Mountains of the Canadian Cordillera and Vancouver Island coalesced and eroded these deposits leaving a series of banks in the central troughs that are remnants of these preglacial sediments (Figure 44.2). Typically, the bank tops

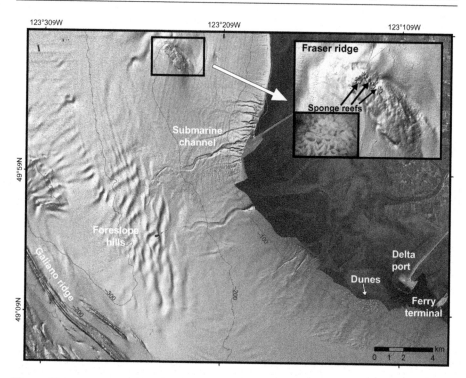

Figure 44.4 Multibeam bathymetry shaded relief image of the prodelta of the southern Fraser Delta showing the submarine channels seaward of the main river channels and the sponge reef located on Fraser Ridge within the delta. Sun angle is from 315° at a 45° azimuth, and the vertical exaggeration is 5 ×. Figure location is shown in Figure 44.1.

range from 10 to 30 m water depth. The major banks of the central Strait of Georgia and the Strait of Juan de Fuca are made up of greater than 80 m of highly stratified sediments composed of well-sorted very fine silt and overlain by an ice contact diamicton [2,10]. On the banks of the Strait of Georgia, linear ridges (constructional fluted tills) formed by the action of the glaciers where obstructions interrupted the southerly flow of glacial ice (Figure 44.5).

Fjords—The BC mainland of the Salish Sea is punctuated by four dominant fjords (Howe, Jervis, Toba, Bute) and several smaller inlets with average water depths of 500 m (Figures 44.1 and 44.2). In addition, one fjord occurs on southeastern Vancouver Island (Sannich Inlet) with an average water depth of 250 m. The fjords range from 40 km in length to over 110 km in length and average 3 km in width. At the head of all the inlets are delta systems where seafloor channels (Figure 44.6) can be mapped as much as 35 km down fjord, except Jervis and Sannich, which do not have submarine channels. Over 25 turbidity current events occur annually in Bute related to increased snow and glacier melt [11]. Turbidity currents have been measured to flow a minimum distance of 26 km and possibly as far as 40–50 km, moving sand down fjord in the mud-dominated environment [12].

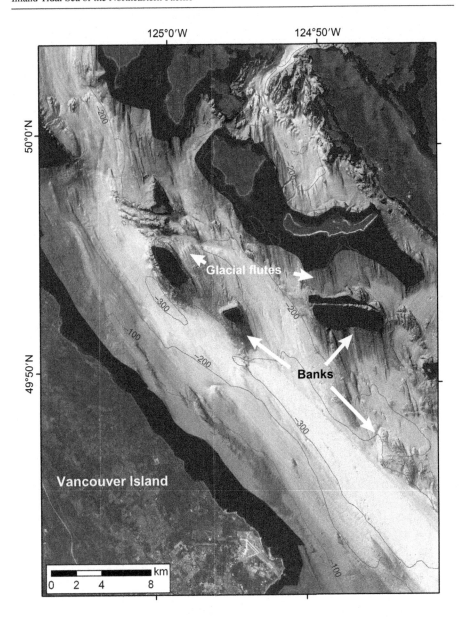

Figure 44.5 Backscatter draped on multibeam shaded relief image showing the fluted banks of the northern Strait of Georgia, suggesting a glacial flow direction to the south to southeast. Areas of high backscatter indicate glacial sediments, while weak backscatter areas represent mud. Sun angle is from 315° at a 45° azimuth, and the vertical exaggeration is 5×. Figure location is shown in Figure 44.1.

Figure 44.6 Multibeam bathymetry relief image of Bute Inlet showing the delta at the head of the fjord and submarine channel extending down fjord. Sun angle is from 315° at a 45° azimuth, and the vertical exaggeration is 5×. Color scale bar represents water depth. Figure location is shown in Figure 44.1. (For interpretation of the references to color in this figure legend, the reader is referred to the web version of this book.)

Mud—Within the Salish Sea, silty-clay sediments occur in the coastal fjords, a portion of central Vancouver Island in coastal inlets, intertidal, and prodelta areas of the Fraser Delta and smaller inlets (fjords) within the San Juan Islands and Puget Sound (Figures 44.1 and 44.2). Intertidal mud flats form a very small part of the greater Salish Sea area but are key migratory bird habitats. Pockmarks are a common feature of submarine mud trough areas, particularly where thick Fraser River muddy sediments exist. For example, English Bay at the entrance to Vancouver Harbor is covered in hundreds of pockmarks (Figure 44.1). In some cases, pockmarks define fault zones, such as the $25 km^2$ field of pockmarks in the northeastern sector of the southern Strait of Georgia that are aligned along the Fraser Delta Fault [13]. Pockmarks vary considerably in size, with the largest being 300 m in diameter and 15 m deep and more commonly 50 m in diameter and 1–2 m deep and occur over the entire depth range of the Salish Sea in muddy sediments.

Sedimentary dunes—Subaqueous dunes are common throughout the Salish Sea (Figure 44.2), particularly within the interconnecting channels (Figure 44.3) and on the Fraser Delta foreslope (Figure 44.4) [8,14]. Each area has been modified as a result of the sediment transport dynamics in a meso- to macrotidal environment with superimposed estuarine induced current flow. The other primary characteristic controlling the environments is the overall bathymetric morphology that restricts flow within the basins [14]. In most cases, the dunes occur within narrowing basins that close at shallow sills or occur at a point of morphological change within the basin, such as on the southern Fraser Delta (Figure 44.4). Four areas in particular contain very large subaqueous dune fields at the point where an interconnecting channel meets the larger water bodies (Discovery Passage and Strait of Georgia, Strait of Georgia and Boundary Pass (Figure 44.3), Haro Strait and Strait of Juan de Fuca, Admiralty Inlet (Puget Sound) and Strait of Juan de Fuca (Figure 44 .1)). These subaqueous dunes have wavelengths between 100 and 300 m, dune heights up to 28 m [14]. In addition, in the central trough of the southern Strait of Georgia over an area of $>60 km^2$, a ridge and swale morphology 20 m high and greater than 5 km long (Foreslope Hills (Figure 44.4)), are interpreted to be deep-water sediment waves (very large subaqueous dunes) [15].

Sponge reefs—Siliceous sponges form reefs on Fraser Ridge, an isolated glacial bank just off the Fraser Delta [16] (Figure 44.4), along the eastern approaches to the Gulf Islands, on a glacial moraine within the fjords and in areas of low-relief glacial deposits in central Strait of Georgia [16,17]. The Fraser Ridge complex has formed in an area where sedimentation rates adjacent to the reef site are greater than 2 cm per year. The Fraser Ridge reefs consist of roughly circular interconnected mounds up to 14 m in height and 200 m in diameter, found in water depths of 150–190 m and restricted to the top and flanks of an isolated bank in the midst of the rapidly expanding Holocene prodelta (Figure 44.4). The reefs that form on low-relief glacial banks of the central strait and within the fjords occur in 120–210 m water depth and reach up to 14 m in height, and are somewhat like dunes in shape and aspect ratio with a wavelength of 30–100 m. The western Galiano Ridge reefs occur on a rocky reef in 95–105 m water depth and reach up to 4 m in height (Figure 44.5).

Biological Communities

No quantitative studies of the biological communities exist for each geomorphic feature [18]. There is considerable descriptive information pertaining to benthos of the Salish Sea, however, and based on this qualitative information several aspects of benthic habitat for some geomorphic features can be highlighted.

Generally all the bedrock outcrops of the Salish Sea are habitat for 37 species of rockfish (*Sebastes* spp.) that form part of a long-term fishery. Many of these species (total 13) are considered candidates for "Species of Concern" in the USA and the rockfish Bocaccio (*Sebastes paucispinis*) is considered threatened in Canada [19]. Consequently, fishery closures have been implemented to conserve rocky areas within the Salish Sea (e.g., Rockfish Conservation Areas (Fisheries and Oceans Canada)). In addition, some rocky ridges host sponge reefs, known to be key habitats for rockfish [20].

The muddy areas of the Salish Sea, including the Fraser Delta prodelta, fjords, and inlets have variable invertebrate benthos abundance (Figure 44.2), with those areas influenced by the Fraser River discharge providing apparently more productive substrate for macroinvertebrates than areas of marine deposition [18]. Marine aquaculture is found within the protected muddy inlets of the Salish Sea, such as salmon fish farming in the fjords and shellfish along the coast of eastern Vancouver Island.

The deep-water Pacific sand lance (*Ammodytes hexapterus*) is dependent upon benthic sediment habitats to burrow into; therefore, this species is most often associated with oxygenated, well-sorted, medium- to coarse-grained sand, particularly a grain size of 0.36–1.0 mm (Figure 44.2). Sediments in the tidal-dominated Salish Sea that fall within this critical sediment criteria are normally part of medium to large subaqueous dune systems (Figures 44.3 and 44.4). The sand lance is a vital food source for many species of coastal seabirds, as well as several marine mammals and commercial and sport fish. Specifically, this species is a primary component in the diet of common murres, marbled murrelet, rhinoceros auklets, tufted puffins, Cassin's auklet, and Pacific salmon, all of which are species of concern or endangered in the Salish Sea [19].

Sponge reefs require a raised glacial environment (moraine, flute, bank) to initiate development. They are made up of two species of hexactinosidan sponges, *Aphrocallistes vastus* and *Heterochone calyx* (Figure 44.7) that build a framework of densely packed sponge skeletons, while several other species of hexactinellida and demosponges are accessory fauna [16,17]. Of the seven known reefs in the Salish Sea, three are considered undamaged, two are totally damaged (total loss of all living sponges), and the other two are damaged but potentially recovering [20]. One undamaged reef showed a higher taxonomic richness and abundance of rockfish (*Sebastes* spp.), both adult and juvenile (Figure 44.7), compared to an adjacent damaged reef, suggesting that the healthy reefs may act as refugia for rockfish [20].

Surrogacy

Infauna in the Strait of Georgia was evaluated in terms of water depth, substrate type, organic content of sediments, and sediment textural characteristics based on data collected over a 19-year period [21]. The data were based on 607 grab samples

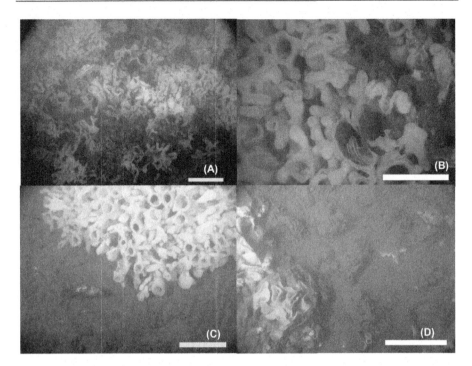

Figure 44.7 Photographs from Galiano Ridge (Figure 44.3) of (A) a juvenile yelloweye rockfish in dense sponges; (B) a juvenile rockfish in *Aphrocallistes vastus*; (C) quillback rockfish beside *Aphrocallistes vastus*; and (D) mechanically damaged sponges. Scale bar in images is 30 cm long.

collected between 1988 and 2006, from areas of the Strait of Georgia and adjacent fjords in natural areas, and 347 samples from areas where anthropogenic discharges occur. Burd et al. (2008) [18] found that neither abundance nor biomass were predictable by particle size, organic content, or water depth. Organic flux, however, was an important indicator of certain biotic factors [21]. For example, areas of high accumulation of sediment, such as the Fraser River, supported the highest macroinfaunal abundance and biomass and were dominated by bivalves. Polychaetes dominated in low organic deposition conditions, particularly where anthropogenic deposition (pulp mills, waste water outfalls) was high [21].

References

[1] R.E. Thomson, Oceanography of the British Columbia coast, Can. Fish. Aquat. Sci. 56 (1981) 1–291.

[2] J.V. Barrie, P.R. Hill, K.W. Conway, K. Iwanowska, K. Picard, Georgia Basin: seabed features and marine geohazards, Geosci. Can. 32 (2005) 145–156.

[3] D.C. Mosher, T.S. Hamilton, Morphology, structure and stratigraphy of the offshore Fraser Delta and adjacent Strait of Georgia, in: J.J. Clague, J.L. Luternauer,

D.C. Mosher, (Eds.), Geology and Natural Hazards of the Fraser River Delta, British Columbia, Geological Survey of Canada, Ottawa, pp. 147–160. (Bulletin 525)

[4] B.S. Hart, T.S. Hamilton, J.V. Barrie, Sedimentation on the Fraser Delta slope and pro-delta, Canada, based on high-resolution seismic stratigraphy, lithofacies and ^{137}Cs fallout stratigraphy, J. Sediment. Res. 68 (1998) 556–568.

[5] S.C. Johannessen, M.C. O'Brien, K.L. Denman, R.W. Macdonald, Seasonal and spatial variations in the source and transport of sinking particles in the Strait of Georgia, British Columbia, Canada, Mar. Geol. 216 (2005) 59–77.

[6] H.G. Greene, J.V. Barrie, Potential Marine Benthic Habitats of the Southern Gulf Islands and San Juan Archipelago, Canada and USA, Map Series 1 to 4, Geological Survey of Canada, Open File 6625, (2010).

[7] P.S. Mustard, The upper cretaceous nanaimo group, Georgia Basin, in: J.W.H. Monger, (Ed.), Geology and Geological Hazards of the Vancouver Region, Southwestern British Columbia, Geological Survey of Canada, pp. 27–95. (Bulletin 481)

[8] P.R. Hill, K. Conway, D.G. Lintern, S. Meule, K. Picard, J.V. Barrie, Sedimentary processes and sediment dispersal in the southern Strait of Georgia, BC, Canada, Mar. Environ. Res. 66 (2008) S39–S48.

[9] J.V. Barrie, R.G. Currie, Human impact and sedimentary regime of the Fraser River Delta, Canada, J. Coastal Res. 16 (2000) 747–755.

[10] A.T. Hewitt, D.C. Mosher, Late quaternary stratigraphy and seafloor geology of eastern Juan de Fuca Strait, British Columbia and Washington, Mar. Geol. 177 (2001) 295–316.

[11] B.D. Bornhold, P. Ren, D.B. Prior, High-frequency turbidity currents in British Columbia fjords, Geo-Mar. Lett. 14 (1994) 238–243.

[12] D.B. Prior, B.D. Bornhold, W.J. Wiseman, D.R. Lowe, Turbidity current activity in a British Columbia fjord, Science 237 (1987) 1330–1333.

[13] J.V. Barrie, P.R. Hill, Holocene faulting on a tectonic margin: Georgia Basin, British Columbia, Canada, Geo-Mar. Lett. 24 (2004) 86–96.

[14] J.V. Barrie, K.W. Conway, K. Picard, H.G. Greene, Large-scale sedimentary bedforms and sediment dynamics on a glaciated tectonic continental shelf: examples for the Pacific margin of Canada, Cont. Shelf Res. 29 (2009) 796–806.

[15] D.C. Mosher, R.E. Thomson, The Foreslope Hills: large scale, fine-grained sediment waves in the Strait of Georgia, British Columbia, Mar. Geol. 192 (2002) 275–295.

[16] K.W. Conway, J.V. Barrie, M. Kruatter, Modern siliceous sponge reefs in a turbid sili-clastic setting: Fraser River delta, British Columbia, Canada, J. Neues Jahrbuch fuer Geol. 6 (2004) 335–350.

[17] K.W. Conway, J.V. Barrie, M. Kruatter, Complex deep-sea habitats: sponge reefs in the Pacific northwest, in: B.J. Todd, H.G. Greene, (Eds.), Mapping the Seafloor for Habitat Characterization, Geological Association of Canada, St. John's. pp. 259–269. (Special paper 47).

[18] B.J. Burd, P.A.G. Barnes, C.A. Wright, R.E. Thomson, A review of subtidal benthic hab-itats and invertebrate biota of the Strait of Georgia, British Columbia, Mar. Environ. Res. 66 (2008) S3–S38.

[19] N.A. Brown, J.K. Gaydos, Species of concern within the Georgia Basin puget sound marine ecosystem: changes from 2002 to 2006, in: Proc. 2007 Georgia Basin Puget Sound Research Conference, 2007, 10 pp.

[20] S.E. Cook, K.W. Conway, B. Burd, Status of the glass sponge reefs in the Georgia Basin, Mar. Environ. Res. 66 (2008) S80–S86.

[21] B.J. Burd, R.W. Macdonald, S.C. Johannessen, A. van Roodselaar, Responses of subtidal benthos of the Strait of Georgia, British Columbia, Canada to ambient sediment conditions and natural and anthropogenic depositions, Mar. Environ. Res. 66 (2008) S62–S79.

45 Cold-Water Coral Distribution in an Erosional Environment: The Strait of Gibraltar Gateway

Ben De Mol[1], David Amblas[1], German Alvarez[1], Pere Busquets[1], Antonio Calafat[1], Miquel Canals[1], Ruth Duran[1], Caroline Lavoie[1], Juan Acosta[2], Araceli Muñoz[3], HERMESIONE Shipboard Party

[1]GRC Geociències Marines, Parc Científic de Barcelona/Departament d'Estratigrafia, Paleontologia i Geociències Marines/Facultat de Geologia/Universitat de Barcelona/Campus de Pedralbes, Barcelona, Spain, [2]Instituto Español de Oceanografiac/Corazon de Maria, Madrid, Spain [3]Tragsa-SGM C/Nuñez de balboa, Madrid, Spain

Abstract

The sill of the Strait of Gibraltar is the morphological and oceanographic gateway between the Atlantic Ocean and the Mediterranean Sea for the post-Messinian period. It was surveyed by multibeam echosounder, TOPAS profiler, and 380 grab samples to study the distribution of cold-water corals. The sill represents an elevated seabed topography characterized in its central part by two main "mounts" separated by E-W oriented depressions. The irregular morphology suggests a rocky aspect, which is confirmed by the TOPAS profiler data showing a thin veneer of sediment or bare rocks. Low sedimentation is enforced by an extremely variable and strong current regime. Overall N-S elevated morphological structures dominate the sill area as an expression of the tectonic stress field and represent Early Pliocene flysch outcrops. Pliocene and/or Quaternary sediments consist mainly of calcareous conglomerates and cold-water coral accumulations, as well as sands and in the deeper part, muddy sediments. Cold-water coral distribution is related to elevated seafloor morphology of various sizes and low-sedimentation environments.

Key Words: cold-water corals, Strait of Gibraltar, erosion, sill, Mediterranean Sea, Atlantic Ocean

Seafloor Geomorphology as Benthic Habitat. DOI: 10.1016/B978-0-12-385140-6.00045-1

Introduction

The sill of the Strait of Gibraltar is the morphological, oceanographical, and ecological gateway between the Atlantic Ocean and the Mediterranean Sea for the post-Messinian crisis period (Figure 45.1). The Messinian crisis was an event during which the Mediterranean Sea partly or nearly completely dried out around 5.96–5.33 Ma [1]. The post-Messinian flooding (Zanclean flood) of the Mediterranean basin resulted in the scouring of the two deep channels (Canal Norte and Sur) in the Strait of Gibraltar; the flooding was very abrupt and did not last for more than 2 years [2].

At the end of the Messinian (Late Miocene) and during the Early Pliocene, the tectonic stress field was dominantly N-S and created pull-apart basins under a transtensional regime, which induced the opening of the Atlantic–Mediterranean gateway at the Strait of Gibraltar. The seabed is composed of synorogenic Betic-Rif clayey flysch overlaid by Pliocene and/or Quaternary calcareous conglomerates and coral accumulations, as well as current transported sand and mud in the deepest parts (Figure 45.2) [3].

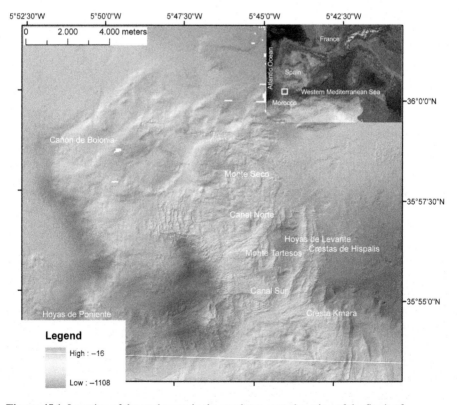

Figure 45.1 Location of the study area in the southern-central section of the Strait of Gibraltar based on a high-resolution (20 m) bathymetric map derived from a compilation of IEO and GRCGM data. The names of seabed features described in the text are indicated.

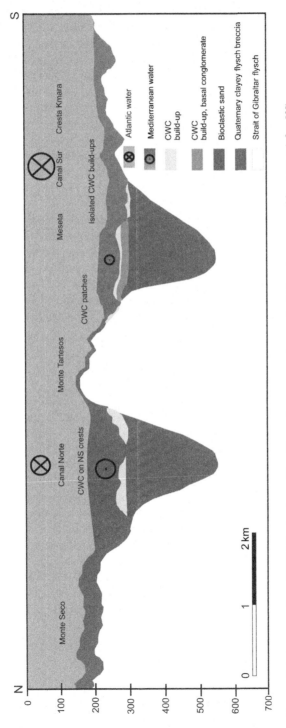

Figure 45.2 Interpretation of TOPAS seismic profile with coral occurrence in a geological and oceanographic context (after [3]).

The oceanographic circulation in the Strait of Gibraltar is characterized by a two-layer system: a surface, eastward, Atlantic water inflow and a deep, westward outflow of saline Mediterranean water, with a variable interface depth of around 100 m [4] (Figure 45.2). This current regime is strongly influenced by tidal currents as well as by wind and atmospheric pressure variations [5]. The interaction of the different oceanographic processes and the seabed topography results in a pulsating upwelling area and enriches the upper illuminated zone with nutrients [6], increasing the biomass of the northwestern part of the Strait of Gibraltar.

The Strait of Gibraltar is one of the busiest maritime zones in the world and is thus affected by invasive species, but the benthic community is poorly studied. Furthermore, the study zone is affected by benthic trawl fisheries on the shelf and near-shelf areas and by the laying of submarine cables. Overall, the naturalness of the study area is considered to be largely unmodified.

This study focuses on a 235 km^2 section of the central-southern part of the sill area of the Strait of Gibraltar, mapped using multibeam sonar (Figure 45.1). The maximum depths within the study area are 500 m in the east and 630 m on the western side. The shallowest parts of the study area are Monte Seco in the north (90 m water depth), Monte Tartesos in the central portion (150 m), and Cresta Kmara in the southeast (200 m). Channel-like depressions separate the elevated areas: Canal Norte separates Monte Seco and Monte Tartesos, and Canal Sur separates Monte Tartesos and Cresta Kmara (Figure 45.1).

Methods

A base map of part of the Gibraltar Strait was made available by the Spanish Ocean Institute (IEO) in a grid resolution of 50 m, obtained aboard R/V *Vizconde de Eza* using a Simrad EM 300 multibeam echosounder (MBES). During the HERMESIONE 2009 cruise, further multibeam surveying was completed and improved the resolution in an area where cold-water coral (CWC) occurrences have been reported [7]. Simrad EM120 (12 kHz) and EM1002 (100 kHz) multibeam sonars were used to map the survey area (235 km^2). Data was processed with Caris Hips and Sips software to a final, tidally corrected (based on port of Tanger tide gauge data) grid at a resolution of 15 m. Four hundred and eight nautical miles of TOPAS sub-bottom profiler data were recorded in a synchronized mode with MBES. Sociedad Española de Estudios para la Comunicación Fija Europa-Africa a través del Estrecho de Gibralta (SECEGSA) conducted an intensive sampling of 380 samples (spacing of about 500 m) collected in 1994. These samples have been analyzed for textural facies and scleractinian coral content [7].

Geomorphological Features and Habitats

The seafloor exhibits a complex geology, manifest as many small peaks and blocks (Figures 45.1 and 45.3). The northern channel is characterized by numerous N-S oriented crests, about 35 m in height. In the southern portion of the survey area, the number of crests is less than in the north, rising 20–25 m above the surrounding seafloor (Figure 45.3). Medium-size topographic irregularities produce a series of elevations and depressions related to the erosional force of the strong W-E currents, generally smoothing the western flanks of all morphological structures.

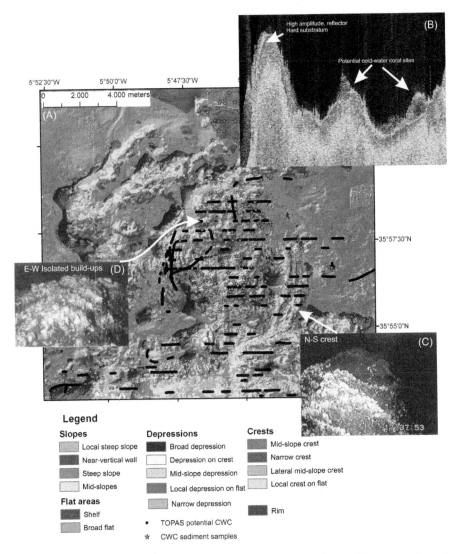

Legend

Slopes
- Local steep slope
- Near-vertical wall
- Steep slope
- Mid-slopes

Flat areas
- Shelf
- Broad flat

Depressions
- Broad depression
- Depression on crest
- Mid-slope depression
- Local depression on flat
- Narrow depression
- • TOPAS potential CWC
- ☆ CWC sediment samples

Crests
- Mid-slope crest
- Narrow crest
- Lateral mid-slope crest
- Local crest on flat
- Rim

Figure 45.3 (A) Morphological classification of the bathymetric dataset (15 m resolution) of the Strait of Gibraltar and the CWC distribution of sediment samples and TOPAS profiles; (B) TOPAS profile illustrating potential coral accumulations; (C) photograph of N-S orientated crest with coral colonization (courtesy of SECEGSA); (D) isolated build-ups with an E-W orientation (courtesy of SECEGSA). (For interpretation of the references to color in this figure legend, the reader is referred to the web version of this book.)

The 10 km long eastern Crestes de Hispalis (Figure 45.1) rises about 50 m above the seafloor and is the largest ridge in the survey area. Shallow knolls having flat summits with minor N-S trending crests are separated by well-delimited depressions.

The rugose morphology suggests that sediment cover is thin or absent over rocky outcrops in the central part of the survey area, which is also consistent with the high-amplitude reflections observed on TOPAS profiles. Within some bathymetric depressions, subsurface reflectors are observed, indicating local sediment accumulations. Diffraction hyperbolas observed above the strong-amplitude reflector are related locally to grab samples containing corals (Figure 45.3). Based on this correlation, the hyperbolas in the TOPAS profiles have been mapped as potential coral accumulations (Figure 45.3). At the eastern and western limits of the surveyed area the seafloor has a smooth topography, indicating that sediment overlies the rocky basement in that area.

A maximum likelihood statistically predicted map of the sediment distribution was created based on sediment samples and interpretations of TOPAS profiles, indicating rocky outcrops (Figure 45.3). Bathymetry-derived morphological grids were analyzed for rugosity, morphological classification (Figure 45.4), and slope. Bare rock is correlated with the morphological classes: crest, steep slope, and near-vertical walls. In the near vicinity of steep slopes and the eastern part of the survey area, gravel is predicted. The largest textural group in the area is a thin veneer of sand. Mud covers the smallest area and is found only in the deepest parts (Figure 45.4). Gravel, rocky outcrops, and a thin layer of sand are suitable for coral settlement [7].

Cold-water coral: Sixteen species of calcareous corals have been identified in water depths between 13 and 443 m by Álverez-Pérez et al. [7]. Azooxanthellate scleractinia corals are the most common at the Gibraltar sill in water deeper than 200 m, with rare occurrences of *Stylasteridae* and *Gorgonacea*. Coral patches colonize hard substrates, rocky outcrops of narrow topographic elevations like rocky crests and, fossilized coral build-ups. The distribution of the corals correlates positively with large and small morphological elevations without any preferential orientation. As the sill region is known as an overall nondepositional to low-sedimentation environment, sedimentation is not limiting the coral distribution. Nutrients in the water column are distributed by a complex combination of oceanographic process interacting with the topography, favoring coral settlement at elevated areas outside the zone of influence of bottom sediment transport. These favorable zones depend upon many local factors, such that no general orientation trend can be observed in spatial pattern of coral colonization.

Coral banks: Reef-forming CWC of *Lophelia pertusa* and to a lesser extent of *Madrepora occulata*, associated with significant quantities of dead fragments, occur at the depths between 180 and 330 m. This zone is located in the deepest part of the Strait of Gibraltar associated with complex seabed morphology (Figures 45.3 and 45.4). Some of the elevated and elongated seabed structures in water depths of around 300 m are reported to be a kind of lithoherm [8,9]. The morphological structures associated with coral occurrences show elongations parallel to the broad N-S structural lineation and E-W current pattern. At present the seafloor at the base of the elevations is composed out of coral rubble, coral colonies, and carbonate sediments

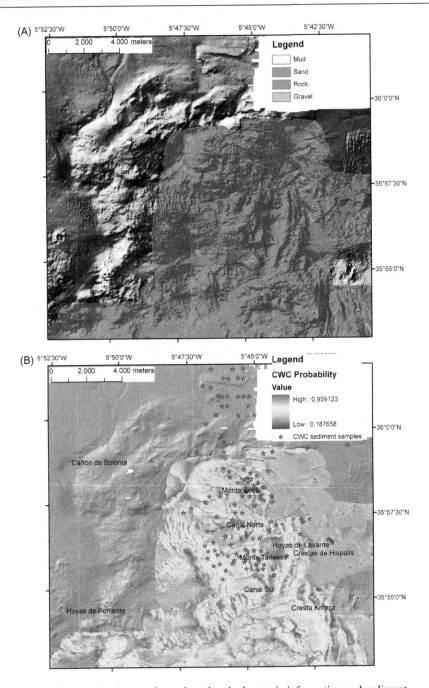

Figure 45.4 (A) Predicted textural map based on bathymetric information and sediment samples; (B) predicted CWC distribution based on a maximum entropy algorithm of morphological parameters and sediment samples indicating CWC occurrences. (For interpretation of the references to color in this figure legend, the reader is referred to the web version of this book.)

(Figures 45.2–45.4). The size of the lithoherms are 250 × 100 m up to 800 × 200 m in plan view. The larger structures show several subdivisions of clusters of smaller entities of coral build-ups.

Three coral settlement types have been recognized. Two of them show a current dominated, E-W elongation, and the third type has a dominant tectonic N-S orientation.

1. *Patches*: On top of a flat area: these are found on the summit and rim of the Monte Seco, a major elevation with a flat summit. The coral accumulations have a reduced height and are not well defined with an E-W orientation (Figure 45.3).
2. *Isolated build-ups*: This type of coral build-up is observed along the southern limits of the Monte Tartesos and at the base of Cresta Kmara, in the deepest part of the Gibraltar sill (300 m) (Figure 45.3). The build-ups are well defined and relatively small (3 km²), separated by small straight moats with an E-W direction that are attributed to current erosion. Isolated build-ups have also been reported on the flanks of Monte Tartesos in the south of the Canal Norte. These coral accumulations are up to 500 m long with exceptional dimension of 1 km.
3. Build-ups on top of N-S orientated crests (Figure 45.3). The colonization of coral on structural N-S oriented crests is the most common form in the Strait of Gibraltar. The coral colonization is directly related to crest dimension. The largest in dimension are observed in the northern part of the Canal Norte, and the smallest in dimension at the western flank of the Cresta Kmara.

A maximum thickness of 40 m is estimated from seismic profiles for the isolated build-ups in the northern part of the Monte Tartesos. In the Canal Sur, 20–30 m thick build-ups have been observed, and in the Meseta area 10 m thick accumulations occur [3].

Surrogacy

The study area includes a zone with elevated structures reported between 180 and 300 m water depths with living *Lophelia pertusa* and *Madrepora oculata* scleractinian corals [8,9]. Based on the Benthic Terrain model scheme [10], a morphological classification has been made of the bathymetric MBES data, indicating several morphological habitats in the area that can cause fragmentation in the main ecosystem (Figure 45.3). This classification is based on the rugosity, slope, curvature, depth, bathymetric positioning index (BPI) with annulus neighborhood of 1,125, 300, 120, and 60 m, and the range of standard deviation of the depth over a distance of 45 m based on a bathymetric grid of 15 m. The predicted textural distribution map is based on the rugosity, slope, and objectively classified morphological zones identified in combination with textural information of the sampled stations used in a maximum-likelihood statistic algorithm provided by ArcGIS. Coral distribution is based on grab samples and the layers of the derived morphological grids (Figures 45.2 and 45.4). These data, in combination with the morphologic descriptive parameters, have been used in a Maximum Entropy algorithm (Figure 45.4B).

In this case study, the textural distribution could not be performed in the same predictive model as the coral distribution, because the same data points have textural and coral presence information rather than actual presence or absence of different classes. For this reason, a presence-only CWC distribution model was applied by using the maximum entropy principal. It states that the least-biased distribution that encodes certain given information is that which maximizes the information entropy. CWC suitable habitats are widely distributed in the sill area at local crests (Figure 45.4B). The occurrence of CWC is in this area not limited by sediment pressure, as in many other sites, but mainly by too strong bottom currents and availability of nutrients.

Acknowledgments

This research was supported by the HERMIONE project, EC contract 226354-HERMIONE, funded by the European Commission's Seventh Framework Programme, a Generalitat de Catalunya "Grups de Recerca Consolidats" grant (2009 SGR 1305), and SMT Inc. via the educational User License for Kingdom Suite interpretation software. Collaboration with Salah Ben Cherifi and his staff at Institut National de Recherches Halieutiques from Casablanca, Morocco, is warmly appreciated.

References

[1] A.C. Campillo, A. Maldonado, A. Mauffret, Stratigraphic and tectonic evolution of the western Alboran Sea: Late Miocene to recent, Geo-Mar. Lett. 12 (1992) 165–172.

[2] D. Garcia-Castellanos, F. Estrada, I. Jiménez-Munt, C. Gorini, M. Fernàndez, J. Vergés, et al., Catastrophic flood of the mediterranean after the messinian salinity crisis, Nature 462 (2009) 778–781.

[3] J.M. Pliego, Open session—the Gibraltar Strait tunnel. An overview of the study process, Tunnel. Underground Space Technol. 20 (2005) 558.

[4] H. Lacombe, C. Richez, in: C.J. Nihoul Jacques (Ed.), Elsevier Oceanography Series Volume, vol. 34, Elsevier, 1982, pp. 13–73.

[5] J.G. Lafuente, J. Delgado, J.M. Vargas, T. Sarhan, M. Vargas, F. Plaza, Low-frequency variability of the exchanged flows through the Strait of Gibraltar during CANIGO, Deep Sea Res. II 49 (2002) 4051–4067.

[6] J.M. Santana-Casiano, M. Gonzalez-Davila, L.M. Laglera, The carbon dioxide system in the Strait of Gibraltar, Deep Sea Res. II 49 (2002) 4145–4161. 10.1016/S0967-0645(02)00147-9.

[7] G. Álvarez-Pérez, P. Busquets, B. De Mol, N.G. Sandoval, M. Canals, J.L. Casamor, Deep-water coral occurrences in the Strait of Gibraltar, in: A. Freiwald, M. Roberts, (Eds.), Cold-Water Corals and Ecosystems, Springer-Verlag, Berlin Heidelberg, 2005, pp. 207–221.

[8] F. Izquierdo, M. Esteras, N. Sandoval, Depósitos coralinos litificados en el Estrecho de Gibraltar, Geogaceta 20 (1996) 401–404.

[9] N. Sandoval, J.L. Sanz, F.J. Izquierdo, Fisiografía y Geología del umbral del Estrecho de Gibraltar, Geogaceta 20 (1996) 343–346.

[10] E.R. Lundblad, D.J. Wright, J. Miller, E.M. Larkin, R. Rinehart, D.F. Naar, et al., A benthic terrain classification scheme for American Samoa, Mar. Geod. 29 (2006) 89–111.

46 Habitat Mapping of a Cold-Water Coral Mound on Pen Duick Escarpment (Gulf of Cadiz)

Lies De Mol[1], Ana Hilário[2], David Van Rooij[1], Jean-Pierre Henriet[1]

[1]Renard Centre of Marine Geology (RCMG), Department of Geology and Soil Science, Ghent University, Gent, Belgium, [2]CESAM and Departamento de Biologia, Campus Universitário de Santiago, Aveiro, Portugal

Abstract

On top of Pen Duick Escarpment on the Moroccan continental margin (southern Gulf of Cadiz), several mound structures were discovered in 2002. Within this case study, habitat mapping of one of these cold-water coral mounds, Beta Mound, was performed using a multibeam echosounder, ROV observations, and boxcore sampling. This habitat mapping revealed the presence of four different habitats: (1) soft (bioturbated) sediment, (2) soft sediment with patchy coral rubble, (3) dense cold-water coral rubble fields, and (4) rock slabs. All cold-water corals are dead except for one living *Dendrophyllia cornigera*. The coral rubble consists mainly of dead *Dendrophyllia*, *Lophelia pertusa*, and *Madrepora oculata*. On top of the coral rubble, numerous crinoids were observed, while the soft sediment is mostly colonized by several species of soft corals. Sponges, squat lobsters, echinoids, and holothurians were also observed. The fish *Helicolenus dactylopterus* was found in all the four habitats.

Key Words: Gulf of Cadiz, habitat mapping, mounds, cold-water corals, *Dendrophyllia, Lophelia pertusa*

Introduction

The Pen Duick Escarpment is a 6 km long, NW-SE oriented, 80–125 m high escarpment with a southwest-facing slope of 8–12°. It is located within the El Arraiche mud volcano field on the Moroccan continental margin (southern Gulf of Cadiz) in water depths of 550–650 m (Figure 46.1). Pen Duick Escarpment represents the southeastern leg of Renard Ridge, one of the two submarine ridges within the mud volcano field. These ridges, the result of extensional tectonics, are characterized by

Seafloor Geomorphology as Benthic Habitat. DOI: 10.1016/B978-0-12-385140-6.00046-3

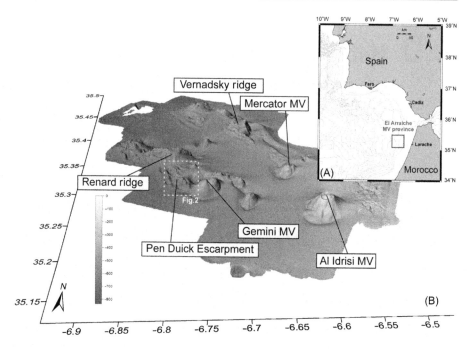

Figure 46.1 (A) Location of the El Arraiche mud volcano field on the Moroccan margin (GEBCO bathymetry, contour lines every 250 m); (B) three-dimensional bathymetric map of the mud volcano field, with the location of the largest mud volcanoes (Mercator, Gemini, and Al Idrisi mud volcanoes), the two ridges (Renard and Vernadsky ridge), and the Pen Duick Escarpment (after [3]).

large rotated blocks bound by lystric faults [1]. On top of Pen Duick Escarpment, 15 mound structures were identified during the R/V *Belgica* CADIPOR cruise in 2002 [2,3]. All mounds are influenced by the North Atlantic Central Water (NACW), with a bottom current speed of $10 \, \text{cm} \, \text{s}^{-1}$ at 1 m above the seafloor [4].

The naturalness of Pen Duick Escarpment is uncertain, as the area is frequently used for fishing activities. However, on the escarpment itself no human impact (e.g., trawl marks and lost fishing gear) has been noticed, although trawling vessels were observed during several cruises to the area (mostly near the mud volcanoes). Along the northwestern African coast, fishing activities are not strictly monitored and therefore the ecosystems on top of the mounds might be impacted; this remains unknown.

Since the discovery of the Pen Duick Escarpment in 2002, several cruises were organized to unveil the morphology and subsurface architecture of these mounds within the framework of the European Science Foundation (ESF) EuroDIVERSITY MiCROSYSTEMS project and the EC FP6 HERMES and EC FP7 HERMIONE projects. Bathymetric data were collected during the R/V *Marion Dufresne* MD169 cruise (2008) using a hull-mounted Thomson SeaFalcon 11 "dual mode" multi-beam echosounder over a survey area of 60 km², resulting in a 20 × 20 m grid [4]. Remotely Operated Vehicle (ROV) observations as well as boxcore sampling were made during the R/V *Belgica* 07/13 and 09/14b cruises.

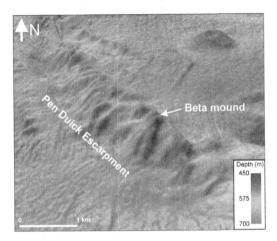

Figure 46.2 Three-dimensional multibeam bathymetric map of Pen Duick Escarpment [4] showing the location of Beta Mound (20 × 20 m grid size).

Geomorphic Features and Habitats

The mounds on top of Pen Duick Escarpment are elongated and have a maximum length of about 500 m, with a maximum height of 60 m [3]. One of these mound structures is Beta Mound, which is located on the southern part of Pen Duick Escarpment (Figure 46.2). Beta Mound is a north-south elongated feature up to 40 m high, 500 m long to 300 m wide, with two summits. The northernmost summit is located at a water depth of 520 m, and the southernmost summit at a water depth of 515 m. The western flank of the mound has an average slope gradient of 15°, while on the eastern flank an average value of 12° is measured. The average slope gradient of the northern flank varies between 12° and 15°, while the southern slope is much steeper with an average slope of 19°. In between the two summits the slope values vary between 8° and 10°.

ROV observations were obtained during four dives using Ghent University's ROV *Genesis* (Table 46.1). ROV *Genesis* is a sub-Atlantic Cherokee-type ROV with an operational survey depth down to 1,300 m. A color camera, recorded with and without navigational overlay, and a black-and-white camera were used to observe the seafloor. In addition, high-resolution still images were obtained with a Canon Powershot camera. A positioning accuracy of about 2–3 m was reached using the Ultra Short Base Line (USBL) IXSEA Global Acoustic Positioning System (GAPS). Two parallel laser beams set 10 cm apart were used as a scale during the seabed observations. The processing and the qualitative interpretation of the dives were performed using the Ocean Floor Observations Protocol (OFOP) version 3.2.0c [5]. Based on the observations, a number of habitats characteristic for the study area were identified. Each habitat was given a color code and integrated into ArcGIS 9.1, resulting in a habitat interpretation map.

Four different habitats were recognized: (i) soft (bioturbated) sediment, often colonized by octocorals (Anthozoa, Octocorallia) (Figure 46.3A), (ii) soft sediment with a patchy distribution of sediment-clogged dead cold-water coral rubble (Figure 46.3B), (iii) dense cold-water coral rubble fields varying in size between 10 and 50 m

Table 46.1 Metadata of Four ROV Surveys Performed at Beta Mound During the R/V *Belgica* 07/13 and 09/14b Cruises

Dive	Start Track				End Track			
	Latitude	Longitude	Depth (m)	Time	Latitude	Longitude	Depth (m)	Time
B07-07	35°17.663'N	6°47.209'W	574	08:45	35°17.740'N	6°47.265'W	543	09:59
B09-06	35°17.692'N	6°47.249'W	512	08:13	35°17.767'N	6°47.255'W	520	10:22
B09-07	35°17.744'N	6°47.257'W	526	07:05	35°17.719'N	6°47.273'W	520	08:26
B09-08	35°17.769'N	6°47.308'W	533	10:01	35°17.586'N	6°47.298'W	548	11:24

Figure 46.3 ROV images of the four different facies observed on Beta Mound: (A) soft sediment with a sea pen; (B) soft sediment with a patchy distribution of coral rubble colonized by crinoids; (C) dense coral rubble field colonized by mainly crinoids and sponges; (D) rock slabs colonized by sponges, octocorals, and crinoids.

in diameter and locally covered with a very thin layer of soft sediment (Figure 46.3C), and (iv) rock slabs (Figure 46.3D) probably made up of authigenic carbonates [6,7]. The cold-water coral rubble fields in facies 3 can be either open-coral frameworks or buried graveyards, depending on the thickness of the soft-sediment layer on top of the rubble. Based on our observations, these coral rubble fields can be considered as coral plates, indicating an important step in the transition from reef to mound.

Facies 1 (soft sediment) is mostly present on the eastern flank of the mound and in between the two summits (Figure 46.4), covering about 20% of the dive track. Facies 2 (patchy corals) comprises 35% of the dive track and mostly occurs on the flanks of the mound, while facies 3 (coral rubble fields) is concentrated on the top of the mound and the western flank. Facies 3 covers the largest part of the dive track, 44%. Based on these observations, the cold-water corals tend to prefer the western flank and the top of the mound rather than the eastern flank. This might indicate a SW-NE current direction as living cold-water corals prefer areas with strong topographically guided bottom currents [8]. Finally, facies 4 (rock slabs) covers only 1% of the dive track.

A total of three boxcore samples (Table 46.2; Figure 46.5) were retrieved on top of Beta Mound, with site selection based on the ROV observations. The location of

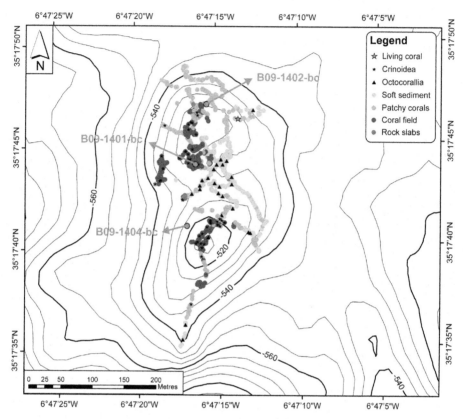

Figure 46.4 Facies interpretation map of Beta Mound. The locations of the boxcore samples are marked in blue. (For interpretation of the references to color in this figure legend, the reader is referred to the web version of this book.)

Table 46.2 Metadata of Three Boxcores Collected During the R/V *Belgica* 09/14b Cruise

Core Number	Latitude	Longitude	Water Depth (m)	Recovery (cm)
B09-1401-bc	35°17.734′N	6°47.272′W	526	13
B09-1402-bc	35°17.783′N	6°47.260′W	528	5–20
B09-1404-bc	35°17.684′N	6°47.282′W	524	30

the samples was accurately determined using the GAPS USBL system. Afterwards, subsamples were analyzed for grain size distribution with a Malvern Mastersizer 2000 (Marine Biology Section, Ghent University). All three boxcore samples indicate a sandy mud seafloor with an average mean grain size of 21 μm (coarse silt). All samples are poorly sorted and have a unimodal, symmetrical to coarse skewed grain size distribution. The sediment color varies from pale brown to light olive brown at

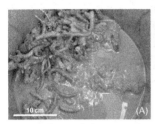

Figure 46.5 Pictures of the three boxcore samples on top of Beta Mound: (A) B09-1401-bc; (B) B09-1402-bc; (C) B09-1404-bc. (For interpretation of the references to color in this figure legend, the reader is referred to the web version of this book.)

the surface to grayish brown at a depth of about 15–20 cm, indicating a change to a more reducing environment.

Biological Communities

Cold-water corals (Scleractinia or stony corals) are widely distributed along the European Atlantic continental margin [9]. One of the most common scleractinian species is *Lophelia pertusa*, which forms bush-like colonies up to several meters across. In contrast, *Madrepora oculata* is much more fragile and forms smaller colonies 30–50 cm in height. A less common species in cold-water coral reefs is the yellow coral *Dendrophyllia* (*Dendrophyllia alternata* and *Dendrophyllia cornigera*), which forms small colonies about 15 cm in height. *Desmophyllum cristagalli*, a solitary coral (the other three are colonial), is also observed on the reef. On Beta Mound, only dead cold-water corals were observed, except for one living coral thicket of *Dendrophyllia cornigera* (Figure 46.6F) on the eastern flank of the northernmost summit at a water depth of 528 m. The coral rubble is dominated by crushed dendrophylliid corals (Figure 46.6A and B), while at some locations well-preserved remnants of small bushes of *Lophelia pertusa* are present (Figure 46.6C–E). This pattern is also observed in the boxcore samples, where the surface and the top 10 cm were completely filled with *Dendrophyllia* corals. From a depth of about 10–15 cm, *Lophelia pertusa*, *Madrepora oculata*, and *Desmophyllum cristagalli* fragments appear. Dendrophylliid corals indicate relatively stable and warm conditions, while *Lophelia pertusa* and *Madrepora oculata* occur during glacial periods with long-term stable currents and temperatures [10].

Cold-water corals provide hard-bottom substrate, refuge, and nursery for a large variety of species and are therefore considered important biodiversity hotspots [7]. On Beta Mound, several invertebrate and fish species are living in the three-dimensional framework formed by the dead cold-water coral rubble. The megafauna of the coral rubble patches or fields is dominated by crinoids (Echinodermata, Crinoidea), which cover most of the surface of the rubble. Poriferans and crustaceans such as squat lobsters (Decapoda, Galatheidae) are also commonly found on the coral rubble areas. In contrast, the megafauna of the soft-sediment habitat

Figure 46.6 ROV images presenting the high biodiversity on cold-water coral mounds: (A and B) zoom on a semiburied coral rubble field colonized by crinoids and sponges; (C and D) well-preserved coral reef fragments (*Lophelia pertusa*) with heights up to 60 cm, also intensively colonized by crinoids; (E) detail of a large coral reef fragment with a sea urchin and a high amount of crinoids; (F) a living *Dendrophyllia cornigera*.

is characterized by the presence of high densities of octocorals, including sea pens (Penatulacea) and gorgonians (Gorgonacea). Crinoids are often found on top of dense gorgonian bushes. Poriferans, sea urchins (Echinodermata, Echinoidea), and holothurians (Echinodermata, Holothuroidea) are also observed in these areas. The fish *Helicolenus dactylopterus* is present on all habitats.

Surrogacy

No statistical analyses have been carried out on this data set to examine relationships between physical surrogates and benthos.

Acknowledgments

The authors would like to thank the shipboard parties of the R/V *Belgica* 07/13 and 09/14 cruises for their help and cooperation during the cruises. This work was financially supported by the ESF FP6 HERMES project "Hotspot Ecosystem Research on the Margins of European Seas" (GOCE-CT-2005-511234-1), the ESF FP7 HERMIONE project "Hotspot Ecosystem Research and Man's Impact on European Seas" (contract number 226354), and the ESF EuroDIVERSITY MiCROSYSTEMS project (05_EDIV_FP083-MICROSYSTEMS). We are grateful to Prof. Dr. M. Vincx (Marine Biology Section, Ghent University) for allowing us to use the Malvern Mastersizer 2000. A. Hilário was supported by the FCT grant SFRH/BPD/22383/2005. L. De Mol acknowledges the support of the Agency for Innovation by Science and Technology in Flanders (IWT).

References

[1] J. Flinch, Tectonic Evolution of the Gibraltar Arc. PhD Thesis, Rice University, Houston, Texas, 1993.
[2] P. Van Rensbergen, D. Depreiter, B. Pannemans, G. Moerkerke, D. Van Rooij, B. Marsset, et al., The El Arraiche mud volcano field at the Moroccan Atlantic slope, Gulf of Cadiz, Mar. Geol. 219 (2005) 1–17.
[3] A. Foubert, D. Depreiter, T. Beck, L. Maignien, B. Pannemans, N. Frank, et al., Carbonate mounds in a mud volcano province off north-west Morocco: key to processes and controls, Mar. Geol. 248 (2008) 74–96.
[4] D. Van Rooij, D. Blamart and the MiCROSYSTEMS shipboard scientists, Cruise Report MD169 MiCROSYSTEMS, Brest (FR)-Algeciras (ES), 15–25 July 2008. ESF EuroDIVERSITY MiCROSYSTEMS internal report, 2008.
[5] E. Huetten, J. Greinert, Software controlled guidance, recording and post-processing of seafloor observations by ROV and other towed devices: The software package OFOP. Geophys. Res. Abs. 10 (2008), EGU2008-A-03088.
[6] R. León, L. Somoza, T. Medialdea, F.J. González, V. Díaz-del-Río, M.C. Fernández-Puga, et al., Sea-floor features related to hydrocarbon seeps in deepwater carbonate-mud mounds of the Gulf of Cádiz: from mud flows to carbonate precipitates, Geo-Mar. Lett. 27 (2007) 237–247.
[7] L. Maignien, D. Depreiter, A. Foubert, J. Reveillaud, L. De Mol, P. Boeckx, et al., Anaerobic oxidation of methane in a cold-water coral carbonate mound from the Gulf of Cadiz. Int. J. Earth Sci. (in press), doi: 10.1007/s00531-010-0528-z.
[8] A. Freiwald, J.H. Fossa, A. Grehan, T. Koslow, J.M. Roberts, Cold-water Coral Reefs. UNEP-WCMC Report, Cambridge, UK, 2004.

[9] A. Freiwald, J.M. Roberts, Cold-Water Corals and Ecosystems, Springer-Verlag, Berlin Heidelberg, 2005.

[10] C. Wienberg, D. Hebbeln, H.G. Fink, F. Mienis, B. Dorschel, A. Vertino, et al., Scleractinian cold-water corals in the Gulf of Cádiz—first clues about their spatial and temporal distribution, Deep Sea Res. I 156 (2009) 1873–1893.

47 Habitats at the Rockall Bank Slope Failure Features, Northeast Atlantic Ocean

Janine Guinan[1], Yvonne Leahy[2], Thomas Furey[3], Koen Verbruggen[1]

[1]INFOMAR Integrated Mapping for the Sustainable Development of Ireland's Marine Resource, Marine and Geophysics Programme, Geological Survey of Ireland, Beggars Bush, Haddington Road, Dublin 4, Ireland, [2]Department of Arts, Heritage and the Gaeltacht, National Parks and Wildlife Service, Custom House, Flood Street, Galway, Ireland, [3]Marine Institute, Renville, Oranmore, Co. Galway, Ireland

Abstract

Underwater video data were acquired along the eastern flank of the Rockall Bank as part of a research survey to acquire information on the extent and distribution of Annex 1 geogenic reef (in accordance with the European Commission's Habitats Directive 92/43/EEC). The Rockall Bank forms part of the Rockall Plateau located in the Northeast Atlantic Ocean, west of Ireland, occurring in water depths from 200 m and extending to 3,000 m in the Rockall Trough, the adjacent deep-water basin. Previous studies examining the geomorphology of the eastern Rockall Bank have described the presence of extensive broadscale (>10 km) "slope failure features, e.g., scarps and channels." Geomorphic features identified included seabed scarps, overhangs and horizontal ledges, vertical rock walls, and pinnacles. Benthic habitats identified were characterized by (i) the framework reef-building cold-water coral *Lophelia pertusa* occurring on overhangs, boulders, and drop-stone pavements, (ii) gorgonians, encrusting sponges, soft corals, and black corals occurring on pinnacles and overhangs, and (iii) corallimorphs, desmospongiae, encrusting sponges, and scleractinia occurring on vertical rock walls and horizontal ledges.

Key Words: Geogenic reef; seabed geomorphology; underwater video data; benthic habitat; Rockall Bank Northeast Atlantic Ocean

Introduction

This case study presents the preliminary findings of research undertaken to provide information on the extent and distribution of Annex 1 geogenic reef habitat, pertaining

Seafloor Geomorphology as Benthic Habitat. DOI: 10.1016/B978-0-12-385140-6.00047-5

to offshore morphological features such as escarpments, submarine canyons and channels, and so on (Table 47.1), under the European Commission Habitats Directive
(Council Directive 92/43/EEC on the conservation of natural flora and fauna) in the
Irish Exclusive Economic Zone (EEZ). The Rockall Bank forms part of the Rockall
Plateau located in the Northeast Atlantic Ocean, ~370 km west of Ireland trending
in a northeast–southwest direction. Gravity and magnetic data suggest the Plateau is
continental in composition and thickness, and its present isolation began in Mesozoic
time with the formation of the Rockall Trough. During the Lower Tertiary, spreading
had migrated to or commenced at the present Reykjanes Ridge axis, thus separating
the Plateau from Greenland. The Bank is bounded to the east by the Rockall Trough,
and the region provides a conduit for saline water masses originating in the south and
transiting northwards into the Nordic Seas [1]. Deep-water circulation across the area
involves a number of complex water masses. Cold, dense Norwegian Sea deep water
overflows into the northern trough, mixing with Labrador Sea water and Antarctic

Table 47.1 Geogenic Subhabitats Associated with Geomorphological Features in the Irish
Offshore Region

Geomorphological Features	Annex 1 Geogenic Reef Habitat Types
Submarine canyons and channels	Vertical rock walls
	Horizontal ledges
	Overhangs
	Pinnacles
	Gullies
	Ridges
	Sloping/flat bedrock
	Boulder fields
	Cobble fields
Escarpments	Vertical rock walls
	Horizontal ledges
	Overhangs
Boulder/cobble/drop-stone pavements	Boulder fields
	Cobble fields
Sloping/flat bedrock	Vertical walls
	Horizontal ledges
	Overhangs
	Pinnacles
	Gullies
	Ridges
Ridges	Vertical walls
	Horizontal ledges
	Overhangs
Trenches	Vertical rock walls
	Horizontal ledges
	Overhangs

Bottom water to form North Atlantic deep water, which flows southwards along the eastern Rockall Bank [2]. Erosional features along the eastern margin of the Rockall Bank are a result of counterclockwise thermohaline currents, which have resulted in a major contourite drift build-up, the Feni Drift [3]. Sea temperatures in the area vary between 7°C and 15°C at the sea surface and between 5.5°C and 7.5°C in deep waters.

In 2009, the Irish national research vessel R.V. *Celtic Explorer* carried out underwater video surveys at several locations in the Northeast Atlantic Ocean, including sites at the Rockall Bank in water depths ranging between 800 and 1,700 m (Figure 47.1). Underwater video transect data were acquired using a combination of remote operated vehicle and drop-frame camera systems at seabed features considered potential geogenic reef habitat, e.g., canyons and escarpments (Table 47.1). The multibeam bathymetry data (acquired using Kongsberg Maritime EM120 and EM1002 hull-mounted systems) were collected as part of the Irish National Seabed Survey (INSS) deep-water mapping survey of Ireland's territorial seabed (2000–2005) carried out on behalf of the Geological Survey of Ireland [4].

Geomorphic Features and Habitats

A digital elevation model of the entire offshore area has been developed based on performing a hydrological network analyses on the INSS bathymetric data set to

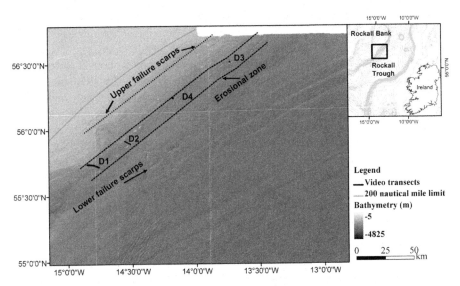

Figure 47.1 Shaded relief bathymetric map incorporating the interpretation map of the Rockall Bank Mass Flow with the regions of failure from Elliott et al. [6]. Locations of four video transects (D1–D4) acquired from the eastern Rockall Bank, Northeast Atlantic Ocean, during a survey onboard the *R.V. Celtic Explorer* in 2009 are indicated. Inset map showing study area location.

identify canyon, escarpment, and channel features [5]. A close examination of the multibeam bathymetry for the region (gridded at 170 m) reveals the complex morphology of the seabed at the eastern margin of the Rockall Bank. Slope failure features along the margin have been described in previous studies [6–10]. Unnithan et al. [11] first described the Rockall Bank Mass Flow (RBMF) at the eastern flank where large-scale episodic slope failure is thought to have occurred, resulting in features such as scarps, channels, debrites, and turbidites. Several morphological zones have been identified based on the integration of multibeam bathymetry with sidescan sonar data and multichannel seismic profiles to investigate the external and internal morphology of the failure mass (Figure 47.1) [6].

Previously, the seabed of the Rockall Bank has been reported to be colonized by discrete patches of the ahermatypic coral *Lophelia pertusa*, commonly found in depths between 130 and 4,000 m [12]. The fauna associated with the reef-forming *L. pertusa* is highly diverse and rich [13]. The three-dimensional structures are home to a variety of fish fauna; these fish associations have been found to be largely depth related [14]. Carbonate mounds are a feature on the eastern margin of the bank, with the Logachev mounds found in depths of between 550 and 1,200 m. Their shape is thought to be directly affected by bottom currents, showing a preferred orientation parallel to flow direction [15,16]. Sidescan sonar data has shown that glaciagenic coarse material, suitable for the establishment of coral colonies, is common on the upper slopes. Similarly boulders, stone, and gravel fragments provide the hard substrate required for colony settlement [12]. The presence of pinnacles and scarps as found in the area of mounds on Rockall Bank is associated with local flow acceleration and provides a favorable environment for suspension feeders, corals in particular [17,18].

At the study area, video transect D1 was acquired from the morphological region of the RBMF referred to as the "upper failure scarp zone," while video transects D2–D4 were acquired from the "erosional zone." The erosional zone is thought to be a continuation of a depression formed between the Feni Drift and the flanks of the Rockall Bank to the south based on an interpretation of the regional-scale bathymetry. The region is characterized by seabed pinnacles in the southern zone, with numerous rounded headwall scarps observed in the seismic data acquired from the area. The broadscale geomorphic features at the study area comprise seabed scarps, while the fine-scale features comprise pinnacles, overhangs, and horizontal ledges.

Seabed scarps: At the southern limit of the study area, transect D1 traverses a seabed scarp located in the western region of the upper failure scarps zone. Video data was acquired over a 6-km transect trending east to west, in water depths between 835 and 1,700 m. Transect D4 targets a seabed scarp feature in the erosional zone midway along the study area in water depth 1,600 m.

Overhangs and horizontal ledges: The most extensive overhang and horizontal ledge features were observed to the north of the study area along transects D3 and D4. It was estimated that the overhangs extend up to 200 m and in places associated with blocks of consolidated sediment.

Vertical rock walls: Vertical walls were associated with transects D2 and D3 in water depths ranging between 400 and 1,700 m. Vertical relief along the transects was estimated at 1–3 m.

Pinnacles: Seabed pinnacles were associated with transects D1 and D2. The relief of the pinnacles is estimated at ~2 m. The biology associated with the pinnacles and other seabed features is described below.

Biological Communities

Video Data Real-Time Acquisition and Postprocessing

Underwater video observations in real-time were logged using Ocean Floor Observation Protocol (OFOP) software, a package developed to facilitate real-time visual observations of video data acquired during the deployment of ROVs (remotely operated vehicle) and TV-sled tows (http://ofop.texel.com). OFOP integrates "button" files providing the data logger with a list of geomorphological and biological feature classes that can be used to characterize and identify habitat types observed in the video. OFOP reads a variety of position data and formats, including data from the GAPS underwater navigation system. Postprocessing of the video data was carried out with the video data replayed through OFOP and observations logged to a new protocol. The objective of the video data acquisition on the survey was to record presence data for the many different species and geomorphological features occurring at each study site. These records were logged continuously to OFOP, with observations logged on average every 5–10 s. Information from both the real-time and postprocessing protocols provided a detailed summary of the observations recorded. Faunal groupings were left at high taxonomic level for real-time logging in the expectation that this would achieve consistent identification by all scientists throughout the cruise.

To the south of the study area, at transect D1, the seabed is characterized by hard substrate with pebbles, cobbles, boulders, and exposed rock (Figure 47.2A). Cold-water corals are widespread over the area, with 90% of images reviewed recording a combination of dead and living coral. The reef-building coral *Lophelia pertusa* was observed along with small colonies of *cf. Solemosmilia variabilis*. Live solitary corals (possibly *Flabellum* sp.) were present. Several species of gorgonians were present, with sponges (encrusting, desmospongiae and glass sponge) abundant at pinnacles along the transect (Figure 47.3A). Further north, transect D2 covered a distance of 3.5 km in water depths ranging between 1,400 and 1,700 m (Figure 47.2B). The seabed was characterized by vertical walls and cliffs with a series of horizontal ledges covered in a mud veneer with extensive areas of dead coral. In general, fauna on the hard substrate was dominated by encrusting sponges, desmospongiae, shrimp, corallimorphs, and *Stichopathes*-type species. Coral rubble was observed on the vertical walls, with black corals, soft corals, and scleractinia observed at the top of pinnacles (Figure 47.3B). Approximately 40% of images reviewed recorded geogenic reef habitat.

To the north of the study area, transect D3 traversed 500 m from a starting depth of 1,450 m to the end of the transect at 1,650 m (Figure 47.2C). The transect began in an area of flat seabed characterized by burrowed, soft sediment before traversing an overhang where corallimorphids, ophiuroids, gorgonians, and several species of encrusting

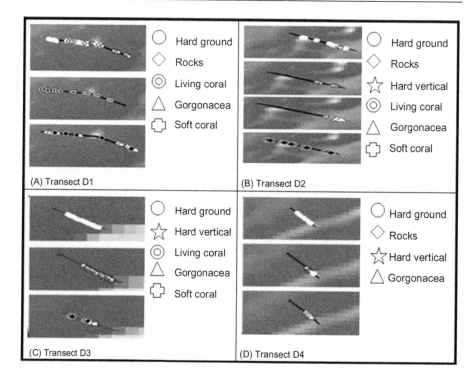

Figure 47.2 Ocean Floor Observation Protocol (OFOP) software was used to record real-time visual observations of video data. OFOP reads a variety of positional data and formats, including data from the underwater navigation system. The maps (A–D) were derived using OFOP and show the geogenic and biogenic features identified from underwater video characterizing transects D1–D4.

sponge were observed in addition to dead and living corals, the living coral inhabited by brisingids (Figure 47.3D). Further along the transect, the terrain was characterized by overhanging blocks of consolidated sediment (Figure 47.3B). Below this, an area of dead coral framework draped in sediment and dominated by encrusting sponges and ophiuroids was observed. A section of the overhang was dominated by gorgonians, encrusting sponges (several species), corallimorphids, and ophiuroids with extensive coral framework following the exposed region, with living *Lophelia pertusa*. This section of the transect was characterized by a high biological diversity with the following taxa observed in the coral framework: corallimorphids, *Anthomastus* sp., gorgonians, encrusting and desmospongiae sponges, mobile crinoids, black corals, brisingids, asteroids, glass sponges, ascidians, and bamboo corals. The remainder of the transect was characterized by vertical hard ground, with occasional sediment-draped terraces characterized by encrusting sponges, corallimorphids, gorgonians, echiuran, and ophiuroids (Figure 47.3C). Geogenic reef habitat was identified from ~75% of the images reviewed.

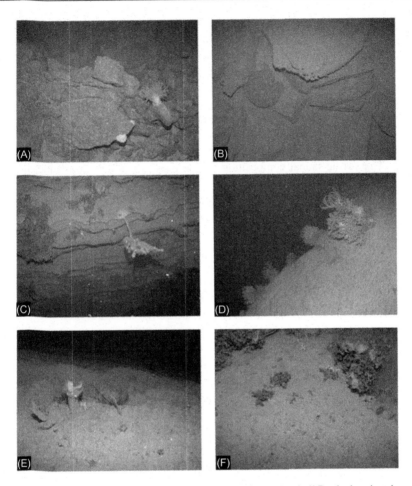

Figure 47.3 Underwater digital stills imagery acquired at the Rockall Bank showing the biological and geological features of the study area. (A) Numerous ophiuroids and at least four species of sponges present on hard substrate, along with the Actinarian anthozoan. (B) Numerous small tubes visible on the overhangs; a number of sponges and the Alcyonacean anthozoan *Anthomastus* sp. were observed. (C) Among the coral framework on this steep slope numerous ophiuroids and sponges were noted, and a number of dead bivalve shells *cf. Acester* sp. (D) High biological diversity occurs around the *Lophelia* reef. On the surrounding hard ground, two species of encrusting sponge were observed, along with ophiuroids and numerous small tubes, while the coral framework itself bears brisingid sp., sponges, and ophiuroids. (E) A rich diversity of faunal types occurs on the irregular terrain, including a variety of sponges, crinoids, and ophiuroids as well as cup corals, a hydroid, and asteroid. (F) Sponges and ophiuroids were observed occurring on the coral rubble.

Midway along the study area, transect D4 focused on an escarpment feature running parallel to the slope (Figure 47.2D). The transect occurred over a distance of 1.5 km from a starting depth of 1,600 m crossing the escarpment. The seabed here was characterized by soft sediment with occasional cerianthids, desmospongia, and crinoids observed. Hard substrate (mainly drop stones and patches of exposed hard ground) were colonized by solitary corals, crinoids, anemones, and sponges (Figure 47.3F). In deeper water, the seabed was characterized by a mud overlay and patches of exposed hard ground. Small terraces and inclined hard grounds were observed over a distance of ~200 m. Boulders and rock outcrops with overhangs were present. The sessile faunal composition on the hard ground and overhangs included several species of desmospongiae, at least two species of encrusting sponges, and the glass sponge (*Aphrocallistes* sp). Three species of antipatharian corals, solitary corals and anemones, ascidians, and stylasteroid hydrozoans were observed. The motile fauna comprises ophiuroidea, asteroidea, echinoidea, and crionoidea. The seabed below the escarpment was characterized by soft substrate (Figure 47.3E). Geogenic reef was reported from ~40% of images reviewed.

Surrogacy

To date, no statistical analyses have been carried out on this data set to examine relationships between physical surrogates and benthos.

Future Work

It is proposed that the underwater video data presented in this case study may assist further investigation of the geomorphology of the RBMF, with future surveys planned to acquire cores from the region. Terrain analyses using topographic algorithms would further define the complexity of the seabed at the study sites. Statistical analyses of the images would provide additional information on the relationship between the benthic fauna and substrate.

Acknowledgments

The authors wish to acknowledge the expert contributions of those who participated in the 2009 survey. We acknowledge the professionalism of the captain and crew of the *R.V. Celtic Explorer* during CE0915, and the survey and administrative support received from Advanced Mapping Services and Research Vessel Operations at the Marine Institute. We thank the staff of the INFOMAR programme at the Geological Survey of Ireland for their valuable contributions, survey assistance, and data support throughout the project. Images from the 2009 Offshore Geogenic Reef Mapping Project are provided courtesy of the Department of the Environment, Heritage and Local Government, and the Marine Institute and Geological Survey of Ireland as part of INFOMAR.

References

[1] A.L. New, D. Smythe-Wright, Aspects of circulation in the Rockall trough, Cont. Shelf Res. 21 (2001) 777–810.

[2] J.A. Howe, M.S. Stoker, K.J. Woolfe, Deep-marine seabed erosion and gravel lags in the northwestern Rockall Trough, North Atlantic Ocean, J. Geol. Soc. Lond. 158 (2001) 427–438.

[3] M.S. Stoker, R.J. Hoult, T. Nielsen, B.O. Hjelstuen, J.S. Laberg, P.M. Shannon, et al., Sedimentary and oceanographic responses to early Neogene compression on the NW European margin, Mar. Pet. Geol. 22 (2005) 1031–1044.

[4] Geological Survey of Ireland, Report of the Survey in Zone 3 of the Irish National Seabed Survey, Volume 1: Describing the Results and the Methods Used, GOTECH, Dublin, 2002.

[5] B. Dorschel, A.J. Wheeler, X. Monteys, K. Verbruggen, Atlas of the Deep-Water Seabed: Ireland, Springer International, Dordrecht, Heidelberg, London, New York, 2010.

[6] G.M. Elliott, P.M. Shannon, P.D.W. Haughton, L.K. Øvrebo, The Rockall Bank Mass Flow: collapse of a moated contourite drift onlapping the eastern flank of Rockall Bank, west of Ireland, Mar. Geol. 27 (2010) 92–107.

[7] D. Evans, Z. Harrison, P.M. Shannon, J.S. Laberg, T. Nielsen, S. Ayers, et al., Palaeoslides and other mass failures of Pliocene to Pleistocene age along the Atlantic continental margin of NW Europe, Mar. Pet. Geol. 22 (2005) 1131–1148.

[8] J.-C. Faugères, E. Gonthier, F. Grousset, J. Poutiers, The Feni Drift: the importance and meaning of slump deposits on the Eastern slope of the Rockall Bank, Mar. Geol. 40 (1981) 49–57.

[9] R.D. Flood, C.D. Hollister, P. Lonsdale, Disruption of the Feni sediment drift by debris flows from Rockall Bank 32 (1979) 311–334.Mar. Geol. 32 (1979) 311–334.

[10] D.G. Roberts, Slumping on the eastern margin of the Rockall Bank, North Atlantic Ocean, Mar. Geol. 13 (1972) 225–237.

[11] V. Unnithan, P.M. Shannon, K. McGrane, P.W. Readman, A.W.B. Jacob, R. Keary, et al., Slope instability and sediment redistribution in the Rockall Trough: constraints from GLORIA, in: P.M. Shannon, P.D.W. Haughton, D.V. Corcoran (Eds.), The Petroleum Exploration of Ireland's Offshore Basins, vol. 188, Geological Society, London, 2001, pp. 439–454. Special Publication

[12] J.B. Wilson, 'Patch' development of the deep-water coral *Lophelia pertusa* (L.) on Rockall Bank, J. Mar. Biol. Assoc. UK 59 (1979) 165–177.

[13] A. Jensen, R. Frederiksen, The fauna associated with the bank forming deepwater coral *Lophelia pertusa* (Scleractinia) on the Faroe shelf, Sarsia 77 (1992) 53–69.

[14] M. Costello, M. McCrea, T. Lundälv, L. Jonsson, B. Bett, et al., Role of cold-water *Lophelia pertusa* coral reefs as fish habitat in the NE Atlantic, in: A. Freiwald. André; Roberts, J. Murray (Eds.), Cold-Water Corals and Ecosystems, Springer, Berlin, Heidelberg, New York, 2005, pp. 771–805.

[15] N.H. Kenyon, A.M. Akhmetzhanov, A.J. Wheeler, T.C.E. van Weering, H. de Haas, M.K. Ivanov, Giant carbonate mud mounds in the southern Rockall Trough, Mar. Geol. 195 (1–4) (2003) 5–30.

[16] N.H. Kenyon, M.K. Ivanov, A.M. Akmetzhanov, Cold water carbonate mounds and sediment transport on the Northeast Atlantic margin, IOC Technical Series, UNESCO, Paris, 1998, p. 179.

[17] A. Genin, P.K. Dayton, F.N. Spiess, Corals on seamount peaks provide evidence of current acceleration over deep-sea topography, Nature 322 (1986) 59–61.

[18] H.E. Huppert, K. Bryan, Topographically generated eddies, Deep Sea Res. 23 (1976) 655–679.

48 Evaluating Geomorphic Features as Surrogates for Benthic Biodiversity on Australia's Western Continental Margin

Franziska Althaus[1], Alan Williams[1], Rudy J. Kloser[1], Jan Seiler[1,2], Nicholas J. Bax[1]

[1]CSIRO Wealth from Oceans Flagship, Hobart Marine Laboratories, Hobart, Tasmania, Australia, [2]University of Tasmania, Tasmania, Australia

Abstract

We analyzed 75,349 video frames to compare megabenthos assemblages between four types of geomorphic features on Australia's western continental margin (2,000 km, ~100–1,000 m depths): undifferentiated shelf and slope, canyons, and one peak. These features were evaluated for their surrogacy potential in the context of an ecologically based, hierarchical habitat classification scheme. On this margin, characterized by few geomorphic feature types, megabenthos assemblages differed markedly between provinces—subdivisions of the marine environment determined by regional scale oceanography and differences in fauna—and between bathomes (depth zones); however, they showed weak relationships with geomorphic features. We conclude that, while some geomorphic features have high potential to act as surrogates for biodiversity at intermediate spatial scales, a hierarchical context is necessary to define and validate them within a larger, biogeographical context.

Key Words: hierarchical habitat classification, epifauna, video data, Western Australia

Introduction

Australia's western continental margin extends over ~2,000 km, from subtropical to temperate latitudes (~18–35°S; Figure 48.1). The regional oceanography overlying the deep continental shelf and slope area (~100–1,000 m depths) is profoundly influenced by the southward-flowing Leeuwin current (LC) to depths of ~300 m, and by a northward-flowing counter-current, the Leeuwin Undercurrent (LUC) below those depths. The LC is characterized by warm, low-salinity, low-productivity waters,

Seafloor Geomorphology as Benthic Habitat. DOI: 10.1016/B978-0-12-385140-6.00048-7

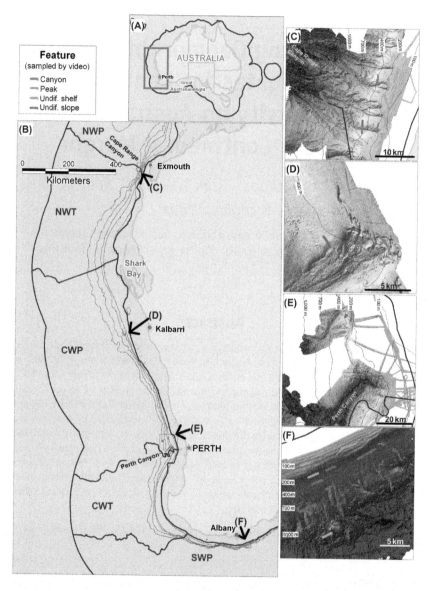

Figure 48.1 (A) Map of Australia indicating the sampling area. (B) Location map of Australia's western coast showing the biogeographic provinces (North, Central, and South Western Province—NWP, CWP, SWP), Transition Zones (Northern and Central Western Transition Zones—NWT, CWT), bathymetry contours (100, 200, 400, 700, and 1,000 m) associated with the bathome structure identified by Last et al. [1], and the location of video samples coded by geomorphic feature type (see legend). Insets (C–F) show high-resolution sun-illuminated swath bathymetry identified on (B) with boxes (grid size: 20 m; sun-illumination. 315° azimuth and 30° elevation). (C) Canyons off Exmouth (in the upper reaches of Cape Range Canyon). (D) A canyon head off Kalbarri. (E) The Perth Canyon and its tributary canyon to the north. (F) The slope off Albany with a small volcanic peak (sampled).

whereas the LUC is characterized by colder more oxygenated waters [2]. Based on the national bioregionalization of Australia [3] Australia's western continental margin extends over three biogeographic provinces and two transition zones (Figure 48.1). There are few types of broad-scale (>10 km) (*sensu* Harris et al. [4]) geomorphic features in depths <1,000 m described by Geoscience Australia [3]. These comprise the Perth Canyon, the Cape Range, and Montebello Canyon complexes (off Exmouth), two terraces below ~500 m (off Shark Bay and off the southwest corner of Australia); the remaining areas are classified as undifferentiated shelf and undifferentiated slope [3]. Additional high-resolution bathymetry from multibeam sonar (MBS) revealed other, mostly smaller, canyons incising the slope [5]. The southern margin (western Great Australian Bight; Figure 48.1A) is characterized by a series of slope-incising canyons, both broad- and small scale (> and <10 km, respectively; *sensu* Harris et al. [4]), listed by Heap and Harris [6]; these include the Vancouver, Kalgan, and Mermaid Canyons. A small-scale (~7 km base area) raised feature, termed volcanic peak by Harris et al. [4], also exists on the southern margin.

The relationships of megabenthos diversity to habitat heterogeneity at multiple spatial scales ranging from oceanographic features to local distributions of substrate types was examined by Williams et al. [5] based on catch samples. The identity of taxa comprising these samples is detailed by McEnnulty et al. [7]. Here, we used images and mapping data collected for the same project on the National Facility *RV Southern Surveyor* in 2005 (survey SS200507) to describe and examine the distribution of sessile epifauna in relation to geomorphic features. The role of these features as ecological habitats and their potential as surrogates for biodiversity were examined in the context of a hierarchical habitat classification scheme used by Last et al. [8] to define bioregions for marine management planning in Australian waters [3].

Geomorphic Features and Habitats

For the purposes of this paper, we refer only to the four classes of geomorphic features of both broad and small scale (*sensu* Harris et al. [4]) mapped with MBS and sampled during the SS200507 survey in 2005:

Undifferentiated shelf: Continental shelf from the coast to depths of ~200 m; samples were taken at ~100 m and at ~200 m. The width and steepness of the shelf varies from very narrow (~10 km) in the north, near Exmouth, to wide (>50 km) in the central west (Figure 48.1B). For the remainder of this document this class is referred to as shelf.

Undifferentiated slope: Continental slope, sampled between 200 and 1,000 m depth. The width and steepness of the slope along Australia's western coastline covaries with the shelf, being narrow and steep (~10–20 km) in the north and south (western Great Australian Bight - GAB), and wider and more gently sloping in the central west (Figure 48.1B). For the remainder of this document this is referred to as slope.

Canyons: Perth Canyon (Figure 48.1E) is the most prominent canyon in the region, with dimensions comparable to the Grand Canyon of the USA: >80 km long and ~10 km wide, it reaches a depth of 1,200 m about 10 km from the canyon head located at the 200 m contour. Video samples were taken only in its upper reaches, and in a smaller adjacent canyon (>34 km thalweg, 2–4 km width) to the north (Figure 48.1E). Video data were also taken from a system of small canyons located in the upper reaches of the Cape Range Canyon off Exmouth (Figure 48.1C), from a canyon head off Kalbarri (Figure 48.1D) and from one canyon in the western GAB. All canyons are of similar length and width as the canyon adjacent to the Perth Canyon (>30 km thalweg, 2–4 km width).

Peak: A volcanic peak situated in mid-slope depths (500 m contour) off Albany. It is a small-scale feature (*sensu* Harris et al. [4]) with a base circumference of ~10 km, an area of 7 km^2, an elevation of 180 m, and a depth at peak of 880 m. One successful video tow was taken on the lower reaches of this feature (~994–1,022 m; Figure 48.1F).

Within the ecologically based hierarchical habitat classification scheme of Last et al. [8] employed here, geomorphic features, regardless of their scale, represent habitats at the third level of the hierarchy. They are mappable within higher level subdivisions of the marine environment determined by regional scale oceanography and differences in fauna (biogeographic provinces at scales of 10,000 km^2—level 1) [8], as well as depth-related physical factors such as temperature and oxygen (bathomes at scales of 100–1,000 km^2—level 2) [8]. Habitats defined by substratum types (biotopes—terrains of hard, soft, and mixed substrata inferred from MBS; Figure 48.2) [9] are nested within geomorphic features as level 4 (scales of tens of meters to kilometers) [8]. Our analysis of habitats is based on 70 video transects with 75,349 individual, consecutive but nonoverlapping video frames (1 s intervals, area ~30 m^2), geolocated and assigned to a habitat type within the hierarchical scheme (Table 48.1).

Biological Communities

Associations of epibenthic megafauna with habitats along Australia's western margin were evaluated quantitatively summarizing the presence/absence scores of 11 fauna types (scored following Williams et al. [10]) in the video frames into percentage occurrences of dominant fauna for each sample. The scored fauna types were: sessile fauna absent, bioturbators, anemones, seapen, stalked sponge, ascidians, bryozoa, crinoids, sponges, coral, and coral (reef); all scored at low or high abundance (</>10% cover within each video frame). Each sample represented a unique combination of province, bathome, geomorphic feature, and biotope (Table 48.1). These samples were analyzed using multivariate analyses of similarity (ANOSIM) and nonmetric multidimensional scaling (MDS) to determine if there were significant differences between provinces, bathomes, and/or features [11]. Transition zones (biotones) exist between provinces but were not as intensively sampled, and thus were excluded from most statistical analyses.

Biotopes: The shelf and shelf-break areas (~80–250 m depths) of Australia's western coast sampled during our surveys were dominated by hard substrates; >50% of the sampled video frames were classified as hard in each of the three provinces (Figure 48.2) as well as in the Central Western Transition zone (CWT; Table 48.1).

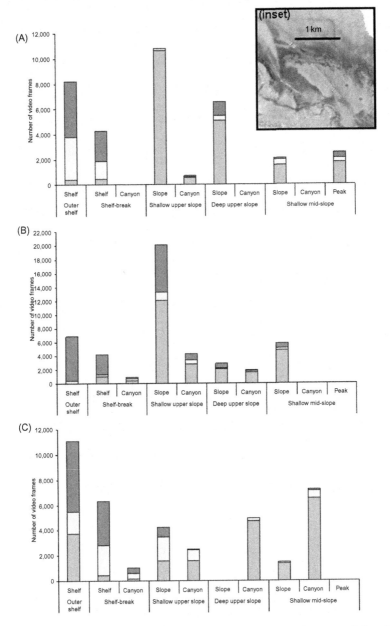

Figure 48.2 Classification of video frames into primary biotopes (substrates as inferred from MBS: hard—red; intermediate—yellow; soft—green; [9]), nested within the habitat classification hierarchy of Last et al. [8]; separate plots are shown for the three biogeographic provinces off Australia's west coast (Figure 48.1B): (A) SWP, (B) CWP, and (C) NWP. The inset gives an example of a video transect classified into primary biotopes overlaid on a MBS backscatter map, showing the correspondence of the classified hard substrates with the dark gray shading representing intense backscatter. (For interpretation of the references to color in this figure legend, the reader is referred to the web version of this book.)

Table 48.1 Habitat Classification Following Last et al. [8] of the Areas of Australia's West Coast Sampled During the National Facility
RV Southern Surveyor Voyage SS200507

Level 1: Biogeographical Province	Level 2: Bathome	Bathome Depth Range	Level 3: Geomorphical Unit (Feature)	Level 4: Primary Biotope					
				Soft		Intermediate		Hard	
				Number of Frames	%	Number of Frames	%	Number of Frames	%
Southern Province	Outer shelf (~100 m)	80–150 m	Shelf	385	4.7	3372	41.1	4448	54.2
	Shelf-break (~200 m)	150–250 m	Shelf	439	10.3	1427	33.3	2413	56.4
	Shallow upper slope (~400 m)	250–500 m	Slope	10,662	98.4	137	1.3	33	0.3
			Canyon	556	78.2	68	9.6	87	12.2
	Deep upper slope (~700 m)	500–800 m	Slope	5,115	78.0	344	5.2	1099	16.8
	Shallow mid-slope (~1,000 m)	800–1,100 m	Slope	1,554	74.2	429	20.5	111	5.3
			Peak	1,784	70.3	289	11.4	465	18.3
Central Western Transition	Outer shelf (~100 m)	80–150 m	Shelf	66	3.7	311	17.4	1406	78.9
	Shallow upper slope (~400 m)	250–500 m	Slope	4,504	99.9	6	0.1	0	0.0
Central Western Province	Outer shelf (~100 m)	80–150 m	Shelf	61	0.9	377	5.5	6446	93.6
	Shelf-break (~200 m)	150–250 m	Shelf	1,006	23.9	271	6.4	2941	69.7
			Canyon	415	44.3	337	36.0	184	19.7
	Shallow upper slope (~400 m)	250–500 m	Slope	12,024	59.6	1225	6.1	6921	34.3
			Canyon	2,845	65.9	574	13.3	901	20.9
	Deep upper slope (~700 m)	500–800 m	Slope	2,167	75.6	93	3.2	606	21.1
			Canyon	1,667	86.5	158	8.2	103	5.3
	Shallow mid-slope (~1,000 m)	800–1100 m	Slope	4,844	83.4	247	4.3	715	12.3
North Western Transition	Shallow upper slope (~400 m)	250–500 m	Slope	5,126	99.7	13	0.3	0	0.0
North Western Province	Outer shelf (~100 m)	80–150 m	Shelf	3,763	33.9	1717	15.5	5624	50.6
	Shelf-break (~200 m)	150–250 m	Shelf	421	6.7	2416	38.3	3471	55.0
			Canyon	1,59	15.5	452	43.9	418	40.6
	Shallow upper slope (~400 m)	250–500 m	Slope	1,580	37.4	1921	45.5	723	17.1
			Canyon	1,604	64.6	821	33.1	58	2.3
	Deep upper slope (~700 m)	500–800 m	Canyon	4,733	95.7	212	4.3	0	0.0
	Shallow mid-slope (~1,000 m)	800–1,100 m	Slope	1,354	91.7	122	8.3	0	0.0
			Canyon	6,515	90.1	591	8.2	121	1.7
Total				75,349	56.8	17,930	13.5	39294	29.6

Soft substrates made up a large fraction (34%) of outer shelf samples only in the North Western Province (NWP). On the slope (below ~200 m) mostly soft substrates (>60%) were observed, regardless of province or feature. Only in the shallow upper slope of the Central Western Province (CWP) did we find a high percentage of hard substrate (34%; Figure 48.2).

SWP biological communities: Biological communities on the shelf of the SWP had high occurrences of corals (>29% of video frames; Figure 48.3A). A high proportion of the hard substrates was devoid of fauna (35%), while sponges were relatively common (34%) on intermediate substrates. Bioturbators (>55%) dominated the few video frames from soft substrates. The shelf-break had an extensive cover of sponges (>87%). Anemones and bioturbators were the most common communities of the mostly soft substrates on the shallow upper slope (>31% and >18%, respectively), with the exception of the few observations of hard substrate in the canyon where 49% of video frames showed no sessile fauna. Sponges, bioturbators, and anemones were most prevalent on the deep upper-slope bathome (>17%, >16%, and >13%, respectively). The shallow mid-slope bathome in the south showed a high level of bioturbators on the slope (>79%), while the peak had anemones and corals (>17% and >10%, respectively) in some areas, although >19% of video frames on the peak showed no sessile fauna (Figure 48.3A).

CWP biological communities: Sponge communities clearly covered the mostly hard substrates of the shallowest two bathomes (80–250 m) in the CWP, regardless of feature: 44% of video frames on the outer shelf—hard; >82% on the shelf-break. The few soft substrates on the outer shelf were characterized by the absence of sessile fauna (>82%). With increasing depth, sponge communities were replaced by bioturbators, particularly on the soft substrates of the slope bathomes (>51%); although canyons showed higher proportions of video frames characterized by sponges on the deep upper slope (68%). Other noteworthy faunal types were seapens in soft substrates within the upper-slope bathome (14%) and crinoids in the mid-slope bathome (>38%; Figure 48.3B).

NWP biological communities: In all bathomes of the NWP, bioturbators were most prevalent (>33% of frames). Sponge communities were recorded commonly (34%) on the hard substrates of the outer shelf bathome, as well as in the canyons at mid-slope depths (>21%). Seapens made up 59% of the fauna type recorded on soft substrates in the canyons at shelf-break depths. A stalked glass sponge (Figure 48.4) was the most common fauna type in the soft and intermediate substrates of the canyon of the deep upper-slope bathome (>33%; Figure 48.3C).

A selection of *in situ* images of the biological communities within each of the five bathomes and biogeographic provinces is shown in Figure 48.4, with a description of each image in Table 48.2. The association between biotopes, as inferred from MBS, and the dominant fauna, as scored from video, may not always match expectations— e.g., bioturbators on hard substrates (Figure 48.3). Associations developed at fine scales here require the interpretation of data from multiple sampling devices (typically acoustics and video) at different scales (here ~400 m^2 and ~30 m^2, respectively) and resolutions [9]. This multiple-scale interpretation may therefore lead to apparent mismatches

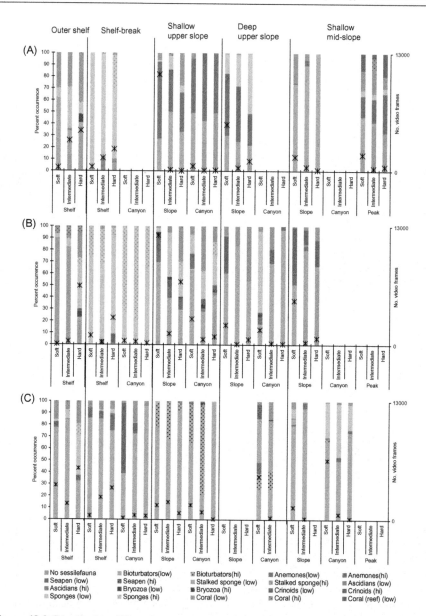

Figure 48.3 Distribution of faunal types scored in video (abundance score high/low: >/<10% area cover per image), as percentage occurrence of video frames in bathomes, geomorphic features, and biotopes [8] on the western Australian margin for the three biogeographic provinces (A) SWP, (B) CWP, and (C) NWP (Figure 48.1B). The asterisks, related to the secondary y-axis, show the number of video frames scored in each category.

(A) South Western Province   (B) Central Western Province   (C) Northern Western Province
Outer shelf

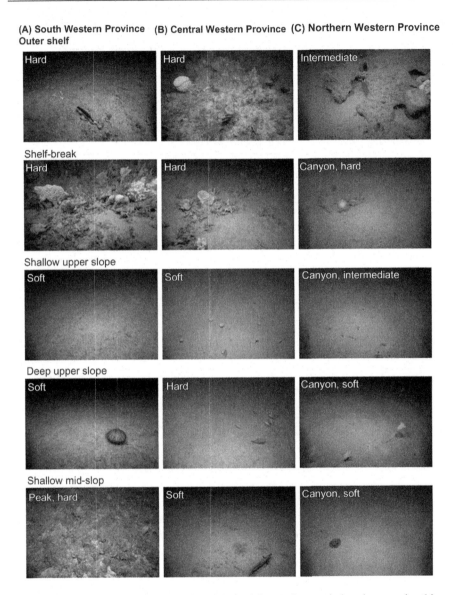

Figure 48.4 Seabed images illustrating a variety of fine-scale associations between benthic megafauna and physical habitats in each of the five bathomes of the three biogeographic provinces present off Australia's west coast (Figure 48.1B). The geomorphic feature (other than shelf/slope) and biotope inferred from MBS are indicated in each image; Table 48.2.

as shown in the video frame of Figure 48.4, where the substrate of CWP deep upper slope appears soft in the image but was identified as hard from the acoustic data.

Transition zones biological communities: The transition zones between provinces were less thoroughly sampled. Sponges, bryozoans, and corals occurred in

Table 48.2 Habitat Descriptions to Accompany Figure 48.4

(A) South Western Province	(B) Central Western Province	(C) North Western Province
Outer shelf		
Subcropping rock with low, encrusting sponges and bryozoa and a compound ascidian	Outcropping rock with diverse community of sponges, corals, and bryozoa	Outcropping rock with large sponges attached to the hard substrate
Shelf-break		
Outcropping rock with diverse community dominated by sponges	Subcropping rock with attached sponges	Small outcropping boulder providing an attachment point for an anemone; a fish
Shallow upper slope		
Soft, muddy sediments with a layer of detritus; cucumber fish	Soft, muddy sediments with sparsely distributed sponges	Bioturbated muddy sediments, some shrimps
Deep upper slope		
Soft, muddy sediments with a layer of detritus; pancake urchin	Muddy sediments (possibly forming a veneer over hard rock) with a seapen; some fish	Bioturbated muddy sediments with a stalked glass sponge
Shallow mid-slope		
Rocky substrate covered in orange solitary and purple colonial corals, also visible a glass sponge cup	Bioturbated muddy sediments with some hydroids; an eel	Soft, muddy sediments; a shrimp and an octopus

Descriptions in Each Cell Correspond to the Photograph in the Equivalent Cell in Figure 48.4.

all biotopes of the CWT zone shelf (all >23%); this bathome was not sampled in the North Western Transition zone (NWT). Bioturbators were most common on the mostly soft or intermediate substrates of the shallow upper-slope bathome of both the CWT and the NWT (>52% and >85%, respectively); anemones were also common in this bathome of the CWT (33%).

Statistical analyses: Two-way crossed analysis of similarity (ANOSIM) of the faunal data set identified a significant difference between bathomes across provinces (and transition zones) ($R = 0.533$, $p = 0.001$), and between provinces across bathomes ($R = 0.556$, $p = 0.001$) (Table 48.3). Pairwise tests showed that the shelf differed from all slope bathomes (except the deepest; $R = 0.6$), while the shelf-break differed from all slope bathomes ($R > 0.6$). The shelf and shelf-break were also very different from each other ($R = 0.53$). Pairwise tests between provinces showed that the NWP was clearly different from both other provinces ($R > 0.61$); the difference between the SWP and the CWP was not as prominent ($R = 0.38$). The transition zones were not intensively enough sampled to warrant further analytical treatment

Table 48.3 Summary of the Results of a Series of Two-Way Crossed Analysis of Similarities (ANOSIM) Tests

(a) All provinces and transition zones

Province (x Bathome): $R = 0.556$, $p = 0.001$
Pairwise tests

	SWP	SWT	CWP	CWT
SWT	0.10			
CWP	0.38	0.40		
CWT	0.90	1.00	0.40	
NWP	0.64	0.61	0.71	−0.03

Bathome (x Province): $R = 0.533$, $p = 0.001$
Pairwise tests

	Shelf	Shelf-edge	S upper slope	D upper slope
Shelf-edge	0.53			
S upper slope	0.61	0.68		
D upper slope	0.90	0.90	0.35	
S mid-slope	0.15	0.61	0.35	0.42

All provinces and transition zones

Feature (x Province): $R = 0.198$, $p = 0.03$
Pairwise tests

	Shelf	Slope	Canyon
Slope	0.53		
Canyon	0.09	0.01	
Peak	0.42	0.41	0.78

Feature (x Bathome): $R = 0.196$, $p = 0.04$
Pairwise tests

	Shelf	Slope	Canyon
Slope	Undef		
Canyon	0.08	0.12	
Peak	Undef	0.73	1.00

(b) South Western Province (SWP)

Bathome (x Feature): $R = 0.863$, $p = 0.001$
Pairwise tests

	Shelf	Shelf-edge	S upper slope	D upper slope
Shelf-edge	1.00			
S upper slope	Undef	Undef		
D upper slope	Undef	Undef	0.26	
S mid-slope	Undef	Undef	1.00	1.00

Feature (x Bathome): $R = 0.593$, $p = 0.04$
Pairwise tests

	Shelf	Slope	Canyon
Slope	Undef		
Canyon	Undef	0.19	
Peak	Undef	1.00	Undef

(Continued)

Table 48.3 (continued)

(c) Central Western Province (CWP)

Bathome (x Feature): $R = 0.535$, $p = 0.001$
Pairwise tests

Feature (x Bathome): $R = 0.049$, $p = 0.371$
Pairwise tests not applicable

	Shelf	Shelf-edge	S upper slope	D upper slope
Shelf-edge	**0.78**			
S upper slope	Undef	**1.00**		
D upper slope	Undef	**0.74**	0.28	
S mid-slope	Undef	Undef	0.44	0.22

(d) North Western Province (NWP)

Bathome (x Feature): $R = 0.446$, $p = 0.001$
Pairwise tests

Feature (x Bathome): $R = 0.187$, $p = 0.11$
Pairwise tests not applicable

	Shelf	Shelf-edge	S upper slope	D upper slope
Shelf-edge	0.30			
S upper slope	Undef	0.04		
D upper slope	Undef	**0.92**	**0.50**	
S mid-slope	Undef	Undef	0.44	0.39	**1.00**

Notes: (a) comparing provinces/transition zones, bathomes, and features over the entire coast off western Australia (Figure 48.1B); and (b–d) separate tests comparing features and bathomes within the three provinces. The R and p values for the overall test are given, along with the R-values for the pairwise tests where these were appropriate (i.e., overall test was significant); R-values referred to in the text are bolded. Notation for upper-slope bathome: S—shallow, D—deep.

of these data. There was a significant difference between features crossed by province or bathome, however R-values were low, indicating little explanatory power ($R = 0.198$, $p = 0.03$ and $R = 0.196$, $p = 0.04$); the main difference was between peak and all other features (pairwise $R > 0.4$) and for the feature across province ANOSIM between undifferentiated shelf and undifferentiated slope (pairwise $R > 0.4$).

To test for differences of faunal compositions between geomorphic features, we looked at each province separately. MDS plots of the Bray–Curtis dissimilarity between samples showed the separation of the shelf and shelf-break bathomes from the slope bathomes in the SWP and CWP; this separation was not observed in the NWP (Figure 48.5). Geomorphic feature types were not a consistent classification for clustering faunal composition in any of these ordination plots. Peak and canyon samples in the SWP were the only clear groupings of geomorphic features, and these were not all distinct from other geomorphic feature types. This result is confirmed by separate ANOSIM analyses for each of the three provinces (Table 48.3). While the results for bathomes across geomorphic features were significant ($p = 0.001$) with R-values of 0.863, 0.535, and 0.446 from south to north, the analysis of geomorphic features across bathomes yielded only one significant result ($R = 0.593$, $p = 0.004$) for the SWP. The pairwise test showed that the result stemmed from the difference between the undifferentiated slope and the peak in the mid-slope bathome ($R = 1$).

Surrogacy

Our data, and a previous study using catch data [5], show the degrees of association between epibenthic megafauna and seabed habitats along Australia's western continental margin (~100 to ~1,000 m depths) are both taxon and scale dependent. Surrogacy relationships therefore need to be specified in terms of both taxonomic resolution and spatial scale. Within the hierarchy of an ecologically based habitat classification scheme [8], and with fauna resolved as 11 dominant types on three categories of substratum, our data showed a consistently strong relationship with biogeographic provinces (scales of 10,000 km^2 [8]); with bathomes (depth zones at scales of 100–1,000 km^2 [8]); and, as shown by Williams et al. [5], indirectly, with substratum types (scales of tens of meters to kilometers) [8]. Relationships are variable at the intermediate scales defined by geomorphic units using the IHO classification of 53 types. Only peak—a small-scale (*sensu* Harris et al. [4]) geomorphic feature—was strongly differentiated from other types. The contrasts between geomorphic features may be underestimated in this study by low density of epifaunal sampling and the low taxonomic resolution provided by video-based definition of dominant fauna. However, weak delineation of the undifferentiated shelf and undifferentiated slope geomorphic features—the majority of the shallow western margin (<1,000 m depths)—and the often large spatial scales of individual features [6], also results in them adding little surrogacy potential to that provided at the higher hierarchical levels of bathomes and provinces [12]. Conversely, strong relationships between fauna and substrata (as inferred from MBS) are apparent at the fine

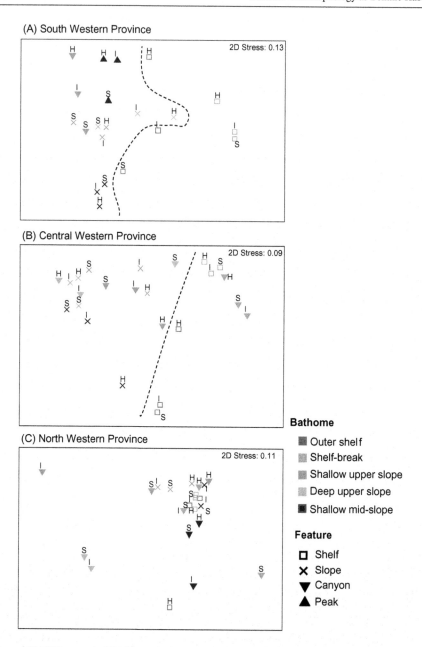

Figure 48.5 Nonmetric MDS ordinations of the faunal composition of aggregated video data for each of the three biogeographic provinces off Australia's west coast (Figure 48.1B). The habitat classification of the samples is identified by color (Bathome), symbol (Feature), and label (Primary Biotope, inferred from MBS: H—hard, I—intermediate, S—soft substrate); the dashed lines show the main separation of groups in the plots. (For interpretation of the references to color in this figure legend, the reader is referred to the web version of this book.)

scales (meters) seen in images (Figure 48.4), and the patchy terrains (10s m to km) they form (biotopes, *sensu* [8]). These observations show the potentially high utility of geomorphic features to act as surrogates for biodiversity at intermediate spatial scales, but that a hierarchical context is necessary to define and validate them. It is necessary to have information at coarser scales (e.g., the depth zones in which features exist) to provide ecologically meaningful context, while information at finer scales (e.g., biotopes represented within the feature) provides ecologically meaningful detail.

References

[1] P. Last, V. Lyne, G. Yearsley, D. Gledhill, M. Gomon, T. Rees, et al., Department of the Environment and Heritage and CSIRO Marine Research, Validation of national demersal fish datasets for the regionalisation of the Australian continental slope and outer shelf (>40 m depth), 2005, 99pp.

[2] M. Fieux, R. Molcard, R. Morrow, Water properties of the Leeuwin current and eddies off Western Australia, Deep Sea Res. I 52 (2005) 1617–1635.

[3] A. Heap, P.T. Harris, A. Hinde, M. Woods, Geoscience Australia, Benthic marine bioregionalisation of Australia's Exclusive Economic Zone—Report to the national Oceans Office on the development of a National benthic marine bioregionalisation in support of regional marine planning, 2005.

[4] P.T. Harris, S.L. Nicol, T.J. Anderson, A.D. Heap, Habitats and benthos of a deep-sea marginal plateau, Lord Howe Rise, Australia, in: P.T. Harris (Ed.), Seafloor Geomorphology as Benthic Habitat. pp 777–789.

[5] A. Williams, F. Althaus, P.K. Dunstan, G.C.B. Poore, N.J. Bax, R.J. Kloser, et al., Scales of habitat heterogeneity and megabenthos biodiversity on an extensive Australian continental margin (100–1100 m depths), Mar. Ecol. 31 (2010) 222–236.

[6] A.D. Heap, P.T. Harris, Geomorphology of the Australian margin and adjacent seafloor, Aust. J. Earth Sci. 55 (2008) 555–585.

[7] F. McEnnulty, K. Gowlett-Holmes, A. Williams, F. Althaus, J. Fromont, G.C.B. Poore, et al., The deepwater megabenthic invertebrates on the western continental margin of Australia (100–1500 m depths): composition, distribution and novelty. Record West. Aust. Mus. 80 (2011) 1–191.

[8] P.R. Last, V.D. Lyne, A. Williams, C.R. Davies, A.J. Butler, G.K. Yearsley, A hierarchical framework for classifying seabed biodiversity with application to planning and managing Australia's marine biological resources, Biol. Conserv. 143 (2010) 1675–1686.

[9] R.J. Kloser, J. Penrose, A. Butler, Multi-beam backscatter measurements used to infer seabed habitats, Continent. Shelf Res. 30 (2010) 1772–1782.

[10] A. Williams, C. Gardener, F. Althaus, B. Barker, D. Mills, Understanding shelf-break habitat for sustainable management of fisheries with spatial overlap. Final report to the FRDC, Project no. 2004/066. 2009, 254pp.

[11] K.R. Clarke, Non-parametric multivariate analyses of changes in community structure, Aust. J. Ecol. 18 (1993) 117–143.

[12] A. Williams, N.J. Bax, R.J. Kloser, F. Althaus, B. Barker, G. Keith, Australia's deepwater reserve network: implications of false homogeneity for classifying abiotic surrogates of biodiversity, ICES J. Mar. Sci. 66 (2009) 214–224.

49 Habitats of the Chella Bank, Eastern Alboran Sea (Western Mediterranean)

Claudio Lo Iacono[1], Eulàlia Gràcia[1], Rafael Bartolomé[1], Enrique Coiras[2], Juan Jose Dañobeitia[1], Juan Acosta[3]

[1]Unidad de Tecnología Marina, Consejo Superior de Investigaciones Científicas (UTM-CSIC), Barcelona, Spain, [2]NATO Undersea Research Centre (NURC), La Spezia, Italy, [3]Instituto Español de Oceanografía (IEO), Madrid, Spain

Abstract

This study is the first characterization of the geomorphic features and benthic habitats of the Chella Bank, a flat-topped volcanic peak situated in the Eastern Alboran Sea, Western Mediterranean. The Chella Bank, also named "Seco de los Olivos," occurs along the upper slope of the Almeria Margin, showing a subcircular shape and covering a surface area of $100\,km^2$ within a depth range of 70–700 m. High-resolution swath bathymetric mapping reveals three main large-scale morphological features on Chella Bank: the flat subhorizontal top and two main ridges, located to the west and to the east of the bank-top. Video tracks acquired in the area showed the occurrence of macrobenthic communities such as gorgonian assemblages (*Callogorgia verticillata*, *Viminella flagellum*), small patches of living cold-water corals (*Madrepora oculata*), and sponges (*Fakelia ventilabrum*). An automatic classification of multibeam data has been tested on the area as a predictive habitat mapping method, using backscatter characteristics (intensity, texture) and depth measurements as surrogate descriptors of the habitats recognized on the Chella Bank.

Key Words: Swath bathymetry, Seamounts, Geomorphology, Habitat mapping, Alboran Sea, Western Mediterranean

Introduction

This study of the Chella Bank ($36°31'N$, $2°52'E$) in the NE Alboran Sea (Figure 49.1) integrates swath bathymetry and video images. The Alboran Sea is located in the SW

Seafloor Geomorphology as Benthic Habitat. DOI: 10.1016/B978-0-12-385140-6.00049-9

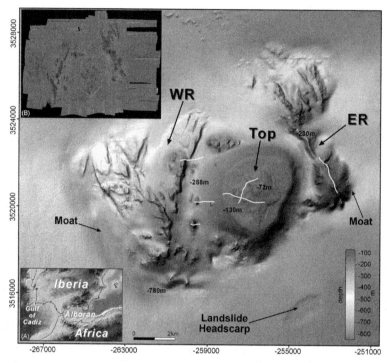

Figure 49.1 Multibeam bathymetric map of the Chella Bank based on a 10m grid. WR: western ridge; ER: eastern ridge. White lines: ROV and deep-towed video camera tracks. (A) Location map showing the Alboran Sea and the survey area, outlined in red; (B) backscatter image of the area (2 m resolution). (For interpretation of the references to color in this figure legend, the reader is referred to the web version of this book.)

Mediterranean, along the westernmost sector of the Alpine orogenic belt (Betic-Rif Cordilleras), and corresponds to a back-arc domain developed during the Neogene as a consequence of the collision between the African and European continental plates [1,2]. The Alboran Sea includes several clusters of seamounts and volcanic banks are present [3,4]. Amongst them, the Chella Bank, located on the western slope of the Almeria Margin, is the geomorphological expression of middle Miocene to Pleistocene calc-alkaline K-rich volcanism [5] (Figure 49.1). The Chella Bank displays a subcircular shape within a depth range of 70–700m and covers an area of up to 100km^2 [6] (Figure 49.1). It consists of a subhorizontal flat top and two ridges previously named as AW (Afloramiento Oeste–West Volcanic Exposure) and ANE (Afloramiento Nor-Este–Northeast Volcanic Exposure), respectively [4]. Previous studies in the area described the Chella Bank as a potential habitat for cold-water corals due to the presence of a rough seafloor on the top of the bank formed by carbonate build-ups [4,5]. Even though the composition of the Chella Bank and the seamounts in the Alboran Sea is volcanic [4,7], some banks are partially comprised of metasediments and amphibole gabbro rocks [8]. Eastward currents in the area are a consequence of the periodic formation of the anticyclonic Eastern Alboran Gyre (EAG) [9]. Nevertheless, a reliable estimation of the local hydrographic setting should take into account the local seafloor topography and the proximity to the coast of the Chella Bank. The Chella Bank is subjected to intense bottom trawling and fishing-line activities due to the presence of many commercial fishery species on its top. Some of the benthic habitats of the area, such as cold-water corals, are endangered by the fishery activities and frequently appear in a poor state of conservation. The Chella Bank is one of the study areas of the LIFE-INDEMARES Project (www.indemares.es), which aims to contribute to the protection and sustainable use of the marine biodiversity in the Spanish seas.

Geomorphic Features and Habitats

The Chella Bank shows a subcircular shape with a diameter of 14 km in EW direction and 10 km in NS direction. The height of the bank ranges from 300 to 550 m. The depth of its base ranges from 370 m at the northern flank to 680 m at the southern flank (Figure 49.1). The flanks of the Chella Bank are very steep, with an average slope of 25° and a maximum slope of 48° along isolated areas. The base of the bank is partially surrounded by circular moats generated by deep-water currents. The most incised moat, up to 600 m deep, is located at the southwestern base of the bank, showing higher backscatter values than those of the surrounding area (Figure 49.1). Slope failure features, such as landslide headscarps, occur in front of the bank, at minimum depths of 800 m (Figure 49.1).

The drafting of a 10 m resolution bathymetric map using Fledermaus, Global Mapper, and Surfer programs allowed us to describe and measure the geomorphic features found in the area. Sediments sampled in this area were analyzed with a large-diameter settling tube for the coarse-grained fraction (>50 μm) and a Sedigraph 5000D for the fine-grained fraction (<50 μm).

Three main large-scale geomorphic features (100 m scale) can be distinguished on the bank: a flat subhorizontal top and two main ridges, located to the west and to the east of the top (Figure 49.1). Acoustic backscatter characteristics and depth measurements along the observed geomorphic features were automatically fused and then classified and ground-truthed using video transects of known biological communities as a ground-truth, with the aim of mapping and distinguishing the most prevalent habitats observed in the area.

The top of the Chella Bank: The top of the Chella Bank has a subcircular shape and covers a total surface area of 7.6 km² (Figure 49.2). The area is composed of an irregular central area showing a rough seafloor on a flat and subhorizontal seafloor (Figure 49.2). The irregular central area occurs in water depths from 76 to 118 m and displays local relieves of 5–40 m above the level of surrounding seafloor. It covers a surface area of 2.2 km², representing 28% of the total area of the top. Backscatter images of this sector alternate between highly reflective stripes, mainly present along the borders and correlated to rocky outcrops in the video images, and a central low-reflective facies consisting of sandy sediments. Along the northwestern region of the central rough area, two subhorizontal terraces, slightly dipping toward the northeast, occur in water depths of 100 and 75 m, respectively (Figure 49.2).

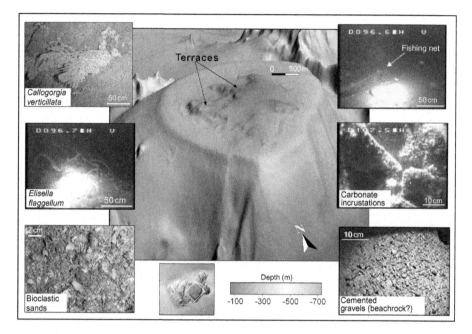

Figure 49.2 Three-dimensional multibeam image of the top of Chella Bank. Still images have been extracted from the video tracks.

The area surrounding the rough central sector consists of a smooth, low-reflective subhorizontal seafloor, with depths ranging between 130 and 115 m, corresponding to coarse sands and gravels, entirely composed of bioclasts (Figure 49.2).

The western ridge: The western ridge of the Chella Bank covers a depth range of 160–620 m and a surface area of 16 km² (Figure 49.3). The ridge exhibits a NS orientation in the southernmost area of the ridge and bifurcates toward the north in a northwest and a northeast direction (Figure 49.3). The flanks have an average slope of 8°, with maximum values of 17°. The western ridge has up to 25 crests and subcircular peaks, with a depth range of 200–500 m, coinciding with areas of high reflectivity. The surface areas of the individual peaks and crests range from 0.05 to 0.80 km², covering 34% of the entire ridge area. Peaks rise 15–35 m above the level of surrounding seafloor and are 100–300 m in diameter, and crests rise 15–40 m above the level of surrounding seafloor and are 1–3.4 km long. Most peaks and crests are orientated NW-SE or NS, coinciding with the orientation of normal and strike slip active fault systems observed to the north and to the west of the study area [10,11] (Figure 49.3).

The western ridge is connected to the top of the Chella Bank through a saddle 2 km wide and 5 km long, striking NS (Figure 49.3). Up to 12 peaks and crests are present along the saddle, with depths between 200 and 300 m (Figure 49.3). They are 10–70 m above the level of surrounding seafloor, and their flanks have an average slope of 15–20°. Peaks are up to 100 m in length and crests are 250–500 m long. These features are thought to be volcanic, although available data do not reveal their nature. In this region, a 400 m long, 20 m deep circular depression was mapped contiguous to a 250 m long, 20 m high peak (Figure 49.3).

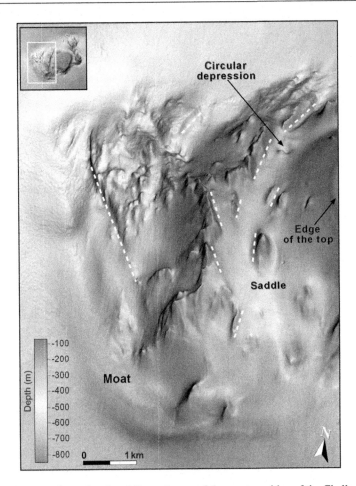

Figure 49.3 Three-dimensional multibeam image of the western ridge of the Chella Bank. Dashed white lines indicate the main orientation of the crests which coincide with the orientation of the active fault lineations observed in the Almeria Margin [10]. (For interpretation of the references to color in this figure legend, the reader is referred to the web version of this book.)

The eastern ridge: The eastern ridge of the Chella Bank develops over a depth range of 100–480 m along a NW-SE direction and covers a surface area of 10 km^2 (Figure 49.4). It consists of two main banks, located in the southern and northern regions of the ridge, connected through a linear, slightly sinuous 2 km long, 400 m wide structure (Figure 49.4). This linear structure rises up to 60 m above the level of surrounding seafloor and occurs in a water depth of 300 m. The southern bank is 200 m in elevation above the level of surrounding seafloor and shows a subconical shape, covering a surface area of 4 km^2. Its flanks have an average slope of 15–20°. The base along the eastern side is 470 m deep, with a 700 m long, 40 m deep erosive moat (Figure 49.4). Based on video images, the southern bank consists of basaltic slabs. The

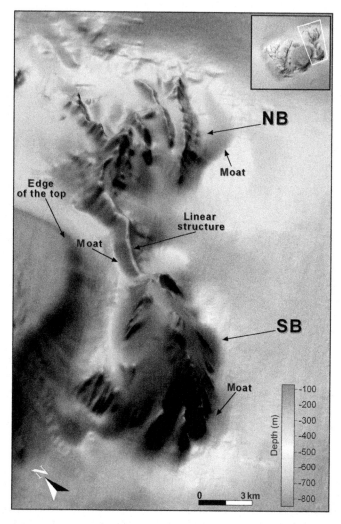

Figure 49.4 Three-dimensional multibeam image of the eastern ridge of the Chella Bank.
NB: Northern Bank; SB: Southern Bank.

northern bank covers a surface area of 3.5 km² and is 150–350 m deep. It consists of
four main NW-SE trending crests and a single NE-SW oriented crest. The crests are
1–2 km long and rise 20–30 m above the level of surrounding seafloor (Figure 49.4).

Biological Communities

Different biological communities associated with the studied geomorphic features
have been recognized in the video images.

The top of the Chella Bank: Video tracks along the terraces on top of the bank reveal the occurrence of rough rocky outcrops extensively covered by carbonate incrustations, and subhorizontal surfaces composed of cemented gravel (probably beachrocks) (Figure 49.2). These rocky outcrops, partly corresponding to carbonate relict bioherms, provide the substrate for extensive deep-sea gorgonian assemblages, mostly *Callogorgia verticillata* and *Viminella flaggellum*. Small patches of living stony corals (*Madrepora oculata*) are also present in the area (Figure 49.2). Video tracks show the habitat to be seriously damaged by constant trawling and longline fishing activity. Analysis of the composition of sand samples collected in the area reveal the presence of planktonic foraminifera, shells, mollusks (especially bivalves), echinoid spines, and bryozoans remains [6] (Figure 49.2).

Peaks and crests: Peaks and crests found along the western and eastern ridges of the Chella Bank are composed of hard massive bedrock. These features, with very steep subvertical walls, occur typically below 200 m water depth and correspond to habitats dominated mostly by the presence of hat-shaped glass sponges, probably *Asconema setubalense*. Spotty sandy sediment veneers are also found in these areas. Additionally, the presence of dead cold-water corals (mainly *Madrepora oculata* and *Lophelia pertusa*) and of related coral debris along the western and eastern ridges of the Chella Bank have recently been confirmed through the collection of rock samples [12] and video images [13].

The occurrence of suspension feeder organisms such as cold-water corals and gorgonians may be favored by bottom currents interacting with the rough local topography [14]. The bottom current flow approaching the rough peaks and upwelling to the top of a seamount can reach a velocity almost twice the mean values registered along the slope region with a flat seafloor [14]. Accelerated currents can increase the supply of planktonic food lifted from deep to surface waters, and induce a major recruitment and growth of passive suspension feeders able to exploit suspended particles as a potential food source [15].

Surrogacy

The local topographic gradient, the intensity and the textural variations of the acoustic backscatter were chosen as descriptive elements of the physical environment along the Chella Bank. Once compensated for the angular incidence and weighted in relation to the different insonified areas along the swaths, multibeam data were classified automatically [16]. A supervised approach was used for the automatic classification of the data [17]. The classifier was trained on multibeam samples calibrated with video images, which were used to ground-truth and to determine the class of interest. Based on the results of video analysis, the visited seafloor areas were set to one of the three following classes: gorgonians and patchy cold-water corals, sponges and patchy cold-water corals, and sediments. This resulted in a ground-truthed multibeam training set of 368 samples. Segmentation and classification of multibeam data was performed using Matlab's linear classifier [18], which resulted in the final class

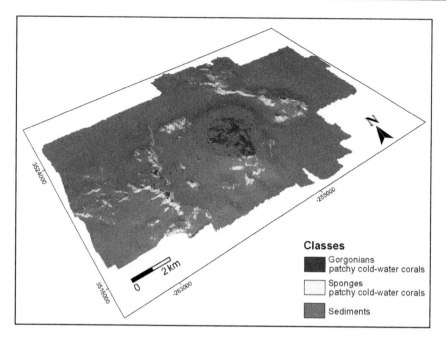

Figure 49.5 Automatic classification of the main macrobenthic communities on the Chella Bank.

map shown in Figure 49.5. Predictive mapping indicate that the gorgonians, cold-water corals, and related fauna occur at the top of the seamounts and along the crests of the ridges between 100 and 200 m (hard substrate with medium intensity and rich texture acoustic returns, flat seafloor). Spotty sponges are present on steep relieves (hard substrate with high-intensity, low-texture acoustic returns, deeply sloping). The third class generally refers to sediments (on soft substrate with low acoustic returns, flat or gently sloping). At the top of the bank, sediments correspond to coarse bioclastic sands plus epibenthic organisms and could also include spotty patches of coral rubble. To strengthen our results, the classification approach has also been validated by 10-fold cross-validation of the ground-truthed training data [19], which resulted in a reasonable 12% classification error. Onsite observations during forthcoming surveys in the area will be necessary to corroborate classification results. Apart from this validation, the new information will serve as additional training data for more robust estimation of habitat distribution in the area.

Acknowledgments

We thank the R/V *Garcia del Cid* and *Sarmiento de Gamboa* crews and the UTM technicians for their kind assistance during the video surveys. We acknowledge the Spanish National Projects EVENT (CGL2006-12861-C02-02) and SHAKE (CGL2011-30005-C02-02) of the Spanish Ministry of Science and Innovation (MICINN), the Acción Complementaria

EVENT-SHELF (CTM2008-03346-E) of the MICINN, and the Grup de Recerca de la Generalitat de Catalunya B-CSI (2009 SGR 146). We are grateful to J.M. Gili, C. Orejas, and A. Gori (ICM-CSIC, Barcelona) for their advice during video analysis, and to Susana Diez (UTM-CSIC) for her assistance in multibeam file conversion. We finally thank Montse Demestre (ICM-CSIC) and Xavier Monteys (GSI-Ireland) for providing the ROV and the digital still camera.

References

[1] J.F. Dewey, M.L. Helman, E. Turco, D.H.W. Hutton, S.D. Knott, Kinematics of the western Mediterranean, Geol. Soc. Lond. 45 (1989) 265–283.

[2] M.C. Comas, V. García-Dueñas, M.J. Jurado, Neogene tectonic evolution of the Alboran Sea from MCS data, Geo-Mar. Lett. 12 (1992) 157–164.

[3] J.D. Milliman, Y. Weiler, D.J. Stanley, Morphology and Carbonate sedimentation on Shallow Banks in the Alboran Sea, in: D.J. Stanley, (Ed.), The Mediterranean Sea: A Natural Sedimentation Laboratory. Dowden, Huthcinson and Ross, Stroudsburg, Pennsylvania (USA), pp. 241–259.

[4] A. Munòz, M. Ballesteros, I. Montoya, J. Rivera, J. Acosta, E. Uchupi, Alborán Basin, southern Spain—Part I: geomorphology, Mar. Petrol. Geol. 25 (2008) 59–73.

[5] S. Duggen, K. Hoernle, H. van de Bogaard, Magmatic evolution of the Alboran region: the role of subduction in forming the western Mediterranean and causing the Messinian Salinity Crisis, Earth Planet. Sci. Lett. 218 (2004) 91–108.

[6] C. Lo Iacono, E. Gràcia, S. Diez, G. Bozzano, X. Moreno, JJ. Dañobeitia, B. Alonso, Seafloor characterization and backscatter variability of the Almería Margin (Alboran Sea, SW Mediterranean) based on high-resolution acoustic data, Mar. Geol. 250 (2008) 1–18.

[7] J. Alvarez-Marrón, Pliocene to Holocene structure of the eastern Alborán Sea (western Mediterranean), in: R. Zahn, M.C. Comas, A. Klaus (Eds.), Proceedings of Ocean Drilling Program, Scientific Results, 1999, 161, 345–555. College Station, Texas.

[8] K. Hoernle, Ostatlantik Mittlelmeer-Schwarzes Meer. Part 1. Cruise No. M51, Leg 1. 12 September–15 October 2001. Warnemunde-Málaga. Meteor-Beriche 03-1, 2003, 38 pp.

[9] A. Viudez, J. Tintoré, Time and space variability in the eastern Alboran Sea from March to May 1990, J. Geophys. Res. 100 (C5) (1995) 8571–8586.

[10] E. Gràcia, R. Pallàs, J.I. Soto, M. Comas, X. Moreno, E. Masana, P. Santanach, et al., Active faulting offshore SE Spain (Alboran Sea): implications for earthquake hazard assessment in the Southern Iberian Margin, Earth Planet. Sci. Lett. 241 (2006) 734–749.

[11] D. Stich, J. Batlló, J. Morales, R. Maciá, D. Savka, Source parameter of the M_w = 6.1 1910 Adra earthquake (southern Spain), Geophys. J. Int. 155 (2003) 539–546.

[12] M. Taviani, the CORTI and COBAS Shipboard Teams, Coral mounds of the Mediterranean Sea: results of EUROMARGINS Cruises CORTI and COBAS, 2nd EUROMARGINS Conference, Barcelona 11–13 November, 2004.

[13] D.G. Masson, C. Brendt, 3D seismic acquisition over mud volcanoes in the Gulf of Cadi and submarine landslides in the Eivissa Channel, western Mediterranean Sea. R.R.S. Charles Darwin Cruise 178, 14 March – 11 April. Cruise Report No.3, 2006, p. 39.

[14] A. Genin, P.K. Dayton, P.F. Lonsdale, F.N. Spiess, Corals on seamounts peaks provide evidence of current acceleration over deep-sea topography, Nature 322 (1986) 3.

[15] J.M. Gili, R. Coma, Benthic suspension feeders: their paramount role in littoral marine food webs, TREE 13 (1998) 8.

[16] E. Coiras, C. Lo-Iacono, E. Gràcia, J.J. Dañobeitia, J.L. Sanz, Automatic segmentation of multi-beam data for predictive mapping of benthic habitats. Submitted to IEEE—Geosci. Remote Sens. Lett., 2010.

[17] E. Coiras, D. Williams, Approaches to automatic seabed classification: Pattern Recognition, In-Teh Publications, Croatia, 2009. pp. 461–472.

[18] W.J. Krzanowski, Principles of Multivariate Analysis, Oxford University Press, New York, NY, 1988. pp. 563.

[19] R.O. Duda, P.E. Hart, D.G. Stork, Pattern Classification, John Wiley & Sons, New York, NY, 2001. pp. 654.

50 Habitat Heterogeneity in the Nazaré Deep-Sea Canyon Offshore Portugal

Veerle A.I. Huvenne[1], Abigail D.C. Pattenden[1], Douglas G. Masson[1], Paul A. Tyler[2]

[1]National Oceanography Centre, Southampton, UK, [2]School of Ocean and Earth Science, University of Southampton, European Way, Southampton, UK

Abstract

Submarine canyons act as main transport pathways between the continental shelf and the deep sea, and provide a range of habitats for a diverse fauna. This study presents Nazaré Canyon, one of the main submarine canyons offshore Portugal. The canyon is over 200 km long, ranges from 50 to >4,800 m water depth, and is in places up to 1,600 m deep. It is not connected to a major river system: sediment input is through the interception of alongshore transport. The canyon was surveyed using shipborne and ROV-borne multibeam, 30 kHz sidescan sonar, cores, photographs, and video footage. High terrain heterogeneity at a medium (km) scale may be the main cause of the observed patchiness in epibenthic megafauna communities and high β-diversity. Community structure (assemblage similarity) is influenced by depth and substratum type, as is α-diversity. Megafauna abundance is significantly correlated with substratum type and local heterogeneity (over 10–100 s of meters), but not with depth. The thalweg axis is generally devoid of fauna due to the repeated disturbance and high current velocities associated with gravity flows.

Key Words: Submarine canyon, NE Atlantic, Portuguese Margin, habitat heterogeneity, nested mapping, ROV, epibenthic megafauna, diversity

Introduction

The Portuguese continental margin is intersected by several large deep-sea canyons, providing pathways for sediment transport from the shelf to the Iberia and Tagus Abyssal Plains. One of the most remarkable examples is Nazaré Canyon (Figure 50.1), cutting deeply into the shelf offshore the town of Nazaré. The total length of this canyon is over 200 km, while locally it is up to 1,600 m deep [1]. It runs from *ca.* 50 m water depth (mwd) at the canyon head to >4,800 mwd at the canyon mouth. Nazaré Canyon is not directly connected to any major river system; instead its location is

Seafloor Geomorphology as Benthic Habitat. DOI: 10.1016/B978-0-12-385140-6.00050-5

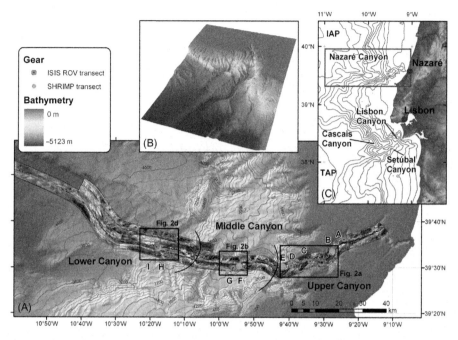

Figure 50.1 (A) Overview map of Nazaré Canyon, including locations of ground-truthing sites (ROV and SHRIMP transects, indicated with letters) and the extent of the maps presented in Figure 50.2. (B) 3D image of Nazaré Canyon, looking ENE, 5× vertical exaggeration. (C) Regional location map illustrating the main submarine canyons offshore Portugal. IAP, Iberia Abyssal Plain; TAP, Tagus Abyssal Plain.

determined by a reactivated Hercynian fault [2]. Capture of sediment transported along the shelf appears to be the main input process, with fine-grained terrigenous sediment that can be traced back to Portuguese rivers predominating [3]. Further down-canyon sediment transport is caused by a combination of tidal currents, response to wind forcing over the shelf and gravity driven surges as a result of winter storms [4,5]. Occasional large-scale turbidity currents may be caused by slope failures in the canyon head or along the shelf break, potentially related to earthquake activity [6]. The ocean-ography of the area is dominated by seasonal processes: northerly winds during spring and summer cause upwelling, while the system reverses during autumn and winter [7,8]. The salinity maximum of the northward-flowing Mediterranean Outflow Water is found at ∼1,200 mwd and forms an important density boundary affecting the forma-tion of nepheloid layers (layers in the water column with increased turbidity, gener-ally due to a high suspended sediment load) and internal waves [5]. Human impacts in Nazaré Canyon are less extensive than in canyons further south that are located close to large population centers (Lisbon, Cascais, and Setúbal Canyons, Figure 50.1). Litter found in the Nazaré Canyon consists mainly of lost nets and fishing gear [9].

The study of Nazaré Canyon started within the framework of several EU-funded projects, such as EuroSTRATAFORM, HERMES, and HERMIONE. This coordi-nated research and survey effort resulted in a large data set, including shipborne multibeam data (100 m × 100 m grid cells); TOBI 30 kHz sidescan sonar imagery (Towed Ocean Bottom Instrument, 6 m × 6 m grid cells); ROV-based multibeam (recorded with a Simrad SM2000 system on the Remotely Operated Vehicle (ROV) ISIS, 0.5 m × 0.5 m grid cells), video, photo, and sample materials; towed video and photographic records (SHRIMP vehicle); CTD (Conductivity Temperature, Depth probe) casts; and piston, gravity, and megacores. This large and varied data set allows a nested habitat mapping approach. In particular, the ROV-based bathymetry data provide unique insights into the canyon morphology and local heterogeneity at the spatial scale of the biological communities, while the TOBI side scan sonar allowed cost-effective, medium- to high-resolution coverage of the entire canyon.

Geomorphic Features and Habitats

The overall geomorphology of Nazaré Canyon has been described in detail by Lastras et al. [1], based on a thorough analysis of the shipborne bathymetry and TOBI sidescan sonar data. These authors divide the canyon into three sections, following the approach by de Stigter et al. [4] (Figures 50.1 and 50.2).

Upper Nazaré Canyon: The upper Nazaré Canyon stretches in a west–southwesterly direction from the canyon head at *ca.* 50 mwd down to its intersection with the shelf break, where the canyon reaches *ca.* 2,000 mwd. The TOBI data, in addition to cross-sectional bathymetry profiles, show a slightly sinuous, steep-walled (20–30°) V-shaped valley incised by gullies on the northern flank and bound by scarps and rock outcrops on the southern flank (Figure 50.2A).

Middle Nazaré Canyon: The middle Nazaré Canyon still consists of a V-shaped valley, although broader and slightly less steep (10–20° walls), and trending roughly westwards. Large, well-developed gully systems cut the canyon flanks. The course of the thalweg displays a number of large meanders with perched terraces and a narrow thalweg channel at the bottom (~400 m wide and ~140 m deep). This morphology cannot be interpreted from shipborne bathymetry alone and has been demonstrated in detail only when ROV-based bathymetry data became available (Figure 50.2B and C). The terraces can presently be interpreted as inner levees with high sedimentation rate [10], although thalweg erosion, for example during the last glacial, most probably played a role in their initial formation.

Lower Nazaré Canyon: The lower Nazaré Canyon extends for *ca.* 85 km west–northwestwards from the point where the narrow V-shaped canyon opens to a broader, flat-bottomed channel, at *ca.* 4,050 mwd. The channel is bound by depositional levees; the northern levee is especially well developed into a ridge. A secondary thalweg channel, about 40 m deep and 400 m wide, meanders across the broad channel floor. Another, similar, thalweg channel can be distinguished close to the northern flank of the lower Nazaré Canyon (Figure 50.2D and E). Variable backscatter patterns on the TOBI imagery indicate the presence of bedforms both within the thalweg channel and across the broad channel floor [1,6].

Where adequate sidescan sonar coverage is available, different sedimentary environments can be recognized, reflecting the depositional regime, and by extension,

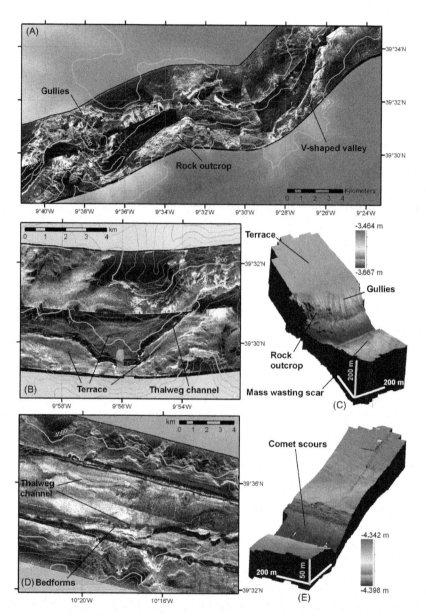

Figure 50.2 (A) Detail of upper canyon morphology, including TOBI sidescan sonar data (high backscatter represented by lighter colors) and contour lines based on shipborne multibeam bathymetry (500-m spacing). (B) Detail of middle canyon morphology, including TOBI sidescan sonar, shipborne bathymetry (100-m spacing), and ROV-borne bathymetry (colored patch illustrates the limited extent of this data set). (C) 3D image of middle canyon thalweg morphology, based on ROV-borne multibeam bathymetry. No vertical exaggeration. (D) Detail of lower canyon morphology, illustrating the broad channel with secondary thalweg channels. Includes TOBI sidescan sonar imagery and bathymetric contours based on shipborne multibeam data (100-m spacing). Small colored patch indicates location and extent of ROV-borne bathymetry. (E) 3D image of lower canyon secondary thalweg channel morphology based on ROV-borne multibeam data. Vertical exaggeration: 2. (For interpretation of the references to color in this figure legend, the reader is referred to the web version of this book.)

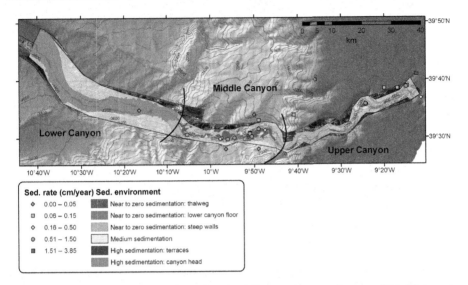

Figure 50.3 Sedimentary environments in Nazaré Canyon, as proxy for potential habitats and identified through visual interpretation of the bathymetry, sidescan sonar, video, and core data. The extent of the mapped area is limited to the coverage of the TOBI data, avoiding areas of far-range noise. Core locations used for the interpretation of sedimentation rates are indicated and are based on Masson et al. [10] and de Stigter et al. [4].

the physical environment at the seabed (current regime, substratum, nutrient input, etc.). The map in Figure 50.3 therefore presents the potential habitats that can be delineated in Nazaré Canyon.

Near to zero net sedimentation, including erosional areas (44% of surface area)

- *Thalweg floor*: The Nazaré thalweg floor is formed by the bottom of the V-shaped valley in the upper canyon and continues as secondary thalweg channel further down. ROV video data and high-resolution bathymetry show a variety of bedforms ranging from ripples to sand waves and comet scours around boulders (Figures 50.2 and 50.4H). Attempts to (push-) core the sediments all failed due to their lack of cohesion, indicating they are composed of well-sorted sands. Additionally, several large blocks and boulders are strewn across the thalweg floor, which most probably originate from local landsliding and debris flow activity. The relatively fresh appearance of these deposits and the general lack of mud deposition in the thalweg channel suggest that the seabed is reworked regularly, potentially as a result of tidal currents constrained by the thalweg morphology [10]. Hence the overall net sedimentation in this environment is considered negligible.
- *Steep walls*: Throughout the canyon, several areas with slopes >40° can be recognized, especially in the finer-scale data sets (Figure 50.2). They include rock outcrops, slide headwalls, mass wasting scars, and steep gullies. They mainly occur in the upper canyon and along the secondary thalweg channel in the middle and lower canyons. Landsliding and gully erosion appear to be the dominant processes, exposing both semiconsolidated sediments and outcrops of hard rock.
- *Lower canyon floor*: Deposition in the lower canyon is reduced by an order of magnitude relative to the middle canyon (see below). Oliveira et al. [5], citing core studies from Mougenot [11], mention 0.01–0.02 cm/year on the lower canyon levees, while de Stigter

et al. [4] find similar values from ^{210}Pb measurements in the lower channel. The latter, however, may be artificially high due to bioturbation of ^{210}Pb into the seabed, and hence represent a maximum value. Long-term sedimentation rates based on piston core interpretation are in the order of 0.001–0.003 cm/year. Hence, the lower canyon floor is also mapped as near to zero net sedimentation.

Areas with extremely high sedimentation rates (5% of surface area)

- *Canyon head*: High sediment accumulation occurs at specific locations in the canyon head, where shelf spillover and along-shelf transported sediments are trapped by the local bathymetry [4,13]. Sedimentation rates up to 3.8 cm/year have been measured in piston cores and seem to be consistent on timescales of 10–100s of years. It is, however, possible that such temporary depocenters become unstable due to oversteepening and increased pore pressure because of the high sedimentation rate, and that they form a source for infrequent but larger turbidity currents throughout the canyon [6,14].
- *Terraces*: The perched terraces in the middle canyon are the main centers of increased sediment deposition. Sedimentation rates of 0.5–1.4 cm/year are measured in several terrace cores [13]. The sediments mainly consist of heavily bioturbated muds, although occasional sand layers are also present. The origin of this mid-canyon depocenter is attributed by Masson et al. [10] to the rain-out of intermediate nepheloid layers (INLs) that may form through the lofting of either tidally resuspended sediments or weak turbidity currents developed from shelf nepheloid layers. As sediments deposit from weakly turbid downcanyon flows, the flows lose their downward momentum and may detach from the seabed. This appears to happen preferentially in the middle canyon, and may also explain the lack of regular fine-grained sedimentation in the lower canyon.

Medium sedimentation (51% of surface area)

- By default, the remaining areas accumulate intermediate amounts of mainly fine-grained sediment. They include less steep canyon walls plus steeper slopes that accumulate a large amount of sediment but fail regularly rendering the net sedimentation to a moderate level [13]. Sedimentation rates are in the order of 0.2 cm/year in the upper and middle canyons, which is still considerably more than the background sedimentation at similar depths along the Iberian continental slope (0.05–0.15 cm/year) [4].

Biological Communities

Using a combination of ROV video records (split in 15 s sections) and still images, plus camera stills from the towed SHRIMP vehicle, we identified the epibenthic megafauna in Nazaré Canyon, based on the work of Pattenden et al. [15]. Unfortunately, due to problems with the laser scales, only transects B, C, E, F, and H could be quantified in terms of species density.

The biological communities within Nazaré Canyon vary according to the local environmental conditions [16]. Depth is a major discriminating factor, but the sedimentary regime, defining substratum and seabed disturbance, appears also to be important. For a full species list, see Pattenden et al. [15].

Thalweg floor: Throughout the canyon, the thalweg floor is practically barren in terms of megafauna. This is most probably due to the high level of disturbance related to the regular tidal activity and occasional turbidity currents. In addition, the absence of fine-grained cohesive material may prohibit the formation of burrows and limits the availability of organic carbon (which is preferentially associated with the finer fractions) [13]. Koho et al. [17] also identified the high levels of disturbance in the canyon axis as the main reason for a strongly reduced total standing stock of benthic foraminifera compared to the canyon walls and terraces.

Upper canyon: Five transects were carried out in the upper Nazaré Canyon, resulting in the identification of a total of 82 nominal species. Sample rarefaction curves (cumulative number of species with increasing study area) indicate that the shallower parts of the upper canyon may be most species-rich (Figure 50.5). However, due to the high patchiness of the benthic communities, and the resulting high variances, significant differences could only be identified between site F in the middle canyon and sites B and C in the upper canyon.

Exposed rock surfaces are dominated by sessile fauna and suspension feeders such as brachiopods (lamp shells, displaying the highest abundances), (encrusting) sponges, ascidians (sea squirts), scleractinian corals (*Lophelia pertusa*), and soft corals (Figure 50.4E). Gullies and ridges cut through semiconsolidated sediment are home to sabellid

Figure 50.4 Typical fauna and habitats in Nazaré Canyon. Scale bar equals to *ca*. 10 cm. Upper canyon: (A) actinarian anemone (unidentified); (B) brisingid sea star (unidentified); (C) echinoid *Calveriosoma hystrix*; (D) stylasterene coral (unidentified); (E) soft coral *Gersemmia* sp.; middle canyon: (F) stalked crinoid *Anachalypsicrinus nefertiti*; lower canyon: (G) xenophyophores *Reticulammina cerebreformis* sp. nov. [12]; (H) ripples in the thalweg channel.

worms, cerianthid anemones, and brisingid sea stars (Figure 50.4A and B). Areas with high accumulation of soft sediments are dominated by sabellid worms, quill worms, echiuran worms, holothurians (sea cucumbers), and anemones. A variety of infauna may also be present, but cannot be identified from the video data. Finally, steeper sedimented slopes, with medium sediment accumulation, are home to large numbers of the xenophyophore *Syringammina*, stalked sponges, holothurians, and brisingids.

Middle canyon: Two transects crossed the middle canyon. A striking difference was observed between the high-sedimentation terraces, characterized by sabellid worms, holothurians, and a variety of burrowing infauna, versus the thalweg walls which provided a holdfast for stalked crinoids (sea lilies: e.g., *Anachalypsicrinus nefertiti*, Figure 50.4F), the anemone *Anthomastus* sp., and another white anemone species. In addition, ophiurids (brittle stars) and echinoids (sea urchins) were found slightly further away from the thalweg channel.

Overall, species numbers were much lower (15 species) in comparison with the upper canyon, a trend that was confirmed by the sample rarefaction curves (Figure 50.5).

Lower canyon: Two transects were carried out within the lower canyon, but they only covered a secondary thalweg channel plus its slopes and part of the broad canyon floor. No ground-truthing of the levees was carried out. A maximum of 19 species was encountered, including two new species of xenophyophore present on the canyon floor (Figure 50.4G) [12]. An additional cerianthid anemone was also found

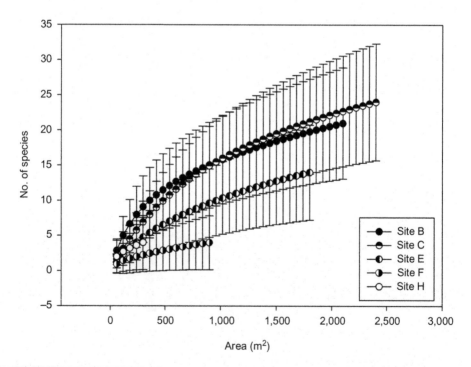

Figure 50.5 Sample-based rarefaction curves illustrating the species density at each of the transects where laser scale measurements were possible.
After Ref. [15].

in this low-sediment input community. The thalweg channel walls were mainly barren, although a few pennatulids (sea pens) and cerianthid anemones were found, while some stalked crinoids were attached to conglomerate boulders in the flank of the secondary thalweg channel.

Surrogacy

Pattenden [15] used the biological community data to investigate the potential relationships of megafauna abundance, α-diversity (within-site species diversity, measured with the Shannon index), and community structure (the set of species forming the faunal assemblage at each site) with environmental parameters such as depth, substratum type, and substratum heterogeneity. The ROV and SHRIMP transects were split in 100-m depth bins, and the parameters measured in five photo/video subsamples taken at random in each depth bin were averaged. Each subsample consisted of sufficient photographs or 15-s video sections to cover *ca.* 60 m². Substratum type was identified from the video/photo data as soft, mixed, or hard (exposed rock) and quantified as 1, 2, and 3, respectively. The heterogeneity at local scale (10–100s of meters) was estimated as the standard deviation of these substrata within the subsamples and depth bins.

Megafauna abundance showed a positive correlation with substratum type and heterogeneity (Table 50.1), indicating higher abundances on rocky substrata and in mixed and heterogeneous environments. Abundance did not decrease significantly with depth, a relation often observed on continental slopes [18]. Pattenden [15] suggests that the lateral advection of particulate organic matter within the canyon, in addition to the vertical input, may be the reason for the sustained high abundances.

Both substratum type and depth have an influence on megafauna diversity (Table 50.1), with more diverse communities occurring at shallower depths and on

Table 50.1 Spearman's Rank Correlations between the Biotic and Abiotic Variables Measured in Nazaré Canyon ($n = 125$)

Correlation Coefficient P-Value	Depth	Substratum Type	Substratum Heterogeneity	Megafauna Abundance	Shannon Diversity Index
Substratum type	−0.656* 0.000				
Substratum heterogeneity	−0.387* 0.000	0.418* 0.000			
Megafauna abundance	−0.105 0.244	0.355* 0.000	0.289* 0.001		
Shannon diversity index	−0.319* 0.000	0.235* 0.008	0.091 0.315	0.583* 0.000	
Percentage filter feeders	−0.0435 0.669	0.155 0.125	−0.658 0.517	−0.0865 0.394	−0.0143 0.888

*Significant result at 0.05 level.

rocky or mixed substrata. Substratum heterogeneity, at the scale measured here, did not have a significant correlation with diversity. Although sites with high heterogeneity seem to carry a high diversity, substratum type may override heterogeneity, with homogeneous rocky substrata supporting some of the most diverse communities. Also the homogeneous soft, but steep substrata of the lowermost transect in the upper canyon (transect E, semiconsolidated sediment) host a very diverse fauna [15].

A similar pattern was observed in relation to community structure, using the BIOENV routine of the Primer software package, which calculates the set of environmental variables that seem to have the highest influence on community (dis-)similarity between sites (see Ref. [15] for details). As demonstrated above, depth is the main discriminating factor, followed to some extent by substratum type. Heterogeneity, at the scale quantified here, again did not show a significant influence. Also, when species were assigned to functional groups depending on feeding type (filter feeders versus nonfilter feeders), no statistically significant correlations were found between the percentage filter feeders and either depth, substratum type, or local heterogeneity (Table 50.1). The amount of species overlap between different sites is very low (even in the upper canyon only 30% of species were encountered in more than one transect, and none were common between all five transects). This high β-diversity (between-site species turnover) suggests that heterogeneity at a medium spatial scale (1,000s of meters) may be the more important factor. Due to the complexity of the terrain, the patchiness in the benthic communities is very high, which could not be captured sufficiently within the nine video/photo transects available for this study. The vastness and medium-scale heterogeneity of the system form a difficult challenge toward the ground-truthing of the mapped (potential) habitats and the identification of biological communities. More work is needed to establish the relationships between abiotic surrogates and deep-sea canyon communities, although the example of Nazaré Canyon presented here already gives a first insight into the spatial distribution of species abundance and diversity.

Acknowledgments

The authors would like to thank the captains, crews, and scientific parties of the cruises DIS297, CD179, and JC010 for their support at sea, and the ISIS team for the excellent ROV work. Veit Hühnerbach (NOCS) took on the task of processing the TOBI data, and Galderic Lastras (Universitat de Barcelona) and Alexandra Morgado (Instituto Hidrografico, Lisbon) helped out with the bathymetry and TOBI data integration. The data were collected and analyzed within the framework of the EU FP6 and FP7 projects HERMES (Hotspot Ecosystem Research on the Margins of European Seas, contract number GOCE-CT-2005-511234) and HERMIONE (Hotspot Ecosystem Research and Man's Impact ON European Seas, grant agreement number: 226354).

References

[1] G. Lastras, R.G. Arzola, D.G. Masson, R.B. Wynn, V.A.I. Huvenne, V. Huehnerbach, et al., Geomorphology and sedimentary processes in the Central Portuguese submarine canyons, western Iberian margin, Geomorphology 103 (2009) 310–329.

[2] J. Vanney, D. Mougenot, Un canyon sous-marin du type "gouf": le Canhão da Nazaré (Portugal), Oceanol. Acta 13 (1990) 1–14.

[3] R.G. Arzola, Controls on Sedimentation in Submarine Canyons: Nazaré, Setúbal and Cascais Canyons, West Iberian Margin, Ph.D. Thesis, University of Southampton, Southampton, 2008.

[4] H.C. de Stigter, W. Boer, P.A. de Jesus Mendes, C. César Jesus, L. Thomsen, G.D. van den Bergh, et al., Recent sediment transport and deposition in the Nazaré Canyon, Portuguese continental margin, Mar. Geol. 246 (2007) 144–164.

[5] A. Oliveira, A.I. Santos, A. Rodrigues, J. Vitorino, Sedimentary particle distribution and dynamics on the Nazaré Canyon system and adjacent shelf (Portugal), Mar. Geol. 246 (2007) 105–122.

[6] R.G. Arzola, R.B. Wynn, G. Lastras, D.G. Masson, P.P.E. Weaver, Sedimentary features and processes in the Nazaré and Setúbal submarine canyons, west Iberian margin, Mar. Geol. 250 (2008) 64–88.

[7] J. Vitorino, A. Oliveira, J.M. Jouanneau, T. Drago, Winter dynamics on the northern Portuguese shelf. Part 1: Physical processes, Prog. Oceanogr. 52 (2002) 123–153.

[8] P. Relvas, E.D. Barton, J. Dubert, P.B. Oliveira, A. Peliz, J.C.B. da Silva, et al., Physical oceanography of the western Iberia ecosystem: latest views and challenges, Prog. Oceanogr. 74 (2007) 149–173.

[9] G. Mordecai, P.A. Tyler, D.G. Masson, V.A.I. Huvenne, Litter in submarine canyons off the west coast of Portugal, Deep Sea Res. Part II 58 (2011) 2489–2496 doi:10.1016/j.dsr2.2011.08.009.

[10] D.G. Masson, V.A.I. Huvenne, H.C. de Stigter R.G. Arzola, T.P. Le Bas, Sedimentary processes in the middle Nazaré Canyon: the importance of small-scale heterogeneity in defining the large-scale canyon environment, Deep Sea Res. II 58 (2011) 2369–2387 doi:10.1016/j.dsr2.2011.04.003.

[11] D. Mougenot, Geologie de la Marge Portugaise, Documentos técnicos do Instituto Hidrográfico, 1989, 259 pp.

[12] A.J. Gooday, A. Aranda da Silva, J. Pawlowski, Xenophyophores (Rhizaria, Foraminifera) from the Nazaré Canyon (Portuguese margin, NE Atlantic), Deep Sea Res. Part II 58 (2011) 2401–2419 doi:10.1016/j.dsr2.2011.04.005.

[13] D.G. Masson, V.A.I. Huvenne, H. de Stigter, G.A. Wolff, K. Kiriakoulakis, R.G. Arzola, et al., Efficient burial of carbon in a submarine canyon, Geology 38 (9) (2010) 831–834.

[14] T.A. Okey, Sediment flushing observations, earthquake slumping, and benthic community changes in Monterey Canyon head, Cont. Shelf Res. 17 (1997) 877–897.

[15] A.D.C. Pattenden, The influence of submarine canyons on the strucure and dynamics of megafaunal communities. PhD Thesis, University of Southampton, Southampton, 2008.

[16] P. Tyler, T. Amaro, R. Arzola, M.R. Cunha, H. de Stigter, A.J. Gooday, et al., Europe's Grand Canyon: Nazare submarine canyon, Oceanography 22 (2009) 48–57.

[17] K.A. Koho, T.J. Kouwenhoven, H.C. de Stigter, G.J. van der Zwaan, Benthic foraminifera in the Nazaré Canyon, Portuguese continental margin: sedimentary environments and disturbance, Mar. Micropaleontol. 66 (2007) 27–51.

[18] M.A. Rex, R.J. Etter, J.S. Morris, J. Crouse, C.R. McClain, N.A. Johnson, et al., Global bathymetric patterns in standing stock and body size in the deep-sea benthos, Mar. Ecol. Prog. Ser. 317 (2006) 1–8.

51 Banks, Troughs, and Canyons on the Continental Margin off Lofoten, Vesterålen, and Troms, Norway

Lene Buhl-Mortensen[1], Reidulv Bøe[2], Margaret F.J. Dolan[2], Pål Buhl-Mortensen[1], Terje Thorsnes[2], Sigrid Elvenes[2], Hanne Hodnesdal[3]

[1]Institute of Marine Research, Nordnes, Bergen, Norway, [2]Geological Survey of Norway, Trondheim, Norway, [3]Norwegian Hydrographic Service, Stavanger, Norway

Abstract

The continental margin off Lofoten, Vesterålen, and Troms comprises a great variation of marine landscapes, ranging from 0 to 3,000 m in depth largely influenced by previous glaciations and oceanographic processes governed by three different water masses. The biological communities identified with multivariate analyses of results from visual surveys were strongly correlated with water mass distribution, sediment type, and various topographic indices. Broadscale geomorphic features such as canyons and troughs have a significant influence on the distribution of sediment types and biological communities through their modification of current patterns. Currents also influence the distribution of geomorphic features such as sand waves and coral reefs. At a broadscale, the largest faunistic change occurs between the warm Atlantic water and the cold intermediate water mass at depths around 700 m.

Key Words: Canyons, sand waves, coral reefs, Arctic fauna, biodiversity, marine landscapes, seabed sediments

Introduction

The continental margin offshore Lofoten–Vesterålen–Troms (LVT) (North Norway) described here is in the area 68°N–70.1°N and 10°E–19°E. Marine landscapes and faunal diversity on this part of the Norwegian margin have earlier been described by Thorsnes et al. [1] and Buhl-Mortensen et al. [2]. This area is in near-pristine condition, and the main human activity is fisheries. In certain parts of the continental

Seafloor Geomorphology as Benthic Habitat. DOI: 10.1016/B978-0-12-385140-6.00051-7

Figure 51.1 Bathymetry offshore Lofoten–Vesterålen–Troms, North Norway.

shelf, the impact from fisheries is substantial; however, the area described here is affected by fishing only to a very limited extent.

The continental shelf comprises shallow banks and plains alternating with deeper troughs (Figure 51.1) formed during the last glaciations. Massive diamictic sediments are found on the continental shelf [3], indicating that ice streams advanced through fjords onto the continental shelf and to the shelf edge during the last (late Weichselian) glaciation, which reached a maximum slightly before 18,000 ^{14}C BP.

The continental slope is incised by many large deep canyons that have developed over a span of several hundred thousand years or more. In this part of Norway, the continental shelf is very narrow (10–90 km), and outside Andøya, the canyon Bleiksdjupet occurs only around 10 km from land. Deglaciation along this margin took place from ~15,000 ^{14}C BP on the outer shelf, according to the dated onset of

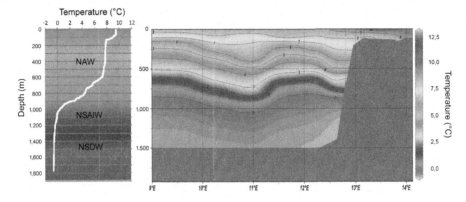

Figure 51.2 Representative temperature profile and bathymetrical distribution of water masses off Lofoten and Vesterålen. Three water masses occur offshore in this area: North Atlantic Water (NAW) above a transition zone of Norwegian Sea Arctic Intermediary Water (NSAIW). The Norwegian Sea Deep Water (NSDW) occurs deeper than ~1,000 m. The depth of the transition zones vary through the year and long tidal waves can occur. The graphs show the spring/summer situation.

glacimarine and then open-marine sedimentation. At 13,600 ^{14}C BP, the ice margin was located along the present coastal area off Vesterålen [4].

The oceanography of the area is influenced by three major water masses [5] (Figure 51.2). Near the coast, the northward-flowing Norwegian Coastal Current comprises the low-salinity Norwegian Coastal Water with variable temperature. This water overlies the Norwegian Atlantic Current with North Atlantic Water (NAW). The NAW extends down to about 500–600 m and is part of the relatively warm and saline North Atlantic Current. Below this depth, two cold-water masses are present: the Norwegian Sea Arctic Intermediate Water (NSAIW) and the Norwegian Sea Deep Water (NSDW). NSAIW has temperatures between −0.5°C and 0.5°C, whereas the NSDW typically has temperatures from −0.5°C to −1.1°C. The border between these two water masses occurs typically at around 1,000 m depth off the Norwegian coast in the Norwegian Sea.

The descriptions of geomorphology and benthic habitats presented here are based on results from the MAREANO (Marine AREA database for Norwegian coast and sea areas) mapping program (see www.mareano.no) [6]. MAREANO includes acquisition of multibeam bathymetry and acoustic backscatter data together with a comprehensive, integrated biological and geological sampling program (underwater video, boxcorer, multicorer, grab, hyperbenthic sled, and beam trawl).

Geomorphic Features and Habitats

Banks and Plains

Banks and plains are important geomorphic elements on the continental shelf off LVT. From north to south, they are Malangsgrunnen, Sveinsgrunnen, Vesterålsbankene

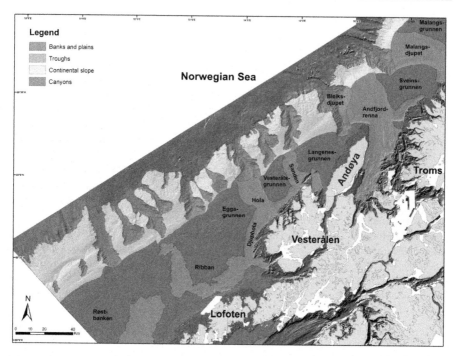

Figure 51.3 3D color shaded bathymetry offshore Lofoten–Vesterålen–Troms, North Norway. Shallow coastal waters are shaded white, water depths on the banks and plains are 40–200 m (light blue), troughs are 200–300 m deep (mid-blue) while depths fall away as deep as 3,000 m (dark blue) at the bottom of the continental slope. (For interpretation of the references to color in this figure legend, the reader is referred to the web version of this book.)

(Langnesgrunnen, Nordgrunnen, Eggagrunnen), Ribban, and Røstbanken (Figure 51.3). The first mentioned are clearly defined banks, while Ribban and Røstbanken are wider plains. Water depths on banks and plains vary from ∼40–200 m, with the shallow-est banks in the north (Figure 51.1). Banks and plains cover 8,600 km² in this part of the Norwegian offshore. They are flat, overall, while bank margins have slopes of up to 20°. Bank and plain surfaces generally slope in an offshore direction. This is partly a consequence of postglacial isostatic uplift of Scandinavia. Banks and plains exhibit numerous large and small geomorphic features.

Iceberg ploughmarks created at the final stages of the last glaciation, when float-ing icebergs scraped the seabed, may be many kilometers long, several hundred meters wide, and more than 10 m deep.

Small moraine ridges created during intermittent halts and readvances of the ice margin may be several meters high, tens of meters wide, and several kilometers long.

Larger, more regional moraine ridges: The shallowest bank areas were dry land or close to sea level at the end of the last glaciation, and moraine ridges are locally reworked by waves and currents. These processes have formed beach bars and cur-rent lineations. Current lineation may be several kilometers long.

Small sand waves and sand ripples are common in sandy areas of the banks. Where currents are strong, there are also examples of gravel waves.

Banks and plains are usually covered by a coarse-grained erosional lag that varies in thickness from a few centimeters to a few decimeters. The erosional lag has developed over a long time period due to wave action (immediately after the last glaciation, when sea level was lower) and current reworking of the underlying tills and moraines deposited during the glaciations. Sandy gravel is the most common sediment type, while gravel with cobbles and boulders dominate the shallowest areas and elevated moraine ridges. On bank slopes, gravelly sand is also common. In one area on the shelf outside Lofoten, crystalline bedrock is exposed at the seafloor.

Troughs

Glacial troughs separate banks and plains on the continental shelf offshore LVT. From north to south, they are Malangsdjupet, Andfjordrenna, Sanden, Hola, and two unnamed troughs outside Lofoten. Some troughs, like Andfjordrenna, are direct continuations of fjords. In addition, some NE–SW-trending troughs occur immediately off the coast, e.g., Djuphola, outside Vesterålen. Water depths in the troughs vary between 200 and 300 m in the southwest, where they separate wide plains. Toward the northeast, trough water depths increase to ~500 m. Troughs cover around 5,400 km^2 in this part of the Norwegian continental shelf. They are generally overdeepened, with thresholds near the shelf edge. Some troughs also have a threshold (elevated bedrock or moraine ridge areas) in their middle or inner parts. Thresholds near the shelf edge comprise till interpreted as glacial grounding zone wedges [7], while thresholds in the middle and inner parts of troughs are made of sedimentary bedrock units that are resistant to glacial erosion. Moraine ridges are locally present within the troughs and along their margins. A glaciotectonic hill in the trough east of Røstbanken is formed from a block of sedimentary bedrock eroded and left by the glaciers.

Hydrography of Troughs

Troughs with a width of a few kilometers are able to break the dynamic balances that otherwise force the broadscale water flow to follow bathymetric contours. They are often also able to sustain enhanced turbulence levels and vertical mixing rates. The Coriolis force leads to a landward-flowing current on the southern side of troughs and a seaward-flowing current on the northern sides. Where ridges transverse the troughs, as in Malangsdjupet, this leads to sedimentation both on the inside of the ridge and immediately outside.

Coral reefs: The troughs exhibit numerous large and small geomorphic features, of which the most spectacular occur in Hola (Figure 51.4). Around 330 coral reefs occur in the southeastern part of the trough, in water depths of 200–270 m. The reefs can be regarded both as geomorphic features and biogenic structures. More details about the reefs are provided in the Biological Communities section. In the southern part of the coral reef area, erosional scours, up to 12 m deep and 1.5 km long, occur behind (north of) coral reefs, due to strong bottom currents [8].

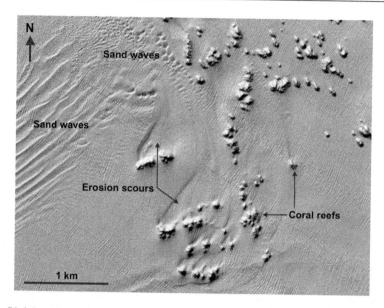

Figure 51.4 Longitudinal erosional scours created by strong bottom currents behind coral reefs in the Hola trough.

Sand waves and sand wave fields: These occur in several troughs, but are especially well developed in Hola. Straight-crested and barchan-type sand waves up to 3 km long and 7 m high are interpreted to form by tidal and geostrophic currents with bottom current speeds reaching 0.7–0.8 m/s [8,9].

Ridge/groove structures: These represent megascale glacial lineations and are also common. The largest ridges are 8–16 m high, 200–700 m wide, many kilometers long, and spaced a few hundred meters apart [7]. They are interpreted to have formed under fast-flowing ice streams.

Pockmarks: These are common in areas of fine-grained sediments, e.g., in the trough between Røstbanken and Ribban. They are typically a few meters deep and 30–60 m in diameter. Pockmarks show that gas or fluids leak from deeper geological strata. In areas of coarse-grained seabed sediments, pockmarks cannot develop; however, we have observed bacterial mats and carbonate crusts on sandy seabed in Hola, indicating leakages from the subsurface [10].

Sediment types: The most common sediment types in the troughs are sand and gravelly sand, but gravelly sandy mud covers large areas of several troughs. Sandy mud occurs in some troughs, e.g., east of Røstbanken. Till/moraine material is common along bank margins and on moraine ridges not covered by finer-grained sediments, e.g., in the outer parts of troughs.

Canyons and Continental Slope

The continental slope along this part of the Norwegian margin covers 8,900 km², of which canyons comprise 3,400 km². Water depths increase from a few hundred meters to 3,000 m over a distance of only 30–40 km, and the average slope gradient is around 7° (9° in canyons and 5° between canyons; local slope gradients reach 60°). The slope is incised by 10 large and several small canyons, some of them being cut more than 500 m into the Pleistocene and Tertiary sediments [11]. Some major canyons on the slope start where glacial troughs on the shelf end. Some canyons are finger- or pear-shaped; others comprise more complex forms. Both erosion from fluid-flow processes and sliding have been important for their formation. Most of the canyons extend to the shelf break, but some end in the upper slope, indicating a retrogressive development through time.

The largest canyon is Bleiksdjupet, which is up to 1 km deep, 10 km wide, 30 km long, with local slopes up to 30°. Channels occur at the lower end of some of the canyons; in Bleiksdjupet, the Lofoten Basin Channel can be traced from the uppermost reaches of the canyon over a distance of more than 40 km out in the Lofoten Basin. At irregular intervals, turbidity currents transport sediments along the channels either due to sliding and mass movements or due to sediment-laden ocean currents going down the canyons. Canyon margins are often steep and irregular, and subvertical walls with rock fall activity occur where there has been erosion and sliding. Submarine fans and blocky slide deposits occur seaward of several of the canyons.

There is a gradual transition from coarse-grained sediments on the upper continental slope to more fine-grained sediments on the lower slope. Sandy gravel dominates the shelf edge and the uppermost slope. This is followed by gravelly sand, which dominates on terraces on the uppermost slope. On the middle and lower slopes, there is predominance of gravelly muddy sand, but the sediment type varies depending on depositional process.

Canyons exhibit great variation in sediment type, and one side of a canyon may be very different from the other. Gravelly sand is common in the upper parts, while gravelly muddy sand and gravelly sandy mud are common lower down. Sand and gravelly sand are common in channels. Consolidated sediments or sedimentary bedrock crop out in steep slopes.

The Lofoten contourite drift occurs on the continental slope in the southwestern part of the area. Fine-grained sediments are deposited on the slope by the contour-parallel North Atlantic Current; however, sedimentation rates are low, and only *ca.* 1 m of sediments have been deposited over the past 10,000 years. Many slides have occurred on this part of the continental slope. The largest, the Trænadjupet Slide, occurred around 4,000 years ago, but several slides may be younger. Sedimentation rates are generally low along this part of the Norwegian continental margin.

Biological Communities

Banks and Plains

Iceberg ploughmarks: The topography of iceberg ploughmarks modifies the local current pattern, resulting in both erosional and depositional areas. The erosion areas, located around the edges of the ploughmarks, represent hard substrata suitable for settlement of sessile colonial invertebrates such as sponges and sea octocorals. The bottoms of the ploughmarks is commonly inhabited by lampshells (Brachiopoda). Their shells accumulate in the ploughmark depressions and often cover large parts of the seabed.

Sponge communities: Fields with large sponges (*Geodia* spp., *Aplysilla sulfurea*, *Stryphnus ponderosus*, and *Steletta* sp.) occur at the banks and plains mainly at depths between 200 and 300 m (Figure 51.5). The substrate beneath the sponges consists of a mixture of sponge spicules and mud. Sponges are known to increase biodiversity by providing microhabitats for mobile species [12]. A high number of associated species was commonly observed in the sponge fields.

Small moraine ridges: A rich fauna was found on moraine ridges. This habitat is difficult to sample with grab and other bottom-sampling gears. In general, habitat complexity increases with an increasing amount of stones in the sediment. The fauna of these ridges have many similarities with the edges of ploughmarks, but many places such as ridges are colonized by large colonial invertebrates such as gorgonian corals. This is probably caused by acceleration of currents over the ridges.

Troughs

Sand waves: These occur in the Hola area as well as in other troughs off LVT. In general, the currents are strong and the fauna is poor in these areas with sand in motion (Figure 51.6). The only sessile organisms are found in small areas of exposed gravel between the waves here, underlying substrates are exposed.

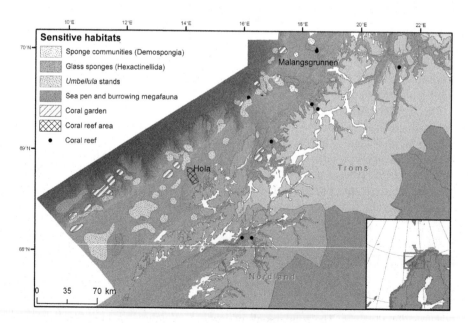

Figure 51.5 Distribution of sensitive habitats off Lofoten, Vesterålen, and Troms.

Sea pens and burrowing megafauna: In muddy parts of the troughs, the OSPAR-listed habitat "Seapens and burrowing megafauna" was found (Figure 51.5). Here the holoturian *Stichopus tremulus* occurred together with the lobster *Nephrops norvegicus* close to its northern limit of distribution. The sea pens in this habitat were *Funiculina quadrangularis*, *Virgularia mirabilis*, *Pennatula phosforea*, and *Kophobelemnon stelliferum*. *Funiculina quadrangularis* commonly occurred in areas with higher currents and muddy sand. Often, the brittle star *Asteronyx loveni* was found on this sea pen. *Pennatula phosforea* and *Kophobelemnon stelliferum* seem to be more closely associated with finer sediments and weaker currents. These were typically found closer to the coast, whereas *Virgularia mirabilis* occurred both offshore and nearshore.

Pockmarks: Pockmarks in the Hola trough contained hard substrata in the form of autogenic carbonates (Figure 51.6). White bacteria mats were observed close to the carbonates. There are no examples of special chemotrophic species within these

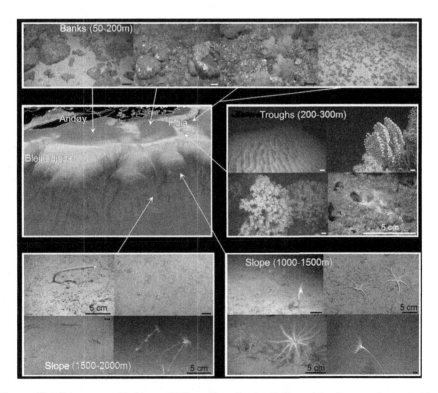

Figure 51.6 Biotopes and habitats off Vesterålen. On the shallowest banks, moraines provide substratum for calcareous red algae, whereas on the edges of deeper banks cauliflower corals (*Drifa glomerata*) may dominate. Sand waves, coral reefs (*Lophelia pertusa*), and pockmarks occur in the Hola trough. The two groups of inserted pictures from the slope illustrate difference in fauna related to changing sediment type with increasing depth. At the deepest locations tube building polychaetes and the sea lily *Rhizocrinus lofotensis* dominate. Scale bars are 10 cm where numbers are not provided.

pockmarks, but the carbonate crusts in this area may have had a function as seeding substratum for the coral reefs.

Coral reefs: They occur in different marine landscapes (troughs, ridges, and shelf break). In the trough north of Malangsgrunnen (Figure 51.5), coral reefs occur at local elevations and on a glacial ridge crossing the trough. These are more or less circular in outline and resemble most of the Norwegian coral reefs built by the colonial stone coral *Lophelia pertusa*. In the Hola trough off Vesterålen, a large cluster of coral reefs occur on a relatively level seabed (Figure 51.4). These are elongated, with only a living part at the end facing the main currents flowing from the coast toward the shelf break. The seabed is relatively level with small gravelly patches within a mainly sandy environment. One active gas seep was discovered in the outer part of the reef area, but no signs of seepage or carbonate crusts were found within the central and denser reef area. However, such substratum may still have been important as a settlement substrate for coral larvae.

In total, there are around 330 reefs in the area. These are 31–334 m long, 27–114 m wide, and 4–17 m high. Most of the reefs are smaller than 100 m in longest direction, whereas only 81 reefs are longer than 200 m. Chains of coral reefs may be up to 500 m long. The shape of the reefs changes from circular to elongate with increasing size or age. Larger reefs normally consist of a relatively small (20 m × 20 m) living up-current front and a longer "tail" of coral debris. Signs of erosion were common around the reef front and in some cases also behind the reefs. Seabed topography has no direct influence on the coral distribution, but it influences the environment by modifying the hydrodynamic setting. Local topographic features such as peaks and ridges induce accelerated currents and are favorable locations for reef growth. The Hola trough probably induces strong currents, bringing nutrient-rich water from local production at the shelf. Within the range of the coral's temperature and salinity tolerance, the combination of hard-bottom substrate for coral larvae settlement and relevant food transport rates are probably more important than the local topography of the seabed.

Canyons and Continental Slope

Canyons: They represent shelf incisions that provide important oceanographic processes and flow paths that allow deeper, slope-derived water to advect closer to shore than normal cross-shelf mixing would allow. Movement of water is a major structuring force of the environment controlling the ecosystem by its influence on sediment resuspension and deposition and transport of nutrients. Examples of typical species and habitats in Bleiksdjupet canyon are presented in Figure 51.7. The sea pen *Umbellula encrinus* is common in canyons and may occur with relative high densities, with just a few meters between the individual colonies.

Sand and gravelly sand are common in channels. Here, high densities of small red sea anemones were commonly observed, together with solitary hydroids.

Consolidated sediments or sedimentary bedrock cropping out in steep slopes represent hard substrates suitable for settlement of octocorals and glass sponges.

Upper slope. It represents a highly diverse landscape with a great diversity of substrata and species. Gravelly bottom in moraines on the shelf break are often

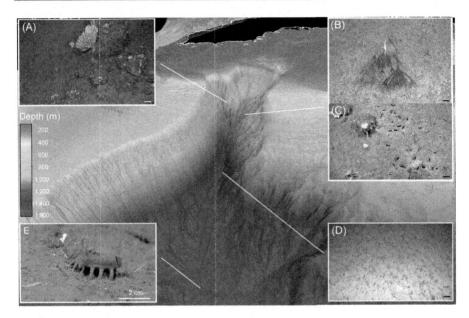

Figure 51.7 Bleiksdjupet canyon is one of Europe's largest submarine canyons. (A) Outcropping consolidated sediments, rock, and bubblegum coral (*Paragorgia arborea*) at 650 m depth. (B) The sea pen, *Umbellula encrinus* (at 850 m depth). (C) Burrows from infaunal megafauna, scattered cobble, and soft corals (Nephtheidae). (D) Sandy bottom in the thalweg with small red sea anemones (1,130 m depth). (E) Fine-grained deep sea mud with the holothurians *Elpidia glacialis* (1,890 m depth). Scale bars are 10 cm where numbers are not provided.

characterized by the basket star (*Gorgonocephalus eucnemis*) and cauliflower corals (Nephtheidae). This biotope is often intersected by areas with strong currents and large sand waves. Some locations also support high densities of gorgonians (coral gardens, Figure 51.5).

Lower slope: The substrate of the lower slope typically consists of muddy sand and gravelly sandy mud. The area is characterized by bottom temperatures between −0.5°C and −1.1°C and an arctic fauna. The megafauna at these depths appears to be common for the deep northern parts of the Atlantic and the Norwegian Sea. This fauna was dominated by the holothurians *Elpidia glacialis* and *Kolga hyalina*, the sea lily *Rhizocrinus lofotensis* together with the crustaceans *Bythocaris leucopis* and *Saduria* sp., and the sea urchin *Pourtalesia cf. jeffreysi*. The fauna is not species rich, but specific for the arctic deep water. On the soft bottom on the lower slope, a rich fauna of small crustaceans (Peracarida) were found living on stalks and tubes of other organisms (polychaetes, sea lilies, hydroids, glass sponges, etc.). At great depths, different glass sponges (Hexactinellidae) may occur in relatively high densities. One of the most common species of large glass sponges is *Caulophacus arcticus*, which occurs on hard substrate on the lower slope.

Surrogacy

Methods

Communities or biotopes were identified by systematic observations of biota and substrata carried out in the field [2] and later by analysis of video records [13]. Video records of the seabed were split into sequences representing 200-m long distances. These video sequences were compared with respect to species composition using detrended correspondence analysis (DCA) to identify groups of sequences. The biotopes identified by MAREANO can be described by combining substratum, terrain, and typical taxa. The broadscale topographic features do not represent environmental variables in themselves, but may modify the environment by influencing the currents and can often be related to the distribution patterns of different substrata. The methods for identifying physical biotope descriptors for habitat/biotope prediction have been described in more detail in publications from the MAREANO project [2,13,14].

The faunal composition on the slope is mainly correlated with depth, which is an expression of the pronounced environmental differences associated with the three major water masses in the area. Within these water mass "regions," acoustic backscatter is the second most important "community predictor." This is a proxy for substratum type. Additionally, certain terrain descriptors (slope angle, rugosity, curvature, etc.) explain parts of the faunistic variation not explained by depth and backscatter in multivariate analyses. At Tromsøflaket (north of the area considered here), the best descriptors (including multiscale physical descriptors of the seabed derived from multibeam echosounder data) were identified by using forward selection. Prediction of biotope distribution was performed using a supervised GIS classification with the multibeam-derived physical seabed descriptors with the strongest explanatory ability identified by the DCA. The species diversity of the identified biotopes was described from the content of the bottom samples.

Identification of surrogates is an important component of MAREANO's ongoing work to map the distribution of benthic habitats. These surrogates can serve as predictor variables in models of the distribution of habitats.

Acknowledgments

The authors are grateful for the support provided by the MAREANO program financed by the Norwegian government. We would like to thank Valerie Bellec, Leif Rise, and Odd Harald Hansen for their contributions to this article.

References

[1] I. Thorsnes, L. Erikstad, M.F.J. Dolan, V.K. Bellec, Submarine landscapes along the Lofoten–Vesterålen–Senja margin, northern Norway, Norw. J. Geol. 89 (2009) 5–16.

[2] P. Buhl-Mortensen, L. Buhl-Mortensen, M. Dolan, J. Dannheim, K. Kröger, Megafaunal diversity associated with marine landscapes of northern Norway: a preliminary assessment, Norw. J. Geol. 89 (2009) 163–171.

[3] D. Ottesen, J.A. Dowdeswell, L. Rise, Submarine landforms and the reconstruction of fast-flowing ice streams within a large Quaternary ice sheet: the 2,500 km-long Norwegian–Svalbard margin (57°–80°N), Geol. Soc. Am. Bull. 117 (2005) 1033–1050.

[4] J. Knies, C. Vogt, J. Matthiessen, S.-I. Nam, D. Ottesen, L. Rise, et al., Re-advance of the Fennoscandian Ice Sheet during Heinrich Event 1, Mar. Geol. 240 (2007) 1–18.

[5] B. Hansen, S. Østerhus, North Atlantic–Nordic Seas exchanges, Prog. Oceanogr. 45 (2000) 109–208.

[6] L. Buhl-Mortensen, H. Hodnesdal, T. Thorsnes, Til bunns i Barentshavet, Skipnes Press, Trondheim, Norway (executive summary in English), (2010) 128 pp.

[7] D. Ottesen, C.R. Stokes, L. Rise, L. Olsen, Ice-sheet dynamics and ice streaming along the coastal parts of northern Norway, Quat. Sci. Rev. 27 (2008) 922–940.

[8] R. Bøe, V.K. Bellec, M.F.J. Dolan, P.B. Buhl-Mortensen, L. Buhl-Mortensen, L. Rise, Giant sand waves in the Hola glacial trough off Vesterålen, North Norway, Mar. Geol. 267 (2009) 36–54.

[9] V. Bellec, L. Rise, R. Bøe, T. Thorsnes, Influence of bottom currents on the Lofoten continental margin, North Norway, in: F.L. Chiocci, D. Ridente, D. Casalbore, A. Bosman (Eds.), International Conference on Seafloor Mapping for Geohazard Assessment, 11–13 May 2009, Ischia, Italy. Rendiconti Online, Società Geologica Italiana, vol. 7, Extended Abstracts, 2009, pp. 155–157.

[10] S. Chand, L. Rise, V. Bellec, M. Dolan, R. Bøe, T. Thorsnes, et al., Active Venting System Offshore Northern Norway, EOS 89 (29) (2008) 261–262.

[11] L. Rise, V. Bellec, R. Bøe, T. Thorsnes, The Lofoten–Vesterålen continental margin, North Norway: canyons and mass-movement activity, in: F.L. Chiocci, D. Ridente, D. Casalbore, A. Bosman (Eds.), International Conference on Seafloor Mapping for Geohazard Assessment, 11–13 May 2009, Ischia, Italy. Rendiconti Online, Societa Geologica Italiana, Extended Abstracts, vol. 7, 2009, pp. 79–82.

[12] A.B. Klitgaard, The fauna associated with outer and upper slope sponges (Porifera, Demospongiae) at the Faroe Islands, Northeastern Atlantic, Sarsia 80 (1995) 1–22.

[13] P. Buhl-Mortensen, M. Dolan, L. Buhl-Mortensen, Prediction of benthic biotopes on a Norwegian offshore bank using a combination of multivariate analysis and GIS classification, ICES J. Mar. Sci. 66 (2009) 2026–2032.

[14] M.F.J. Dolan, P.B. Mortensen, T. Thorsnes, L. Buhl-Mortensen, V. Bellec, R. Bøe, Developing seabed nature-type maps offshore Norway: initial results from the MAREANO programme, Norw. J. Geol. 89 (2009) 17–28.

52 Distribution of Hydrocorals Along the George V Slope, East Antarctica

Alexandra L. Post[1], Philip E. O'Brien[1],
Robin J. Beaman[2], Martin J. Riddle[3],
Laura De Santis[4], Stephen R. Rintoul[5]

[1]Marine and Coastal Environment Group, Geoscience Australia,
Canberra, ACT, Australia, [2]School of Earth and Environmental Sciences,
James Cook University, Cairns, QLD, Australia, [3]Environmental
Protection and Change, Australian Antarctic Division, Channel
Highway, Kingston, TAS, Australia, [4]Instituto Nazionale di
Oceanografia e Geofisica Sperimentale, Sgnoico, Trieste, Italy, [5]CSIRO
Marine and Atmospheric Research, Hobart, TAS, Australia

Abstract

Dense coral–sponge communities on the upper continental slope (570–950 m) off
George V Land, east Antarctica, have been identified as Vulnerable Marine Ecosystems.
We propose three main factors governing their distribution on this margin: (i) their
depth in relation to iceberg scouring, (ii) the flow of organic-rich bottom waters, and
(iii) their location at the head of shelf-cutting canyons. Icebergs scour to depths of
500 m in this region, and the lack of such disturbance is a likely factor allowing the
growth of rich benthic ecosystems. In addition, the richest communities are found in
the heads of canyons that receive descending plumes of Antarctic bottom water formed
on the George V shelf, which could entrain abundant food for the benthos. The canyons
harboring rich benthos are also those that cut the shelf break. Such canyons are known
sites of high productivity in other areas due to strong current flow and increased mixing
with shelf waters, and the abrupt, complex topography.

Key Words: hydrocorals, Vulnerable Marine Ecosystem, submarine canyons, continental
slope, oceanographic processes

Introduction

The George V margin is located in east Antarctica, in the region of the Mertz Glacier
Tongue (Figure 52.1). The continental shelf break along this part of the margin occurs

Seafloor Geomorphology as Benthic Habitat. DOI: 10.1016/B978-0-12-385140-6.00052-9

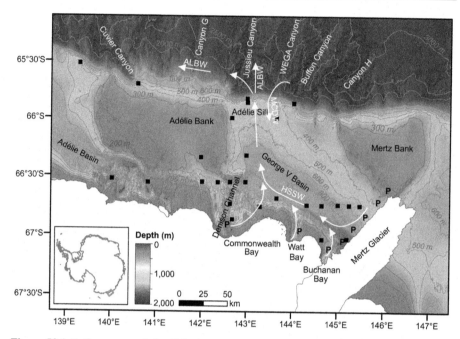

Figure 52.1 Bathymetry and simplified bottom circulation over the George V shelf with stations shown by black squares. The bathymetry grid is at 250 m resolution [1], and a simplified oceanographic circulation is shown [2,3,4] HSSW: high-salinity shelf water; MCDW: modified circumpolar deep water; ALBW: Adélie Land bottom water. P marks the extent of the Mertz polynya along the coast and western edge of the Mertz Ice Tongue. The inset map shows the location of the study area highlighted by the red box. (For interpretation of the references to color in this figure legend, the reader is referred to the web version of this book.)

at an average depth of 500 m, extending to the abyssal plain below 3,500 m [1]. The continental slope is cut by a series of large canyons, of which the Cuvier and Jussieu Canyons appear to reach the shelf break [5].

The region is a zone of bottom-water production formed from the highly saline shelf waters produced in coastal polynyas, which sink and flow through the George V Basin (also known as the Mertz-Ninnis trough) and off the shelf via the Adélie Sill [2]. The ice-free waters of coastal polynyas support a relatively long growing season, with phytoplankton production commencing soon after the spring equinox [6]. These waters therefore have high productivity compared to shelf areas with extensive sea ice cover.

Fishing within the CCAMLR managed region is restricted to depths below 550 m [7]. Two regions along the shelf break off the George V shelf have recently been declared Vulnerable Marine Ecosystems due to the occurrence of dense hydrocoral/sponge communities and are now closed to bottom fishing [8].

The RSV *Aurora Australis* was used as part of the Collaborative East Antarctic Marine Census (CEAMARC) series of surveys on the George V Land continental shelf and slope in December 2007 to January 2008. This survey focused on benthic and demersal sampling as well as physical and chemical oceanography. This chapter presents results on the distribution of hydrocoral communities as determined from underwater video and still-camera transects, and their relationship to physical parameters of the upper continental slope [9].

Geomorphic Features and Habitats

A bathymetric model based on multibeam swath bathymetry and singlebeam data has been produced with a 250 m grid size for the George V shelf and adjacent continental slope [1]. The key multibeam data set along the continental slope is derived from the Italian PNRA/MOGAM (Morphology and Geology of Antarctic Margins) survey [10]. This revised bathymetry model indicates that the shelf break occurs at depths between 400 and 500 m adjacent to the Mertz and Adélie banks, and up to 660 m in the region of the Adélie Sill [1] (Figure 52.1). Significantly, the average depth of the shelf break is below the depth of iceberg scours in this region, which have been found to depths of 500 m along the eastern flank of Mertz Bank [11]. Grounded icebergs on the George V shelf are sourced from the Ross Ice Shelf, the Ninnis Glacier, and the Cook Ice Shelf, and advected westwards via the westwind drift [12]. While icebergs produced by the Mertz Glacier are relatively small (maximum keel depth of 298 m) and are therefore only likely to ground on the shallowest parts of the banks, those from the Cook Ice Shelf have keel depths of up to 521 m [13] and ground readily along the margins of the Mertz and Adélie banks.

Along the upper slope, two canyons appear to reach the shelf break: the Jussieu Canyon, which connects to the shelf break at 142.5°E; 65.8°S, and the Cuvier Canyon at 140.5°E; 65.7°S. Swath data (Reson SeaBat 8150 multibeam system) reveal that the head of the Jussieu Canyon comprises dendritic branches approximately 5 km wide, which are smooth and flat floored, containing isolated outcrops of partially buried sedimentary ridges. The flat smooth character of the seabed at the head of the canyon is consistent with the occurrence of a sheet flowing bottom current, rather than current focusing in the canyon axis.

Canyons that cut the shelf break have been shown to be effective conduits for phytoplankton and other detritus produced in productive shelf waters, with the material distributed throughout the canyon system by strong tidal and gravity currents [14]. Submarine canyons are therefore associated with an enhanced food supply able to sustain high densities of filter feeders (e.g., anemones, corals, and sponges).

Connection between the shelf and the slope bottom waters in the region of the Adélie Sill is also revealed by oceanographic measurements [2,15]. Bottom salinity and density is relatively high along the slope to the west of the Adélie Sill, whereas relatively low salinity and density occurs toward the east (Figure 52.2). This pattern reflects the production of highly saline and dense bottom waters within the polynyas

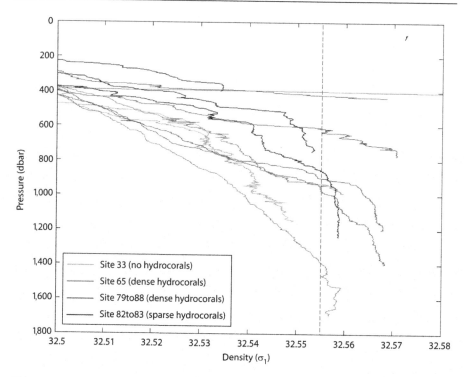

Figure 52.2 Density measured from Conductivity, temperature, depth (CTD) profiles at continental slope sites. The vertical dotted line illustrates the transition to dense waters at depth on transects 79–88, 65, and 82–83. This is consistent with the presence of ALBW along the slope to the west of the Adélie Sill.

along the George V coast and adjacent to the Mertz Glacier, which then flows northwestward through the George V Basin, forming Adélie Land bottom water (ALBW) as is flows off the shelf through the sill [3,15,16] (Figure 52.1). The ALBW then flows down the continental slope via the Jussieu Canyon and is also advected westward along the slope. ALBW likely entrains nutrients and organic matter from the productive shelf area of the Mertz polynya. Evidence for the export of biogenic siliceous sediment off the shelf via ALBW is supported by sedimentological studies linking sediment deposition and bottom-water production throughout the late Quaternary [5,17]. In addition, sediment waves, drifts, and moat features revealed from seismic facies demonstrate the action of bottom current flow across the continental rise during pre-Quaternary times [5,18,19].

Sub-bottom chirp profiles acquired during the MOGAM cruise (with sweep of 2–7 kHz) on the upper slope adjacent to the Jussieu Canyon show a very high-amplitude seabed reflector (Figure 52.3) in water depths of between 580 and 1,300 m. Artificial whitening of the highest amplitude acoustic facies is graphically used to better show the patchy distribution of the highest amplitude sediment reflectance at and below the seafloor. The low penetration of the acoustic signal below the seabed, caused

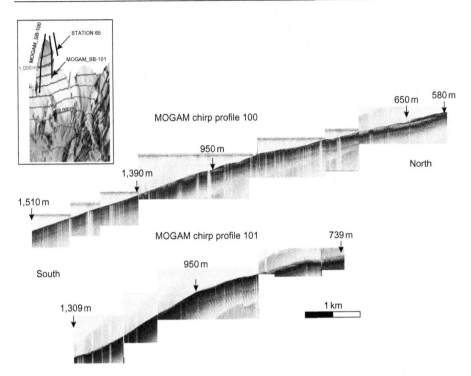

Figure 52.3 Swath bathymetric map and sub-bottom chirp profiles 100 and 101, collected in 2006 during the Italian PNRA/MOGAM (Morphology and Geology of Antarctic Margins) cruise from the continental slope of the George V Land with the R/V *OGS Explora*, adjacent to transect 65. The chirp profiles show a very high-amplitude seabed reflector in water depth between 580 and 1,300 m. Artificial whitening of the highest amplitude acoustic facies is graphically used to better show the patchy distribution of the highest amplitude sediment at and below the seafloor.

by the high reflectivity of the seafloor, suggests that the seabed is composed of coarse and/or compacted sediment. This is consistent with the flow of bottom currents through the upper branches of the Jussieu Canyon, which would erode and resuspend fine-grained material and transport it to deeper water. The high-amplitude acoustic facies observed in the chirp profiles can therefore be interpreted to be caused by the high contrast between the water column and a coarse, hard, biogenically encrusted seabed.

Biological Communities

Detailed video and still footage was obtained along five transects in four areas of the George V continental slope: from east to west these are stations 33, 63, 65, 79–88, and 82–83. In addition, 25 sites were imaged on the continental shelf. Dense communities of hydrocoral (*Errina* sp.) and demosponges (numerous species) occur at stations 65 and 79–88 (Figure 52.4; Table 52.1). The other two areas on the slope

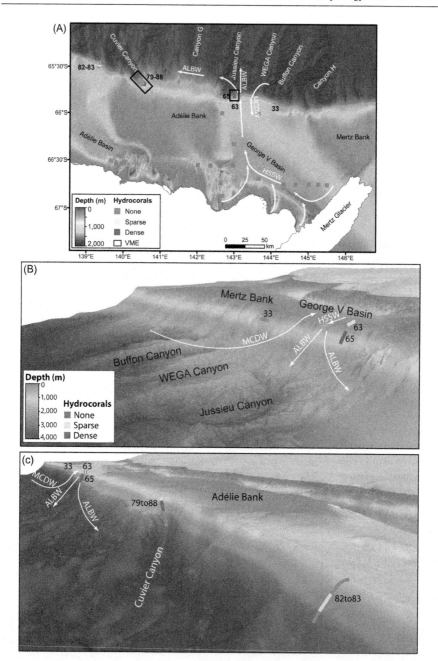

Figure 52.4 Occurrence of *Errina* sp. hydrocoral communities along the George V continental slope (colored squares) with 250 m bathymetry grid and general bottom circulation (based on [2,3,20]) in plan view (A), and three-dimensional views along the continental slope (B and C; 15 × vertical exaggeration). Black outlines on (A) indicate areas proclaimed as Vulnerable Marine Ecosystems. The smooth seafloor along the upper slope in (C) reflects the absence of multibeam bathymetry, though numerous ship tracks allow definition of the shelf break. (For interpretation of the references to color in this figure legend, the reader is referred to the web version of this book.)

Table 52.1 Summary of Slope Transects Sampled on the George V Continental Shelf

Station	Depth Range	Occurrence of *Errina sp.*	Dominant Community	Seafloor Image
82–83	400–1,200 m	Sparse: 600–920 m; Dense: 900 m	Medium-density soft corals and bryozoans	
79–88	435–1,395 m	Sparse: 440–570 m; Dense: 570–950 m	Dense *Errina* sp. and demosponge	
65	650–950 m	Dense: 650–950 m	Dense *Errina* sp. and demosponge	
63	420–430 m	Sparse: 430 m	Medium-density soft corals and gorgonian whips	
33	740–770 m	None	Sparse demosponge	

contain only sparse (station 82–83) or no corals (station 33). Station 63, upslope from station 65, also has only sparse hydrocoral occurrence. All other sites analyzed on the continental shelf contained no hydrocoral occurrences (Figure 52.4).

The mid-slope transect at station 65 crosses depths of approximately 650–950 m. Dense seabed communities occur along the entire transect, dominated by the hydrocoral *Errina* sp. and a diverse assemblage of demosponges (Table 52.1). Station 63, on the upper slope, contains sparse occurrences of *Errina* sp. at 430 m. Along transect 79–88, the hydrocoral *Errina* sp. is present in low densities from 440 to 570 m in a community co-dominated by various demosponge species and soft corals. A high-density hydrocoral–sponge community occurs at depths of 570–950 m along this transect, with highest cover occurring on ridges. No hydrocorals were observed at depths greater than 1,025 m on this transect. At station 82–83, sparse occurrences of the *Errina* sp. corals occur at depths of 600–920 m, with a small area of dense cover at 900 m. At no time is *Errina* sp. the dominant component of the community along transect 82–83, with soft corals, bryozoa, and at times hexactinellids (glass sponges), the dominant taxa. The seabed communities along transect 33 (670–820 m water depth) contain no hydrocorals, with only a very sparse cover of demosponges present.

The sparse occurrence of hydrocorals on transect 63 reflects relatively shallow water depths reaching a maximum depth of 430 m—shallower than the occurrence of dense communities on the other transects. The sparse occurrence of hydrocoral communities at depths shallower than 570 m on the George V shelf likely reflects their slow recovery to disturbance from iceberg scouring, given likely iceberg keel depths of up to 520 m in this region [13]. Transects 82–83 and 33 exhibited only sparse or no hydrocorals, yet encompassed water depths well below the depth of iceberg scouring (400–1,500 and 670–820 m, respectively), and are within the depth range where dense hydrocoral communities occur on other transects.

The two rich coral–sponge communities are located at the head of the Jussieu and Cuvier canyons (Figures 52.3–52.4). The Jussieu Canyon is the main conduit for ALBW exiting the shelf [2,15], and continental slope sites to the west also have water properties indicative of ALBW [3,20] (Figure 52.2). At site 33, to the east, bottom salinity and density is relatively low, consistent with the presence of Ross Sea Bottom Water on this part of the slope [2] (Figure 52.2). The absence of corals at site 33 is likely associated with lower food supply due to the absence of ALBW. In addition, the shelf edge at the head of the Jussieu and Cuvier canyons is indented, indicating that these two canyons cut the continental shelf break [5]. No such indentations are observed at sites 33 and 82–83, suggesting that canyons below these sites do not cut the shelf break. Canyons that cut the shelf break have been shown to be effective conduits for transporting nutrients and detrital material from the productive shelf waters, and the relatively strong current flow and availability of hard substrates also make them prime habitat for filter feeders. Hydrocoral communities sampled in the Ross Sea have also been shown to be located in areas of high current flow on the shelf break and slope [21], further suggesting the strong association between dense hydrocoral communities and oceanographic and morphologic factors along the Antarctic margin.

Surrogacy

Due to the low number of transects containing hydrocorals, no statistical analyses have been carried out to examine the relationship between the hydrocorals and the physical parameters.

Acknowledgments

We would like to thank Captain Ian Moodie, the crew, and scientific the party aboard RSV *Aurora Australis* survey V3 who undertook the sampling for the CEAMARC program. An earlier copy of this manuscript was improved by critical review by Dr. Matthew McArthur and Dr. Scott Nicol (Geoscience Australia) and Dr. David Dowden (NIWA). The work is published with the permission of the Chief Executive Officer, Geoscience Australia.

References

[1] R.J. Beaman, P.E. O'Brien, A.L. Post, L. De Santis, A new high-resolution bathymetry model for the Terre Adélie and George V continental margin, East Antarctica, Antarct. Sci. 23 (1) (2011) 95–103. 10.1017/S095410201000074X.

[2] N.L. Bindoff, S.R. Rintoul, R. Massom, Bottom water formation and polynyas in Adélie Land, Antarctica, Pap. Proc. R. Soc. Tasm. 133 (2000) 51–56.

[3] S.R. Rintoul, On the origin and influence of Adélie Land Bottom Water, in: S. Jacobs R. Weiss(Eds.), Ocean, Ice and Atmosphere: Interaction at the Antarctic Continental Margin, vol. 75, American Geophysical Union, Washington, DC, 1998, pp. 151–171.

[4] R.A. Massom, K.L. Hill, V.I. Lytle, A.P. Worby, M.J. Paget, I. Allison, Effects of regional fast-ice and iceberg distributions on the behaviour of the Mertz Glacier polynya, East Antarctica, Ann. Glaciol. 33 (2001) 391–398.

[5] A. Caburlotto, L. De Santis, C. Zanolla, A. Cemerlenghi, J.K. Dix, New insights into Quaternary glacial dynamic changes on the George V Land continental margin (East Antarctica), Quat. Sci. Rev. 25 (2006) 3029–3049.

[6] R.N. Sambrotto, A. Matsuda, R. Vaillancourt, M. Brown, C. Langdon, S.S. Jacobs, C. Measures, Summer plankton production and nutrient consumption patterns in the Mertz Glacier Region of East Antarctica, Deep Sea Res. II 50 (2003) 1393–1414.

[7] M.L. Stevenson, S.M. Hanchet, A. Dunn, A characterisation of the toothfish fishery in Subareas 88.1 & 88.2 from 1997/98 to 2007/08, in WG-FSA-08/22, CCAMLR, 2008.

[8] CCAMLR, Encounters with vulnerable marine ecosystems in the convention area, in: Commission for the Conservation of Antarctic Marine Living Resources Vulnerable Marine Ecosystems Workshop, Paper WG-EMM-09/8. CCAMLR, La Jolla, CA, USA, 2009.

[9] A.L. Post, P.E. O'Brien, R.J. Beaman, M.J. Riddle, L. De Santis, Physical controls on deep-water coral communities on the George V Land slope, East Antarctica, Antarctic Science 22 (2010) 371–378, doi:10.1017/S095410201000180..

[10] L. De Santis, G. Brancolini, A. Accettella, A. Cova, A. Caburlotto, F. Donda, C. Pelos, F. Zgur, M. Presti, Keystone in a Changing World—Online Proceedings of the 10th International Symposium on Antarctic Earth Sciences, 2007.

[11] P.W. Barnes, R. Lien, Icebergs rework shelf sediments to 500 m off Antarctica, Geology 16 (1988) 1130–1133.

[12] R.A. Massom, Recent iceberg calving events in the Ninnis Glacier region, East Antarctica, Antarct. Sci. 15 (2003) 303–313.

[13] J.A. Dowdeswell, J.L. Bamber, Keel depths of modern Antarctic icebergs and implications for sea-floor scouring in the geological record, Mar. Geol. 243 (2007) 120–131.

[14] E.W. Vetter, P.K. Dayton, Organic enrichment by macrophyte detritus, and abundance patterns of megafaunal populations in submarine canyons, Mar. Ecol. Prog. Ser. 186 (1999) 137–148.

[15] G.D. Williams, N.L. Bindoff, S.J. Marsland, S.R. Rintoul, Formation and export of dense shelf water from the Adélie Depression, East Antarctica, J. Geophys. Res. 113 (2008) CO4039, 10.1029/2007JC004346.

[16] N.L. Bindoff, G.D. Williams, I. Allison, Sea-ice growth and water-mass modification in the Mertz Glacier polynya, East Antarctica, during winter, Ann. Glaciol. 33 (2001) 399–406.

[17] P.T. Harris, G. Brancolini, L. Armand, M. Busetti, R.J. Beaman, G. Giorgetti, M. Presti, F. Trincardi, Continental shelf drift deposit indicates non-steady state Antarctic bottom water production in the Holocene, Mar. Geol. 179 (2001) 1–8.

[18] F. Donda, G. Brancolini, L. De Santis, F. Trincardi, Seismic facies and sedimentary processes on the continental rise off Wilkes Land (East Antarctica): evidence of bottom current activity, Deep Sea Res. II 50 (2003) 1509–1527.

[19] C. Escutia, L. De Santis, F. Donda, R.B. Dunbar, A.K. Cooper, G. Brancolini, S.L. Eittreim, Cenozoic ice sheet history from East Antarctic Wilkes Land continental margin sediments, Global Planet. Change 45 (2005) 51–81.

[20] N.L. Bindoff, M.A. Rosenberg, M.J. Warner, On the circulation of water masses over the Antarctic continental slope and rise between 80 and 150°E, Deep Sea Res. II 47 (2000) 2299–2326.

[21] S.J. Parker, D.A. Bowden, Identifying taxonomic groups as vulnerable to bottom longline fishing gear in the Ross Sea region: Commission for the Conservation of Antarctic Marine Living Resources Vulnerable Marine Ecosystems Workshop, Paper WS-VME-09/8, CCAMLR, La Jolla, CA, USA, 2009.

53 The Cook Strait Canyon, New Zealand: Geomorphology and Seafloor Biodiversity of a Large Bedrock Canyon System in a Tectonically Active Environment

Geoffroy Lamarche[1], Ashley A. Rowden[1], Joshu Mountjoy[1], Vanessa Lucieer[2], Anne-Laure Verdier[1]

[1]National Institute of Water and Atmospheric Research, Wellington, New Zealand, [2]Marine Research Laboratories, TAFI, University of Tasmania, Hobart, TAS, Australia

Abstract

The multibranched Cook Strait Canyon system (41°S–42°S) deeply incises the bedrock between the North and the South islands of New Zealand, cutting approximately 40 km into the continental shelf. Water depths range from 120 m on the shelf to 2,700 m in the Hikurangi Trough. The geomorphology of the canyons reflects a tidally dominated seaway and intense seismotectonic activity associated with the neighboring active Pacific–Australia plate boundary. The study area covers ~4,500 km², including the Cook Strait Canyon (~1,800 km²), areas of continental shelf, and deep sedimentary trough. This study uses Kongsberg EM300 multibeam data, 110 sediment samples, ~100 biological samples, and ~200 benthic taxa identified from seabed images and direct samples. Derivative spatially continuous data sets include geomorphic mapping, quantitative backscatter analysis, and sediment distribution, which enabled us to spatially analyze the geomorphic–sedimentary–biologic relationships to characterize the geohabitat and benthic biodiversity of a large, bedrock-incised submarine canyon system.

Key Words: submarine canyon, subduction, tidal seaways

Introduction

Cook Strait is located between the North and the South islands of New Zealand, forming a 20–60 km wide oceanic passage to east-moving oceanographic currents and atmospheric winds between the Tasman Sea and the southwest Pacific Ocean. Cook Strait is of strategic, public, and economic interest for New Zealand. This study

Seafloor Geomorphology as Benthic Habitat. DOI: 10.1016/B978-0-12-385140-6.00053-0

focuses on the multibranched Cook Strait canyon system, which is carved more than 1,100 m deep into the bedrock of continental shelf [1].

The diverse and complex geomorphology of the Cook Strait canyon system is the result of dynamic, climatic, tectonic, and oceanic forcings that impacted on the region since at least the last postglacial period [2]. The tectonic deformation is associated with the active oblique subduction of the Pacific Plate beneath the Australia Plate along the Hikurangi Trough immediately to the east of the strait. There, major tectonic faults are capable of generating damaging earthquakes. The strait is subject to extreme tidal currents, with peak flows reaching 1.5 m/s and strong oceanographic currents resulting from oceanographic mass transfer between Tasman Sea and Pacific Ocean. This current regime is coupled with a vigorous, locally generated wave climate and strong sediment delivery from North and South island's rivers, which altogether have produced distinctive patterns of sedimentation in the region [1–3]. The strait contains seabed power cables joining the two islands, and important fish stocks and biological resources are all concentrated in the vicinity of the country's capital city. The naturalness of the Cook Strait environment is also potentially impacted by the discharge of treated sewage from Wellington and surrounding urban areas.

This study reports the use of EM300 (30 kHz) multibeam echosounder (MBES) data collected between 2001 and 2005 onboard RV *Tangaroa*, in water depth greater than 100 m. Multichannel and high-resolution seismic reflection data, sediment cores, seafloor grab/dredge samples, and seafloor photographs complement the bathymetry data set (Figure 53.1). The geophysical and geological data provide excellent means for ground-truthing habitat maps generated from multibeam data analysis.

We use the backscatter imagery (Figure 53.2) as an indicator for seafloor geology and microtopography [5], as it relates to grain size and sediment volume scattering. Backscatter imagery emphasizes topographic and geological features not necessarily recognized with conventional surveying. The processing of the raw backscatter signal attenuates the effects due to recording equipment, seafloor topography, and water column, and includes signal calibration, compensation, and speckle noise filtering.

We used published [1,5,6] and unpublished data from over 100 seafloor samples collected since the late 1950s in Cook Strait [7], using gravity corers, piston corers, grabs, dredges, trawls, and sleds to describe the substratum type and biological community composition. Sedimentological analyses include mean grain size analysis obtained using a laser granulometer.

Geomorphic Features and Habitats

During the Last Glacial Maximum, parts of the continental shelf (present-day water depths shallower than 150 m) were emergent as a coastal plain [2]. As sea level rose, a widespread wave-based erosional surface was formed on the shelf. Most of this surface was subsequently blanketed by a wedge of post-last glacial mud, but some areas remain bare of mud, so that the typical erosional surface outcrops at the seafloor. Pervasive semicircular scars along the upper canyon walls attest to their highly unstable condition. The canyons control the routing of sediment pathways from land

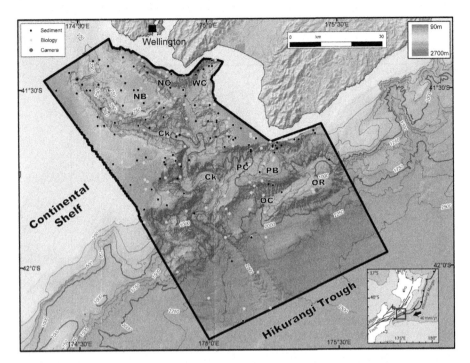

Figure 53.1 Cook Strait bathymetry Digital Terrain Model (DTM) gridded at 25 m showing the location of sediment and biological samples, and camera stations. Pastel colors are from the regional bathymetry of the New Zealand National Institute of Water and Atmospheric Research (NIWA) [4]. NB: Nicholson Bank; NC: Nicholson Canyon, WC: Wairarapa Canyon; Ck: Cook Strait Canyon, PB: Palliser Bank; PC: Palliser Canyon; OR: Opouawe Ridge; OC: Opouawe Canyon. Inset: The Pacific–Australia plate boundary (teeth line) in the New Zealand region. The main faults of the North and South islands are indicated on land. The arrow indicates the Pacific–Australia relative plate motion.

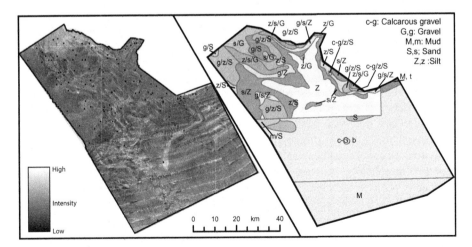

Figure 53.2 Calibrated and compensated backscatter imagery of Cook Strait, with strong intensity in white. Red and black dots are location of sediment samples. Extract of the New Zealand sediment map from Orpin et al. [5] for the study area. (For interpretation of the references to color in this figure legend, the reader is referred to the web version of this book.)

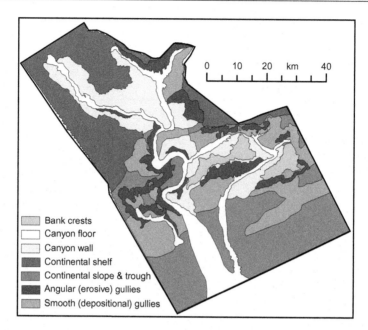

Figure 53.3 Geomorphic Domains of the Cook Strait canyon system.

to the turbidite-filled Hikurangi Trough. The continental shelf and slope are affected by NE-trending structurally controlled ridges. Active strike-slip faults have strong morphological expressions as ENE/WSW-trending scarps [8].

The study area has been segmented into seven geomorphic domains (Figure 53.3) for the purposes of habitat analysis, based on previous work that concerned sedimentary processes in the Cook Strait canyon system [1] (Table 35.1).

Continental shelf: A limited area of continental shelf is included in the study area, constrained by the extent of multibeam bathymetry. The continental shelf is defined as extending to the canyon and continental slope edge, generally between the 130–200 m contour. The seafloor of the continental shelf within the study area is known to be variably characterized by rock outcrop of marine siltstone and mudstone, and sediment ranging from coarse clastics to mud [5–8].

Angular and smoothed gullies: Slope gullies are steep-sided, confined channels that incise into the slopes of canyon walls and tectonically forced bathymetric features. Two classes of gullies are recognized, distinguished by their angular versus smoothed morphology. Angular gullies are interpreted as bedrock incised and predominantly occur in sedimentary (siltstone/mudstone) sequences. Angular gullies are inferred to be active or recently active erosive features. Smoothed gullies are either angular gullies that are draped with sediment or depositional features on the open continental slope. The two gully types are primarily distinguished by local slope gradient, with angular gullies typically exhibiting slopes exceeding 20°.

Canyon walls affected by mass failure: Significant areas of the canyon walls are affected by deep-seated mass failures resulting in a characteristic undulating

Table 53.1 Data Available in the Seven Geomorphic Domains of the Cook Strait canyon System

Domain	Total Area (%)	Total Area (km²)	Number of Sediment Samples	Biological Samples	Camera Stations	Number of Taxa
Continental shelf	19.1	976.4	36	33	0	67
Bank crests	3.2	165.4	0	6	6*	82
Smoothed slope gullies	16.8	858.3	8	5	0	14
Angular slope gullies	9.3	475.5	20	15	4	27
Canyon wall landslides	15.0	769.1	16	19	0	42
Canyon floor	11.9	607.6	23	10	3	25
Continental slope and trough	24.7	1265.9	10	12	0	34

*One camera station and five video/camera transects.

morphology [1]. It is inferred that most mass failure affected areas are draped with fine-grained sediment.

Canyon floor: Canyon floors cover a surprisingly significant proportion of the study area (~12%) and span a large bathymetric depth range, from 250 to 2,700 m. Canyon floor substrate is characterized by sand. From morphological interpretation it is inferred that sedimentary bedrock is exposed in several places (e.g., knickpoints or bathymetric steps).

Bank crests: Bank crests are given their own category for two reasons: (i) they are generally levelled and isolated areas on the otherwise inclined continental slope and subsequently are dominated by deposition (hemipelagic sedimentation); and (ii) cold fluid seepage documented on the Hikurangi Margin predominantly occurs on bank crests [3], including the Opouawe Bank within the study area, a significant seep site with significant biogenic association. Two bank crests occur within the study area and have a combined area of 165 km².

Continental slope and Hikurangi Trough: The continental slope is characterized by a smooth surface and a low slope gradient (<5°), and typically has a soft sediment substrate. The Hikurangi Trough is the main sedimentary basin within which sediment shed from the canyon and slopes systems deposits. It has a low (<1°) to flat slope gradient and is characterized by surface sediment ranging from mud to fine sand [2].

Biological Communities

Whilst approximately 100 fauna samples and tens of photographic images of the seafloor have been collected in the Cook Strait canyon system, no formal description of the benthic communities of the canyon has been forthcoming. The reason for this lies in the generally ad hoc nature of the biological sampling to date. The spatially and temporally sporadic samples have been used primarily to describe the distribution of individual taxa or specific assemblages [9,10]. Generally, samples/images have been obtained using different gear types, and the fauna identified to inconsistent taxonomic levels, which restricts the type of descriptions and comparisons that can be made among different locations. Basic qualitative descriptions of the faunal assemblages found at the different geomorphic habitats are presented below, based on all seabed samples, as well as the results of limited statistical analyses based on a smaller number of samples where taxa are identified to family level. Seafloor images were used when available to illustrate community composition and substratum type.

Continental shelf: Decapod crustaceans are among the conspicuous members of the seabed assemblage of the shelf, particularly the majid crabs (e.g., *Thacanophrys filholi*) and pagurid hermit crabs. Asteroid (e.g., *Odontaster behami*) and ophiuroid (e.g., *Amphipholis squamata*) echinoderms also occur frequently in the samples from the shelf. Pecten bivalves and buccinid gastropods are among the reasonably well-represented mollusk taxa. Polychaetes and other burrowing fauna such as sipunculids and small bivalve species are found in the soft sediments of the shelf.

Angular and smoothed gullies: Echinoids are the dominant faunal echinoderms of the angular gullies, with at least five species (e.g., *Goniocidaris umbraculum* and *Spatangus mathesoni*) occurring at this geomorphic habitat. Asteroids and ophiuroids are also found. Mollusk representatives such as bivalves and gastropods occur frequently in these gullies (Figure 53.4A). Fauna such as sponges, brachiopods, and scleractinian corals (e.g., *Caryophyllia profunda* and *Flabellum apertum*) are found attached to exposed hard substrates in the angular gullies. Few samples have been obtained from the smoothed gullies, but the assemblage composition appears to be similar to that found in the other type of gully habitat.

Canyon walls: The exposed hard substrates of the canyon walls provide suitable habitat for at least four species of scleractinian corals (e.g., *Caryophyllia* spp. and *Goniocorella dumosa*), as well as sponges, ascidians, hydrozoans, brachiopods, and bryozoans. Decapod crustaceans occur relatively frequently in this habitat, in particular galatheid squat lobsters (e.g., *Phylladiorhynchus pusillus*). Serolid isopods also occur among the crustacean fauna. Fauna of soft sediment are also present within the canyon wall geomorphic habitat, including polychaetes, echinoids (e.g., *Brissopsis oldhami*), and bivalves.

Canyon floor: The sandy floor of the canyon appears to have sparse fauna, i.e., relatively few taxa in low frequency of occurrence among samples. However, the few samples and seafloor images that were obtained from this habitat show that ophiuroids can occur in high abundances (Figure 53.4B). Among the more common fauna that occur in samples from this habitat are some small-sized crustaceans such as amphipods (belonging to the Lysianassidae family) and isopods (Serolidae and Cirolanidae), as well as decapods. Pycnogonids (sea spiders) are also reasonably well represented. Burrowing polychaetes (e.g., *Goniada vorax*), echinoids, and bivalves occur in the substrates of the canyon floor.

Bank crests: Apart from a single early sample, the remaining biological samples were taken in 2006 and have been better processed than the majority of samples taken from other geomorphic habitats in the Cook Strait canyon system. This is particularly the case for the identification of the polychaete fauna. As such, the

Figure 53.4 Seafloor photos in the Cook Strait canyon system: (A) ribbed sand and benthic fauna (including gastropods) of an angular gully habitat; (B) sandy substrate and the abundant ophiuroid assemblage of the canyon floor habitat; (C) carbonate structure at a cold seep site and associated fauna (including sibloglinid worms and, vesicomyid bivalves) of the bank crest habitat (Opouawe Bank).
Source: Images: (A,B) NIWA; (C) NOAA/NIWA.

assemblage of the bank crest habitat appears to be taxon rich. A number of cold seeps are located on the Opouawe Bank, characterized by carbonate structures and a particular seep fauna, either associated with the biogenic substrate or the soft sediment surrounding these features [9,10] (Figure 53.4C). Over 20 species of polychaete have been sampled, including members of Siboglinidae typically found at chemosynthetic habitats such as seeps. Other seep-associated taxa are found among the frequently occurring bivalve mollusk fauna (e.g., *Acharax clarificata*). Corals (e.g., *Flabellum apertum*), sponges, actinarians, and hydrozoans are among the fauna found attached to the carbonate structures at the seep sites. Decapod crustaceans, ophiuroid and holothurian echinoderms, and gastropod mollusks occur among the "background" assemblage of the bank crest habitat.

Continental slope and Hikurangi Trough: Although many different taxonomic groups occur on the continental slope area of the Cook Strait canyon system, no particular faunal group or species dominates the assemblage of this habitat. The decapod crustaceans are represented by species of majids, galatheids (e.g., *Munida gracilis*), pagurids, and pasiphaeids. Echinoid, ophiuroid, and asteroid echinoderms are also found. Among the mollusk groups, which include the gastropods and scaphopods, the bivalves are particularly well represented by at least 11 species. The latter include epi- as well as infaunal species (belonging to the orders Nuculoidea and Veneroida). Some fauna associated with hard substrates, such as sponges, gorgonian and scleractinian corals, and hydrozoans, are also found on the continental slope. Only one sample has been obtained from the Hikurangi Trough adjacent to the Cook Strait canyon; it contained an unidentified polychaete and bivalve.

Differences in Taxon Richness among Geomorphic Habitats

It is not possible to undertake robust multivariate statistical analysis of community composition because of the inconsistent levels of taxonomic identification among samples from the different geomorphic habitats. However, it is possible to estimate taxon richness for each habitat using a presence/absence estimator metric [11] for data at the family level. The results of such an analysis (Figure 53.5) revealed that taxon richness is estimated to be lowest for the presumably two most disturbed habitats: angular and smoothed gullies, and the canyon walls affected by mass failure. Estimated taxon richness is highest for the communities of the banks (which include the cold seep sites) and the continental slope. The communities of the predominantly soft sediment canyon floor and continental shelf have intermediate levels of estimated taxon richness.

Surrogacy

The backscatter data have been successfully used as a proxy for seafloor substrate mapping [5], and an object-based image analysis technique was applied to the Cook Strait backscatter data that resulted in a first-level, unsupervised, seabed substrate map [12,13] (Figure 53.6). The classification method is based on textural

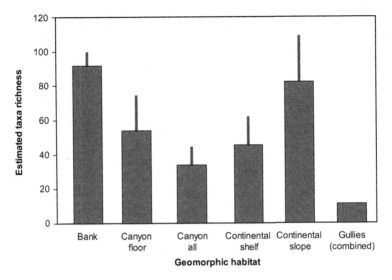

Figure 53.5 Estimated taxa richness (using Chao 2 estimator at $n = 6$ [11]) for the geomorphic habitats of the Cook Strait canyon system. Error bars = 1 Standard Error. Note: Hikurangi Trough habitat is not included because $n < 6$.

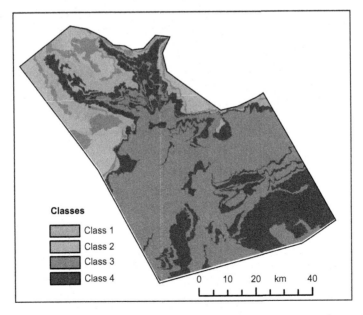

Figure 53.6 Object-based classification of the Cook Strait canyon system study area from backscatter data (Figure 53.2) and bathymetry (Figure 53.1).

image analysis [14] and examines the physical factors influencing seafloor texture at various scales in order to evaluate how differences in backscatter texture indicate differences in substrate and ultimately seabed biotopes. The four classes in the unsupervised classification map have no associated classification value, and further validation will be necessary to correlate classes to metrics of biodiversity. However, initial tests are promising for the production of biodiversity maps that use ecological theory to predict biodiversity from knowledge of seabed substrate heterogeneity, as derived from statistically compensated backscatter data [12]. This method will be applied to Cook Strait in the near future after the sparse historical biological data has been augmented with data from recently collected samples from a range of habitats in the study area. These data, and the analytical approach outlined, provide an excellent opportunity to further develop automated methods to characterize marine habitats over large areas and assess how the classification relates to biodiversity.

Owing to the wide range of water depth, geomorphological expression, and oceanographic conditions, Cook Strait also offers an excellent opportunity to test fundamental hypotheses concerning the influence of habitat type and heterogeneity on patterns of seafloor biodiversity.

References

[1] J.J. Mountjoy, P.M. Barnes, J.R. Pettinga, Morphostructure and evolution of submarine canyons across an active margin: Cook Strait sector of the Hikurangi Margin, New Zealand, Mar. Geol. 260 (1–4) (2009) 45–68.

[2] K.B. Lewis, L. Carter, F.J. Davey, The opening of Cook Strait: interglacial tidal scour and aligning basins at a subduction to transform plate edge, Mar. Geol 116 (1994) 293–312.

[3] P.M. Barnes, G. Lamarche, J. Bialas, S. Henrys, I. Pecher, G. Netzeband, et al., Tectonic and geological framework for gas hydrates and cold seeps on the Hikurangi subduction margin, New Zealand, Mar. Geol. 272 (1–4) (2010) 26–48.

[4] CANZ, Undersea New Zealand, New Zealand Region Physiography, 1:4,000,000, second ed., New Zealand Oceanographic Institute Chart, Miscellaneous Series N. 85, 2008.

[5] G. Lamarche, X. Lurton, A.-L. Verdier, J.-M. Augustin, Quantitative characterization of seafloor substrate and bedforms using advanced processing of multibeam backscatter. Application to the Cook Strait, New Zealand, Continent. Shelf Res. 31 (2011) S93–S109.

[6] L. Carter, Acoustical characterisation of seafloor sediments and its relationship to active sedimentary processes in Cook Strait, New Zealand, N. Z. J. Geol. Geophys. 35 (1992) 289–300.

[7] A. Orpin, L. Carter, A. Goh, K. Mackay, A.-L. Verdier, S.M. Chiswell, P.J.H. Sutton, New Zealand diverse seafloor sediments: NIWA Charts, Miscellaneous Series N. 86, National Institute of Water and Atmospheric Research, Wellington, New Zealand, 2008.

[8] P.M. Barnes, J.-C. Audru, Recognition of active strike-slip faulting from high-resolution marine seismic reflection profiles: Eastern Marlborough fault system, New Zealand, Geol. Soc. Am. Bull. 111 (4) (1999) 538–559.

[9] K.B. Lewis, B.A. Marshall, Seep faunas and other indicators of methane-rich dewatering on New Zealand convergent margins, N. Z. J. Geol. Geophys. 39 (1996) 181–200.

[10] A.R. Baco, A.A. Rowden, L.A. Levin, C.R. Smith, D.A. Bowden, Initial characterization of cold seep faunal communities on the New Zealand Hikurangi margin, Mar. Geol. 272 (2010) 251–259.

[11] A. Chao, Estimating the population size for capture-recapture data with unequal catchability, Biometrics 43 (1987) 783–791.

[12] G. Lamarche, V. Lucieer, A.A. Rowden, A.-L. Verdier, J.-M. Augustin, X. Lurton, Submarine Substrate and Biodiversity Mapping using Multiscale Analysis of Bathymetric and Backscatter data—Examples from Cook Strait and the Kermadec Ridge, New Zealand, Pacific Science Association (Ed.). 2009: 11th Pacific Science Inter-Congress, Tahiti—French Polynesia, 111.

[13] V. Lucieer, G. Lamarche, Unsupervised fuzzy classification and object-based image analysis of multibeam data to map deep water substrates, Cook Strait, New Zealand, Continent. Shelf Res. 31 (2011) 1236–1247.

[14] V.L. Lucieer, Object-oriented classification of sidescan sonar data for mapping benthic marine habitats, Int. J. Remote Sens. 29 (3) (2008) 905–921.

54 The Ascension–Monterey Canyon System: Habitats of Demersal Fishes and Macroinvertebrates Along the Central California Coast of the USA

Mary Yoklavich[1], H. Gary Greene[2]

[1]National Marine Fisheries Service, Southwest Fisheries Science Center, Santa Cruz, CA, USA, [2]Tombolo Institute, Eastsound, WA, USA

Abstract

The Ascension–Monterey canyon system, located 160 km south of San Francisco, is the largest seafloor physiographic feature on the continental margin of California. This system includes six submarine canyons: Ascension, Año Nuevo, Cabrillo, Soquel, Monterey, and Carmel. We focus attention on the heads of these canyons, which provide diverse habitats of rocky, steep relief surrounded by low-relief sand and mud in water depths from 20 to 800 m. The rock exposures are manifested locally as ledges, crevices, and overhangs that afford refuge to fishes. Rockfishes (*Sebastes* spp.) dominate fish assemblages in the canyon heads and occur in all types of habitats. Fish assemblages in these canyons discriminate based on depth and substratum. Common macroinvertebrates include brittle stars (*Ophiacantha* spp.), sessile sea cucumbers (*Psolus squamatus*), fragile pink urchins (*Allocentrotus fragilis*), crinoids (*Florometra serratissima*), and spot prawns (*Pandalus platyceros*) occurring on mixed mud, rock, and cobbles. Some large invertebrates [e.g., various sponges, gorgonians, sea pens, and basket stars (*Gorgonocephalus eucnemis*)] may be components of habitat for demersal fishes in these canyons.

Key Words: Submarine canyon, demersal, habitat, rockfish, species assemblage, seafloor

Introduction

The Ascension–Monterey canyon system (Figure 54.1), located in the vicinity of Monterey Bay about 160 km south of San Francisco, is the largest seafloor physiographic feature on the continental margin of California. This feature is located within

Seafloor Geomorphology as Benthic Habitat. DOI: 10.1016/B978-0-12-385140-6.00054-2

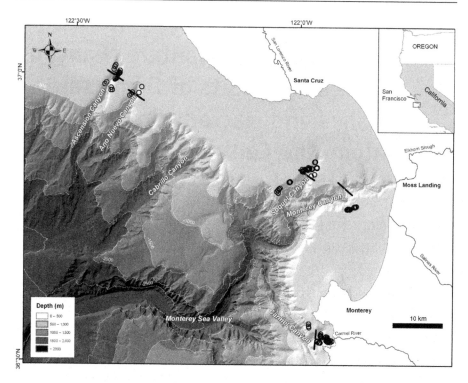

Figure 54.1 Ascension–Monterey canyon system, located off central California, includes Ascension, Año Nuevo, Cabrillo, Soquel, Monterey, and Carmel canyons. Circles indicate locations of visual surveys of organisms and associated habitats using *Delta* submersible. Solid lines are cross-section depth profiles (see Figure 54.2). Bathymetric data were collected with a Kongsberg Simrad EM-300™ multibeam echosounder and processed at 15-m grid cell size (*source*: MBARI Digital Data Series No. 3, available from cdrom@mbari.org).

the tectonically deformed 100-km-wide boundary between the Pacific and North American plates, represented by the San Andreas Fault system. These plates have a relative slip rate of about 8 cm per year [1], resulting in the deformation and fracturing of seafloor rocks. Earthquakes, common along this fault system, often produce landslides on the steep walls of the canyons [2].

The Ascension–Monterey canyon system includes six moderate-to-large size canyons: from north to south, Ascension, Año Nuevo, Cabrillo, Soquel, Monterey, and Carmel (Figure 54.1). Ascension, Año Nuevo, and Cabrillo canyons are located along the exposed open coastline and represent the northern segment of this system. Soquel, Monterey, and Carmel canyons are partially or entirely located inside Monterey and Carmel Bays and represent the southern segment [2]. These two segments intersect at about 3,800 m depth to become the Monterey Sea Valley.

In this case study, we focus on the heads of these submarine canyons, spanning a total area of 180 km². The head of Ascension canyon covers 24 km² and ranges

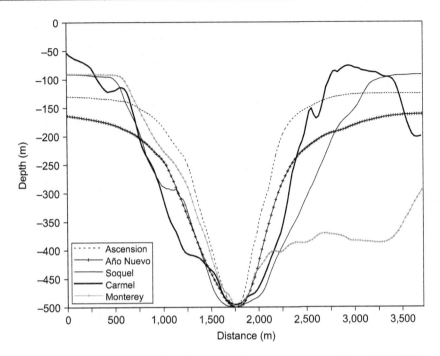

Figure 54.2 Depth profiles across the heads of five submarine canyons at a nadir of 500 m (see Figure 54.1 for location of profiles).

in depth from 120 to 800 m. The Año Nuevo canyon head is 17 km² in area and 118–836 m water depth. Cabrillo canyon comprises nine small heads (62 km² total area) that just notch the distal part of the continental shelf at about 500 m depth and continue to ~1,200 m. The head of Monterey canyon is about 50 km² in area and 20–780 m water depth. Soquel canyon head spans 19 km² and 90–650 m depth. Carmel canyon head is only about 8 km² ranging from 33 to 502 m water depth.

From a comparison of five cross-canyon profiles at 500 m depth (Figure 54.2), the head of Ascension canyon is steep, narrow, and intersects the shelf in 130 m of water. The head of nearby Año Nuevo canyon is not as steep as Ascension canyon and has a rim at 160 m depth. The headward part of Monterey canyon is more or less V-shaped, but with several slump meanders in its thalweg, and meets the shelf at about 90 m. The head of Soquel canyon also cuts the northern Monterey Bay shelf at a depth of about 90 m and is broader in profile at the 500-m isobath than all but Monterey canyon. The cross-canyon profile of the head of Carmel canyon at 500 m depth intersects the shelf in 50 m of water.

The heads of these canyons provide habitats of rugged, often rocky, steep relief in an ocean floor environment of flat or gently sloping unconsolidated sediment. Additionally, the morphologic expression of these canyon heads often serves to intercept cold nutrient-rich waters from the deep sea and deliver them to coastal areas where marine life of the entire food chain (from zooplankton, to fishes, to whales, pinnipeds, and turtles) feed. Some of these canyon heads also can be conduits that transport sediment and marine debris (natural and otherwise) from shallow nearshore to deep offshore ecosystems, or act as natural sinks for sediment, their walls often coated with mud, drift kelp, and organic material composed of dead pelagic organisms coming from the sunlit waters above.

Monterey and Carmel canyons extend to within 450–500 m of the shoreline, while the heads of the other four canyons notch the distal part of the continental shelf (Cabrillo canyon) or extend to the midshelf region (Ascension, Año Nuevo, and Soquel canyons). All of these canyon heads are located close to the ports of Santa Cruz, Moss Landing, and Monterey and have long been recreationally and commercially fished using a variety of gear types that include trawl, longline, traps, gillnets, and hook and line [3]. The walls in the heads of the Ascension–Monterey canyon system include exposed plutonic, metamorphic, and sedimentary rocks that serve both as substrate for growth of sessile invertebrates and as refugia for many economically valuable species. On the US west coast, all demersal fish species that have been officially designated as overfished (due to substantially diminished population levels) occur in seafloor habitats of the canyon heads off central California. In addition, the likelihood of habitat destruction is much greater in these deep areas than in nearshore habitats due to the predominance of potentially damaging fishing gears. In a recent study off southern and central California, the highest densities of derelict fishing gear were found in the heads of Monterey and Soquel canyons [4]. These human activities can greatly alter deep-water habitats [5] and community structure of demersal species [6]. In 2007, most of Soquel canyon and parts of Monterey and Carmel canyons were included in marine protected areas designed to protect these communities and habitats [7].

Geomorphic Features and Habitats

The geologic processes and resultant physiography associated with the Ascension–Monterey canyon system have been described by Greene et al. [2]. Potential seafloor habitats in this area have been reported by Greene et al. [8] and Copps et al. [9]. These geomorphic and habitat descriptions are largely based on bathymetric data collected for the Monterey Bay Aquarium Research Institute (MBARI, Digital Data Series No. 3, available from cdrom@mbari.org) in 1998. An area of 16,780 km² and 30–3,700 m depth was surveyed, including all heads of the Ascension–Monterey canyon system. The data were gridded to 4–5% of water depth using an overall 15-m grid cell to produce digital bathymetric models or artificially illuminated relief images. Habitat interpretations have been verified locally using the submersible *Delta* and MBARI's remotely operated vehicles (ROVs) *Ventana* and *Tiburon*.

Several potential habitat types in the headward parts of the Ascension–Monterey canyon system have been identified based on megahabitat types (i.e., the continental shelf and slope) and induration (i.e., soft sediments and rock; Figure 54.3). The most prolific habitat types in all the canyon heads are sedimentary shelf and slope canyon walls, which comprise soft hummocky unconsolidated materials ranging from mud

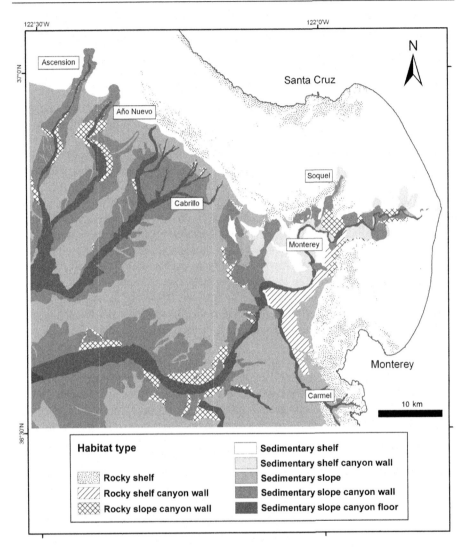

Figure 54.3 Seafloor habitat map of Ascension–Monterey canyon system [8].

to sand and gravel. These sediments also carpet the canyon floors, with grain size generally becoming finer down canyon. All six canyons contain consolidated and differentially eroded sedimentary rocks that are exposed within the walls (i.e., the rocky shelf and slope canyon walls identified in Figure 54.3). These rock exposures extend for tens of meters and are manifested locally as ledges, small caves, crevices, and overhangs, which provide relief and refuge for fishes and other organisms.

The heads of Ascension, Año Nuevo, and Cabrillo canyons are far removed from the wave and current action of the littoral zone, are inactive, and do not transport significant amounts of sediment to the deep sea. The headward parts of Ascension and Año Nuevo canyons are draped primarily with Quaternary fine-grain sediments of mud and silt and have Pliocene sedimentary rocks locally exposed along the walls. Cabrillo canyon head is mainly eroded into Quaternary unconsolidated sediment, although local outcrops of upper Tertiary sedimentary bedrock may occur.

Monterey canyon dwarfs the other five canyons and is the dominant geomorphologic feature in the region. It is 170 km long (470 km from head to distal end of its submarine fan, including its sea valley), with a maximum water depth of 3,800 m, a maximum rim-to-floor relief of 1,700 m, and a maximum rim-to-rim width of 12 km. Its size is comparable to that of the Grand Canyon of the Colorado River. Monterey canyon is deeply incised in the continental shelf, bifurcating Monterey Bay, with the head located near the beach at Moss Landing and bringing 500 m of water depth within 0.5 km of shore. The head intercepts the littoral drift and is the southern terminus of a major sediment transport cell that initiates at the Golden Gate, some 160 km to the north where the San Francisco Bay Estuary and Sacramento River spill into the Pacific Ocean [10]. This active canyon transports an estimated 200,000 km^3 of sand to the deep sea each year [11]. The lower walls include exposed upper Tertiary sedimentary rocks, and the upper walls comprise locally exposed Pleistocene terrestrial and deltaic sediments covered with Quaternary unconsolidated material. Recurring wall failures and mass wasting, often stimulated by earthquakes, simultaneously destroy and construct seafloor habitats.

Soquel canyon, a tributary to Monterey canyon, is not active and presently is being filled; it intersects Monterey canyon at 1,000 m. The head displays subtle relief indicative of slumping, while the base of the walls expose Pliocene sedimentary bedrock. Coarse-grain lag deposits (gravels, pebbles, and cobbles) representative of the last low-stand of sea level in this region occur along the rim of this canyon, as well as along the rim of Monterey canyon.

Carmel is an active canyon with three heads, one of which extends almost to shore. This canyon intersects Monterey canyon at 2,000 m depth. The two canyon heads in Carmel Bay are the southern terminations of the Monterey Bay littoral drift cell, which extends from the head of Monterey canyon some 22 coastal km to the north. The upper part of Carmel canyon is eroded into Cretaceous granite and constantly receives coarse-to-fine-grain arkosic sand. Granite crops out along most of the eastern wall, with some exposures locally covered with sand. Cretaceous sandstone crops out along the basal parts of the eastern canyon walls, which may be locally covered with Quaternary unconsolidated sediment. With the exception of isolated rock falls, few landslides occur in this canyon.

Biological Communities

The abundance of demersal fishes and their associated seafloor habitats (as defined by substratum type and depth) have been characterized for five of the canyon heads within the Ascension–Monterey canyon system using quantitative visual transect methods

Table 54.1 Median and Range of Depth (m) and Number of Quantitative Visual Transects Conducted from the Occupied Submersible *Delta* in the Heads of Five Canyons within the Ascension–Monterey Canyon System

Location	Median Depth (Range) (m)	Number of Transects
Ascension	235 (180–320)	26
Año Nuevo	251 (170–273)	6
Soquel	180 (95–305)	81
Monterey	140 (110–179)	16
Carmel	194 (88–305)	32

($n = 161$ transects) conducted from the occupied submersible *Delta* (Table 54.1). While objectives of these surveys were primarily focused on fish assemblages in rocky and mixed rock and mud habitats, at least 30% of the transects occurred on steep walls covered with soft sediment.

Rockfishes (*Sebastes* spp.) dominated the canyon assemblages; for example, half of the 52 species and 77% of the total number of demersal fishes surveyed at 95–305 m depth in Soquel canyon were rockfishes [12]. Members of this diverse group were found in all habitats. Relatively small (40 cm total length) species [e.g., stripetail (*S. saxicola*) and greenstriped (*S. elongatus*) rockfishes] (Figure 54.4A) occurred mostly on mud with some low-relief substrata (cobble, pebbles), and larger (60–90 cm total length) species of rockfishes [e.g., bocaccio (*S. paucispinus*), cowcod (*S. levis*), yelloweye (*S. ruberrimus*), and redbanded (*S. babcocki*)] (Figure 54.4B and C) were associated with high-relief structure (e.g., vertical rock crevices, undercut ledges, boulder fields). Common nonrockfish species included Dover sole (*Microstomus pacificus*; Figure 54.4D), Pacific hagfish (*Eptatretus stoutii*), and unidentified species of poachers (family Agonidae) on soft sediments, and lingcod (*Ophiodon elongatus*; Figure 54.4C) in mixed and rock habitats. Species richness and diversity have been reported for fish assemblages from various substratum types in Soquel canyon [12]. The most diverse groups of evenly distributed species composition occurred in habitats of broken rock, cobbles, and pebbles interspersed with soft mud and in habitats of rock exposures with moderate vertical fractures and rock ledges. The lowest species diversity was found in uniform mud and boulder habitats, wherein a single species dominated each assemblage [i.e., stripetail rockfish in mud and pygmy rockfish (*S. wilsoni*) in boulders].

Macroinvertebrate assemblages and associated seafloor habitats (as defined by substratum type and depth) have also been characterized from the video transects conducted in the most northern (Ascension) and southern (Carmel) canyon heads (Bianchi, Tissot, and Yoklavich; unpublished data). Additionally, habitat of the economically valuable spot prawn (*Pandalus platyceros*; Figure 54.4E) has been described from the video transects located in the headward part of Carmel canyon [13]. Brittle stars (*Ophiacantha* spp.), unidentified hermit crabs, sessile sea cucumbers (*Psolus squamatus*), fragile pink urchins (*Allocentrotus fragilis*), and crinoids (*Florometra*

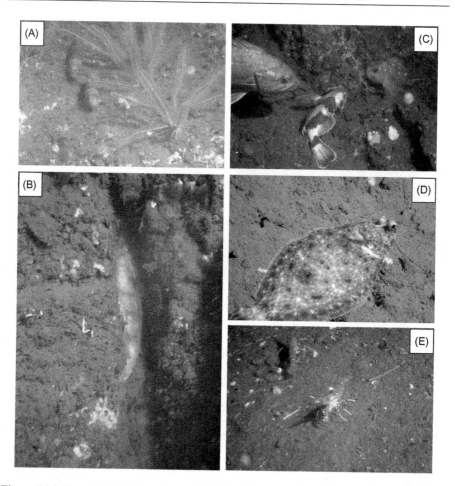

Figure 54.4 Representative fishes and macroinvertebrates in important seafloor habitats
in the heads of the Ascension–Monterey canyon system: (A) young greenstriped (*Sebastes
elongatus*) and greenspotted (*S. chlorostictus*) rockfishes in crinoid (*Florometra serratissima*)
field and shell hash at 110 m depth on edge of Soquel canyon (photo: S. Untiedt); (B) cowcod
(*S. levis*) in rock crevice at 220 m depth in Soquel canyon (photo: J. Blaine); (C) lingcod
(*Ophiodon elongatus*) and vermilion (*S. miniatus*), greenspotted, and redbanded (*S. babcocki*)
rockfishes on rock outcrop at 200 m depth in Soquel canyon (photo: M. Yoklavich); (D) Dover
sole (*Microstomus pacificus*) on steep mud slope at 300 m depth in Monterey canyon
(photo: T. Laidig); and (E) spot prawns (*Pandalus platyceros*) on steep slope of coarse sand
at 190 m depth in Carmel canyon (photo: M. McCrea).

serratissima; Figure 54.4A) occurred in habitats of mixed mud, rock, and cobbles
and were the dominant species in transects from Ascension canyon. Spot prawns
were the most abundant macroinvertebrate species observed during transects conducted in Carmel canyon. Highest spot prawn densities were found in habitats of
sediment (coarse sand and mud) mixed with boulders, cobbles, or gravel at depths

of 150–300 m [13]. Hard granitic rock that is locally covered with sand, particularly along the shoreward rim of Carmel canyon, provides a mixed sand-rock habitat often occupied by spot prawns. Brittle stars and crinoids were abundant primarily on moderate- to high-relief rock in Carmel canyon, and squat lobsters (*Munida quadrispina*) were abundant on both high-relief rock and steep sediment slopes. Some species of macroinvertebrate [e.g., various sponges, gorgonians, sea pens, and basket stars (*Gorgonocephalus eucnemis*)] may be components of habitat for demersal fishes in these canyons.

Surrogacy

Demersal fish assemblages in deep water on the central California coast have been distinguished primarily by depth, substratum type, and seafloor slope [12,14,15]. Using the 161 visual transect samples collected with the occupied submersible *Delta*, canonical discriminant analysis was conducted of fish densities constrained by the five heads within the Ascension–Monterey canyon system (Figure 54.5A). Canonical variate 1 (the horizontal axis) discriminates between Soquel and Monterey canyons and the other three canyons (Ascension, Año Nuevo, and Carmel) based on species density. The second canonical variate discriminates between two groups along the vertical axis (i.e., Ascension and Año Nuevo canyons versus Carmel canyon). The structure coefficients (Figure 54.5B) are helpful in interpreting the canonical results in terms of those variables contributing to the analysis. Soquel and Monterey canyons, located inside Monterey Bay and having the shallowest median sample depths (Table 54.1), had similar species profiles in contrast to the other three canyon locations on axis 1 of Figure 54.5B. Ascension and Año Nuevo canyons, both located on the outer coast and having the deepest median depths, had similar species profiles in contrast to that of Carmel canyon (inside Carmel Bay, and with an intermediate median sample depth) on axis 2 of Figure 54.5B. In addition to differences in depth, transects in Ascension and Año Nuevo canyons largely comprised deep fine-sediment habitats with few rock outcrops. Major habitats sampled in Monterey and Soquel canyons were vertical rock walls with crevices, overhangs, small and large ledges, talus slopes, and boulder fields interspersed with soft mud. Significantly more boulder habitat was sampled in Carmel canyon. These differences in both depth and substratum are reflected in the species assemblages (Figure 54.5B) that represent the three broad groupings of sample transects from the five canyon heads (Figure 54.5A), as identified in the canonical discriminant analysis.

Acknowledgments

We acknowledge the assistance of several colleagues with data collection, particularly L. Browne, G. Cailliet, R. Lea, M. Love, D. Sullivan, and R. Starr. We thank the personnel from the RVs *Jolly Roger, Cavalier*, and *Pt. Sur* and from *Delta Oceanographics* for providing suitable survey vehicles and support platforms. We thank D. Watters and C. Syms for

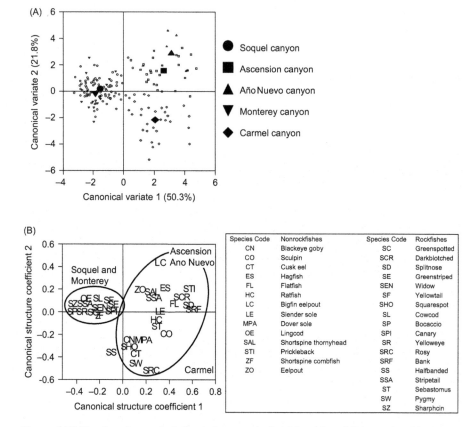

Figure 54.5 Results of canonical discriminant analysis of densities of fish species: (A) the first and second canonical variates discriminate among the five canyons based on species densities of 161 visual transect samples (small open symbols are coefficients for each transect at the five canyons; large solid symbols are mean coefficients at each of the five canyons); and (B) species compositions that contribute to the canyon groupings in (A), based on the structure coefficients of the canonical correlation among samples.

assistance with figure preparation and data analyses. T. Laidig, D. Watters, G. Lamarche, and an anonymous reviewer provided helpful comments on earlier drafts. This study was partially supported by NOAA's National Undersea Research Program, West Coast and Polar Regions Undersea Research Center, University of Alaska Fairbanks (grant nos. UAF-92-0063 and UAF-93-0036). Underwater photographs were taken during surveys supported by the California Ocean Protection Council.

References

[1] T. Atwater, Implications of plate tectonics for the Cenozoic evolution of western North America, Geol. Soc. Am. Bull. 81 (1970) 3513–3536.

[2] H.G. Greene, N.M. Maher, C.K. Paull, Physiography of the Monterey Bay National Marine Sanctuary and implications about continental margin development, Mar. Geol. 181 (2002) 55–82.

[3] M.S. Love, Subsistence, commercial, and recreational fisheries, in: L.G. Allen, D.J. Pondella, II, M.H. Horn, (Eds.), The Ecology of Marine Fishes: California and Adjacent Waters, University of California Press, Berkeley, 2006, pp. 567–594.

[4] D.L. Watters, M. Yoklavich, M. Love, D. Schroeder, Assessing marine debris in deep seafloor habitats off California, Mar. Pollut. Bull. 60 (2010) 131–138.

[5] P.W. Barnes, J.P. Thomas (Eds.), Benthic habitats and the effects of fishing, AFS Symposium 41, Bethesda, MD, 2005.

[6] M.S. Love, M.M. Yoklavich, Deep rock habitats, in: L.G. Allen, D.J. Pondella, II, M.H. Horn, (Eds.), The Ecology of Marine Fishes: California and Adjacent Waters, University of California Press, Berkeley, 2006, pp. 253–266.

[7] R. Starr, M. Yoklavich, Monitoring MPAs in deep water off central California: 2007 IMPACT submersible baseline survey, California Sea Grant College Program Publication T-067, 2008, pp. 1–22.

[8] H.G. Greene, J.J. Bizzarro, I. Herbert, M. Erdey, H. Lopez, L. Murai, et al., Essential fish habitat characterization and mapping of California continental margin, Moss Landing Marine Laboratories Technical Publication Series 2003-01, 2003, 29 pp. (2 CDs).

[9] S. Copps, M. Yoklavich, G. Parkes, W. Wakefield, A. Bailey, H.G. Greene, et al., Applying marine habitat data to fishery management on the US west coast: initiating a policy-science feedback loop, in: B. Todd, H.G. Greene (Eds.), Mapping the Seafloor for Habitat Characterization, Geological Association of Canada, Special Paper 47, 2007, pp. 451–462.

[10] T.C. Best, G.B. Griggs, A sediment budget for the Santa Cruz littoral cell, California, in: R.H. Osborne, (Ed.), From Shoreline to Abyss, Society of Economic Paleontologists and Mineralogists, Tulsa, OK, 1991, Special Publication 46, pp. 35–50.

[11] S.L. Eittreim, J.P. Xu, M. Noble, B.D. Edwards, Towards a sediment budget for the northern Monterey Bay continental shelf, Mar. Geol. 181 (2002) 235–248.

[12] M. Yoklavich, H.G. Greene, G. Cailliet, D. Sullivan, R. Lea, M. Love, Habitat associations of deep-water rockfishes in a submarine canyon: an example of a natural refuge, Fish. Bull. 98 (2000) 625–641.

[13] K.L. Schlining, The Spot Prawn (*Pandalus platyceros* Brandt 1851) Resource in Carmel Submarine Canyon, CA: Aspects of Fisheries and Habitat Associations, M.S. Thesis, California State University, Stanislaus, 1999, 53 pp.

[14] M.M. Yoklavich, G.M. Cailliet, R.N. Lea, H.G. Greene, R. Starr, J. deMarignac, et al., Deepwater habitat and fish resources associated with the Big Creek Ecological Reserve, CalCOFI Rep. 43 (2002) 120–140.

[15] T.E. Laidig, D.L. Watters, M.M. Yoklavich, Demersal fish and habitat associations from visual surveys on the central California shelf, Estuarine Coastal Shelf Sci. 83 (2009) 629–637.

55 A Study of Geomorphological Features of the Seabed and the Relationship to Deep-Sea Communities on the Western Slope of Hatton Bank (NE Atlantic Ocean)

Miriam Sayago-Gil[1], Pablo Durán-Muñoz[2], F. Javier Murillo[2], Víctor Díaz-del-Río[1], Alberto Serrano[3], L. Miguel Fernández-Salas[1]

[1]Instituto Español de Oceanografía, Centro Oceanográfico de Málaga, Puerto Pesquero s/n, Fuengirola, Málaga, Spain, [2]Instituto Español de Oceanografía, Centro Oceanográfico de Vigo, Subida al RadiofaroVigo, Pontevedra, Spain, [3]Instituto Español de Oceanografía, Centro Oceanográfico de Santander, Promontorio San Martín, Santander, Spain

Abstract

The paper provides the connection between the seabed morphology and the invertebrate benthic species communities along the western upper-middle slope of Hatton Bank between 600 and 2,000 m water depth, an important area for deep-sea fisheries. This work is supported by the *ECOVUL/ARPA* project (Instituto Español de Oceanografía), which was carried out using multibeam bathymetry, high-resolution seismic profiles, benthos samples, bottom trawl, and longline samples; the results were added to the information from observers on commercial trawlers. Two main domains were interpreted: outcrop (nondepositional area) and contourite drift (depositional area). Most species associated with hard substrata belong to the phylum Cnidaria, being mainly cold-water corals. Benthic communities are sparse where the seabed comprises mobile sediments (drift) in deeper water, while they are common closer to the top of the bank (outcrop) in shallower water, where the substrate is hard.

Key Words: Hatton Bank, mounds, ridge, cold-water corals, benthic communities, Hatton Drift, NE Atlantic Ocean

Seafloor Geomorphology as Benthic Habitat. DOI: 10.1016/B978-0-12-385140-6.00055-4

Introduction

The study area, located between 600 and 2,000 m water depth along the western upper-middle slope of Hatton Bank, has a sinuous bathymetric planform. Hatton Bank is one of the three shallow-water banks (with George Bligh Bank and Rockall Bank) that form the Rockall Plateau (Figure 55.1). The Rockall Plateau represents a volcanic continental margin with the continental-ocean transition located beneath the lower western slope of Hatton Bank [1]. The structure of Hatton Bank has been described as a bedrock surface composed of flood basalts [2] dated as late Palaeocene, although in some areas younger rocks (Mesozoic) may be found [3].

The western slope of Hatton Bank is mainly dominated by a contourite drift (Hatton Drift, [4]), which is a sediment deposit created (or substantially reworked) by the action of bottom currents [5]. Sediment transport along the slope is to the northeast and comprises both fine-grained suspended material in locally generated nepheloid layers and cohesionless, sand-size bedload material [6]. The main features of the drift (morphology, sedimentation rates, and thickness) have been influenced by bottom currents since early Eocene times [7]. The western slope of Hatton Bank is influenced by a branch of the Labrador Sea Water that meets with the Iceland–Scotland Overflow Water to comprise the Deep Northern Boundary Current [8], which flows along the Hatton Bank forming a part of the cyclonic circulation system within the Iceland Basin [9].

Different geomorphologic features have been interpreted on the top of Hatton Bank (~500 m deep), such as iceberg ploughmarks, terraces, circular depressions, and isolated topographic highs (pinnacles and carbonate mounds), which are ideal for biological communities. The mounds on Hatton Bank bear a strong resemblance

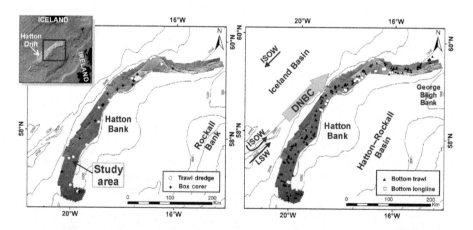

Figure 55.1 Study area location (multibeam bathymetry) along the western slope of Hatton Bank on regional bathymetric contours. Location of samples—trawl dredges, boxcorers, bottom trawls, and bottom longlines—take on the study area. Present-day bottom-current circulation (arrows). ISOW. Iceland–Scotland Outflow Water; LSW: Labrador Sea Water; DNBC: Deep Northern Boundary Current.

(although being smaller in size) to features reported in the Porcupine Seabight [10] and at the southern end of Rockall Bank [11]. They were shown to be bioclastic accumulations sustained by the growth of cold-water corals. Small mounds on the crest of Hatton Bank are principally comprised of *Lophelia pertusa* based on previous researches [12] and verified by this study.

The western slope of Hatton Bank is an important area for deep-sea fisheries. Benthic surveys are revealing the increasing extent to which bottom trawling is altering deep-sea coral habitats [13]. This is a particular concern because reefs take centuries to millennia to develop [14]. The impact of demersal fisheries on reefs, seamounts, and the associated deep-water coral, *Lophelia pertusa*, in NE Atlantic waters is discussed in papers by several authors [12,14–19].

This work has been carried out using different data sets (Figure 55.1): bathymetric data (as well as backscatter) was collected by multibeam echosounder (EM300) covering a total area of 18,760 km² and processed at a grid resolution of 50 × 50 m. Likewise, a network of about 1,120 km of very high-resolution seismic profiles was collected with a parametric Topas PS018 echosounder. A total of 35 sediment/benthos samples were collected: 13 with a boxcorer (0.25 m² × 0.35 m), and 22 with a trawl dredge. A total of 38 fishing trawls (Lofoten net) were carried out during *Ecovul-Arpa* cruises. In addition, 268 bottom longline sets and 230 bottom trawls from science-industry cooperative surveys were used for this project. Data sets were managed and integrated within a GIS. All information was collected under the *Ecovul-Arpa* multidisciplinary project, which focused on the study of the relationship between benthic and demersal communities (which could be influenced by anthropogenic disturbance) and seabed geology in order to increase the knowledge base supporting the designation of Marine Protected Areas.

Geomorphic Features and Habitats

Two major seabed domains can be distinguished (Figure 55.2): (1) outcrop, described as a nondepositional area corresponding to the top of the bank or area on the slope adjacent to the top; and (2) drift, a depositional area in which a contourite deposit (Hatton Drift) has been developed on the western slope of the bank. The limit between both domains is located at approximately 1,100 m water depth (except in the ridges area where the outcrop is extended, reaching 1,600 m water depth).

Outcrop: Characterized by an uneven surface (originating from tectonic activity and erosion) with ridges and escarpments trending predominantly WE. In some locations it may be covered by a thin deposit of sediments (up to 20 ms Two-Way Travel Time (TWTT)). Seafloor gradients can locally reach 40°, and acoustic backscatter intensity is typically high (> −20 dB). The study area does not cover the entire outcrop (e.g., the top of the bank is not considered as the objective of the current project). Consequently, all interpretations given in this work correspond solely to this specific part of the outcrop.

A series of parallel ridges (Figure 55.2) (basalt scarps probably associated with faults in depth) were identified in relation to the outcrop (adjacent to the top of the bank)

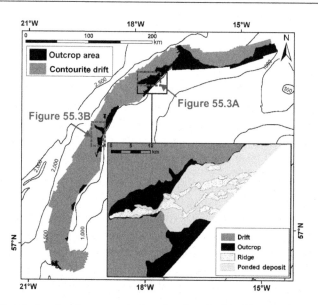

Figure 55.2 Sketch showing the two main domains described in this work, outcrop area and contourite drift. A detail of the outcrop area showing the ridges and ponded deposits is zoomed. Location of Figure 55.3 selected areas are also indicated.

above 1,600 m water depth. These ridges are long, narrow, spaced 5 km apart, and segmented into 2–7 km long sections. Ridges are 5–45 m high, with sloping sides showing gradients up to 17°. Dozens of small mounds, probably carbonate reefs (based on other authors' studies [12]), have been identified superimposed on the ridges (Figure 55.3), extending 10–25 m above the basalt substrate, some reaching a few hundred meters in width. Some mounds show symmetrical features, while others are asymmetrical (which may be in relation to clusters of mounds). Below these mounds, the seismic signal is often chaotic and sometimes opaque, possibly as a consequence of sound attenuation by the mound composition. The existence of buried mounds, which can be observed in seismic profiles, suggests variation in the strength of the contour current or mound growth rate over time. The ridges are associated with ponded deposits that infill hollows between ridges and drape the surrounding uneven surface. These deposits (up to 50 ms (TWTT) thickness) result from the along- and down-slope movement of sediment, where the ridges are acting as barriers trapping sediment. Boxcorer samples as well as trawl dredges show that this material is a mixture of sediment (from the Hatton Drift) plus coral rubble (probably derived from the adjacent ridges and mounds).

Hatton Drift: Onlaps onto the outcrop defined above (Figure 55.2) and comprises mainly fine or very fine sediments. It follows the trend of the slope and has gradients of 0–3° (Figure 55.3), although it can reach 30° locally (associated to specific morphologies such as furrows, for example). The acoustic backscatter intensity is moderate–low (generally < -20 dB). The drift deposit generally increases in thickness basinward (>400 ms (TWTT) down slope) and onlaps upslope with well defined stratified layers. In places, this deposit can be observed covering outcrop areas. Analyses of the samples recovered

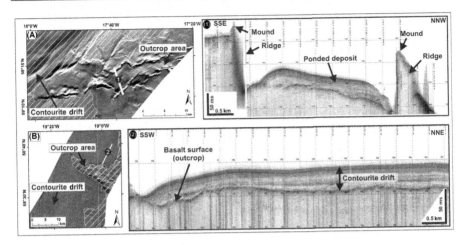

Figure 55.3 Bathymetry detail of the outcrop area showing ridges (A) and a drift detail (B). (1) Topas seismic profile collected on the outcrop area, showing mounds on top of the ridges (this interpretation is supported by the extended morphology studies; [24]) as well as the ponded deposit generated between ridges. (2) Topas seismic profile displaying the contourite drift attached to the outcrop.

from the contourite drift show that the present-day seabed sediments in the area are mainly muddy sands. This information was used to calibrate the backscatter data to generate morphological sketches of the study area according to sediment type. These maps were used for locating benthic and demersal communities.

The morphological features presented do not necessarily reflect present-day conditions but may have been associated with past current events, in accordance with earlier interpretations [20].

Biological Communities

The biological data collected have permitted the identification of deep-sea communities in relation to the two major geomorphologic domains described earlier: Hatton Bank outcrop and Hatton Drift (Table 55.1).

As a result of this study, the Hatton Bank outcrop is characterized to be a suitable substratum for settlement of cold-water corals (Figure 55.4). Live stony corals (*Lophelia pertusa*, *Solenosmilia variabilis*, and *Madrepora oculata*) were identified in this domain. Other identified species were cup corals (genus *Caryophyllia* sp., *Desmophyllum* sp., and *Stephanocyathus* sp.), bamboo corals (*Acanella* sp.), seafans (*Acanthogorgia* sp., *Callogorgia verticillata*, *Primnoa resedaeformis*, and the Family Plexauridae), soft corals (*Capnella florida* and the family Nephtheidae), black corals (*Stichopathes* sp., *Parantipathes* sp., and Antipatharia indet.), and lace corals (family Stylasteridae). The outcrop is occasionally covered by drift sediments forming a suitable soft substrate for seapens (*Pennatula* sp., *Anthoptilum murrayi*, and *Halipteris* sp.). Fragile and small

Table 55.1 Invertebrate Presence in the Two Main Domains Described in this Work (D: drift; O: outcrop)

Taxa		D	O	Taxa		D	O	Taxa		D	O
Porifera	Porifera indet		+	Annelida	Polychaeta indet		+	Mollusca	Buccinidae	+	+
	Geodiidae	+			Aphroditidae	+			Bivalvia indet		+
	Axinellidae	+			Eunicidae	+			Pectinidae		+
	Euplectellidae		+		Terebellidae				Verticordiidae	+	
	Aphrocallistidae		+	Chelicerata	Colossendeidae	+			Sepiolidae	+	
Cnidaria	Anthozoa indet	+	+	Crustacea	Crustacea indet		+		Octopotheutidae	+	+
	Gorgonacea indet		+		Scalpellidae	+	+		Histioteuthidae	+	
	Acanthogorgiidae		+		Balanomorpha indet	+	+		Ommastrephidae	+	+
	Isididae		+		Aegidae	+			Cranchiidae	+	
	Paragorgiidae	+			Aristeidae	+			Opisthoteuthidae	+	
	Plexauridae	+	+		Pasiphaeidae	+			Cirroteuthidae	+	
	Primnoidae		+		Nematocarcinidae	+			Octopodidae	+	
	Pennatulacea indet	+	+		Crangonidae	+		Bryozoa	Bryozoa indet		+
	Anthoptilidae	+	+		Oplophoridae	+		Brachiopoda	Brachiopoda indet		+
	Halipteridae	+	+		Glyphocrangonidae	+		Echinodermata	Asteroidea indet		+
	Pennatulidae	+	+		Nephropidae	+			Astropectinidae	+	+
	Nephtheidae	+			Polychelidae	+			Goniasteridae	+	
	Actiniaria indet	+			Chirostylidae		+		Poraniidae	+	
	Actinernidae	+	+		Parapaguridae	+	+		Solasteridae	+	

Hormathiidae

Epizoanthidae

Antipatharia indet

Antipathidae

Caryophylliidae

Oculinidae

Flabellidae

Hydrozoa indet

Stylasteridae

Lithodidae

Galatheidae

Majidae

Pisidae

Geryonidae

Pterasteridae

Zoroasteridae

Ophiuroidea indet

Asteronychidae

Gorgonocephalidae

Asteroschematidae

Echinoidea indet

Cidaridae

Echinothuridae

Echinidae

Holoturoidea indet

Psolidae

Stichopodidae

Synallactidae

Psychropotidae

Elasipodida indet

Crinoidea indet

Ascidacea indet

Chordata

Based on the cooperative surveys (bottom trawl and bottom longline).

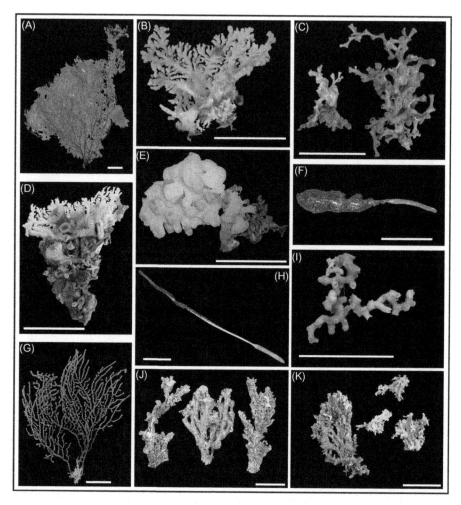

Figure 55.4 Examples of several species of benthic communities collected on the study area.
(A) Antipatharia; (B) lace coral: stylasteridae; (C) stony coral (*Lophelia pertusa*); (D) photo
showing the high biodiversity of this kind of habitat: Scleractinia, Stylasteridae, cup coral
(*Desmophyllum* sp.); (E) glass sponge (*Aphrocallistes* sp.); (F) pennatulacean (*Anthoptilum
murrayi*); (G) gorgonian: Plexauridae, settled on *Madrepora oculata*; (H) pennatulacean
(*Halipteris* sp.); (I) stony coral (*Solenosmilia variabilis*); (J) stony coral (*Enallopsammia
rostrata*); (K) stony coral (*Madrepora oculata*). Scale bars are 10 cm.

glass sponges (*Aphrocallistes* sp. and Porifera indet.) were also observed. Moreover,
high biodiversity was found associated with dead fragments of corals. Based on the fishing effort and the morphology of the area, outcrop communities would not have been
strongly influenced by fishing, at least compared to the sedimentary surrounding areas.
The importance of the ponded deposits areas, in terms of biodiversity, lies in the biogenic rubble consisting of dead degraded coral, shells, and plates of balanomorph

barnacles mixed with mud/sand from the drift. At least in terms of macrofauna the outcrop is often much richer and more diverse than the surrounding sedimentary areas. The number of species associated with dead structures and the ecological importance is unknown at this stage. From an ecological perspective, dead coral can provide habitat, feeding grounds, recruitment, and nursery areas for a range of deep-water organisms [12,21]. Gorgonians, actinarians, and sponges (different species of sponges grow on the dead coral skeletons; [22]) are conspicuous and abundant. Invertebrate fauna included Porifera (*Aphrocallistes* sp., *Phorbas* sp., and *Craniella* sp.) and encrusting sponges (*Hymedesmia* sp., *Forcepia* sp., and *Myxilla* sp.), Cnidaria (Scleractinia, Octocorallia, and Hydrozoa), Echinodermata, Polychaeta, Crustacea, Mollusca, Bryozoa, Brachiopoda, and Tunicata [23].

Surveys carried out on Hatton Drift indicate occasional catches of cold-water corals, such as small live pieces and coral rubble of stony corals (*Solenosmilia variabilis*), gorgonians, and antipatharians. This fact suggests scarce and patchy distribution of cold-water corals on the drift. The patches are commonly associated with areas around outcrops and generally are related to hauls carried out accidentally over (or very close to) the outcrop area.

Table 55.1 shows a drift community composed of a higher number of species (but with lower abundances) than in the outcrop. The methodology used in every domain—bottom trawl or bottom longline—may affect to the results. Based on researches carried out on the study area [23], the greater biomass was deeper but diversity was less, whereas the shallower area of the Hatton Drift was dominated by hauls with less biomass and more diversity.

Cnidarians associated with soft bottoms, such as seapens and cup corals (*Flabellum* sp.), were also observed. It is noteworthy to indicate that the habitat is not a "wild environment" because it may have been disturbed by deep-water trawl fishing.

Based on this work, we can conclude that the shallow hard substrate (outcrop area) acts as a suitable platform for cold-water corals to become established and develop into mounds (as for example on top of the ridges), while cold-water coral communities are currently sparse in deeper water where the seabed comprises mobile sediments (Drift).

Acknowledgments

This work was supported by the *Ecovul/Arpa* project. Our most sincere thanks to the positive institutional support of the Instituto Español de Oceanografía and Secretaría General del Mar, Spain.

References

[1] G.S. Kimbell, J.D. Ritchie, H. Johnson, R.W. Gatliff, Controls on the structure and evolution of the NE Atlantic margin revealed by regional potential field imaging and 3D modelling, in: A.G. Doré, B.A. Vining, (Eds.), Petroleum Geology: North-West Europe and Global Perspectives, Geological Society of London, London (2005) pp. 933–945.

[2] L.K. Smith, R.S. White, N.J. Kusznir, Structure of the Hatton Basin and adjacent con-
     tinental margin, in: A.G. Doré, B.A. Vining, (Eds.), Petroleum Geology: North-West
     Europe and Global Perspectives, Geological Society of London, London (2005) pp.
     947–956.
[3] D.M. McInroy, K. Hitchen, M.S. Stoker, Potencial Eocene and Oligocene stratigraphic
     traps of the Rockall Plateau, NE Atlantic Margin, In: M.R. Allen, G.P. Goffey, R.K.
     Morgan, I.M. Walker, (Eds.), The Deliberate Search for the Stratigraphic, Geological
     Society of London Special Publications, London (2006) 254: 247–266. Trap.
[4] W.F. Ruddiman, Sediment redistribution on the Reykjanes Ridge: seismic evidence,
     Geol. Soc. Am. Bull. 83 (1972) 2039–2062.
[5] M. Rebesco, Contourites, In: R.C. Selley, L.R.M. Cocks, I.R. Plimer (Eds.),
     Encyclopedia of Geology, vol. 4, Elsevier, Oxford, 2005, pp. 513–527.
[6] I.N. McCave, P.F. Lonsdale, C.D. Hollister, W.D. Gardner, Sediment transport over the
     Hatton and Gardar contourite drifts, J. Sediment. Petrol. 50 (4) (1980) 1049–1062.
[7] D.A.V. Stow, J.A. Holbrook, Hatton Drift contourites, northeast Atlantic, Deep Sea
     Drilling project LEG 81, University of Edinburgh Report no 25, 1984, pp. 695–699.
[8] M.S. McCartney, Recirculating components to the deep boundary current of the northern
     North Atlantic, Prog. Oceanogr. 29 (1992) 283–383.
[9] J.F. Read, CONVEX-91: water masses and circulation of the Northeast Atlantic subpolar
     gyre, Progr. Oceanogr. 48 (2001) 461–510.
[10] W. Bailey, P.M. Shannon, J.J. Walsh, V. Unnithan, The spatial distribution of faults and
     deep sea carbonate mounds in the Porcupine Basin, offshore Ireland, Mar. Petrol. Geol.
     20 (2003) 509–522.
[11] T.C.E. Van Weering, H. de Haas, H.C. de Stigter, H. Lykke-Andersen, I. Kouvaev,
     Structure and development of giant carbonate mounds at the SW and SE Rockall Trough
     margins, NE Atlantic Ocean, Mar. Geol. 198 (2003) 67–81.
[12] J.M. Roberts, L.A. Henry, D. Long, J.P. Hartley, Cold-water coral reef frameworks,
     megafauna communities and evidence for coral carbonate mounds on the Hatton Bank,
     north east Atlantic, Facies 54 (2008) 297–316.
[13] J.M. Hall-Spencer, A.D. Rogers, A.J. Davies, A. Foggo, Deep-sea coral distribution on
     seamounts, oceanic islands, and continental slopes in the Northeast Atlantic, Bull. Mar.
     Sci. 81 (2007) 135–146.
[14] J.M Hall-Spencer, V. Allain, J.H. Fossa, Trawling damage to Northeast Atlantic ancient
     coral reefs, Proc. R. Soc. Lond. Ser. B: Biol. Sci. (2002) 507–511.
[15] J.M. Roberts, S.M. Harvey, P.A. Lamont, J.D. Gage, J.D. Humphery, Seabed photogra-
     phy, environmental assessment and evidence for deep-water trawling on the continental
     margin west of the Hebrides, Hydrobiologia 441 (2000) 173–183.
[16] J.H. Fossa, P.B. Mortensen, D.M. Furevik, The deep-water coral Lophelia pertusa in
     Norwegian waters: distribution and fishery impacts, Hydrobiologia 471 (2002) 1–12.
[17] A. Freiwald, J.M. Roberts, Cold-water corals and ecosystems. Erlangen Earth
     Conference Series, Springer Verlag, 2005, 1243 pp.
[18] A. Davies, J.M. Roberts, J.M. Hall-Spencer, Preserving deep-sea natural heritage:
     emerging issues in offshore conservation and management, Biol. Conserv. 138 (2007)
     299–312.
[19] ICES, Report of the ICES/NAFO Joint Working Group on Deep-water Ecology
     (WGDEC), 22–26 March 2010, ICES CM 2010/ACOM:26, 2010, 161 pp.
[20] L. Due, H.M. Van Aken, L.O. Boldreel, A. Kuijpers, Seismic and oceanographic evi-
     dence of the present day bottom water dynamics in the Lousy Bank–Hatton Bank area,
     NE Atlantic, Deep Sea Res. I 53 (2006) 1729–1741.

[21] J.K. Reed, A.N. Shepard, C.C. Koenig, K.M. Scanlon, R.G. Gilmore, Mapping, habitat characterization, and fish surveys of the deep-water Oculina coral reef Marine Protected Area: a review of historical and current research, in: A. Freiwald, J.M. Roberts, (Eds.), Cold-Water Corals and Ecosystems, Springer, Berlin, 2005, pp. 443–465.

[22] M.L. Hovland, Deep Water Coral Reefs. Unique Biodiversity Hot-Spots, Springer, Germany, 2008. pp. 278

[23] P. Durán-Muñoz, M. Sayago-Gil, J. Cristobo, S. Parra, A. Serrano, V. Díaz del Rio, et al., Seabed mapping for selecting cold-water coral protection areas on Hatton Bank, Northeast Atlantic", ICES J. Mar. Sci. 66 (2009) 2013–2025.

[24] M. Sayago-Gil, D. Long, K. Hitchen, V. Díaz-del-Río, L.M. Fernández-Salas, P. Durán-Muñoz, Evidence for current-controlled morphology along the western slope of Hatton Bank (Rockall Plateau, NE Atlantic Ocean), Geo-Mar. Lett. 30 (2010) 99–111.

56 Seafloor Habitats and Benthos of a Continental Ridge: Chatham Rise, New Zealand

Scott D. Nodder, David A. Bowden, Arne Pallentin, Kevin Mackay

National Institute of Water and Atmospheric Research (NIWA) Ltd, Kilbirnie, Wellington, New Zealand

Abstract

Chatham Rise is a submarine continental ridge that extends eastward from New Zealand into the Southwest Pacific Ocean and lies beneath the highly productive Subtropical Front. It is >160,000 km^2 in area and spans water depths of <50 to >2,000 m with emergent islands at its easternmost end (Chatham Islands). The north-facing flanks are generally steeper (3–10°) than those to the south (<3°). The crest is essentially flat-lying at ~350–450 m depth, with locally irregular topography and several shallower large banks. Isolated groups of volcanic seamounts are present on the flanks of the rise. Glauconite- and phosphorite-rich muddy sands dominate the crests and upper flanks, with carbonate-rich foraminiferal sands at deep sites on the southern flank and hemipelagic muds on the deep central and westernmost, north-facing flanks. Strong current flows at depths to 1,000 m are indicated by rippled sands on the northeastern and southern flanks. The area supports major commercial fisheries. Benthic communities are diverse and locally highly abundant, ranging from shallow reef fauna on the banks to soft-sediment assemblages dominated by infauna and deposit feeders, with extensive cold-water coral stands on some seamounts. Initiatives to map benthic fauna distributions using remote-sensed data are in development, but have met with limited success to date.

Key Words: Bathyal, deep sea, benthic communities, seamounts, biodiversity, soft sediments, Subtropical Front, Southwest Pacific

Introduction

Chatham Rise is a submarine ridge extending ~1,500 km due east of the South Island, New Zealand (Figure 56.1). It consists of continental basement rocks (Mesozoic graywackes) overlain by Mid- to Late Cretaceous marine sedimentary rocks and Paleocene–Miocene chalks and greensands, with sporadic Pliocene–Pleistocene intraplate volcanic edifices, especially along its northern and

Seafloor Geomorphology as Benthic Habitat. DOI: 10.1016/B978-0-12-385140-6.00056-6

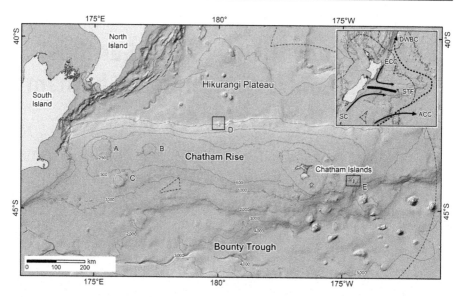

Figure 56.1 Chatham Rise, showing locations of: (A) Mernoo Bank; (B) Veryan Bank; (C) Reserve Bank; (D) Graveyard seamounts complex; (E) Andes seamounts complex. Inset shows the approximate position of the Subtropical Front (STF, black bar) formed by convergence of the Southland Current (SC, lower arrow) and East Cape Current (ECC, upper arrow). The Antarctic Circumpolar Current (ACC) and Deep Western Boundary Current (DWBC, thick dashed line) are also shown. Bathymetric contours are in meters [3]. The narrow dashed lines show the extent of the New Zealand 200-nautical-mile Exclusive Economic Zone.

eastern margins [1,2]. The rise is clearly delineated by the 2,000-m isobath with its 130 km wide, essentially flat crestal region generally at 350–450 m below sea level [3]. Several raised bank areas are prominent along the length of the crest (e.g., Mernoo Bank, Reserve Bank, Veryan Bank), with the Chatham Islands rising above sea level at the eastern end of the rise. The northern flank is typically steeper (3–10°) than the southern flank (<3°). Pelagic calcareous oozes dominate the Bounty Trough and lower southern flanks of the rise [4], becoming increasingly characterized by glauconitic muddy sands, with abundant phosphate nodules, on the crest and upper slopes [1,5]. Terrigenous sediments dominate the western crest and northern flanks of the rise where the ridge flanks intersect the Hikurangi Plateau [4,6].

The Chatham Rise underlies and partially constrains the highly productive Subtropical Front (STF, insert in Figure 56.1) [7,8]. The STF is formed by the interaction between warm, highly saline, macronutrient-limited subtropical surface waters that are transported southward along the eastern continental margin of the North Island (East Cape Current) and colder, less saline, macronutrient-rich, micronutrient-poor sub-Antarctic surface waters associated with the flows along the eastern South Island (Southland Current) [6]. The STF on the Chatham Rise is characterized by high water column biological production [8] and associated high benthic biomass [9,10], and it provides the basis of New Zealand's largest deep-water fisheries [e.g., peak annual tonnages of about 30,000 t and 70,000 t in the early 1980s

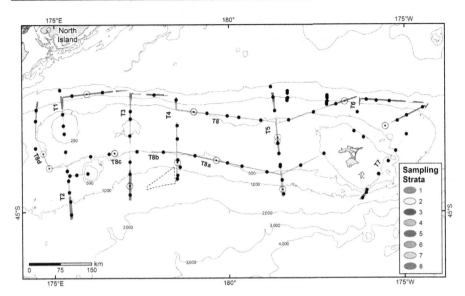

Figure 56.2 Ocean Survey 20/20 sampling across Chatham Rise in 2006–2007: multibeam acoustic transects (individual transects labeled with T prefix), *a priori* environmental classes (coded in legend; see text for details about how these sampling strata were derived), and seabed sampling sites (black dots). The dashed line shows the intersection of New Zealand's 200-nautical-mile Exclusive Economic Zone boundaries.

and 1998 for orange roughy (*Hoplostethus atlanticus*) and hoki (*Macruronus novaez-elandiae*), respectively] [11]. Seabed habitats are impacted by bottom trawling across most parts of the rise, with major fisheries for orange roughy, oreos (*Allocytus niger* and *Pseudocyttus maculatus*), and alfonsino (*Beryx splendens*) [12] on pinnacles and slope areas along the northern and eastern flanks, hoki over water depths of 500–800 m on the flanks, and scampi (*Metanephrops challengeri*), hake (*Merluccius australis*), and ling (*Genypterus blacodes*) on the crest of the rise. Before the development of deep-water fisheries in the 1970s, Chatham Rise supported very large populations of orange roughy. These populations are now considerably reduced and larger aggregations, particularly in spawning season, are known to persist only on a few seamounts [13,14]. Consistent with high fishing activity over recent decades, trawl marks are ubiquitous seafloor features on soft sediments of the rise, especially along the southwestern and northeastern upper flanks.

In the 1980s, extensive survey work was undertaken in crestal regions with the main aim of characterizing the phosphate nodule resource on the rise [1,5]. A program of biological oceanographic research, including benthic studies, focused on the Chatham Rise and STF through the 1990s and early 2000s [10,15]. Since 2001, there has been a series of studies of benthic invertebrate assemblages on Chatham Rise seamounts, primarily the Graveyard and Andes complexes [16–18]. In August 2006 and April–May 2007, the Chatham Rise was surveyed as part of New Zealand's Ocean Survey 20/20 (OS 20/20) initiative (Figure 56.2). Bathymetric data were collected using a 30-kHz Simrad EM300 multibeam sonar system and processed with 25-m grid resolution over a series of transects designed to capture environmental and fishing intensity gradients. Eight environmental strata were defined along these transects by statistical classification based on acoustic parameters (backscatter, rugosity, depth) and remotely sensed and modeled oceanographic data [sea surface temperature (SST) amplitude, chlorophyll *a* amplitude and phase, seabed current speed] [19]. At 100 sites distributed across the eight predefined environmental strata, seabed samples for fauna and sediments were collected to characterize habitat diversity and its relationship to benthic biodiversity. The primary sampling gears used were a towed underwater camera system with video and still image cameras [20] and an epibenthic sled (25 mm mesh). At a subset of 17 sites, a multicorer was deployed to sample infauna (bacteria, meiofauna, macrofauna) and sediments (%mud/sand/gravel, %organic matter, chloroplastic pigments, %carbonate).

Geomorphic Features and Habitats

The main geomorphic features of the Chatham Rise are the smoothly sloping flanks (predominantly north- and south-facing), the flat-lying, though locally topographically irregular crestal region (including several prominent banks), and isolated groups of volcanic peaks or seamounts (Figure 56.1), with occasional sea valleys found on the flanks. Except for the emergent Chatham Islands, the rise deepens generally to the east, eventually merging with the abyssal plain of the Southwest Pacific Basin. Other than the OS 20/20 multibeam transects described above, the only detailed mapping of specific geomorphic features on Chatham Rise has concentrated on two seamount complexes: the Graveyard complex on the northern flank and the Andes complex on the eastern end of the rise [16,21]. Based primarily on information from OS 20/20 and these seamounts studies, the main feature types on the rise are described below.

Flanks: The steep northern flanks extend down to water depths of 3,000 m on the Hikurangi Plateau (Figure 56.1). Evidence for slope instability and/or current scour are present along various sections of the northern flank (unpublished NIWA 3.5 kHz and multibeam data [22]), while strong currents along the flanks to the north and east of the Chatham Islands are inferred from deep-sea photographs collected in 2007 (Figure 56.3A). The less steep southern flank drops down into the Bounty Trough at water depths of 2,500–2,800 m (Figure 56.1). Persistent eastward currents of up to 26 cm/s at 500 m above the seafloor have been observed on the southern flank over a 1-year mooring deployment [23] and are associated with areas of scour and erosion, particularly at 1,000–1,200 m depths (unpublished NIWA 3.5 kHz data). Near-bed currents (1 m above seafloor) on the southern flank are typically between 5 and 15 cm/s and have a strong tidal component [24]. Slopes on the eastern margin of the rise descend to abyssal depths [3] and are modified by seabed currents associated with the Deep Western Boundary Current [25].

Crest and banks: The crest of the Chatham Rise is relatively flat-lying between water depths of 350–450 m, is emergent at the Chatham Islands, and has occasional large banks that rise to within 30 m of the sea surface (Mernoo Bank) [3] (Figure 56.1).

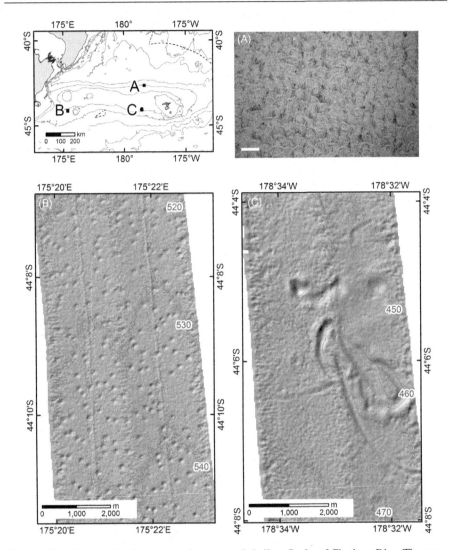

Figure 56.3 Geomorphic features on the crest and shallow flanks of Chatham Rise. The map in the top left panel shows the locations of: (A) Bottom photograph of current-swept sandy substrate at 1,000 m depth on the northeastern flank of Chatham Rise. Scale bar shows 20 cm. (B) Fields of pockmarks indicating fluid escape at 500–600 m depth west of Veryan Bank. (C) Iceberg scour marks at 450 m depth on the southeastern crest. Individual scours are *ca.* 10 m deep, and pockmarks are *ca.* 5 m deep.

On Mernoo Bank, there are exposed Mesozoic rocks and cobbles with carbonate sands, whereas Veryan Bank is a volcanic knoll and Reserve Bank consists predominantly of medium to fine glauconitic sands. The local topography of the crest is highly irregular, with multiple hummocks and swales, many of which are sediment-filled [5]. The crest and associated banks are also swept by locally strong, tidally driven currents of

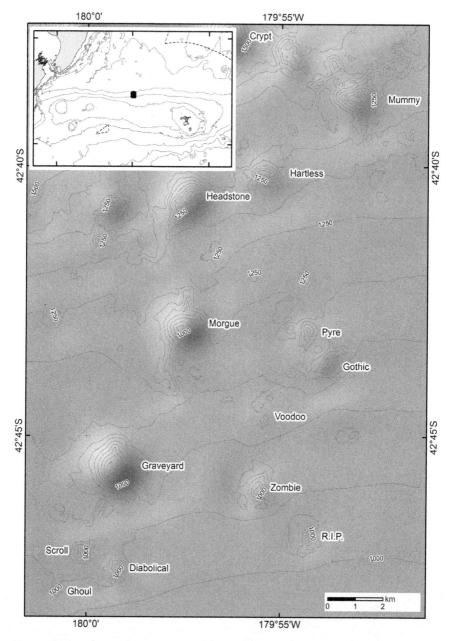

Figure 56.4 The Graveyard seamounts complex on the northern flank of Chatham Rise.

up to 25 cm/s [24,26,27]. In 2007, an extensive field of shallow (~2- to 5-m deep) pockmarks was discovered in soft muddy sediments at 500–600 m depths south of Mernoo Bank, indicating fluid and/or gas escape (Figure 56.3B). Remnants of inferred last-glacial iceberg scour marks are ubiquitous at water depths of 400–500 m along the

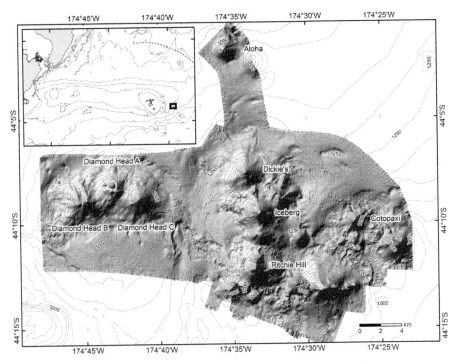

Figure 56.5 The Andes seamounts complex at the eastern end of Chatham Rise.

southern side of the crest, reaching maximum scour depths of 10 m below seafloor over horizontal scales of 200–500 m (Figure 56.3C). Previous geological studies on the rise identified iceberg dropstones [1], which is consistent with the more recent multibeam mapping results.

Pinnacles: Isolated groups of volcanic pinnacles occur around the Chatham Rise [2] and have been mapped in detail at two locations [3]. In the Graveyard complex (Figure 56.4), there are at least 15 small volcanic pinnacles across an area of 190 km², ranging in elevation from 50 to 300 m in basal water depths of 1,000–1,600 m Figure 56.4. In the Andes complex (Figure 56.5), there are at least 11 named peaks across an area of 150 km², ranging in elevation from 100 to 800 m in basal water depths of 1,200–1,500 m (Figure 56.5). Other volcanic features are known to exist elsewhere on the flanks of the rise (primarily through anecdotal reports from fishing vessels), but have yet to be mapped. There are likely to be others that are yet to be discovered.

Sea valleys and troughs: Because the entire Chatham Rise has not been mapped in detail, there is only limited information about the areal extent of sea valleys on the rise flanks. Gullies and larger valley-like features are apparent on the northeastern flank (~1,000 m depth), while in central areas of the southern flank in water depths of 1,000–1,200 m several incised features have been reported, some of which are associated with the deep oreo fishery grounds (*RV Tangaroa* master, personal

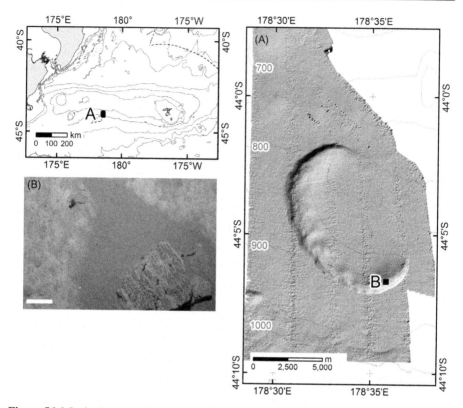

Figure 56.6 Incised trough-like feature in 900 m depth on the central southern flank of Chatham Rise (A). Seabed photograph taken in April 2007 (B) shows accumulation of phytodetritus in the feature. Scale bar shows 20 cm.

communication). One interesting trough-like feature, discovered and mapped in detail in 2007, has an unusual semicircular form with a steep southern bank that is incised 50–70 m into the seabed slope. At the time of surveying (April 2007), this curved trough was observed to have a thick accumulation of phytodetritus (degraded marine algal material) on the trough floor, with evidence of white bacterial mats and no epifauna evident in photographs (Figure 56.6). This valley is now believed to be a giant gas expulsion feature [28].

Sediments: In 2007, a total of 69 new sediment samples were collected on Chatham Rise using a multicorer (17 sites) and pipe dredges attached to beam trawls and epibenthic sleds (52 sites) (Figure 56.2). Together with archived sediment samples (grab, core, dredge) collected since the 1950s [4,6] and observations from OS 20/20 video transects, these data indicate a predominance of muddy, organic-rich, and terrigenous-influenced sediments in western areas of the Chatham Rise (Figure 56.7). High carbonate (80–90%), foram-rich silty sands are found on the deep southern flanks. Glauconite-rich, moderately carbonate, muddy sands cover the upper flanks (<1,500 m) and crest. Phosphorite nodules occur commonly in crestal sediments (<500 m), formed authigenically around "core" material, such as pebbles,

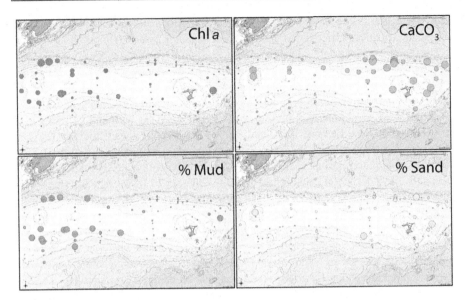

Figure 56.7 Chatham Rise sediments: relative proportions of chlorophyll *a* (Chl *a*, top left, symbols scaled from 0.002 to 0.17 μg/g dry weight), calcium carbonate (CaCO₃, top right, 7–90%), mud (% mud, <63 μm, lower left, 37–94%), and sand and gravel (% sand, >63 μm, lower right, 5–65%) in sediments from multicorer and pipe dredge samples collected during the Ocean Survey 20/20 voyage in 2007.

shells, and occasionally cetacean bones [1]. The phosphatization is believed to have occurred ~12–7 million years ago [1], although the high flux of organic material to the seafloor [23,24] and moderately strong currents across the rise [26,27] are possibly also responsible for maintaining reducing conditions at the seafloor, and perhaps the associated precipitation of authigenic glauconite.

Biological Assemblages

During the 2007 OS 20/20 research voyage, 115 1-hour video transects were collected and analyzed for the presence and abundance of benthic invertebrate fauna. Epibenthic sled samples were taken at the same sites and were used to confirm identifications of taxa seen in the video, as well as for generating independent estimates of abundance and diversity. The average (±SD) swept area of seabed per video transect was 2,359 ± 644 m², and all fauna and bioturbation marks *ca.* >5 cm were recorded from each transect, together with continuous categorical descriptions of primary substrate type (bedrock, boulders, muddy sediment, etc.) using methods described in Bowden et al. [29]. All faunal abundances were standardized to numbers per 1,000 m² of seabed. Detailed analyses of variations in assemblage structure and function, including modeling of distributions in relation to environmental gradients, are in progress, but here broad variations in assemblage structure are described and the more abundant and conspicuous taxa, characteristic of each area, are noted.

The diverse habitats across Chatham Rise, ranging from shallow banks on the crest to bathyal sediments at the base and seamounts on the flanks, support a wide range of benthic faunal assemblages (Figure 56.8). At <50m deep at its shallowest, Mernoo Bank is within the photic zone and supports benthic macroalgae, fish, and diverse assemblages of sessile and mobile fauna normally associated with coastal rocky reef sites (Figure 56.8A). Reserve Bank and Veryan Bank, both toward the western end of the rise, are deeper, and although some coralline algae persist here (Figure 56.8B), assemblages are more similar to those on hard substrate continental shelf areas at similar depths. Most of the remainder of the rise crest consists of sediment substrates populated by mobile fauna, with conspicuous examples including scampi (*Metanephrops challengeri*), squat lobsters (*Munida gracilis*), several crab species, quill worms (*Hyalinoecia longibranchiata*), urchins (*Parametia peloria* and other spatangids) (Figure 56.8C), and asteroids. Sediment cover is often thin and where underlying rock or phosphorite nodules are exposed, sponges, stylasterid hydrocorals, and other sessile suspension-feeding fauna occur. Most sessile fauna on the crest are relatively sparsely distributed and of small size, but large (>1 m) hexactinellid sponges (*Hyalscus* sp.) are locally common in the center and toward the western end of the rise (Figure 56.9).

The flanks of the rise are heterogeneous at all scales, with habitats ranging from soft sediments, rich in organic material, to current-scoured sands and high-relief, exposed bedrock. Benthic assemblages vary considerably with habitat, and there are pronounced differences in assemblage composition from north to south and from west to east. The southern flank has generally higher numbers of taxa and abundances, with the highest taxon richness of benthic fauna occurring around the southwestern end of the rise at upper slope depths (*ca.* 200–500m). Characteristic fauna on the southern slopes include anemones, cidaroid urchins, the regular urchin *Gracilechinus multidentatus*, predatory gastropods (Buccinidae, Volutidae), and sponges (*Neoaulaxinia persicum, Geodia regi*). At depths below 1,000m on the southern flank, soft-sediment assemblages are dominated over large areas by highly abundant populations of the brittle star *Ophiomusium lymani* (Figure 56.8G). Elsewhere at these depths on the southern flanks, holothurians are the most conspicuous fauna. The northern flanks, by contrast, are characterized by shrimps (*Campylonotus* sp., *Nematocarcinus* sp.), pagurid crabs, cnidarians (including pennatulaceans), soft corals (*Anthomastus* sp.), anemones, gorgonians (Isididae, *Radicipes* sp.), xenophyophores (foraminifera), and holothurians (*Enypniastes eximia*).

Seamounts in the Graveyard complex on the northern flank of the rise, and the Andes complex on the southeastern flank, support dense stands of cold-water corals (mostly *Solenosmilia variabilis*; Figure 56.8E and F), with associated invertebrate assemblages that differ markedly from assemblages on the surrounding flanks of the rise [30]. On many of the larger seamount features, these coral stands have been greatly reduced by bottom trawling [18,21]. Several smaller hills remain unfished, however, and some of these, together with one of the fished hills (Morgue), have been protected from fishing since 2001.

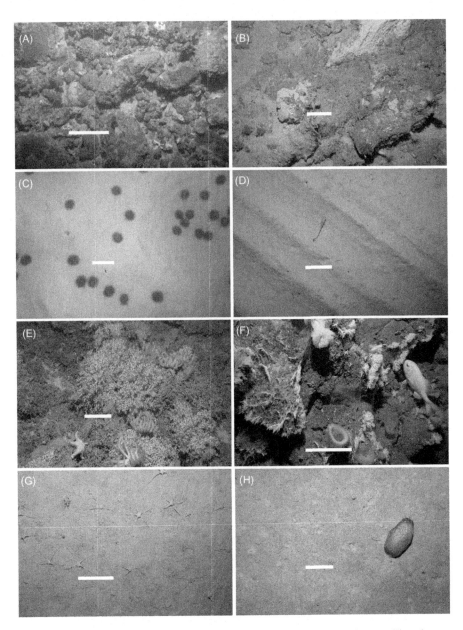

Figure 56.8 Benthic habitats on Chatham Rise: (A) sponges and encrusting coralline algae at 58 m on Mernoo Bank; (B) sponges, encrusting coralline algae, and anemones at 150 m on Veryan Bank; (C) urchins (*Parametia peloria*) at 420 m on soft sediment on the eastern crest; (D) rat-tail (*Coelorinchus innotabilis*) on soft sediment with trawl marks at 510 m on the southwest flank; (E) cold-water coral (*Solenosmilia variabilis*) and seastars (Brisingidae and *Hippasteria phyringia*) at 1,000 m in the Graveyard seamounts complex; (F) orange roughy (*Hoplostethus atlanticus*), stylasterid hydrocorals, and sponges at 800 m in the Graveyard seamounts complex; (G) dense population of brittle stars (*Ophiomusium lymani*) on soft sediments at 1,250 m on the southwestern flank; (H) large holothurian (*Benthodytes incerta*) on soft sediments at 1,200 m on the southeastern flank. Scale bar shows 20 cm.

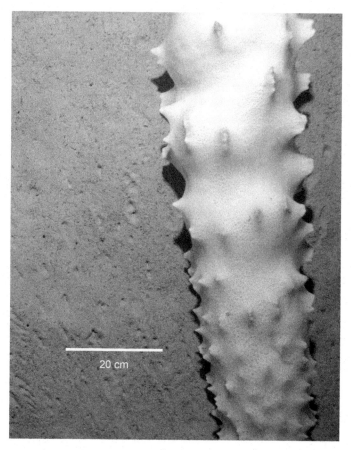

Figure 56.9 Hexactinellid sponge (*Hyalscus* sp.) on the southern flank of Chatham Rise at 600 m.

Surrogacy

Observed distributions of benthic faunal assemblages during OS 20/20 have been used to assess the utility of a Marine Environment Classification (MEC) scheme that was developed in 2006 [19] using only remotely sensed and modeled oceanographic layers (e.g., SST gradient, SST amplitude, winter SST, depth, seabed slope, mean annual solar radiation, tidal currents). In parallel with these assessments, continuous distributions of individual species and full benthic assemblage composition across Chatham Rise have been modeled in relation to environmental gradients (NIWA, unpublished data). These methods are in early stages of development and validation but, to date, the utility of the existing environmental classification schemes and models for predicting benthic distributions appears to be limited to coarse spatial and taxonomic scales. Thus, while even the earliest iteration of the MEC distinguishes between depth ranges, and to some extent between the northern and southern and eastern and western sectors of Chatham Rise, the degree to which these environmental classes correspond with observed faunal distributions at finer scales is limited.

Acknowledgments

The support of the Ocean Survey 20/20 Steering Committee and the funding agencies (Ministry of Fisheries, Land Information New Zealand, Department of Conservation, and NIWA) is gratefully acknowledged. Thanks also to the officers and crew of the RV *Tangaroa*, NIWA technical and scientific staff, and other personnel from NZ institutions who participated in the planning and implementation stages of the Ocean Survey 20/20 Chatham–Challenger project.

References

[1] D.J. Cullen, The submarine phosphate resource on central Chatham Rise, DFMS Report 2 (ISSN 0113-2210), Division of Marine and Freshwater Science (now NIWA), Department of Scientific and Industrial Research, Wellington, New Zealand (1987) 22 pp.

[2] R.A. Wood, P.B. Andrews, R.H. Herzer, R.A. Cook, N. de B. Hornibrook, R.H. Hoskins, et al., Cretaceous–Cenozoic geology of the Chatham Rise region, South Island, NZGS Basin Studies 3, New Zealand Geological Survey (now GNS Science), Department of Scientific and Industrial Research, Wellington, New Zealand (1989) 76 pp.

[3] K.A. Mackay, B.A. Wood, M.R. Clark, Chatham Rise Bathymetry, National Institute of Water and Atmospheric Research Ltd, Wellington, New Zealand (2005) NIWA Miscellaneous Chart Series 82.

[4] J.S. Mitchell, L. Carter, J.C. McDougall, New Zealand Regional Sediments (1:6 000 000), New Zealand Oceanographic Institute (now NIWA), Department of Scientific and Industrial Research, Wellington, New Zealand (1989) NZOI Miscellaneous Chart Series 67.

[5] H.-R. Kudrass, U. von Rad, Geology and some mining aspects of the Chatham Rise phosphorite: a synthesis of SONNE-17 results, Geol. Jahrb. D65 (1984) 233–252.

[6] A. Orpin, L. Carter, A. Goh, E. Mackay, A. Pallentin, A.-L. Verdier, et al., New Zealand's Diverse Seafloor Sediments, National Institute of Water and Atmospheric Research Ltd, Wellington, New Zealand (2008) NIWA Miscellaneous Chart Series 86.

[7] P. Sutton, Detailed structure of the Subtropical Front over Chatham Rise, east of New Zealand, J. Geophys. Res. 106 (C12) (2001) 31045–31056.

[8] R.J. Murphy, M.H. Pinkerton, K.M. Richardson, J.M. Bradford-Grieve, P.W. Boyd, Phytoplankton distributions around New Zealand derived from SeaWiFS remotely-sensed ocean colour data, N. Z. J. Mar. Freshwater Res. 35 (2001) 343–362.

[9] P.K. Probert, D.G. McKnight, Biomass of bathyal macrobenthos in the region of the Subtropical Convergence, Chatham Rise, New Zealand, Deep Sea Res. Part I 40 (1993) 1003–1007.

[10] S.D. Nodder, C.A. Pilditch, P.K. Probert, J.A. Hall, Variability in benthic biomass and activity beneath the Subtropical Front, Chatham Rise, SW Pacific Ocean, Deep Sea Res. Part I 50 (2003) 959–985.

[11] K.J. Sullivan, P.M. Mace, N.W.M. Smith, M.H. Griffiths, P.R. Todd, M.E. Livingston, et al. (Comps.), Report from the Fishery Assessment Plenary, May 2005: Stock Assessments and Yield Estimates (2005) 792 pp (Unpublished report held in NIWA library, Wellington).

[12] M.R. Clark, R.L. O'Driscoll, Deepwater fisheries and aspects of their impact on seamount habitat in New Zealand, J. Northwest Atl. Fish. Sci. 31 (2003) 441–458.

[13] M. Clark, Fisheries for orange roughy (*Hoplostethus atlanticus*) on seamounts in New Zealand, Oceanol. Acta 22 (1999) 593–602.

[14] M.R. Clark, O.F. Anderson, R. Francis, D.M. Tracey, The effects of commercial exploitation on orange roughy (*Hoplostethus atlanticus*) from the continental slope of the Chatham Rise, New Zealand, from 1979 to 1997, Fish. Res. 45 (2000) 217–238.

[15] J.M. Bradford-Grieve, P.W. Boyd, F.H. Chang, S. Chiswell, M. Hadfield, J.A. Hall, et al., Pelagic ecosystem structure and functioning in the Subtropical Front region east of New Zealand in austral winter and spring 1993, J. Plankton Res. 21 (1999) 405–428.

[16] A.A. Rowden, S. O'Shea, M.R. Clark, Benthic biodiversity of seamounts on the north-west Chatham Rise, Marine Biodiversity & Biosecurity Report 2, Ministry of Fisheries, Wellington, New Zealand (2002), 21 pp.

[17] M.R. Clark, D.A. Bowden, S.J. Baird, R. Stewart, Effects of fishing on the benthic biodiversity of seamounts of the "Graveyard" complex on the northern Chatham Rise, New Zealand Aquatic Environment & Biodiversity Report 46, Ministry of Fisheries, Wellington, New Zealand (2010) 40 pp.

[18] A. Williams, T.A. Schlacher, A.A. Rowden, F. Althaus, M.R. Clark, D.A. Bowden, et al., Seamount megabenthic assemblages fail to recover from trawling impacts, Mar. Ecol. 31 (2010) 183–199.

[19] T.H. Snelder, J.R. Leathwick, K.L. Dey, A.A. Rowden, M.A. Weatherhead, G.D. Fenwick, et al., Development of an ecologic marine classification in the New Zealand region, Environ. Manage. 39 (2006) 12–29.

[20] P. Hill, Designing a deep-towed camera vehicle using single conductor cable, Sea Technol. December (2009) 49–51.

[21] M.R. Clark, A.A. Rowden, Effect of deepwater trawling on the macro-invertebrate assemblages of seamounts on the Chatham Rise, New Zealand, Deep Sea Res. Part I 56 (2009) 1540–1554.

[22] P.M. Barnes, Mid-bathyal current scours and sediment drifts adjacent to the Hikurangi deep-sea turbidite channel, eastern New Zealand: evidence from echo character mapping, Mar. Geol. 106 (1992) 169–187.

[23] S.D. Nodder, L.C. Northcote, Episodic particulate fluxes at southern temperate mid-latitudes (42–45°S) in the Subtropical Front region, east of New Zealand, Deep Sea Res. Part I 48 (2001) 833–864.

[24] S.D. Nodder, G.C.A. Duineveld, C.A. Pilditch, P.J. Sutton, P.K. Probert, M.S.S. Lavaleye, et al., Focusing of phytodetritus deposition beneath a deep-ocean front, Chatham Rise, New Zealand, Limnol. Oceanogr. 52 (2007) 299–314.

[25] L. Carter, I.N. McCave, Development of sediment drifts approaching an active plate margin under the SW Pacific Deep Western Boundary Current, Paleoceanography 9 (1994) 1061–1085.

[26] R.A. Heath, Observations on Chatham Rise currents, N. Z. J. Mar. Freshwater Res. 17 (1983) 321–330.

[27] S.M. Chiswell, Acoustic Doppler Current Profiler measurements over the Chatham Rise, N. Z. J. Mar. Freshwater Res. 28 (1994) 167–178.

[28] B. Davy, Pecher, R. Wood, L. Carter, K. Gohli, Gas escape features off New Zealand: evidence of massive release of methane from hydrates, Geophys. Res. Letters 37 (2010), L211309, doi:10.1029/2010GL045184.

[29] D.A. Bowden, S. Schiaparelli, M.R. Clark, G.J. Rickard, A lost world? Archaic crinoid-dominated assemblages on an Antarctic seamount, Deep Sea Res. Part II (2010) doi:10.1016/j.dsr2.2010.09.006.

[30] A.A. Rowden, T.A. Schlacher, A. Williams, M.R. Clark, R. Stewart, F. Althaus, et al., A test of the seamount oasis hypothesis: seamounts support higher epibenthic megafaunal biomass than adjacent slopes, Mar. Ecol. 31 (2010) 95–106.

57 Habitats and Benthos of a Deep-Sea Marginal Plateau, Lord Howe Rise, Australia

Peter T. Harris, Scott L. Nichol, Tara J. Anderson
Andrew D. Heap

Marine and Coastal Environment Group, Geoscience Australia, Canberra, ACT, Australia

Abstract

Lord Howe Rise is a deep-sea marginal plateau located in the Coral Sea and Tasman Sea, ~125,000 km² in area and 750–1,200 m in water depth. An area of the western flank of northern Lord Howe Rise covering ~25,500 km² was mapped and sampled by Geoscience Australia in 2007 to characterize the deep-sea environments and benthic habitats. Geomorphic features in the survey area include ridges, valleys, plateaus, and basins. Smaller superimposed features include peaks, moats, holes, polygonal furrows, scarps, and aprons. The physical structure and biological composition of the seabed were characterized using towed video and sampling of epifaunal and infaunal organisms. These deep-sea environments are dominated by thick, depositional, soft sediments (sandy mud), with local outcrops of volcanic rock and mixed gravel–boulders. Ridge, valley, and plateau environments were moderately bioturbated, but few organisms were directly observed or collected. Volcanic peaks were bathymetrically complex hard-rock structures that supported sparse distributions of suspension feeders (e.g., cold-water corals and glass sponges) and associated epifauna (e.g., crinoids and brittle stars). Isolated outcrops along the sloping edge of one ridge also supported similar assemblages, some with high localized densities of coral-dominated assemblages.

Key Words: Deep sea, plateau, epibenthic, cold-water corals, multibeam sonar, Coral Sea, Tasman Sea

Introduction

Lord Howe Rise is a marginal plateau located in the Coral Sea and Tasman Sea, composed mainly of continental fragments that detached from the eastern margin of

Seafloor Geomorphology as Benthic Habitat. DOI: 10.1016/B978-0-12-385140-6.00057-8

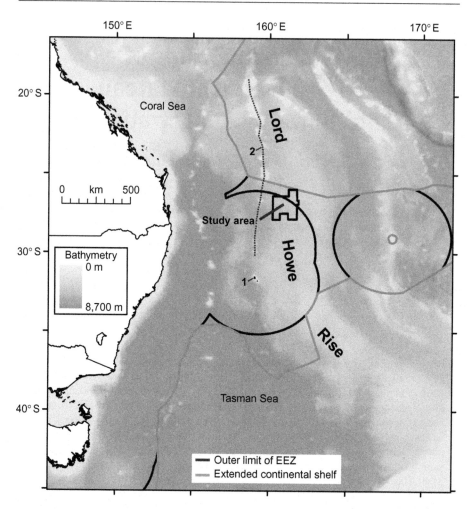

Figure 57.1 Location map of study area on the Lord Howe Rise, eastern Australian margin. The outer limit of the exclusive economic zone (EEZ) and extended continental shelf of Australia: 1—Lord Howe Island; 2—dotted line represents Lord Howe seamount chain.

continental Australia during the late Jurassic and Cretaceous [1]. Lord Howe Rise is an extensive feature of the South Pacific Ocean, spanning ~2,800 km in latitude (19°S to 43°S) and 450–650 km wide. The crest of the Lord Howe Rise lies around 750–1,200 m below sea level and it is surmounted by small volcanic islands and seamounts (i.e., Lord Howe Island and Lord Howe seamount chain), while the 2,000-m isobath outlines the base of the plateau (Figure 57.1). The oceanic environment is part of the lower-bathyal biome having intermediate surface primary production, low dissolved oxygen in bottom waters, and mean bottom water temperatures of between 1.7°C and 2.3°C [2]. The region is influenced by eddies shed from the East Australia current [3].

The Lord Howe Rise benthic environment is in a near-pristine condition with "low" human impact according to Halpern et al. [4]. The region is affected by shipping, limited fishing, and laying of telecommunications cables but is otherwise considered to be near pristine. In October–December 2007, part of the western flank of northern Lord Howe Rise was mapped and sampled by Geoscience Australia using the New Zealand research vessel *Tangaroa* [5]. Bathymetric data were collected

Table 57.1 Data Acquired at Sampling Stations during Geoscience Australia Survey TAN0713 [5]

Geomorphic Feature/Unit	Station	Video	Still Images	Surface Sediments	Infauna	Epifauna
Ridge	04	✓		✓	✓	
Volcanic peak	05	✓	✓			✓
Ridge	06			✓		✓
Valley	07					✓
Ridge	08	✓	✓	✓		✓
Valley	09	✓	✓			
Ridge	10	✓	✓			
Plateau	11	✓	✓			✓
Ridge	12	✓		✓		
Ridge	13	✓	✓	✓	✓	
Ridge	14	✓	✓	✓		✓
Ridge	15	✓	✓	✓		✓
Ridge	16	✓	✓	✓	✓	
Volcanic peak	17	✓	✓			✓
Ridge	18	✓	✓	✓	✓	
Volcanic peak	19	✓	✓			✓
Ridge	20	✓	✓	✓		✓
Ridge	21	✓	✓	✓	✓	
Ridge	22	✓	✓	✓		✓
Ridge	23	✓	✓	✓	✓	
Ridge	24	✓	✓	✓	✓	
Ridge	25	✓	✓	✓		
Ridge	26	✓	✓	✓	✓	
Ridge	27	✓	✓	✓	✓	
Plateau	29	✓	✓	✓	✓	
Ridge	30	✓	✓	✓	✓	✓
Volcanic peak	31	✓	✓	✓	✓	
Volcanic peak	32	✓	✓	✓		✓
Volcanic peak	33	✓	✓	✓		✓
Plateau	42	✓		✓		
Plateau	43	✓	✓	✓		
Plateau	44	✓	✓	✓		
Plateau	45	✓	✓	✓		
Plateau	46	✓	✓	✓		

using a 30-kHz Simrad multibeam sonar system over a survey area of ~25,500 km^2 and gridded to a spatial resolution of 50 m. Seabed habitats and associated assemblages were observed using underwater towed video at 32 stations, while benthic organisms (epifauna and infauna) were collected from 14 and 12 stations, respectively (Table 57.1). Surface sediment samples were collected by grab, boxcore, and piston core at 28 representative stations across the survey area (Table 57.1; Figure 57.2).

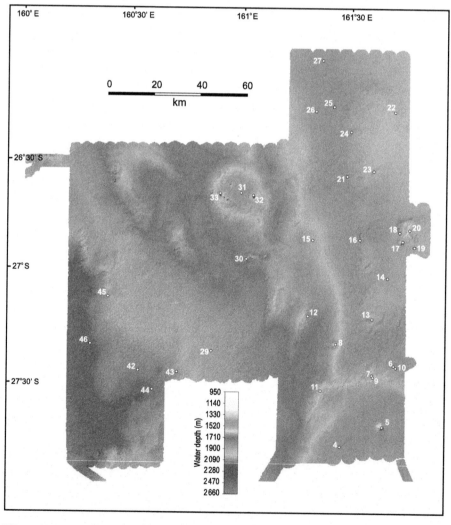

Figure 57.2 Multibeam sonar bathymetric map of northern Lord Howe Rise based on a 50-m grid, showing station locations from Geoscience Australia survey TAN0713.

Geomorphic Features and Habitats

Bathymetry across the western flank of the Lord Howe Rise plateau is characterized by a trend of increasing water depth toward the southwest, with a gentle regional gradient of ~0.3° (Figure 57.2). Within this area, large-scale geomorphic features are broadly oriented north–south, tending to a northwest–southeast alignment across the western part of the survey area. Overall, the relief of the area is on the scale of tens of meters. However, local bathymetric anomalies exist where peaks, valleys, and scarps introduce hundreds of meters of relief to the seafloor [6].

The classification of seabed geomorphic features used here is based on feature names and definitions from the International Hydrographic Organization [7], with additional terms for other small-scale features taken from the literature [8] (Table 57.2). Geomorphic features were identified using bathymetric profiles drawn

Table 57.2 Definitions for Geomorphic Features, Units, and Elements Mapped within the Lord Howe Rise Survey Area

Geomorphic Feature	Definition
Plateau	A flat or nearly flat elevation of considerable areal extent, dropping off abruptly on one or more sides.
Geomorphic unit	
Basin	A depression in the seabed, more or less equidimensional in plan and of variable extent.
Ridge	An elongated narrow elevation of varying complexity having steep sides.
Valley	A relatively shallow, wide depression, the bottom of which usually has a continuous gradient.
Geomorphic element	
Apron	A gently dipping surface, underlain primarily by sediment, at the base of any steeper slope.
Escarpment (scarp)	An elongated, characteristically linear, steep slope separating horizontal or gently sloping sectors of the seabed in nonshelf areas.
Hole	A small local depression, often steep sided, in the seabed.
Moat	An annular depression that may not be continuous, located at the base of many seamounts, oceanic islands, and other isolated elevations.
Peak	A prominent elevation either pointed or of a very limited extent across the summit.
Polygonal furrows	A network of cracks in the seabed forming a polygonal pattern. Cracks can be straight to curved, with patterns, lengths, depth, and width showing great variation.

Note: All definitions are from the International Hydrographic Organization standardized list of undersea features [6], except polygonal furrows taken from Goudie [8].

in IVS 3D Fledermaus software (version 6). Boundary contours were then mapped as polygons in ArcGIS, from which geomorphic features, units, and elements were interpreted (Figure 57.3). In this study, the Lord Howe Rise plateau is the primary geomorphic feature upon which geomorphic units are mapped; these include ridges, valleys, and basins. Geomorphic elements are superimposed on these units and include peaks, moats, holes, polygonal furrows, scarps, and aprons [6].

Ridges: These are the most extensive geomorphic units in the survey area, covering 12,700 km^2 (Figure 57.3). The eastern sector of the mapped area is occupied by

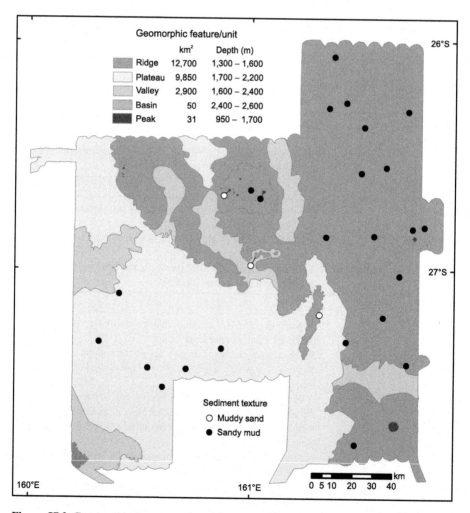

Figure 57.3 Geomorphic features, units, and volcanic peaks of northern Lord Howe Rise, with surface areas and water depth ranges indicated. Texture of surface sediments at sample stations is also shown.

the largest ridge, which extends 150 km north to south and is 30–50 km wide, with slightly convex slopes of up to 0.5°. Locally, holes and broad depressions introduce topographic variability of 50–100 m to the ridge crest across distances of 10–60 km. North of latitude 27.1°S and within the central to western part of the survey area, the seafloor is characterized by a relatively complex terrain of a multiple ridges and valleys. Here the ridges are associated with volcanic peaks that generally sit on the mid- to lower-ridge slopes.

Valleys: These are formed across water depths ranging from 1,600 m in the east to 2,400 m in the west of the survey area and cover 2,900 km². The deepest valley in the survey area is located in the far southwest, in 2,000–2,400 m water depth (Figure 57.3). This valley extends ~30 km in an east–west direction, widening from 6 km at the headwall to 16 km at the mouth. The slope valley floor decreases from 20° at the headwall to ~1.5° along the upper reaches and ~0.5° along the lower reaches and thalweg. Elsewhere, valleys are of similar dimensions (30–45 km long and 3.5–14 km wide), with gradients of 0.01–0.03°.

Plateaus: Low-gradient plateaus cover 9,850 km² of the survey area in water depths ranging from 1,700 to 2,200 m (Figure 57.3). The most extensive area of plateau occupies 8,365 km² in the southwest sector of the mapped area. Here the seabed slopes gently to the southwest at ~0.2° between 1,900 and 2,100 m water depth. The central-north sector is also occupied by a small low-relief plateau that covers <600 km² and is bordered by ridges.

Volcanic peaks: The 16 volcanic peaks mapped within the survey area cover 31 km² and range in height from 65 to 450 m (Figure 57.3). The two largest peaks are located in the shallowest water depth (1,400 m) near the eastern margin of the survey area and rise to 950 and 1,020 m water depth, respectively. Of the 16 volcanic peaks, 13 are clustered into three groups on ridges located in the western, central, and eastern sectors of the mapped area. All peaks have conical shapes, with slope gradients of 10–30°, and a moat at their base that is up to 50 m deep (Figure 57.4).

Basin: The southwest corner of the survey area captures a small section of steepening seafloor that extends from 2,400 to 2,600 m water depth. This is the edge of the Middleton Basin, which extends westward from this point [5].

Sediments: Grain size properties of the 28 sediment samples were determined by sieve separation of the gravel, sand, and mud fractions and by laser granulometry on the combined mud and sand fractions, using a Malvern Mastersizer 2000. Carbonate content of sediments was measured by acid digestion of a bulk subsample. Twenty-five of these samples were classified as sandy mud, and the other three as muddy sand (Figure 57.3). Mean grain size ranges from medium to very coarse silt (9–47 μm), and all samples are very poorly sorted. Bulk carbonate content ranges from 85% to 94%, incorporating forams and other nannofossils that have formed stiff dewatered deposits. Samples were collected from ridges, peaks, holes, the main plateau, and a valley, with slightly coarser grained sediments (muddy sands) occurring on peaks, small ridges, and holes. However, this is not a consistent pattern, as other peaks, ridges, and holes are characterized by sandy mud. Overall, there is no clear relationship between sediment type and geomorphic setting within the sampled area [6].

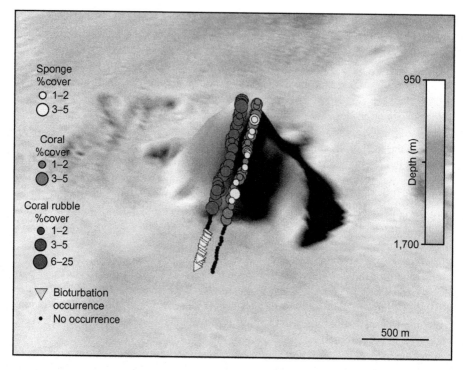

Figure 57.4 A typical volcanic peak located on a ridge in the central-north sector of the mapped area. This peak is 250 m high with slopes of up to 30°. The distribution of key taxa across the peak and adjacent seabed are plotted for station 33 (depth range 1,360–1,610 m). "No occurrence" denotes the absence of these key taxa. Data are presented for each 15-second video frame along the video transect.

Biological Communities

Assessments of benthos in this study are based primarily on underwater video, supplemented by higher-resolution still images and taxonomic identification of collected epifauna and infauna [9,10]. Seabed habitats and benthic assemblages were characterized in real-time using Characterization of the Benthos and Ecological Diversity (C-BED), a three-tiered characterization scheme that quantifies substratum cover, bedform/relief, and the presence of macroorganisms and percent cover of key taxa at 30-second intervals along each transect, whereby the seabed at each 30-second location is assessed for a period of 15 s (details provided in Refs. [9,10]).

Over the entire survey area, substrata was dominated by homogeneous soft sediments that comprised 84% of the seabed, with 12% volcanic outcrops, and the remaining 4% comprising a range of mixed habitats with gravels or boulders. Seabed relief was generally flat, with rare occurrences of low- (9%) and moderate- (2%) relief habitats and sand waves (1%). Overall, these environments supported sparse biological assemblages. Bioturbation marks (e.g., burrows (53%), tracks (38%), mounds (32%)) were the most common signs of life (61% of all basin locations had bioturbation marks), while sessile organisms, such as cold-water corals, sponges, crinoids, and brittle stars, occurred sparsely on rocky substrata (Figure 57.5). Motile species including shrimp and prawns (15%), fishes (8%), and jellyfish (5%) were also regularly recorded but were sparsely distributed and never abundant. The relationships between the physical structure of the seabed and biological assemblages are examined in detail in Anderson et al. [10]. Three benthic environments (peaks, ridges, and plateaus) were identified from these analyses as important predictors of biological patterns and are summarized below.

Volcanic peaks (rocky outcrops): These are dominated by homogeneous hard substrata (71% of all locations within this geomorphic class), with a mixture of rock and soft sediments and homogeneous soft sediments (14 and 15%, respectively) occurring mostly at the base of peaks. Although volcanic outcrops are bathymetrically distinct in the multibeam images (Figure 57.2), at fine scales, these peaks were characterized by low or flat (82% combined) relief habitats (Figure 57.5A and B). The rocky substrata of peaks had surprisingly few attached or associated organisms, and almost no dense habitat-forming taxa. Suspension feeders, such as cold-water corals and sponges, were commonly seen (55 and 18% of all locations, respectively) (Figure 57.5), but were present only in low densities (mean 4.5 and 1%, respectively). Cold-water corals and sponges provided structure for other species, such as brittle stars (e.g., *Ophiocreas oedipus*, *Ophiophycis john*, *Asteroschema tubiferum*) and crinoids (Figure 57.5A–D). Dead coral rubble (mostly Enallopsammia) was also common on the upper slopes of peaks (mean 5.3%, range 0–55%). Motile species, including shrimp/prawns, fish, and jellyfish, were only sporadically recorded (10, 2, and 1%, respectively).

Ridges (sediment covered): They are composed of flat soft sediments (92% of all ridge locations) with small (~3–70 m) isolated low-relief rock outcrops (8% of ridge locations) that occurred along the flanks where the slope was >10°. Soft sediments were bioturbated (79%) by combinations of burrows (60%), tracks (42%), mounds (41%), and characteristic rosettes and crater rings (8.2 and 4.8%, respectively; e.g., Figure 57.5G). Very few epifaunal organisms were directly observed or collected in these soft-sediment environments. The isolated rock ridges supported patchy but diverse cold-water coral and sponge assemblages (mean 7%, range 0–80%; Figure 57.5C), characterized by bamboo (*Keratoisis*) and golden corals, the latter entwined with brittle stars (*Asteroschema tubiferum*). Dead coral rubble was also patchy (mean 7%, range 0–50%). Motile species, such as shrimp and prawns (18%), fishes (10%), and jellyfish (7%), were occasionally recorded. Infaunal diversity and abundance were also low in ridge environments, with 3–25 species/taxa per station (Figure 57.6). Standardized species richness values indicate that Station 31 (1,518 m), which is immediately adjacent to a peak, had the highest number of infaunal species. Although infaunal sample sizes are low, there appears to be no direct relationship between species richness and either depth or distance to rocky outcrop (Figure 57.6).

Plateaus (soft sediments): They are composed of flat soft sediments (all plateau locations), bioturbated by burrows (86%), tracks (95%), and mounds (62%), with

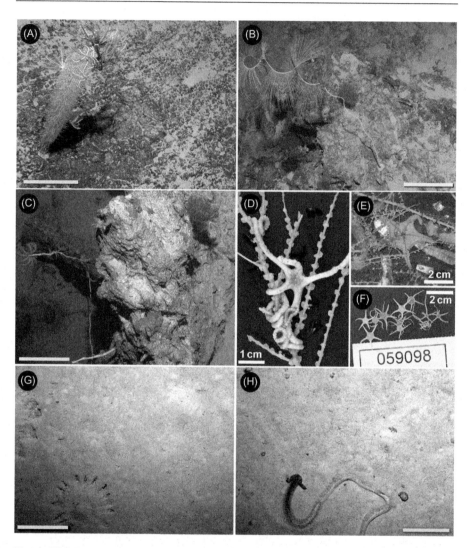

Figure 57.5 Photographs taken on hard and soft substrata of the Lord Howe Rise, deep-sea plateau. Filter feeders on hard substrata: (A) glass sponge on volcanic peak with brittle stars and crinoids (station 33: 1,351 m); (B) spiraled gorgonian (Chrysogorgiidae: *Iridogorgia*), family on volcanic peak (station 19: 1,395 m); (C) cold-water corals with brittle stars and crinoids on rocky ridge (station 30: 1,566 m); (D) brittle star (Asteroschematidae) entwined in a Golden coral (*Callogorgia*) (station 32: 1,594 m); (E) sea star (*Circeaster arandae*) and bamboo corals (station 6: 1,532 m); (F) collection of soft-sediment associated brittle stars (station 11: 1,572 m); (G) ridge sediments with "crater-ring" marks (station 13: 1,437 m); (H) plateau sediments with acorn worm (Hemichordata) (station 29: 1,939 m). Scale bars are 20 cm unless otherwise defined.

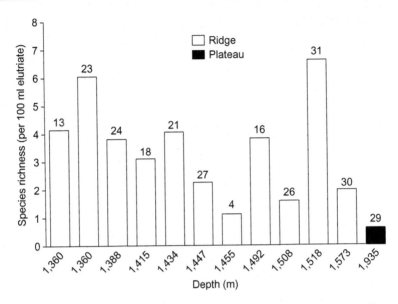

Figure 57.6 Species richness of infauna by station in relation to water depth and ridges and plateaus across the study area. Species richness is calculated as a standardized measure of species number per 10 ml of elutriate. Station numbers are indicated above each histogram bar.

characteristic acorn worms (phylum Hemichordata, class Enteropneusta) and their spiral and meandering trails (Figure 57.5H) [10,11].

Except for the sporadic occurrences of gorgonian whips, seapens, and hydroids (6% of locations), soft-sediment brittle stars (Figure 57.5F), motile species (shrimp/prawns, fishes, and jellyfish—20, 5, and 1%, respectively), and very few epifaunal organisms were directly observed or collected from the plateau environment. Of the six plateau stations, station 29 (1,940 m) was the only one sampled for infauna, due to the difficulties in successfully collecting boxcore samples at depths >2,000 m. This station supported the lowest number of infaunal species (Figure 57.6), although it is unclear if this reflects any geomorphic differences.

Concluding Remarks

In some deep-sea environments, rocky substrates support dense assemblages of suspension feeders, such as corals and sponges [12]. In this study, however, organisms were only sparsely recorded on rocky outcrops, with large areas of rock uncolonized. Epifaunal and infaunal diversity and abundances were low in soft-sediment habitats. While the array of cold-water corals and sponges recorded on rocky outcrops of both ridge and peak environments was diverse, epifaunal abundance was surprisingly low. Factors known to influence faunal diversity and abundance include nutrients, oxygen, organic content, and trace element levels [13]. However, it remains unclear which of these factor(s) may be limiting epifaunal abundances and infaunal abundance and diversity across this region of Lord Howe Rise.

Acknowledgments

This study was undertaken as part of the Offshore Energy Security Programme (2007–2011) funded by the Commonwealth Government of Australia. We thank the crew of RV *Tangaroa* for their technical support during survey TAN0713; Melissa Fellows (Geoscience Australia) who helped with production of the geomorphic map; and the taxonomic experts who identified biological specimens from this collection, with special thanks to Tim O'Hara (asteroids and ophiuroids), Rachel Przeslawski (infauna), Phil Allderslade, and Ron Thresher (cold-water corals). Reviews from Lene Buhl-Mortensen and M. Sayago-Gil are gratefully acknowledged. This chapter was produced with the support of funding from the Australian Government's Commonwealth Environment Research Facilities (CERF) program and is a contribution of the CERF Marine Biodiversity Hub. This chapter is published with permission of the Chief Executive Officer, Geoscience Australia.

References

[1] J.B. Willcox, J. Sayers, Geological framework of the central Lord Howe Rise (Gower Basin) region with consideration of its petroleum potential, Geoscience Australia Record 2002/11, Canberra, 2002.

[2] P.T. Harris, T. Whiteway, High Seas Marine Protected Areas: benthic environmental conservation priorities from a GIS analysis of global ocean biophysical data, Ocean Coast. Manage. 52 (2009) 22–38.

[3] P.J. Mulhearn, Variability of the East Australia Current over most of its depth and a comparison with other regions, J. Geophys. Res. 93 (C11) (1988) 13925–13929.

[4] B.S. Halpern, S. Walbridge, K.A. Selkoe, C.V. Kappel, F. Micheli, C. D'Agrosa, et al., A global map of human impact on marine ecosystems, Science 319 (2008) 948–952.

[5] A.D. Heap, M. Hughes, T. Anderson, S. Nichol, R. Hashimoto, J. Daniell, et al., Shipboard Party, Seabed environments of the Capel–Faust Basins and Gifford Guyot, Eastern Australia—post survey report, Geoscience Australia Record 2009/22, Canberra, 2009, 166 pp.

[6] S.L. Nichol, A.D. Heap, J. Daniell, High resolution geomorphic map of a submerged marginal plateau, northern Lord Howe Rise, east Australian margin, Deep Sea Res. Part II, (2011), 58(2011) 889–898. doi:10.1016/j.dsr2.2010.10.045.

[7] IHO, Standardization of Undersea Feature Names: Guidelines Proposal form Terminology, International Hydrographic Organisation and International Oceanographic Commission, Monaco, 2001.

[8] A.S. Goudie, Encyclopedia of Geomorphology, Routledge, London, 2004.

[9] T.J. Anderson, G.R. Cochrane, D.A. Roberts, H. Chezar, G. Hatcher, A rapid method to characterize seabed habitats and associated macro-organisms, in: B.J. Todd, H.G. Greene, (Eds.), Mapping the Seafloor for Habitat Characterization, Geological Association of Canada , Special Paper 47, pp. 71–79.

[10] T.J. Anderson, S.L. Nichol, C. Syms, R. Przeslawski, P.T. Harris, Deep sea bio-physical variables as surrogates for biological assemblages, an example from the Lord Howe Rise, Deep Sea Res. Part II, 2011, 58: 970–978, doi:10.1016/j.dsr2.2010.10.xxx.

[11] T.J. Anderson, R. Przeslawski, M. Tran, Distribution, abundance and trail characteristics of acorn worms at Australian continental margins, Deep Sea Res. Part II, 2011, 58: 979–991, doi:10.1016/j.dsr2.2010.10.xxx.

[12] A.D. Rogers, The biology of seamounts, Adv. Mar. Biol. 30 (1994) 305–350.

[13] L.A. Levin, R.J. Etter, M.A. Rex, A.J. Gooday, C.R. Smith, J. Pineda, et al., Environmental influences on regional deep-sea species diversity, Annu. Rev. Ecol. Syst. 32 (2001) 51–93.

58 Seamounts, Ridges, and Reef Habitats of American Samoa

Dawn J. Wright[1], Jed T. Roberts[2], Douglas Fenner[3], John R. Smith[4], Anthony A.P. Koppers[5], David F. Naar[6], Emily R. Hirsch[7], Leslie Whaylen Clift[8], Kyle R. Hogrefe[9]

[1]Department of Geosciences, Oregon State University, Corvallis, OR, USA, [2]Oregon Department of Geology and Mineral Industries, Portland, OR, USA, [3]Department of Marine and Wildlife Resources, American Samoa Government, Pago Pago, AS, USA, [4]Hawaii Undersea Research Laboratory, University of Hawaii, Honolulu, HI, USA, [5]College of Oceanic and Atmospheric Sciences, Oregon State University, Corvallis, OR, USA, [6]College of Marine Science, University of South Florida, St. Petersburg, FL, USA, [7]Geospatial Consulting Group International, Alexandria, VA, USA, [8]Coastal Marine Resource Associates, Honolulu, HI, USA, [9]US Geological Survey, Alaska Science Center, Anchorage, AK, USA

Abstract

We present the geomorphology of the Eastern Samoa Volcanic Province, covering 28,446 km², and depths ranging from ~50 to 4,000 m. A new compilation of available multibeam data reveals 51 previously undocumented seamounts, and delineates major submarine rift zones, eruptive centers, and volcanic plateaus. Moving from a regional to local scale, and with regard to specific coral reef habitats, we report the results of three *Pisces V* submersible dives to the submerged flanks of Tutuila, with overall objectives of species identification of deep water fish and invertebrates (32 species of invertebrates and 91 species of fish identified, 9 new records), determining the base of extensive live bottom (i.e., coral cover of 20% and greater) as well as relations to any prior benthic terrain classifications at 100 m and deeper.

Key Words: bathymetry; geomorphology; submarine volcanism; seamounts; American Samoa; South Pacific; coral reefs; coral reef habitat

Seafloor Geomorphology as Benthic Habitat. DOI: 10.1016/B978-0-12-385140-6.00058-X

Introduction

The Samoan volcanic lineament in the southwest Pacific Ocean extends from the large subaerial islands of Savai'i and Upolu (independent nation of Samoa) in the west to the small island of Ta'u (American Samoa) in the east. Hart et al. [1] in addressing the longstanding debate as to whether or not the Samoan volcanic lineament is plume driven, along with the direction of the lineament's age progression, established a Western Samoa Volcanic Province (WSVP) and an Eastern Samoan Volcanic Province (ESVP).

With regard to geomorphology, we focus at the broad regional scale on the ESVP, which is comprised of the American Samoa islands of Tutuila, Aunu'u, Ofu, Olosega, and Ta'u, the large submarine volcanoes known as Vailulu'u, South Bank (renamed "Papatua" by Hart et al. [1]), 2% Bank (renamed "Tulaga" by Hart et al. [1]), Southeast Bank (renamed "Malumalu" by Hart et al. [1]), Malulu, Soso, and Tama'i; as well as Rose Atoll (aka Motu O Manu Atoll) and numerous smaller guyots and submarine seamounts (aka volcanic knolls; Figure 58.1). It should also be

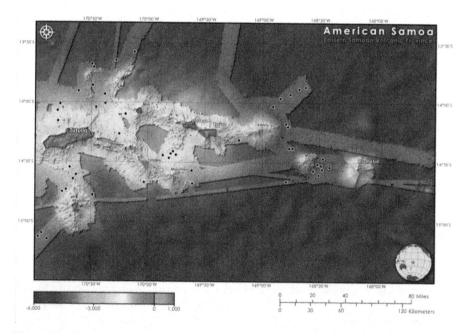

Figure 58.1 Bathymetry of American Samoa and the broader ESVP [2]. The resolution of the multibeam bathymetry is 200 m, but for visual continuity, a 1 km grid of bathymetry derived from the satellite altimetry of Sandwell and Smith [3] is used as a backdrop (this backdrop is present in Figures 58.2–58.4 as well). Filled circles indicate the distribution of 51 previously unidentified seamounts, several to be contributed to the Seamount Catalog [4]. Map projection for this and all other figures is Mercator, and map geodetic datum is WGS-84.

noted that Rose Atoll is a highly eroded edifice with stellate morphology, suggesting that it is substantially older than the other volcanoes of the American Samoa province [5]. It is therefore not considered a part of the ESVP and is perhaps the product of either volcanism in the Cook-Austral region or ancient ridge origin and plate transport. However, it is included in the discussion and map compilation described below by virtue of its location. In addition, Swains Atoll, within the unincorporated US territory of American Samoa, is located approximately 320 km to the north, far beyond the extent of the ESVP.

Seamounts, guyots, and knolls, with the fluid flow, nutrient supply, and modification to local circulation patterns they provide, are all extremely important habitats (e.g., corals, invertebrates, benthic fish, sea turtles, and sharks) and may include some of the richest biological "hotspots" in the oceans [6–8]. In the American Samoa region, the most well known and spectacular example thus far is Vailulu'u seamount at the eastern end of the Samoan archipelago, particularly with the hydrothermal vents discovered at its summit and rapidly growing Nafanua volcanic cone within the summit crater/caldera [9–11]. Indeed, seamounts have become priority habitats under the Convention on Biological Diversity [12], and in 2004 the Oslo–Paris Convention for the Protection of the Marine Environment of the Northeast Atlantic included seamounts on a list of priority habitats in need of protection, along with other deep-sea reefs, sponge aggregations, and hydrothermal vent sites on the Mid-Atlantic Ridge [13].

At a more local scale, and with regard to specific coral reef habitats, this project focuses on areas just offshore of the island of Tutuila, which is home to the Fagatele Bay National Marine Sanctuary (FBNMS). Geologically, the shallow flanks of Tutuila, and of the Samoan islands in general, are characterized by outcrops of basalt and limestone, biogenic and volcanic silt, sand and gravel, calcareous pavements, and calcareous ooze [14]. Many of the "cookie cutter" bays that are found along the southern coast of Tutuila, such as Fagatele Bay, Larsen's Bay, Pago Pago Harbor, and Faga'itua, are thought to be the result of volcanic collapse and erosion [15]. Then, as the island subsided due to crustal loading, large portions of these eroded valleys were flooded by the sea.

A particular emphasis is on understanding the mesophotic reefs in these areas, mesophotic meaning lower light levels at 30–150 m in the transition from euphotic to dysphotic. Knowledge of American Samoan mesophotic reefs is very limited, yet they include the deepest reefs—and those most untouched by humans in the archipelago—thus helping to delineate what unimpacted coral reefs are like in the territory [16–18]. Knowledge of what natural unimpacted reefs are like in the territory is very important for gauging the impacts humans have had on reefs at this location. Unfortunately, there are no good examples of shallow unimpacted reefs around Tutuila, where human impacts are greatest. Mesophotic reefs around Tutuila have the potential to be some of the least impacted reefs around Tutuila, and thus information on them is extremely valuable for determining the baseline, and the goal for reef management and conservation.

Regional Scale Geomorphic Features

Seamounts: Previous studies of the Samoan islands have reported regional bathymetry as predicted from satellite altimetry [1]. Multibeam bathymetry surveys have been restricted to local areas covered by one or two cruises [19,20] that primarily supported studies of the geochemical signature and age progression of western portion of the Samoan volcanic province 1,20–22. This case study introduces the first regional scale, multicruise, multibeam bathymetry of the eastern portion of the Samoan lineament, providing an overview of the bathymetric setting west of Savai'i and Upolu, and clearer implications of the seamount trails and other volcanic knolls revealed therein. Although we are using the IHO [23] definition of seamount in this book, it is useful to note that definitions of what constitute a "seamount" do vary, especially in light of a growing multidisciplinary seamount biogeosciences community where knolls are included in the definition (as discussed in more detail in Ref. [24]).

Figure 58.1 shows a new bathymetric compilation of American Samoa based on multibeam sonar data available from 14 cruises from 1984 through 2006, covering an area of 28,446 km^2. The map also shows the locations of 51 previously undocumented seamounts and volcanic knolls. Figure 58.2 summarizes the geomorphic interpretation for the region by delineating major rift zones, subaerial and submarine eruptive centers, volcanic plateaus, and the outlines of the most prominent seamounts and volcanic knolls. Particularly of note are the many small seamounts prevalent throughout ESVP from Ofu-Olosega westward, especially on the northern flank of South Bank and stretching from Ofu-Olosega across the inter-rift valley to the 2% Bank-Southeast Bank (aka Tulaga-Malumalu) saddle.

In the current global census of seamounts, only 200 have been sampled, and in no systematic fashion [7], but future studies based on this current study hold promise for exploring a possible relationship between seamount shape and habitat. Topographic/bathymetric position index (TPI/BPI) is important in a vertical sense (e.g., species richness along a vertical biodiversity gradient, as discussed in Refs. [25–27]). Further studies (beyond the scope of this case study) will test the hypothesis if seamount size and shape (in a more horizontal cross-sectional sense) bear any relationship to species diversity and richness, especially with geomorphic and hydrodynamic processes influencing marine ecological communities on a range of scales [28,29]. Below we focus on some of the largest seamounts, but the reader is referred to Roberts [30] for more extensive description and geomorphic analysis of undersea volcanic features throughout the entirety of the ESVP, as well as shape and distribution analyses of seamounts and a discussion of the age progressions of volcanic lineaments.

Tutuila: The large seamounts, guyots, knolls, and breaching islands of the ESVP demonstrate complex eruptive patterns. Perhaps the most intricate is the Tutuila complex, composed of five separate volcanic centers ([15]; Figure 58.2) and representing the largest structure in the ESVP, with a volume of 4,957 km^3 (as calculated using sources from Ref. [4]. Tutuila is unique with its highly elongate primary rift zone that trends 70° [15], as opposed to the rest of the Samoan chain trending at 110°. Indeed, en echelon lineaments both to its east and west delineate that primary rift zone trend (110°), with the island itself marking an interruption in that dominant rift

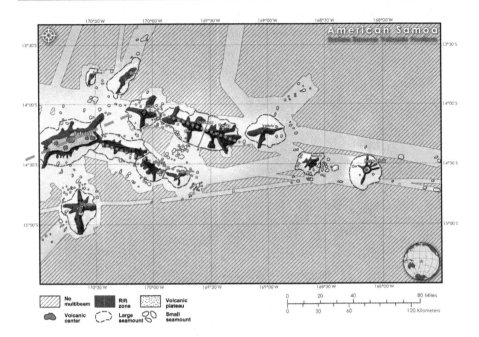

Figure 58.2 Geomorphic interpretation of major volcanic features of the islands of American Samoa and the broader ESVP, based on the multibeam bathymetry of Figure 58.1. Major rift zones are shown in black, subaerial and submarine eruptive centers in red, volcanic plateaus in stipple, outlines of the most prominent large seamounts in dashed circles, and small seamounts in small open circles. Cross-hatching indicates no multibeam bathymetry coverage. Large dashed line shows the northeasterly volcanic rift trend of Tutuila, which differs markedly from the southeasterly trends of the two major seamount chains directly to the east (shorter dashed lines). (For interpretation of the references to color in this figure legend, the reader is referred to the web version of this book.)

direction. Tutuila may in fact be an extension of the North Fiji Fracture Zone far to the east, which also trends 70° [31].

The morphology of Tutuila exhibits several highly incised secondary rift zones radiating away from the primary trend (Figure 58.3). Protruding slightly from the southwestern corner of the island is a rift oriented 20°. Reinstating the primary en echelon trend, a massive rift protrudes at 110° and connects to 2% Bank (aka Tulaga). A third rifting system extends from the northeastern corner of the island in a 30° trend. The linear nature of these features implies structural guidance of volcanism by fault or fracture zone. The island flanks are in a stage of advanced erosion, exhibiting numerous slope failures and incised rifts (Figure 58.3). Sparse populations of small seamounts occupy the western flank, as well as the northern and southern flanks, which are in line with the primary rift of South Bank.

South Bank: South Bank (aka Papatua Guyot) is the largest isolated edifice in the ESVP. Though it has not been radiometrically dated, it is probably at least as old as Tutuila, based on its location in the ESVP. The summit of South Bank sits very near

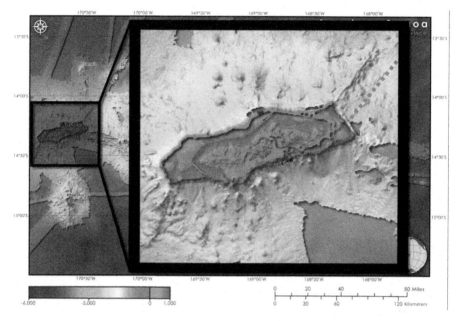

Figure 58.3 Detailed bathymetry of Tutuila Island, based on the 200 m multibeam bathymetry of Figure 58.1, as well as available nearshore multibeam bathymetry at 2–5 m resolution. The morphology of Tutuila exhibits several highly incised secondary rift zones radiating away from the primary trend, a primary one of which extends from the northeastern corner of the island in a 30° trend (dashed line). The linear nature of these features implies structural guidance of volcanism by fault or fracture zone.

to sea level. It likely breached in the past and has since been eroded by wave action to produce a flat summit surface. South Bank has two perpendicular rifting zones trending nearly in line with the four cardinal directions (Figure 58.4). Though it is probably at least 1 million years old [1], its northern and southwestern flanks show relatively little evidence of slope failure and are superimposed with small seamounts. It possesses an emerging stellate morphology, though it is not nearly as developed as on Tutuila or Northeast Bank. The shield-building stage for South Bank is not easily attributable to a plume source, based on its divergent location and anomalous primary rift trending at N0°E, though Hart et al. [1] suggest that decompressional melting due to slab–plume interactions could account for the location of South Bank.

Northeast Bank: Northeast Bank (renamed "Muli" by Hart et al. [1]) is the second largest isolated edifice in the ESVP. It is partially connected to the Ofu-Olosega complex by a deep saddle and exhibits a near-stellate morphology, a testament to its highly eroded state and once circular shape (Figure 58.5). Its flat summit lies within 100 m of sea level and therefore we speculate that, like South Bank, it may have breached sea level at some point in the past. Northeast Bank has two primary rift zones trending at 30° and 120°. Its flanks are smooth and largely vacant of small

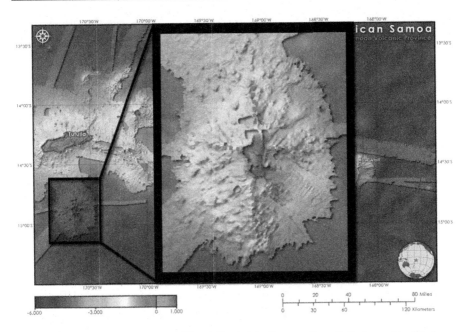

Figure 58.4 Detailed bathymetry of South Bank (aka Papatua Guyot), the largest isolated edifice in the ESVP. South Bank has two perpendicular rifting trends nearly in line with the four cardinal directions (dashed lines).

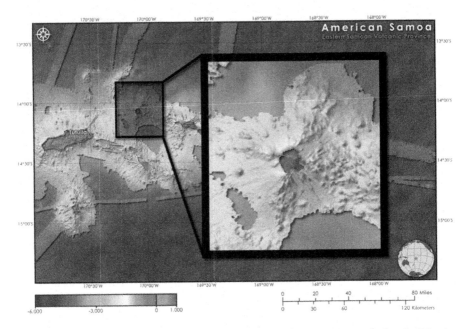

Figure 58.5 Detailed bathymetry of the Northeast Bank, the second largest isolated edifice in the ESVP. Northeast Bank has two primary rift zones trending at 30° and 120°.

seamounts. The exception is its eastern rift forming the saddle with the Ofu-Olosega complex, which is interspersed with small seamounts.

Local Scale Geomorphic Features and Habitats

Reefs: In terms of biological surveys in the study area, the NOAA Coral Reef Ecosystem Division (CRED) of the Pacific Islands Fisheries Science Center (PIFSC) surveyed the flanks of Tutuila and the Manu'a Islands in 2004 for management of benthic habitats associated with coral reefs [32]. The following year, the Hawaii Undersea Research Laboratory (HURL) initiated another survey of the Tutuila near-shore, this time accompanied by submersible dives again aimed at documenting the characteristics of the benthic habitat [33]. HURL extended its benthic habitat survey-ing to Rose (Motu O Manu) Atoll as well [34].

Towed camera and scuba surveys in the area are ongoing, but deeper submers-ible surveys into the mesophotic, dysphotic, and aphotic zones are rarer. As such, we provide examples here from HURL cruise KOK0510 (Figure 58.6; [36]). The cruise consisted of three *Pisces V* submersible dives to the submerged flanks of Tutuila, American Samoa, specifically the coral reef platform of Taema Bank, and the submerged caldera forming Fagatele Bay and Canyon, with overall objectives

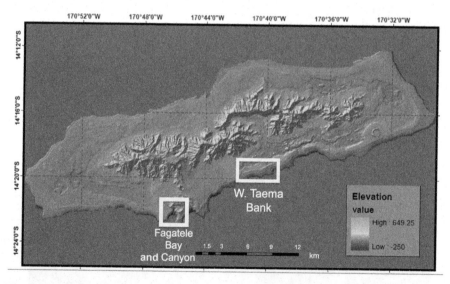

Figure 58.6 Coastal terrain model of Tutuila, American Samoa with surrounding bathymetry, after Hogrefe et al. [35]. Rectangles show the locations of the two dive sites of Cruise KOK0510: West Taema Bank offshore of south-central Tutuila (Dives P5–648 and P5–640) and Fagatele Bay and Canyon to the southwest (Dive P5–649). The coastal resolution coastal terrain model was developed at 5 m resolution from a USGS digital elevation model, the multibeam bathymetry of NOAA PIFMC, and nearshore bathymetry derived from IKONOS 4 m satellite imagery [35].

of species identification of fish and invertebrates (32 species of invertebrates and 91 species of fish identified, 9 new records) and determining the base of extensive live bottom (i.e., coral cover of 20% and greater). In addition and where possible, we sought to ground-truth previous benthic terrain classifications at 100 m and deeper that had been derived from bathymetric position index and rugosity analyses in GIS. Lundblad et al. [37] describe these methods in complete detail and the resulting classifications specifically for American Samoa.

Taema Bank and Fagatele Bay and Canyon (Figure 58.6) were chosen as primary dive sites due to the occurrence of previous shallow (≤150 m) multibeam surveys in the area (especially by Oregon State University, OSU, and University of South Florida, USF, as described in Refs. [38–40]), their importance for coral monitoring and protection, and for safety. Indeed, at both sites the water is suitably deep for the safe navigation of a 68 m long research vessel needing to track a submersible almost directly below it for shallow dives of 500 m or less.

Taema Bank: Taema Bank (Figures 58.6 and 58.7) is a long, narrow, submarine platform located ~3 km off the south central coast of Tutuila. It is ~3 km long by 30 m wide, rising ~30 m above a surrounding seafloor, averaging 50–100 m in depth [38]. Because the platform is largely flat and fairly smooth, it is interpreted as an ancient reef terrace that may have once experienced wave erosion at sea level.

Figure 58.7 shows BPI "zone" and "structure" maps for West Taema Bank, created from 1 m resolution multibeam bathymetry data of Lundblad et al. [37]. BPI is a scale-dependent index representing a grid cell's location within a seascape relative to its local surroundings. Lundblad et al. [37] defines a "zone" as a coarse-scale surficial characteristic of the seafloor that combines slope with a coarse-scale BPI, to delineate large crests or ridges, valleys, basins, plains, and slopes. "Structures" are finer scale classifications resulting from the combination of bathymetry and slope, with both coarse- and fine scales of BPI. Therefore, structures include categories

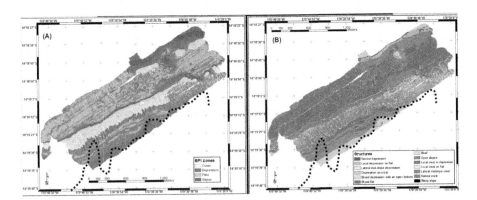

Figure 58.7 (A) BPI "zone" map of West Taema Bank, created from 1 m resolution multibeam bathymetry data, with classifications based on the scheme of Lundblad et al. [37]. Dashed line shows the smoothed trackline of *Pisces V* Dive P5–648. (B) Classification of same West Taema Bank bathymetry into "structures," with same submersible track overlain.

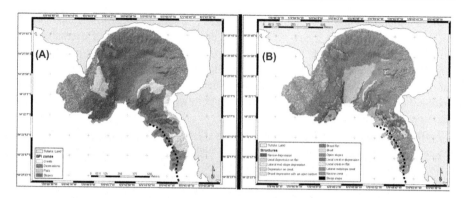

Figure 58.8 (A) BPI "zone" map of FBNMS, created from 1 m resolution multibeam bathymetry data, with classifications based on the scheme of Lundblad et al. [37]. Dashed line shows smoothed track of *Pisces V* Dive P5–649. (B) Classification of same FBNMS bathymetry into "structures," with same submersible track overlain.

such as narrow depressions or grooves, narrows crests or ridges, local depressions/ crests on plains, lateral mid-slope depressions or grooves, lateral mid-slope crests or ridges, open slopes, shelves, and broad flats [37]. Zone and structure classifications of Taema Bank (Figure 58.7) are based on the scheme of Lundblad et al. [37] for American Samoa.

Taema Bank is made up of mostly flats and slopes in terms of zones, but there are very distinct crests and depressions throughout (Figure 58.7A). Open slopes surround the crests of the bank along with the broad flats. As the shelf reaches an escarpment near a series of broad flats, there is also a series of spurs and grooves. These pervasive features are given structure classifications of narrow crests, lateral mid-slope depressions, and lateral mid-slope crests (Figure 58.7B). The open slopes lead down to broad depressions with open bottoms on both sides of Taema Bank. Submersible observations on Dives P5–648 and P5–650 visually confirmed these classifications and noted that the shelf contains stretches of colonized pavement covered with a veneer of sand [41]. Within the fringing lateral mid-slope depressions and crests on the open slopes are colonized pavement and hummocky bottom with low relief and ~5% sand cover.

Fagatele Bay and Canyon: Fagatele Bay (Figures 58.6 and 58.8), and its continuation deeper offshore as a canyon, is the result of an ancient caldera that collapsed and subsided, causing the seaward rim to be breached by the ocean and flooded [38]. The fringing coral reef is indeed of continuing interest and concern in this US federal marine sanctuary that has essentially recovered from a near-devastating infestation of crown-of-thorns starfish in the late 1970s [42]. The bay was also affected by hurricanes in 1990, 1991, 2004, and 2005; and a coral bleaching event occurred in 1994, possibly due to high sea-surface temperatures from an El Niño [43]. The live coral cover has recovered well from near-total destruction, and populations of small benthic fish still thrive, particularly surgeonfish, damselfish, and angelfish [44,45].

The submerged caldera that is Fagatele Bay dramatically slopes downward into a canyon (canyon has not yet been classified with regard to habitat). At around 20 m open slopes dominate, which further descend toward the broad depressions with open bottoms (Figure 58.8A). Open slopes above and on the edge of the narrow depressions in the center of the bay suggest prior seafloor subsidence, resulting also in a noteworthy ridge at the east center of the bay (~14°22″1′S, 170°45″52′W; Figure 58.8B). It is classified as a narrow crest with fringing lateral mid-slope crests and depressions. The edge of the narrow crest deepens so dramatically in some places that there is a narrow strand of steep slopes around it. Steep slopes are also seen at the edge of other narrow crests and lateral mid-slope features throughout the bay. The area that appears most complex, containing a diverse combination of BPI zones, as well as high to medium-high rugosity, is in the southeast portion of the bay, which extends to depths safe enough for submersible observations. Submersible observations on Dive P5–649 visually confirmed the presence of narrow mid-slope depressions and lateral mid-slope crests, while noting also the presence of several small box canyons cut into the southeast wall that were not detected in the original terrain classification. Of note also is a transition at ~185–209 m depth from old carbonate reef to a basalt layer, and another carbonate layer before transitioning to sediment at ~235 m.

Biological Communities

Pisces V submersible Dive P5–648 consisted of a video and photographic survey up the southwest wall of Taema Bank, noting 36 m as the depth at which the main corals extend to (base of main reef on bank) on a fairly consistent basis. It then proceeded to a deeper, safer contour of interest for sub/ships operations (down to 110 m, below significant surface wave surge), following it to the east along the south side of the bank, making observations of biota and physical structure.

The dive followed the 110 m depth contour for ~7 km in the broad depression/open bottom habitat class, and noted a significant assemblages of gorgonian corals (*Iciligorgia*), sea fans (e.g., *Annella reticulata*, *Melithaea*), whip corals (*Cirrhipathes*), and sea cucumbers (*Holothuria edulus*, *Thelonota anax*). A transition was noted from west to east of sea fans in the east having crinoids attached to their tops, and with three-armed, feathery brittlestars. There were also alternating "provinces" of barren, sloping calcareous (*Halimeda*) algae, sand plains, to slopes cut by deep crevices in calcareous conglomerate blocks to sea fans assemblages. In this same habitat class, various species of groundfish (e.g., greeneye or *Chlorophthalmus priridens*, orange sea toad or *Chaunax fimbriatus*, and the black-botched stingray or *Taeniura meyeni*; Figure 58.9) congregated in high rugosity, carbonate rubble piles, which may have been created by the fish as habitat.

Dive P5–649 consisted of a video and photographic survey around the edge of Fagatele Bay and Canyon, starting from the southwest corner and reaching the far southeast portion of the national marine sanctuary. Most of the dive scaled both southwest and southeast walls of the bay and canyon; hence, the predominant habitat classes were narrow (vertical) depression and lateral mid-slope crest. Main species observed included the harlequin grouper (*Cephalopholis*; Figure 58.9), the bigeye (*Heteropriacanthus cruentatus*), sea fans (*Annella reticulata*), gorgonian corals (*Iciligorgia*), and "doughboy"

Figure 58.9 Photographs of new records for American Samoa as observed on Dives P5–648 and P5–649 from the *Pisces V* submersible, to Taema Bank, and Fagatele Bay and Canyon: (A) black-blotched stingray (*Taeniura meyeni*), ~2 m wide, 110 m depth on calcareous (*Halimeda algae*) sand and high-rugosity carbonate rubble, broad depression/open bottom habitat class, Taema Bank; (B) underside of "lounge cushion" sea star, recovered from 57 m depth, broad flat habitat class, Taema Bank—returned to ocean; (C) harlequin grouper (*Cephalopholis polleni*), 93 m depth, west wall of Fagatele Canyon, lateral mid-slope crest habitat class; (D) doughboy sea star (*Choriaster granulatus*) with orange and purple sea fans (*Annella reticulata, Melithaea*), 90 m depth, west wall of Fagatele Canyon, lateral mid-slope crest habitat class; (E) batfish (*Ogcocephalidae*), 247 m depth on calcareous sand, east Fagatele Canyon floor, broad depression/open bottom habitat class; (E) base of west wall of Fagatele Canyon, 230 m, showing clear contact between basalt flow overlying carbonate province. http://dusk.geo.orst.edu/djl/samoa/hurl/

sea stars (*Choriaster granulatus*). At the floor of the canyon, in the broad depression/ open bottom habitat class, a notable discovery was the batfish (*Ogcocephalidae*) and the black-blotched stingray (Figure 58.9), as well as sightings of the deep water grouper (*Epinephelus timorensis*) among many other fish species.

Dive P5–650 returned to Taema Bank to investigate the full extent of the sheer carbonate wall encountered on P5–648, and hence was below the depth of prior benthic terrain classifications. Upon finding the base of the wall at 440 m, a video and photographic survey proceeded north from that point along that contour, where a preponderance of galatheid and hermit crabs, urchins, shrimp, a soft corals, sea cucumbers, and small stars were noted. Farther up the bank at ~115 m (broad flat habitat class), a large province of foraminifers in calcareous sands was noted: genus *Cycloclypea*, the largest in the world.

All three dives were extremely successful, with a cumulative bottom time of 18 h and identification at both sites of 32 species of invertebrates and 91 species of fish, at least nine of which are "new records" for American Samoa (Figure 58.9). The base of extensive live bottom for Taema Bank (coral cover of 20% and greater) was identified at a depth of 36 m. Alternating sections of carbonate reef and basalt were observed at ~185–220 m depth along both the east and west walls of Fagatele Canyon, and large, grooved, mass-wasting scarps were noted at ~300–400 m depth near the base of the south central wall of Taema Bank. No evidence of eutrophication or slurry from Pago Pago harbor was seen on the south side of Taema Bank. Complete species lists, dive track maps, and cruise report are available in Wright [41].

Surrogacy

At this time, no statistical analyses have been carried out on these data sets to examine relationships between physical surrogates and benthos. We await additional video and photographic surveys.

Acknowledgments

We thank captains, crew, and scientific personnel aboard the several ships who have been responsible for obtaining bathymetric data throughout the Eastern Samoa region.

Special thanks to Nancy Daschbach, former manager of the Fagatele Bay National Marine Sanctuary, for providing travel funding over the years via the NOAA National Marine Sanctuary Program. Thanks also to Joyce Miller, Emily Lundblad Hirsch, and Scott Ferguson, NOAA Coral Reef Ecosystem Division/Pacific Islands Benthic Habitat Mapping Center/ Pacific Fisheries Science Center, as well as to Peter Craig of the National Park of American Samoa for helpful discussions. The manuscript benefited greatly from the comments of an anonymous reviewer and Robin Beaman of James Cook University. This work was supported in part by the US NOAA-National Undersea Research Program Hawai'i Undersea Research Laboratory and the US National Science Foundation.

References

[1] S.R. Hart, M. Coetzee, R.K. Workman, J. Blusztajn, K.T.M. Johnson, J.M. Sinton, et al.,
     Genesis of the Western Samoa seamount province: age, geochemical fingerprint and tec-
     tonics, Earth. Planet. Sci. Lett. 227 (2004) 37–56.

[2] D.J. Wright, Fagatele Bay National Marine Sanctuary GIS Data Archive. http://dusk.geo.
     orst.edu/djl/samoa, 2010 (last accessed 10 August 2010).

[3] D.T. Sandwell, W.H.F. Smith, Marine gravity anomoly from Geosat and ERS 1 satellite
     altimetry, J. Geophys. Res. 102 (B5) (1997) 10,039–10,054.

[4] A.A.P. Koppers, Seamount Catalog, Seamount Biogeosciences Network, http://earthref.
     org/cgi-bin/er.cgi?s = sc-s0-main.cgi, 2010 (last accessed 10 August 2010).

[5] N.C. Mitchell, Transition from circular to stellate forms of submarine volcanoes,
     J. Geophys. Res. 106 (B2) (2001) 1987–2003.

[6] H. Staudigel, A.A.P. Koppers, J.W. Lavelle, T.J. Pitcher, T.M. Shank, Mountains in the
     sea, Oceanography 23 (1) (2010) 18–19.

[7] P. Etnoyer, J. Wood, T.C. Shirley, Box 12: How large is the seamount biome?,
     Oceanography 23 (1) (2010) 206–209.

[8] T.J. Pitcher, T. Morato, K.I. Stocks, M.R. Clark, Box 6: Seamount Ecosystem
     Evaluation Framework (SEEF): A tool for global seamount research and data synthesis,
     Oceanography 23 (1) (2010) 123–125.

[9] H. Staudigel, S.R. Hart, A.A.P. Koppers, C. Constable, R. Workman, M. Kurz, et al.,
     Hydrothermal venting at Vailulu'u Seamount: the smoking end of the Samoan chain,
     Geochem. Geophys. Geosyst. 5 (2004) Q02003, 10.1029/2003GC000626.

[10] H. Staudigel, S.R. Hart, A. Pile, B.E. Bailey, E.T. Baker, S. Brooke, et al., Vailulu'u sea-
     mount, Samoa: life and death on an active submarine volcano, Proc. Natl. Acad. Sci.
     U. S. A. 103 (17) (2006) 6448–6453.

[11] A.A.P. Koppers, H. Staudigel, S.R. Hart, C. Young, J.G. Konter, Spotlight 8: Vailulu'u
     seamount, 14°12.96'S, 169°03.46'W, Oceanography 23 (1) (2010) 164–165.

[12] Convention on Biological Diversity, Azores Scientific Criteria and Guidance for Identifying
     Ecologically or Biologically Significant Marine Areas and Designing Representative
     Networks of Marine Protected Areas in Open Ocean Waters and Deep Sea Habitats.
     Secretariat of the Convention on Biological Diversity, Montréal, Canada, 2009, p. 12.

[13] T. Morato, T.J. Pitcher, M.R. Clark, G. Menezes, F. Tempera, F. Porteiro, et al., Box 12:
     Can we protect seamounts for research? A call for conservation, Oceanography 23 (1)
     (2010) 190–199.

[14] N.F. Exon, Offshore sediments, phosphorite and manganese nodules in the Samoan
     region, Southwest Pacific, South Pacific Marine Geological Notes, SOPAC Technical
     Report, Suva, Fiji (1982) 103–120.

[15] H.T. Stearns, Geology of the Samoan Islands, Geol. Soc. Am. Bull. 55 (1944)
     1279–1332.

[16] D. Fenner, The state of the coral reef habitat in American Samoa: Proceedings of the
     NOAA Reef Fisheries Workshop, NOAA Coral Reef Task Force, Pago Pago, American
     Samoa, 2008, 74 pp.

[17] D. Fenner, M. Speicher, S. Gulick, G. Aeby, S.C. Aletto, P. Anderson, et al., The state of
     coral reefs ecosystems of American Samoa, in: J.E. Waddell, A.M. Clarke, (Eds.), The State
     of Coral Reef Ecosystems of the United States and Pacific Freely Associated States: 2008,
     NOAA Technical Memorandum NOS NCCOS 73, NOAA/NCCOS Center for Coastal
     Monitoring and Assessment's Biogeography Team, Silver Spring, MD, 2008, pp 307–351.

[18] D. Fenner, Annual Report for 2008 of the Territorial Coral Reef Monitoring Program
      for American Samoa, Benthic Section. Department of Marine and Wildlife Resources,
      American Samoa, 2009.
[19] S.R. Hart, H. Staudigel, A.A.P. Koppers, J. Blusztajn, E.T. Baker, R. Workman, et al.,
      Vailulu'u undersea volcano: the New Samoa, Geochem. Geophys. Geosys. 1 (12) (2000).
[20] A.A.P. Koppers, J.A. Russell, M.G. Jackson, J. Konter, H. Staudigel, S.R. Hart, Samoa
      reinstated as a primary hotspot trail, Geology 36 (6) (2008) 435–438.
[21] M.G. Jackson, S.R. Hart, A.A.P. Koppers, H. Staudigel, J. Konter, J. Blusztajn, et al.,
      The return of subducted continental crust in Samoan lavas, Nature 448 (2007) 684–687.
[22] R.K. Workman, J.M. Eller, S.R. Hart, M.G. Jackson, Oxygen isotopes in Samoan lavas:
      confirmation of continent recycling, Geology 36 (7) (2008) 551–554.
[23] International Hydrographic Organisation, Standardization of Undersea Feature Names:
      Guidelines Proposal form Terminology. International Hydrographic Organisation and
      International Oceanographic Commission, Monaco, 2001.
[24] H. Staudigel, A.A.P. Koppers, J.W. Lavelle, T.J. Pitcher, T.M. Shank, Box 1: Defining
      the word "seamount", Oceanography 23 (1) (2010) 20–21.
[25] A. Guisan, S.B. Weiss, A.D. Weiss, GLM versus CCA spatial modeling of plant species
      distribution, Plant Ecol. 143 (1999) 107–122.
[26] P. Iampietro, R. Kvitek, Quantitative seafloor habitat classification using GIS terrain
      analysis: effects of data density, resolution, and scale: Proceedings of the 22nd Annual
      ESRI User Conference, ESRI, San Diego, CA, 2002.
[27] E. Lundblad, D.J. Wright, J. Miller, E.M. Larkin, R. Rinehart, T. Battista, et al.,
      Classifying benthic terrains with multibeam bathymetry, bathymetric position and rugos-
      ity: Tutuila, American Samoa, Mar. Geod. 29 (2) (2006) 89–111.
[28] S.A. Levin, The problem of pattern and scale in ecology, Ecology 73 (1992) 1943–1967.
[29] R.H. Karlson, H.V. Cornell, T.P. Hughes, Coral communities are regionally enriched
      along an oceanic biodiversity gradient, Nature 429 (2004) 867–870.
[30] J. Roberts, The Marine Geomorphology of American Samoa: Shapes and Distributions
      of Deep Sea Volcanics. M.S. Thesis, Oregon State University, Corvallis, OR, http://
      marinecoastalgis.net/jed07, 2007 (last accessed 10 August 2010).
[31] G.P.L. Walker, P.R. Eyre, Dike complexes in American Samoa, J. Volcanol. Geotherm.
      Res. 69 (1995) 241–254.
[32] Pacific Islands Benthic Habitat Mapping Center, Benthic Habitat Mapping. http://www.
      soest.hawaii.edu/pibhmc/pibhmc_mapping.html, 2010 (last accessed 10 August 2010).
[33] D.J. Wright, Report of HURL Cruise KOK0510: Submersible Dives and Multibeam
      Mapping to Investigate Benthic Habitats of Tutuila, American Samoa. Unpublished
      Technical Report, NOAA's Office of Undersea Research Submersible Science Program,
      Hawai'i Undersea Research Lab, Honolulu, HI, 2005.
[34] J.R. Smith, R.B. Dunbar, F.A. Parrish, First multibeam mapping of deep-water habitats
      in the U.S. Line Islands, EOS. Trans. Am. Geophys. Union 86 (36) (2006).
[35] K.R. Hogrefe, D.J. Wright, E.J. Hochberg, Derivation and integration of shallow-water
      bathymetry: implications for coastal terrain modeling and subsequent analyses, Mar.
      Geod. 31 (4) (2008) 299–317.
[36] D.J. Wright, E.R. Lundblad, D. Fenner, L. Whaylen, J.R. Smith, Initial results of sub-
      mersible dives and multibeam mapping to investigate benthic habitats of Tutuila,
      American Samoa, EOS. Trans. Am. Geophys. Union 86 (36) (2006).
[37] E. Lundblad, D.J. Wright, J. Miller, E.M. Larkin, R. Rinehart, T. Battista, et al., A benthic
      terrain classification scheme for American Samoa, Mar. Geod. 29 (2) (2006) 89–111.

[38] D.J. Wright, B.T. Donahue, D.F. Naar, Seafloor mapping and GIS coordination at America's remotest national marine sanctuary (American Samoa), in: D.J. Wright, (Ed.), Undersea with GIS, ESRI Press, Redlands, CA, 2002, pp. 33–63.

[39] NOAA Coral Reef Ecosystem Division, Benthic Habitat Mapping and Characterization. http://crei.pifsc.noaa.gov/hmapping, 2004 (last accessed 10 August 2010).

[40] R. Brainard, J. Asher, J. Grove, J. Helyer, J. Kenyon, F. Mancini, J. Miller, et al., Coral Reef Ecosystem Monitoring Report for American Samoa: 2002–2006. NOAA Pacific Islands Fisheries Science Center, Coral Reef Ecosystem Division, Honolulu, HI, 2008, 495 pp.

[41] D.J. Wright, Cruise KOK0510 —R/V Ka'imikai-o-Kanaloa. http://dusk.geo.orst.edu/djl/samoa/hurl, 2005 (last accessed 10 August 2010).

[42] C.E. Birkeland, R.H. Randall, R.C. Wass, B. Smith, S. Wilkins, Biological Assessment of the Fagatele Bay National Marine Sanctuary. NOAA Technical Memorandum, Fagatele Bay National Marine Sanctuary, Pago Pago, American Samoa, 1987, 232 pp.

[43] A.L. Green, C.E. Birkeland, R.H. Randall, Twenty years of disturbance and change in Fagatele Bay National Marine Sanctuary, American Samoa, Pac. Sci. 53 (4) (1999) 376–400.

[44] P. Craig, Temporal spawning patterns for several surgeonfishes and wrasses in American Samoa, Pac. Sci. 52 (1998) 35–39.

[45] R. Pyle, B.P. Bishop Museum Exploration and Discovery: The Coral-Reef Twilight Zone, Fagatele Bay National Marine Sanctuary, 14–18 May 2001. Bernice Pauahi Bishop Museum, Honolulu, HI, http://www2.bishopmuseum.org/PBS/samoatz01, 2001 (last accessed 10 August 2010).

59 Mapping Condor Seamount Seafloor Environment and Associated Biological Assemblages (Azores, NE Atlantic)

Fernando Tempera[1], Eva Giacomello[1], Neil C. Mitchell[2], Aldino S. Campos[3], Andreia Braga Henriques[1], Igor Bashmachnikov[4], Ana Martins[1], Ana Mendonça[1], Telmo Morato[1], Ana Colaço[1], Filipe M. Porteiro[1], Diana Catarino[1], João Gonçalves[1], Mário R. Pinho[1], Eduardo J. Isidro[1], Ricardo S. Santos[1], Gui Menezes[1]

[1]Departamento de Oceanografia e Pescas, Universidade dos Açores, Rua Prof. Dr. Frederico Machado, 4, Horta, Açores, Portugal, [2]School of Earth, Atmospheric and Environmental Sciences, University of Manchester, Manchester, UK, [3]Portuguese Task Group for the Extension of the Continental Shelf (EMEPC), Rua Costa Pinto, Paço de Arcos, Portugal, [4]Instituto de Oceanografia, Faculdade de Ciências da Universidade de Lisboa, Campo Grande, Lisboa, Portugal

Abstract

Condor seamount is a linear volcano located in the Azores (northeast Atlantic), 35 km in length, 2–6 km wide, and of varied seafloor morphology. A scientific observatory devoted to research on seamount ecosystem structure and functioning has been established on Condor, secured by a temporary fishing closure. Multiple projects have contributed to this observatory by targeting the seamount with snapshots and long-term deployments of moored, satellite-based, and shipborne technologies. This chapter presents a brief characterization of the seamount's seafloor environment by focusing on the multibeam bathymetry data and a series of video, oceanographic, and fishery surveys. A classification based upon the bathymetric position index is presented to characterize the landscape composition of the seamount. Habitats of conservation importance,

Seafloor Geomorphology as Benthic Habitat. DOI: 10.1016/B978-0-12-385140-6.00059-1

such as coral gardens and deep-sea sponge aggregations, are documented. A qualitative zonation of the benthic assemblages based on the video surveys is presented along with dominant fish and crustacean catch data for comparable depth strata.

Key Words: Condor seamount, scientific observatory, Azores, seamount geomorphology, landscape classification, habitat mapping, benthic zonation

Introduction

Seamounts are among the most common topographic features in the world ocean. Depending on their particular morphological traits, they can also be referred to as banks, knolls, guyots, mounds, or hills [1]. The total global area of the seamount biome has recently been estimated as 28.8 million km^2 [2]. It is known that these seafloor elevations are hotspots for pelagic biodiversity [3] and hold fishery resources of increasing value [4] as well as fragile habitats [5] that have been suffering impacts from destructive fishing practices like trawling [6]. Understanding how deep-sea habitat-building species like corals and sponges distribute at fine scales over the complex topography of individual seamounts is therefore critical information to design usage zonation schemes.

Condor is a ridge seamount located about 17 km to the WSW of Faial Island in the archipelago of the Azores (Portugal, northeast Atlantic) (Figure 59.1A). For decades, it has been targeted by local artisanal fisheries, mostly using bottom-tending longlines and handlines. However, the lack of bottom trawling has maintained the benthic environment of Condor in a relatively good state, with exploratory visual surveys revealing well-preserved deep-sea biotopes of conservation importance such as coral gardens and deep-sea sponge aggregations within the range of fisheries activities [8,9].

Taking advantage of the accessibility of this site for research, as well as its fair conservation status [10], local fishermen associations, regional administrative bodies, and the University of the Azores have recently agreed a 2-year moratorium on fishing over this seamount (Figure 59.1B). This temporary no-fishing area has allowed the secure installation of a scientific observatory devoted to improving our understanding of seamount ecological structure and functioning from the sea surface down to the seafloor.

Projects CONDOR, CORALFISH, CORAZON, and HERMIONE all have Condor seamount as a focus research site and conducted both sporadic and continuous observations there. A suite of remote sensing technology (e.g., multibeam sonar, EK500 sonar, Acoustic Doppler Current Profiler, satellite imagery), as well as *in situ* sensors (such as CTDs, i.e., Conductivity Temperature Depth profilers), have been used. Several deployments have aimed at providing continuous multiannual time series such as the Ecological Acoustic Recorders, biotelemetry listening posts, and current-meter arrays. Collections of physical samples have been conducted using a water-sampling rosette, grabs, midwater trawls, traps, and longlines. Two Remotely-Operated Vehicles (ROVs) have been used to collect video imagery and voucher specimens.

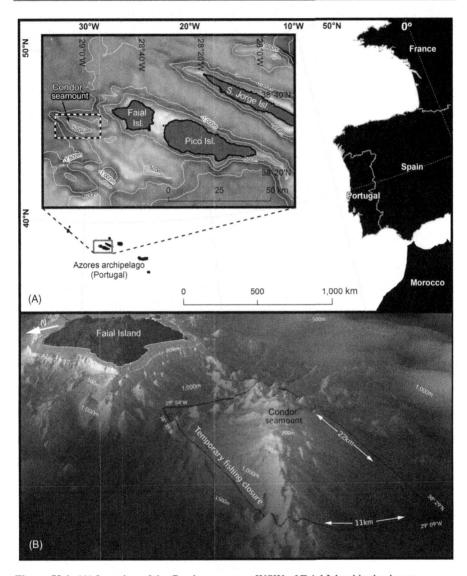

Figure 59.1 (A) Location of the Condor seamount WSW of Faial Island in the Azores archipelago (Portugal, NE Atlantic). (B) Perspective of the Condor seamount relief showing the limits of the temporary no-fishing area allowing the installation of a scientific observatory. *Source*: Graphics: F. Tempera ©*Imag*DOP. Bathymetry data credits: EMEPC, DOP-UAz, Project STRIPAREA/J.Luís/UAlg-CIMA, Ref. [7].

Bathymetry and backscatter data, along with ground-truthing information provided by video and biological surveys, were gathered to produce a qualitative zonation of the Condor seamount benthic assemblages.

Surveys

Multibeam bathymetry and backscatter data were collected from multiple surveys conducted on the seamount and its vicinity. The main surveys were conducted in 2008 and 2010, using a 12-kHz Kongsberg-Simrad EM120 multibeam sonar. Both types of data were acquired using the Seafloor Information System (SIS) package, and all procedures followed the International Hydrographic Organization standards for Survey Order 2. The data were processed using CARIS HIPS, based on the Combined Uncertainty and Bathymetric Estimator (CUBE) algorithm, and binned at a 50-m resolution. A complementary grid for the seamount vicinity was derived from the 2006 STRIPAREA data set acquired using a 30-kHz Simrad EM300 swath bathymetry sonar. The two grids were combined in ArcGIS. Acoustic backscatter mosaics also produced in CARIS were used along with the general geomorphological interpretation and seafloor imagery to infer seabed composition.

Seafloor imagery was collated from multiple surveys that in total visited 25 different locations on the seamount. The multiple platforms used included (i) a customized 2,000-m rated drop-down camera (2006), (ii) the 6,000-m rated ROV *Luso* (2008, 2010), and (iii) the 300-m rated ROV *SP* (2009, 2010). The footage collated was visualized to identify the depth strata and substrate types occupied by assemblages of conspicuous habitat-building species.

Three oceanographic cruises were carried out, including CTD profiling of conductivity, temperature, oxygen, and turbidity along the water column. Four current-meter moorings were deployed over the seamount in 2009 and 2010 (Figure 59.2A).

Longline and shrimp-trap sets were conducted in 2009 and 2010 on the seamount to characterize fish and shrimp assemblages, respectively (Figure 59.2A). Nine longline sets corresponding to 31,790 hooks were conducted along four radial transects that covered from the seamount summit to 1,200 m depth. Additionally, four sets using a sturdier type of longline and totaling 4,292 hooks were conducted roughly parallel to bathymetry between 1,150 and 1,350 m depth. Finally, seven shrimp-trap sets using a total of 883 traps were conducted on different sectors of the seamount between 185 and 1,000 m depth.

Geomorphic Features and Habitats

Geomorphology: The multibeam bathymetry resolved Condor's morphology in unprecedented detail (Figure 59.2). The seamount is revealed to be an elongated ridge with a major semi-axis of 35 km and a minor semi-axis of 20 km at the base. Its outer edge, delineated using the farthest convex bathymetric contours that radiated from the main ridge, are roughly outlined by the 1,700-m depth contour along the edge farthest from Faial Island. Its shoreward edge adjoins the Faial Island slope at depths between 1,500 and 1,000 m. A planar area of 431.61 km^2 is occupied by the seamount. It has an estimated surface area of 446.87 km^2 and a surface to area ratio of ~1.04.

The WNW-ESE orientation of the seamount's major axis is parallel with the main volcanic ridges in the area south of the Terceira rift, including the Faial–Pico complex [7]. Its morphology most likely resulted from the accumulation of lavas

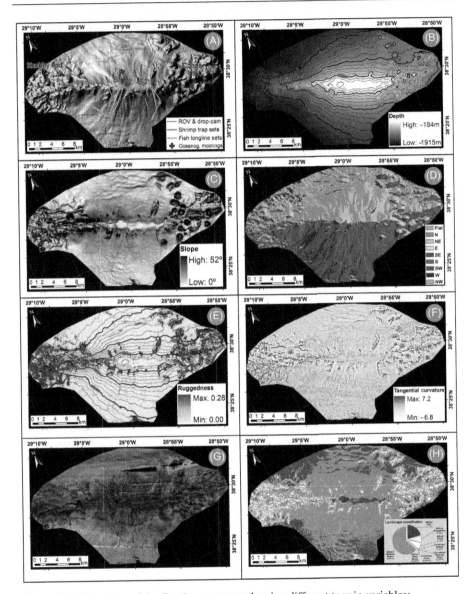

Figure 59.2 Plan view of the Condor seamount showing different terrain variables: (A) shaded-relief view showing ground-truthed locations and labels of geomorphological sectors highlighted in the description; (B) depth; (C) slope; (D) aspect; (E) ruggedness; (F) tangential curvature; (G) backscatter mosaic; and (H) landscape classification.

along a linear volcano, i.e., one produced by multiple superimposed cones or ridges that erupted from volcanic dykes oriented along its main axis. The orientations of dykes may be controlled by the oblique opening (relative to the plate boundary) of the European and African plates here [11,12].

Terrain variables and landscape classification: The seamount's extent was used as a mask for the subsequent extraction of terrain parameters from the 50-m resolution bathymetry grid. Slope, aspect, and curvature were derived using the ArcGIS Spatial Analyst extension (Figure 59.2B–D). Ruggedness (Figure 59.2E) was calculated using Sappington et al.'s Vector Ruggedness Measure [13] and the respective ArcGIS 9.2 script (available at http://arcscripts.esri.com/details.asp?dbid=15423).

A landscape classification was derived using the benthic terrain modeler tool for ArcGIS [14]. A broadscale neighborhood range of 450 m was applied along with a customized dictionary that integrated classical bionomical depth zones, slope classes, and topographical categories (Table 59.1). Bathyal moderate slopes (i.e., open inclined ground slanting between 5° and 45°) were the predominant class, composing 63.5% of the seamount landscape (Figure 59.2H).

Seamount summit: The seamount has a flat and broad summit area in the west (S1, Figure 59.2A), and an elongated, relatively flat but narrower area of deeper elevation in the central and eastern sectors (S2, Figure 59.2A). In the acoustic backscatter data, the summit region exhibits high backscatter (Figure 59.2G) that is shown by seafloor imagery to correspond to large rocky seafloor outcrops, boulders, and gravels (Figure 59.3A and B).

The flat shape of Condor's summit, along with the well-rounded boulders observed in the seafloor footage, further suggests that surf was previously at this level during periods of lowered sea level. Because the shallowest point of the seamount at 184 m depth is below maximum depth of sea-level lowstands commonly considered to have occurred during the Pleistocene [15], we suggest the seamount summit has probably subsided since it formed.

Seamount slope: The flanks of the seamount are generally smooth, and seabed imagery here has revealed both unconsolidated sediments and areas where the sediment surface has been cemented into 10- to 15-cm thick plates (Figure 59.3E–G).

Sediment composition on the northern slope of the seamount was predominantly bioclastic down to a depth of 965 m, where it shifted into a predominantly volcaniclastic black sand.

A downslope-oriented shallow ridge and gully morphology, together with observations of fragmented surficial plates, suggest that some gravity-driven downslope sediment movements occur. These are probably facilitated by the steep slope of the seafloor (ranging from 10° to 20°) and the moderate seismic activity that occurs in the region. The slope gullies are commonly floored by large-wavelength shallow dunes that are probably shaped during the gravity flows (Figure 59.2G).

Seamount ends: The small volcanic cones that dominate the eastern seamount end and the knobby terrain found to the east of the main massif clearly confer a distinct geomorphology to the seamount extremities. In Hawaii, the contrast between the rugged knobby morphology of Puna Ridge and smooth submarine slope adjacent to the SE Rift Zone of Kilauea has been interpreted as being created by submarine erupted lava in the former and to break up of lava extruded above sea level reaching the shoreline and interacting with surf in the latter [16]. We suspect that a similar difference in eruption style has controlled the different morphological textures of the center and ends of Condor seamount, in turn providing substrates for biology of a very different nature between them.

Table 59.1 Parameterization of the Benthic Terrain Modeler Classification Scheme

Zone	BPI Parameter		Slope (°)		Depth (m)	
	Lower Bound	Upper Bound	Lower Bound	Upper Bound	Lower Bound	Upper Bound
Flats						
Infralittoral shelf	−100	100		5	−50	
Circalittoral shelf	−100	100		5	−200	−50
Bathyal flat	−100	100		5	−2,000	−200
Abyssal rise			0.5	5		−2,000
Abyssal plain	−100	100		0.5		−2,000
Slopes						
Infralittoral moderate slope	−100	100	5	45	−50	
Infralittoral steep slope	−100	100	45	70	−50	
Infralittoral cliff	−100	100	70		−50	
Circalittoral moderate slope	−100	100	5	45	−200	−50
Circalittoral steep slope	−100	100	45	70	−200	−50
Circalittoral cliff	−100	100	70		−200	−50
Bathyal moderate slope	−100	100	5	45	−2,000	−200
Bathyal steep slope	−100	100	45	70	−2,000	−200
Bathyal cliff	−100	100	70		−2,000	−200
Abyssal moderate slope	−100	100	5	45		−2,000
Abyssal steep slope	−100	100	45	70		−2,000
Abyssal cliff	−100	100	70			−2,000
Depressions						
Infralittoral depression		−100			−50	
Circalittoral depression		−100			−200	−50
Bathyal depression		−100			−2,000	−200
Abyssal depression		−100				−2,000
Crests						
Infralittoral crest	100				−50	
Circalittoral crest	100				−200	−50
Bathyal crest	100				−2,000	−200
Abyssal crest	100					−2,000

The seabed in the seamount extremities is probably bedrock or rock talus, an inference supported by the high backscatter (Figure 59.2G). The shape and disposition of volcanic cones observed to the east of the ridge suggest that they likely originated from hydromagmatic eruptions from localized vents leading to piles of volcaniclastic debris [17]. The NW end is characterized by a likely recent series of anastomosing volcanic ridges apparently extruded from an abundance of fissures.

Oceanographic measurements: Data from three campaigns have revealed oceanographic patterns over the Condor seamount that differs from the surrounding ocean. In

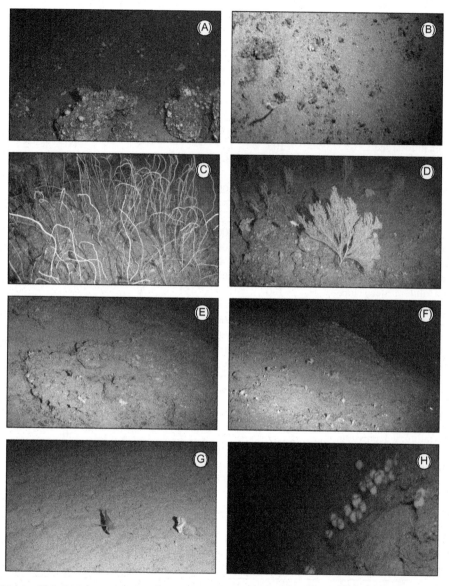

Figure 59.3 Different substrates and biological communities observed on the Condor seamount. (A, B) Mixed and sedimentary substrates identified on the northern side of the S1 sector (depth: ~200 m). (C, D) Coral gardens (*Viminella flagellum*, *Dentomuricea* sp., and a large primnoid gorgonian) on the southeastern side of the S1 sector (depth: ~200–270 m). (E, F) Sedimentary seafloor with surficial plates dominated by small encrusting sponges (depth: ~400 m). (G) Sedimentary area on the northern flank (depth: 600 m). (H) *Pheronema carpenteri* (Hexactinellida) aggregation on the "cones" sector (depth: ~730 m).
Source: Image credits: A–G, ©Greenpeace/Gavin Newman; H, EMEPC.

the upper 100 m layer, i.e., close to the level of the summit, a checkered pattern of temperature and salinity fields has been observed that consists of two upwelling centers (with higher turbidity levels) entwined with two downwelling centers. The centers of the anomalies were situated 5–7 km away from the seamount summit and collectively rotate around the seamount in an anticyclonic direction following a tidal period. Such structures typically originate when a part of the energy associated with the barotropic tide is trapped at topographic rises in the form of internal tidal waves [18]. Information provided by the two moorings deployed north and south of the summit confirmed a mean anticyclonical flow pattern around the seamount. Such quasiclosed circulation patterns (Taylor caps) form when the seamount is subject to a mean flow of moderate intensity, but may also be produced by nonlinearity of a periodic tidal flow [18,19].

Tidal forcing is also responsible for periodic changes in thermocline depth over the seamount. Temperature anomalies, combined with data from the tide gauge on Faial tidal station (http://oceano.horta.uac.pt/azodc/tidegauge.php), suggest that the thermocline was rising when water level was falling and vice versa. CTD profiles have further revealed that the upper mixed layer may extend down to 200 m depth during the cold season, therefore intersecting the seamount summit and possibly influencing its biological assemblages.

Biological Communities

Video surveys have so far extended from 185 to 1,097 m depth, while fishing gear sets reached 1,350 m depth (Figure 59.2A). Locations on the seamount summit, flanks, and extremities containing various substrate types have been sampled in the process. A preliminary vertical zonation of Condor seamount's benthic assemblages is presented in Figure 59.4. The nekto-benthic species of commercial interest dominating the fishing sets are also identified for comparable depth strata and accompanied by their catch-per-unit-effort (CPUE; individuals per 1,000 hooks for fish; individuals per 100 traps for shrimp).

Two habitats of conservation importance are documented: coral gardens and deep-sea sponge aggregations. The densest coral gardens were recorded on the seamount summit down to 287 m depth and were dominated by gorgonians (*Viminella flagellum* and *Dentomuricea* sp.) in association with tall hydrarians (cf. *Polyplumaria flabellata*) (Figure 59.3C and D). Variations were observed in the relative abundance of the species: while *Viminella flagellum* and cf. *Polyplumaria flabellata* were recorded mainly on hard grounds and occurred down to 480 m depth, *Dentomuricea* sp. seemed more abundant in seafloor areas exhibiting a sediment veneer [8] and was not observed much beyond the break between the planated summit and the seamount slope.

Sponge aggregations were prevalent between 720 and 860 m depth and were dominated by the bird's nest sponge *Pheronema carpenteri* (Hexactinellida), which colonized both consolidated and unconsolidated substrates [2] (Figure 59.3E–G). Some strata were apparently poor in tall epibenthic species, e.g., 380–720 m and 965–1,097 m in unconsolidated sediments and 480–714 m in consolidated substrates.

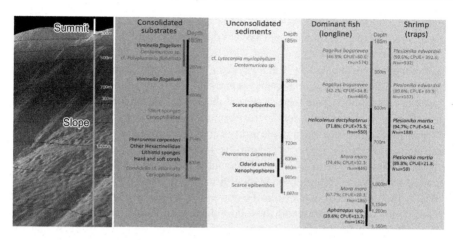

Figure 59.4 Qualitative zonation of benthic assemblages according to substrate nature and depth. The variation of dominant nekto-benthic species (fish and shrimps) along the depth strata is shown on the right, including the respective percentage of total catch in number of individuals, the catch-per-unit-effort (CPUE), and the total number of individuals caught.

A rocky ridge explored on the northern flank between 714 and 837 m depth held exuberant epibenthic assemblages comprising a variety of habitat-building sponges (hexactinellids and lithistids) and corals (primnoid gorgonians, alcyonaceans, and stylasterids). The deepest coral gardens were found between 968 and 1,001 m depth and were dominated by the gorgonian *Candidella* cf. *imbricata* and small yellow Caryophyllidae.

Surrogacy

The investigation of the statistical relationships between physical surrogates and benthos, as well as the association of fish with the biologically structured habitats, is ongoing. After systematic biological observations are extracted from the fishery surveys and seafloor imagery, colocated terrain and oceanographic information will be associated with them and used to develop predictive models of species and assemblages distributions or habitat suitability. Exploiting terrain-based characteristics as environmental proxies is considered a necessity in cases like Condor, where fine-scale validated models of near-bottom oceanographic conditions (currents, temperatures, and biogeochemical parameters) are lacking and regional models provide only coarse oceanographic information. Although these geomorphological characteristics are not substitutes for quality oceanographic data, they may nevertheless be effective in capturing the variability induced by both oceanographic exposure and seafloor nature.

Acknowledgments

IMAR-DOP/UAz is Research and Development Unit no. 531 and LARSyS-Associated Laboratory no. 9 funded by the Portuguese Foundation for Science and Technology (FCT) through pluriannual and programmatic funding schemes (OE, FEDER, POCI2001, FSE) and by the Azores Directorate for Science and Technology (DRCT). Surveys have been supported by projects CORALFISH (FP7 ENV/2007/1/213144), CONDOR (EEA Grants PT0040/2008), and CORAZON (FCT/PTDC/MAR/72169/2006) whilst data analysis has been also supported by projects Hermione (EC/FP7-226354) and MeshAtlantic (AA- 10/1218525/BF). A. Braga-Henriques is funded by a FRCT doctoral grant (ref. M3.1.2/F/016/2008). Further acknowledgements are due to the Portuguese Task Group for the Extension of the Continental Shelf (EMEPC) for conducting a multibeam opportunistic survey of the seamount and sharing video imagery of ROV *Luso* dives as well as providing the ROV team and equipment during the CORALFISH 2010 cruise, Joaquim Luís for sharing a bathymetry grid derived from the 2006 STRIPAREA multibeam survey, and Greenpeace for sharing the drop-down camera footage from the Azores leg of the 2006 "Defending our Oceans" Expedition.

References

[1] IHO, Standardization of Undersea Feature Names: Guidelines Proposal Form Terminology, fourth ed., IHO and IOC, International Hydrographic Bureau, Monaco, 2008.

[2] P.J. Etnoyer, J. Wood, T.C. Shirley, How large is the seamount biome?, Oceanography 23 (2010) 206–209.

[3] T. Morato, S.D. Hoyle, V. Allain, S.J. Nicol, Seamounts are hotspots of pelagic biodiversity in the open ocean, Proc. Natl. Acad. Sci. USA 107 (21) (2010) 9707–9711.

[4] M.R. Clark, V.I. Vinnichenko, J.D.M. Gordon, G.Z. Beck-Bulat, N.N. Kukharev, A.F. Kakora, Large-scale distant-water trawl fisheries on seamounts, in: T.J. Pitcher, T. Morato, P.J.B. Hart, M.R. Clark, N. Haggan, R.S. Santos, (Eds.), Seamounts: Ecology, Fisheries and Conservation, Blackwell, Oxford, 2007, pp. 361–399.

[5] J. Hall-Spencer, A. Rogers, J. Davies, A. Foggo, Deep-sea coral distribution on seamounts, oceanic islands, and continental slopes in the Northeast Atlantic, Bull. Mar. Sci. 81 (Suppl. 1) (2007) 135–146.

[6] M.R. Clark, J.A. Koslow, Impacts of fisheries on seamounts, in: T.J. Pitcher, T. Morato, P.J.B. Hart, M.R. Clark, N. Haggan, R.S. Santos, (Eds.), Seamounts: Ecology, Fisheries and Conservation, Blackwell, Oxford, 2007, pp. 413–441.

[7] N. Lourenço, J.M. Miranda, J.F. Luís, A. Ribeiro, L.A. Mendes Victor, J. Madeira, et al., Morpho-tectonic analysis of the Azores Volcanic Plateau from a new bathymetric compilation of the area, Mar. Geophys. Res. 20 (3) (1998) 141–156.

[8] A. Braga-Henriques, F. Cardigos, G. Menezes, O. Ocaña, F.M. Porteiro, F. Tempera, et al., Recent observations of cold-water coral communities on the "Condor de Terra" seamount, Azores, Program and Book of Abstracts of the 41st European Marine Biology Symposium, Cork (Ireland), Poster 116, 2006, pp. 79.

[9] F. Tempera, A. Henriques, F. Porteiro, A. Colaço, G. Menezes, Constraining the geomorphological setting of deep-sea biotopes using high resolution bathymetry at the Condor de Terra seamount, Program and Book of Abstracts of the ICES International Symposium on Issues Confronting the Deep-Sea, Horta (Azores, Portugal), Poster E27, 2009, pp. 89–90. http://www.turangra.com/deepocean/AbstractBook.pdf.

[10] T.J. Pitcher, T. Morato, P.J.B. Hart, M.R. Clark, N. Haggan, R.S. Santos, The depths of ignorance: an ecosystem evaluation framework for seamount ecology, fisheries and conservation, in: T.J. Pitcher, T. Morato, P.J.B. Hart, M.R. Clark, N. Haggan, R.S. Santos, (Eds.), Seamounts: Ecology, Fisheries and Conservation, Blackwell, Oxford, 2007, pp. 476–488.

[11] J.F. Luís, J.M. Miranda, A. Galdeano, P. Patriat, J.C. Rossignol, L.A.M. Victor, The Azores triple junction evolution since 10 Ma from an aeromagnetic survey of the Mid-Atlantic Ridge, Earth Planet. Sci. Lett. 125 (1994) 439–459.

[12] P.R. Vogt, W.Y. Jung, The Terceira Rift as hyperslow, hotspot dominated oblique spreading axis: a comparison with other slow spreading plate boundaries, Earth Planet. Sci. Lett. 218 (2004) 77–90.

[13] J.M. Sappington, K.M. Longshore, D.B. Thomson, Quantifying landscape ruggedness for animal habitat analysis: a case study using bighorn sheep in the Mojave Desert, J. Wildl. Manage. 71 (5) (2007) 1419–1426.

[14] E. Lundblad, D.J. Wright, J. Miller, E.M. Larkin, R. Rinehart, D.F. Naar, et al., A benthic terrain classification scheme for American Samoa, Mar. Geod. 29 (2006) 89–111.

[15] R. Bintanja, R.S.W. van de Wal, J. Oerlemans, Modelled atmospheric temperatures and global sea levels over the past million years, Nature 437 (2005) 125–128.

[16] D.K. Smith, L.S.L. Kong, K.T.M. Johnson, J.R. Reynolds, Volcanic morphology of the submarine Puna Ridge, Kilauea Volcano, in: E. Takahashi, P.W. Lipman, M.O. Garcia, J. Naka, S. Aramaki, (Eds.), Evolution of Hawaiian Volcanoes: Recent Progress in Deep Underwater Research, American Geophysical Union, Washington D.C., 2002, pp. 125–142, AGU Monograph 128.

[17] D.A. Clague, J.G. Moore, G.R. Reynolds, Formation of flat-topped volcanic cones in Hawaii, Bull. Volcanol. 62 (2000) 214–233.

[18] D. Codiga, C.C. Eriksen, Observations of low-frequency circulation and amplified subinertial tidal currents at Cobb Seamount, J. Geophys. Res. 102 (C10) (1997) 22993–23007.

[19] M. White, I. Bashmachnikov, J. Aristegui, A. Martins, Physical processes and seamount productivity, in: T.J. Pitcher, T. Morato, P.J.B. Hart, M.R. Clark, N. Haggan, R.S. Santos, (Eds.), Seamounts: Ecology, Conservation and Management, Blackwell, Oxford, 2007, pp. 65–84.

60 Cold-Water Coral Colonization of Alboran Sea Knolls, Western Mediterranean Sea

Ben De Mol[1], David Amblas[1], Antonio Calafat[1], Miquel Canals[1], Ruth Duran[1], Caroline Lavoie[1], Araceli Muñoz[2], Jesus Rivera[3], HERMESIONE, DARWIN CD178, and COBAS Shipboard Parties

[1]GRC Geociències Marines, Parc Cientific de Barcelona, Departament d'Estratigrafia, Paleontologia i Geociències Marines, Facultat de Geologia, Universitat de Barcelona, Campus de Pedralbes, Barcelona, Spain, [2]Tragsa-SGM C/ Nuñez de balboa, Madrid, Spain, [3]Instituto Español de Oceanografía, C/ Corazon de Maria, Madrid, Spain

Abstract

Knolls are prominent features in the Alboran Sea, Western Mediterranean. Multibeam bathymetry and backscatter data, combined with TOPAS parameteric profiles, surficial sediment sampling, and video transects for ground-truthing revealed a large relict cold-water coral habitat. This habitat consists mainly of dead and living *Dendrophyllia* sp. and, to a lesser extent, *Lophelia pertusa* and *Madrepora occulata* associated with giant deep-sea oysters *Neopycnodonte zibrowii* sp., and gorgonians. The knolls display flat shoaling tops ranging from 70 m (Chella Bank) down to 430 m (La Herradura knoll) depth and rise ~400–600 m above the surrounding seafloor. Sediment cover on the knolls is in general thin and mainly consists of bioclastic coarse sandy sediment. The slope of knoll flanks is up to 70°, with stepped ridges and near-vertical walls colonized with corals. At the summit of the knolls, cold-water corals are observed.

Key Words: cold-water corals, Alboran Sea, Mediterranean Sea, knolls, ridges, deep sea

Introduction

The Alboran Sea is the westernmost basin of the Mediterranean Sea and represents a basin 350 km long and 150 km wide, with water depths between 0 and 2,000 m (Figure 60.1). It corresponds to a Neogene back-arc basin resulting from African/Euroasian plate convergence [1]. The Alboran Sea is divided into two sub-basins,

Seafloor Geomorphology as Benthic Habitat. DOI: 10.1016/B978-0-12-385140-6.00060-8

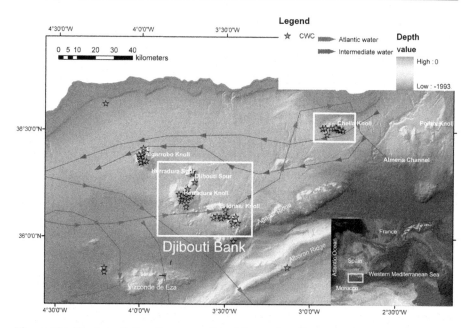

Figure 60.1 Shaded relief seafloor map of the Alboran Sea. Green stars represent CWC occurrences in the area. Present-day surface and intermediate water circulation in the Alboran Sea are indicated with arrows. Djibouti Bank and Chella knoll have been surveyed in detail and are marked with white rectangles that correspond to Figures 60.2–60.4, respectively. (For interpretation of the references to color in this figure legend, the reader is referred to the web version of this book.)

the Western Alboran Basin, located between the Strait of Gibraltar and the volcanic Alboran Ridge, and the Eastern Alboran Basin, extending eastwards from the Alboran Ridge. The two detailed studies in this paper are located in the Western Alboran Basin, which is the tectonically inverted block of the Djibouti Bank and Chella knoll (also known as Seco de los Olivos) [2] (Figure 60.1).

The Alboran Sea is characterized by highly dynamic and variable water masses that make it one of the most productive areas in the oligotrophic Mediterranean Sea. Hydrodynamic features occur across all time and length scales, such as tidal motion, strong baroclinic jets, large-scale gyres, mesoscale eddies, upwelling regions, and frontal zones, all with important implications on the dynamics of plankton and benthic ecosystems [3]. The strong surface inflow of Atlantic water through the Strait of Gibraltar, known as the Atlantic Jet, maintains two semipermanent anticyclonic gyres consisting of a mixture of different proportions of Mediterranean and Atlantic waters that change in sympathy with tidal cycles [4].

The knolls in the Alboran Sea are affected by benthic trawl fisheries, evidenced by trawl marks and lost fishing gear observed in the study area. Overall, the naturalness of the study area is considered to be modified.

A database of cold-water coral (CWC) occurrences has been collected based on samples and towed video (SCHRIMP) transects during the COBAS 2004, Darwin

178 2006, and HERMESIONE 2009 cruises, and from the scientific literature [5,6]. Background bathymetric data collected aboard RV *Vizconde de Eza* using a Simrad EM 300 multibeam sonar was used to create a bathymetric grid of 100 m resolution over a depth range of from 0 to 2,000 m. High-resolution swath bathymetry and backscatter data were collected over two knolls at the Djibouti Bank (El Idrissi and Herradura) over a depth range of 200–1,700 m and an area of 1,240 km^2, using a SIMRAD EM120 and an EM1002 at depths shallower than 650 m, during the HERMESIONE cruise carried out by the University of Barcelona in September–October 2009. Bathymetry and backscatter data were processed with Caris Hips and Sips software to a grid size of 15 m. Simultaneously with the multibeam sonar survey, 840 nautical miles of TOPAS PS18 sub-bottom profiles have been acquired to investigate the upper sedimentary layer. Thirty-six sediment grabs and one boxcore were collected from Djibouti Bank for textural analysis, along with several rock dredge samples from Algorrobo knoll. At the Chella Bank, two transects (3.5 nautical miles total length) with the SCHRIMP video camera system have been collected in combination with eight rock dredges and three sediment grabs.

Geomorphic Features and Habitats

The topography of the Alboran Basin seafloor is characterized by pinnacles, knolls, banks, ridges, and troughs as a direct expression of the Pliocene-Quaternary compressive tectonic regime (Figure 60.1). The origin of the larger morphological features has been ascribed to uplifted basement, while smaller accentuated structures have been linked to Miocene volcanism. Vizconde de Eza Bank, Djibouti Bank, and Herradura knoll are built-up by metasedimentary and gabbroic rock related to Miocene rifting. Chella knoll and probably most adjacent pinnacles are the morphological expression of middle Miocene to Pleistocene calc-alkaline volcanic eruptions [7]. The majority of the Holocene sediments consist of hemipelagic mud from fluvial and aeolian sources, as well as *in situ* production of shallow marine carbonates [8–11].

Djibouti Bank and Algorrobo and Chella knolls display flat tops, shoaling to 207 m (El Idrissi knoll), 273 m (Herradura knoll), 422 m (Djibouti Spur), 293 m (Algorrobo knoll), and 70 m (Chella knoll), and rise about 400–600 m above the surrounding seafloor. In general, the knoll flanks show a steep middle section up to 70° at 300–400 m water depth. In plan view, the knolls have a semicircular shape with diameters of 10–13 km on Djibouti Bank and 4 km for the central part of Chella Bank (9 km if side crests are included). Algorrobo knoll is located about 40 km and El Idrissi knoll about 65 km from the coast (Figure 60.1). Chella knoll is 20 km from the nearest coast, but only 9 km from the Almeria shelf break. Pinnacles southeast of El Idrissi knoll and south of Herradura knoll are much more irregular in shape and display N-S oriented fabrics of post-Pliocene neotectonic origin (Figures 60.1 and 60.2). All knolls display sediment-draped and rocky flanks with complex and sharp slope gradient changes, surrounded by depressions that resemble ~50 m deep moats. Sediment distribution on the flanks of the knolls and the asymmetric development of the moats illustrates the local current regime and sediment availability at the various

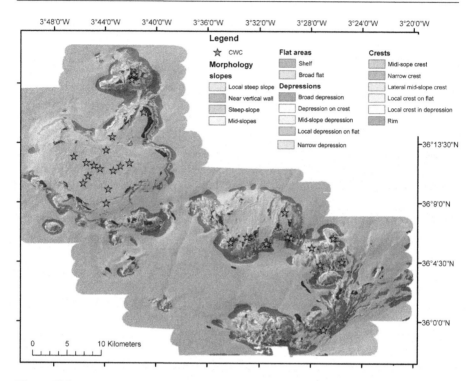

Figure 60.2 Morphological classification of the Djibouti Bank area, based on a 15 m grid swath bathymetry and the distribution of sampled CWC.

knoll locations, with no general trend. Scarps related to mass wasting processes have been observed along the flanks of most knolls, locally leading to near-vertical walls. The summits of the knolls are decorated by about 15–30 m high crests and blocks with a main N-S to NE-SW trend (Figure 60.2). The Chella Bank complex consists of a main circular knoll, flanked to the east and west by crests with a N-S to NE-SW orientation surrounded by 50 m deep elongated depressions.

Textural Map

The statistically maximum-likelihood predicted textural distribution map is based on bulk grain size analysis of the sediment samples, in standard textural classes (gravel sand, muddy sand, and mud), rock, and derivate multibeam data products such as slope, rugosity, backscatter, and a nonsubjectively morphological classification based on set of quantifiable parameters such as slope of the seafloor, rugosity, and bathymetric position index (BPI) of the Djibouti Bank area (Figure 60.3). Rock exposures have been statistically correlated with near-vertical walls at the mid-slope flanks of the knolls and along local crests on the knoll summits. The most prominent outcrops are on the southern flanks of the Herradura knoll and Djibouti spur and the crests at

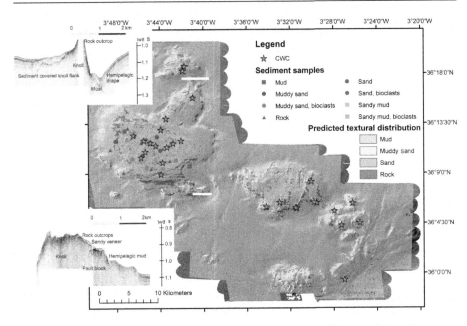

Figure 60.3 Statistically predictive textural map and textural classification of the sediment samples based for ground-truthing in Djibouti Bank study area. The textural predicted map has been calculated using morphological parameters and backscatter data in a maximum-likelihood algorithm (ArcGIS). TOPAS profiles, indicated with a white line, are used to control the predicted textural distribution. Sediment seems to cover the flanks of knolls and crests in the area, with pronounced moats at the foot of the knoll attributed to scouring by bottom currents. (For interpretation of the references to color in this figure legend, the reader is referred to the web version of this book.)

the summit of the El Idrissi knoll. Rocky outcrops are limited to about 4% of the survey area due to thick sediment draping (Figure 60.3). Sand and muddy sand occur in small patches in pockets within the elevated rocky outcrops at the base of the summit crests and knoll flank; in total this textural class represents 21% of the predicted sediment area (Figure 60.3). Mud is the most prevalent sediment type in the area, representing 75% of the survey area. It is found on local flat parts of the summits and the flat basin floor. The same textural–morphological trend has been observed at the Chella Bank based on video transects and sediment samples (Figure 60.4). The video observations revealed sandy sediments near the base of the crests, and fine sand and fossil coral fragments (mainly *Dendrophyllia* sp.) on the flanks. (Table 60.1)

Biological Communities

The CWC database of the Western Mediterranean sea holds 67 (sampling stations) in the Alboran Sea, of which 29 are on the Djibouti Bank (Figures 60.1 and 60.2) and 16 on Chella Bank (Figure 60.4). Video observations combined with the sediment

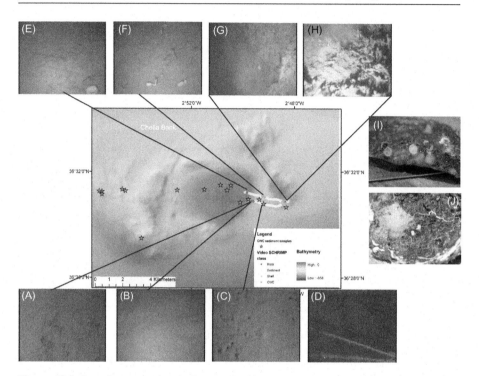

Figure 60.4 Overview map of the different sediment facies observed on the SHRIMP video system. Sediment surrounding the Chella knoll is composed out of densely bioturbated mud, with coral fragments near the base of the knoll and clearly observable trawl marks (A–C). In the vicinity of rock outcrops, wires and fishing gear have been found (D). The flanks of the knoll are characterized by a thin veneer of sediment covering underlying rocks, which are colonized by benthic fauna (E). Common species are sponges, in particular glass sponges and CWC fragments. Near the steepest flanks of the knoll more rocky outcrops are observed, and Fe-Mn-stained oyster shells colonized by CWC *Madrepora occulata* and *Lophelia pertusa*. (H, J) are common. Juvenile corals have been observed on shell fragments and rocky outcrops (I).

samples show extensive cover of relict *Dendrophyllia* sp. and *Desmophyllium* sp. coral fragments, with lesser amounts of *Madrepora occulata* and *Lophelia pertusa*. Pebbles, boulders, and steep rocky slopes are colonized also by a variety of benthic species, including glass sponges, gorgonians, crustaceans, and mollusks. The highest density of fossil CWC fragments is associated with dead giant oyster shells, *Neopycnodonte zibrowii* sp. [12]. In this survey, juvenile living coral settlements have been observed on shell fragments of *Neopycnodonte zibrowii* sp. on the steep flanks of Chella Bank. In general, the basin seafloor consists of bioturbated muddy sediments without large quantities of macro- or megabenthic communities (Figure 60.4), which limits the coral habitat to steep rocky outcrops and crests. These surfaces provide the corals with a hard substratum to settle on and protection against fine-grained sedimentation. The living corals located at Chella Bank are associated

Table 60.1 Results of the Multivariance Statistical Analysis of Textural Data with Acoustic Data

Multivariate Analysis							
Effect	Wilks Lambda	Ho Df: null hypothesis degree of freedom	Error Df: degree of freedom for error	*F*: F-value	Prob.: Probability		
Class	0.597	50	377	0.903	0.663		
Classification Table							
	Pred. Group (Std)						
Act. Group	Rock	Sand	Sandy Mud	Mud	No Class	Correctly Classified	
Rock	12	5	4	4	0	0.440	
Sand	0	6	1	6	3	0.375	
Sandy Mud	1	0	13	1	1	0.813	
Mud	0	1	0	3	1	0.600	
No Class	0	1	1	1	3	0.500	

with the boundary between the two water masses (Modified Atlantic Water and the Levantine Intermediate Water) characterized by a salinity and temperature contrast.

Surrogacy

The objective of the mapping surveys in the Alboran Sea was to obtain a probability distribution map of relict and present-day CWC ecosystems (Figure 60.6). The statistically derived relations between morphological parameters, sediment texture, and CWC occurrences on the Djibouti Bank complex and the Chella knoll have been extrapolated over the entire basin at a lower resolution in order to plan forthcoming ground-truthing surveys (Figures 60.2–60.5). Rock dredges have been deployed on knoll flanks as a first approach to identify CWC area on the knolls. For the statistical analysis between morphological parameters and CWC occurrences, the midpoint of the dredging transect has been used. Boxcores have been used for grain size analysis and CWC localization in flat areas, such as the knoll summits and at the base of the knoll slopes.

Morphological parameters correlating with CWC occurrences on the Chella Bank and Djibouti Bank are similar, suggesting that the physical habitat requirements for CWC in the Alboran Sea can predicted by morphological classifications (Figures 60.2–60.5). A benthic terrain classification dictionary has been developed to classify all morphological structures on the Alboran knolls (Figure 60.2) that can fraction the main ecosystem, with the Benthic Terrain model scheme [13]. This classification is based on the broad-scale BPI 3,000m to detect the large structural zones as flanks, knolls, depressions, and channels, and the fine-scale BPI of 100 detecting ridges and depressions on flanks and tops of the knolls, based on a bathymetric data set of 15m cell size.

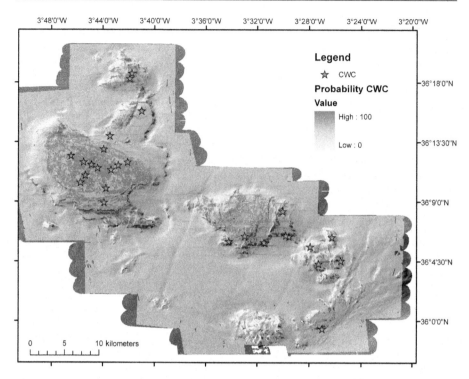

Figure 60.5 Predictive suitability CWC habitat map at the Djibouti Bank, based on morphological parameters and backscattering derived from multibeam data and sediment sampling. The map illustrates the distribution of CWC at local crests and steep sections of the knoll flanks. Fossil coral fragments in sandy sediment are observed at the base of local elevations and the foot of the knoll flank.

Seventy-five percent of the sample stations have been taken into account for statistical analysis (Discriminant Function Coefficients) between CWC, textural class, and morphological parameters and backscatter data in the Djibouti Bank. The outcome of this analysis has been used to calculate the different factors in an ArcGIS environment based on the bathymetry-derived parameter grids. In a second step, a signature file of covariances between the introduced factors and the textural classes and CWC occurrences has been calculated, allowing identification of the different classes with the maximum-likelihood statistic tool in ArcGIS. Maximum-likelihood statistics allow the prediction of normal distribution of classes of multiple data sets and incorporate the idea of positive spatial autocorrelation, in which the occurrence of a class in one area is expected to be more likely if the class occurs in many surrounding cells.

Statistical relations were derived between morphological parameters, sediment texture, and CWC occurrences. Classification was carried out by Discriminant Function Coefficients on acoustic backscatter and bathymetry-derived parameters including rugosity, slope, curvature, BPIs (calculated at 3,000 and 100 m radius) [12], and standard deviation of depth. The CWC habitat maps of 75% maximum-likelihood

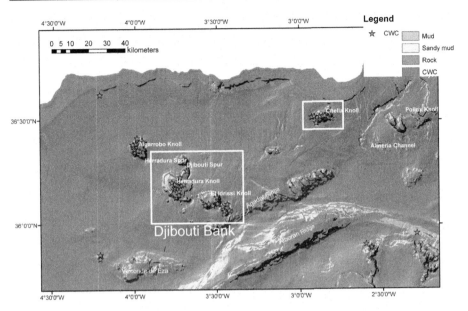

Figure 60.6 Predicted textural and CWC distribution of the Alboran Basin, based on bathymetric information and relationship between the morphological surrogates for CWC distribution, Djibouti Bank and Chella knoll (marked with a white rectangular). CWC seems to be associated with knoll flanks and ridges, and with steep flanks in canyons and fault walls as well. The latter areas have not been verified with ground-truthing surveys. (For interpretation of the references to color in this figure legend, the reader is referred to the web version of this book.)

probability (Figure 60.3), predict that fossil and living *Scleractinina* CWC are found on steep slopes (more than 18°) and on convex structures (crests) with little sediment cover (Figure 60.5). The good correlation between the observed and predicted CWC locations at the small-scale survey (15 m) show that bathymetry-derived parameters are valuable as high-resolution surrogates for CWC habitats. Nevertheless, on the basin-scale predictive CWC maps with a resolution of 100 m (Figure 60.6), the small morphological elevations on the summit and flanks that are identified as suitable habitat for CWC are out of the detection range for morphological analysis. The broad-scale map demonstrates that the highest probability for CWC occurrences is located near the upper flanks of the knolls, steep slopes in canyons, and fault-related elevations and ridges. Both canyons and fault walls have not yet been verified with observation data.

Conclusion

Detailed mapping of selected knolls revealed that morphology parameters are good predictors of CWC habitats in the basin. Most CWC observations are of fossil fragments associated with steep slope at mid-water depths and rocky outcrops. CWC have been related to knoll crests, near-vertical walls, and escarpments on the knoll flanks. The relict CWC habitat is associated with fossil *Neopycnodonte zibrowii* sp. at the knoll flanks, which in turn provides a substrate for new CWC settlements. Colonization in the past and present does not correspond to any particular side of the knoll flanks; hence, it appears there is no enhanced source of nutrients nor any increased pressure related to sedimentation from any particular direction in the study area. Upscaling from detailed habitat maps of knolls to basin-wide predictive habitat maps reveals various potential CWC habitats that are not yet surveyed and might lead to new discoveries of relict and living CWC ecosystems in the basin. In the future, more precisely positioned ground-truthing surveys (ROV and video transect), in combination with oceanographic data sets, might distinguish relict from living CWC ecosystems in the basin.

Acknowledgments

This research was supported by the HERMIONE project, EC contract 226354-HERMIONE, funded by the European Commission's Seventh Framework Programme, and a Generalitat de Catalunya "Grups de Recerca Consolidats" grant (2009 SGR 1305), and SMT Inc. via the educational User License for Kingdom Suite interpretation software.

References

[1] L. Lonergan, & White, N. Origin of the Betic-Rif mountain belt. Tectonics 16, 504–522, 10.1029/96tc03937 (1997).
[2] A. Muñoz, M. Ballesteros, I. Montoya, J. Rivera, J. Acosta, E. Uchupi, Alborán Basin, southern Spain—Part I: geomorphology, Mar. Petrol. Geol. 25 (2008) 59–73.
[3] N. Skliris, & J.-M. Beckers, Modelling the Gibraltar Strait/Western Alboran Sea ecohydrodynamics. Ocean Dynam. 59, 489–508 (2009).
[4] D. Macias, C.M. García, F. Echevarría Navas, A. Vázquez-López-Escobar, M. Bruno Mejías, Tidal induced variability of mixing processes on Camarinal Sill (Strait of Gibraltar): a pulsating event, J. Mar. Syst. 60 (2006) 177–192. 10.1016/j.jmarsys.2005.12.003.
[5] K. Hoernle, A. Agouzouk, B. Berning, T. Buchmann, S. Christiansen, S. Duggen, et al., METEOR Cruise No. M51, Leg 1 VULKOSA: KIEL, GEOMAR, 2001, p. 1–94.
[6] J.G. Herrera, J.T. Vázquez, T. García, N. López, J. Canoura, D. Palomino, et al., DEEPER 0908 Instituto Español de Oceanografía, 2008.
[7] S. Duggen, K. Hoernle, P. van den Bogaard, C. Harris, Magmatic evolution of the Alboran region: the role of subduction in forming the western Mediterranean and causing the Messinian Salinity Crisis, Earth Planet. Sci. Lett. 218 (2004) 91–108.
[8] B. Alonso, A. Maldonado, Plio-Quaternary margin growth patterns in a complex tectonic setting: Northeastern Alboran SeaGeo-Mar. Lett., 12
[9] G., Ercilla, B. Alonso, & J. Baraza, Sedimentary evolution of the northwestern Alboran Sea during the Quaternary. Geo-Mar. Lett. 12, 144–149 (1992).

[10] J. Fabres, A. Calafat, A. Sanchez-Vidal, M. Canals, S. Heussner, Composition and spatiotemporal variability of particle fluxes in the Western Alboran yre, Mediterranean Sea, J. Mar. Syst. 33–34 (2002) 431–456.

[11] P. Masqué, J. Fabres, M. Canals, J.A. Sanchez-Cabeza, A. Sanchez-Vidal, I. Cahco, et al., Accumulation rates of major constituents of hemipelagic sediments in the deep Alboran Sea: a centennial perspective of sedimentary dynamics, Mar. Geol. 193 (2003) 207–233.

[12] M. Wisshak, M. López Correa, S. Gofas, C. Salas, M. Taviani, J. Jakobsen, et al., Shell architecture, element composition, and stable isotope signature of the giant deep-sea oyster *Neopycnodonte zibrowii* sp. n. from the NE Atlantic, Deep Sea Res. I 56 (2009) 374–407.

[13] E.R. Lundblad, D.J. Wright, J. Miller, E.M. Larkin, R. Rinehart, D.F. Naar, et al., A benthic terrain classification scheme for American Samoa, Mar. Geod. 29 (2006) 89–111.

61 Fluid Venting Through the Seabed in the Gulf of Cadiz (SE Atlantic Ocean, Western Iberian Peninsula): Geomorphic Features, Habitats, and Associated Fauna

José L. Rueda, Víctor Díaz-del-Río, Miriam Sayago-Gil, Nieves López-González, Luis M. Fernández-Salas, Juan T. Vázquez

Instituto Español de Oceanografía, Centro Oceanográfico de Málaga, Puerto Pesquero s/n, Fuengirola (Málaga), Spain

Abstract

The Gulf of Cádiz, enclosed between the Iberian Peninsula and Morocco, is characterized by deposits more than 2.4 km in thickness, forming major gravitational gliding nappes along the slope. Compressional stress produces hydrocarbon-rich fluid venting that generates numerous seepage-related structures (chimneys, slabs, pavements). These kinds of hard substrates are formed in a matrix of different types of soft bottoms, promoting an increase in the bottom complexity and the diversity of different habitat types. The fauna from stable hard bottoms is the most biodiverse, including porifera (*Asconema setubalense*), cold-water corals (*Madrepora oculata*), gorgonians (*Callogorgia verticillata*), and mollusks (*Asperarca nodulosa, Neopycnodonte zibrowii*). In soft bottoms, the fauna includes solitary corals (*Flabellum chunii*), gorgonians (*Isidella elongata*), mollusks (*Colus gracilis, Ranella olearia, Thyasira* spp.), commercial crustaceans (*Nephrops norvegicus*), and frenulate polychaetes. In mixed bottoms, the black coral *Leiopathes glaberrima*, the echinoid *Cidaris cidaris*, and the crustacean *Bathynectes maravigna* are common components.

Key Words: Iberian Margin, upper slope, leaking gases, mud volcanoes, pockmarks, chimneys, benthic communities, hard bottoms, black corals, mollusks

Introduction

The Gulf of Cadiz is located in the SW Iberian Peninsula and can be considered as the structural front of the Betic-Rifian Arc, located at the westernmost tectonic belt of the

Alpine–Mediterranean convergence zone between the African and Eurasian plates. This front is an olistostrome, comprising a large volume of shale and salt diapirism [1]. It extends from the Spanish and Moroccan margins toward the Atlantic abyssal plain, reaching a sedimentary thickness of more than 4 km. Gravity spreading and sliding along the upper continental slope provides upward migration routes to fluid expulsion toward the seabed surface, mainly methane and fine sediments [2–4].

A variety of seep-related geomorphic features has been described in this region, such as mud volcanoes (depths between ∼300 and 3,000 m), carbonate mounds, pockmarks, brine pools, and gas-related sediments [5]. These structures contain precipitates in the form of slabs, pavements, and chimneys, generated by aggregation of carbonate cement resulting from microbial oxidation of gas emissions [6,7]. These increase the complexity of the soft bottoms where they are formed, promoting a mixture of hard- and soft-bottom fauna. On stable hard substrates, corals, gorgonians, and sponges increase this complexity even more, resulting in a higher number of species than in adjacent soft bottoms. In adjacent soft bottoms, fisheries for different demersal and benthic species occur, although no detailed information is available on their impact upon mud volcano communities.

The oceanographic setting is characterized by the exchange of water masses between the Atlantic Ocean and the Mediterranean Sea through the Straits of Gibraltar. The exchange involves the near-bottom, highly saline, warm Mediterranean outflow water (MOW) and the inflow at the sea surface of less saline and colder North Atlantic superficial water (NASW). The MOW exits the Strait and flows northward as a contour current past the Cape of San Vicente (SW Portugal) at depths between 400 and 1,500 m, with a mean velocity of *ca.* 46 cm/s, salinity values from 36.5‰ to 37.42‰ and temperatures between 12°C and 13.7°C [7].

Scientific data were collected on board the RV *Cornide de Saavedra* and the RV *Hesperides* in four expeditions between 1999 and 2001, covering an area of 8,500 km^2 and a depth range of 300–2,000 m. Swath bathymetry was collected by means of multibeam echosounder Simrad EM12S-120 system (13 kHz and 81 beams) achieved with a GPSD (Differential Global Positioning System) and processed at a grid resolution of 50 m. Very-high-resolution seismic profiles were obtained with TOPAS (Topographic Parametric Sounding) (15/18 kHz), and high-resolution sections were collected with Sparker (3,500/7,000 J). Sampling methods included a rectangular benthic-type dredge, a boxcorer, and a standard gravity core (3 m in length) for collecting samples from selected mud volcanoes. A set of more than 1,000 photographs has been taken in some sites related to fluid flow structures using a deep-sea benthos camera.

Geomorphic Features and Habitats

Mud volcanoes, pockmarks, and other fluid escape structures are common along the Iberian margin of the Gulf of Cadiz. They are interpreted as indicators of gas-rich, overpressured sediments occurring at different depths (300–1,200 m) [5]. Mud volcanoes are conduits for fluid venting and consequent carbonate precipitation within the sediments or at the seafloor. Some of the widespread shallow fluid venting on the seafloor is attributed to the local destabilization of gas-hydrate-rich sediments due

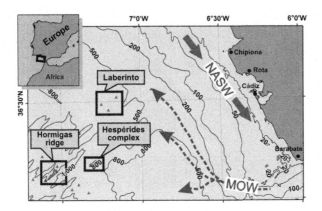

Figure 61.1 Location map of the study area on the southwestern Iberian Margin displaying three representative fluid venting areas: Laberinto, Hormigas Ridge, and Hespérides complex. MOW, Mediterranean outflow water; NASW, North Atlantic superficial water.

to MOW warming [4]. Three major sites of seepage-related cluster structures can be identified along the Iberian upper continental slope [3,7,8] (Figure 61.1) that are representative of the fluid venting phenomena in three different geological environments: contouritic drifts, megasliding deposits (on the continental margin), and diapiric ridges.

El Laberinto: This region is located on the upper slope (Figure 61.1) and comprises a cluster of separate, circular volcanoes outcropping over an extensive contouritic drift, covering 201 km^2 (Figure 61.2). The volcanoes are surrounded by rings of subcircular pockmarks, formed as a product of degassing diapirs. Notable characteristic features include three mud volcanoes (Anastasya, Tarsis, and Pipoca) together with an adjacent prominent diapiric ridge and contouritic channels. Anastasya mud volcano has a regular cone geometry (diameter ~1.5 km) covering 7.5 km^2 and located at a depth of 452 m, with a height of 80 m above the level of surrounding seafloor. Sediments are plastic, mousse-like muds that are methane saturated and draped by a thin layer of recent medium-grained sand. Tarsis mud volcano has a regular geometry (diameter ~1 km) covering 4.5 km^2, occurring at 762 m depth, with a height of 60 m. It has a 50-m deep pockmark depression circling the cone. The volcano is entirely surrounded by a sheeted contourite drift that is densely burrowed by decapods. Pipoca mud volcano (diameter ~1.5 km) covers 6.6 km^2 and displays a mud flow lobe flowing downslope and placed between two MOW moats.

Hesperides Complex: This complex is representative of megasliding deposits and consists of a cluster of five single-cone mud volcanoes, covering *ca.* 7 km^2 (680–730 m depth) [3] (Figure 61.3). The complex is affected by a MOW flow filament that descends along a channel, warming the deposits and thereby triggering the destabilization of hydrates and generating several pockmarks. Three cone morphotypes include: (a) twin circular and regular cones, (b) twin slightly flat cones with smooth surfaces and irregular contour lines, and (c) elongated single cone with strongly irregular and asymmetric hillsides. Scattered over the complex are carbonate chimneys and crusts providing hard substrata for benthic fauna. A large pockmark of 2.1 km^2, with a

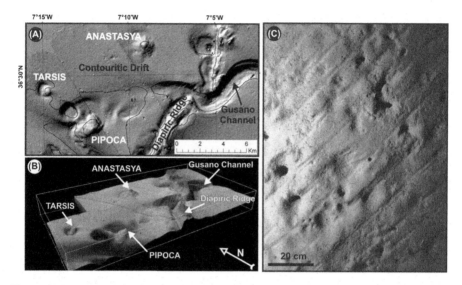

Figure 61.2 Main geomorphic features around El Laberinto: (A) bathymetric map displaying three main mud volcanoes; (B) 3D view of the previous figure; (C) adjacent contouritic soft bottoms with trawling marks and decapod burrows.

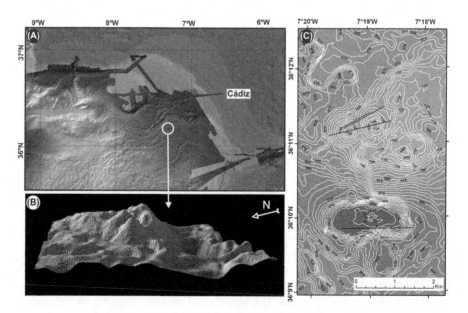

Figure 61.3 Main geomorphic features around Hespérides Complex: (A) location on the regional bathymetry; (B) 3D view of the volcano; (C) bathymetry of the area displaying several cones and the pockmark (southward). Continuous lines represent dredge samples collected in the complex.

maximum depth of 150 m, next to the complex contains bottom mud deposits without evidence of gas-enriched sediments or hard structures.

Hormigas Ridge: The ridge is formed by a set of single mud volcanoes on top of several small hills from the deepest sector of one of the diapiric ridges, covering 45 km^2 (890–935 m depth) and containing numerous carbonate chimneys, crusts, and slabs. The largest mound is the Cornide mud volcano (diameter ~1.2 km) (Figure 61.4), covering 0.5 km^2 (935 m depth) and located next to the Cadiz Channel. It is conformed by a twin flat mound with a height of 35 m below the diapir summit and with a cliff of *ca.* 235 m above the channel. The crest of the cone yielded large amounts of brown carbonate crusts and chimneys, and a polygenic matrix breccia covered by sandy sediments with a strong H$_2$S smell.

Geomorphic features related to fluid flow and availability of hard bottoms (e.g., chimneys, slabs) is influenced by the erosion patterns that affect each volcano. Therefore, different habitat types occur in association with different volcanoes. For example, exhumed chimneys and slabs can be seen in those volcanoes where strong deep currents have eroded the cone that originally enclosed these structures. Soft bottoms, from coarse sands to mud, generally occur adjacent to the volcanoes or in the pockmarks. Three major types of substrate for benthic and demersal fauna are as follows.

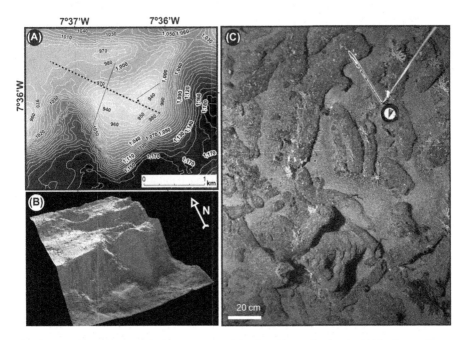

Figure 61.4 Main geomorphic features around the Cornide mud volcano: (A) bathymetric map; (B) 3D view displaying the twin flat mound; (C) extensive chimney fields promoting the establishment of hard-bottom fauna. Continuous lines represent dredge samples; dashed line is a photographic transect.

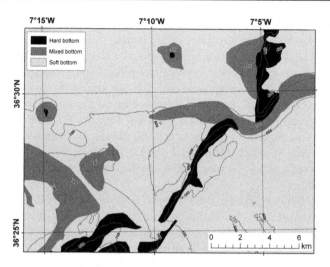

Figure 61.5 Map showing different types of substrates (hard, mixed, and soft) and habitats in El Laberinto.

Stable hard bottoms: These include crusts, pavements, and slabs as well as carbonate chimneys that are generally not covered by mud or sand. These types of bottoms are normally located in eroded volcanoes (Hespérides and Cornide), close to the summit or at the side of the volcano. This is a rare type of substrate compared to soft bottoms (Figures 61.5–61.7) and represents an ideal habitat for colonial organisms (e.g., corals, sponges).

Mixed bottoms: These include scattered chimneys and crusts that occur in a matrix of sandy mud to muddy sediments. This mixture of substrates generally occurs within mud volcanoes, scattered between the stable hard substrates or at the base of the volcano, where chimneys and crusts are deposited (Hespérides and Córnide) (Figures 61.6 and 61.7). This substrate is less abundant than soft bottoms but exhibits similar coverage to rocky bottoms, especially in the El Laberinto and Cornide mud volcanoes (Figures 61.5 and 61.6). A mixed fauna occupies this habitat.

Soft bottoms: These are mainly contouritic sands and muds occurring between the volcanoes and in the pockmark deposits. Soft-bottom types have the greatest spatial coverage and are composed of different sediment types (coarse sands, fine sands, and mud). Fisheries of commercial species in the vicinity of some mud volcanoes (El Laberinto) occur in these types of bottoms.

Biological Communities

Benthic assemblages were studied from samples collected with a rectangular benthic-type dredge in both mud volcanoes and pockmarks. Underwater photography (five locations in 2001) was also used as a nondestructive sampling method for fauna and for characterizing the different types of bottoms (slabs and crusts,

Figure 61.6 Map showing different types of substrates (hard, mixed, and soft) and habitats in Hesperides.

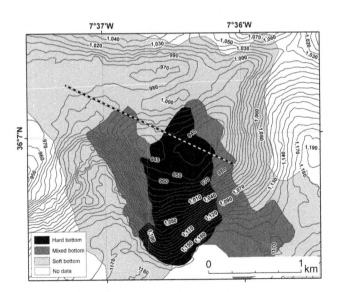

Figure 61.7 Map showing different types of substrates (hard, mixed, and soft) and habitats in Cornide, as well as a photographic transect.

chimneys, and soft bottoms) using the image analysis software J Microvision v1.27. In photos from different locations and bottom types (hard, mixed, and soft), sessile (porifera, octocorallia, and antipatharia) and mobile faunistic groups (echinoderms) were also quantified in framed areas of $4\,m^2$ ($n = 7$ photos per bottom type).

As mentioned before, different bottom types occur on mud volcanoes and mounds, making three major types of habitat with different faunistic assemblages.

Stable hard bottoms (large crusts, chimneys, and slabs): The fauna includes several species of sponges and cnidarians: the sponge *Asconema setubalense*; colonial (*Dendrophyllia cornigera, Lophelia pertusa*, and *Madrepora oculata*) and solitary (*Caryophyllia* spp.) scleractinian corals; and octocorallia (*Callogorgia verticillata* and two unidentified holaxonian gorgonians) [9] (Figure 61.8). Porifera and gorgonians display much higher densities (10–50 colonies $\cdot 4\,m^{-2}$) in stable hard bottoms than in mixed bottoms (10–12 colonies $\cdot 4\,m^{-2}$) or in soft bottoms (<5) (Figure 61.9). Densities were also different between mud volcanoes, reaching a maximum at Cornide (one of the deepest volcanoes in the study area). Regarding polychaetes, serpulids (four spp.) are an abundant group, especially under chimneys and crusts, as well as the brachiopods *Novocrania anomala, Megerlia* sp., and *Terebratulina retusa*. Bivalves represent the main molluscan group inhabiting chimneys/slabs and include *Spondylus gussonii, Asperarca nodulosa* (highly common in crevices), *Limopsis angusta* or the recently described giant deep-sea oyster *Neopycnodonte zibrowii*, a long-lived mollusc with an estimated lifespan of several centuries [10]. Mobile organisms such as echinoderms display densities lower than the aforementioned

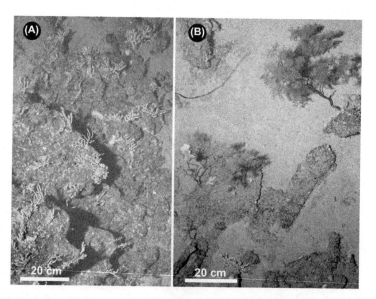

Figure 61.8 Photographs taken at Cornide mud volcano showing two different types of habitats in mud volcanoes (900–1,000 m depth): (A) stable hard bottoms with different gorgonian species and the echinoid *Cidaris cidaris*; (B) mixed bottoms with colonies of the black coral *Leiopathes glaberrima* and the decapod *Bathynectes maravigna*.

fauna, with values up to 4.6 individuals $\cdot$ 4 m^{-2} for large echinoids and *ca.* 1 individual $\cdot$ 4 m^{-2} for asteroids (Figure 61.9). The echinoid *Cidaris cidaris* is one of the top abundant echinoderms, with higher values in predominantly rocky bottoms (2–4 individuals $\cdot$ 4 m^{-2}) than in soft ones (<2 individuals $\cdot$ 4 m^{-2}) and in Arcos mud volcano compared to other volcanoes. The decapods *Monodaeus couchi* and *Pleisonika* spp., as well as the fishes *Beryx decadactylus* and *Chimaera monstrosa*, are other mobile components that occur in lower abundances.

Mixed bottoms: Here, different fauna are associated with the different substrata, but the presence of soft sediment that could cover the chimneys and crusts may represent a disadvantage for the establishment and persistence of colonial organisms on the hard structures (Figure 61.8). Nevertheless, some large colonial species are common in this type of habitat, such as black corals (*Leiopathes glaberrima*) and the gorgonian *C. verticillata*. Black corals (mostly *L. glaberrima*) display lower densities (up to 7 colonies $\cdot$ 4 m^{-2}) than porifera and gorgonians, with maximum values generally on chimneys and crusts from mixed bottoms. Black corals showed different densities in some mud volcanoes, such as those within Hormigas ridge, where numbers were highest on Cornide volcano (3.3 colonies $\cdot$ 4 m^{-2}) and lowest on Arcos volcano (0.3 colonies $\cdot$ 4 m^{-2}). The echinoid *C. cidaris* and the asteroid *Peltaster placenta* also occur together with the decapod *Bathynectes maravigna*, the bivalves *Limopsis angusta*, *L. aurita*, *Pseudamussium septemradiatum*, and the brachiopod *Gryphus vitreus*.

Matrix of soft bottoms (sand, mud): The faunistic assemblage includes the solitary coral *Flabellum chunii* and the bamboo coral *Isidella elongata*, the gastropods *Colus gracilis*, *Euspira fusca*, and *Ranella olearia*, the bivalves *Bathyarca philippiana*, *Parvicardium minimum*, *Atrina pectinata*, *Cuspidaria* spp., and the chemosymbiotic *Thyasira* spp., the crustaceans *Nephrops norvegicus*, *Parapenaeus longirostris*, *Munida intermedia*, and *Goneplax rhomboides*, different species of

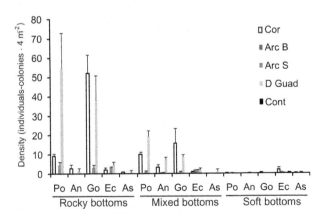

Figure 61.9 Densities of sessile and mobile fauna in different types of habitats from different mud volcanoes estimated using underwater photography. An, Antipatharia; Arc S, Arcos summit; Arc, Arcos base; As, asteroids; Cor, Cornide; Cont, countouritic bottoms; D Guad, Guadalquivir Diapiric ridge; Ec, echinoids; Go, gorgonians; Po, porifera. Mean ± standard error.

frenulates (Polychaetes, Siboglinidae), and the fishes *Bathysolea profundicola, Helicolenus dactylopterus, Leucoraja* spp., and *Callionymus reticulatus,* among others. In the vicinity of the shallow fluid venting areas (Laberinto) are commercial trawling fisheries of the Norway lobster *Nephrops norvegicus* and the rose shrimp *Parapenaeus longirostris,* as well as of the fishes *Merluccius merluccius, Lophius budegassa,* and *Micromesistius poutassou.* These fisheries represent the main threat for the maintenance of the aforementioned benthic assemblages. Nevertheless, porifera, gorgonians, black corals, and echinoderms displayed very-low-density values compared to the other two habitats occurring in mud volcanoes.

Surrogacy

No surrogacy approaches have been carried out in this study.

Acknowledgments

Thanks are due to all those colleagues who participated in data acquisition during the cruises performed on board RV *Cornide de Saavedra,* RV *F. de P. Navarro,* and BIO Hespérides between 1994 and 2009. Thanks are due to captains and crews as well. This work is supported by INDEMARES/CHICA Project (LIFE07/NAT/E/000732).

References

[1] A. Maldonado, L. Somoza, L. Pallarés, The Betic orogen and Iberian–African boundary in the Gulf of Cadiz: geological evolution (Central North Atlantic), Mar. Geol. 155 (1999) 9–43.

[2] J. Baraza, G. Ercilla, Gas-charged sediments and large pockmark like features on the Gulf of Cadiz slope (SW Spain), Mar. Petrol. Geol. 13 (1996) 253–261.

[3] L. Somoza, V. Díaz del Río, R. León, M. Ivanov, M.C. Fernández-Puga, J.M. Gardner, et al., Seabed morphology and hydrocarbon seepage in the Gulf of Cadiz mud volcano area: acoustic imagery, multibeam and ultrahigh resolution seismic data, Mar. Geol. 195 (2003) 153–176.

[4] V. Díaz-del-Río, L. Somoza, J. Martínez-Frías, M.P. Mata, A. Delgado, F.J. Hernández-Molina, et al., Vast fields of hydrocarbon-derived carbonate chimneys related to the accretionary wedge/olistostrome of the Gulf of Cádiz, Mar. Geol. 195 (2003) 177–200.

[5] L.M. Pinheiro, M. Ivanov, N.H. Kenyon, V. Magalhaes, L. Somoza, J. Gardner, Euromargins-MVSEIS Team, et al., Structural control of mud volcanism and hydrocarbon-rich fluid seepage in the Gulf of Cadiz: recent results from the TTR-15 cruise, in: J. Mascle, D. Sakellariou, F. Briand (Eds.), Fluid Seepages/Mud Volcanism in the Mediterranean and Adjacent Domains, vol. 29, pp. 53–58. CIESM Workshop Monographs

[6] C. Martín-Puertas, M.P. Mata, M.C. Fernández-Puga, V. Díaz del Río, J.T. Vázquez, L. Somoza, A comparative mineralogical study of gas-related sediments of the Gulf of Cadiz, Geo-Mar. Lett. 27 (2–4) (2007) 223–235

[7] R. León, L. Somoza, T. Medialdea, F.J. González, V. Díaz-del-Río, M.C. Fernández-Puga, et al., Sea-floor features related to hydrocarbon seeps in deepwater carbonate-mud mounds of the Gulf of Cádiz: from mud flows to carbonate precipitates, Geo-Mar. Lett. 27 (2007) 237–247.

[8] S. Gofas, J.L. Rueda, M.C. Salas, V. Díaz-del-Río, A new record of the giant deep-sea oyster *Neopycnodonte zibrowii* in the Gulf of Cadiz (SW Iberian Peninsula), Mar. Biodivers. Rec. 3 (2010) e72.

[9] V. Díaz-del-Río, D. Palomino, J.T. Vázquez, J. Rueda, L.M. Fernández-Salas, N. López-González, Nuevas evidencias de enlosados y chimeneas carbonatadas en el campo de volcanes de fango de El Laberinto (Golfo de Cádiz, SO de la Península Ibérica), in: F. Rodríguez, J. Gallastegui, G. Flor Blanco, J. Martín Llaneza, (Eds.), Nuevas Contribuciones al Margen Ibérico Atlántico, Universidad de Oviedo, Oviedo, (2009), pp. 297–300.

[10] M. Wisshak, M. López Correa, S. Gofas, C. Salas, M. Taviani, J. Jakobsen, et al., Shell architecture, element composition, and stable isotope signature of the giant deep-sea oyster *Neopycnodonte zibrowii* sp. n. from the NE Atlantic, Deep Sea Res. Part 1 56 (2009) 374–407.

62 Habitats of the Su Su Knolls Hydrothermal Site, Eastern Manus Basin, Papua New Guinea

Yannick C. Beaudoin[1], Samantha Smith[2]

[1]UNEP/GRID- Arendal School of Geosciences, University of Sydney, Australia, [2]Nautilus Minerals Inc., Milton, QLD 4064, Australia

Abstract

Su Su Knolls, located in the eastern Manus Basin, Bismarck Sea, Papua New Guinea, is comprised of three volcanic domes cresting between 1,150 and 1,520 m below sea level. The main geomorphic features examined for this study are hydrothermally generated mounds associated with the development of polymetallic seafloor massive sulfide occurrences. The main habitat types are soft (sediment) and hard (rock) substrates, which can be further categorized by whether or not they are under the influence of current hydrothermal activity. Biological activity and types of animals present are primarily affected by proximity to active venting, with those associated with hydrothermal vents having a chemosynthetic base to their food chain. Predictive mapping using bathymetry alone would have difficulty distinguishing between substrates either near or away from active venting. Although additional indirect observation methods would add to any evaluation, direct observation of the seafloor is needed to conclusively distinguish between habitats.

Key Words: marine minerals, hydrothermal, vent fauna, Su Su Knolls, Solwara, polymetallic sulphides, chemosynthetic

Introduction

The Manus Basin (Figure 62.1) is a rapidly opening (~10 cm/yr) back-arc basin located between directionally opposed fossil and active subduction zones [1]. The basin, as a whole, tectonically lies within the zone of convergence between the Australian and Pacific plates and is composed of two smaller fragments: the North and South Bismarck microplates. Typical seafloor spreading is observed in the central section along the Manus Spreading Center, while the eastern part of the basin is characterized by extensional rifting. The extensional rifting and formation of submarine volcanic dome complexes and hydrothermal mounds in the eastern Manus Basin has led to the formation of the Su Su Knolls, the geomorphic features that are the subject of this case study.

The Su Su Knolls (Figure 62.2) consist of a series of three andesite to dacite volcanic domes, 1.0–1.5 km in diameter, with crests 1,150–1,520 m below sea level [1]. The

Seafloor Geomorphology as Benthic Habitat. DOI: 10.1016/B978-0-12-385140-6.00062-1

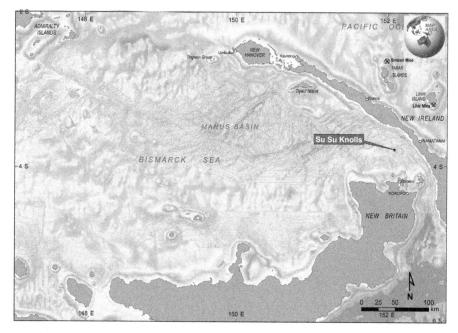

Figure 62.1 Regional map of the Manus Basin showing the location of Su Su Knolls.
Source: Nautilus Minerals.

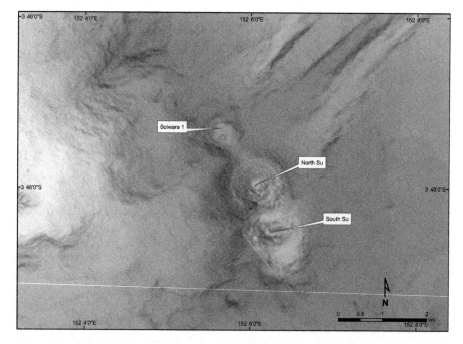

Figure 62.2 High-resolution multibeam bathymetry map of the Su Su Knolls hydrothermal sites.
Source: Nautilus Minerals.

three volcanic domes form a north-northwest trend at a maximum height of ~550 m above the surrounding ocean floor. The highest point of Su Su Knolls is marked by the summit of an active subsea volcano, North Su. Two sites known to exhibit hydrothermal venting, Suzette (also known as Solwara 1) and South Su, lie on the northwest and southeast flanks of North Su, respectively. Hydrothermal vents sometimes host communities of organisms fueled by the chemicals dissolved in the vent fluids. At actively venting locations, chemosynthetic microorganisms form the base of the food chain, supporting diverse animals such as gastropods (snails) and barnacles. At sites where venting has ceased, the animal community make-up is different, with hard substrate animal assemblages often resembling those found on deep seamounts.

In terms of human impacts on the Su Su Knolls site, traditional fisheries operations do not take place at these depths; the main human activity has been local-scale scientific and commercial exploration investigations (including scientific and exploration drilling). Combined with the extensive recent high-resolution multibeam bathymetric mapping conducted over Solwara 1 and South Su (Figure 62.2) and thorough investigations of the hydrothermal vent biota that dominates the setting, these sites provide a reference for examining links between geomorphology, heat flux, and chemosynthetic habitats.

Geomorphic Features and Habitats

The defining physical characteristics of the specific hydrothermal sites at the Su Su Knolls examined in this study include the seafloor massive sulphide mounds of Solwara 1 and South Su, located respectively about 1 km to the northwest and southeast of the central North Su active submarine volcano [2]. In contrast to the more typical approach used to categorize photosynthesis-based benthic habitats according to seafloor morphology (e.g., seamounts, rises, deep-sea canyons), an examination of chemosynthetic communities must consider the main energy source of the lowest trophic level of the system. For the specific biological communities included in this case study, that energy source is the chemicals contained in the hydrothermal fluids present. Although regional morphology has a large-scale role in identifying likely areas with the potential to host chemosynthetic-based communities linked to hydrothermal activity (e.g., active spreading center versus deep abyssal plain), a more firm determination can only be accomplished in combination with the inclusion of an analysis of the physical and temporal parameters that define the hydrothermal venting systems themselves. Taking this into account, five main categories of seafloor habitats best describe the distribution of such communities along the Su Su Knolls:

1. Active and intermittently active hydrothermal vents, as defined by the presence of vent-dependent macrofauna and/or visible signs of venting of warm = water fluids in the immediate vicinity of the site.
2. Hard surfaces (volcanic and hydrothermal rocks) close to vents, with potential for peripheral influence of chemotrophic energy source, but without vent-dependent species or visible signs of venting.

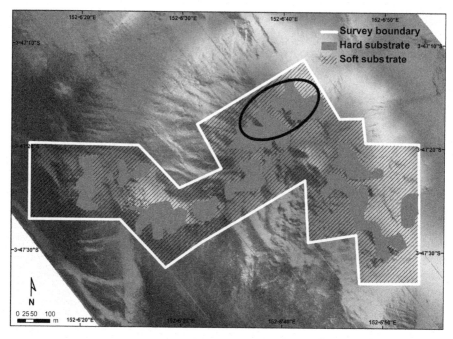

Figure 62.3 Map showing hard- and soft-substrate zones of Solwara 1.
Source: Nautilus Minerals.

3. Hard surfaces remote from vent influences.
4. Unconsolidated sediments (soft substrate) close to hot vent fluids, with diffuse venting or peripheral influence of chemotrophic energy source.
5. Unconsolidated sediments (soft substrate) remote from hot vent fluids.

Proper characterization of the aforementioned habitats can only be effectively achieved with visual observations of the seafloor. This is the only method that can currently accurately confirm, at a site scale, whether a given hydrothermal site is actively venting. Predictive habitat mapping is therefore at present restricted to classification according to the two underlying substrates: hard substrate and soft substrate. These substrate zones, as defined at Solwara 1, are shown in Figure 62.3. For this study, ground-truthing methods included visual surveys using a remotely operated vehicle (ROV) fitted out with high-resolution video cameras, and direct sampling. The visual surveys covered the entire Solwara 1 area, with each animal (or benthic community) and each hard or soft substrate occurrence being logged and recorded.

The hard substrates of Solwara 1 and South Su primarily consist of conical chimney-like structures that form when metal-rich hot fluids emitted from the Earth's interior are released into cold seawater (the ambient seawater temperature at these depths is typically ~2 °C). The metals precipitate out of hot hydrothermal solutions (up to 120 °C), and metal-rich deposits form on the seafloor, resembling spires or chimneys.

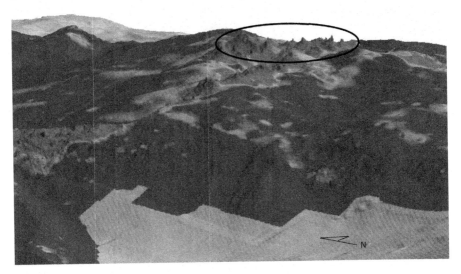

Figure 62.4 3D perspective view of Solwara 1 derived from high-resolution bathymetric mapping. Red indicates location of mineralized zones. Circled area indicates the location of a chimney field, a defining feature of the hard substrate often associated with hydrothermal venting.
Source: Nautilus Minerals.

Figure 62.4 demonstrates how these features appear with detailed bathymetric mapping. Other hard substrates present include blocks/outcrops of volcanic rock.

The soft substrate areas of Solwara 1 and South Su are comprised primarily of dark grey volcaniclastic sandy silts and silty sands [3] and minor hemipelagic muds. It has been proposed that violent hydrothermal eruptions at North Su are the dominant source of the volcaniclastic sediments at Su Su Knolls [4].

Benthic Communities–Hard Substrates

Active hydrothermal vents tend to support a higher abundance of fauna, although of lower diversity, compared to dormant vents and the surrounding deep-sea environments [5]. Where there are active hydrothermal areas, one of the main contributing factors to the biology present is the chemistry of the fluid being emitted.

The benthic communities around vents typically, but not always, include species of tube worms, gastropods, bivalves, and crustaceans. Many of these animals are dependent on the microorganisms that break down hydrogen sulphide in the venting fluid and cannot exist apart from the influence of vent activity.

Benthic communities in the immediate vicinity of the hydrothermal vents are based on chemosynthesis [5], where energy is derived from chemical nutrients rather than sunlight. However, at Solwara 1, South Su, and surrounding areas, there are also hydrothermal vents that are now dormant; these host a different suite of organisms. These

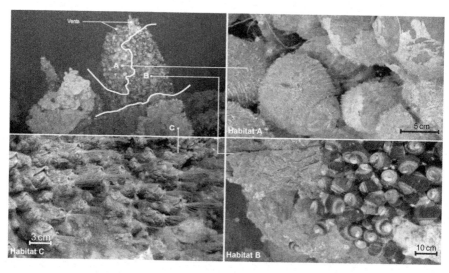

Figure 62.5 Active (venting) hard substrate sites at both Solwara 1 and South Su were dominated by three habitat zones defined by their biomass-dominant species: Habitat A, made up by the gastropod *Alviniconcha* sp. (closest to the vent); Habitat B, made up by the gastropod *Ifremeria nautilei* (middle zone); and Habitat C, made up by the barnacle *Eochionelasmus ohtai* (outer zone, furthest from the vent) [6,7].
Source: Images courtesy of Nautilus Minerals.

"dormant-site" animals are unlikely to be endemic to inactive sulphide systems–at least some have been observed on other hard substrates in the deep sea, such as seamounts [6].

Actively venting sites – the active vent sites at South Su and Solwara 1 share three main habitat zones made up by the gastropods *Alviniconcha* sp. (closest to the vent) and *Ifremeria nautilei* (middle zone), and the barnacle *Eochionelasmus ohtai* (outer zone, furthest from the vent) (Figure 62.5) [6,7]. The species of mussel, *Bathymodiolus manusensis,* and the vestimentiferan tube worms (*Alaysia* sp. and *Arcovestia ivanovi*) were observed only at South Su, where they were associated with the outer barnacle zone. These species are shown in Figure 62.6.

These zones also provide habitat for other species, including crabs, squat lobsters, shrimps, limpets, and polychaete worms that live in association with the biomass-dominant gastropods and barnacles [6].

The species density and diversity at both Solwara 1 and South Su was low for all of the habitat zones when compared with other vent systems worldwide, such as the East Pacific Rise, where the vestimentiferan tube worm habitats (characteristic of vents) are typically more extensive and provide habitats for much greater numbers of epifaunal animals [6].

Dormant sites – the most conspicuous and characteristic species colonizing inactive sites [6] were:

- Suspension-feeding bamboo corals *Keratoisis* sp. (Figure 62.7A)
- Stalked barnacles *Vulcanolepas parensis* (Figure 62.7B)

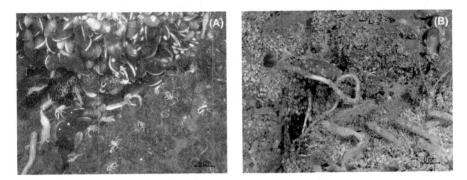

Figure 62.6 Deep-sea (A) mussels (*Bathymodiolus manusensis*) and (B) Vestimentiferan tube worms (*Arcovestia ivanovi*) found at South Su, but not at Solwara 1 [6].
Source: Images courtesy of Nautilus Minerals.

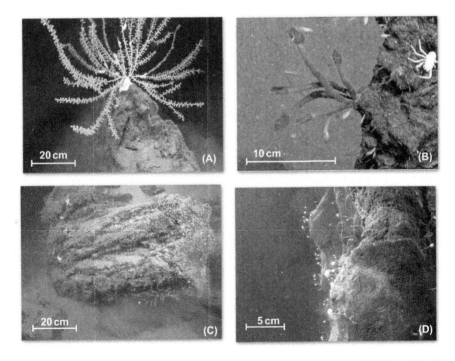

Figure 62.7 (A)bamboo coral (*Keratoisis* sp.) found at dormant sites at both Solwara 1 and South Su [6]; (B) stalked barnacles (*Vulcanolepas parensis*) found at dormant sites at both Solwara 1 and South Su [6]; (C) hydroids found at dormant sites at both Solwara 1 and South Su [6]; (D) carnivorous sponges (*Abyssocladia sp.*) found at dormant sites at both Solwara 1 and South Su [6].
Source: All images courtesy of Nautilus Minerals.

- Hydroids (Figure 62.7C)
- Carnivourous sponges *Abyssocladia* sp. (Figure 62.7D)

Overall, coral and sponge areas were the prominent biogenic features of inactive hard substrates. Assemblages are similar to those observed at seamounts, where dense corals and sponges rely on enhanced delivery of particulate food in flow regimes associated with topographic relief [7]. There is evidence that the bamboo corals near hydrothermal systems obtain some of their organic sulphur from nearby chemotrophic sources [6]. At Solwara 1, they were observed on both sulphide and dacite substrata, which may indicate different species or, more likely, different growth forms in areas of greater exposure to chemotrophic energy (i.e., they can use but are not dependent on this energy) [7].

Benthic Communities–Soft Substrates

At Su Su Knolls, the faunal densities for both active and inactive sediments was low compared to other sites in the eastern Pacific [7,8]. The low numbers of animals made statistical comparisons difficult, and differentiation of active and dormant sites was more complex due to the heterogeneity of the sediments. Variations included color, physical appearance, and biological activity. The dominant taxa included tanaids (crustaceans) and nucleonid bivalves, and most animals were found in the top 2 cm of sediment, with none found more than 5 cm below the seawater-sediment interface [8].

Additional Considerations

Defining Habitats

From bathymetric mapping alone, both active and dormant hard substrate "habitats" could show similar geomorphic characteristics, with distinctive chimney-like structures rising from the seafloor (Figure 62.4). Other parameters, such as venting indicators derived from "plume mapping" and water chemistry analyses, would need to be considered to give a best estimate of which animals might be present at a particular site. Without visual observations of the seafloor, it would be difficult to distinguish either a hard or soft substrate area under the influence of hydrothermal venting from one that is not.

Regional Variability

In addition, it is important to keep in mind that while there are typical "vent animals," not all animals are found at every hydrothermal vent site worldwide. As this paper has shown, even sites 2 km away from one another can exhibit differences in faunal make-up (i.e., there are tube worms and mussels at South Su that are not present at Solwara 1). Some animals, however, are wider spread with, for example, some biomass-dominant species (*Ifremeria nautilei*, *Munidopsis lauensis*) having ranges that extend across all regional back-arc basins, while other species (*Alviniconcha* sp 1 and 2, *Arcovestia ivanovi*) inhabit some basins but not all, and some species (*Bathymodiolus manusensis*) are so far only observed in the Manus Basin [7].

Local-scale Variability

Adding to the lack of clarity regarding observations of chimney features on the seafloor, the natural environment may be variable, and the locations and strength of venting may change. Thus, a habitat map made for one site in one year may not match that made the next year or the year after. This issue of "variable vent activity" was observed at the Solwara 1 site, and the impact this has on the benthic communities, is discussed below.

Areas studied at Solwara 1 in 2006 and visited again in 2007 and 2008 with an ROV fitted with high-resolution cameras revealed that the distribution and level of venting activity at Solwara 1 is variable [7]. There is also evidence of this on a macro scale as observed through year to year variability in regional "plume" mapping [9].

As mentioned above, some animals are dependent on the venting and during the various research campaigns, patches of apparently dead gastropods (snails) were observed in several areas where venting had ceased (Figure 62.8), though sample collection would be necessary to confirm this. These patches may indicate where venting had switched off and/or the movement of snails in search of venting fluid.

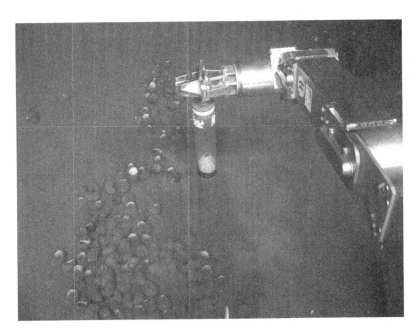

Figure 62.8 Map showing locations of dead gastropod (snail) observations at Solwara 1 [7]. Insert photo shows a patch of presumed dead gastropods observed while acquiring a push-core sample.
Source: Images courtesy of Nautilus Minerals.

These observations are important when characterizing the seafloor habitats of the Su Su Knolls, as they indicate a dynamic system wherein it is difficult to assign the terms "active" or "dormant" to any particular site.

Thus, the presence of vent-dependent species can be used to define locations of current vent activity, because these species cannot exist apart from their chemotrophic energy source. Accordingly, the absence of vent-dependent species may indicate a currently dormant habitat in terms of hydrothermal activity, which may be associated with peripheral areas where there may be intermittent activity or partial (e.g., down-current) influence [7].

References

[1] R.A. Binns, S.D. Scott, J.B. Gemmell, K.A.W. Crook, shipboard party, The SuSu Knolls hydrothermal field, eastern Manus basin, PNG, EOS, Transactions American Geophysical Union 78 (1997) F772.

[2] M. Tivey, W. Bach, J. Seewald, M. Tivey, D. Vanko, Hydrothermal systems in the Eastern Manus Basin: Fluid chemistry and magnetic structure as guides to seafloor processes. Cruise report prepared for Nautilus Minerals Inc., Milton, Queensland, 2006.

[3] E. Hrischeva, S. Scott, Quality including trace elements of sediments from the SuSu Knolls, Manus Basin, Bismarck Sea, Papua New Guinea. Report to Nautilus Minerals Inc., Milton, Queensland, 2008 (Report available from www.cares.nautilusminerals.com)

[4] R.A. Binns, Eastern Manus Basin, Papua New Guinea: Guides for volcanogenic massive sulphide exploration from a modern seafloor analogue, in: T.F. McConachy, B.I.A. McInnes, (Eds.), Copper-zinc massive sulphide deposits in Western Australia, CSIRO Exploration and Mining, Australia, 2004.

[5] C.L. Van Dover, The ecology of hydrothermal vents, Princeton University Press, Princeton, New Jersey, USA, 2000.

[6] P. Collins, C. Logan, M. Mungkaje, R. Jones, K. Yang, C.L. Van Dover, Characterization and comparison of macrofauna at inactive and active sulphide mounds at Solwara 1 and South Su (Manus Basin). Report to Nautilus Minerals Inc., Milton, Queensland, 2008. (Report available from www.cares.nautilusminerals.com).

[7] Coffey Natural Systems and Nautilus Minerals Niugini Limited, Environmental Impact Statement. Solwara 1 Project. Queensland, Australia, 2008. (Report available from www .cares.nautilusminerals.com).

[8] L.A. Levin, F. Mendoza, T. Konotchick, Macrofauna of active and inactive hydrothermal sediments from Solwara 1 and South Su, Manus Basin, Papua New Guinea. Report to Nautilus Minerals Inc., Milton, Queensland, 2008 (Report available from www.cares .nautilusminerals.com).

[9] J.M. Parr, R.A. Binns, Report on the PACMANUS-III cruise, RV "Franklin", Eastern Manus Basin, Papua New Guinea: Exploration and mining Report 345R, CSIRO Division of Exploration and Mining. Australia, 1997.

63 Southern Kermadec Arc– Havre Trough Geohabitats and Biological Communities

Richard J. Wysoczanski, Malcolm R. Clark

National Institute of Water and Atmospheric Research (NIWA), Wellington, New Zealand

Abstract

The Kermadec Arc and Havre Trough are a volcanic arc and back-arc basin system extending north from New Zealand to the Tonga Arc system at 25° 36'S. The region consists of the bounding Colville and Kermadec Ridges, the Kermadec Arc front and the Havre Trough. The southern Kermadec Arc (the focus of this study together with the southern Havre Trough) features volcanoes up to 2,800 m high and active hydro-thermal venting systems. The southern Havre Trough includes basins > 4,000 mbsl. Over 150 fish species (mostly pelagic sharks, tunas, billfishes, and bramids) have been identified, although the fish fauna below 600 mbsl is essentially unknown. Benthic invertebrate assemblages are dominated by echinoderms, cnideria, and arthropods. Hydrothermal vents occur on many of the seamounts, and their communities consist of over 20 invertebrate species. Bathymodiolid mussels are the predominant species at many of the venting sites, with stalked barnacles often associated with black-smoker vents, and alvinocaridid shrimps common at diffuse venting sites. Stratovolcanoes in the south host large beds of an endemic bathymodiolid mussel *Gigantidas gladius*.

Key Words: Kermadec Arc, Havre Trough, oceanic subduction margin, sea-mounts, basins, hydrothermal vents, biodiversity

Introduction

Convergent margins, where oceanic lithosphere is subducted back into the mantle, form the key interface for large-scale element recycling amongst the ocean, crust, mantle, and atmosphere. Here, sediments and oceanic crust are recycled into the Earth's mantle. Conversely, magmas and fluids sourced from the Earth's mantle trans-fer magma, fluid, and gas to the crust, oceans, and atmosphere through volcanism. Importantly, for oceanic habitats this volcanic activity provides nutrients and ele-ments for diverse biological communities and ore deposits (e.g., Cu, Pb, Zn, Au, Ag) concentrated at submarine hydrothermal vents.

Seafloor Geomorphology as Benthic Habitat. DOI: 10.1016/B978-0-12-385140-6.00063-3

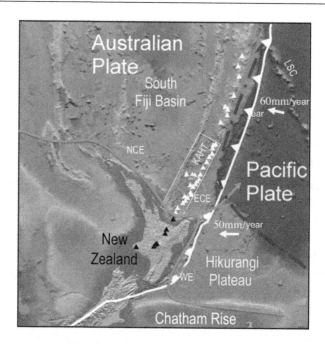

Figure 63.1 Location map of the Kermadec Arc–Havre Trough featuring volcanoes of the
Kermadec Arc (white triangles) and the continental Taupo Volcanic Zone (black triangles).
The white rectangle is the area covered in this study. Ocean currents are the Deep Western
Boundary Current (DWBC: blue) and the East Auckland Current (EAC: orange). ECE = East
Cape Eddy; NCE = North Cape Eddy; WE = Wairarapa Eddy; LSC = Louisville Seamount
Chain. (For interpretation of the references to color in this figure legend, the reader is referred
to the web version of this book.)

The Kermadec Arc volcanic chain and associated Havre Trough back-arc (Figure
63.1) is a type example of an oceanic arc, where the Pacific Plate is being subducted
beneath the Australian Plate [1]. The Kermadec Arc–Havre Trough (KAHT) system
extends southward from 25° 36'S, where subduction of the Louisville Seamount
Chain (LSC) marks the boundary between the northern Kermadec Arc and the Tonga
Arc (Figure 63.1). The southern termination is at 36°40', where the arc undergoes a
dramatic transition from an oceanic to a continental arc. South of ~36°S the KAHT
structure is further complicated by subduction of the Hikurangi Plateau, which is
twice the thickness of normal oceanic crust subducting to the north. The KAHT rises
above sea level as subaerial islands (Raoul, Macauley, and Curtis Islands) and isolated rocks, and reached depths in excess of 4,000 mbsl.

The KAHT is the latest in a succession of volcanic arcs and back-arc basins that
have been established in the SW Pacific since Cretaceous times. At *ca*. 6 Ma, the
proto-Colville-Kermadec arc began rifting, resulting in the opening of the Havre
Trough [2,3]. The present day Colville and Kermadec Ridges are, respectively, the
western and eastern margins of the rifted arc. The present day Kermadec arc was

established by 2 Ma. The principal tectonic sections of the arc/back-arc system, from east to west, are: (i) the Miocene Kermadec Ridge and fore-arc; (ii) the post-Pliocene volcanic arc; (iii) the *ca.* 6 Ma Havre Trough back-arc basin; and (iv) the remnant Miocene Colville Ridge [1].

The ocean current system in this region is dominated at depth (>2,000 m) by the Deep Western Boundary Current, which flows northward around the Chatham Rise, along the northern margin of the Hikurangi Plateau and northward between the Kermadec Ridge and Kermadec Trench, with inflow into the KAHT [4] (Figure 63.1). Ocean surface waters are sourced by the East Australian Current, which flows eastward from Australia across the Tasman Sea, and around North Cape as the East Auckland Current. Ocean currents are further complicated by the anticyclonic warm East Cape Eddy, which flows around the southern Kermadec Ridge, and the North Cape Eddy, northwest of North Cape. Mean transport of the EAC is 9 Sv, with at least an additional 10 Sv circulation in the three eddies [4].

Individual volcanoes of the Kermadec Arc front have been mapped during a number of research cruises involving New Zealand (e.g., NIWA, GNS Science, and universities) and international institutes (e.g., JAMSTEC, NOAA, and WHOI) in the last few years [5,6]. By contrast, the Havre Trough is only partially mapped, with scant multibeam data north of 32°S. As there is little or no sampling or surveying coverage for much of the northern Havre Trough, this study will focus on the southern KAHT (SKAHT) (Figure 63.2), although reference to the northern KAHT (NKAHT) will be made where data are available.

A bathymetric map of the SKAHT (Figure 63.2) has been collated from multiple data sources (including 12 kHz and 30 kHz multibeam data) gridded to a 200 m mesh by Wysoczanski et al. [1]. Backscatter data were also reported in many of the sources, although the quality of the data is highly variable [1]. The SKAHT is characterized by the bounding Colville and Kermadec Ridges, a frontal arc with basins between many seamounts, and a trough consisting of a basal plateau at 2,500 mbsl and basins in excess of 4,000 mbsl. To the east of the arc, the Kermadec–Tonga Trench, which reaches depths of 10,047 m, marks the site of subduction of the Pacific Plate.

Geological and biological sampling of the SKAHT has been conducted for over 50 years using numerous research vessels and sampling techniques. Records of over 300 biological samples, 400 rock dredges, 100 sediment samples, and 30 core sample sites sampled by NIWA alone attest to the scientific interest in the SKAHT. Rock and sediment sampling has generally been restricted to the arc [5] and back-arc regions [1,7], although sediment cores have been collected and analyzed from the fore-arc region, between the Kermadec Trench and Ridge [8]. A summary of the geology and tectonics of the SKAHT is reported by Smith and Price [9] and Wysoczanski et al. [1].

There have been 10 dedicated surveys to describe the benthic biodiversity of seamounts, or groups of seamounts, of the SKAHT, with a number of additional opportunistic sampling stations. These surveys have focused on benthic fauna, in particular invertebrates, for which data on physical specimens is available from the NIWA Invertebrate Collection for 280 stations (comprising 1267 individual taxonomic records).

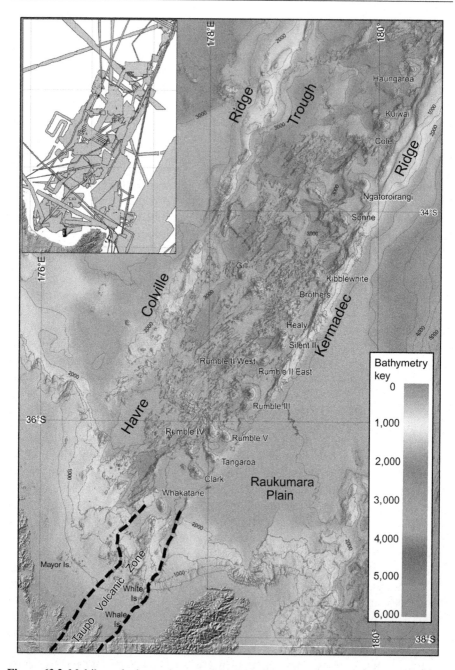

Figure 63.2 Multibeam bathymetric map of the KAHT discussed in this study. Inset shows the extent of multibeam coverage.

Source: Modified after Wysoczanski et al. [1].

Geomorphic Features and Habitats

The Kermadec Arc consists of numerous (>30) volcanic edifices (seamounts and knolls) of both basaltic-andesite stratovolcanoes and silicic caldera complexes. These constructs are generally spaced at 30–50 km intervals and limited to a thin (30 km) NNE trending band west of the Kermadec Ridge. A small number of stratovolcanoes occur in the back-arc region, including close to the Colville Ridge (e.g., Gill volcano). Many of the arc front volcanoes have highly active hydrothermal vent systems with rich biological communities discovered on research cruises in the last decade [5,6]. The Havre Trough consists of a number of basins extending to >4,000 mbsl. Basins also occur along the arc front, separating many of the arc volcanoes.

The geomorphology of the KAHT changes abruptly at ~32°S. Here, the frontal arc converges with the Kermadec Ridge, which together with the Colville Ridge dramatically widens. The Havre Trough is also more sediment filled and has significantly shallower water depths of generally <3,000 mbsl. The region south of this change in geomorphology is hereafter referred to as the southern KAHT, although the distinction between southern, middle, and northern sectors of the KAHT differs in the literature [1,6,9].

The main geomorphic features of the SKAHT are briefly described later. The areal extent of each feature in the SKAHT is given in percent (of total area covered) in Table 63.1 and in the descriptions later.

Table 63.1 Areal Extent (%) of Geomorphic Features of the KAHT

Geomorphic Region	Surface Area (%)
Havre Trough	
Basins (<2,500 m)	27.7
Basal platform (>2,500 m)	34.1
Seamounts	0.5
Kermadec Arc	
Seamounts	4.1
Basins (<2,500 m)	3.7
Ridges (>2500 m)	
Colville Ridge	16.3
Kermadec Ridge	13.6
Total	100.0

Geomorphic Feature	Surface Area (%)
Seamounts	4.6
Basins	31.4
Platforms	34.1
Ridges	29.9
Total	100.0

Ridges: The Colville and Kermadec Ridges are ~20–30 km wide, and have asymmetric profiles, with the troughward side steeper and more eroded than the outward-facing slope, which more gently grades into the plains either side of the SKAHT [10]. The Colville Ridge is slightly larger in area (16.3%) than the Kermadec Ridge (13.6%), and together they make up a significant proportion of the SKAHT (29.9%). The ridges rise above the 2,500 m base of the SKAHT plateaus and ocean plains on either side of the SKAHT and do not reach above sea level. Little is known of their composition due to a lack of sampling, although they are thought to comprise volcanic rock and volcaniclastic sediment and may contain ore deposits [10].

Plateaus and knolls: More than half of the Havre Trough consists of a plateau at approximately 2,500 mbsl, cut by basins and surmounted by linear ridges and knolls. The depth of this basal plateau is coincident with the depth of the ocean plains either side of the SKAHT. The composition of the plateau is uncertain. It may consist of postrifting magmatic activity, or it may be a remnant of the proto-Kermadec Arc that has been stretched by rifting. It displays low backscatter reflectivity, suggesting that it is either sediment covered or consists of old, weathered volcanic rock. Locally, the plateau rises above 2,500 mbsl in the form of linear ridges and knolls that strike orthogonal (045°) to the arc and ridge front (030°). The high reflectivity of these regions, camera observations, and comparisons with the Izu-Bonin arc of Japan suggest that these are young magmatic features, possibly even recent [1]. The most striking area of ridges and knolls on the basal plateau is at the cross-arc chain centred at 36°S, where arc magmatism extends from the arc front to the Colville Ridge [11,12]. The plateau, including areas of ridges and knolls, is the largest (34.1%) geomorphic region of the SKAHT.

Basins: Nearly half of the Havre Trough and the Kermadec Arc front consists of basins that reach depths greater than 4,000 m. The basins make up 31.4% of the entire SKAHT, and generally strike 045°, parallel to the Havre Trough plateau and orthogonal to the arc front. The exception is a central pull-apart basin (unnamed) centred at 34.5°S − 178.7°E, which trends parallel to the arc front. Unlike the majority of other basins, the pull-apart basin shows weak backscatter reflectivity. Visual images and samples of the Ngatoroirangi Rift (immediately SW of Ngatoroirangi Seamount and centered at 33°32′S–179°33′E) collected during two *Shinkai 6500* dives in 2006 suggest the basins are floored by young magmatism, which may reflect present-day rifting or nascent disorganized seafloor spreading [1].

Seamounts: Despite the small area they cover (4.6% of total surface area), seamounts are one of the most conspicuous studied features of the SKAHT. All of them are volcanoes that are currently active or presumed dormant. The majority of seamounts define the Kermadec Arc, although a few (0.5% total surface area) occur in the back-arc region. Some of these are paired with, and 10 km west of, arc-front volcanoes (e.g., Rumble IV west of Rumble V), with the off-arc volcanoes thought to be the older of the pair [11]. In the SKAHT, all but two volcanoes are cone-shaped stratovolcanoes. Brothers and Healy Seamounts, however, are calderas: circular depressions at the volcano summit formed by collapse of the underlying magma chamber [13] (Figure 63.3).

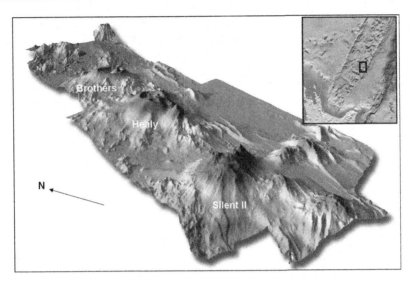

Figure 63.3 Multibeam three-dimensional image of Kermadec arc seamounts (with 5× vertical exaggeration), including the caldera volcanoes Brothers and Healy, and the stratovolcano Silent II. Note the inter-arc basins north of Brothers and west of Silent II, and the flat basal plateau at 2,500 mbsl, SE of Healy.
Source: Figure courtesy of NIWA.

The SKAHT seamounts do not reach sea level. Rumble III comes closest to the sea surface, with a summit at 220 mbsl. It is also the highest at 2,780 m, rising from 3,000 mbsl: the maximum depth of the base of the stratovolcanoes. By contrast, the silicic calderas have low relief (~1,000 m) and the deepest summits (1,350 mbsl for Brothers, 1,100 mbsl for Healy).

Importantly, particularly in terms of biological communities, two-thirds of the arc-front seamounts in the SKAHT are host to active hydrothermal venting [14]. High-temperature black-smoker venting has been particularly well studied at Brothers volcano. In other areas (e.g., Rumble II), venting is characterized by lower temperatures and is more diffuse [15–17]. Other seamounts in the volcanic arc may contain inactive hydrothermal fields. No evidence for venting has yet been found in the back-arc.

The change in morphology of the KAHT north of 32°S indicates that the areal extent of the basal plateau and ridges are substantially greater, and basins significantly smaller than in the SKAHT. Furthermore, the incidence of hydrothermal venting increases to the north (87% in the mid-Kermadec Arc) [14].

Biological Communities

Biological research has focused on the arc volcanoes of the southern region, and biodiversity is, in general, poorly known. Some groups are reasonably well described, such as bryozoans, for which there are records of 256 species, at least 38 of which are currently known only from the Kermadec region. Decapods (crabs, lobsters, and shrimps) number at least 88 species in 70 genera and 30 families. However, it is difficult at present to undertake broad-based analyses of marine biodiversity, so only a general description is given here, focusing on biological communities in the area of the SKAHT, with only a few notes on biodiversity further north.

Overall biodiversity, all habitats: A total of 428 "species" (identified as named species or known to be a unique specimen) of benthic invertebrate are recorded in the NIWA database, from seven phyla (Table 63.2). The most speciose group are echinoderms, especially brittlestars. Cnideria (mostly corals) and arthropods

Table 63.2 A summary of taxonomic group (Phylum level, then further divided by Order, Class or Family) and number of species held in the NIWA National Invertebrate Collection database

Taxonomic Group (Phylum/Order-Family)	Number of Species
Porifera	
Demospongiae	13
Hexactinellida	23
Cnideria	
Anthozoa (black and stony corals)	38
Hydrozoa	10
Octocorallia (seapens, soft corals, gorgonian corals)	35
Bryozoa	34
Brachiopoda	1
Mollusca	
Bivalvia	9
Cephalopoda (squids, octopuses)	2
Gastropoda	32
Scaphopoda	1
Arthropoda	
Amphipoda	8
Decapoda (shrimps, crabs, lobsters, etc.)	52
Isopoda	11
Mysidacea	1
Mysida	1
Pantopoda (sea spiders)	3
Pedunculata (barnacles)	12
Sessilia	2
Echinodermata	
Asteroidea (sea stars)	34
Crinoidea (feather stars)	6
Echinoidea (sea urchins)	14
Holothuroidea (sea cucumbers)	7
Ophiuroidea (brittlestars)	79

Source: "Specify" as at April 2010.

(primarily decapods) are also frequently recorded. In part this is due to these groups having received most attention from taxonomists, and hence other taxa may be found to have more species when examined in the future by experts. More details of the invertebrate fauna are given in the seamounts section later.

A total of about 150 fish species have been reported in the region, but this is believed to be a small proportion of the likely fauna in the general KAHT. The fish fauna of depths greater than 600 mbsl is virtually unknown.

Pelagic fish communities are dominated by sharks, tunas, billfishes, and bramids. Most of these are tropical or subtropical species that migrate southward towards mainland New Zealand as the water warms in summer and autumn. The main sharks are blue (*Prionace glauca*), shortfin mako (*Isurus oxyrinchus*), and thresher (*Alopias vulpinus*) [18]. A commercial line fishery has occurred for yellowtail amberjack (*Seriolla grandis*) [19].

Demersal fish communities are dominated by large predatory fishes, particularly wreckfish (*Polyprion americanus*) and bluenose warehou (*Hyperoglyphe antarctica*). Both have been commercially fished. Hapuku wreckfish (*P. oxygeneios*) and northern spiny dogfish (*Squalus griffini*) are probably at the northern edge of their ranges. Many tropical or subtropical fishes are at the southern end of their ranges, including an endemic moray eel (*Anarchias supremus*) [18]. The Kermadec spiny dogfish (*Squalus raoulensis*) was also recently described from three Raoul Island specimens in the NKAHT and is believed to be endemic.

Ridge fauna: There has been little sampling on the Kermadec and Colville Ridges in the SKAHT. It is likely that the general biological composition and communities of the ridges is similar to that of the non-hydrothermal vent areas of the arc volcanoes. Several dredge samples have been taken on the Colville Ridge, which contained specimens of Bryozoa and bamboo coral.

Most sampling on the Kermadec Ridge has occurred near the Kermadec Islands in the northern KAHT. This northern shallow habitat is notable for its subtropical fauna, and four bryozoan genera (*Kermadecazoon, Reginelloides, Tenthrenulina,* and *Zygopalme*) are thought to be endemic to that part of the Kermadec Ridge (D. Gordon, NIWA, personal communication).

Plateau and knoll fauna: These features of the SKAHT have rarely been sampled. It is likely that general biological composition is similar to that of seamounts (see section below).

Basin fauna: The biodiversity of the deeper areas in the region is poorly known. There has been no direct dredge or trawl sampling in the abyssal depths of the KAHT. Several submersible dives at depths of 2,500–3,500 mbsl were carried out by *Shinkai 6500* in 2006 [1].

Fish species observed in still photographs included cusk eels of the family Ophiidae (potentially of the genera *Halcomycteronus* or *Spectrunculus*), *Bathymicrops breviana-lis*, and a synaphobranchid (A Stewart, Museum of New Zealand Te Papa Tongarewa; P. McMillan, NIWA, personal communications). Holothurians dominated images of soft-sediment seafloor around 3,000 mbsl, while a more diverse assemblage of bamboo corals, primnoid corals, and stalked sponges were recorded on rocky substrate at 3,500 mbsl. A baited camera frame was deployed in a depth just over 7,000 mbsl at the southern end of the Kermadec Trench in 2009, and images obtained of the snailfish *Notoliparis kermadecensis*.

Seamount fauna: The faunal composition on seamounts in the SKAHT varies according to depth of the seamount, its dominant substrate type, and whether hydrothermal venting occurs. A full analysis of all SKAHT data has yet to be completed, pending taxonomic resolution of several invertebrate groups. Nevertheless, it is possible to give a general account of seamount benthic community composition, based largely on direct samples but supplemented with observations from photographic data collected from drop cameras, ROVs, and submersibles.

Invertebrate assemblages on hard substrate are characterized by the presence of gorgonian corals, echinoids, ophiuroids, alcyonaceans, gastropods, and asteroids. The echinoids, such as *Dermechinus horridus*, can form dense patches on rocky ridges along the flanks of the seamounts (Figure 63.4). Coarse substrate is dominated by ophiuroids, asteroids, gastropods, and anemones, with the first three groups also frequent in images of soft substrate [19]. Biodiversity levels can vary considerably between seamounts. Rumble V (with 31 species/sled tow) has up to three times as many benthic invertebrate species as Rumble III (12 species) and Brothers (eight species) [20]. Species richness estimates also differ between a number of seamounts at the southern end of the arc extending into the Bay of Plenty (the northern, offshore sector of the Taupo Volcanic Zone) [21], although the assemblage composition may be similar [19].

Figure 63.4 A "clump" of the urchin *Dermechinus horridus*, common on boulders and exposed rock substrates on ridges and seamounts. The image view is approximately 3 m width, 2.5 m height.
Source: Photo courtesy of NOAA-GNS-NIWA.

In the NKAHT, examination of faunal composition in video images from submersible dives during 2005 [22] indicated that assemblage composition differed between Macaulay, Giggenbach, and Wright seamounts, partly explained by the different summit depths of these seamounts [19]. Depth is widely recognized as an important driver of faunal composition [23].

Schools of demersal fish have been observed during submersible dives on a number of seamounts in the SKAHT. These include bluenose, alfonsino (*Beryx decadactylus*) (Figure 63.5), and pink maomao (*Caprodon longimanus*). Occasional grouper occur near the summit of shallower seamounts.

Numerous seamounts in the SKAHT have active hydrothermal vents. Over 20 species of invertebrates have been identified from ROV and submersible observations of these vents. Many of the vent communities are dominated by bathymodiolid mussels, but there are considerable differences in faunal composition between the seamounts. The black-smoker vent communities on Brothers volcano are dominated by the stalked barnacle *Vulcanolepis osheai*, while on the more diffuse venting region on the cone are large swarms of blind alvinocaridid shrimps, including two recently described species, *Alvinocaris longirostris* and *A. niwa* [24]. Galatheid species include *Munidopsis lauensis* and *M. sonne*. Amongst vestimentiferan tube worms are two species that are new records for New Zealand, *Oasisia fujikurai* and *Lamellibrachia juni*, although these only occur in scattered clumps. In contrast to the barnacle dominance on Brothers, Rumble III and Rumble V seamounts host large beds of an endemic (to the KAHT) bathymodiolid mussel *Gigantidas*

Figure 63.5 Alfonsino (*Beryx decadactylus*) and a single roughy (*Hoplostethus* sp.) near the summit of Rumble V seamount. The image size is approximately 2.5 m width, 2 m height. *Source:* Image courtesy of NOAA-GNS-NIWA.

Figure 63.6 The predatory sea star *Rumbleaster eructans* on beds of the mussel *Gigantidas gladius* at a site of hydrothermal venting on Rumble III seamount. Image view is approximately 2.5 m width, 2 m height.
Source: Photo courtesy of NIWA.

gladius, as well as *Bathymodiolus* spp. [25,26]. Beds of *Gigantidas* often have concentrations of the predatory sea star *Rumbleaster eructans* (Figure 63.6). Shallower seamounts in the NKAHT, including Macauley and Giggenbach, have different species of *Bathymodiolus* [27], beds of a lucinid clam *Bathyaustriella thionipta*, and xenograpsid vent crabs *Gandalfus puia* and *Xenograpsus ngatama* [28]. The species of *Bathymodiolus* appears to change northwards along the ridge, as several southern seamounts have mainly *Bathymodiolus* (sp. "NZ1") [27], and Monowai seamount in the NKAHT just outside the EEZ comprises predominantly *B. manusensis*.

As well as the vent assemblages differing between individual sites, larger regional-scale trends include only scattered pockets of tube worms on southern seamounts of the SKAHT, becoming more abundant northwards. The reverse is the case for stalked barnacles, which are not found in northern vent sites. Also, the species composition of mussels changes, with *Gigantidas gladius* found mainly in the south and the species of *Bathymodiolus* changing northwards along the arc. In comparison with vents north of New Zealand, there is a notable absence of provannid gastropods [29].

Few fish have been observed in close association with hydrothermal venting in the SKAHT. Known species include the zoarcid vent fish *Pyrolycus moelleri* at Brothers seamount, and high densities of a small flatfish (*Symphurus thermophilus*) near sulfur-rich rocks on Rumble III (as well as Macaulay seamount in the NKAHT) [30].

Surrogacy

Research is currently being carried out on the association between benthic invertebrate faunal composition and the nature of the substrate and geological feature type. However, at this stage no statistical analyses have been carried out.

Acknowledgments

We would like to thank Kareen Schnabel (NIWA) for extracting invertebrate data from the NIC database, and Malcolm Francis (NIWA) and Andrew Stewart (Te Papa) for information on some of the Kermadec fishes. Detailed reviews from Claudio Lo Iacono and an anonymous reviewer helped to improve the manuscript. This work was in part funded by Foundation for Research Science and Technology (New Zealand), contracts C01X0808 and C01X0906.

References

[1] R.J. Wysoczanski, E. Todd, I.C. Wright, M.I. Leybourne, J.M. Hergt, C. Adam, et al., Backarc rifting, constructional volcanism and nascent disorganised spreading in the southern Havre Trough backarc rifts (SW Pacific), J. Volcanol. Geotherm. Res. 190 (2010) 39–57.

[2] E. Ruellan, J. Delteil, I.C. Wright, T. Matsumoto, From rifting to active spreading in the Lau Basin–Havre Trough backarc system (SW Pacific): locking/unlocking induced by seamount chain subduction, Geochem. Geophys. Geosys. 4 (5) (2003). 10.1029/2001GC000261.

[3] N. Mortimer, P.B. Gans, J.M. Palin, S. Meffre, R.H. Herzer, D.N.B. Skinner, Location and migration of Miocene-Quaternary volcanic arcs in the SW Pacific region, J. Volcanol. Geotherm. Res. 190 (2010) 1–10.

[4] D. Roemmich, P. Sutton, The mean and variability of ocean circulation past northern New Zealand: determining the representativeness of hydrographic climatologies, J. Geophys. Res. C 103 (1998) 13,041–15,054.

[5] I.C. Wright, T.J. Worthington, J.A. Gamble, New multibeam mapping and geochemistry of the 30°–35°S sector, and overview, of southern Kermadec arc volcanism, J. Volcanol. Geotherm. Res. 149 (2006) 263–296.

[6] I.J. Graham, A.G. Reyes, I.C. Wright, K.M. Peckett, I.E.M. Smith, R.J. Arculus, Structure and petrology of newly discovered volcanic centers in the northern Kermadec–southern Tofua arc, South Pacific Ocean, J. Geohys. Res. B 113 (2008). 10.1029/2007JB005453.

[7] R.J. Wysoczanski, I.C. Wright, J.A. Gamble, E.H. Hauri, J.F. Luhr, S.M. Eggins, et al., Volatile contents of Kermadec Arc–Havre Trough pillow glasses: fingerprinting slab-derived aqueous fluids in the mantle sources of arc and back-arc lavas, J. Volcanol. Geotherm. Res. 152 (2006) 51–73.

[8] J.A. Gamble, J. Woodhead, I.C. Wright, I.E.M. Smith, Basalt and sediment geochemistry and magma petrogenesis in a transect from oceanic island arc to rifted continental margin arc: the Kermadec–Hikurangi margin, SW Pacific, J. Petrol. 37 (1996) 1523–1546.

[9] I.E.M. Smith, R.C. Price, The Tonga-Kermadec arc and Havre-Lau back-arc system: their role in the development of tectonic and magmatic models for the western Pacific, J. Volcanol. Geotherm. Res. 156 (2006) 315–331.

[10] I. Wright, Morphology and evolution of the remnant Colville and active Kermadec arc Ridges south of 33°30'S, Mar. Geophys. Res. 19 (1997) 177–193.

[11] I.C. Wright, L.M. Parson, J.A. Gamble, Evolution and interaction of migrating cross-arc volcanism and backarc rifting: an example from the southern Havre Trough (35°20' − 37°S), J. Geohys. Res. B 101 (1996) 22,071–22,086.

[12] E. Todd, J.B. Gill, R.J. Wysoczanski, M. Handler, I.C. Wright, J.A. Gamble, Sources of constructional cross-chain volcanism in the southern Havre Trough: new Insights from HFSE and REE concentration and isotope systematics, Geochem. Geophys. Geosys. 11 (2010) 4, 10.1029/2009GC002888.

[13] I.C. Wright, J.A. Gamble, Southern Kermadec submarine caldera arc volcanoes (SW Pacific): caldera formation by effusive and pyroclastic eruption, Mar. Geol. 161 (1999) 207–227.

[14] C.E.J. de Ronde, E.T. Baker, G.J. Massoth, J.E. Lupton, I.C. Wright, R.J. Sparks, et al., Submarine hydrothermal venting activity along the mid-Kermadec arc, New Zealand: large-scale effects on venting, Geochem. Geophys. Geosys. 8 (2007) Q07007, 10.1029/2006GC001495.

[15] I.C. Wright, C.E.J. de Ronde, K. Faure, J.A. Gamble, Discovery of hydrothermal sulfide mineralization from southern Kermadec arc volcanoes (SW Pacific), Earth Planet Sci. Lett. 164 (1998) 335–343.

[16] C.E.J. de Ronde, E.T. Baker, G.J. Massoth, J.E. Lupton, I.C. Wright, R.A. Feely, et al., Intraoceanic subduction-related hydrothermal venting, Kermadec volcanic arc, New Zealand, Earth Planet Sci. Lett. 193 (2001) 359–369.

[17] C.E.J. de Ronde, M.D. Hannington, P. Stoffers, I.C. Wright, R.G. Ditchburn, A.G. Reyes, et al., Evolution of a submarine magmatic-hydrothermal system: Brothers volcano, southern Kermadec arc, New Zealand, Econ. Geol. 100 (2005) 1097–1133.

[18] M.P. Francis, D.P. Gordon, S.T. Ahyong, L.H. Griggs, Marine biodiversity of the Kermadec region. NIWA Client Report WLG2009-20, 2009, 27 pp. (unpublished report available from NIWA, Private Bag, Wellington).

[19] J. Beaumont, A.A. Rowden, M.R. Clark, Deepwater biodiversity of the Kermadec Islands Coastal Marine Area. NIWA Client Report WLG2009-55, 2009, 126 p. (unpublished report available from NIWA, Private Bag, Wellington).

[20] A.A. Rowden, M.R. Clark, S. O'Shea, D.G. McKnight, Benthic biodiversity of seamounts on the southern Kermadec volcanic arc. Marine Biodiversity Biosecurity Report No. 3, 2003, 23 p.

[21] A.A. Rowden, M.R. Clark, Benthic biodiversity of seven seamounts at the southern end of the Kermadec Arc. Aquatic Environment and Biodiversity Report No. 62, 2010, 31p.

[22] S. Merle, R. Embley, W. Chadwick, New Zealand American Submarine Ring of Fire 2005 Kermadec Arc Submarine Volcanoes. Unpublished report can be made available through NIWA, Private Bag 14901, Wellington, New Zealand, 2005.

[23] R.S. Carney, Zonation of deep biota on continental margins, Oceanogr. Mar. Biol. Annu. Rev. 43 (2005) 211–278.

[24] W.R. Webber, A new species of *Alvinocaris* (Crustacea: Decapoda: Alvinocaridae) and new records of alvinocarids from hydrothermal vents north of New Zealand, Zootaxa 444 (2004) 1–26.

[25] M.R. Clark, S. O'Shea, Hydrothermal vent and seamount fauna from the southern Kermadec Ridge, New Zealand, International Ridge-Crest Research: Biological Studies 10 (2001) 14–17.

[26] R. Von Cosel, B.A. Marshall, Two new species of large mussels (Bivalvia: Mytilidae) from active submarine volcanoes and a cold seep off the eastern North Island of New Zealand, with description of a new genus, The Nautilus 117 (2) (2003) 31–46.

[27] P.J. Smith, S.M. McVeagh, Y. Won, R.C. Vrijenhoek, Genetic heterogeneity among New Zealand species of hydrothermal vent mussels (Mytilidae: Bathymodiolus), Mar. Biol. 144 (2004) 537–545.

[28] S. Ahyong, Deepwater crabs from seamounts and chemosynthetic habitats off eastern New Zealand (Crustacea: Decapoda: Brachyura), Zootaxa 1708 (2008) 2–78.

[29] A.A. Rowden, M.R. Clark, Macrobenthic Ecology, in: New Zealand American Submarine Ring of Fire 2005 Kermadec Arc Submarine Volcanoes. (Compiled by Merle, S., Embley, R., and Chadwick, W.). Unpublished report can be made available through NIWA, Private Bag 14901, Wellington, New Zealand, 2005.

[30] T.A. Munroe, J. Hashimoto, A new Western Pacific tonguefish (Pleuronectiformes: Cynoglossidae): the first Pleuronectiform discovered at active hydrothermal vents, Zootaxa 1839 (2008) 43–59.

Part III

Synthesis

64 Geohab Atlas of Seafloor Geomorphic Features and Benthic Habitats: Synthesis and Lessons Learned

Peter T. Harris[1], Elaine K. Baker[2]

[1]Marine and Coastal Environment Group, Geoscience Australia, Canberra, ACT, Australia
[2]UNEP/GRID- Arendal School of Geosciences, University of Sydney, Australia

Abstract

This chapter presents a broad synthesis and overview based on the 57 case studies included in Part 2 of this book and on questionnaires completed by the authors. The case studies covered areas of seafloor ranging from 0.15 km^2 to over 1,000,000 km^2 (average of 26,600 km^2) and a broad range of geomorphic feature types. The mean depths of the study areas ranged from 8 to 2,375 m, with about half of the studies on the shelf (depth $<$ 120 m) and half on the slope and at greater depths. Mapping resolution ranged from 0.1 to 170 m (mean of 13 m). There is a relatively equal distribution of studies across the four naturalness categories: near pristine ($n = 17$), largely unmodified ($n = 16$), modified ($n = 13$), and extensively modified ($n = 10$). In terms of threats to habitats, most authors identified fishing ($n = 46$) as the most significant threat, followed by pollution ($n = 12$), oil and gas development ($n = 7$), and aggregate mining ($n = 7$). Anthropogenic climate change was viewed as an immediate threat to benthic habitats by only three authors ($n = 3$).

Water depth was found to be the most useful surrogate for benthic communities in the most studies ($n = 17$), followed by substrate/sediment type ($n = 14$), acoustic backscatter ($n = 12$), wave-current exposure ($n = 10$), grain size ($n = 10$), seabed rugosity ($n = 9$), and bathymetric/topographic position index (BPI/TPI) ($n = 8$). Water properties (temperature, salinity) and seabed slope are less useful surrogates. Multiple analytical methods were used to identify surrogates, with ARC GIS being by far the most popular method (23 out of 44 studies that specified a methodology).

Of the many purposes for mapping benthic habitats, four stand out as preeminent: (1) to support government spatial marine planning, management, and decision making; (2) to support and underpin the design of marine protected areas (MPAs); (3) to conduct scientific research programs aimed at generating knowledge of benthic ecosystems and seafloor geology; and (4) to conduct living and nonliving seabed resource assessments for economic and management purposes. Out of 57 case studies, habitat mapping was intended to be part of an ongoing monitoring program in 24 cases, whereas the

Seafloor Geomorphology as Benthic Habitat. DOI: 10.1016/B978-0-12-385140-6.00064-5

mapping was considered to be a one-off exercise in 33 cases. However, out of the 33 one-off cases, the authors considered that their habitat map would form the baseline for monitoring future changes in 24 cases. This suggests that governments and regulators generally view habitat mapping as a useful means of measuring and monitoring change. In terms of the perceived clients and users of habitat maps, most authors considered industry, marine conservation, and the scientific community to be the primary users of habitat maps. However, the overwhelming majority of habitat surveys were funded by government or government-funded agencies/institutions ($n = 49$), with only minor funding from private industry ($n = 7$) or nongovernment organizations ($n = 4$).

A gap analysis (i.e., geomorphic features and habitats not included in the case studies) illustrates that whereas shelf and slope habitats are well represented in the case studies, estuarine and deltaic coastal habitats plus deep ocean (abyssal–hadal) environments were described in only a few case studies. Geographically, about half of the case studies were from waters around Western Europe, while the margins of the continents of Africa, Asia, and South America were not represented in any case study. Given the intense pressures facing benthic habitats and broad regional differences in ecosystems, species, and habitats, future case studies from these regions should be specifically sought for future editions of the Atlas.

Key Words: Naturalness, surrogates, spatial marine planning, habitat classification schemes, habitat mapping clients, environmental data

Introduction

This chapter presents a broad synthesis and overview based on the 57 case studies presented in Part 2 of this volume. Authors were asked to prepare their case studies using a template that specified required information. To be accepted, case studies had to contain both geomorphic and biologic data, contain a clear description of at least one geomorphic feature type, describe the oceanographic setting, and provide an assessment of the naturalness of the environment. The spatial comparison of biological data with spatial physical data is a key element of every case study. Authors were given the opportunity to describe surrogacy relationships and methods used to identify and quantify them.

On submitting their final papers, authors completed a questionnaire[1] to provide specific details of their study area and other issues ranging from surrogates to data storage. Based on the content of the case studies, together with the authors' responses to the questionnaire, the following topics are considered in the present synthesis of information:

1. study area attributes, including naturalness and anthropogenic threats;
2. surrogates used (and not used) and spatial analysis techniques;
3. the use of published habitat classification schemes;
4. the purpose of habitat mapping (sectors and clients);
5. consultation in habitat survey planning;
6. data storage and accessibility;

---

[1] The questionnaire is reproduced in Appendix 1 of this chapter.

7. gap analysis;
8. best practices for habitat mapping.

Not all parts of the template, nor every question in the questionnaire, are applicable to every case study, and there is more than one possible response to questions in some cases (e.g., multiple mapping methods were used, several or no surrogacy methods used, multiple stakeholders, multiple funding sources). Nevertheless, the authors' responses provide an insight from the perspective of the GeoHab community[2] on the purpose, scope, and value of habitat mapping for the benefit of the broader community.

In the following, we present the responses from the case study authors as a series of graphs and statistical statements. These are followed by our interpretation of the findings with some possible explanations. Unless otherwise stated, the interpretations and explanations provided are from the authors of this chapter and are not necessarily the views of the case study authors.

Attributes of Case Studies

The case studies covered areas of seafloor ranging from $0.15 \mathrm{km}^2$ to over $1,000,000 \mathrm{km}^2$ (average of $26,600 \mathrm{km}^2$) and a broad range of geomorphic feature types. The mean depths of the study areas ranged from 8 to $2,375 \mathrm{m}$, with about half of the studies on the shelf (depth $<120 \mathrm{m}$) and half on the continental slope and at greater depths (depth $>120 \mathrm{m}$). Mapping resolution ranged from 0.1 to $170 \mathrm{m}$ (mean of $13 \mathrm{m}$).

The naturalness of the environment examined in each case study was described by the authors in their own terms. For the purposes of this discussion, we assigned each study to one of four categories: near pristine, largely unmodified, modified, and extensively modified. The results (Figure 64.1) illustrate that there is a good distribution of studies among the four naturalness categories. As might be expected, the modified and extensively modified case studies tended to be located in shallow water depths, on the shelf or in the coastal zone, whereas largely unmodified and near-pristine environments were located in more remote deep-water (slope and abyssal) areas.

In describing the naturalness of study areas, authors specified the main anthropogenic threats that they perceived as having an immediate impact upon the habitats they studied. In several cases, habitat mapping was specifically undertaken in order to manage specific pressures from one or more human uses. The case studies in this book identified the impacts of fishing as the most significant threat to benthic habitats (Figure 64.2). Fishing has four times the number of citations as the second most perceived threat (pollution). Pollution includes the effects of marine debris, which was the most commonly cited form of pollution. Oil and gas development, aggregate mining, and coastal development are perceived as also being among the more significant threats (Figure 64.2).

---

[2] The "GeoHab community" for the purposes of this discussion encompasses all participants in past and future GeoHab conferences and publications (including this book) and those engaged in habitat-mapping activities.

Anthropogenic climate change is viewed as an immediate threat to benthic habitats by only three authors (Chapters 21, 24, and 34; Figure 64.2). This is not necessarily evidence that climate change is not a significant threat to the marine environment in the longer term, but it does indicate that the experts working in the field of habitat mapping are currently documenting the harmful consequences of other human pressures (particularly fishing and pollution) on the condition and status of the marine environment. Future impacts of anthropogenic climate change will

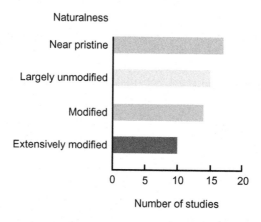

Figure 64.1 Graph showing the naturalness of environments in the case studies.

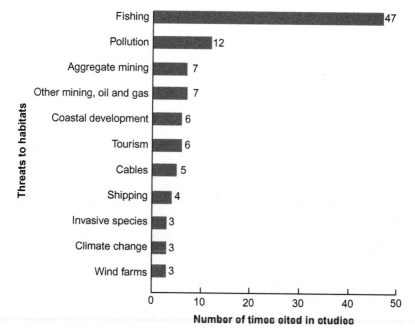

Figure 64.2 Graph showing the anthropogenic pressures cited in the case studies.

be in addition to damage already suffered by marine habitats due to the cumulative impacts of pressures (fishing, pollution, mining, etc.) that have occurred over the past century (see Chapter 3).

Surrogates Used (and Not Used) and Spatial Analysis Techniques

Three questions posed in the questionnaire were related to the variables that were mapped in each study area and their usefulness as surrogates for the occurrence of benthic biota. The extent to which a particular surrogate performed was generally assessed via various statistical tests and multivariate analysis techniques. The most commonly measured variables in the case studies were (in decreasing order of citation frequency): depth (measured in all studies), sediment grain size, acoustic backscatter, seabed slope, substrate or seabed type, and rugosity (Figure 64.3). Each of these parameters is cited by over 20 authors as having been measured as part of the case study.

However, not all measured variables were found to be *useful* surrogates for the occurrence of benthic organisms. This was determined using a range of analytical methods to identify surrogates and their predictive power. ArcGIS is by far the single most popular tool (23 out of 44 studies that specified a methodology). However, a

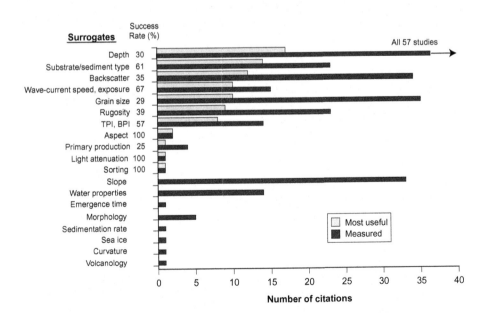

Figure 64.3 Comparison of the number of times variables were measured (blue) versus variables found to be the most useful surrogates for the occurrence of biota (yellow). Water depth was measured in all 57 case studies. (For interpretation of the references to color in this figure legend, the reader is referred to the web version of this book.)

number of different multivariate analysis methods are also used (24 studies), often in parallel with GIS. Of the various multivariate analysis software packages, PRIMER software and its incorporated tools were most commonly used to find relationships between physical and biological data, although other software was used to a lesser extent (see Table 5.2 in Chapter 5).

The most useful surrogate for benthic communities was found to be water depth, followed by substrate/sediment type, acoustic backscatter, wave-current exposure, grain size, seabed rugosity, and BPI/TPI. It is interesting to note the variability in the "success rate" of surrogates (i.e., the number of times a parameter was found to be useful divided by the times it was measured). Some surrogates were measured in only a few studies but were found to be useful in every case [100% success rate; e.g., aspect (Chapter 29), light attenuation (Chapter 24), and sorting (Chapter 42); Figure 64.3]. Wave-current speed was measured in 15 studies and was found to be a useful surrogate in 10 of them (67% success rate), giving this parameter a higher success rate than depth, substrate type, rugosity, TPI/BPI, backscatter, or grain size (Figure 64.3).

There are a number of possible explanations as to why there is a discrepancy between the variables measured versus those considered by authors to be the most useful surrogates (Figure 64.3). Indirect variables such as depth and backscatter correlate with several direct variables, which possibly explain their overall good performance as surrogates. For example, backscatter correlates with sediment grain size, seabed rugosity, and substrate type, which are direct variables that individually performed well as surrogates (Figure 64.3). Some variables, such as emergence time, sea-ice presence, and light attenuation, are not broadly applicable to all marine habitats; they apply only to a small subset. Hence, they were not given an equal opportunity in this survey, which covers all of the case studies.

It may also be argued that different surrogates will be more useful in some environments than in others. For example, water properties such as salinity are more likely to be important surrogates in estuaries than in the open marine environment (and only five of the case studies in this book were from estuarine or fjord environments; Chapters 11, 12, 18, 19, and 20). In the open marine environment, water properties vary most significantly over broad spatial and temporal scales, so unless the case study encompasses a broad area (or data are collected over a long time series) variability in water properties may not be large enough to register as a useful surrogate.

Some data sets that have already been collected by different agencies over a period of time prior to a survey may be freely available at no extra cost and can be incorporated into the surrogacy analysis. Examples are commercial fisheries data, oceanographic time series observations, and modeled ocean currents, waves, and tides. These data can supplement the measurements and observations collected during the habitat mapping survey, but they may not have been collected at an appropriate resolution or in exactly the right location, thereby affecting their performance as a surrogate.

It must be accepted, however, that some variables are simply not good surrogates for the occurrence of biota. Overall, the survey results suggest that water properties and seabed slope are not particularly useful in the environments studied; these variables were measured in more than 15 and 30 case studies (respectively) but not reported as being a useful surrogate in any of them.

Use of Published Classification Schemes

GeoHab has not adopted or endorsed any particular habitat classification scheme, but much effort has gone into the design of various schemes, as reviewed in Chapter 4. Among the several advantages of using standardized classification schemes is that they enable comparisons to be made between different studies, by providing a standard framework and terminology which prompts the user for information through the course of the study. Also, their intrinsic predictive power of describing the relationships between physical habitats and their associated biological communities is useful for analysis, interpretation, and dissemination of results.

The part of the questionnaire dealing with habitat classification schemes was completed by all 57 authors. Several authors used more than one scheme (Chapters 18, 26, 28, 29, and 40). At least one scheme was used in 21 of the case studies, but no scheme was used in 36 studies (i.e., most of the case studies, 36 out of 57, did not use any previously published habitat classification scheme; Figure 64.4). Most schemes were applied in only one or two studies. No single scheme was used by more than five studies. The scheme of Greene et al. [1] is the most cited.

Several schemes listed in Figure 64.4 were developed by national or regional science agencies, and these were sometimes used in the case studies carried out in the relevant jurisdiction. However, the rate of uptake and utilization of schemes is problematic. For example, although 24 studies were carried out in European waters, only four of those studies applied the EUNIS habitat scheme (Figure 64.4).

There may have been some confusion among the GeoHab community as to what constitutes a classification scheme. Some classification schemes nominated by

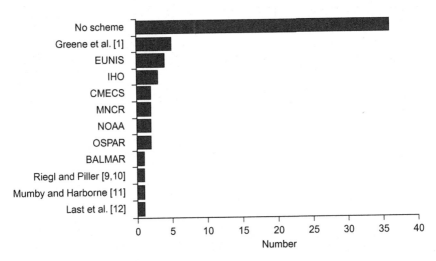

Figure 64.4 List of different habitat classification schemes and the number of times they were used in the case studies. The schemes are as follows: Greene et al. [1]; EUNIS [2]; IHO [3]; CMECS [4]; MNCR [5]; NOAA [6]; OSPAR [7]; BALMAR [8]; Riegl and Piller [9,10]; Mumby and Harborne [11]; and Last et al. [12].

authors in their questionnaires might be better described as definitions of terms and nomenclature. For example, the IHO [3] report is simply a list of geomorphic feature names and broad generic descriptions, rather than an actual classification scheme (with decision rules, etc.). Several authors cited the IHO definitions but did not nominate it as a habitat classification scheme in the questionnaire. The OSPAR [7] list of threatened species and habitats is similarly not an actual classification scheme.

Apart from not fully understanding classification schemes, there may be other reasons why authors generally avoided using them. The extra effort required to apply a published classification scheme may come at too high a price for projects that are on a limited budget. The advantages of using a scheme are mostly only realized downstream in a project, where comparison of the results with other studies and communication become a priority. It is clear that further investigation is needed to fully understand the low rate of habitat classification scheme usage among GeoHab scientists.

The Purpose of Habitat Mapping (Sectors and Clients)

As was reviewed in Chapter 1, there are many purposes for mapping benthic habitats, but four of these stand out as being preeminent: (1) to support government spatial marine planning, management, and decision making; (2) to support and underpin the design of MPAs; (3) to conduct scientific research programs aimed at generating knowledge of benthic ecosystems and seafloor geology; and (4) to conduct living and nonliving seabed resource assessments for economic and management purposes (see Table 1.1 in Chapter 1). Only the final category (conduct living and nonliving seabed resource assessments) is directly relevant to an economic outcome; all others are related to government management and planning (Figure 64.5).

One of the commercial purposes of habitat mapping specified in the list (Figure 64.5) relates to field testing of equipment. It is an interesting observation that seafloor mapping technology has developed over a number of decades through a close partnership between scientists and engineers. Habitat mapping is inextricably linked to developments in marine technology (especially sonar technology), and advances made in one field almost always lead to advances in the other.

Habitat mapping was intended to be part of an ongoing monitoring program in 24 out of 57 cases (two case studies specified that part of their purpose was an ongoing monitoring program; Chapters 10 and 18). By contrast, habitat mapping was considered to be a one-off exercise in 33 case studies. This statistic is consistent with the usual government funding of marine research, where support can often be obtained for a one-off survey to collect new data about a particular environment or ecosystem process. Getting support for ongoing environmental monitoring is more difficult to obtain. However, out of the 33 one-off cases, 24 reported that their habitat map would form the baseline for monitoring future changes. This suggests that governments and regulators generally view habitat mapping as a useful means of measuring and monitoring environmental change, even where a decision to support ongoing monitoring has not been taken.

There appears to be a disconnect between the stated purpose of habitat mapping in most case studies (i.e., government management), and the end users and clients

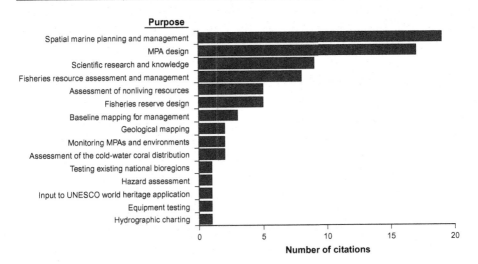

Figure 64.5 Purposes of habitat mapping specified by case study authors and the number of times they were cited.

perceived by case study authors to be the main stakeholders (industry). Most authors considered marine conservation to be the biggest single user (Figure 64.6), followed by the fishing industry, government regulators, the scientific community, the tourism industry, navigation, other industry (deep-sea minerals, wind farms, etc.), oil and gas industry, and aggregate mining. Grouping all industry sectors together shows that most authors believe their mapping work is more relevant to industry in a general sense than to either conservation or government and science applications (Figure 64.6).

Although the GeoHab scientists considered their data to be useful to industry (Figure 64.6), the overwhelming majority of habitat surveys were funded by government or government-funded agencies/institutions ($n = 49$), with only minor funding from private industry ($n = 7$) or nongovernment organizations ($n = 4$). The role of government agencies as a catalyst for encouraging investment and interest from industry may be a key to understanding this apparent contradiction. For example, government agencies which deal with the management of natural resources often have a role of collecting environmental data to support and encourage industry investment in resource development; hence a government-funded survey may actually have industry as a primary purpose. The multiple potential uses of habitat mapping information may also be an explanation, as the data collected by one government agency to encourage industry investment might also be used by a different agency to make decisions about zoning and conservation.

Consultation in Habitat Survey Planning

Habitat mapping surveys are complex and expensive enterprises, requiring extensive planning and consultation to ensure successful outcomes. Consultation with all

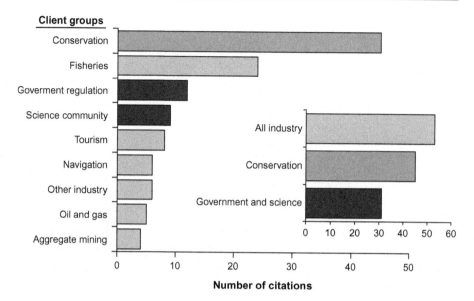

Figure 64.6 Users and clients of habitat mapping specified by case study authors and the number of times they were cited. Inset shows that industry-related users grouped together actually form the largest single group.

stakeholders is essential. Stakeholders comprise the survey participants (including the ship's crew) and may also include industry practitioners, government managers, members of the scientific community, and (where relevant) representatives of indigenous or traditional users of the environment being studied. The latter group was consulted mainly in shallow water surveys along the coast or inner shelf areas.

In general terms, the GeoHab science community has clearly embraced consultation, with most case studies consulting with at least three of the five broad stakeholder groups listed in Figure 64.7. All five stakeholder groups were consulted in eight case studies and four groups were consulted in 10 case studies. All case studies consulted with at least one stakeholder group. In the two studies that consulted only one group, it was the government that was engaged in planning the surveys.

Data Storage and Accessibility

The data collected during a habitat mapping survey is composed of many types, including electronic data (multibeam sonar, underwater video or camera data, current and water property measurements, modeled data, etc.) together with physical samples (sediments and biota) that must first be analyzed before numerical or classification type data are obtained. Government-funded agencies or institutions commonly have an obligation to store the data and samples they collect (for a period of time at least), which is consistent with the response to the question, "Is the habitat mapping

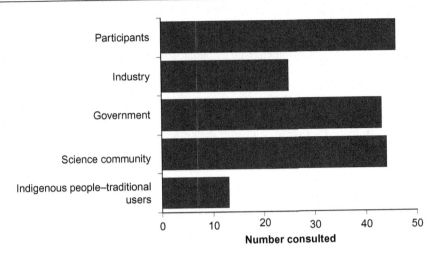

Figure 64.7 Stakeholder groups consulted in preparation of habitat mapping surveys specified by case study authors and the number of times they were cited. A total of 54 case studies responded to the question about stakeholder consultation.

data stored or saved for future use?" Out of 54 responses, 45 (83%) said yes and 9 (17%) said no.

As mentioned above, the majority of habitat surveys are funded by government or government-funded agencies/institutions, which have an obligation to the taxpayers to make their data and survey results freely available and accessible. Indeed, it is commonly a justification for collecting the data in the first place that it will be used by the wider community for more than one purpose (map once, use many ways). This perception is confirmed by the responses received from the question as to "whether the habitat mapping data collected for each case study had been used for any other (additional) purpose by a different group?" Out of 54 responses, 36 (67%) said yes and 18 (33%) said no. However, when it comes to the question of making the data and information available over the Internet, a different response is received. Out of 57 responses to this question, 22 authors (39%) reported that at least some part of the data they collected was accessible over the Internet, whereas 35 authors (61%) reported that none of their data were accessible via the Internet. There are several reasons why most agencies do not make their data web accessible.

In the first place, government agencies often hold confidential information and so are unable to make all of their data web accessible. The data may be kept confidential for commercial reasons. A two- to five-year moratorium on making data public is often imposed to allow scientists involved with the data collection time to publish their findings. The cost of implementing a security system that distinguishes between confidential and nonconfidential data may prohibit some agencies from making any data accessible over the Internet. On the subject of cost, the funding provided to agencies to conduct surveys to collect data may not include the added costs of storing the data and making it web accessible.

Another reason why most agencies do not make their habitat data accessible over the Internet is related to the nature of digital data itself. Simply putting raw data online is rarely useful. If parameters and standards are not published with the data, or if infrastructure (server space and a fast network connection) is not available, it becomes difficult to understand and use the data. In the case of multibeam sonar, data collected at sea are first cleaned of noise and artifacts to produce a processed data file. In order to view raw or processed files, specialized software is needed to create a georeferenced bathymetric map of the seabed. Unless the user owns the correct software, the data files are unreadable.

Data volume is another reason. The files comprising a week-long multibeam sonar survey can be very large (several terabytes), well beyond the scope of what can reasonably be delivered over a standard Internet connection. Instead, agencies commonly make a gridded map product (at an appropriate resolution) available over the Internet. Interested persons may then view the map online and inquire directly to the agency about access to the processed multibeam data files if this is necessary. In fact, several authors noted that their data were partially available in a decimated form, as in the case of providing a gridded bathymetry product.

A final possible explanation as to why most agencies do not make their habitat data accessible over the Internet is that there is no great demand for the raw data; most of the general public are content with the published reports, gridded map products, and papers that are produced from the survey. Although it may be true that scientists need access to raw data to do their work, they can usually get that access by directly contacting colleagues at the agency that hosts the data. Politicians and managers want information, explanations, and advice from scientists (not raw data); they want to know that the advice they receive is backed up by adequate data and peer-reviewed papers, but they rarely get involved with the technical details of how the information was reduced and interpreted. Hence, the cost of building and maintaining a web-based data delivery system to serve up raw or processed digital data may simply not be warranted.

Gap Analysis and Priorities for the Future of Habitat Mapping: Geographic Areas, Geomorphic Features, and Environmental Variables

A gap analysis has been carried out based on: (1) the geographic distribution of case studies; (2) the geomorphic features encompassed by case studies; and (3) the environmental variables considered by case studies.

Geographically, about half of the case studies were from waters around Western Europe, with the remainder scattered around North America and the periphery of the Indian and Pacific Oceans (Figure 64.8). There are a few case studies from the northern margins of Africa (Mediterranean and Red Seas), but none from the Atlantic margin of the continent. The margins of Asia and South America were not represented in any case study (there is no case study from the South Atlantic Ocean). Given the intense pressures facing benthic habitats and broad regional differences in ecosystems, species, and habitats, future case studies from these regions should be specifically sought for future editions of the Atlas.

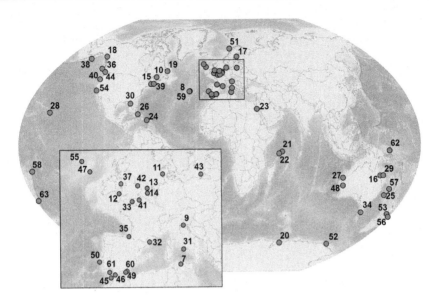

Figure 64.8 Map showing the distribution of case studies. The numbers refer to case study chapters. About half of the case studies are from waters around Western Europe, with the remainder scattered around North America and the Austral-Asian region; few are located in Africa, and there are none in Asia or South America.

In terms of geomorphic features, shelf and slope habitats were well represented in the case studies (Figure 64.9). Sandbanks, sand waves, coral reefs, canyons, and glaciated shelves are particularly well covered. The case studies mainly focused on a single geomorphic feature type (36 out of 57 studies); however, 18 case studies focused on two feature types, and three studies focused on three or more types. Surprisingly (given their accessibility and overall importance to humans), estuarine and deltaic coastal habitats are poorly represented in the case studies. In particular, wave-dominated estuaries and all types of deltas seem to have been avoided. This may be a consequence of the shallow depths of these environments, which greatly restricts boat access and hence the application of sonar mapping technologies (but not LIDAR or other remote sensing techniques) and conducting marine fieldwork in general (e.g., video tows and sampling).

The other obvious gap in geomorphic features is deep ocean (abyssal–hadal) environments, which were described in only a few case studies. Given that around 50% of the earth is located at abyssal depths, this knowledge gap is globally significant and represents a major challenge to the habitat mapping community. Some obvious reasons why the abyssal environment is avoided include: (1) the high cost of operating deep-water survey vessels; (2) the abyssal seafloor is largely located on the high seas and therefore falls outside the responsibility of national science agencies; and (3) the perception of these environments as being remote and hence not heavily impacted by human activities.

The final aspect of the gap analysis relates to physical attributes of the environment that are likely to affect habitats but were not featured in the case studies for

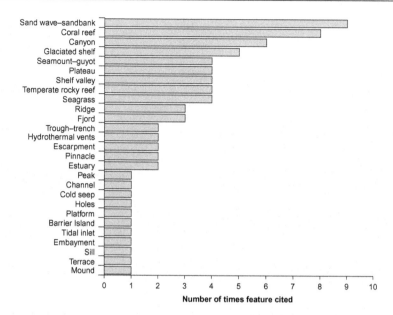

Figure 64.9 Graph showing geomorphic features described in the case studies and the number of times each feature was cited as a major focus for the study ($n = 82$). Some studies considered more than one type of feature.

various reasons. A general observation is that most habitat mapping studies report a snapshot view in which a range of single observations are made over an area in order to produce a set of spatially complete, mostly temporally static, data layers that lend themselves to multivariate analysis techniques. But habitats are composed of many spatially and temporally dynamic elements that cannot be measured by a single observation. Often it is the variability of a physical attribute that characterizes habitats (e.g., the variation in temperature and salinity in an estuary, the frequency of storm disturbances to particular coastal and shelf habitats, or the migration rate of continental shelf dunes). The dynamic aspects of habitats include their connectivity in relation to adjacent habitats as sources or sinks of larvae and colonizers and the biological processes of predator–prey relationships, trophic levels, and gene flow. These temporally dynamic elements seem to be overlooked in many case studies.

It is, of course, very expensive to collect a long time series of measurements needed to accurately represent the dynamic conditions of many environments. This is particularly a problem when the process is episodic in nature or has a long periodicity ("long" in terms of the duration of a typical research program may be only a few years!). However, the geologic record can often assist in such cases, by providing *in situ* environmental time series information. Sediment cores were collected as part of some case studies (Chapters 46 and 50), and changes in the down-core chemistry of sediments, fossil assemblages, sediment accumulation rates, or the physical character of sediment deposits (grain size, presence/absence of primary structures, etc.) are all useful sources of information about changes in the environment [13,14]. Coral core records from the Great Barrier Reef document changes in river sediment supply since European settlement [15], for example. The resolution of sediment records is of course a function of accumulation and sediment mixing rates, and deposits suitable for palaeoenvironmental analysis may not always be available. However, where they occur, the integration of such geologic time series data sets into the characterization of habitats can provide insights into their naturalness, classification, and broader understanding of their temporal variability (and stability).

Numerical modeling is another tool that can provide assistance in some cases where dynamic aspects of the environment are a primary factor characterizing habitats. Such is the case, for example, in many tide-dominated shelf and estuarine habitats and on tropical shelves influenced by storm events (hurricanes, typhoons, and cyclones). For example, repeat multibeam surveys combined with current measurements and modeling were used to understand the migration of large sand waves in Torres Strait, Australia, in relation to sea grass dieback [16]. Two case studies (Chapters 33 and 41) included modeled wave and current information in their analyses of habitat characterization, and both were located on the macrotidal, west European continental shelf. It is also important to note that wave and current energy are generally found to be among the most useful surrogates by the case study authors (Figure 64.3).

Discussion: Best Practices for Benthic Habitat Mapping

The case studies in this volume, and the responses to the questionnaires by authors reported in this chapter, provide valuable insights into what constitutes best practice for conducting a habitat mapping program (Figure 64.10). In the first instance, it is clear that thorough consultation and planning of habitat mapping surveys is standard practice among GeoHab scientists. Planning for most surveys will involve consultation with the scientists directly involved, industry practitioners holding a stake in the area to be surveyed, relevant government agencies, and the broader science community. In cases where indigenous people or traditional users have an interest in the proposed survey area, they are normally consulted by GeoHab scientists in designing the survey.

The main objectives of conducting most habitat mapping surveys are to take stock of what benthos and habitats exist in a given area, measure the condition of the environment and its dependent communities, and try to identify trends in overall condition (or health). Therefore, understanding and quantifying the naturalness of the habitats present in an area is essential in order to establish baselines and measure change. This may be a challenge in areas where undisturbed reference sites are unavailable. Identifying trends in condition (improving or declining) may be accomplished through repeated mapping surveys and/or by taking a time series of measurements of previously identified surrogates. Several of the case studies described modified or extremely modified environments, yet theirs are the first systematic assessment of the condition and status of that particular habitat. The commencement of a habitat mapping and monitoring program in areas heavily impacted by human activities still requires establishing a baseline of relative naturalness so that improvements in

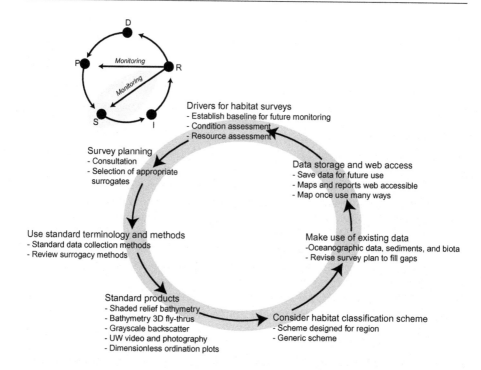

Figure 64.10 Series of steps involved in the conduct of a benthic habitat mapping survey to achieve best practice. Habitat mapping and monitoring conform to the "state" part of the DPSIR framework (yellow shading in insert at top left). (For interpretation of the references to color in this figure legend, the reader is referred to the web version of this book.)

condition can be measured and reported. Such improvements may become the performance indicators to justify the establishment of MPAs or fisheries reserves, for example.

GeoHab scientists represent a range of disciplines, and it has been an objective of this book to try to bring clarity to the terminology used for describing habitats and the benthic marine environment. The glossary to this volume contains around 200 definitions contributed by the authors and adopted by mutual consent. While it remains a work in progress, the glossary provides a useful reference for habitat mapping terminology.

Standard methods and protocols for marine data collection are well established for most data types (sediment grain size, taxonomy, water properties, depth, etc.). The case studies in this book also describe some new developments in video classification (Chapters 46 and 47) and automated methods for analyzing and classifying seabed morphology based on multibeam bathymetry and acoustic backscatter data (Chapters 8, 34, 36, and 58; see also technologies listed in Table 1.2, Chapter 1). The different methodologies described in a number of case studies to identify surrogacy relationships are also among some of the more recent and exciting developments in habitat mapping science.

In terms of habitat data products, there appears to be convergence in the use of 3D imagery and false-colored grids for representing bathymetry data. Fly-through video movies can be generated from bathymetry data sets, providing a powerful communication device, but are of no use in hard-copy-only literature. Maps of sediment type, grain size, and geomorphic features are also a common theme among the case studies. Dimensionless ordination plots are a popular way of displaying relationships between biological observations and physical surrogates. Although these data presentation methods are not standardized, there is some consistency among the case studies in their occurrence, which allows some comparisons to be made.

GeoHab authors seem less certain about adopting any standard classification scheme in their studies. Most case studies in this book did not utilize any scheme in spite of the advantages for comparison of results and communication. The jury is still out as to whether any one scheme will become a habitat mapping standard in the years ahead.

Selecting which biological and physical variables to measure is perhaps one of the most important decisions to be made at the outset of a habitat mapping program. It is obvious that not all variables are equally useful as surrogates (Figure 64.3). However, different variables will be more useful surrogates in particular habitats; no single variable is equally useful as a surrogate in all habitats. Since collecting and analyzing each data type has a direct cost to any program, science teams planning habitat mapping surveys would be well advised to review the literature to optimize the list of variables to be measured.

The collected data should be archived and stored; this is standard practice for most government agencies (Figure 64.10). Where the data were collected using public funds, storage and archiving is generally a requirement. However, even in cases where data have been collected using private resources (where confidentiality may be a requirement), thought should be given to data storage and archiving lest the data be lost.

In terms of making data accessible, it is the common practice of most GeoHab scientists to provide access to reports and publications describing the data over the Internet. This appears to be a high priority for stakeholders. Making people aware that the habitat mapping work has been completed and providing access to summary maps and reports is generally more important than making large volumes of raw digital data accessible over the Internet. If a moratorium is imposed on the release of the data (e.g., to allow scientists time to publish their findings), the length should be kept to as short a time as practicable and for no longer than 5 years.

One final and very important best practice for habitat mapping is to make use of existing information and to make available for reuse data collected during a habitat mapping survey. Using existing data sets, such as commercial fisheries data, oceanographic observations, modeled wave and currents information, museum biological data, and so on, is a cost-effective means of supplementing and adding value to newly collected habitat survey data. Making available data collected for reuse by other groups is a policy adopted by many government departments in relation to any and all government-funded programs. For habitat mapping science, this policy translates into "Map once, use many ways."

Acknowledgments

Thanks to Geoffroy Lamarche (New Zealand Institute of Water and Atmospheric Science) for helpful comments and suggestions on an earlier draft of this chapter. This work was produced with the support of funding from the Australian Government's Commonwealth Environment Research Facilities (CERF) program and is a contribution of the CERF Marine Biodiversity Hub and UNEP/GRID Arendal. This chapter is published with the permission of the Chief Executive Officers of Geoscience Australia and UNEP/GRID Arendal.

Appendix 1 Questionnaire Completed by All Case Study Authors

Name of first author:
Contact e-mail:
Case study number:
Case study title:

Science Questions:
What was the size (spatial area, km^2) of your study site? _____km^2

What was the water depth range of your study site (from most shallow to deepest)? _____m to ____m.

What was the maximum resolution (minimum distance in meters between soundings) of acoustic mapping in your study? _____m.

Apart from geomorphology, what variables were measured in your study to characterize habitats? (Tick the relevant boxes)

- Sediment grain size
- Sediment composition (carbonate content, TOC, etc.)
- Acoustic backscatter
- Water properties (temperature, salinity, dissolved oxygen, etc.)
- Primary production
- Slope
- Rugosity
- Topographic position index (TPI)
- Current strength or bed shear stress
- Wave characteristics or wave-induced bed stress
- Other (specify) _____

Which of the above (if any) did you find to be the most useful surrogates for biota (communities, species, etc.)? _____

Did you use any spatial analysis techniques in your study (e.g., multivariate analysis, GIS tools)? If so please specify _____

Did you use any previously published classification scheme in your study (e.g., EUNIS)? If yes, please specify here: _____

Socioeconomic Questions:

What was the main user group that your study was serving? (Tick the relevant boxes)

- Fisheries
- Conservation
- Oil-gas
- Aggregate mining
- Navigation
- Tourism
- Other (specify) _____

What was the main purpose of the survey (e.g., input to spatial marine planning, marine protected area design, fisheries reserve, seabed resource assessment)?

_____

After the survey was completed, was the data used for a different purpose by another group? If so, please explain: _____

Who paid the costs for your study? (Government research body, private sector, other)

_____

Was your study a one-off study or was it part of an ongoing monitoring program? (Circle one): One-off study/ongoing monitoring

If it was a one-off study, is there any current plan to use the outcome of your study to establish a baseline for future assessment? (Circle one): YES/NO

Who was consulted in planning your survey? (Tick the relevant boxes)

- Survey participants
- Industry practitioners
- Government
- Science community
- Traditional users (indigenous people)
- Other (specify) _____

Are the data collected in your study stored (or planned to be stored) in a national database? (Circle one): YES/NO

Are the data collected in your study currently web accessible? (Circle one): YES/NO

References

[1] H.G. Greene, M.M. Yoklavich, R.M. Starr, V.M. O'Connell, W.W. Wakefield, D.E. Sullivan, et al., A classification scheme for deep seafloor habitats, Oceanol. Acta 22 (1999) 663–678.

[2] C.E. Davies, D. Moss, The EUNIS Classification, European Environment Agency, http://www.eea.europa.eu/data-and-maps/data/nationally-designated-areas-national-cdda-3/eunis-habitat-classification.

[3] IHO, Standardization of Undersea Feature Names: Guidelines Proposal Form Terminology, International Hydrographic Organisation and International Oceanographic Commission, Bathymetric Publication No. 6., fourth ed., IHO, Monaco, 2008.

[4] C.J. Madden, D.H. Grossman, A framework for a coastal/marine ecological classification standard (CMECS), in: B.J. Todd, G. Greene, (Eds.), Mapping the Seafloor for Habitat Characterisation, Geological Association of Canada, St. Johns, NL, 2007, pp. 185–210.

[5] D.W. Connor, J.H. Allen, N. Golding, L.K. Howell, L.M. Lieberknecht, K.O. Northen, et al., The Marine Habitat Classification for Britain and Ireland Version 04.05, JNCC, Peterborough, http://jncc.defra.gov.uk/page-1584, 2004.

[6] NOAA National Centers for Coastal Ocean Science, Shallow-water benthic habitats of American Samoa, Guam, and the Commonwealth of the Northern Mariana Islands, NOAA Technical Memorandum NOS NCCOS 8, Silver Spring, MD, http://ccma.nos.noaa.gov/products/biogeography/us_pac_terr/index.htm, 2005.

[7] OSPAR, Quality Status Report 2010, OSPAR, London, http://qsr2010.ospar.org/en/index.html, 2010, 176 pp.

[8] Z. Al-Hamdani, J. Reker, U. Alanen, J.H. Andersen, J. Bendtsen, U. Bergström, et al. (Eds.), Towards marine landscapes in the Kattegat and Baltic Sea, BALANCE Interim Report No. 10, 2007, 118 pp.

[9] B. Riegl, W.E. Piller, Distribution and environmental control of coral assemblages in northern Safaga Bay (Red Sea, Egypt), Facies 36 (1997) 141–162.

[10] B. Riegl, W.E. Piller, Coral frameworks revisited—reefs and coral carpets in the northern Red Sea, Coral Reefs 18 (1999) 241–253.

[11] P.J. Mumby, A.R. Harborne, Development of a systematic classification scheme of marine habitats to facilitate regional management and mapping of Caribbean coral reefs, Biol. Conserv. 88 (2) (1999) 155–163.

[12] P.R. Last, V.D. Lyne, A. Williams, C.R. Davies, A.J. Butler, G.K. Yearsley, A hierarchical framework for classifying seabed biodiversity with application to planning and managing Australia's marine biological resources, Biol. Conserv. 143 (2010) 1675–1686.

[13] A. McMinn, G.M. Hallegraeff, P. Thomson, A. Jenkinson, H. Heijnis, Cyst and radionucleotide evidence for the recent introduction of the toxic dinflagellate *Gymnodinium catenatum* into Tasmanian waters, Mar. Ecol. Prog. Ser. 161 (1997) 165–172.

[14] K. Morris, J.C. Butterworth, F.R. Livens, Evidence for the remobilization of Sellafield waste radionuclides in an intertidal salt marsh, West Cumbria, U.K., Estuarine Coast. Shelf Sci. 51 (2000) 613–625.

[15] M.T. McCulloch, S. Fallon, T. Wyndham, E. Hendy, J.M. Lough, D. Barnes, Coral record of increased sediment flux to the inner Great Barrier Reef since European settlement, Nature 421 (2003) 727–730.

[16] J. Daniell, P.T. Harris, M. Hughes, M. Hemer, A. Heap, The potential impact of bedform migration on seagrass communities in Torres Strait, Northern Australia, Cont. Shelf Res. 28 (2008) 2188–2202.

Glossary

Abiotic Without life.

Abyss The great depths of the oceans, usually considered to be depths of 2,000–6,000 m, a region of low temperatures, high pressure, and absence of sunlight.

Abyssal hills Tract, sometimes extensive, of low (100–500 m) elevations on the deep seafloor.

Abyssal plain An extensive, flat, gently sloping or nearly level region at abyssal depths.

Abyssopelagic Depths in the water column greater than 4,000 m.

Algae The simplest plants; may be single celled (such as diatoms) or quite large (such as sea weeds). Live in salt or fresh water.

Alpha (α) diversity The species diversity measured at the local scale comparing two sites from different bioregions. See also *Biodiversity*.

Anadromous Fish that live in the ocean mostly, and breed in fresh water. Contrast with *Catadromous*.

Angle of repose The angle attained on a slope comprised of unconsolidated sediments after avalanching, typically around 30–40° for quartz sand, depending on porosity and packing. The angle of repose is common in nature, occurring on the lee sides of migrating bedforms and among continental slope sediment deposits.

Aphotic zone The deepest part of the water column, where light does not penetrate.

Apron Gently dipping featureless surface, underlain primarily by sediment, at the base of any steeper slope.

Archipelago A group of islands.

Aspect The azimuth (compass bearing) that slopes are facing.

Assemblage A neutral substitute for "community" but implying no necessary interrelationships among species; also called species assemblage.

Atoll An annular *coral reef* enclosing a lagoon in which there are no basement rock features other than reefs and islets composed of carbonate reef material.

AUV Autonomous underwater vehicle.

Bank Elevation over which the depth of water is relatively shallow but normally sufficient for safe surface navigation.

Barforms Sandbanks, sandbars; elongate features comprised of unconsolidated sediments formed within or along the margins of tidal channels and shelf seas.

Basin Depression, characteristically in the deep seafloor, more or less equidimensional in plan and of variable extent.

Bathyal (also *Bathypelagic*) The dark, deep part of the water column (1,000–4,000 m) below the euphotic (well-lighted) zone and mesopelagic (poorly lighted) zone but above the abyssopelagic zone.

Bathymetry Study and mapping of seafloor elevations and the variations in water depth; the topography of the seafloor.

Bathypelagic zone See *Bathyal*.

Bedforms Ripples and dunes; wave-like, flow-transverse features formed of unconsolidated sediments on the seabed.

Bench A level or gently sloping plane; a shelf-like area of rock with steeper slopes above and below.

Benthic Associated with the seafloor.

Benthos The collection of organisms living on or within the seabed.

Beta (β) diversity The species diversity measured between two or more sites within a bioregion. See also *Biodiversity*.

Biocoenosis (also biocoenose or biocenose) "Living community"; formulated in 1877 by Karl Möbius, it is a concept synonymous with the modern term "ecosystem."

Biodiversity Biological diversity; used by marine conservationists most commonly with respect to species diversity. Biodiversity can also encompass habitat or community diversity and genetic diversity within a species. With respect to measuring species diversity at increasingly broad spatial scales, alpha-diversity refers to species diversity measured at the local scale comparing specific sites from different bioregions; beta-diversity refers to species diversity comparing sites from within a bioregion; and gamma-diversity refers to species diversity comparing different bioregions measured at a number of sites.

Biogenic In geology, sediments or sedimentary rocks of biological origin, including in particular calcium carbonate and biogenic silica (opal).

Biogeography The study of the spatial and temporal distribution of species.

Bioherm Mound-shaped deposits of rock and sediment produced by marine organisms. Coral reefs and *Halimeda* banks are well-known examples.

Biome A high level in the ecological classification of associations between organisms; marine biomes are commonly considered to include the intertidal, neritic, and bathypelagic.

Bioregionalization A spatial representation depicting the boundaries of areas represented by each level in a hierarchical classification of habitat types (provinces, biomes, habitat complexes, etc.)

Biotone A zone of transition between core provinces. Biotones are not simply "fuzzy" boundaries but represent unique transition zones between the core provinces.

Biotope (ecotope) Both the abiotic and biotic elements of habitats; physical habitats and their associated biota.

Biotope complex (also biotope network) Spatially and biologically related and connected groups of biotopes.

Canyon (also submarine canyon) A relatively narrow, deep depression with steep sides, the bottom of which generally has a continuous slope, developed characteristically on most continental slopes. Larger than a gully, canyons typically extend over a depth range exceeding 1,000 m and are incised >100 m into the slope.

Catadromous Fish that live mostly in fresh water, but breed in the ocean. Contrast with *Anadromous*.

Cay A small, low-elevation, sandy island formed on the surface of a coral reef.

CBD (United Nations) Convention on Biological Diversity.

Circalittoral Subtidal zone defined by EUNIS on the bases of light penetration and presence/absence of red algae. The circalittoral zone can be split into two subzones; upper circalittoral (foliose red algae present but not dominant) and lower circalittoral (foliose red algae absent). The exact depth at which the circalittoral zone begins is directly dependent on the intensity of light reaching the seabed, and hence is dependent on turbidity.

Cliff A vertical or near-vertical rock exposure; a tall steep rock face. See also *Escarpment*.

CMECS Coastal/Marine Ecological Classification Standard of the United States National Oceanic and Atmospheric Administration (NOAA).

Community A group of species that generally are assumed to be interdependent (though this is often not demonstrated). The term can be used in a variety of hierarchies. Communities at larger scales can be progressively subdivided, such as spatially, taxonomically, and trophically, to finer scales.

Continental margin The submerged prolongation of continental land mass consisting of the seabed and subsoil of the continental shelf, slope, and rise, but not the deep ocean floor.

Continental rise A gentle slope rising from the oceanic depths towards the foot of a continental slope

Continental shelf A zone adjacent to a continent (or around an island) and extending from the low-water line to a depth at which there is usually a marked increase of slope towards oceanic depths

Continental slope Located seaward from the shelf edge to the upper edge of a continental rise or the point where there is a general reduction in slope.

Contourite (also contourite drift) Sediments deposited by bottom currents that flow generally along a bathymetric contour, usually along the lower continental slope or rise. See also *Drift deposits*.

Coral reef Reefs of tropical waters developed through biotic processes dominated by corals and calcareous algae. In geology, sedimentary features, comprised of macroscopic skeletal framework, built by the interaction of organisms and their environment, that have synoptic relief and whose biotic composition differs from that found on and beneath the surrounding seafloor. See also *Reef*.

Coral reef ridge Long, narrow elevation with steep sides composed of live or dead coral.

Deep In oceanography, an obsolete term that was generally restricted to depths greater than 6,000 m.

De Geer moraines (also called "washboard" moraines) (a) Parallel, regularly spaced ridges that are oriented transverse to ice movement in a general sense and that collectively resemble a washboard; (b) a subglacial feature formed by the periodic recession and re-advance of a glacier, which pushes previously deposited ground moraine into a ridge.

Delta Seaward prograding sediment body deposited at the mouth of a river.

Demersal species A fish (also called groundfish), cephalopod, or crustacean that lives on or near the seabed.

Drift deposits Elongate, flow-transverse sediment bodies, commonly deposited at the foot of continental slopes and associated with contour-following bottom currents (see also *Contourites*); commonly found at the mouths of submarine canyons and upon submarine fan systems.

Drumlin An elongated, whale-shaped hill formed by glacial ice acting on underlying unconsolidated till or ground moraine. The long axis of drumlins is usually aligned parallel with the movement of the ice, with its steeper slope facing into the direction of glacial movement. Drumlins are typically kilometers in length, less than 50 m in height, and hundreds of meters in width.

Dune Sedimentary bedforms larger than ripples, greater than 0.6 m in wavelength and greater than around 10 cm in height. Dunes are mostly asymmetrical in profile, with a gentle up-current stoss slope and a steeper down-current lee slope, which may be at the angle of repose of the sediment. Dune crestlines may be either linear (two dimensional) or nonlinear (three dimensional, barchan-shape) in plan view. Large dunes may have smaller dunes superimposed upon them.

Dysphotic zone The part of the water column, below the euphotic zone, that receives low levels of sunlight insufficient to support plant growth.

Ecosystems Short for ecological systems. Functional units that result from the interactions of abiotic and biotic components; a combination of interacting, interrelated parts that form a unitary whole. All ecosystems are "open" systems in the sense that energy and matter are transferred in and out. The Earth as a single ecosystem constantly converts solar energy into myriad organic products, and has increased in biological complexity over geologic time.

Ecotone A transition zone between two ecologically distinct regions, that may contain specialized species adapted to the gradient in environmental conditions that occurs there.

Ecotope (biotope) Both the abiotic and biotic elements of habitats; physical habitats and their associated biota; a geographical unit homogeneous within limits for the most important hydraulic, morphological, and physicochemical environmental factors that are relevant for biota.

Endemic Species only known to occur in one location.

Epibenthic Living on the surface of the seabed.

Epifauna Animals that live on the surface of the seabed or upon other benthic animals or plants.

Epiflora Plants that live on the surface of the seabed or upon other benthic animals or plants.

Epipelagic Depth in the water column of less than 200 m.

Escarpment Elongated and comparatively steep (sometimes vertical) slope separating flat or gently sloping areas. See also *Cliff*.

Esker An elongate, ridge-shaped feature deposited by a glacier.

Estuary A drowned river valley that receives sediment from both landward and seaward sources.

EUNIS European Nature Information System habitat classification system, pertaining to land, freshwater, and marine habitats, including European seas.

Euphotic zone The upper part of the water column that receives sufficient light to allow plant growth.

Eutrophic Of or relating to waters rich in plant nutrients that support high rates of plant growth. If the flux of organic matter exceeds the rate at which it can be oxidized (such as occurs in some enclosed water bodies or basins), the water and sediments may become anoxic and dominated by sulfide-reducing bacteria.

Fan (also submarine fan) Relatively smooth, fan-like, depositional feature normally sloping away from the outer termination of a canyon or canyon system.

Fetch The unobstructed distance of ocean over which wind or waves can travel.

Filter feeding In zoology, a form of food procurement in which food particles or small organisms are randomly strained from water. Passive filter feeding (sessile animals that rely on currents for food delivery) is found primarily among the small- to medium-sized invertebrates and active filter feeding (mobile animals) occurs in a few large vertebrates (e.g., flamingos, baleen whales).

Fjord (also spelled "fiord") Long, narrow, deep inlet of the sea between steep slopes formed by glacial erosion.

Gamma (γ) diversity Species diversity comparing different bioregions measured at a number of sites. See also *Biodiversity*.

Geomorphic feature Spatially discrete elements of Earth's surface (including the ocean floor) attributed to a particular formative process. Examples are dunes, seamounts, atolls, and fjords, formed by wind (or water) currents, volcanism, coral reef growth, and glaciers, respectively.

Geomorphology The scientific study of the shape of Earth's surface (including the ocean floor) and the formative processes causing them.

Geodiversity The natural range (diversity) of geological (rocks, minerals, fossils), geomorphological (landforms, processes), and soil (sediment) features. It includes their assemblages, relationships, properties, interpretations, and systems (Gray, 2004).

GIS Acronym for Geographic Information System. GIS is computer software for the capture, storage, analysis, and presentation of spatial data. In the simplest terms, GIS is the merging of cartography, statistical analysis, and database technology.

GOODS Global Open Ocean and Deep Sea habitats classification developed by the United Nations Convention on Biological Diversity (Agnostini et al., 2008).

Gravel Sediment grains larger than 2 mm. Includes granules (≥ 2, < 4 mm), pebbles (≥ 4, < 64 mm), cobbles (≥ 64, < 256 mm), and boulders (≥ 256 mm).

Gully Straight, shallow channel formed in relatively high seafloor-slope settings; smaller than a submarine canyon.

Guyot Seamount having a comparatively smooth flat top formed by wave erosion and sometimes coral reef growth.

Habitat Physically distinct areas of seabed associated with suites of species (communities or assemblages) that consistently occur together. See also *Potential habitat*.

Hadal Pertaining to depths of the ocean greater than 6,000 m.

Hermatypic corals Reef-building corals; coral species that produce an aragonite skeleton, which accretes over time into limestone coral reef deposits.

Holdfast A root-like structure that anchors aquatic sessile organisms, such as seaweed, other sessile algae, stalked crinoids, benthic cnidarians, and sponges, to the substrate.

Hole Local depression, often steep sided, of the seafloor.

Holocene Period of geologic time spanning the last 11,700 years from the end of the Pleistocene period (and the end of the last ice age) to the present.

Hypsometric curve An empirical cumulative distribution function of elevations and/or depths. A curve that illustrates the relative portion of surface area occurring at different depths or elevations within a specified region.

Infauna Animals that live within sediments.

Infralittoral Zone of the seabed below the low-water mark characterized by sufficient light to allow plant growth.

Intermediate disturbance hypothesis (IDH) A hypothesis postulating that the diversity of species is controlled by the frequency of disturbances capable of removing biota from a patch of seabed. The greatest diversity is expected to occur in environments that experience an intermediate frequency of disturbance in relation to the time required for the community to recover.

Intertidal The area (zone) between extreme high-water mark and extreme low-water mark that is submerged at high tide and exposed at low tide.

Invertebrate An animal without a backbone or spinal column (i.e., not vertebrate).

Island biogeography (theory of) The number of species present on an island is proportional to island area and its distance offshore from the mainland. Small remote islands generally have poorer species diversity than large mainland-proximal islands.

IUCN World Conservation Union; formerly known as the International Union for the Conservation of Nature and Natural Resources, from which the acronym derives.

Kilohertz (kHz) In marine mapping, a term used to describe the frequency of sound, equal to 1,000 cycles per second (Hertz). 100 kHz is equal to 100,000 Hz, etc.

Knoll Relatively small (500–1,000 m tall) isolated elevation of a rounded shape; a small seamount.

Lagoon A shallow marine (sometimes brackish to hypersaline) coastal water body, receiving little if any fluvial input, separated from the sea by a restricted inlet usually having a sill.

Levee Sediment embankment deposited along the sides of rivers or submarine canyons.

Limestone Sedimentary rock comprised of $> 50\%$ calcium carbonate.

Lithoherm A type of deep-water reef composed of surface hardened layers of lithified sandy carbonate sediments supporting a diverse array of benthic fauna, including deep-water corals.

Littoral The intertidal zone, located between upper and lower extreme tide marks and exposed during lowest tides. The supralittoral (supratidal) zone is located adjacent to and above the littoral zone. The sublittoral (subtidal) zone is located adjacent to and below the littoral zone.

Macroalgae Seaweeds (algae) visible to the naked eye.

Macrofauna Benthic organisms retained on a 0.3 mm (or larger) sieve.

Marine Protected Area (MPA) Defined by the IUCN as "any area of intertidal or subtidal terrain, together with its overlying water and associated flora, fauna, historical and cultural features, which has been reserved by law or other effective means to protect part or all of the enclosed environment."

Marine spatial planning A process that brings together multiple users of the ocean—including energy, industry, government, conservation, and recreation—to make informed and coordinated decisions about how to use marine resources sustainably based on a spatial framework (maps). Marine protected areas and fisheries reserves are examples of marine spatial planning devices.

Megabenthic community Community comprising megafauna, benthic organisms that can be seen by the naked eye or be detected in underwater photographs.

Megafauna Animals that can be seen by the naked eye; also defined as animals larger than 100 pounds (45 kg) in weight.

Megaripple Type of sedimentary bedform. The term "megaripple" has largely been replaced with "dune" in the marine geology literature. Megaripples are a type of small dune (small sand wave). They are generally asymmetric in profile and move downstream as the upstream slope is eroded and sediment is deposited on the downstream or lee slope. Megaripples may reverse their asymmetry over a 12 h tidal cycle. Their crestlines may be either linear (two dimensional) or nonlinear (three dimensional or barchan-shaped) in plan view.

Meiofauna Animals in the size range 0.063–0.5 mm.

Mesopelagic Depths between 200 and 1,000 m.

Metapopulation A population that exists in a connected complex of spatially discrete habitat fragments.

Microalgae Phytoplankton; small plants visible under a microscope, such as diatoms. The definition includes benthic algae.

Microfauna Small, mostly microscopic animals (less than 0.063 mm), such as protozoa, nematodes, and small arthropods.

Moraine A glacial deposit comprised of till (unconsolidated glacial debris) found in currently glaciated and formerly glaciated regions. Different types of moraine are deposited beneath (ground moraine), alongside (lateral moraine), or at the end (terminal moraine) of a glacier.

Mound A rounded feature of bathymetrically positive relief that rises above the level of surrounding seabed. See also *Bioherm*.

Mud Sediment grains smaller than 0.0625 mm in size. Includes silt (between 0.0625 and 0.004 mm) and clay (<0.004 mm).

Mud volcano Product of mud diapirism, in which a more mobile and ductily deformable material is forced into brittle, overlying rocks and erupts onto the seafloor, forming a volcano-shaped (cone-shaped) deposit.

Nappe In geology, a large sheet-like body of rock that has slumped more than 2 km down the continental slope.

Naturalness In conservation, the extent to which a habitat or environment has been modified from its natural state by human activities and impacts. Commonly used descriptive terms for naturalness are near-pristine, largely unmodified, modified, and extensively modified.

Nekton An aquatic organism, such as fish and krill (euphausiids), that can swim powerfully enough to move against currents (contrast with *Plankton*).

Neritic Pertaining to the water column overlying the continental shelf.

Niche The range of environmental variables (such as temperature, salinity, and nutrients) within which a species can exist and reproduce. The preferred (or fundamental) niche is the one in which the species performs best in the absence of competition or interference from extraneous factors.

Ocean Biogeographic Information System (OBIS) A database originally created to host biological data collected as part of the Census of Marine Life. In 2009, OBIS was adopted by the Intergovernmental Oceanographic Commission of UNESCO, as an activity under its International Oceanographic Data and Information Exchange (IODE) program http://www.iobis.org/.

Olistostrome A sedimentary deposit composed of a chaotic mass of heterogeneous material, such as blocks and mud, known as olistoliths, that accumulates as a semifluid body by submarine gravity sliding or slumping of the unconsolidated sediments.

Palimpsest Sediment that exhibits attributes of a previous depositional environment, but also attributes of the modern environment.

Pediment A plain of eroded bedrock (which may or may not be covered with a thin veneer of alluvium) in an arid climatic zone, developed between mountain and basin areas.

Pelagic Of, relating to, or living in the water column of seas and oceans (as distinct from benthic).

Phanerogam Plants that produce seeds (also known as spermatophytes).

Phanerogam meadow Extended or patchy areas of seagrass, characteristic of shallow, tropical to temperate seas, colonized by the phanerogam plants *Posidonia oceanica* (L.) and/or *Cymodocea nodosa* (Ucria).

Phytoplankton Microscopic free-floating algae that drift in sunlit surface waters.

Physiognomy The apparent characteristics, outward features, or appearance of ecological communities.

Pinnacle High tower- or spire-shaped pillar of rock or coral, alone or cresting a summit. It may extend above the surface of the water, and may or may not be a hazard to surface navigation.

Plankton Small or microscopic aquatic plants and animals that are suspended freely in the water column; they drift passively and cannot move against the horizontal motion of the water (contrast with nekton that are capable of horizontal movement). Planktic animals (zooplankton) include small protozoans and the eggs and larvae of larger animals. Some migrate vertically in the water column each day (diel vertical migration). Planktic plants are phytoplankton and include diatoms, cyanobacteria, dinoflagellates, and coccolithophores.

Plateau In geography, an elevated plain, tableland, or flat-topped region of considerable extent. In oceanography, a bathymetric feature of broad extent elevated above the level of surrounding seafloor with a more or less flat top and steep sides.

Platform coral reef Flat or nearly flat area of live or dead coral reef located in the photic zone, elevated above the level of surrounding seafloor and dropping off in depth abruptly on one or more sides.

Pleistocene Period of geologic time extending from the end of Pliocene period around 2.6 million years ago up to the beginning of the Holocene period, 11,700 years ago.

Pockmarks Craters in the seabed, usually one to tens of meters but as large as a few hundred meters in diameter, formed by the expulsion of gas and/or fluid from sediments.

Potential habitat Physically distinct areas of seabed that may potentially be associated with suites of species (communities or assemblages) that consistently occur together.

Polynya From the Russian word for "lake"; an area of open water surrounded by sea ice.

Precautionary principle The guiding ecological principle that maintains only activities that have been demonstrated not to damage ecological resources be permitted. Too often, however, activities are permitted until it has been demonstrated that they are harmful.

Quaternary Period of geologic time extending from the end of Pliocene period around 2.6 million years ago up to the present; collective term for the Holocene and Pleistocene periods.

Raster Data type consisting of rows and columns of cells, with each cell storing a single value. A raster image is based on a grid of data.

Recruitment The influx of new members into a population by either reproduction or immigration.

Reef Rock (including coral limestone and other rock types) lying at or near the sea surface that may constitute a hazard to surface navigation.

Reef promontory A convex curve in a horizontal reef that occurs at the edge of a shelf creating a protrusion in both vertical and horizontal directions, much like the corner of a table.

Relaxation time The time required for species and populations to adjust to changed environmental conditions.

Relict Sediments (including reef limestone and other bioherms) that were originally deposited under different environmental conditions than those occurring today. See also *Palimpsest*.

Ribbed moraines Group of irregularly subparallel, locally branching, generally smoothly rounded and arcuate ridges that are convex in the downstream direction of a glacier but that curve upstream adjacent to eskers. They are most common in the continental ice sheets and are abundant in the Arctic.

Ridge (a) Long, narrow elevation with steep sides; (b) long, narrow elevation often separating ocean basins; (c) linked major mid-oceanic mountain systems of global extent.

Ripples (ripple marks) Sedimentary bedforms smaller than dunes, less than 0.6 m in wavelength and less than around 10 cm in height. Ripple marks are asymmetrical in profile, with a gentle up-current stoss slope and a steeper down-current lee slope at the angle of repose of the sediment. Wave-formed ripple marks have a symmetrical, almost sinusoidal profile. Ripple crestlines may be either linear (two dimensional) or nonlinear (three dimensional) in plan view.

Rogen moraine See *Ribbed moraine*.

Roughness Measure of variation in depth (bed elevation) of the seabed relative to a specified length scale; in GIS, the largest intercell difference in bed elevation between a central grid cell and its surrounding cells (contrast with *Topographic Position Index*).

ROV Acronym for remotely operated vehicle.

Rugosity In GIS, the ratio of the surface area to the planar area across the neighborhood of a central grid point. See also *Roughness*.

Saddle Broad pass, resembling in shape a riding saddle, in a ridge or between contiguous seamounts.

Sand Sediment grains between 0.0625 and 2 mm in size.

Sandbank (or tidal current ridge) Type of sedimentary barform. Sandbanks are elongate sediment deposits common in macrotidal environments, ranging from 100s of meters to many kilometers in length, 10s to 100s of meters in width, and rising from the seabed to the intertidal zone. Sandbank crests are commonly exposed at low tide. They are aligned with their long axis subparallel to the main current.

Sand wave Type of flow-transverse sedimentary bedform. The term "sand wave" has largely been replaced with "dune" in the marine geology literature. Sand waves are generally >1 m in height, >10 m in wavelength (crest–crest spacing), and typically they have smaller dunes superimposed. They are found in most tidal bays and estuaries and on the seabed wherever currents exceed a speed of around 1 m/s. Sand waves are generally asymmetric in profile and move downstream as the upstream slope is eroded and sediment is deposited on the downstream or lee slope. Their lee slopes are generally around 10°, much less steep than the angle of repose of sand. Sand waves are so large that they do not normally reverse their asymmetry over a 12 h tidal cycle. Their crestlines may be either linear (two dimensional) or nonlinear (three dimensional or barchan-shaped) in plan view.

Seagrass Flowering plants from one of four plant families (*Posidoniaceae*, *Zosteraceae*, *Hydrocharitaceae*, or *Cymodoceaceae*), all in the order Alismatales (in the class of monocotyledons), which grow in marine (brackish to fully saline) environments.

Seamount Large isolated elevation, greater than 1,000 m in relief above the seafloor, characteristically of volcanic origin and conical form.

Seascape The marine version of "landscape," comprised of suites of habitats that consistently occur together.

Shoal Offshore hazard to surface navigation that is composed of unconsolidated material.

Shelf A terrace. See also *Continental shelf.*

Shelf edge (also shelf break, shelf drop-off) Marked by a seaward increase in seabed slope that occurs at the seaward limit of the continental shelf and the upper limit of the continental slope.

Shelf sea Shallow body of water located on a continental shelf.

Siliciclastic Sediments or sedimentary rocks that are almost exclusively silica bearing, either as forms of quartz or other silicate minerals.

Sill Seafloor barrier of relatively shallow depth restricting water movement between adjacent basins. In oceanography, the sill depth signifies the depth of water that a water mass must achieve in order to pass between basins.

Slope Any inclined surface, generally less steep than a cliff or escarpment. See also *Continental slope.*

Sonar Acronym derived from the phrase "sound navigation and ranging"; method or equipment for determining water depth by underwater sound and also the presence, location, or nature of objects in the sea and the character of the seafloor.

Species richness Also "species diversity"; the number of species that occurs in an area. See also *Biodiversity.*

Spermatophytes Plants that produce seeds (also known as phanerogams).

Spur and groove A form of coral reef structure that typically occurs on deep fore-reef environments; characterized by a series of parallel carbonate ridges, aligned with prevailing winds, and built by hermatypic corals and interspersed with sand channels.

Strata Plural of "stratum"; in geology, layers (or beds) of rock or sediment. The term "strata" is used also with reference to the components comprising a stratified classification system.

Subduction zone Adjacent to active plate margins, a place where ocean crust collides with and is subducted beneath continental crust to create a ridge and ocean trench complex.

Sublittoral (subtidal) Zone located adjacent to and below the littoral zone, i.e., below the level of extreme low tide.

Substrate The surface a plant or animal lives upon. The substrate can include biotic or abiotic materials. For example, encrusting algae that lives on a rock can be substrate for another animal that lives on top of the algae.

Substratum In geology, layers (beds or strata) of unconsolidated sediment or lithified sedimentary rock found exposed at the seafloor or beneath overlying layers (beds or strata). In biology, see *Substrate.*

Supralittoral (supratidal) Zone located adjacent to and above the littoral zone, i.e., above the level of extreme high tide.

Surrogate A measurable entity that will represent, or substitute for, a more complex element of biodiversity that is more difficult to define or measure.

Taxonomy The practice and science of classification. Biological classification (sometimes known as "Linnaean taxonomy") includes the prediction, discovery, description, and (re) defining of taxa. It uses taxonomic ranks, including, among others, (in order) kingdom, phylum, class, order, family, genus, species.

Terrace Relatively flat horizontal or gently inclined surface, sometimes long and narrow, which is bounded by a steeper ascending slope on one side and by a steeper descending slope on the opposite side.

Terrigenous Derived from the land, as in terrigenous sediment.

Tidal current ridge See *Sandbank*.

Till Unsorted, unstratified mixtures of clay, silt, sand, and gravel (up to, but not including, boulder size); the usual composition of a moraine.

Topographic Position Index (TPI) (also Bathymetric Position Index, BPI) Difference in seabed depth between a central grid cell and the mean of its surrounding cells.

Trench Long, narrow, characteristically V-shaped in cross-section, very deep, asymmetrical depression of the seafloor, with relatively steep sides (contrast with *Trough*).

Trophic level The position of an organism in the food chain or "food pyramid," determined by the number of transfers of energy that occur between the nonliving energy source and that level.

Trough Long depression of the seafloor, characteristically flat bottomed, steep sided, and normally shallower than a trench.

Tsunami A very fast-moving oceanic wave initiated by an underwater disturbance such as an earthquake, volcanic eruption, or seafloor slump (Japanese for "harbor wave").

Turbidity current Dense mixture of suspended sediment and water that flows down-slope under the influence of gravity. Normally constrained to the continental slope and attributed to the formation of submarine canyons.

Upwelling An oceanographic process by which water rises from the lower depths upwards into shallow surface waters.

Valley (also "sea-valley") Relatively shallow, wide depression, the bottom of which usually has a continuous gradient. This term is generally applied to features found on the continental shelf and is not used for features on the continental slope that have canyon-like characteristics for a significant portion of their extent (which are instead submarine canyons).

Vicariance The subdivision of a continuously distributed biota by climatic and/or geological events. The separation or division of a group of organisms by a geographic barrier, such as a mountain or a body of water, resulting in differentiation of the original group into new varieties or species.

Water mass A volume of water that has defined salinity and temperature characteristics; water masses characterize oceanic and coastal areas and might be considered to be analogs of the major climatic regions of terrestrial environments.

Zooplankton Small, sometimes microscopic, animals that drift in the ocean; protozoa, crustaceans, jellyfish, and other invertebrates that drift at various depths in the water column are zooplankton.